EMISSION SPECTRA

CONTINUOUS SPECTRUM (Incandescent solids or liquids and incandescent gases under high pressure give continuous spectra) INCA...

7500 7000 6500 6000 5500 5000 4500 4000 Å

BRIGHT LINE SPECTRA (Incandescent or electrically excited gases under low pressure give bright line spectra)

MERCURY

SODIUM

HELIUM

HYDROGEN

7500 7000 6500 6000 5500 5000 4500 4000 Å

Adapted from the SPECTRUM CHART, Welch Scientific Company

Useful Conversions and Combinations

$$1 \text{ eV} = 1.602 \times 10^{-19} \text{ J}$$

$$1 \text{ cal} = 4.184 \text{ J}$$

$$1 \text{ u} = 931.5 \text{ MeV}/c^2$$

$$1 \text{ MeV}/c^2 = 1.073 \times 10^{-3} \text{ u} = 1.783 \times 10^{-30} \text{ kg}$$

$$1 \text{ Å} = 10^{-10} \text{ m} = 0.1 \text{ nm}$$

$$1 \text{ fm} = 10^{-15} \text{ m}$$

$$1 \text{ in} = 2.540 \text{ cm}$$

$$1 \text{ mi} = 1609 \text{ m}$$

$$hc = 1.240 \times 10^3 \text{ eV} \cdot \text{nm} = 1.986 \times 10^{-25} \text{ J} \cdot \text{m}$$

$$\hbar c = 1.973 \times 10^2 \text{ eV} \cdot \text{nm} = 1.97.3 \text{ MeV} \cdot \text{fm}$$

$$k_B T = 0.02525 \text{ eV at } T = 300 \text{ K}$$

$$ke^2 = e^2/4\pi\epsilon_0 = 1.440 \text{ eV} \cdot \text{nm}$$

$$1 \text{ barn} = 10^{-28} \text{ m}^2$$

$$1 \text{ curie} = 3.7 \times 10^{10} \text{ decays/s}$$

$$1 \text{ MeV}/c = 5.344 \times 10^{-22} \text{ kg} \cdot \text{m/s}$$

Modern Physics

Second Edition

Modern Physics

Second Edition

RAYMOND A. SERWAY
James Madison University

CLEMENT J. MOSES
Utica College of Syracuse University

CURT A. MOYER
Clarkson University

Saunders Golden Sunburst Series

SAUNDERS COLLEGE PUBLISHING
Harcourt Brace College Publishers

Fort Worth   Philadelphia   San Diego   New York   Orlando   Austin
San Antonio   Montreal   Toronto   London   Sydney   Tokyo

Text Typeface: New Baskerville
Composition: Progressive Information Technologies
Vice President/Publisher: John Vondeling
Development Editor: Susan Dust Pashos
Senior Project Editor: Sally Kusch
Copy Editors: Linda Davoli, Mary Patton
Manager of Art and Design: Carol Bleistine
Illustration Supervisor: Sue Kinney
Art and Design Coordinator: Kathleen Flanagan
Text Designer: Rebecca Lemna
Cover Designer: Carol Bleistine
Text Artwork: Rolin Graphics
Manager of Photos and Permissions: Dena Digilio-Betz
Senior Vice President, Director of EDP: Tim Frelick
Senior Production Manager: Charlene Squibb
Vice President/Director of Marketing: Margie Waldron

Cover Credit: Painting by Andrzej Dudzinski/photograph © Photo Researchers, Inc.

Printed in the United States of America

MODERN PHYSICS, 2nd edition

ISBN 0-03-001547-2

Library of Congress Catalog Card Number: 96-70012

90123456   032   1098765

About the Authors

Raymond A. Serway (Ph.D., Illinois Institute of Technology) is Professor of Physics at James Madison University. He received the Madison Scholar award at James Madison University in 1990, as well as the Distinguished Teaching Award and the Tau Delta Kappa Teaching Excellence Award at Clarkson University, where he taught for thirteen years. He was Guest Scientist at the IBM Research Laboratory in Zurich, Switzerland, where he worked with K. Alex Müller, 1987 Nobel prize recipient; he was also a visiting scientist at Argonne National Laboratory, where he collaborated with Sam Marshall. He is the author of the best-selling texts *Physics for Scientists and Engineers,* 4th ed., *Principles of Physics,* and (with Jerry Faughn) *College Physics,* 4th ed.

Clement Moses (Ph.D., State University of New York at Binghamton) is Associate Professor of Physics at Utica College of Syracuse University, where he has taught since 1982. He was born and brought up in Utica, New York, and holds an A.B. from Hamilton College and an M.S. from Cornell University in addition to his Ph.D. His research work, both in industry and university settings, has been in condensed matter physics and has dealt with defects in solids, the development of improved solar cells, and the dynamics of atoms at surfaces. Recently, he spent a sabbatical year at the University of Genova, Italy, engaged in surface physics research.

Curt A. Moyer (Ph.D., State University of New York at Stony Brook) has been a member of the Physics Department at Clarkson University since 1974. He has written numerous research articles in the fields of condensed matter physics and surface science. Dr. Moyer is an experienced teacher and an advocate for the uses of computers in education. He prepared the MP DESKTOP software that accompanies this text.

Preface

This book is intended as a modern physics text for science majors and engineering students who have already completed an introductory calculus-based physics course. The contents of this text may be subdivided into two broad categories: an introduction to the theories of relativity, quantum and statistical physics (Chapters 1 through 9) and applications of elementary quantum theory to molecular, solid-state, nuclear, and particle physics (Chapters 10 through 15).

OBJECTIVES

Throughout this book our underlying objectives have been threefold:

1. To provide simple, clear, and mathematically uncomplicated explanations of physical concepts and theories of modern physics;
2. to clarify and show support for these theories through a broad range of current applications and examples. In this regard, we have attempted to answer questions such as: What holds molecules together? How do electrons tunnel through barriers? How do electrons move through solids? How can currents persist indefinitely in superconductors?
3. to enliven and humanize the text with brief sketches of the historical development of 20th century physics, including anecdotes and quotations from the key figures as well as interesting photographs of noted scientists and original apparatus.

COVERAGE

Topics.   The material covered in this book is concerned with fundamental topics in modern physics with extensive applications in science and engineering. Chapter 1 presents an introduction to the special theory of relativity and includes an essay on general relativity. Chapters 2 through 4 present an historical and conceptual introduction to early developments in quantum theory, including a discussion of the experiments that show the necessity of quantum theory. Chapters 5 through 9 are an introduction to the real "nuts and bolts" of quantum mechanics, covering the Schrödinger equation, tunneling phenomena, the hydrogen atom, multielectron atoms, and a survey of statistical mechanics. The remainder of the book consists, for the most part, of applications of the theory set forth in earlier chapters to more specialized areas of modern physics. In particular, Chapter 10 discusses the physics of molecules, Chapter 11 is an introduction to the physics of solids, Chapter 12 (new to this edition) is entirely devoted to magnetism and superconductivity, Chapters 13 and 14 cover nuclear physics and methods of obtaining energy from nuclear reactions, and Chapter 15 treats the subjects of particle physics and cosmology.

CHANGES TO THE SECOND EDITION

The second edition contains two major changes from the first edition: *First,* this edition has been extensively rewritten in order to clarify difficult concepts, aid understanding, and bring the text up to date with the rapidly developing technical applications of quantum physics. Artwork and the order of presentation of certain topics have also been revised to help in this process. Typically, each chapter contains about five new worked examples, some followed by student exercises, and five to ten new end-of-chapter questions and problems. Several timely new essays have been added, and all other essays have been revised and updated. Finally, the *Suggestions for Further Reading* at the end of each chapter have been revised in many cases, featuring more recent references.

Second, this edition contains many computer demonstration examples (MP DESKTOP) developed for this text by one of the authors, Curt Moyer. We think these simulations clarify, enliven, and complement the analytical solutions presented in the text. Many chapters contain several strategically placed examples and about five end-of-chapter computer problems that use MP DESKTOP, specifically designed to clarify or extend text developments. In our opinion, learning is enhanced by the use of this software, but those who wish to skip over the computer examples and related end-of-chapter problems may do so without interrupting the logical flow of the text.

Specific changes by chapter in this second edition are as follows:

Chapter 1 has been extensively rewritten. New sections on relativistic collisions, the relativistic Doppler shift, and mass as a form of energy have been added. The concept of speed-dependent mass has been eliminated, and a revised essay on general relativity by Clifford M. Will is included.

Chapter 2 contains a revised and clarified treatment of the photoelectric effect, as well as a more concise treatment of the Rayleigh-Jeans and Planck blackbody laws.

Chapter 3 has been reorganized so that the empirical results on spectral lines are presented before the Bohr theory describing these lines. There is also a completely rewritten section on Bohr's correspondence principle (Section 3.4), and a new section on the Franck-Hertz experiment (Section 3.5).

Chapter 4 contains a most important new section, introducing our unique simulation package, the MP DESKTOP, which is applied to modeling wave groups. This chapter also contains a clarified treatment of wave groups and a new section on Fourier integrals (Section 4.4).

Chapter 5 has many new examples using the MP DESKTOP to find bound state energies and wavefunctions for a variety of one-dimensional potentials.

Chapter 6 now includes revised and clarified explanations of key points in tunneling phenomena, including more details on imposing boundary conditions. New examples dealing with transmission resonances and the use of MP DESKTOP to study transmission through barriers are included. Simulations of ammonia inversion and nuclear decay show the power of MP DESKTOP to depict the time evolution of nonstationary states.

Chapter 7 offers a more detailed explanation of the solution to the Schrödinger equation for atomic hydrogen. The MP DESKTOP is used to illustrate electron distributions and the directional character of hydrogen wavefunctions.

Chapter 8 features a new section on x-ray spectra and Moseley's law (Section 8.7), which is a natural extension of Section 8.6 on the periodic table. This chapter also offers an interesting application of the MP DESKTOP in finding the energy and wavefunctions of screened alkali atoms.

Chapter 9 has a new section on Einstein's theory of specific heat and phonons. This section shows how Einstein broadened quantization to include atomic vibrations in solids as well as light. This chapter also features an exciting new essay by Steven Chu dealing with laser cooling of atoms and recent observations of Bose condensation of rubidium atoms.

Chapter 10 contains new MP DESKTOP examples dealing with diatomic molecules. Both the rotation-vibration transition energies and wavefunctions of diatomic molecules are determined by the software.

Chapter 11 has been substantially reorganized compared to the first edition. Superconductivity has been moved to Chapter 12, and a section dealing with lasers (mainly semiconductor lasers) has been added. This is warranted because the rather dated first edition chapter on lasers has been deleted. Chapter 11 also features two interesting essays: a new essay on the history of the invention of the laser by Sam Marshall and an updated essay on solar cells by John Meakin.

Chapter 12 is entirely new in this second edition. This chapter features an up-to-date survey of superconductivity, including high-temperature superconductivity, as well as descriptions of many superconducting devices. There is also a good introduction to magnetism in matter, which is needed to understand superconductivity. The chapter concludes with an informative essay on superconducting devices by Clark A. Hamilton.

Chapter 13 includes clarification and expanded discussion of the neutron, the exchange force (the glue that holds the nucleus together), the energetics of alpha decay, and the liquid-drop model of the nucleus.

Chapter 14 presents updated fusion results with extensively rewritten sections on both tokomak and laser fusion schemes. New information has been added on radiation detectors in Section 14.10. There is also some reorganization of material in this chapter, primarily the inclusion of nuclear reactions, which more logically fit here rather than in Chapter 13.

Chapter 15 has been extensively rewritten and reorganized. The biggest change in this chapter is in the integration of cosmology with elementary particles. This provides a link between the very small and the very large and a cohesive flow that was missing in the first edition. There has also been extensive clarification and expansion of the subjects of antiparticles, resonance particles, energy considerations in particle production (new Section 15.7), the eightfold way (new Section 15.8), quarks (Section 15.9), and quantum chromodynamics (Section 15.10). Chapter 15 also contains a rewritten section on cosmology with an entirely new derivation of the spectrum emitted by a receding black body. The chapter concludes with an essay by Melissa Franklin and David Kestenbaum on the discovery of the top quark. The essay explains how the most complicated machines of the 20th century—particle accelerators and detectors—work.

FEATURES OF THIS TEXT

Computer Examples and Problems.   Many chapters contain several examples and four to five new problems requiring the use of our unique simulation software, MP DESKTOP. A diskette containing the MP DESKTOP software package is packaged with each text. MP DESKTOP is a sophisticated interactive learning tool with considerable flexibility and scope. Using MP DESKTOP, students can compose matter-wave packets and study their time evolution, find stationary state energies and wavefunctions, and determine the probability for particle transmission and reflection from nearly any potential well or barrier. The software requires an IBM PC (or close compatible) with a VGA display and color monitor. (A 486 processor with a clock speed of at least 50 MHz is recommended for some exercises.)

Style.   We have attempted to write this book in a style that is clear, logical, and succinct, yet somewhat informal and relaxed in the hope that students will find the text appealing and enjoyable to read. All new terms have been carefully defined, and we have tried to avoid jargon.

Worked Examples.   A large number of worked examples of varying difficulty are presented as an aid in understanding concepts. In many cases, these examples will serve as models for solving some end-of-chapter problems. The examples are set off with colored bars for ease of location, and most examples are given titles to describe their content.

Exercises Following Examples.   As an added feature, many of the worked examples are followed immediately by exercises with answers. These exercises are intended to make the textbook more interactive with the student, and to test immediately the student's understanding of key concepts and problem-solving techniques. The exercises represent extensions of the worked examples and are numbered in case the instructor wishes to assign them for homework.

Problems and Questions.   An extensive set of questions and problems is included at the end of each chapter. Most of the problems are listed by section topic. Answers to all odd-numbered problems are given at the end of the book. Problems span a range of difficulty and more challenging problems have colored numbers. Most of the questions serve to test the student's understanding of the concepts presented in that chapter, and many can be used to motivate classroom discussions.

Units. The international system of units (SI) is used throughout the text. Occasionally, where common usage dictates, other units are used (such as the angstrom, Å, and cm^{-1}, so commonly used by spectroscopists), but all such units are carefully defined in terms of SI units.

Chapter Format. Each chapter begins with a **preview,** which includes a brief discussion of chapter objectives and content. **Marginal notes** set in color are used to locate important concepts and equations in the text. Important statements are *italicized* or **highlighted,** and important equations are set in a colored box for added emphasis and ease of review. Each chapter concludes with a **summary** which reviews the important concepts and equations discussed in that chapter.

In addition, many chapters contain **special topic sections** which are clearly marked **optional.** These sections expose the student to slightly more advanced material either in the form of current interesting discoveries or as fuller developments of concepts or calculations discussed in that chapter. Many of these special topic sections will be of particular interest to certain student groups such as chemistry majors, electrical engineers, and physics majors.

Guest Essays. An important feature of this text is the inclusion of up-to-date and interesting material in the form of essays by guest authors. These essays cover a wide range of topics and are intended to convey an insider's view of exciting current developments in modern physics. Furthermore, the essay topics represent extensions and/or applications of the material discussed in the chapter in which they are included. Some of the essay topics covered are recent developments in general relativity, the scanning tunneling microscope, superconducting devices, the history of the laser, laser cooling of atoms, solar cells, and how the top quark was detected.

Mathematical Level. Students using this text should have completed a comprehensive one-year calculus course, as calculus is used throughout the text. However, we have made an attempt to keep physical ideas foremost so as not to obscure our presentations with overly elegant mathematics. Most steps are shown when basic equations are developed, but exceptionally long and detailed proofs which interrupt the flow of physical arguments have been placed in appendices.

Appendices and Endpapers. The appendices in this text serve several purposes. End-of-chapter appendices consist of lengthy derivations of key equations. These appendices have been placed at the end of certain chapters to avoid interrupting the main flow of arguments. Other appendices located at the end of the book contain physical constants, conversion factors, a table of atomic masses, a list of Nobel prize winners and their achievements, and instructions for using MP DESKTOP. The endpapers inside the front cover of the book contain important physical constants and standard abbreviations of units used in the book for quick reference, while a periodic table is included in the rear cover endpapers.

Ancillaries. The ancillaries available with this text include an Instructor's Solutions Manual, consisting of solutions to all problems in the text, and a set of overhead transparencies. Saunders College Publishing may provide such complimentary instructional aids and supplements or supplement packages to those adopters qualified under their adoption policy. Please contact your sales representative for more information. If as an adopter or potential user you receive supplements you do not need, please return them to your sales representative or send them to:

Attn: Returns Department
Troy Warehouse
465 South Lincoln Drive
Troy, MO 63379

TEACHING OPTIONS

As noted earlier, the text may be subdivided into two basic parts: Chapters 1 through 9, which contain an introduction to relativity, quantum physics, and statistical physics, and Chapters 10 through 15, which treat applications to molecules, the solid state, superconductivity, nuclear physics, elementary particles, and cosmology. It is suggested that the first part of the book be covered sequentially. However, Chapter 1 on relativity may actually be covered at any time because $E^2 = p^2c^2 + m^2c^4$ is the only formula from this chapter which is essential for subsequent chapters. Chapters 10 through 15 are independent of one another and can be covered in any order with one exception: Chapter 14 on Nuclear Physics Applications should follow Chapter 13 on Nuclear Structure.

A traditional sophomore or junior level modern physics course for science, mathematics, and engineering students should cover most of Chapters 1 through 9 and several of the remaining chapters, depending on student major. For example, an audience consisting mainly of electrical engineering students might cover most of Chapters 1 through 9 with particular emphasis on tunneling and tunneling devices in Chapter 6, the Fermi-Dirac distribution in Chapter 9, semiconductors, superconductivity, and superconducting devices in Chapters 11 and 12, and particle detectors in Chapter 14. Chemistry and chemical engineering majors could cover most of Chapters 1 through 9 with special emphasis on atoms in Chapter 8, classical and quantum statistics in Chapter 9, and molecular bonding and spectroscopy in Chapter 10. Mathematics and physics majors should pay special attention to the unique development of operator methods and the concept of sharp and fuzzy observables introduced in Chapter 5. The deep connection of sharp observables with classically conserved quantities and the powerful role of sharp observables in shaping the form of system wavefunctions is developed more fully in Chapter 7.

Our experience with this material has shown that there is more material contained in this book than can be covered in a standard three-credit-hour course. For this reason, one has to "pick-and-choose" from topics in the second part of the book as noted earlier. However, the text can also be used in a two-semester sequence with some supplemental material, such as one of many monographs on relativity, and selected readings from other sources in the areas of solid state, nuclear, and elementary particle physics. These suggested readings are listed at the end of each chapter.

ACKNOWLEDGMENTS

We wish to thank the users of the first edition who freely shared with us their comments and criticisms. We remain indebted to the reviewers of the first edition. In preparing this second edition we owe a special debt of gratitude to Robert Bauman, Paul Finkler, Edward Gibson, M. Y. Han, Sam Marshall, and K. To Yue for their enormously helpful and constructive reviews and suggestions. We thank Sam Marshall for his contributions to the manuscript and the solutions manual. We thank Alice Presley for checking solutions to end-of-chapter problems and for her work in preparing the Instructor's Manual. In addition, we wish to thank all of the following reviewers for their helpful comments and suggestions for improving the manuscript and the software:

Stephen G. Alexander, Miami University

Randall Caton, Christopher Newport University

Don Chodrow, James Madison University

Sumner P. Davis, University of California, Berkeley

Paul Finkler, University of Nebraska—Lincoln

Cheng-Ming Fou, University of Delaware

Richard Gass, University of Cincinnati

Edward F. Gibson, California State University, Sacramento

William Graham, University of Pennsylvania

M. Y. Han, Duke University

Randy Harris, University of California, Davis

Roger Judge, University of California, San Diego

Sanford Kern, Colorado State University

Arnold Kleinstein, WorldView Software

Peter D. Loly, University of Manitoba

Richard Madey, Hampton University

S. Douglas Marcum, Miami University

Edward Mooney, Jr., Youngstown State University

Warren F. Perger, Michigan Technological University

William Pong, University of Hawaii at Manoa

Richard A. Saenz, California Polytechnic State University, San Luis Obispo

Leo Sartori, University of Nebraska—Lincoln

Thor F. Stromberg, New Mexico State University

K. To Yue, Emory University

We are grateful to the many students at Utica College, Clarkson University, and James Madison University who provided helpful comments and criticisms of the manuscript while it was tested in the classroom. We thank the professional staff at Saunders College Publishing for their fine work during the development and production of this text, especially Sally Kusch, Susan Dust Pashos, Carol Bleistine, Dena Digilio-Betz, Tim Frelick, Charlene Squibb, John Vondeling, and Marjorie Waldron. We thank Linda Davoli and Mary Patton for copy editing the manuscript, and Suzon O. Kisler for her unerring and valuable reference work. We also wish to acknowledge valuable discussions with many people, including Don Chodrow, Kevin Giovanetti, John R. Gordon, William Ingham, Christopher Moses, Joe Rudmin, and Dorn Peterson. We thank all of the authors of the essays included in this text for their valuable contributions and cooperation: Steven Chu, Melissa Franklin, Roger A. Freedman, Clark A. Hamilton, Paul K. Hansma, David Kestenbaum, Sam Marshall, John Meakin, and Clifford M. Will. Finally, we thank all of our families for their patience and continual support.

Raymond A. Serway
James Madison University
Harrisonburg, VA 22807

Clement Moses
Utica College of Syracuse University
Utica, NY 13502

Curt A. Moyer
Clarkson University
Potsdam, NY 13699

July 1996

Contents Overview

Contents

MP DESKTOP
Examples and Problems

Chapter 10

Chapter 11

1. A. Piccard	11. L. Brillouin	21. I. Langmuir
2. E. Henriot	12. P. Debye	22. M. Planck
3. P. Ehrenfest	13. M. Knudsen	23. M. Curie
4. E. Herzen	14. W.L. Bragg	24. H.A. Lorentz
5. Th. de Donder	15. H.A. Kramers	25. A. Einstein
6. E. Schroedinger	16. P.A.M. Dirac	26. P. Langevin
7. E. Verschaffelt	17. A.H. Compton	27. C.E. Guye
8. W. Pauli	18. L.V. de Broglie	28. C.T.R. Wilson
9. W. Heisenberg	19. M. Born	29. O.W. Richardson
10. R.H. Fowler	20. N. Bohr	

The "architects" of modern physics. This unique photograph shows many eminent scientists who participated in the fifth international congress of physics held in 1927 by the Solvay Institute in Brussels. At this and similar conferences, held regularly from 1911 on, scientists were able to discuss and share the many dramatic developments in atomic and nuclear physics. This elite company of scientists includes fifteen Nobel prize winners in physics and three in chemistry. *(Photograph courtesy of AIP Niels Bohr Library)*

1

Relativity

Chapter Outline

At the end of the 19th century, scientists believed that they had learned most of what there was to know about physics. Newton's laws of motion and his universal theory of gravitation, Maxwell's theoretical work in unifying electricity and magnetism, and the laws of thermodynamics and kinetic theory were highly successful in explaining a wide variety of phenomena.

However, at the turn of the 20th century, a major revolution shook the world of physics. In 1900 Planck provided the basic ideas that led to the formulation of the quantum theory, and in 1905 Einstein formulated his brilliant special theory of relativity. The excitement of the times is captured in Einstein's own words: "It was a marvelous time to be alive." Both ideas were to have a profound effect on our understanding of nature. Within a few decades, these theories inspired new developments and theories in the fields of atomic physics, nuclear physics, and condensed matter physics.

Although modern physics has been developed during this century and has led to a multitude of important technological achievements, the story is still incomplete. Discoveries will continue to evolve during our lifetime, many of which will deepen or refine our understanding of nature and the world around us. It is still a "marvelous time to be alive."

1.1 SPECIAL RELATIVITY

Light waves and other forms of electromagnetic radiation travel through free space at the speed $c = 3.00 \times 10^8$ m/s. As we shall see in this chapter, the speed of light is an upper limit for the speeds of particles and mechanical waves.

Most of our everyday experiences and observations deal with objects that move at speeds much less than that of light. Newtonian mechanics and early ideas on space and time were formulated to describe the motion of such objects, and this formalism is very successful for describing a wide range of phenomena. Although Newtonian mechanics works very well at low speeds, it fails when applied to particles whose speeds approach that of light. Experimentally, one can test the predictions of Newtonian theory at high speeds by accelerating an electron through a large electric potential difference. For example, it is possible to accelerate an electron to a speed of $0.99c$ by using a potential difference of several million volts. According to Newtonian mechanics, if the potential difference (as well as the corresponding energy) is increased by a factor of 4, then the speed of the electron should be doubled to $1.98c$. However, experiments show that the speed of the electron—as well as the speeds of all other particles in the universe—always remains less than the speed of light, regardless of the size of the accelerating voltage. Because it places no upper limit on the speed that a particle can attain, Newtonian mechanics is contrary to modern experimental results and is therefore clearly a limited theory.

In 1905, at the age of only 26, Albert Einstein published his *special theory of relativity*. Regarding the theory, Einstein wrote,

> The relativity theory arose from necessity, from serious and deep contradictions in the old theory from which there seemed no escape. The strength of the new theory lies in the consistency and simplicity with which it solves all these difficulties, using only a few very convincing assumptions. . . .[1]

Although Einstein made many important contributions to science, the theory of relativity alone represents one of the greatest intellectual achievements of the 20th century. With this theory, one can correctly predict experimental observations over the range of speeds from $v = 0$ to speeds approaching the speed of light. Newtonian mechanics, which was accepted for over 200 years, is in fact a limiting case of Einstein's special theory of relativity. This chapter gives an introduction to the special theory of relativity which deals with the analysis of physical events from coordinate systems moving uniformly with respect to one another. A discussion of general relativity that describes events for coordinate systems undergoing general or accelerated motion is presented in an interesting essay written by Clifford Will at the end of the chapter.

In this chapter we show that the special theory of relativity is based on two basic postulates:

1. The laws of physics are the same in all reference systems which move uniformly with respect to one another. That is, basic laws such as $\Sigma \boldsymbol{F} = d\mathbf{p}/dt$ have the same mathematical form for all observers moving at constant velocity with respect to one another.

---

[1] A. Einstein and L. Infeld, *The Evolution of Physics,* New York, Simon and Schuster, 1961.

2. The speed of light in vacuum is always measured to be 3×10^8 m/s, and the measured value is independent of the motion of the observer or of the motion of the source of light. That is, the speed of light is the same for all observers moving at constant velocities.

Special relativity covers phenomena such as the slowing down of clocks and the contraction of lengths in moving reference frames as measured by an observer in another reference frame. We also discuss the relativistic forms of momentum and energy and some consequences of the famous mass-energy equivalence formula, $E = mc^2$.

Although it is well known that relativity plays an essential role in theoretical physics, it also has practical applications, including the design of accelerators and other devices that use high-speed particles. Note that these devices simply will not work if designed according to Newtonian mechanics! We shall have occasion to use the outcomes of relativity in some subsequent chapters of this text.

1.2   THE PRINCIPLE OF RELATIVITY

To describe a physical event, it is necessary to establish a frame of reference, such as one that is fixed in the laboratory. You should recall from your studies in mechanics that Newton's laws are valid in *all* inertial frames of reference. Because an inertial frame of reference is defined as one in which Newton's first law is valid, one can say that *an inertial frame of reference or an inertial system is a system in which a free body is not accelerating.* Furthermore, any system moving with constant velocity with respect to an inertial system must also be an inertial system. There is no preferred frame of reference. This means that the laws that describe an experiment performed in a vehicle moving with uniform velocity will be identical to the laws that describe the same experiment performed in another reference frame.

Inertial frame of reference

According to the **principle of Newtonian relativity,** the laws of mechanics must be the same in all inertial frames of reference. For example, if you perform an experiment while at rest in a laboratory, and an observer in a passing truck moving with constant velocity performs the same experiment, the observations of the person in the moving truck must agree with your observations. Specifically, in the laboratory or in the truck a ball thrown up rises and returns to the thrower's hand. Moreover, both events are measured to take the same time in the truck or in the laboratory, and Newton's second law may be used in both frames to compute this time. Although these experiments may look different to different observers (see Fig. 1.1, in which the Earth observer sees a different path for the ball), all observers agree on the validity of Newton's laws and principles such as conservation of energy and conservation of momentum. This implies that no experiment involving the laws of mechanics can detect any difference between the two inertial frames. The only thing that can be detected is the relative motion of one frame with respect to the other. That is, the notion of *absolute* motion through space is meaningless, as is the notion of a preferred reference frame.

(a)

(b)

Figure 1.1 (a) The observer in the truck sees the ball move in a vertical path when thrown upward. (b) The Earth observer views the path of the ball as a parabola.

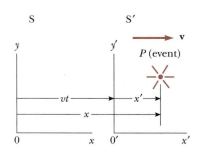

Figure 1.2 An event occurs at a point P. The event is observed by two observers in inertial frames S and S′, in which S′ moves with a velocity **v** relative to S.

Galilean transformation of coordinates

Suppose that some physical phenomenon such as a light bulb flash, which we call an event, occurs in an inertial system. The event's location and time of occurrence can be specified by the coordinates (x, y, z, t). We would like to be able to transform the space and time coordinates of the event from one inertial system to another moving with uniform relative velocity. This is accomplished by using the so-called *Galilean transformation,* which owes its origin to Galileo.

Consider two inertial systems S and S′, as in Figure 1.2. The system S′ moves with a constant velocity **v** along the xx' axes, where **v** is measured relative to the system S. Clocks in S and S′ are synchronized; hence the origins of S and S′ coincide at $t = t' = 0$. We assume that an event occurs at the point P. An observer in the system S would describe the event with space-time coordinates (x, y, z, t), whereas an observer in system S′ would use (x', y', z', t') to describe the same event. As we can see from Figure 1.2, these coordinates are related by the equations

$$
\begin{aligned}
x' &= x - vt \\
y' &= y \\
z' &= z \\
t' &= t
\end{aligned}
\tag{1.1}
$$

These equations constitute what is known as a **Galilean transformation of coordinates.** Note that the fourth coordinate, time, is *assumed* to be the same in both inertial systems. That is, within the framework of classical mechanics, all clocks run at the same rate regardless of their velocity, so that the time at which an event occurs for an observer in S is the same as the time for the same event in S′. Consequently, the time interval between two successive events should be the same for both observers. Although this assumption may seem obvious, it turns out to be *incorrect* when treating situations in which v is comparable to the speed of light. In fact, this point represents one of the most profound differences between Newtonian concepts and the ideas contained in Einstein's theory of relativity.

Now suppose two events are separated by a distance dx and a time interval dt as measured by an observer in S. It follows from Equations 1.1 that the corresponding displacement dx' measured by an observer in S′ is given by $dx' = dx - v\, dt$, where dx is the displacement measured by an observer in S. Because $dt = dt'$, we find that

$$\frac{dx'}{dt} = \frac{dx}{dt} - v$$

or

$$u'_x = u_x - v \qquad (1.2)$$

Galilean addition law for velocities

where u_x and u'_x are the instantaneous velocities of the object relative to S and S′, respectively.

This result, which is called the **Galilean addition law for velocities** (or Galilean velocity transformation), is used in everyday observations and is consistent with our intuitive notions of time and space. However, we will soon see that these results from mechanics lead to serious contradictions when applied to electromagnetic waves.

The Speed of Light

It is natural to ask whether the concept of Newtonian relativity and the Galilean addition law for velocities in mechanics also apply to electricity, magnetism, and optics. Recall that Maxwell in the 1860s showed that the speed of light in free space was given by $c = (\mu_0 \epsilon_0)^{-1/2} = 3.00 \times 10^8$ m/s. Physicists of the late 1800s thought that light waves (like sound and water waves) required a definite medium in which to move called the *ether*[2] and that the speed of light was c only in a special, absolute frame at rest with respect to the ether. In any other frame moving at speed v relative to the absolute ether frame, the Galilean addition law was expected to hold. Thus, the speed of light in this other frame was expected to be $c - v$ for light traveling in the same direction as the frame, $c + v$ for light traveling opposite to the frame, and in between these two values for light moving in an arbitrary direction with respect to the moving frame.

Because the existence of a preferred, absolute ether frame would show that light was similar to other classical waves (in requiring a medium) and that Newtonian ideas of an absolute or special coordinate frame were true, considerable importance was attached to establishing the existence of the ether frame. Because the speed of light is enormous, experiments involving light traveling in media moving at then attainable laboratory speeds had not been capable of detecting small changes of the size of $c \pm v$ prior to the late 1800s. Scientists of the period, realizing that the Earth moved rapidly around the Sun at 30 km/s, shrewdly decided to use the Earth itself as the moving frame in an

---

[2] It was proposed by Maxwell that light and other electromagnetic waves were waves in a luminiferous ether, which was present everywhere, even in empty space, but which allowed planets and other massive objects to pass freely through it without resistance.

Figure 1.3 If the velocity of the ether wind relative to the Earth is **v**, and **c** is the velocity of light relative to the ether, the speed of light relative to the Earth is (a) $c + v$ in the downwind direction, (b) $c - v$ in the upwind direction, and (c) $(c^2 - v^2)^{1/2}$ in the direction perpendicular to the wind.

attempt to improve their chances of detecting these small changes in light velocity.

From our point of view of observers fixed on Earth, we may say that we are stationary and that the absolute ether frame containing the medium for light propagation moves past us with speed v. Determining the speed of light under these circumstances is just like determining the speed of an aircraft in a moving air current or wind, and consequently we speak of an "ether wind" blowing through our apparatus fixed to the Earth. A direct method for detecting an ether wind would be to measure its influence on the speed of light determined by an apparatus fixed in a frame of reference on Earth. If **v** is the velocity of the ether relative to the Earth, then the speed of light should have its maximum value, $c + v$, when propagating downwind as shown in Figure 1.3a. Likewise, the speed of light should have its minimum value, $c - v$, when propagating upwind as in Figure 1.3b, and some intermediate value, $(c^2 - v^2)^{1/2}$, in the direction perpendicular to the ether wind as in Figure 1.3c. If the Sun is assumed to be at rest in the ether, then the velocity of the ether wind would be equal to the orbital velocity of the Earth around the Sun, which has a magnitude of about 3×10^4 m/s compared to $c = 3 \times 10^8$ m/s. Thus, the change in the speed of light would be about 1 part in 10^4 for measurements in the upwind or downwind directions, and changes of this size should be detectable. However, as we show in the next section, all attempts to detect such changes and establish the existence of the ether (and hence of the absolute frame) proved futile!

1.3 THE MICHELSON-MORLEY EXPERIMENT

The most famous experiment designed to detect small changes in the speed of light was performed in 1887 by Albert A. Michelson (1852–1931) and Edward W. Morley (1838–1923).[3] We should state at the outset that the outcome of the experiment was *negative*, thus contradicting the ether hypothesis. The experiment was designed to determine the velocity of the Earth with respect to the hypothetical ether. The experimental tool these pioneers used was the Michelson interferometer shown in Figure 1.4. One of the arms of the interferometer was aligned along the direction of the motion of the Earth through the ether. The Earth moving through the ether would be equivalent to the ether flowing past the Earth in the opposite direction with speed v as shown in Figure 1.4. This ether wind blowing in the opposite direction should cause the speed of light measured in the Earth's frame of reference to be $c - v$ as it approaches the mirror M_2 in Figure 1.4 and $c + v$ after reflection. The speed v is the speed of the Earth through space, and hence the speed of the ether wind, and c is the speed of light in the ether frame. The two beams of light reflected from M_1 and M_2 would recombine, and an interference pattern consisting of alternating dark and bright bands, or fringes, would be formed.

During the experiment, the interference pattern was observed while the interferometer was rotated through an angle of 90°. This rotation would change the speed of the ether wind along the direction of the arms of the interferometer. The effect of this rotation should have been to cause the fringe pattern to shift slightly but measurably. Measurements failed to show any change in the

[3] A. A. Michelson and E. W. Morley, *Am. J. Sci.* 134:333, 1887.

M₁

Arm 2 L Ether wind
              $\mathbf{v}$

Source

M₀ Arm 1

M₂

L

Telescope

Figure 1.4   Diagram of the Michelson interferometer. According to the ether wind concept, the speed of light should be $c - v$ as the beam approaches mirror M_2 and $c + v$ after reflection.

interference pattern! The Michelson-Morley experiment was repeated by other researchers under various conditions and at different times of the year when the ether wind was expected to have changed direction and magnitude, but the results were always the same: *no fringe shift of the magnitude required was ever observed.*[4]

The negative results of the Michelson-Morley experiment not only contradicted the ether hypothesis, they also meant that it was impossible to measure the absolute velocity of the Earth with respect to the ether frame. However, as we shall see in the next section, Einstein offered a postulate for his theory of relativity that places quite a different interpretation on the null results. In later years when more was known about the nature of light, the idea of an ether that permeates all of space was relegated to the ash heap of worn-out concepts. Light is now understood to be a phenomenon that *requires no medium for its propagation.* As a result, the idea of an ether in which these waves could travel became unnecessary.

Details of the Michelson-Morley Experiment

To understand the outcome of the Michelson-Morley experiment, let us assume that the interferometer shown in Figure 1.4 has two arms of equal length L. First consider the beam traveling parallel to the direction of the ether wind, which is taken to be horizontal in Figure 1.4. According to Newtonian mechanics, as the beam moves to the right, its speed is reduced by the wind and its speed with respect to the Earth is $c - v$. On its return journey, as the light beam moves to the left downwind, its speed with respect to the Earth is $c + v$. Thus, the time of travel to the right is $L/(c - v)$, and the time of travel to the

---

[4] From an Earth observer's point of view, changes in the Earth's speed and direction in the course of a year are viewed as ether wind shifts. In fact, even if the speed of the Earth with respect to the ether were zero at some point in the Earth's orbit, six months later the speed of the Earth would be 60 km/s with respect to the ether, and one should find a clear fringe shift. None has ever been observed, however.

left is $L/(c + v)$. The total time of travel for the round trip along the horizontal path is

$$t_1 = \frac{L}{c + v} + \frac{L}{c - v} = \frac{2Lc}{c^2 - v^2} = \frac{2L}{c}\left(1 - \frac{v^2}{c^2}\right)^{-1}$$

Now consider the light beam traveling perpendicular to the wind, which is the vertical direction in Figure 1.4. Because the speed of the beam relative to the Earth is $(c^2 - v^2)^{1/2}$ in this case (see Fig. 1.3c) then, the time of travel for each half of this trip is $L/(c^2 - v^2)^{1/2}$, and the total time of travel for the round trip is

$$t_2 = \frac{2L}{(c^2 - v^2)^{1/2}} = \frac{2L}{c}\left(1 - \frac{v^2}{c^2}\right)^{-1/2}$$

Thus, the time difference between the light beam traveling horizontally and the beam traveling vertically is

$$\Delta t = t_1 - t_2 = \frac{2L}{c}\left[\left(1 - \frac{v^2}{c^2}\right)^{-1} - \left(1 - \frac{v^2}{c^2}\right)^{-1/2}\right]$$

Because $v^2/c^2 \ll 1$, this expression can be simplified by using the following binomial expansions after dropping all terms higher than second order:

$$(1 - x)^n \approx 1 - nx \qquad \text{(for } x \ll 1)$$

In our case, $x = v^2/c^2$, and we find

$$\Delta t = t_1 - t_2 \approx \frac{Lv^2}{c^3} \tag{1.3}$$

The two light beams start out in phase and return to form an interference pattern. Let us assume that the interferometer is adjusted for parallel fringes and that a telescope is focused on one of these fringes. The time difference between the two light beams gives rise to a phase difference between the beams, producing an interference pattern when they combine at the position of the telescope. A difference in the pattern should be detected by rotating the interferometer through 90° in a horizontal plane, such that the two beams exchange roles (Fig. 1.6). This results in a net time difference of twice that given by

Figure 1.5 Albert A. Michelson (1852–1931). A German-American physicist, Michelson invented the interferometer and spent much of his life making accurate measurements of the speed of light. In 1907 he was the first American to be awarded the Nobel prize, which he received for his work in optics. His most famous experiment, conducted with Edward Morley in 1887, implied that it was impossible to measure the absolute velocity of the Earth with respect to the ether. Subsequent work by Einstein in his special theory of relativity eliminated the ether concept by assuming that the speed of light has the same value in all inertial reference frames. *(SPL/Photo Researchers)*

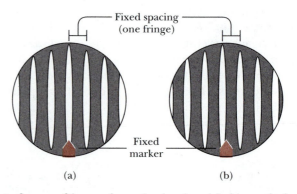

(a)   (b)

Figure 1.6 Interference fringe schematic showing (a) fringes before rotation and (b) expected fringe shift after a rotation of the interferometer by 90°.

Equation 1.3. Thus, the path difference corresponding to this time difference is

$$\Delta d = c(2\,\Delta t) = \frac{2Lv^2}{c^2}$$

The corresponding fringe shift is equal to this path difference divided by the wavelength of light, λ, because a change in path of 1 wavelength corresponds to a shift of one fringe.

$$\text{Shift} = \frac{2Lv^2}{\lambda c^2} \tag{1.4}$$

In the experiments by Michelson and Morley, each light beam was reflected by mirrors many times to give an increased effective path length L of about 11 m. Using this value, and taking v to be equal to 3×10^4 m/s, the speed of the Earth about the Sun, gives a path difference of

$$\Delta d = \frac{2(11\ \text{m})(3 \times 10^4\ \text{m/s})^2}{(3 \times 10^8\ \text{m/s})^2} = 2.2 \times 10^{-7}\ \text{m}$$

This extra distance of travel should produce a noticeable shift in the fringe pattern. Specifically, using light of wavelength 500 nm, we find a fringe shift for rotation through 90° of

$$\text{Shift} = \frac{\Delta d}{\lambda} = \frac{2.2 \times 10^{-7}\ \text{m}}{5.0 \times 10^{-7}\ \text{m}} \approx 0.40$$

The instrument used by Michelson and Morley had the capability of detecting a shift in the fringe pattern as small as 0.01 fringe. However, *they detected no shift in the fringe pattern*. Since then, the experiment has been repeated many times by various scientists under various conditions, and no fringe shift has ever been detected. Thus, it was concluded that one cannot detect the motion of the Earth with respect to the ether.

Many efforts were made to explain the null results of the Michelson-Morley experiment and to save the absolute ether frame concept and the Galilean addition law for the velocity of light. Because all these proposals have been shown to be wrong, we consider them no further here and turn instead to an auspicious proposal made by George F. Fitzgerald and Hendrik A. Lorentz. In the 1890s, Fitzgerald and Lorentz tried to explain the null results by making the following ad hoc assumption. They proposed that the length of an object moving at speed v would contract along the direction of travel by a factor of $\sqrt{1 - v^2/c^2}$. The net result of this contraction would be a change in length of one of the arms of the interferometer such that no path difference would occur as it was rotated.

Never in the history of physics were such valiant efforts devoted to trying to explain the absence of an expected result as those directed at the Michelson-Morley experiment. The stage was set for the brilliant Albert Einstein, who solved the problem in 1905 with his special theory of relativity.

1.4 POSTULATES OF SPECIAL RELATIVITY

In the previous section we noted the impossibility of measuring the speed of the ether with respect to the Earth and the failure of the Galilean addition law for speeds near that of light. In 1905, Albert Einstein (Fig. 1.7) proposed a theory that boldly removed these difficulties and at the same time completely altered our notion of space and time.[5] Einstein based his special theory of relativity on two postulates.

1. **The Principle of Relativity:** All the laws of physics are the same in all inertial reference frames.

Postulates of special relativity

2. **The Constancy of the Speed of Light:** The speed of light in vacuum has the same value, $c = 3.00 \times 10^8$ m/s, in all inertial frames, regardless of the velocity of the observer or the velocity of the source emitting the light.

Figure 1.7 Albert Einstein playing his beloved fiddle, 1941. *(Hansel Micth, Life Magazine © Time, Inc.)*

[5] A. Einstein, "On the Electrodynamics of Moving Bodies," *Ann. Physik* 17: 891, 1905. For an English translation of this article and other publications by Einstein, see the book by H. Lorentz, A. Einstein, H. Minkowski, and H. Weyl, *The Principle of Relativity*, Dover, 1958.

The first postulate asserts that *all* the laws of physics, those dealing with electricity and magnetism, optics, thermodynamics, mechanics, and so on, will be the same in all coordinate frames moving with constant velocity relative to one another. This postulate is a sweeping generalization of Newton's concept of relativity that refers only to the laws of mechanics. From an experimental point of view, Einstein's principle of relativity means that any kind of experiment (measuring the velocity of light for example) performed in one laboratory at rest must give the same result when performed in another laboratory moving at constant velocity relative to the first one. Hence, no preferred inertial system exists, and it is impossible to detect absolute motion.

Note that postulate 2, the principle of the constancy of the speed of light, is consistent with postulate 1: If the speed of light was not the same in all inertial frames but was c in only one, it would be possible to distinguish between inertial frames, and one could identify a preferred, absolute frame in contradiction to postulate 1. Postulate 2 also does away with the problem of measuring the speed of the ether by essentially denying the existence of the ether and boldly asserting that light always moves with speed c with respect to any inertial observer whatsoever.

BIOGRAPHY

ALBERT EINSTEIN
(1879–1955)

Albert Einstein, one of the greatest physicists of all time, was born in Ulm, Germany. As a child, Einstein was very unhappy with the discipline of German schools and completed his early education in Switzerland at age 16. Because he was unable to obtain an academic position following graduation from the Swiss Federal Polytechnic School in 1901, he accepted a job at the Swiss Patent Office in Berne. During his spare time, he continued his studies in theoretical physics. In 1905, at the age of 26, he published four scientific papers that revolutionized physics. One of these papers, which won him the Nobel prize in 1921, dealt with the photoelectric effect. Another was concerned with Brownian motion, the irregular motion of small particles suspended in a liquid. The remaining two papers were concerned with what is now considered his most important contribution of all, the special theory of relativity. In 1915, Einstein published his work on the general theory of relativity, which relates gravity to the structure of space and time. One of the remarkable predictions of the theory is that strong gravitational forces in the vicinity of very massive objects cause light beams to deviate from a straight-line path. This and other predictions of the general theory of relativity have been experimentally verified (see the essay in this chapter by Clifford Will).

Einstein made many other important contributions to the development of modern physics, including the concept of the light quantum and the idea of stimulated emission of radiation, which led to the invention of the laser 40 years later. However, throughout his life he rejected the probabilistic interpretation of quantum mechanics when describing events on the atomic scale, in favor of a deterministic view. He is quoted as saying "God does not play dice with nature." This comment is reputed to have been answered by Niels Bohr, one of the founders of quantum mechanics, with "Don't tell God what to do!"

In 1933, Einstein left Germany under the power of the Nazis and spent his remaining years at the Institute for Advanced Study in Princeton, New Jersey. He devoted most of his later years to an unsuccessful search for a unified theory of gravity and electromagnetism.

Photo courtesy of AIP/Niels Bohr Library

Although the Michelson-Morley experiment was performed before Einstein published his work on relativity, it is not clear that Einstein was aware of the details of the experiment. Nonetheless, the null result of the experiment can be readily understood within the framework of Einstein's theory. According to his principle of relativity, the premises of the Michelson-Morley experiment were incorrect. In the process of trying to explain the expected results, we stated that when light traveled against the ether wind its speed was $c - v$, in accordance with the Galilean addition law for velocities. However, if the state of motion of the observer or of the source has no influence on the value found for the speed of light, one will always measure the value to be c. Likewise, the light makes the return trip after reflection from the mirror at a speed of c, and not with the speed $c + v$. Thus, the motion of the Earth should not influence the fringe pattern observed in the Michelson-Morley experiment, and a null result should be expected.[6]

If we accept Einstein's theory of relativity, we must conclude that relative motion is unimportant when measuring the speed of light. At the same time, we must alter our common sense notions of space and time and be prepared for some rather bizarre consequences.

1.5 CONSEQUENCES OF SPECIAL RELATIVITY

Almost everyone who has dabbled even superficially with science is aware of some of the startling predictions that arise because of Einstein's approach to relative motion. As we examine some of the consequences of relativity in this section, we shall find that they conflict with our basic notions of space and time. We restrict our discussion to the concepts of length, time, and simultaneity, which are quite different in relativistic mechanics and Newtonian mechanics. For example, we will find that *the distance between two points and the time interval between two events depend on the frame of reference in which they are measured.* That is, *there is no such thing as absolute length or absolute time in relativity.* Furthermore, *events at different locations that occur simultaneously in one frame are not simultaneous in another frame moving uniformly past the first.*

Before we discuss the consequences of special relativity, we must first understand how an observer in an inertial reference frame describes an event. We define an event as an occurrence described by three space coordinates and one time coordinate. In general, different observers in different inertial frames would describe the same event with different space-time coordinates.

The reference frame used to describe an event consists of a coordinate grid and a set of clocks situated at the grid intersections, as shown in Figure 1.8 in two dimensions. It is necessary that the clocks be synchronized. This can be

---

[6] Perhaps at this point you have rightly concluded that the Galilean transformation is incorrect; that is, it does not keep all the laws of physics in the same form for a frame at rest and one in uniform motion. The correct coordinate and time transformations that preserve the form of all physical laws in two coordinate systems moving uniformly with respect to each other are called Lorentz transformations. These are derived in Section 1.6. Although the Galilean transformation preserves the form of Newton's laws in two frames moving uniformly with respect to each other, Newton's laws of mechanics are limited laws that are only valid for low speeds. In general, Newton's laws must be replaced by Einstein's relativistic laws of mechanics, which hold for all speeds and are invariant under the Lorentz transformations.

Figure 1.8 In relativity, we use a reference frame consisting of a coordinate grid and a set of synchronized clocks.

accomplished in many ways with the help of light signals. For example, suppose an observer at the origin with a master clock sends out a pulse of light at $t = 0$. The light pulse takes a time r/c to reach a second clock, situated a distance r from the origin. Hence, the second clock will be synchronized with the clock at the origin if the second clock reads a time r/c at the instant the pulse reaches it. This procedure of synchronization assumes that the speed of light has the same value in all directions and in all inertial frames. Furthermore, the procedure concerns an event recorded by an observer in a specific inertial reference frame. Clocks in other inertial frames can be synchronized in a similar manner. An observer in some other inertial frame would assign different space-time coordinates to events, using another coordinate grid with another array of clocks.

Simultaneity and the Relativity of Time

A basic premise of Newtonian mechanics is that a universal time scale exists that is the same for all observers. In fact, Newton wrote that "Absolute, true, and mathematical time, of itself, and from its own nature, flows equably without relation to anything external." Thus, Newton and his followers simply took simultaneity for granted. In his special theory of relativity, Einstein abandoned this assumption. According to Einstein, *a time interval measurement depends on the reference frame in which the measurement is made.*

Einstein devised the following thought experiment to illustrate this point. A boxcar moves with uniform velocity, and two lightning bolts strike the ends of the boxcar, as in Figure 1.9a, leaving marks on the boxcar and ground. The marks left on the boxcar are labeled A' and B'; those on the ground are labeled A and B. An observer at O' moving with the boxcar is midway between A' and B', and a ground observer at O is midway between A and B. The events recorded by the observers are the light signals from the lightning bolts.

The two light signals reach the observer at O at the same time, as indicated in Figure 1.9b. This observer realizes that the light signals have traveled at the same speed over equal distances. Thus, observer O concludes that the events at A and B occurred simultaneously. Now consider the same events as viewed by the observer on the boxcar at O'. By the time the light has reached observer O, observer O' has moved as indicated in Figure 1.9b. Thus, the light signal

Figure 1.9   Two lightning bolts strike the ends of a moving boxcar. (a) The events appear to be simultaneous to the stationary observer at O, who is midway between A and B. (b) The events do not appear to be simultaneous to the observer at O', who claims that the front of the train is struck *before* the rear.

from B' has already swept past O', but the light from A' has not yet reached O'. According to Einstein, *observer O' must find that light travels at the same speed as that measured by observer O.* Therefore, observer O' concludes that the lightning struck the front of the boxcar before it struck the back. This thought experiment clearly demonstrates that the two events, which appear to O to be simultaneous, do not appear to O' to be simultaneous. In other words,

> two events that are simultaneous in one frame are in general not simultaneous in a second frame moving with respect to the first. That is, simultaneity is not an absolute concept, but one that depends on the state of motion of the observer.

At this point, you might wonder which observer is right concerning the two events. The answer is that *both are correct,* because the principle of relativity states that *there is no preferred inertial frame of reference.* Although the two observers reach different conclusions, both are correct in their own reference frame because the concept of simultaneity is not absolute. This, in fact, is the central point of relativity—any uniformly moving frame of reference can be used to describe events and do physics. However, observers in different inertial frames of reference will always measure different time intervals with their clocks and different distances with their meter sticks. Nevertheless, they will both agree on the forms of the laws of physics in their respective frames, because these laws must be the same for all observers in uniform motion. It is the alteration of time and space that allows the laws of physics (including Maxwell's equations) to be the same for all observers in uniform motion.

Time Dilation

The fact that observers in different inertial frames always measure different time intervals between a pair of events can be illustrated in another way by a vehicle moving to the right with a speed v, as in Figure 1.10a. A mirror is fixed to the ceiling of the vehicle, and observer O', at rest in this system, holds a laser a distance d below the mirror. At some instant the laser emits a pulse of light directed toward the mirror (event 1), and at some later time, after reflecting from the mirror, the pulse arrives back at the laser (event 2). Observer O'

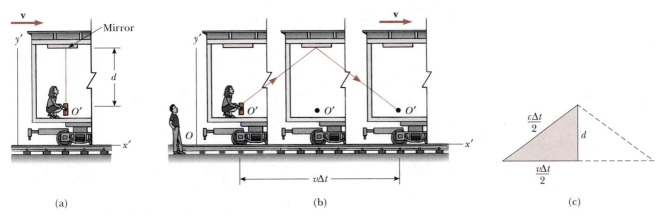

Figure 1.10   (a) A mirror is fixed to a moving vehicle, and a light pulse leaves O' at rest in the vehicle. (b) Relative to a stationary observer on Earth, the mirror and O' move with a speed v. Note that the distance the pulse travels measured by the stationary observer on Earth is greater than $2d$. (c) The right triangle for calculating the relationship between Δt and $\Delta t'$.

carries a clock, C', which she uses to measure the time interval $\Delta t'$ between these two events. Because the light pulse has the speed c, the time it takes to travel from O' to the mirror and back can be found from the definition of speed:

$$\Delta t' = \frac{\text{distance traveled}}{\text{speed of light}} = \frac{2d}{c} \tag{1.5}$$

This time interval $\Delta t'$—measured by O', who, remember, is at rest in the moving vehicle—requires only a *single* clock, C', in this reference frame.

Now consider the same set of events as viewed by an observer O in a different frame (Fig. 1.10b). According to this observer, the mirror and laser are moving to the right with a speed v. The sequence of events would appear entirely different as viewed by this observer. By the time the light from the laser reaches the mirror, the mirror will have moved a distance $v\,\Delta t/2$, where Δt is the time for the light pulse to travel from O' to the mirror and back as measured by the second observer O. In other words, the second observer concludes that, because of the motion of the clock, if the light is to hit the mirror, it must leave the laser at an angle with respect to the vertical direction. Comparing Figures 1.10a and 1.10b, we see that the light must travel farther in the second frame than in the first frame.

According to the second postulate of relativity, both observers must measure c for the speed of light. Because the light travels farther in the second frame, it follows that the time interval Δt measured by an observer in the second frame is longer than the time interval $\Delta t'$ measured by an observer in the first frame. To obtain a relationship between Δt and $\Delta t'$, it is convenient to use the right triangle shown in Figure 1.10c. The Pythagorean theorem gives

$$\left(\frac{c\,\Delta t}{2}\right)^2 = \left(\frac{v\,\Delta t}{2}\right)^2 + d^2$$

Solving for Δt gives

$$\Delta t = \frac{2d}{\sqrt{c^2 - v^2}} = \frac{2d}{c\sqrt{1 - v^2/c^2}} \tag{1.6}$$

Because $\Delta t' = 2d/c$, we can express Equation 1.6 as

Time dilation

$$\Delta t = \frac{\Delta t'}{\sqrt{1 - (v^2/c^2)}} = \gamma \, \Delta t' \tag{1.7}$$

where $\gamma = (1 - v^2/c^2)^{-1/2}$ This result says that the time interval Δt measured by the observer moving with respect to the clock is *longer* than the time interval $\Delta t'$ measured by the observer at rest with respect to the clock *because γ is always greater than unity.* This effect is known as time dilation.

A moving clock runs slower

The time interval $\Delta t'$ in Equation 1.7 is called the proper time. In general, **proper time** is defined as *the time interval between two events as measured by an observer who sees the events occur at the same point in space.* In our case, observer O' measures the proper time. That is, *proper time is always the time measured by an observer moving along with the clock.*

Because the time between ticks of a moving clock, $\gamma(2d/c)$, is observed to be longer than the time between ticks of an identical clock at rest, $2d/c$, one often says, *"A moving clock runs slower than a clock at rest by a factor of γ."* This is true for ordinary mechanical clocks as well as for the light clock just described. In fact, we can generalize these results by stating that *all physical processes, including chemical reactions and biological processes, slow down relative to a stationary clock when they occur in a moving frame.* For example, the heartbeat of an astronaut moving through space would keep time with a clock inside the spaceship, but both the astronaut's clock and her heartbeat appear slow to an observer, with another clock, in any other reference frame. The astronaut would not have any sensation of life slowing down in her frame.

Time dilation is a very real phenomenon that has been verified by various experiments. For example, muons are unstable elementary particles, which have a charge equal to that of an electron and a mass 207 times that of the electron. Muons can be produced by the collision of cosmic radiation with atoms high in the atmosphere. Muons have a lifetime of only 2.2 μs when measured in a reference frame at rest with respect to them. If we take 2.2 μs as the average lifetime of a muon and assume that its speed is close to the speed of light, we would find that these particles could travel a distance of about 650 m before they decayed. Hence, they could not reach the Earth from the upper atmosphere where they are produced. However, experiments show that a large number of muons *do* reach the Earth. The phenomenon of time dilation explains this effect (see Fig. 1.11a). Relative to an observer on Earth, the muons have a lifetime equal to $\gamma\tau$, where $\tau = 2.2$ μs is the lifetime in a frame of reference traveling with the muons. For example, for $v = 0.99c$, $\gamma \approx 7.1$ and $\gamma\tau \approx 16$ μs. Hence, the average distance traveled as measured by an observer on Earth is $\gamma v\tau \approx 4700$ m, as indicated in Figure 1.11b.

Figure 1.11 (a) Muons traveling with a speed of 0.99c travel only about 650 m as measured in the muons' reference frame, where their lifetime is about 2.2 μs. (b) The muons travel about 4700 m as measured by an observer on Earth. Because of time dilation, the muons' lifetime is longer as measured by the Earth observer.

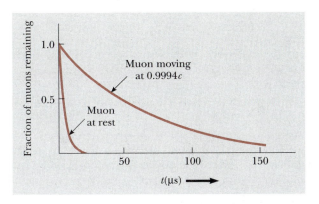

Figure 1.12   Decay curves for muons traveling at a speed of $0.9994c$ and for muons at rest.

In 1976, experiments with muons were conducted at the laboratory of the European Council for Nuclear Research (CERN) in Geneva. Muons were injected into a large storage ring, reaching speeds of about $0.9994c$. Electrons produced by the decaying muons were detected by counters around the ring, enabling scientists to measure the decay rate, and hence the lifetime of the muons. The lifetime of the moving muons was measured to be about 30 times as long as that of the stationary muon (see Fig. 1.12), in agreement with the prediction of relativity to within two parts in a thousand.

The results of an experiment reported by Hafele and Keating provided direct evidence for the phenomenon of time dilation.[7] The experiment involved the use of very stable cesium beam atomic clocks. Time intervals measured with four such clocks in jet flight were compared with time intervals measured by reference atomic clocks located at the U.S. Naval Observatory. To compare these results with the theory, many factors had to be considered, including periods of acceleration and deceleration relative to the Earth, variations in direction of travel, and the weaker gravitational field experienced by the flying clocks compared with the Earth-based clock. Their results were in good agreement with the predictions of the special theory of relativity and can be explained in terms of the relative motion between the Earth's rotation and the jet aircraft. In their paper, Hafele and Keating report the following: "Relative to the atomic time scale of the U.S. Naval Observatory, the flying clocks lost 59 ± 10 ns during the eastward trip and gained 273 ± 7 ns during the westward trip. . . . These results provide an unambiguous empirical resolution of the famous clock paradox with macroscopic clocks."

[7] J. C. Hafele and R. E. Keating, "Around the World Atomic Clocks: Relativistic Time Gains Observed," *Science*, July 14, 1972, p. 168.

EXAMPLE 1.1   What Is the Period of the Pendulum?

The period of a pendulum is measured to be 3.0 s in the rest frame of the pendulum. What is the period of the pendulum when measured by an observer moving at a speed of 0.95c with respect to the pendulum?

Solution   In this case, the proper time is equal to 3.0 s. Instead of the observer moving at 0.95c, we can take the equivalent point of view that the observer is at rest and that the pendulum is moving at 0.95c past the stationary observer. Hence the pendulum is an example of a moving

clock. Because a moving clock runs slower than a stationary clock by γ, Equation 1.7 gives

$$T = \gamma T' = \frac{1}{\sqrt{1 - (0.95c)^2/c^2}}\ (3.0\text{ s})$$

$$T' = (3.2)(3.0\text{ s}) = 9.6\text{ s}$$

That is, a moving pendulum slows down or takes longer to complete a period.

Length Contraction

We have seen that measured time intervals are not absolute, that is, the time interval between two events depends on the frame of reference in which it is measured. Likewise, the measured distance between two points depends on the frame of reference. The **proper length** of an object is defined as *the length of the object measured by someone who is at rest with respect to the object.* You should note that proper length is defined similarly to proper time, in that proper time is the time between ticks of a clock measured by an observer who is at rest with respect to the clock. The length of an object measured in a reference frame in which the object is moving is always less than the proper length. This effect is known as **length contraction.**

To understand length contraction quantitatively, let us consider a spaceship traveling with a speed v from one star to another, and two observers: one on Earth and the other in the spaceship. The observer at rest on Earth (and also assumed to be at rest with respect to the two stars) measures the distance between the stars to be L', where L' is the proper length. According to this observer, the time it takes the spaceship to complete the voyage is $\Delta t = L'/v$. What does an observer in the moving spaceship measure for the distance between the stars? Because of the time dilation, the space traveler measures a smaller time of travel: $\Delta t' = \Delta t/\gamma$. The space traveler claims to be at rest and sees the destination star as moving toward the spaceship with speed v. Because the space traveler reaches the star in the time $\Delta t'$, he or she concludes that the distance, L, between the stars is shorter than L. This distance measured by the space traveler is given by

$$L = v\,\Delta t' = v\frac{\Delta t}{\gamma}$$

Because $L' = v\,\Delta t$, we see that $L = L'/\gamma$ or

Length contraction

$$L = L'\left(1 - \frac{v^2}{c^2}\right)^{1/2} \tag{1.8}$$

where $(1 - v^2/c^2)^{1/2}$ is a factor less than one. This result may be interpreted as follows:

If an object has a length L' when it is at rest, then when measured by an observer moving with relative speed of v in a direction parallel to its length, it will appear to contract to the length L, where $L = L'(1 - v^2/c^2)^{1/2}$.

Note that the length contraction takes place only along the direction of motion. For example, suppose a stick moves past a stationary Earth observer with a speed v as in Figure 1.13. The length of the stick as measured by an observer in the frame attached to it is the proper length L', as illustrated in Figure 1.13a. The length of the stick, L, as measured by the Earth observer in the stationary frame is shorter than L' by the factor $(1 - v^2/c^2)^{1/2}$. Note that length contraction is a symmetric effect: If the stick were at rest on Earth, an observer in the moving frame would also measure its length to be shorter by the same factor $(1 - v^2/c^2)^{1/2}$.

It is important to emphasize a particular type of experiment for which $L = vt$. Let us return to the example of the decaying muons moving at speeds close to the speed of light. An observer in the muon's reference frame would measure the proper lifetime, whereas an Earth-based observer measures the proper height of the mountain in Figure 1.11. In the muon's reference frame, there is no time dilation, but the distance of travel is observed to be shorter when measured in this frame. Likewise, in the Earth observer's reference frame, there is time dilation, but the distance of travel is measured to be the proper height of the mountain. Thus, when calculations on the muon are performed in both frames, one sees the effect of "offsetting penalties," and the outcome of the experiment is the same! Note that proper length and proper time are measured in *different* reference frames.

If an object in the shape of a box passing by could be photographed, its image would show length contraction, but its shape would also be distorted. This is illustrated in the computer-simulated drawings shown in Figure 1.14 for a box moving past an observer with a speed $v = 0.8c$. When the shutter of the camera is opened, it records the shape of the object at a given instant of time. Because light from different parts of the object must arrive at the shutter at the same time (when the photograph is taken), light from more distant parts of the object must start its journey earlier than light from closer parts. Hence, the photograph records different parts of the object at different times. This results in a highly distorted image, which shows horizontal length contraction, vertical curvature, and image rotation.

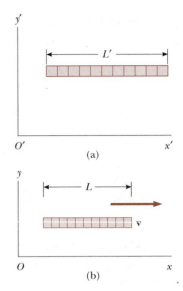

Figure 1.13 A stick moves to the right with a speed v. (a) The stick as viewed in a frame attached to it. (b) The stick as seen by an observer who sees it move past her at v. Any inertial observer finds that the length of a meter stick moving past her with speed v is less than the length of a stationary stick by a factor of $(1 - v^2/c^2)^{1/2}$.

Figure 1.14 Computer-simulated photographs of a box (a) at rest relative to the camera and (b) moving at a speed $v = 0.8c$ relative to the camera.

EXAMPLE 1.2   The Contraction of a Spaceship

A spaceship is measured to be 100 m long while it is at rest with respect to an observer. If this spaceship now flies by the observer with a speed of $0.99c$, what length will the observer find for the spaceship?

Solution   From Equation 1.8, the length measured as the spaceship flies by is

$$L = L' \sqrt{1 - \frac{v^2}{c^2}} = (100 \text{ m}) \sqrt{1 - \frac{(0.99c)^2}{c^2}} = 14 \text{ m}$$

Exercise 1   If the ship moves past the observer at $0.01000c$, what length will the observer measure?

Answer   99.99 m.

EXAMPLE 1.3   How High Is the Spaceship?

An observer on Earth sees a spaceship at an altitude of 435 m moving downward toward the Earth at $0.970c$. What is the altitude of the spaceship as measured by an observer in the spaceship?

Solution   The moving observer in the ship finds the altitude to be

$$L = L' \sqrt{1 - \frac{v^2}{c^2}} = (435 \text{ m}) \sqrt{1 - \frac{(0.970c)^2}{c^2}} = 106 \text{ m}$$

EXAMPLE 1.4   The Triangular Spaceship

A spaceship in the form of a triangle flies by an observer at $0.950c$. When the ship is measured by an observer at rest with respect to the ship (Fig. 1.15a), the distances x and y are found to be 50.0 m and 25.0 m, respectively.

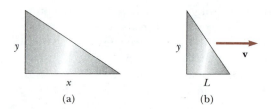

Figure 1.15   (Example 1.4) (a) When the spaceship is at rest, its shape is as shown. (b) The spaceship appears to look like this when it moves to the right with a speed v. Note that only its x dimension is contracted in this case.

What is the shape of the ship as seen by an observer who sees the ship in motion along the direction shown in Figure 1.15b?

Solution   The observer sees the horizontal length of the ship to be contracted to a length of

$$L = L' \sqrt{1 - \frac{v^2}{c^2}}$$

$$= (50.0 \text{ m}) \sqrt{1 - \frac{(0.950c)^2}{c^2}} = 15.6 \text{ m}$$

The 25-m vertical height is unchanged because it is perpendicular to the direction of relative motion between the observer and the spaceship. Figure 1.15b represents the shape of the spaceship as seen by the observer who sees the ship in motion.

O P T I O N A L | ## THE TWIN PARADOX

An intriguing consequence of time dilation is the so-called clock or twin paradox. Consider an experiment involving a set of identical 20-year-old twins named Speedo and Goslo. The twins carry with them identical clocks that have been synchronized. Speedo, the more adventuresome of the two, sets out on an epic journey to planet X, which is 10 lightyears from Earth. Furthermore, his spaceship is capable of reaching a speed of $0.500c$ relative to the inertial frame of his twin brother. After reaching planet X, Speedo becomes homesick and impetuously sets out on a return trip to Earth at the same high speed he had attained on the outbound journey. On his return, Speedo is shocked to discover that many things have changed during his absence. To Speedo, the most significant change is that his twin brother Goslo *has aged more than he* and is now 60 years of age. Speedo, on the other hand, has aged by only 34.6 years. We see that the bodily processes continued for a longer period of time for one twin than for the other. The accelerated twin moved from one reference frame to another, so his accumulated time was less.

At this point, it is fair to raise the following question—Which twin is the traveler and which twin would really be the younger of the two? From Speedo's perspective, it is he who is at rest while Goslo is on a high-speed space journey. To Speedo, it is Goslo and the Earth that have raced away on a 17.3-year journey and then headed back for another 17.3 years. This leads to an apparent contradiction. Which twin will have developed the signs of excess aging?

To resolve this apparent paradox, recall that special relativity deals with inertial frames of reference moving with respect to one another at *uniform speed*. However, the trip situation is not symmetric. Speedo, the space traveler, must experience acceleration during his journey. As a result, his state of motion is not always uniform, and consequently Speedo is not in an inertial frame. He cannot regard himself to always be at rest and Goslo to be in uniform motion. Hence Speedo cannot simply apply time dilation to Goslo's motion, because to do so would be an incorrect application of special relativity. Therefore there is no paradox.

The conclusion that Speedo is in a noninertial frame is inescapable. We may diminish the length of time needed to accelerate and decelerate Speedo's spaceship to a negligible interval by using very large and expensive rockets and claim that he spends all but a negligible amount of time coasting to planet X at $0.500c$ in an inertial frame. However, to return to Earth, Speedo must slow down, reverse his motion, and return in a different inertial frame, one which is moving uniformly toward the Earth. At the very best, Speedo is in two different inertial frames. The important point here is that even when we idealize Speedo's trip, it consists of motion in two different inertial frames and a very real lurch as he hops from the outbound ship to the returning Earth shuttle. In other words, only Goslo remains in a single inertial frame, and so only he can correctly apply the simple time dilation formula to Speedo's trip. Thus, Goslo finds that instead of aging 40 years $(20\,\text{lightyears}/0.500c)$, Speedo actually ages only $(\sqrt{1 - v^2/c^2})(40\,\text{years})$, or 34.6 years. Clearly, Speedo spends 17.3 years going to planet X and 17.3 years returning in agreement with our earlier statement.

The result that Speedo ages 34.6 years while Goslo ages 40 years can be confirmed in a very direct experimental way by Speedo if we use the special theory of relativity but take into account the fact that *Speedo's trip takes place in two different inertial frames.* In yet another flight of fancy, suppose that Goslo celebrates his birthday each year in a flashy way, sending a powerful laser pulse to inform his twin by light flash that Goslo is another year older and wiser. Because Speedo is in an inertial frame on the outbound trip in which the Earth appears to be receding at $0.500c$, the flashes occur at a rate of one every

$$\frac{1}{\sqrt{1 - (v^2/c^2)}}\,\text{year} = \frac{1}{\sqrt{1 - [(0.500c)^2/c^2]}}\,\text{year} = 1.15\,\text{years}$$

This occurs because moving clocks run slower. Also, because the Earth is receding, each successive flash must travel an additional distance of $(0.500c)(1.15\,\text{years})$ between flashes. Consequently, Speedo observes flashes to arrive with a total time between flashes of 1.15 years $+ (0.500c)(1.15\,\text{years})/c = 1.73$ years. The total number of flashes seen by Speedo on his outbound voyage is therefore $(1\,\text{flash}/1.73\,\text{years})(17.3\,\text{years}) = 10$ flashes. This means that Speedo views the Earth clocks to run more slowly than his own on the outbound trip because he observes 17.3 years to have passed for him while only 10 years have passed on Earth.

On the return voyage, because the Earth is racing toward Speedo with speed $0.500c$, successive flashes have less distance to travel, and the total time Speedo sees between the arrival of flashes is drastically shortened: 1.15 years $-$ $(0.500)(1.15\,\text{years}) = 0.577$ year/flash. Thus, during the return trip, Speedo sees $(1\,\text{flash}/0.577\,\text{year})(17.3\,\text{years}) = 30$ flashes in total. In sum, during his 34.6 years of travel, Speedo receives (10 + 30) flashes, indicating that his twin has aged 40 years. Notice that there has been no failure of special relativity for Speedo as long as we take his *two* inertial frames into account. On both the outbound and inbound trips Speedo correctly judges the Earth clocks to run slower than his own, but on the return trip his rapid movement toward the light flashes more than compensates for the slower rate of flashing.

The Relativistic Doppler Shift

Figure 1.16 "I love hearing that lonesome wail of the train whistle as the frequency of the wave changes due to the Doppler effect."

Another important consequence of time dilation is the shift in frequency found for light emitted by atoms in motion as opposed to light emitted by atoms at rest. A similar phenomenon, the mournful drop in pitch of the sound of a passing train's whistle, known as the Doppler effect, is quite familiar to most cowboys (Fig. 1.16). The Doppler shift for sound is usually studied in introductory physics courses and is especially interesting because motion of the source with respect to the medium of propagation can be clearly distinguished from motion of the observer. This means that in the case of sound we can distinguish the "absolute motion" of frames moving with respect to the air, which is the medium of propagation for sound.

Light waves must be analyzed differently from sound, because light waves require no medium of propagation and no method exists of distinguishing the motion of the light source from the motion of the observer. Thus, we expect to find a different formula for the Doppler shift of light waves, one that is only sensitive to the *relative* motion of source and observer and that holds for relative speeds of source and observer approaching c.

Consider a source of light waves at rest in frame S, emitting waves of frequency f and wavelength λ as measured in S. We wish to find the frequency f' and wavelength λ' of the light as measured by an observer fixed in frame S', which is moving with speed v toward S as shown in Figure 1.17a and b. In general, we expect f' to be greater than f if S' approaches S because more wave crests are crossed per unit time, and we expect f' to be less than f if S' recedes from S. In particular, consider the situation from the point of view of an observer fixed in S' as shown in Figure 1.18. This figure shows two successive wave fronts (color) emitted when the approaching source is at positions 1 and 2, respectively. If the time between the emission of these wavefronts as measured in S' is T', during this time front 1 will move a distance cT' from position 1. During this same time, the light source will advance a distance vT' to the left of position 1, and the distance between successive wave fronts will be measured in S' to be

$$\lambda' = cT' - vT' \tag{1.9}$$

Because we wish to obtain a formula for f' (the frequency measured in S') in terms of f (the frequency measured in S), we use $f' = c/\lambda'$ to obtain

$$f' = \frac{c}{(c - v)T'} \tag{1.10}$$

To eliminate T' in favor of T, note that T is the proper time; that is, T is the time between two events (the emission of successive wave fronts) that occur at the same place in S, and consequently,

(a)

(b)

Figure 1.17   (a) A light source fixed in S emits wave crests separated in space by λ and moving outward at speed c as seen from S. (b) What wavelength λ' is measured by an observer at rest in S'? S' is a frame approaching S at speed v along $x - x'$.

$$T' = \frac{T}{\sqrt{1 - (v^2/c^2)}}$$

Substituting for T' in Equation 1.10 and using $f = 1/T$ gives

$$f' = \frac{\sqrt{1 - (v^2/c^2)}}{1 - (v/c)} f \qquad (1.11)$$

or

$$f' = \frac{\sqrt{1 + (v/c)}}{\sqrt{1 - (v/c)}} f \qquad (1.12)$$

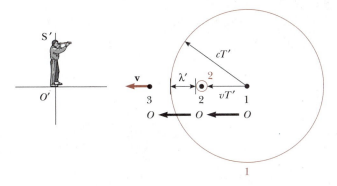

Figure 1.18   The view from S'. 1, 2, and 3 (in black) show three successive positions of O separated in time by T', the period of the light as measured from S'.

For clarity this expression is often written

$$f_{obs} = \frac{\sqrt{1 + (v/c)}}{\sqrt{1 - (v/c)}} f_{source} \qquad (1.13)$$

where f_{obs} is the frequency measured by an observer approaching a light source, and f_{source} is the frequency as measured in the source's rest frame.

Equation 1.13 is the relativistic Doppler shift formula, which, unlike the Doppler formula for sound, depends only on the relative speed v of the source and observer and holds for relative speeds as large as c. Equation 1.13 agrees with physical intuition in predicting f_{obs} to be greater than f_{source} for an approaching emitter and receiver. The expression for the case of a receding source is obtained by replacing v with $-v$ in Equation 1.13.

Although Christian Johann Doppler's name is most frequently associated with the effect in sound, he originally developed his ideas in an effort to understand the shift in frequency or wavelength of the light emitted by moving atoms and astronomical objects. The most spectacular and dramatic use of the Doppler effect has occurred in just this area in explaining the famous red shift of absorption lines (wavelengths) observed for most galaxies. (A galaxy is a cluster of millions of stars.) The term *red shift* refers to the shift of known absorption lines toward longer wavelengths, that is, toward the red end of the visible spectrum. For example, lines normally found in the extreme violet region for a galaxy at rest with respect to the Earth are shifted about 100 nm toward the red end of the spectrum for distant galaxies—indicating that these distant galaxies are rapidly *receding* from us. The American astronomer Edwin Hubble used this technique to confirm that most galaxies are moving away from us and that the Universe is *expanding*. (For more about the expanding Universe see Chapter 15, Section 12.)

EXAMPLE 1.5   Determining the Speed of Recession of the Galaxy Hydra

The light emitted by a galaxy contains a continuous distribution of wavelengths because the galaxy is composed of millions of stars and other thermal emitters. However, some narrow gaps occur in the continuous spectrum where the radiation has been strongly absorbed by cooler gases in the galaxy (Fig. 15.19). In particular, a cloud of ionized calcium atoms produces very strong absorption at 394 nm for a galaxy at rest with respect to the Earth. For the galaxy Hydra, which is 2 billion lightyears away, this absorption is shifted to 475 nm. How fast is Hydra moving away from the Earth?

Solution For an approaching source and observer, $f_{obs} > f_{source}$ and $\lambda_{obs} < \lambda_{source}$ because $f_{obs}\lambda_{obs} = c = f_{source}\lambda_{source}$. In the case of Hydra $\lambda_{obs} > \lambda_{source}$, so Hydra must be receding and we must use

$$f_{obs} = \frac{\sqrt{1 - (v/c)}}{\sqrt{1 + (v/c)}} f_{source}$$

Substituting $f_{obs} = c/\lambda_{obs}$ and $f_{source} = c/\lambda_{source}$ into this equation gives

$$\lambda_{obs} = \frac{\sqrt{1 + (v/c)}}{\sqrt{1 - (v/c)}} \lambda_{source}$$

Finally, solving for v/c, we find

$$\frac{v}{c} = \frac{\lambda_{obs}^2 - \lambda_{source}^2}{\lambda_{obs}^2 + \lambda_{source}^2}$$

or

$$\frac{v}{c} = \frac{(475 \text{ nm})^2 - (394 \text{ nm})^2}{(475 \text{ nm})^2 + (394 \text{ nm})^2} = 0.185$$

Therefore, Hydra is receding from us at $v = 0.185c = 5.54 \times 10^7$ m/s.

1.6   THE LORENTZ TRANSFORMATION

We have seen that the Galilean transformation is not valid when v approaches the speed of light. In this section, we shall derive the correct transformation equations that apply for all speeds in the range of $0 \le v < c$. This transformation, known as the Lorentz transformation, was developed by Hendrik A. Lorentz (1853–1928) in 1890 in connection with electromagnetism. However, its real significance in a physical theory transcending electromagnetism was first recognized by Einstein.

The Lorentz transformation is a set of formulas that relates the space and time coordinates of two inertial observers moving with a relative speed v. We have already seen two examples of the Lorentz transformation in the time dilation and length contraction formulas. The Lorentz transformation equations provide a more formal, concise, and almost mechanical method of solution that is well suited to further derivations such as the transformation of velocities between moving frames.

We start our derivation of the Lorentz transformation by noting that a reasonable guess about the form of these equations (based on physical intuition) can greatly reduce the algebraic complexity of the derivation. For simplicity, consider the standard frames, S and S′, with S′ moving at a speed v along the $+x$ direction (see Fig. 1.2). A reasonable guess about the dependence of x' on x and t is

$$x' = G(x - vt) \tag{1.14}$$

where G is a dimensionless factor that does not depend on x or t but is some function of v/c such that G is 1 in the limit as v/c approaches 0. The form of Equation 1.14 is suggested by the form of the Galilean transformation, $x' = x - vt$, which we know is correct for low speeds or in the limit as v/c approaches zero. The fact that Equation 1.14 is linear in x and t is also important because we require a single event in S (specified by x_1, t_1) to correspond to a single event in S′ (specified by x_1', t_1'). Assuming that Equation 1.14 is correct, we can write the *inverse Lorentz transformation* for x in terms of x' and t' as

$$x = G(x' + vt') \tag{1.15}$$

This follows from Einstein's postulate of relativity, which requires the laws of physics to have the same form in both S and S′ and where the sign of v has been changed to take into account the difference in direction of motion of the two frames.

Consider a rocket moving with a speed v along the xx' axes as in Figure 1.19. The frame of the rocket S′ is indicated with the coordinates (x', y', z', t'), and an observer in S uses coordinates (x, y, z, t). A flashbulb mounted on the rocket emits a pulse of light at the instant that the origins of the two reference frames coincide.

At the instant the flashbulb goes off and the two origins coincide, we define $t = t' = 0$. The light signal travels as a spherical wave, where the origin of the wavefront is the fixed point O where the flash originated. At some later time, a point such as P on the spherical wavefront is at a distance r from O and a distance r' from O', as shown schematically in Figure 1.19b. According to Einstein's second postulate, the speed of light should be c for both observers. Hence, the distance to the point P on the wavefront as measured by an observer

Figure 1.19 A rocket moves with a speed v along the xx' axes. (a) A pulse of light is sent out from the rocket at $t = t' = 0$ when the two systems coincide. (b) Coordinates of some point P on an expanding spherical wavefront as measured by observers in both inertial systems. (This figure is entirely schematic, and you should not be misled by the geometry.)

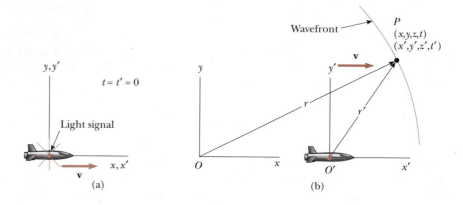

in S is given by $r = ct$, while the distance to the point P as measured by an observer in S' is given by $r' = ct'$. That is,

$$r = ct \tag{1.16}$$

$$r' = ct' \tag{1.17}$$

If we accept Einstein's second postulate we must require that the times t and t' taken for the light to reach P be *different*. This is in contrast to the Galilean transformation, where $t = t'$.

Because the motion of S' is along the xx' axes, it follows that the y and z coordinates measured in the two frames are always equal. That is, they are unaffected by the motion along x, and therefore $y = y'$ and $z = z'$.

To simplify matters consider the points x, x' along the x axes at which the spherical light wave crosses the horizontal axes. Equations 1.16 and 1.17 become $x = ct$ and $x' = ct'$. Substituting these values into Equations 1.14 and 1.15, we obtain

$$ct = G(ct' + vt') \tag{1.18}$$

$$ct' = G(ct - vt) \tag{1.19}$$

Solving Equation 1.19 for t' and substituting this value into Equation 1.18 gives

$$ct = \frac{G^2}{c}(c + v)(c - v)t \tag{1.20}$$

or

$$G \equiv \gamma = \frac{1}{\sqrt{1 - (v^2/c^2)}} \tag{1.21}$$

The direct transformation is thus $x' = \gamma(x - vt)$, and the inverse transformation is $x = \gamma(x' + vt')$. To get the time transformation (t' as a function of t and x), substitute $x' = \gamma(x - vt)$ into $x = \gamma(x' + vt')$ to eliminate x':

$$x = \gamma[\gamma(x - vt) + vt'] \tag{1.22}$$

or

$$t' = \gamma\left[t + \frac{x}{v}\left(\frac{1}{\gamma^2} - 1\right)\right] \tag{1.23}$$

Because $1/\gamma^2 - 1 = -v^2/c^2$,

$$t' = \gamma\left(t - \frac{vx}{c^2}\right) \qquad (1.24)$$

In summary, the complete transformations between an event found to occur at (x, y, z, t) in S and (x', y', z', t') in S' are

$$x' = \gamma(x - vt) \qquad (1.25)$$

$$y' = y \qquad (1.26)$$

$$z' = z \qquad (1.27)$$

$$t' = \gamma\left(t - \frac{vx}{c^2}\right) \qquad (1.28)$$

Lorentz transformation for S → S'

$$\text{where} \quad \gamma = \frac{1}{\sqrt{1 - (v^2/c^2)}} \qquad (1.29)$$

If we wish to transform coordinates in the S' frame to coordinates in the S frame, we simply replace v by $-v$ and interchange the primed and unprimed coordinates in Equations 1.25 through 1.29. The resulting transformation is given by

$$x = \frac{x' + vt'}{\sqrt{1 - (v^2/c^2)}} = \gamma(x' + vt')$$

$$y = y'$$

$$z = z' \qquad (1.30)$$

$$t = \frac{t' + (v/c^2)x'}{\sqrt{1 - (v^2/c^2)}} = \gamma\left(t' + \frac{vx'}{c^2}\right)$$

Inverse Lorentz transformation for S' → S

In the Lorentz transformation, t depends on both t' and x'. Likewise, t' depends on both t and x. This is unlike the case of the Galilean transformation, in which $t = t'$.

When $v \ll c$, the Lorentz transformation should reduce to the Galilean transformation. To check this, note that as $v \to 0$, $v/c < 1$ and $v^2/c^2 \ll 1$, so that Equations 1.25–1.28 reduce in this limit to the Galilean coordinate transformation equations, given by

$$x' = x - vt \qquad y' = y \qquad z' = z \qquad t' = t$$

EXAMPLE 1.6   Time Dilation Is Contained in the Lorentz Transformation

Show that the phenomenon of time dilation is contained in the Lorentz transformation. A light located at (x_0, y_0, z_0) is turned abruptly on at t_1 and off at t_2 in frame S. (a) For what time interval is the light measured to be on in frame S'? (See Fig. 1.2 for a picture of the two standard frames.) (b) What is the distance between where the light is turned on and off as measured by S'?

Solution   (a) The two events, the light turning on and the light turning off, are measured to occur in the two frames as follows:

	Event 1 (light on)	Event 2 (light off)
Frame S	x_0, t_1	x_0, t_2
Frame S′	$x_1' = \gamma(x_0 - vt_1)$	$x_2' = \gamma(x_0 - vt_2)$
	$t_1' = \gamma\left(t_1 - \dfrac{vx_0}{c^2}\right)$	$t_2' = \gamma\left(t_2 - \dfrac{vx_0}{c^2}\right)$

Note that the y and z coordinates are not affected because the motion of S′ is along x. As measured by S′, the light is on for a time interval

$$t_2' - t_1' = \gamma\left(t_2 - \frac{vx_0}{c^2}\right) - \gamma\left(t_1 - \frac{vx_0}{c^2}\right) = \gamma(t_2 - t_1)$$

Because $\gamma > 1$ and $(t_2 - t_1)$ is the proper time, it follows that $(t_2' - t_1') > (t_2 - t_1)$, and we have regained our previous result for time dilation, Equation 1.7.

(b) Although event 1 and event 2 occur at the same place in S, they are measured to occur at a separation of $x_2' - x_1'$ in S′ where

$$x_2' - x_1' = (\gamma x_0 - \gamma v t_2) - (\gamma x_0 - \gamma v t_1) = \gamma v(t_1 - t_2)$$

This result is reasonable because it reduces to

$$v(t_1 - t_2) \text{ for } v/c \ll 1.$$

Can you explain why $x_2' - x_1'$ is negative?

Exercise 2   Use the Lorentz transformation to derive the expression for length contraction. Note that the length of a moving object is determined by measuring the positions of both ends simultaneously.

Lorentz Velocity Transformation

Let us now derive the Lorentz velocity transformation, which is the relativistic counterpart of the Galilean velocity transformation. Suppose that an object is observed in the S′ frame with an instantaneous speed u_x' measured in S′ given by

$$u_x' = \frac{dx'}{dt'} \tag{1.31}$$

Using Equations 1.25 and 1.28, we have

$$dx' = \frac{dx - v\,dt}{\sqrt{1 - (v^2/c^2)}}$$

$$dt' = \frac{dt - (v/c^2)\,dx}{\sqrt{1 - (v^2/c^2)}}$$

Substituting these into Equation 1.31 gives

$$u_x' = \frac{dx'}{dt'} = \frac{dx - v\,dt}{dt - (v/c^2)\,dx} = \frac{(dx/dt) - v}{1 - (v/c^2)(dx/dt)}$$

But dx/dt is just the velocity component u_x of the object measured in S, and so this expression becomes

Lorentz velocity transformation for S → S′

$$u_x' = \frac{u_x - v}{1 - (u_x v/c^2)} \tag{1.32}$$

Similarly, if the object has velocity components along y and z, the components in S′ are

$$u'_y = \frac{u_y}{\gamma[1 - (u_x v/c^2)]} \quad \text{and} \quad u'_z = \frac{u_z}{\gamma[1 - (u_x v/c^2)]} \quad (1.33)$$

When u_x and v are both much smaller than c (the nonrelativistic case), we see that the denominator of Equation 1.32 approaches unity, and so $u'_x \approx u_x - v$. This corresponds to the Galilean velocity transformations. In the other extreme, when $u_x = c$, Equation 1.32 becomes

$$u'_x = \frac{c - v}{1 - (cv/c^2)} = \frac{c[1 - (v/c)]}{1 - (v/c)} = c$$

From this result, we see that an object moving with a speed c relative to an observer in S also has a speed c relative to an observer in S'—*independent* of the relative motion of S and S'. Note that this conclusion is consistent with Einstein's second postulate, namely, that the speed of light must be c with respect to all inertial frames of reference. Furthermore, the speed of an object can never exceed c. That is, the speed of light is the "ultimate" speed. We return to this point later when we consider the energy of a particle.

To obtain u_x in terms of u'_x, we replace v by $-v$ in Equation 1.32 and interchange the roles of u_x and u'_x. This gives

$$u_x = \frac{u'_x + v}{1 + (u'_x v/c^2)} \quad (1.34)$$

Inverse Lorentz velocity transformation for S' → S

EXAMPLE 1.7   Relative Velocity of Spaceships

Two spaceships A and B are moving in *opposite* directions, as in Figure 1.20. An observer on Earth measures the speed of A to be $0.750c$ and the speed of B to be $0.850c$. Find the velocity of B with respect to A.

Figure 1.20   (Example 1.7) Two spaceships A and B move in *opposite* directions. The velocity of B relative to A is *less* than c and is obtained by using the relativistic velocity transformation.

Solution   This problem can be solved by taking the S' frame as being attached to spacecraft A, so that $v =$ $0.750c$ relative to an observer on Earth (the S frame). Spacecraft B can be considered as an object moving to the left with a velocity $u_x = -0.850c$ relative to the Earth observer. Hence, the velocity of B with respect to A can be obtained using Equation 1.32:

$$u'_x = \frac{u_x - v}{1 - \dfrac{u_x v}{c^2}} = \frac{-0.850c - 0.750c}{1 - \dfrac{(-0.850c)(0.750c)}{c^2}}$$

$$= -0.9771c$$

The negative sign for u'_x indicates that spaceship B is moving in the negative x direction as observed by A. Note that the result is less than c. That is, a body whose speed is less than c in one frame of reference must have a speed less than c in *any other* frame. If the Galilean velocity transformation were used in this example, we would find that $u'_x = u_x - v = -0.850c - 0.750c = -1.600c$, which is greater than c.

EXAMPLE 1.8   The Speeding Motorcycle

Imagine a motorcycle rider moving with a speed of $0.800c$ past a stationary observer, as shown in Figure 1.21. If the

Figure 1.21   (Example 1.8) A motorcyclist moves past a stationary observer with a speed of $0.800c$ and throws a ball in the direction of motion with a speed of $0.700c$ relative to himself.

rider tosses a ball in the forward direction with a speed of $0.700c$ relative to himself, what is the speed of the ball as seen by the stationary observer?

Solution   In this situation, the velocity of the motorcycle with respect to the stationary observer is $v = 0.800c$. The velocity of the ball in the frame of reference of the motorcyclist is $u_x' = 0.700c$. Therefore, the velocity, u_x, of the ball relative to the stationary observer is

$$u_x = \frac{u_x' + v}{1 + (u_x'v/c^2)}$$
$$= \frac{0.700c + 0.800c}{1 + [(0.700c)(0.800c)/c^2]} = 0.9615c$$

Exercise 3   Suppose that the motorcyclist moving with a speed $0.800c$ turns on a beam of light that moves away from him with a speed of c in the same direction as the moving motorcycle. What would the stationary observer measure for the speed of the beam of light?

Answer c.

EXAMPLE 1.9   Relativistic Leaders of the Pack!

Imagine two motorcycle gang leaders racing at relativistic speeds along perpendicular paths from the local pool hall as shown in Figure 1.22. How fast does pack leader Beta recede over Alpha's right shoulder as seen by Alpha?

Solution   Figure 1.22 shows the situation as seen by a stationary police officer located in frame S, who observes

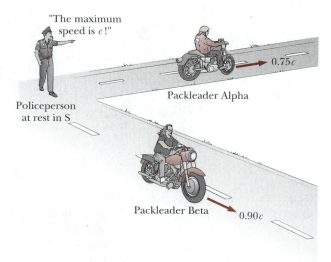

Figure 1.22   (Example 1.9) Two motorcycle pack leaders, Alpha and Beta, blaze past a stationary police officer. They are leading their respective gangs from the pool hall along perpendicular roads.

the following:

Pack leader Alpha	$u_x = 0.75c$	$u_y = 0$
Pack leader Beta	$u_x = 0$	$u_y = -0.90c$

To get Beta's speed of recession as seen by Alpha, we take S′ to move along with Alpha as shown in Figure 1.23, and we calculate u_x' and u_y' for Beta using Equations 1.32 and 1.33:

$$u_x' = \frac{u_x - v}{1 - (u_xv/c^2)} = \frac{0 - 0.75c}{1 - [(0)(0.75c)/c^2]} = -0.75c$$

$$u_y' = \frac{u_y}{\gamma[1 - (u_xv/c^2)]}$$
$$= \frac{\sqrt{1 - [(0.75c)^2/c^2]}(-0.90c)}{1 - [(0)(0.75c)/c^2]} = -0.60c$$

The speed of recession of Beta away from Alpha as observed by Alpha is then found to be less than c as required by relativity.

$$u' = \sqrt{(u_x')^2 + (u_y')^2} = \sqrt{(-0.75c)^2 + (-0.60c)^2} = 0.96c$$

Exercise 4   Calculate the classical speed of recession of Beta from Alpha using a Galilean transformation.

Answer $1.2c$

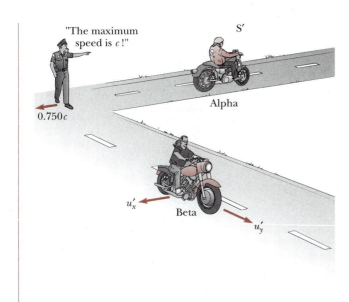

Figure 1.23 (Example 1.9) Pack leader Alpha's view of things.

1.7 RELATIVISTIC MOMENTUM AND THE RELATIVISTIC FORM OF NEWTON'S LAWS

Because Newton's laws are invariant under a Galilean transformation but we know that the correct relativistic transformation is the Lorentz transformation, we must correct or generalize Newton's laws to conform to the Lorentz transformation and the principles of relativity. These generalized definitions of Newton's laws of momentum and energy should reduce to the classical (nonrelativistic) definitions for $v \ll c$.

First, recall that the conservation of momentum states that when two bodies collide, the total momentum remains constant, assuming the bodies are isolated (that is, they interact only with each other). Suppose the collision is described in a reference frame S in which the momentum is conserved. If the velocities in a second reference frame S′ are calculated using the Lorentz transformation and the classical definition of momentum, $\mathbf{p} = m\mathbf{u}$, one finds that momentum *is not* conserved in the second reference frame. However, because the laws of physics are the same in all inertial frames, the momentum must be conserved in all systems. In view of this condition and assuming the Lorentz transformation is correct, we must modify the definition of momentum.

To see how the classical form $\mathbf{p} = m\mathbf{u}$ fails and to determine the correct relativistic definition of $\mathbf{p}$, consider the case of an inelastic collision between two particles of equal mass. Figure 1.24a shows such a collision for two identical particles approaching each other at speed v as observed in an inertial reference frame S. Using the classical form for momentum, $\mathbf{p} = m\mathbf{u}$, the observer in S finds momentum to be conserved as shown in Figure 1.24a. Suppose we now view things from an inertial frame S′ moving to the right with speed v with respect to S. In S′ the new speeds are v_1', v_2' and V' (see Fig. 1.24b). If we use the Lorentz velocity transformation

Momentum is conserved according to S
$$p_{\text{before}} = mv + m(-v) = 0$$
$$p_{\text{after}} = 0$$

Momentum is *not* conserved according to S'
$$p'_{\text{before}} = \frac{-2mv}{1 + v^2/c^2}$$
$$p'_{\text{after}} = -2mv$$

Figure 1.24   (a) An inelastic collision between two equal clay lumps as seen by an observer in frame S. (b) The same collision viewed from a frame S' that is moving to the right with speed v with respect to S.

$$u'_x = \frac{u_x - v}{1 - (u_x v / c^2)} \tag{1.32}$$

to find v'_1, v'_2 and V', and the classical form for momentum, $\mathbf{p} = m\mathbf{u}$, will momentum be conserved according to the observer in S'? To answer this question we first calculate the velocities in S' in terms of the given velocities in S.

$$v'_1 = \frac{v_1 - v}{1 - (v_1 v / c^2)} = \frac{v - v}{1 - (v^2 / c^2)} = 0$$

$$v'_2 = \frac{v_2 - v}{1 - (v_2 v / c^2)} = \frac{-v - v}{1 - [(-v)(v)/c^2]} = \frac{-2v}{1 + (v^2 / c^2)}$$

$$V' = \frac{V - v}{1 - (Vv / c^2)} = \frac{0 - v}{1 - [(0)v/c^2]} = -v$$

Checking for momentum conservation in S', we have,

$$p'_{\text{before}} = mv'_1 + mv'_2 = m(0) + m\left[\frac{-2v}{1 + (v^2/c^2)}\right] = \frac{-2mv}{1 + (v^2/c^2)}$$

$$p'_{\text{after}} = 2mV' = -2mv$$

Thus, in the frame S', the momentum before the collision is not equal to the momentum after the collision, and *momentum is not conserved.*

Conservation of linear momentum is a fundamental result, and it must be valid in all inertial frames. Because we have found that using the relativistically correct velocity transformation and the classical definition of momentum lead to the nonconservation of momentum, the only way to preserve conservation of momentum is to modify the definition of momentum. It can be shown (see

Example 1.15) that momentum is conserved in both S and S', (and indeed in all inertial frames), if we redefine momentum as

$$\mathbf{p} \equiv \frac{m\mathbf{u}}{\sqrt{1 - (u^2/c^2)}} \qquad (1.35)$$

Definition of relativistic momentum

where $\mathbf{u}$ is the velocity of the particle and *m is the proper mass, that is, the mass measured by an observer at rest with respect to the mass.*[8] We use the symbol $\mathbf{u}$ for a particle velocity rather than $\mathbf{v}$, which is used for the relative velocity of two reference frames. Note that when u is much less than c, the denominator of Equation 1.35 approached unity, so that $\mathbf{p}$ approaches $m\mathbf{u}$. Therefore, the relativistic equation for $\mathbf{p}$ reduces to the classical expression when u is small compared with c. Because it is simpler, Equation 1.35 is often written as $\mathbf{p} = \gamma m\mathbf{u}$, where $\gamma = 1/\sqrt{1 - (u^2/c^2)}$. Note that this γ has the same functional form as the γ in the Lorentz transformation, but here γ contains u, the particle speed, and in the Lorentz transformation, γ contains v, the relative speed of the two frames.

The relativistic force $\mathbf{F}$ on a particle whose momentum is $\mathbf{p}$ is defined by the expression

$$\mathbf{F} \equiv \frac{d\mathbf{p}}{dt} \qquad (1.36)$$

where $\mathbf{p}$ is given by Equation 1.35. This expression is reasonable because it preserves classical mechanics in the limit of low velocities and requires the momentum of an isolated system ($\mathbf{F} = 0$) to be conserved relativistically as well as classically. It is left as a problem (Problem 29) to show that the acceleration a of a particle decreases under the action of a constant force, in which case $a \propto (1 - u^2/c^2)^{3/2}$. From this formula note that as the velocity approaches c, the acceleration caused by any finite force approaches zero. Hence, it is impossible to accelerate a particle from rest to a speed equal to or greater than c.

---

[8] In this book we shall always take m to be constant with respect to speed, and we call m the speed invariant mass, or proper mass. Some physicists refer to the mass in Equation 1.35 as the rest mass m_0 and call the term $m_0/\sqrt{1 - (u^2/c^2)}$ the relativistic mass. Using this description, the relativistic mass is imagined to increase with increasing speed. We exclusively use the invariant mass m because we think it is a clearer concept and that the introduction of relativistic mass leads to no deeper physical understanding.

EXAMPLE 1.10   Momentum of an Electron

An electron, which has a mass of 9.11×10^{-31} kg, moves with a speed of $0.750c$. Find its relativistic momentum and compare this with the momentum calculated from the classical expression.

Solution   Using Equation 1.35 with $u = 0.750c$, we have

$$p = \frac{mu}{\sqrt{1 - (u^2/c^2)}}$$

$$= \frac{(9.11 \times 10^{-31} \text{ kg})(0.750 \times 3.00 \times 10^8 \text{ m/s})}{\sqrt{1 - [(0.750c)^2/c^2]}}$$

$$= 3.10 \times 10^{-22} \text{ kg} \cdot \text{m/s}$$

The incorrect classical expression would give

$$\text{Momentum} = mu = 2.05 \times 10^{-22} \text{ kg} \cdot \text{m/s}$$

Hence, the correct relativistic result is 50% greater than the classical result!

EXAMPLE 1.11   An Application of the Relativistic Form of F = dp/dt: The Measurement of the Momentum of a High-Speed Charged Particle

Suppose a particle of mass m and charge q is injected with a relativistic velocity $\mathbf{u}$ into a region containing a magnetic field $\mathbf{B}$. The magnetic force $\mathbf{F}$ on the particle is given by $\mathbf{F} = q\mathbf{u} \times \mathbf{B}$. If $\mathbf{u}$ is perpendicular to $\mathbf{B}$, the force is radially inward, and the particle moves in a circle of radius R with $|\mathbf{u}|$ constant. From Equation 1.36 we have

$$\mathbf{F} = \frac{d\mathbf{p}}{dt} = \frac{d}{dt}(\gamma m\mathbf{u})$$

Because the magnetic force is always perpendicular to the velocity, it does no work on the particle, and hence the speed, u, and γ are both constant with time. Thus, the magnitude of the force on the particle is

$$F = \gamma m \left| \frac{d\mathbf{u}}{dt} \right| \tag{1.37}$$

Substituting $F = quB$ and $|du/dt| = u^2/R$ (the usual definition of centripetal acceleration) into Equation 1.37, we can solve for $p = \gamma mu$. We find

$$p = \gamma mu = qBR \tag{1.38}$$

Equation 1.38 shows that the momentum of a relativistic particle of known charge q may be determined by measuring its radius of curvature R in a known magnetic field, $\mathbf{B}$. This technique is routinely used to determine the momentum of subatomic particles from photographs of their tracks through space.

1.8   RELATIVISTIC ENERGY

We have seen that the definition of momentum and the laws of motion required generalization to make them compatible with the principle of relativity. This implies that the relativistic form of the kinetic energy must also be modified.

To derive the relativistic form of the work-energy theorem, let us start with the definition of work done by a force F and make use of the definition of relativistic force, Equation 1.37. That is,

$$W = \int_{x_1}^{x_2} F \, dx = \int_{x_1}^{x_2} \frac{dp}{dt} \, dx \tag{1.39}$$

where we have assumed that the force and motion are along the x axis. To perform this integration and find the work done on a particle or the relativistic kinetic energy as a function of the particle velocity u, we first evaluate dp/dt:

$$\frac{dp}{dt} = \frac{d}{dt} \frac{mu}{\sqrt{1 - (u^2/c^2)}} = \frac{m(du/dt)}{[1 - (u^2/c^2)]^{3/2}} \tag{1.40}$$

Substituting this expression for dp/dt and $dx = u \, dt$ into Equation 1.39 gives

$$W = \int_{x_1}^{x_2} \frac{m(du/dt) \, u \, dt}{[1 - (u^2/c^2)]^{3/2}} = m \int_0^u \frac{u \, du}{[1 - (u^2/c^2)]^{3/2}}$$

where we have assumed that the particle is accelerated from rest to some final velocity u. Evaluating the integral, we find that

$$W = \frac{mc^2}{\sqrt{1 - (u^2/c^2)}} - mc^2 \tag{1.41}$$

Recall that the work-energy theorem states that the work done by a force acting on a particle equals the change in kinetic energy of the particle. Because the initial kinetic energy is zero, we conclude that the work W is equivalent to the relativistic kinetic energy K, that is,

$$K = \frac{mc^2}{\sqrt{1 - (u^2/c^2)}} - mc^2 \qquad (1.42)$$

Relativistic kinetic energy

This equation is routinely confirmed by experiments using high-energy particle accelerators. At low speeds, where $u/c \ll 1$, Equation 1.42 should reduce to the classical expression $K = \frac{1}{2}mu^2$. We can check this by using the binomial expansion $(1 - x^2)^{-1/2} \approx 1 + \frac{1}{2}x^2 + \cdots$ for $x \ll 1$, where the higher-order powers of x are ignored in the expansion. In our case, $x = u/c$, so that

$$\frac{1}{\sqrt{1 - (u^2/c^2)}} = \left(1 - \frac{u^2}{c^2}\right)^{-1/2} \approx 1 + \frac{1}{2}\frac{u^2}{c^2} + \cdots$$

Substituting this into Equation 1.42 gives

$$K \approx mc^2\left(1 + \frac{1}{2}\frac{u^2}{c^2} + \cdots\right) - mc^2 = \frac{1}{2}mu^2$$

which agrees with the classical result. A graph comparing the relativistic and nonrelativistic expressions is given in Figure 1.25. Note that in the relativistic case, the particle speed never exceeds c, regardless of the kinetic energy. The two curves are in good agreement when $u \ll c$.

It is useful to write the relativistic kinetic energy in the form

$$K = \gamma mc^2 - mc^2 \qquad (1.43)$$

where

$$\gamma = \frac{1}{\sqrt{1 - u^2/c^2}}$$

The constant term mc^2, which is independent of the speed, is called the **rest energy** of the particle. The term γmc^2, which depends on the particle speed,

Figure 1.25 A graph comparing relativistic and nonrelativistic kinetic energy. The speeds are plotted versus energy. In the relativistic case, u is always *less* than c.

is therefore the sum of the kinetic and rest energies. We define γmc^2 to be the **total energy** E, that is,

Definition of total energy

$$E = \gamma mc^2 = K + mc^2 \qquad (1.44)$$

Mass-energy equivalence

The expression $E = \gamma mc^2$ is Einstein's famous mass-energy equivalence equation, which shows that mass is a measure of energy of all forms. Furthermore, this result shows that a small mass corresponds to an enormous amount of energy because c^2 is a very large number. This concept has revolutionized the field of nuclear physics and is treated in detail in the next section.

In many situations, the momentum or energy of a particle is measured rather than its speed. It is therefore useful to have an expression relating the total energy E to the relativistic momentum p. This is accomplished by using the expression $E = \gamma mc^2$ and $p = \gamma mu$. By squaring these equations and subtracting, we can eliminate u (Problem 32). The result, after some algebra, is

Energy-momentum relation

$$E^2 = p^2c^2 + (mc^2)^2 \qquad (1.45)$$

When the particle is at rest, $p = 0$, and so we see that $E = mc^2$. That is, the total energy equals the rest energy. For the case of particles that have zero mass, such as photons (massless, chargeless particles of light) and neutrinos (massless, chargeless particles associated with electron or beta-decay of a nucleus),[9] we set $m = 0$ in Equation 1.45, and we see that

$$E = pc \qquad (1.46)$$

This equation is an *exact* expression relating energy and momentum for photons and neutrinos, which always travel at the speed of light.

Finally, note that because the mass m of a particle is independent of its motion, m must have the same value in all reference frames. On the other hand, the total energy and momentum of a particle depend on the reference frame in which they are measured, because they both depend on velocity. Because m is a constant, then according to Equation 1.45 the quantity $E^2 - p^2c^2$ must have the same value in all reference frames. That is, $E^2 - p^2c^2$ is *invariant* under a Lorentz transformation.

When dealing with electrons or other subatomic particles, it is convenient to express their energy in electron volts (eV), since the particles are usually given this energy by acceleration through a potential difference. The conversion factor is

$$1 \text{ eV} = 1.60 \times 10^{-19} \text{ J}$$

For example, the mass of an electron is 9.11×10^{-31} kg. Hence, the rest energy of the electron is

$$m_e c^2 = (9.11 \times 10^{-31} \text{ kg})(3.00 \times 10^8 \text{ m/s})^2 = 8.20 \times 10^{-14} \text{ J}$$

---

[9] Neutrino, from the Italian, means "little tiny neutral one." Following this practice, neutron should probably be neutrone (pronounced noo-trōn-eh) or "great big neutral one."

Converting this to electron volts, we have

$$m_e c^2 = (8.20 \times 10^{-14}\ \text{J})(1\ \text{eV}/1.60 \times 10^{-19}\ \text{J}) = 0.511\ \text{MeV}$$

where 1 MeV = 10^6 eV. Finally, note that because $mc^2 = 0.511$ MeV, the mass of the electron may be written $m = 0.511$ MeV/c^2, accounting for the practice of measuring particle masses in units of MeV/c^2.

EXAMPLE 1.12   The Energy of a Speedy Electron

An electron has a speed $u = 0.850c$. Find its total energy and kinetic energy in electron volts.

Solution   Using the fact that the rest energy of the electron is 0.511 MeV together with $E = \gamma mc^2$ gives

$$E = \frac{m_e c^2}{\sqrt{1 - (u^2/c^2)}} = \frac{0.511\ \text{MeV}}{\sqrt{1 - [(0.85c)^2/c^2]}}$$

$$= 1.90(0.511\ \text{MeV}) = 0.970\ \text{MeV}$$

The kinetic energy is obtained by subtracting the rest energy from the total energy:

$$K = E - m_e c^2 = 0.970\ \text{MeV} - 0.511\ \text{MeV} = 0.459\ \text{MeV}$$

EXAMPLE 1.13   The Energy of a Speedy Proton

The total energy of a proton is three times its rest energy.
    (a) Find the proton's rest energy in electron volts.

$$\text{Rest energy} = m_p c^2$$

$$= (1.67 \times 10^{-27}\ \text{kg})(3.00 \times 10^8\ \text{m/s})^2$$

$$= (1.50 \times 10^{-10}\ \text{J})(1\ \text{eV}/1.60 \times 10^{-19}\ \text{J})$$

$$= 938\ \text{MeV}$$

    (b) With what speed is the proton moving?

Because the total energy E is three times the rest energy, $E = \gamma mc^2$ gives

$$E = 3m_p c^2 = \frac{m_p c^2}{\sqrt{1 - (u^2/c^2)}}$$

$$3 = \frac{1}{\sqrt{1 - (u^2/c^2)}}$$

Solving for u gives

$$\left(1 - \frac{u^2}{c^2}\right) = \frac{1}{9} \quad \text{or} \quad \frac{u^2}{c^2} = \frac{8}{9}$$

$$u = \frac{\sqrt{8}}{3}\ c = 2.83 \times 10^8\ \text{m/s}$$

    (c) Determine the kinetic energy of the proton in electron volts.

$$K = E - m_p c^2 = 3m_p c^2 - m_p c^2 = 2m_p c^2$$

Because $m_p c^2 = 938$ MeV, $K = 1876$ MeV.
    (d) What is the proton's momentum?

We can use Equation 1.45 to calculate the momentum with $E = 3mc^2$:

$$E^2 = p^2 c^2 + (m_p c^2)^2 = (3m_p c^2)^2$$

$$p^2 c^2 = 9(m_p c^2)^2 - (m_p c^2)^2 = 8(m_p c^2)^2$$

$$p = \sqrt{8}\ \frac{m_p c^2}{c} = \sqrt{8}\ \frac{(938\ \text{MeV})}{c} = 2650\ \frac{\text{MeV}}{c}$$

Note that the unit of momentum is written MeV/c for convenience.

1.9   MASS AS A MEASURE OF ENERGY

The equation $E = \gamma mc^2$ as applied to a particle suggests that even when a particle is at rest ($\gamma = 1$) it still possesses enormous energy through its mass. The clearest experimental proof of the equivalence of mass and energy occurs in nuclear and elementary particle interactions in which both the conversion of mass into energy and the conversion of energy into mass take place. Because of this convertibility from the currency of mass into the currency of energy we can no longer accept the separate classical laws of the conservation of mass and the conservation of energy; we must instead speak of a single unified law, **the conservation of mass-energy.** Simply put, this law requires that *the sum of the mass-energy of a system of particles before interaction must equal the sum of the mass-energy of the system after interaction where the mass-energy of the ith particle is defined as*

Conservation of mass-energy

$$E_i = \frac{m_i c^2}{\sqrt{1 - (u_i^2/c^2)}}$$

To understand the conservation of mass-energy and to see how the new relativistic laws possess more symmetry and wider scope than the classical laws of momentum and energy conservation, we consider the simple inelastic collision treated earlier.

As one can see in Figure 1.24a, *classically* speaking momentum is conserved but kinetic energy is not because the total kinetic energy before collision equals mu^2 (we have replaced the v of Figure 1.24 with u) and the total kinetic energy after is zero. Now consider the same two colliding clay lumps using the relativistic mass-energy conservation law. If the mass of each lump is m, and the mass of the composite object is M, we must have:

$$E_{\text{before}} = E_{\text{after}}$$

$$\frac{mc^2}{\sqrt{1 - (u^2/c^2)}} + \frac{mc^2}{\sqrt{1 - (u^2/c^2)}} = Mc^2$$

or

$$M = \frac{2m}{\sqrt{1 - (u^2/c^2)}} \tag{1.47}$$

Because $\sqrt{1 - (u^2/c^2)} < 1$, the composite mass M is greater than the sum of the two individual masses! What's more, it is easy to show that the mass increase of the composite lump, $\Delta M = M - 2m$, is equal to the sum of the incident kinetic energies of the colliding lumps ($2K$) divided by c^2:

$$\Delta M = \frac{2K}{c^2} = \frac{2}{c^2} \left(\frac{mc^2}{\sqrt{1 - (u^2/c^2)}} - mc^2 \right) \tag{1.48}$$

Thus, we have an example of the conversion of kinetic energy to mass, and the satisfying result that in relativistic mechanics, kinetic energy is not lost in an inelastic collision but shows up as an increase in the mass of the final composite object. In fact the deeper symmetry of relativity theory shows that *both relativistic mass-energy and momentum are always conserved in a collision,* whereas classical methods show that momentum is conserved but kinetic energy is not unless the collision is perfectly elastic. Indeed, as we show in Example 1.15, relativistic momentum and energy are inextricably linked because momentum conservation in all inertial frames only holds if mass-energy conservation (Eq. 1.47) also holds.

EXAMPLE 1.14

(a) Calculate the mass increase for a completely inelastic head-on collision of two 5.0-kg balls each moving toward the other at 1000 miles per hour in opposite directions (the speed of a fast jet plane). (b) Explain why measurements on macroscopic objects have reinforced the *incorrect* beliefs that mass is conserved ($M = 2m$) and that kinetic energy is lost in an inelastic collision.

Solution   (a) $u = 1000$ mi/h $= 450$ m/s

$$\frac{u}{c} = \frac{4.5 \times 10^2 \text{ m/s}}{3.0 \times 10^8 \text{ m/s}} = 1.5 \times 10^{-6}$$

Because $u^2/c^2 \ll 1$, substituting

$$\frac{1}{\sqrt{1 - (u^2/c^2)}} \approx 1 + \frac{1}{2}\frac{u^2}{c^2}$$

in Equation 1.48 gives

$$\Delta M = 2m\left(\frac{1}{\sqrt{1 - (u^2/c^2)}} - 1\right)$$

$$\approx 2m\left(1 + \frac{1}{2}\frac{u^2}{c^2} - 1\right) \approx \frac{mu^2}{c^2}$$

$$= (5.0 \text{ kg})(1.5 \times 10^{-6})^2 = 1.1 \times 10^{-11} \text{ kg}$$

(b) Because the mass increase of 1.1×10^{-11} kg is an unmeasurably minute fraction of $2m$ (10 kg), it is quite natural to assume that the mass remains constant when macroscopic objects suffer an inelastic collision. On the other hand, the change in kinetic energy from mu^2 to 0 is so large (10^6 J) that it is readily measured to be lost in an inelastic collision of macroscopic objects.

Exercise 5   Prove that $\Delta M = 2\Delta K/c^2$ for a completely inelastic collision as stated.

EXAMPLE 1.15

Show that use of the relativistic definition of momentum

$$p = \frac{mu}{\sqrt{1 - (u^2/c^2)}}$$

leads to momentum conservation in both S and S' for the inelastic collision shown in Figure 1.24.

Solution   In frame S:

$$p_{\text{before}} = \gamma m v + \gamma m(-v) = 0$$

$$p_{\text{after}} = \gamma M V = (\gamma M)(0) = 0$$

Hence, momentum is conserved in S. Note that we have used M as the mass of the two combined masses after the collision and allowed for the possibility in relativity that M is not necessarily equal to $2m$.

In frame S':

$$p'_{\text{before}} = \gamma m v'_1 + \gamma m v'_2 = \frac{(m)(0)}{\sqrt{1 - (0)^2/c^2}}$$

$$+ \frac{m}{\{\sqrt{1 - [-2v/1 + (v^2/c^2)]^2}\}(1/c^2)}\left(\frac{-2v}{1 + v^2/c^2}\right)$$

After some algebra, we find

$$\frac{m}{\{\sqrt{1 - [2v/1 + (v^2/c^2)]^2}\}(1/c^2)} = \frac{m(1 + v^2/c^2)}{(1 - v^2/c^2)}$$

and we obtain

$$p'_{\text{before}} = \frac{m(1 + v^2/c^2)}{(1 - v^2/c^2)}\left(\frac{-2v}{1 + v^2/c^2}\right) = \frac{-2mv}{(1 - v^2/c^2)}$$

$$p'_{\text{after}} = \gamma M V' = \frac{M(-v)}{\sqrt{1 - [(-v)^2/c^2]}} = \frac{-Mv}{\sqrt{1 - v^2/c^2}}$$

To show that momentum is conserved in S', we use the fact that M is not simply equal to $2m$ in relativity. As shown above, the combined mass, M, formed from the collision of two particles, each of mass m moving toward each other with speed v, is greater than $2m$. This occurs because of the equivalence of mass and energy, that is, the kinetic energy of the incident particles shows up in relativity theory but appears as another form of energy. Thus, from Equation 1.47 we have

$$M = \frac{2m}{\sqrt{1 - (v^2/c^2)}}$$

Substituting this result for M into p'_{after}, we obtain

$$p'_{\text{after}} = \frac{2m}{\sqrt{1 - (v^2/c^2)}}\frac{-v}{\sqrt{1 - (v^2/c^2)}}$$

$$= \frac{-2mv}{1 - (v^2/c^2)} = p'_{\text{before}}$$

Hence, momentum is conserved in both S and S', provided that we use the correct relativistic definition of momentum, $p = \gamma mu$, and invoke conservation of mass and energy.

The absence of observable effects of the conservation of mass and energy with inelastic collisions of macroscopic objects impels us to look for other areas to test this law, where particle velocities are higher, masses are more precisely known, and forces are stronger than electrical or mechanical forces. This leads us naturally to consider nuclear collisions and decays because nuclear masses can be measured very precisely with a mass spectrometer, nuclear forces are much stronger than electrical forces, and decay products are often produced with extremely high velocities.

Perhaps the most direct confirmation of the conservation of mass and energy occurs in the decay of a heavy radioactive nucleus at rest into several lighter particles that are emitted with large kinetic energies. For such a nucleus of mass **Fission** M undergoing *fission* into particles with masses M_1, M_2, and M_3 and having speeds u_1, u_2, and u_3, conservation of mass and energy requires

$$Mc^2 = \frac{M_1 c^2}{\sqrt{1 - (u_1^2/c^2)}} + \frac{M_2 c^2}{\sqrt{1 - (u_2^2/c^2)}} + \frac{M_3 c^2}{\sqrt{1 - (u_3^2/c^2)}} \qquad (1.49)$$

Because the square roots are all less than one, $M > M_1 + M_2 + M_3$ and the loss of mass, $M - (M_1 + M_2 + M_3)$, appears as energy of motion of the products. This disintegration energy released per fission is often denoted by the symbol Q and can be written for our case as

$$Q = [M - (M_1 + M_2 + M_3)]c^2 = \Delta mc^2 \qquad (1.50)$$

EXAMPLE 1.16   A Fission Reaction

An excited $^{236}_{92}\text{U}$ nucleus decays at rest into $^{90}_{37}\text{Rb}$, $^{143}_{55}\text{Cs}$, and several neutrons, $^{1}_{0}\text{n}$. (a) By conserving charge and the total number of protons and neutrons write a balanced reaction equation and determine the number of neutrons produced. (b) Calculate by how much the combined "offspring" mass is less than the "parent" mass. (c) Calculate the energy released per fission. (d) Calculate the energy released in kilowatt hours when 1 kg of uranium undergoes fission in a power plant which is 40% efficient.

Solution In general an element is represented by the symbol ^A_ZX, where X represents the symbol for the element, A is the number of neutrons plus protons in the nucleus (mass number), and Z is the number of protons in the nucleus (atomic number). Conserving charge and number of nucleons gives:

$$^{236}_{92}\text{U} \longrightarrow \, ^{90}_{37}\text{Rb} + \, ^{143}_{55}\text{Cs} + 3\,^{1}_{0}\text{n}$$

So three neutrons are produced per fission.
(b) The masses of the decay particles are given in Appendix B in terms of atomic mass units, u, where $1\,\text{u} = 1.660 \times 10^{-27} = 931.5\ \text{MeV}/c^2$.

$$\Delta m = M_\text{U} - (M_\text{Rb} + M_\text{Cs} + 3m_\text{n}) = 236.045563\ \text{u}$$
$$- 89.914811\ \text{u} + 142.927220\ \text{u} + (3)(1.008665)\ \text{u}$$
$$= 0.177537\ \text{u} = 2.9471 \times 10^{-28}\ \text{kg}$$

Therefore the reaction products have a combined mass that is about 3.0×10^{-28} kg less than the initial uranium mass.
(c) The energy given off per fission event is just Δmc^2. This is most easily calculated if Δm is first converted to mass units of MeV/c^2. Because $1\,\text{u} = 931.5\ \text{MeV}/c^2$,

$$\Delta m = (0.177537\ \text{u})(931.5\ \text{MeV}/c^2) = 165.4\ \text{MeV}/c^2$$

$$Q = \Delta mc^2 = 165.4\ \frac{\text{MeV}}{c^2}\, c^2 = 165.4\ \text{MeV}$$

(d) To find the energy released by the fission of 1 kg of uranium we need to calculate the number of nuclei, N, contained in 1 kg of ^{236}U.

$$N = \frac{(6.02 \times 10^{23}\ \text{nuclei/mol})}{(236\ \text{g/mol})}(1000\ \text{g})$$
$$= 2.55 \times 10^{24}\ \text{nuclei}$$

The total energy produced, E, is

$$E = (\text{efficiency})NQ$$
$$= (0.40)(2.55 \times 10^{24}\ \text{nuclei})(165\ \text{MeV/nucleus})$$
$$= 1.68 \times 10^{26}\ \text{MeV}$$
$$= (1.68 \times 10^{26}\ \text{MeV})(4.45 \times 10^{-20}\ \text{kWh/MeV})$$
$$= 7.48 \times 10^{6}\ \text{kWh}$$

Exercise 6 How long will this amount of energy keep a 100-W lightbulb burning?
Answer ≈ 8500 years.

Although the simplest case showing the release of nuclear energy is the decay of a heavy unstable element into several lighter elements, the most common case is the one in which the mass of a composite particle is *less than* the sum of the particle masses composing it. By examining Appendix B, you can see that the mass of any nucleus is less than the sum of its component neutrons and protons by an amount Δm. This occurs because the nuclei are stable, bound systems of neutrons and protons (bound by strong attractive nuclear forces), and in order to disassociate them into separate nucleons an amount of energy Δmc^2 must be supplied to the nucleus. This energy or work required to pull a bound system apart, leaving its component parts free of attractive forces and at rest, is called the binding energy, *BE*. Thus, we describe the mass and energy of a bound system by the equation

$$Mc^2 + BE = \sum_{i=1}^{n} m_i c^2 \qquad (1.51)$$

where M is the bound system mass, the m_i's are the free component particle masses, and n is the number of component particles. Two general comments are needed about Equation 1.51. First, it applies quite generally to any type of system bound by attractive forces, whether gravitational, electrical (chemical) or nuclear. For example, the mass of a water molecule is less than the combined mass of two free hydrogen atoms and a free oxygen atom, although the mass difference cannot be directly measured in this case. (The mass difference can be measured in the nuclear case because the forces and the binding energy are so much greater.) Second, Equation 1.51 shows the possibility of liberating huge quantities of energy, *BE*, if one reads the equation from right to left; that is, one collides nuclear particles with enough energy to overcome proton repulsion and fuse the particles into new elements with less mass. Such a process is called *fusion*, one example of which is a reaction in which two deuterium nuclei combine to form a helium nucleus, releasing 23.9 MeV per fusion. We can write this reaction schematically as follows:

Fusion

$$^2_1\text{H} + ^2_1\text{H} \longrightarrow ^4_2\text{He} + 23.9 \text{ MeV}$$

EXAMPLE 1.17

(a) How much lighter is a molecule of water than two hydrogen atoms and an oxygen atom? The binding energy of water is about 3 eV. (b) Find the fractional loss of mass per gram of water formed. (c) Find the total energy released (mainly as heat and light) when one gram of water is formed.

Solution   (a) Equation 1.51 may be solved for the mass difference as follows:

$$\Delta m = (m_\text{H} + m_\text{H} + m_\text{O}) - M_{\text{H}_2\text{O}} = \frac{BE}{c^2} = \frac{3 \text{ eV}}{c^2}$$

$$= \frac{(3.0 \text{ eV})(1.6 \times 10^{-19} \text{ J/eV})}{(3.0 \times 10^8 \text{ m/s})^2} = 5.3 \times 10^{-36} \text{ kg}$$

(b) To find the fractional loss of mass per molecule we divide Δm by the mass of a water molecule, $M_{\text{H}_2\text{O}} = 18u = 3.0 \times 10^{-26}$ kg:

$$\frac{\Delta m}{M_{\text{H}_2\text{O}}} = \frac{5.3 \times 10^{-36} \text{ kg}}{3.0 \times 10^{-26} \text{ kg}} = 1.8 \times 10^{-10}$$

Because the fractional loss of mass per molecule is the same as the fractional loss per gram of water formed, 1.8×10^{-10} g of mass would be lost for each gram of water formed. Because this is much too small a mass to be measured directly, this calculation shows that nonconservation of proper mass does not generally show up as a measurable effect in chemical reactions.

(c) The energy released when one gram of H_2O is formed is simply the change in mass when one gram of water is formed times c^2:

$$E = \Delta mc^2 = (1.8 \times 10^{-13} \text{ kg})(3.0 \times 10^8 \text{ m/s})^2 \approx 16 \text{ kJ}$$

This energy change, as opposed to the decrease in mass, is easily measured, providing another case similar to Example 1.14 in which mass changes are minute but energy changes are easily measured.

1.10 CONSERVATION OF RELATIVISTIC MOMENTUM, MASS, AND ENERGY

So far we have considered only cases of the conservation of mass and energy. By far, however, the most common and strongest confirmation of relativity theory comes from the daily application of relativistic momentum and energy conservation to elementary particle interactions. Often the measurement of momentum (from the path curvature in a magnetic field—see Example 1.11) and kinetic energy (from the distance a particle travels in a known substance before coming to rest) are enough when combined with conservation of momentum, mass, and energy to determine fundamental particle properties like mass, charge, and mean lifetime.

EXAMPLE 1.18   Measuring the Mass of the π^+ Meson

The π^+ meson (also called the pion) is a subatomic particle responsible for the strong nuclear force between protons and neutrons. It is observed to decay at rest into a μ^+ meson (muon) and a neutrino, denoted ν. Because the neutrino has no charge or mass (talk about elusive!), it leaves no track in a bubble chamber. (A bubble chamber is a large chamber filled with liquid hydrogen that shows the tracks of charged particles as a series of tiny bubbles.) However, the track of the charged muon is visible as it loses kinetic energy and comes to rest (Fig. 1.26). If the mass of the muon is known to be 106 MeV/c^2, and the kinetic energy, K, of the muon is measured to be 4.6 MeV from its track length, find the mass of the π^+.

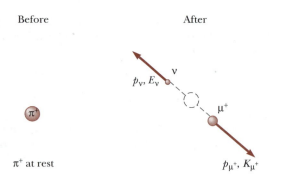

Before   After

p_ν, E_ν

ν

μ^+

π^+

π^+ at rest p_{μ^+}, K_{μ^+}

Figure 1.26   (Example 1.18) Decay of the pion at rest into a neutrino and a muon.

Solution   Conserving energy gives

$$E_\pi = E_\mu + E_\nu$$

Because the pion is stopped when it decays, and the neutrino has zero mass,

$$m_\pi c^2 = \sqrt{(m_\mu c^2)^2 + (p_\mu^2 c^2)} + p_\nu c \qquad (1)$$

Conserving momentum in the decay yields $p_\mu = p_\nu$. Substituting the muon momentum for the neutrino momentum in Equation 1 gives the following expression for the rest energy of the pion in terms of the muon mass and momentum:

$$m_\pi c^2 = \sqrt{(m_\mu c^2)^2 + (p_\mu^2 c^2)} + p_\mu c \qquad (2)$$

Finally, to obtain p_μ from the measured value of the muon's kinetic energy, K_μ, we start with Equation 1.45, $E_\mu^2 = p_\mu^2 c^2 + (m_\mu c^2)^2$, and solve it for $p_\mu^2 c^2$:

$$p_\mu^2 c^2 = E_\mu^2 - (m_\mu c^2)^2 = (K_\mu + m_\mu c^2)^2 - (m_\mu c^2)^2$$
$$= K_\mu^2 + 2K_\mu m_\mu c^2$$

Substituting this expression for $p_\mu^2 c^2$ into Equation (2) yields the desired expression for the pion mass in terms of the measured quantities:

$$m_\pi c^2 = \sqrt{(m_\mu^2 c^4 + K_\mu^2 + 2K_\mu m_\mu c^2}$$
$$+ \sqrt{K_\mu^2 + 2K_\mu m_\mu c^2} \qquad (3)$$

Finally substituting $m_\mu c^2 = 106$ MeV and $K_\mu = 4.6$ MeV into Equation (3) gives

$$m_\pi c^2 = 111 \text{ MeV} + 31 \text{ MeV} \approx 1.4 \times 10^2 \text{ MeV}$$

Thus, the mass of the pion is

$$m_\pi = 140 \text{ MeV}/c^2$$

This result shows why this particle is called a meson; it has a *me*dium mass between the light electron $(0.511 \text{ MeV}/c^2)$ and the heavy proton $(938 \text{ MeV}/c^2)$.

SUMMARY

The two basic postulates of the **special theory of relativity** are as follows:

- The laws of physics must be the same for all observers moving at constant velocity with respect to one another.
- The speed of light must be the same for all inertial observers, independent of their relative motion.

To satisfy these postulates, the Galilean transformations must be replaced by the **Lorentz transformations** given by

$$x' = \gamma(x - vt) \tag{1.25}$$

$$y' = y \tag{1.26}$$

$$z' = z \tag{1.27}$$

$$t' = \gamma\left(t - \frac{v}{c^2}x\right) \tag{1.28}$$

where

$$\gamma = \frac{1}{\sqrt{1 - (v^2/c^2)}} \tag{1.29}$$

In these equations, it is assumed that the primed system moves with a speed v along the xx' axes.

The relativistic form of the **velocity transformation** is

$$u_x' = \frac{u_x - v}{1 - (u_x v/c^2)} \tag{1.32}$$

where u_x is the speed of an object as measured in the S frame and u_x' is its speed measured in the S′ frame.

Some of the consequences of the special theory of relativity are as follows:

- Clocks in motion relative to an observer appear to be slowed down by a factor γ. This is known as **time dilation.**
- Lengths of objects in motion appear to be contracted in the direction of motion by a factor of $1/\gamma$. This is known as **length contraction.**
- Events that are simultaneous for one observer are not simultaneous for another observer in motion relative to the first.

These three statements can be summarized by saying that duration, length, and simultaneity are not absolute concepts in relativity.

The relativistic expression for the **momentum** of a particle moving with a velocity **u** is

$$\mathbf{p} \equiv \frac{m\mathbf{u}}{\sqrt{1 - (u^2/c^2)}} = \gamma m\mathbf{u} \tag{1.35}$$

where γ is redefined in the following equation as

$$\gamma = \frac{1}{\sqrt{1 - (u^2/c^2)}}$$

The relativistic expression for the **kinetic energy** of a particle is

$$K = \gamma mc^2 - mc^2 \tag{1.42}$$

where mc^2 is called the **rest energy** of the particle.

The total energy E of a particle is related to the mass through the expression:

$$E = \gamma mc^2 = \frac{mc^2}{\sqrt{1 - (u^2/c^2)}} \tag{1.44}$$

Another useful expression relates the relativistic momentum to the total energy through the equation

$$E^2 = p^2 c^2 + (mc^2)^2 \tag{1.45}$$

Finally, the law of the conservation of mass and energy states that the sum of the masses and energies of a system of particles before interaction must equal the sum of the masses and the energies of the particles after interaction where the energy of the ith particle is

$$E = \frac{m_i c^2}{\sqrt{1 - (u_i^2/c^2)}}$$

Application of the conservation of mass and energy to the specific cases of (1) the fission of a heavy nucleus at rest and (2) the fusion of several particles into a composite nucleus with less total mass allows us to define (1) the energy released per fission, Q, and (2) the binding energy of a composite system, BE.

SUGGESTIONS FOR FURTHER READING

1. E. F. Taylor and J. A. Wheeler, *Spacetime Physics,* San Francisco, W. H. Freeman, 1963.
2. R. Resnick, *Introduction to Special Relativity,* New York, Wiley, 1968.
3. A. P. French, *Special Relativity,* New York, Norton, 1968.
4. H. Bondi, *Relativity and Common Sense,* Science Study Series, Garden City, N.Y., Doubleday, 1964.
5. J. Bronowski, "The Clock Paradox," *Sci. American,* February, 1963, p. 134.
6. A. Einstein, *Out of My Later Years,* New York, World Publishing, 1971.
7. A. Einstein, *Ideas and Opinions,* New York, Crown, 1954.
8. G. Gamow, *Mr. Tompkins in Wonderland,* New York, Cambridge University Press, 1939.
9. L. Infeld, *Albert Einstein,* New York, Scribner's, 1950.
10. J. Schwinger, *Einstein's Legacy,* Scientific American Library, New York, W. H. Freeman, 1985.
11. R. S. Shankland, "The Michelson-Morley Experiment," *Sci. American,* November 1964, p. 107.
12. R. Skinner, *Relativity for Scientists and Engineers,* New York, Dover Publications, 1982.
13. N. Mermin, *Space and Time in Special Relativity,* Prospect Heights, Illinois, Waveland Press, 1989.

QUESTIONS

1. What two measurements will two observers in relative motion *always* agree on?
2. A spaceship in the shape of a sphere moves past an observer on Earth with a speed 0.5c. What shape will the observer see as the spaceship moves past?
3. An astronaut moves away from Earth at a speed close to the speed of light. If an observer on Earth could make measurements of the astronaut's size and pulse rate, what changes (if any) would he or she measure? Would the astronaut measure any changes?
4. Two identically constructed clocks are synchronized. One is put in an eastward orbit around Earth while the other remains on Earth. Which clock runs slower? When the moving clock returns to Earth, will the two clocks still be synchronized?
5. Two lasers situated on a moving spacecraft are triggered simultaneously. An observer on the spacecraft claims to see the pulses of light simultaneously. What condition is necessary in order that another observer agrees that the two pulses are emitted simultaneously?
6. When we say that a moving clock runs slower than a stationary one, does this imply that there is something physically unusual about the moving clock?
7. When we speak of time dilation, do we mean that time passes more slowly in moving systems or that it simply appears to do so?
8. List some ways our day-to-day lives would change if the speed of light were only 50 m/s.
9. Give a physical argument to show that it is impossible to accelerate an object of mass m to the speed of light, even with a continuous force acting on it.
10. It is said that Einstein, in his teenage years, asked the question, "What would I see in a mirror if I carried it in my hands and ran at the speed of light?" How would you answer this question?
11. Because mass is a measure of energy, can we conclude that a compressed spring has more mass than the same spring when it is not compressed? On the basis of your answer, which has more mass, a spinning planet or an otherwise identical but nonspinning planet?
12. Suppose astronauts were paid according to the time spent traveling in space. After a long voyage at a speed near that of light, a crew of astronauts return and open their pay envelopes. What will their reaction be?
13. What happens to the density of an object as its speed increases, as measured by an Earth observer?

PROBLEMS

1.2 The Principle of Relativity

1. In a laboratory frame of reference, an observer notes that Newton's second law is valid provided it is expressed as $\mathbf{F} = d\mathbf{p}/dt$. Show that it is also valid for an observer moving at a constant speed relative to the laboratory frame.
2. Show that Newton's second law is not valid in a reference frame moving past the laboratory frame of Problem 1 with a constant acceleration.
3. A 2000-kg car moving with a speed of 20 m/s collides with and sticks to a 1500-kg car at rest at a stop sign. Show that because momentum is conserved in the rest frame, momentum is also conserved in a reference frame moving with a speed of 10 m/s in the direction of the moving car.
4. A billiard ball of mass 0.3 kg moves with a speed of 5 m/s and collides elastically with a ball of mass 0.2 kg moving in the opposite direction with a speed of 3 m/s. Show that because momentum is conserved in the rest frame, it is also conserved in a frame of reference moving with a speed of 2 m/s in the direction of the second ball.

1.3 Michelson-Morley Experiment

5. An airplane flying upwind, downwind, and crosswind shows the main principle of the Michelson-Morley experiment. A plane capable of flying at speed c in still air is flying in a wind of speed v. Suppose the plane flies upwind a distance L and then returns downwind to its starting point. (a) Find the time needed to make the round trip and compare it with the time to fly crosswind a distance L and return. Before calculating these times, sketch the two situations. (b) Compute the time difference for the two trips if $L = 100$ mi, $c = 500$ mi/h, and $v = 100$ mi/h.

1.5 Consequences of Special Relativity

6. With what speed will a clock have to be moving in order to run at a rate that is one half the rate of a clock at rest?
7. How fast must a meter stick be moving if its length is observed to shrink to 0.5 m?
8. A clock on a moving spacecraft runs 1 s slower per day relative to an identical clock on Earth. What is the relative speed of the spacecraft? (*Hint:* For $v/c \ll 1$, note that $\gamma \approx 1 + v^2/2c^2$.)
9. A meter stick moving in a direction parallel to its length appears to be only 75 cm long to an observer. What is the speed of the meter stick relative to the observer?
10. A spacecraft moves at a speed of $0.900c$. If its length is L as measured by an observer on the spacecraft, what is the length measured by a ground observer?
11. The average lifetime of a pi meson in its own frame of

reference is 2.6×10^{-8} s. (This is the proper lifetime.) If the meson moves with a speed of $0.95c$, what is (a) its mean lifetime as measured by an observer on Earth and (b) the average distance it travels before decaying, as measured by an observer on Earth?

12. An atomic clock is placed in a jet airplane. The clock measures a time interval of 3600 s when the jet moves with a speed of 400 m/s. How much longer or shorter a time interval does an identical clock held by an observer on the ground measure? (*Hint:* For $v/c \ll 1$, $\gamma \approx 1 + v^2/2c^2$.)

13. An astronaut at rest on Earth has a heartbeat rate of 70 beats/min. What will this rate be when the astronaut is traveling in a spaceship at $0.90c$ as measured (a) by an observer also in the ship and (b) by an observer at rest on the Earth?

14. The muon is an unstable particle that spontaneously decays into an electron and two neutrinos. If the number of muons at $t = 0$ is N_0, the number at time t is given by $N = N_0 e^{-t/\tau}$, where τ is the mean lifetime, equal to 2.2 μs. Suppose the muons move at a speed of $0.95c$ and there are 5.0×10^4 muons at $t = 0$. (a) What is the observed lifetime of the muons? (b) How many muons remain after traveling a distance of 3.0 km?

15. A rod of length L_0 moves with a speed v along the horizontal direction. The rod makes an angle of θ_0 with respect to the x' axis. (a) Show that the length of the rod as measured by a stationary observer is given by $L = L_0[1 - (v^2/c^2) \cos^2 \theta_0]^{1/2}$. (b) Show that the angle that the rod makes with the x axis is given by the expression $\tan \theta = \gamma \tan \theta_0$. These results show that the rod is both contracted and rotated. (Take the lower end of the rod to be at the origin of the primed coordinate system.)

16. *The classical Doppler shift for light.* A light source recedes from an observer with a speed v that is small compared with c. (a) Show that in this case, Equation 1.13 reduces to

$$\frac{\Delta f}{f} \approx -\frac{v}{c}$$

(b) Also show that in this case

$$\frac{\Delta \lambda}{\lambda} \approx \frac{v}{c}$$

(*Hint:* Differentiate $\lambda f = c$ to show that $\Delta \lambda / \lambda = -\Delta f/f$) (c) Spectroscopic measurements of an absorption line normally found at $\lambda = 397$ nm reveal a red shift of 20 nm for light coming from a galaxy in Ursa Major. What is the recessional speed of this galaxy?

17. Calculate, for the judge, how fast you were going in miles per hour when you ran the red light because it appeared Doppler-shifted green to you. Take red light to have a wavelength of 650 nm and green to have a wavelength of 550 nm.

18. (a) How fast and in what direction must galaxy A be moving if an absorption line found at 550 nm (green) for a stationary galaxy is shifted to 450 nm (blue) for A? (b) How fast and in what direction is galaxy B moving if it shows the same line shifted to 700 nm (red)?

19. *Doppler radar.* An important practical application of the Doppler effect is the use of radar to determine the speed of a moving object. In this case the Doppler shift of the electromagnetic radar signal reflected from the moving object is directly proportional to the radial speed of the moving object with respect to the radar transmitter. If a police radar transmitter radiates at 10.0 GHz, calculate the frequency shift observed by the police for a car traveling at (a) 60.0 mi/h and (b) 70.0 mi/h. (c) What measurement precision in frequency is required to distinguish between the two cars? State your answer as a fraction $\Delta f/f$. How can such a high precision be obtained? (1.00 mi/h = 0.447 m/s).

1.6 The Lorentz Coordinate and Velocity Transformations

20. Two spaceships approach each other, each moving with the *same* speed as measured by an observer on the Earth. If their *relative* speed is $0.70c$, what is the speed of each spaceship?

21. An electron moves to the right with a speed of $0.90c$ relative to the laboratory frame. A proton moves to the right with a speed of $0.70c$ relative to the electron. Find the speed of the proton relative to the laboratory frame.

22. An observer on Earth observes two spacecraft moving in the *same* direction toward the Earth. Spacecraft A appears to have a speed of $0.50c$, and spacecraft B appears to have a speed of $0.80c$. What is the speed of spacecraft A measured by an observer in spacecraft B?

23. *Speed of light in a moving medium.* The motion of a medium such as water influences the speed of light. This effect was first observed by Fizeau in 1851. Consider a light beam passing through a horizontal column of water moving with a speed v. (a) Show that if the beam travels in the same direction as the flow of water, the speed of light measured in the laboratory frame is given by

$$u = \frac{c}{n} \left(\frac{1 + nv/c}{1 + v/nc} \right)$$

where n is the index of refraction of the water. (*Hint:* Use the velocity transformation relation, Equation 1.34, and note that the speed of light with respect to the moving frame is given by c/n.) (b) Show that for $v \ll c$, the expression above is in good agreement with Fizeau's experimental result

$$u \approx \frac{c}{n} + v - \frac{v}{n^2}$$

This proves that the Lorentz velocity transformation and not the Galilean velocity transformation is correct for light.

24. An observer in frame S sees lightning simultaneously strike two points 100 m apart. The first strike occurs at $x_1 = y_1 = z_1 = t_1 = 0$ and the second at $x_2 = 100$ m, $y_2 = z_2 = t_2 = 0$. (a) What are the coordinates of these two events in a frame S' moving in the standard configuration at $0.70c$ relative to S? (b) How far apart are the events in S'? (c) Are the events simultaneous in S'? If not, what is the difference in time between the events, and which event occurs first?

25. As seen from Earth, two spaceships A and B are approaching along perpendicular directions. If A is observed by an Earth observer to have velocity $u_y = -0.90c$ and B to have a velocity $u_x = +0.90c$, find the speed of ship A as measured by the pilot of B.

1.7 Relativistic Momentum and the Relativistic Form of Newton's Laws

26. Calculate the momentum of a proton moving with a speed of (a) $0.010c$, (b) $0.50c$, (c) $0.90c$. (d) Convert the answers of (a)–(c) to MeV/c.

27. An electron has a momentum that is 90% larger than its classical momentum. (a) Find the speed of the electron. (b) How would your result change if the particle were a proton?

28. Consider the relativistic form of Newton's second law. Show that when **F** is parallel to **v**,

$$F = m \left(1 - \frac{v^2}{c^2} \right)^{-3/2} \frac{dv}{dt}$$

where m is the mass of an object and v is its speed.

29. A charged particle moves along a straight line in a uniform electric field E with a speed v. If the motion and the electric field are both in the x direction, (a) show that the magnitude of the acceleration of the charge q is given by

$$a = \frac{dv}{dt} = \frac{qE}{m} \left(1 - \frac{v^2}{c^2} \right)^{3/2}$$

(b) Discuss the significance of the dependence of the acceleration on the speed. (c) If the particle starts from rest at $x = 0$ at $t = 0$, find the speed of the particle and its position after a time t has elapsed. Comment on the limiting values of v and x as $t \rightarrow \infty$.

30. Recall that the magnetic force on a charge q moving with velocity **v** in a magnetic field **B** is equal to $q\mathbf{v} \times \mathbf{B}$. If a charged particle moves in a circular orbit with a fixed speed v in the presence of a constant magnetic field, use Newton's second law to show that the frequency of its orbital motion is

$$f = \frac{qB}{2\pi m} \left(1 - \frac{v^2}{c^2} \right)^{1/2}$$

31. Show that the momentum of a particle having charge e moving in a circle of radius R in a magnetic field B is given by $p = 300BR$, where p is in MeV/c, B is in teslas, and R is in meters.

1.8 Relativistic Energy

32. Show that the energy-momentum relationship given by $E^2 = p^2c^2 + (mc^2)^2$ follows from the expressions $E = \gamma mc^2$ and $p = \gamma mu$.

33. A proton moves at a speed of $0.95c$. Calculate its (a) rest energy, (b) total energy, and (c) kinetic energy.

34. An electron has a kinetic energy five times greater than its rest energy. Find (a) its total energy and (b) its speed.

35. Find the speed of a particle whose total energy is 50% greater than its rest energy.

36. A proton in a high-energy accelerator is given a kinetic energy of 50 GeV. Determine the (a) momentum and (b) speed of the proton.

37. An electron has a speed of $0.75c$. Find the speed of a proton which has (a) the same kinetic energy as the electron and (b) the same momentum as the electron.

38. Protons in an accelerator at the Fermi National Laboratory near Chicago are accelerated to an energy of 400 times their rest energy. (a) What is the speed of these protons? (b) What is their kinetic energy in MeV?

39. *How long will the Sun shine, Nellie?* The Sun radiates about 4.0×10^{26} J of energy into space each second. (a) How much mass is released as radiation each second? (b) If the mass of the Sun is 2.0×10^{30} kg, how long can the Sun survive if the energy release continues at the present rate?

40. Electrons in projection television sets are accelerated through a total potential difference of 50 000 V. (a) Calculate the speed of the electrons using the relativistic form of kinetic energy assuming the electrons start from rest. (b) Calculate the speed of the electrons using the classical form of kinetic energy. (c) Is the difference in speed significant in the design of this set in your opinion?

41. As noted in Section 1.8, the quantity $E - p^2c^2$ is an *invariant* in relativity theory. This means that the quantity $E^2 - p^2c^2$ has the same value in all inertial frames even though E and p have different values in different frames. Show this explicitly by considering the following case. A particle of mass m is moving in the $+x$ direction with speed u and has momentum p and energy E in the frame S. (a) If S' is moving at speed v in the standard way, find the momentum p' and energy E' observed in S'. (*Hint:* Use the Lorentz velocity transformation to find

p' and E') Is $E = E'$ and $p = p'$? (b) Show that $E^2 - p^2c^2$ is explicitly equal to $E'^2 - p'^2c^2$.

1.9   Mass as a Measure of Energy

42. A radium isotope decays to a radon isotope ^{222}Rn by emitting an α particle (a helium nucleus) according to the decay scheme ^{226}Ra $\rightarrow$ ^{222}Rn + ^{4}He. The masses of the atoms are 226.0254 (Ra), 222.0175 (Rn), and 4.0026 (He). How much energy is released as the result of this decay?

43. Consider the decay $^{55}_{24}$Cr $\rightarrow$ $^{55}_{25}$Mn + e$^-$, where e$^-$ is an electron. The ^{55}Cr nucleus has a mass of 54.9279 u, and the ^{55}Mn nucleus has a mass of 54.9244 u. (a) Calculate the mass difference in MeV. (b) What is the maximum kinetic energy of the emitted electron?

44. Calculate the binding energy in MeV per nucleon in the isotope $^{12}_6$C. Note that the mass of this isotope is exactly 12 u, and the masses of the proton and neutron are 1.007276 u and 1.008665 u, respectively.

45. The free neutron is known to decay into a proton, an electron, and an antineutrino $\bar{\nu}$ (of zero rest mass) according to

$$\text{n} \longrightarrow \text{p} + \text{e}^- + \bar{\nu}$$

This is called *beta decay,* and will be discussed further in Chapter 13. The decay products are measured to have a total kinetic energy of 0.781 MeV $\pm$ 0.005 MeV. Show that this observation is consistent with the excess energy predicted by the Einstein mass-energy relationship.

1.10   Conservation of Relativistic Momentum, Mass, and Energy

46. An electron having kinetic energy $K = 1.000$ MeV makes a head-on collision with a positron at rest. (A positron is an antimatter particle that has the same mass as the electron but opposite charge.) In the collision the two particles annihilate each other and are replaced by

two γ rays of equal energy, each traveling at equal angles θ with the electron's direction of motion. (Gamma rays are massless particles of electromagnetic radiation having energy $E = pc$.) Find the energy E, momentum p, and angle of emission θ of the γ rays.

47. The K^0 meson is an uncharged member of the particle "zoo" that decays into two charged pions according to K^0 $\rightarrow$ π^+ + π^-. The pions have opposite charges as indicated and the same mass, $m_\pi = 140$ MeV/c^2. Suppose that a K^0 at rest decays into two pions in a bubble chamber in which a magnetic field of 2.0 T is present (see Fig. P1.47). If the radius of curvature of the pions is 34.4 cm, find (a) the momenta and speeds of the pions and (b) the mass of the K^0 meson.

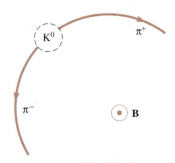

Figure P1.47   A sketch of the tracks made by the π^+ and π^- in the decay of the K^0 meson at rest. The pion motion is perpendicular to **B**. (**B** is directed out of the page.)

48. An unstable particle having a mass of 3.34×10^{-27} kg is initially at rest. The particle decays into two fragments that fly off with velocities of $0.987c$ and $-0.868c$. Find the rest masses of the fragments.

ADDITIONAL PROBLEMS

49. In 1962, when Scott Carpenter orbited Earth 22 times, the press stated that for each orbit he aged 2 millionths of a second less than if he had remained on Earth. (a) Assuming that he was 160 km above Earth in an eastbound circular orbit, determine the time difference between someone on Earth and the orbiting astronaut for the 22 orbits. You will need to use the approximation, $\sqrt{1 - x} \approx 1 - x/2$, for small x. (b) Did the press report accurate information? Explain.

50. The proper length of one spaceship is three times that of another. The two spaceships are traveling in the same direction and, while both are passing overhead, an Earth observer measures the two spaceships to have the same

length. If the slower spaceship is moving with a speed of $0.35c$, determine the speed of the faster spaceship.

51. The pion has an average lifetime of 26.0 ns when at rest. For it to travel 10.0 m, how fast must it move?

52. If astronauts could travel at $v = 0.95c$, we on Earth would say it takes $(4.2/0.95) = 4.4$ years to reach Alpha Centauri, 4.2 lightyears away. The astronauts disagree. (a) How much time passes on the astronauts' clocks? (b) What distance to Alpha Centauri do the astronauts measure?

53. A spaceship moves away from Earth at a speed v and fires a shuttle craft in the forward direction at a speed v relative to the ship. The pilot of the shuttle craft launches

a probe at speed v relative to the shuttle craft. Determine (a) the speed of the shuttle craft relative to Earth, and (b) the speed of the probe relative to Earth.

54. In a nuclear power plant the fuel rods last three years before being replaced. If a 1.0-GW plant operates at 80% capacity for the three years, how much fuel is consumed?

55. An observer in a rocket moves toward a mirror at speed v relative to the reference frame labeled by S in Figure P1.55. The mirror is stationary with respect to S. A light pulse emitted by the rocket travels toward the mirror and is reflected back to the rocket. The front of the rocket is a distance d from the mirror (as measured by observers in S) at the moment the light pulse leaves the rocket. What is the total travel time of the pulse as measured by observers in (a) the S frame and (b) the front of the rocket?

Figure P1.55

56. A physics professor on Earth gives an exam to her students who are on a spaceship traveling at speed v relative to Earth. The moment the ship passes the professor, she signals the start of the exam. If she wishes her students to have time T_0 (spaceship time) to complete the exam, show that she should wait a time (Earth time) of

$$T = T_0\sqrt{\frac{1 - v/c}{1 + v/c}}$$

before sending a light signal telling them to stop. (*Hint:* Remember that it takes some time for the second light signal to travel from the professor to the students.)

57. A yet-to-be-built spacecraft starts from Earth moving at constant speed to the yet-to-be-discovered planet Retah, which is 20 lighthours away from Earth. It takes 25 h (according to an Earth observer) for a spacecraft to reach this planet. Assuming that the clocks are synchronized at the beginning of the journey, compare the time elapsed in the spacecraft's frame for this one-way journey with the time elapsed as measured by an Earth-based clock.

58. Suppose our Sun is about to explode. In an effort to escape, we depart in a spaceship at $v = 0.80c$ and head toward the star Tau Ceti, 12 lightyears away. When we reach the midpoint of our journey from the Earth, we see our Sun explode and, unfortunately, at the same instant we see Tau Ceti explode as well. (a) In the spaceship's frame of reference, should we conclude that the two explosions occurred simultaneously? If not, which occurred first? (b) In a frame of reference in which the Sun and Tau Ceti are at rest, did they explode simultaneously? If not, which exploded first?

59. Two powerless rockets are on a collision course. The rockets are moving with speeds of $0.800c$ and $0.600c$ and are initially 2.52×10^{12} m apart as measured by Liz, an Earth observer, as shown in Figure P1.59. Both rockets are 50.0 m in length as measured by Liz. (a) What are their respective proper lengths? (b) What is the length of each rocket as measured by an observer in the other rocket? (c) According to Liz, how long before the rockets collide? (d) According to rocket 1, how long before they collide? (e) According to rocket 2, how long before they collide? (f) If both rocket crews are capable of total evacuation within 90 min (their own time), will there be any casualties?

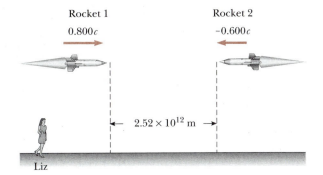

Figure P1.59

60. As measured by observers in a reference frame S, a particle having charge q moves with velocity $\mathbf{v}$ in a magnetic field $\mathbf{B}$ and an electric field $\mathbf{E}$. The resulting force on the particle is then measured to be $\mathbf{F} = q(\mathbf{E} + \mathbf{v} \times \mathbf{B})$. Another observer moves along with the charged particle and measures its charge to be q also but measures the electric field to be $\mathbf{E}'$. If both observers are to measure the same force, $\mathbf{F}$, show that $\mathbf{E}' = \mathbf{E} + \mathbf{v} \times \mathbf{B}$.

Essay

THE RENAISSANCE OF GENERAL RELATIVITY

Clifford M. Will

McDonnell Center for the Space Sciences, Washington University

Despite its enormous influence on scientific thought in its early years, general relativity had become a sterile, formalistic subject by the late 1950s, cut off from the mainstream of physics. Yet by 1970, it had become one of the most active and exciting branches of physics. It took on new roles both as a theoretical tool of the astrophysicist and as a playground for the elementary-particle physicist. New experiments verified its predictions in unheard-of ways and to remarkable levels of precision. One of the most remarkable and important aspects of this renaissance was the degree to which experiment and observation motivated and complemented theoretical advances.

This was not always the case. In deriving general relativity during the final months of 1915, Einstein was not particularly motivated by a desire to account for observational results. Instead, he was driven by purely theoretical ideas of elegance and simplicity. His goal was to produce a theory of gravitation that incorporated both the special theory of relativity, which deals with physics in inertial frames, and the principle of equivalence, the proposal that physics in a frame falling freely in a gravitational field is in some sense equivalent to physics in an inertial frame.

Once he formulated the general theory, however, Einstein proposed three tests. One was an immediate success: the explanation of the anomalous advance in the perihelion of Mercury of 43 arcseconds per century, a problem that had bedeviled celestial mechanicians of the latter part of the 19th century. The next test, the deflection of light by the Sun, was such a success that it produced what today would be called a "media event." The third test, the gravitational red shift of light, remained unfulfilled until 1960, by which time it was no longer viewed as a true test of general relativity.

The turning point for general relativity came in the early 1960s, when discoveries of unusual astronomical objects such as quasars demonstrated that the theory has important applications in astrophysics. After 1960, the pace of research in general relativity and in an emerging field called "relativistic astrophysics" accelerated.

EINSTEIN'S THREE TESTS

Perihelion Shift The explanation of the anomalous perihelion shift of Mercury's orbit was an early triumph of general relativity. This had been an unsolved problem in celestial mechanics for over half a century, since the announcement by Urbain Le Verrier in 1859 that, after the perturbing effects of the planets on Mercury's orbit had been accounted for, there remained in the data an unexplained advance in the perihelion of Mercury. The modern value for this discrepancy is 42.98 arcseconds per cenury, and general relativity accounted for it in a natural way (Fig. E1). Radar measurements of the orbit of Mercury since 1966 have led to improved accuracy, so that the predicted relativistic advance agrees with the observations to about 0.1%.

Deflection of Light Einstein's second test concerns the deflection of light. According to general relativity, a light ray that passes the Sun at a distance d is deflected by an angle $\Delta\theta = 1.75''/d$, where d is measured in units of the solar radius, and the notation $''$ denotes seconds of arc (Fig. E2). Confirmation by the British astronomers Arthur S. Eddington and Crommelin of the bending of optical starlight observed during a total solar eclipse in the first months following World War I helped make Einstein a celebrity. However, those measurements had only 30% accuracy, and succeeding eclipse experiments weren't much better.

The development of long-baseline radio interferometry produced a method for greatly improved determinations of the deflection of light. Coupled with this techno-

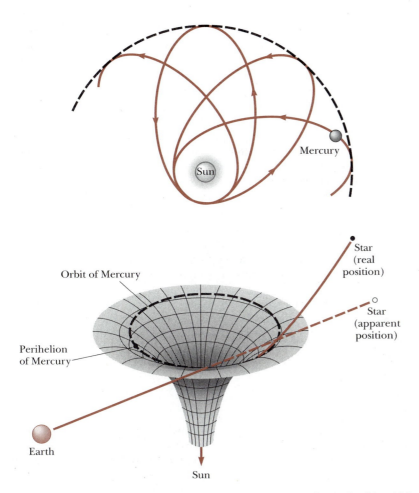

Figure E1   (a) Advance of the perihelion of Mercury. The elliptical orbit of Mercury about the Sun rotates very slowly relative to the system connected with the Sun. General relativity successfully explains this small effect, which predicts that the direction of the perihelion should change by only 43 arcseconds per century. (b) The diagram shows how a two-dimensional surface warped into three dimensions can change the direction of a "straight" line that is constrained to its surface; the warping of space is analogous, although with a greater number of dimensions to consider. The effect is similar to the golfer putting on a warped green. Though the ball is hit in a straight line, we see it appear to curve. (From Jay M. Pasachoff, *Astronomy: From The Earth to the Universe,* 4th ed., Philadelphia, Saunders College Publishing, 1991.)

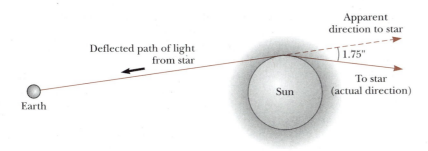

Figure E2   Deflection of starlight passing near the Sun. Because of this effect, the Sun and other remote objects can act as a *gravitational lens.* In his general relativity theory, Einstein calculated that starlight just grazing the Sun's surface should be deflected by an angle of 1.75."

logical advance is a series of heavenly coincidences: each year groups of strong quasars pass near the Sun (as seen from Earth). The idea is to measure the differential deflection of radio waves from one quasar relative to those from another as they pass near the Sun. A number of measurements of this kind occurred almost annually over the period from 1969 to 1975, yielding a confirmation of the predicted deflection to 1.5%. Recent measurements have improved this to 0.1%.

Gravitational Red Shift   Another consequence of Einstein's insight is the gravitational red shift effect, which is a frequency shift between two identical clocks placed at different heights in a gravitational field. For small differences in height h between clocks, the shift in the frequency Δf is given by

$$\frac{\Delta f}{f} = \frac{gh}{c^2}$$

where g is the local gravitational acceleration and c is the speed of light. If the receiver clock is at a lower height than the emitter clock, the received signal is shifted to higher frequencies ("blue shift"); if the receiver is higher, the signal is shifted to lower frequencies ("red shift").

The most precise gravitational red shift experiment performed to date was a rocket experiment carried out in June 1976. A "hydrogen maser" atomic clock was flown on a Scout D rocket to an altitude of 10 000 km, and its frequency compared with a similar clock on the ground using radio signals. After the effects of the rocket's motion were taken into account, the observations confirmed the gravitational red shift to 0.02%.

APPLICATIONS

General relativity has passed every experimental test to which it has been put, and many alternative theories have fallen by the wayside. Most physicists now take the theory for granted and look to see how it can be used as a practical tool in physics and astronomy. For example, gravitational red shift must now be taken into account properly in order for the Global Positioning System (GPS) to perform its navigation function accurately.

Gravitational Radiation   One of these new tools is gravitational radiation, a subject almost as old as general relativity. By 1916, Einstein had succeeded in showing that the field equations of general relativity admitted wave-like solutions analogous to those of electromagnetic theory. For example, a dumbbell rotating about an axis passing at right angles through its handle will emit gravitational waves that travel at the speed of light. They also carry energy away from the dumbbell, just as electromagnetic waves carry energy away from a light source.

In 1968, Joseph Weber initiated a program of gravitational-wave detection using massive aluminum bars as detectors. A passing gravitational wave acts as an oscillating gravitational force field that alternately compresses and extends the bar lengthwise. Currently a dozen laboratories around the world are engaged in building and improving on the basic "Weber bar" detector, striving to reduce noise from thermal, electrical, and environmental sources in order to detect the very weak oscillations produced by a gravitational wave. For a bar of one meter length, the challenge is to detect a variation in length smaller than 10^{-20} meters, or 10^{-5} of the radius of a proton. At the same time, several "laser-interferometric" gravitational-wave observatories are under construction in the United States and Europe. They are bouncing laser beams to monitor variations in length between mirrors spaced several kilometers apart. The observatories are scheduled to be operational around the year 2000.

Although gravitational radiation has not been detected directly, we know that it exists, through a remarkable system known as the binary pulsar. Discovered in 1974 by radio astronomers Russell Hulse and Joseph Taylor, it consists of a pulsar (which is a rapidly spinning neutron star) and a companion star in orbit around each other. Although the companion has not been seen directly, it is also believed to be a neutron star. The pulsar acts as an extremely stable clock, its pulse period of approximately 59 milliseconds drifting by only 0.25 ns/year. By measuring the arrival times of radio pulses at Earth, observers were able to determine the motion of the pulsar about its companion with amazing accuracy. For example, the accurate value for the orbital period is 27 906.980 895 s, and the orbital eccentricity is 0.617 131. Like a rotating dumbbell, an orbiting binary system should emit gravitational radiation and, in the process, lose some of its orbital energy. This energy loss will cause the pulsar and its companion to spiral in toward each other, and the orbital period to shorten. According to general relativity, the predicted decrease in the orbital period is 75.8 μs/year. The observed decrease rate is in agreement with the prediction to better than 0.5%. This confirms the existence of gravitational radiation and the general relativistic equations that describe it. Hulse and Taylor received the Nobel prize in 1993 for this discovery.

Black Holes   The first glimmerings of the black hole idea date to the 18th century, in the writings of a British amateur astronomer, the Reverend John Michell. Reasoning on the basis of the corpuscular theory that light would be attracted by gravity, he noted that the speed of light emitted from the surface of a massive body would be reduced by the time the light was very far from the source. (Michell of course did not know special relativity.) It would therefore be possible for a body to be sufficiently massive and compact to prevent light from escaping from its surface.

Although the general relativistic solution for a nonrotating black hole was discovered by Karl Schwarzschild in 1916, and a calculation of gravitational collapse to a black hole state was performed by J. Robert Oppenheimer and Hartland Snyder in 1939, black hole physics didn't really begin until the middle 1960s, when astronomers confronted the problem of the energy output of the quasars and started to take black holes seriously.

A black hole is formed when a star has exhausted the thermonuclear fuel necessary to produce the heat and pressure that support it against gravity. The star begins to collapse, and if it is massive enough, it continues to collapse until its radius approaches a value called the gravitational radius, or Schwarzschild radius. In the nonrotating spherical case, this radius is $2GM/c^2$, where M is the mass of the star. For a body of one solar mass, the gravitational radius is about 3 km; for a body of the mass of Earth, it is about 9 mm. An observer sitting on the surface of the star sees the collapse continue to smaller and smaller radii, until both star and observer reach the origin $r = 0$, with consequences too horrible to describe. On the other hand, an observer at great distances observes the collapse to slow down as the radius approaches the gravitational radius, a result of the gravitational red shift of the light signals sent outward. The distant observer never sees any signals emitted by the falling observer once the latter is inside the gravitational radius. This is because any signal emitted inside can never escape the sphere bounded by the gravitational radius, called the "event horizon."

In 1974, Stephen Hawking discovered that the laws of quantum mechanics require a black hole to evaporate by the creation of particles with a thermal energy spectrum and to have an associated temperature and entropy. The temperature of a Schwarzschild black hole is $T = hc^3/8\pi k_B GM$, where h is Planck's constant and k_B is Boltzmann's constant. This discovery demonstrated a remarkable connection between gravity, thermodynamics, and quantum mechanics that helped renew the theoretical quest for a grand synthesis of all the fundamental interactions. On the other hand, for black holes of astronomical masses, the evaporation is completely negligible, because for a solar-mass black hole, $T \approx 10^{-6}$ K.

Although a great deal is known about black holes in theory, rather less is known about them observationally. There are several instances in which the evidence for the existence of black holes is impressive, but in all cases it is indirect. For instance, in the x-ray source

Cygnus XI, the source of the x-rays is believed to be a black hole with a mass larger than about six solar masses in orbit around a giant star. The x-rays are emitted by matter pulled from the surface of the companion star and sent into a spiralling orbit around the black hole. Recently, the Hubble Space Telescope detected convincing evidence for a black hole, billions of times the mass of the Sun, in the center of galaxy M87.

Cosmology   Although Einstein in 1917 first used general relativity to calculate a model for the Universe, the subject was not considered a serious branch of physics until the 1960s, when astronomical observations lent credence to the idea that the Universe is expanding from a "Big Bang." In 1965 came the discovery of the cosmic background radiation by Arno Penzias and Robert Wilson. This radiation is the remains of the hot electromagnetic blackbody radiation that once dominated the Universe in its earlier phase, now cooled to 3 K by the subsequent expansion of the Universe. Next came calculations of the amount of helium synthesized from hydrogen by thermonuclear fusion in the very early Universe, around 1000 s after the Big Bang. The amount, approximately 25% by mass, was in agreement with the abundances of helium observed in stars and in interstellar space. This was an important confirmation of the hot Big-Bang picture because the amount of helium believed to be produced by fusion in the interiors of stars is woefully inadequate to explain the observed abundances.

Today, the general relativistic hot Big-Bang model of the Universe has broad acceptance, and cosmologists now focus their attention on more detailed issues, such as how galaxies and other large-scale structures formed out of the hot primordial soup, and on what the Universe might have been like earlier than 1000 s.

Suggestions for Further Reading

Steven W. Hawking, *A Brief History of Time,* New York, Bantam Books, 1988.
Kip S. Thorne, *Black Holes and Time Warps: Einstein's Outrageous Legacy,* New York, W. W. Norton, 1994.
S. Weinberg, *The First Three Minutes: A Modern View of the Origin of the Universe,* 2nd ed., New York, Basic Books, 1993.
C. M. Will, *Was Einstein Right? Putting General Relativity to the Test,* Revised edition, New York, Basic Books, 1993.

2

The Quantum Theory of Light

At the beginning of the 20th century, following the lead of Newton and Maxwell, physicists might have rewritten the biblical story of creation as follows:

In the beginning He created the heavens and the earth—

$$F = G\frac{mm'}{r^2} = ma$$

and He said "Let there be light"—

$$\oint \mathbf{E} \cdot d\mathbf{A} = \frac{Q}{\epsilon_0} \qquad \oint \mathbf{E} \cdot d\mathbf{s} = -\frac{d\Phi_B}{dt}$$

$$\oint \mathbf{B} \cdot d\mathbf{A} = 0 \qquad \oint \mathbf{B} \cdot d\mathbf{s} = \mu_0 I + \epsilon_0\mu_0\frac{d\Phi_E}{dt}$$

Actually, in addition to the twin pillars of mechanics and electromagnetism erected by the giants Newton and Maxwell, there was a third sturdy support for physics in 1900—thermodynamics and statistical mechanics. Classical thermodynamics was the work of many men (Carnot, Mayer, Helmholtz, Clausius, Lord Kelvin). It is especially notable because it starts with two simple propositions and gives solid and conclusive results independent of detailed physical mechanisms. Statistical mechanics, founded by Maxwell, Clausius, Boltzmann,[1] and Gibbs, uses the methods of probability theory to calculate averages and fluctuations from the average for systems containing many particles or modes of vibration. It is interesting that quantum physics started not with a breakdown of Maxwell's or Newton's laws applied to the atom but with a problem of classical statistical mechanics—that of calculating the intensity of radiation at a given wavelength from a heated cavity. The desperate solution to this radiation problem was found by a thoroughly classical thermodynamicist, Max Planck, in 1900. Indeed, it is significant that both Planck and Einstein returned again and again to the simple and general foundation of thermodynamics and statistical mechanics as the only certain bases for the new quantum theory. Although we shall not follow the original thermodynamic arguments completely, we shall see in this chapter how Planck arrived at the correct spectral distribution for cavity radiation by allowing only certain energies for the oscillators in the cavity walls. We shall also see how Einstein extended this quantization of energy to light itself, thereby brilliantly explaining the photoelectric effect. We conclude our brief history of the quantum theory of light with a discussion of the scattering of light by electrons (Compton effect), which showed conclusively that the light quantum carried momentum as well as energy. Finally, we describe the pull of gravity on light in Section 2.7.

2.1 HERTZ'S EXPERIMENTS—LIGHT AS AN ELECTROMAGNETIC WAVE

It is ironic that the same experimentalist who so brilliantly confirmed that the "newfangled" waves of Maxwell actually existed and possessed the same properties as light also caused the downfall of the electromagnetic wave theory as the complete explanation of light. To understand this irony, let us briefly review the theory of electromagnetism developed primarily by the great Scottish physicist James Clerk Maxwell between 1865 and 1873.

Maxwell was primarily interested in the effects of electric current oscillations in wires. According to his theory, an alternating current would set up fluctuating electric and magnetic fields in the region surrounding the original disturbance. Moreover, these waves were predicted to have a frequency equal to the frequency of the current oscillations. *In addition, and most importantly, Maxwell's theory predicted that the radiated waves would behave in every way like light:* electromagnetic waves would be reflected by metal mirrors, would be refracted by dielectrics like glass, would exhibit polarization and interference, and would travel through a vacuum with a speed of 3.0×10^8 m/s. Naturally this led to the unifying and simplifying postulate that light was also a type of Maxwell wave or electromagnetic disturbance, created by extremely high frequency electric oscillators in matter. At the end of the 19th century the precise nature of these charged submicroscopic oscillators was unknown (Planck called them resonators), but physicists assumed that somehow they were able to emit light waves whose frequency was equal to the oscillator's frequency of motion.

---

[1] On whose tombstone is written $S = k_B \log W$, a basic formula of statistical mechanics, attributed to Boltzmann.

Even at this time, however, it was apparent that this model of light emission was incapable of direct experimental verification, because the highest electrical frequencies then attainable were about 10^9 Hz, and visible light was known to possess a frequency a million times higher. But Heinrich Hertz (Fig. 2.1) did the next best thing. In a series of brilliant and exhaustive experiments, he showed that Maxwell's theory was correct and that an oscillating electric current does indeed send out electromagnetic waves that possess every characteristic of light except the same wavelength as light. Using a simple spark gap oscillator consisting of two short stubs terminated in small metal spheres separated by an air gap of about half an inch, he applied pulses of high voltage, which caused a spark to jump the gap and produce a high-frequency electric oscillation of about 5×10^8 Hz. This oscillation, or ringing, occurs while the air gap remains conducting, and charge surges back and forth between the spheres until electrical equilibrium is established. Using a simple loop antenna with a small spark gap as the receiver, Hertz very quickly succeeded in detecting the radiation from his spark gap oscillator, even at distances of several hundred meters. Moreover, he found the detected radiation to have a wavelength of about 60 cm, corresponding to the oscillator frequency of 5×10^8 Hz. (Recall that $c = \lambda f$, where λ is the wavelength and f is the frequency.)

In an exhaustive tour de force, Hertz next proceeded to show that these electromagnetic waves could be reflected, refracted, focused, polarized, and made to interfere—in short, he convinced almost every leading physicist of the period that Hertzian waves and light waves were one and the same. The classical model for light emission was an idea whose time had come. It spread like wildfire. The idea that light was an electromagnetic wave radiated by submicroscopic electric oscillators was applied in rapid succession to the transmission of light through solids, to reflection from metal surfaces, and, perhaps most impressively, to the newly discovered Zeeman effect. In 1896, Pieter Zeeman discovered that a strong magnetic field changes the frequency of the light emitted by a glowing gas. In an impressive victory for Maxwell, it was found that Maxwell's equations correctly predicted (in most cases) the change of vibration of the electric oscillators and hence the change in frequency of the light emitted. Maxwell, with Hertz behind the throne, reigned supreme, for he had united the formerly independent kingdoms of electricity, magnetism, and light! (See Fig. 2.2.)

Figure 2.1 Heinrich Hertz (1857–1894), an extraordinarily gifted German experimentalist. (*The Bettmann Archive*)

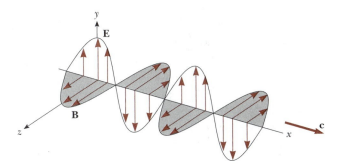

Figure 2.2   A light or radio wave far from the source according to Maxwell and Hertz.

A terse remark made by Hertz ends our discussion of his brilliant confirmation of the electromagnetic wave nature of light. In describing his oscillator he notes that "it is essential that the pole surfaces of the spark gap should be frequently repolished" to ensure reliable operation of the spark.[2] Apparently this result was initially quite mysterious to Hertz. In an effort to resolve the mystery, he later carefully investigated this side effect and concluded that it was the ultraviolet light from the initial spark acting on a clean metal surface that caused current to flow more freely between the poles of the spark gap. In the process of verifying the electromagnetic wave theory of light, Hertz had discovered the photoelectric effect, a phenomenon that would ultimately undermine the priority of the wave theory of light and establish the particle theory of light on an equal footing.

2.2 BLACKBODY RADIATION

The tremendous success of Maxwell's theory of light emission immediately led to attempts to apply it to a long-standing puzzle about radiation—the so-called "blackbody" problem. The problem is to predict the radiation intensity at a given wavelength emitted by a glowing solid at a specific temperature. Instead of launching immediately into Planck's solution of this problem, let us develop a feeling for its importance to classical physics by a quick review of its history.

Thomas Wedgwood, Charles Darwin's relative and a renowned maker of china, seems to have been the first to note the universal character of all heated objects. In 1792, he observed that all the objects in his ovens, regardless of their chemical nature, size, or shape, became red at the same temperature. This crude observation was sharpened considerably by the advancing state of spectroscopy, so that by the mid-1800s it was known that glowing solids emit continuous spectra rather than the bands or lines emitted by heated gases. (See Fig. 2.3.) In 1859, Gustav Kirchhoff proved a theorem as important as his circuit loop theorem when he showed by arguments based on thermodynamics that for any body in thermal equilibrium with radiation[3] the emitted power is proportional to the power absorbed. More specifically,

$$e_f = J(f, T)A_f \tag{2.1}$$

where e_f is the power emitted per unit area per unit frequency by a particular heated object, A_f is the absorption power (fraction of the incident power absorbed per unit area per unit frequency), and $J(f, T)$ is a universal function (the same for all bodies) that depends only on f, the light frequency and T, the absolute temperature of the body. A *black body* is defined as an object that

Figure 2.3 Emission from a glowing solid. Note that the amount of radiation emitted (the area under the curve) increases with increasing temperature.

Black body

---

[2] H. Hertz, *Ann. Physik* (Leipzig), 33:983, 1887.

[3] An example of a body in equilibrium with radiation would be an oven with closed walls at a fixed temperature and the radiation within the oven cavity. To say that radiation is in thermal equilibrium with the oven walls means that the radiation has exchanged energy with the walls many times and is homogeneous, isotropic, and unpolarized. In fact, thermal equilibrium of radiation within a cavity can be considered to be quite similar to the thermal equilibrium of a fluid within a container at constant temperature—both will cause a thermometer in the center of the cavity to achieve a final stationary temperature equal to that of the container.

absorbs all the electromagnetic radiation falling on it and consequently appears black. It has $A_f = 1$ for all frequencies and so Kirchhoff's theorem for a black body becomes

$$e_f = J(f, T) \qquad (2.2)$$

Equation 2.2 shows that the power emitted per unit area per unit frequency by a black body depends only on temperature and light frequency and not on the physical and chemical makeup of the black body, in agreement with Wedgwood's early observation.

Because absorption and emission are connected by Kirchhoff's theorem, we see that a black body is also an ideal radiator. In practice a small opening in any heated cavity, such as a port in an oven, behaves like a black body because such an opening traps radiation (Fig. 2.4). If the direction of the radiation is reversed in Figure 2.4, the light emitted by a small opening is in thermal equilibrium with the walls, because it has been absorbed and reemitted many times.

The next important development in the quest to understand the universal character of the radiation emitted by glowing solids came from the Austrian physicist Josef Stefan (1835–1893) in 1879. He found experimentally that the total power per unit area emitted at all frequencies by a hot solid, e_{total}, was proportional to the fourth power of its absolute temperature. Thus, Stefan's law may be written as

$$e_{\text{total}} = \int_0^\infty e_f \, df = \sigma T^4 \qquad (2.3)$$

where e_{Total} is the power per unit area emitted by the black body, e_f is the power per unit area per unit frequency emitted by the black body, T is the absolute temperature of the body, and σ is the Stefan-Boltzmann constant, given by $\sigma = 5.67 \times 10^{-8} \text{ W} \cdot \text{m}^{-2} \cdot \text{K}^{-4}$. A body that is not an ideal radiator will obey the same general law but with a coefficient, a, less than 1:

$$e_{\text{total}} = a\sigma T^4 \qquad (2.4)$$

Only five years later another impressive confirmation of Maxwell's electromagnetic theory of light occurred when Boltzmann derived Stefan's law from a combination of thermodynamics and Maxwell's equations.

Figure 2.4 The opening to the cavity inside a body is a good approximation of a black body. Light entering the small opening strikes the far wall, where some of it is absorbed but some is reflected at a random angle. The light continues to be reflected, and at each reflection a portion of the light is absorbed by the cavity walls. After many reflections essentially all of the incident energy is absorbed.

Stefan's law

EXAMPLE 2.1 Stefan's Law Applied to the Sun

Estimate the surface temperature of the Sun from the following information. The Sun's radius is given by $R_s = 7.0 \times 10^8$ m. The average Earth-Sun distance is $R = 1.5 \times 10^{11}$ m. The power per unit area (at all frequencies) from the Sun is measured at the Earth to be 1400 W/m². Assume that the Sun is a black body.

Solution For a black body, we take $a = 1$, so Equation 2.4 gives

$$e_{\text{total}}(R_s) = \sigma T^4 \qquad (2.5)$$

Because we have the total power per unit area at the Earth,

$e_{\text{total}}(R)$, we need the connection between $e_{\text{total}}(R)$ and $e_{\text{total}}(R_s)$. This comes from the conservation of energy:

$$e_{\text{total}}(R_s) \cdot 4\pi R_s^2 = e_{\text{total}}(R) \cdot 4\pi R^2$$

or

$$e_{\text{total}}(R_s) = e_{\text{total}}(R) \cdot \frac{R^2}{R_s^2}$$

Using Equation 2.5 we have

$$T = \left[\frac{E(R) \cdot R^2}{\sigma R_s^2} \right]^{1/4}$$

or

$$T = \left[\frac{(1400 \text{ W/m}^2)(1.5 \times 10^{11} \text{ m})^2}{(5.67 \times 10^{-8} \text{ W/m}^2 \cdot \text{K}^4)(7.0 \times 10^8 \text{ m})^2} \right]^{1/4}$$

$$= 5800 \text{ K}$$

EXAMPLE 2.2   The Surface Temperature of the Sun Revisited!

As can be seen in Figure 2.3, *the wavelength marking the maximum power emission of a black body*, $\lambda_{\max}$, *shifts toward shorter wavelengths as the black body gets hotter.* Even this simple effect of $\lambda_{\max} \propto T^{-1}$ was not definitely established experimentally until about 20 years after Kirchhoff's seminal paper had started the search to find $J(f, T)$. In 1893, Wilhelm Wien proposed a general form for the blackbody distribution law that gave the correct experimental behavior of $\lambda_{\max}$ with temperature. This law is called *Wien's displacement law* and may be written:

$$\lambda_{\max} T = 2.898 \times 10^{-3} \text{ m} \cdot \text{K} \qquad (2.6)$$

where $\lambda_{\max}$ is the wavelength in meters corresponding to maximum intensity, and T is the absolute temperature. Assuming that the peak sensitivity of the human eye (which occurs at about 500 nm—blue-green light) coincides with $\lambda_{\max}$ for the Sun (a black body), we can calculate the Sun's surface temperature:

$$T = \frac{2.898 \times 10^{-3} \text{ m} \cdot \text{K}}{500 \times 10^{-9} \text{ m}} = 5800 \text{ K}$$

Thus we have good agreement between measurements made at all wavelengths (Example 2.1) and at the maximum intensity wavelength.

Exercise 1   How convenient that the Sun's emission peak is at the same wavelength as our eyes' sensitivity peak. Can you account for this?

Spectral energy density of a black body

So far, the power radiated per unit area per unit frequency by the black body, $J(f, T)$ has been discussed. *However, it is more convenient to consider the spectral energy density, or energy per unit volume per unit frequency within the black body cavity,* $u(f, T)$. For light in equilibrium with the walls, the power emitted per square centimeter of opening is simply proportional to the energy density of the light in the cavity. Because the cavity radiation is isotropic and unpolarized, one can average over direction to show that the constant of proportionality between $J(f, T)$ and $u(f, T)$ is $c/4$, where c is the speed of light. Thus

$$J(f, T) = u(f, T)c/4 \qquad (2.7)$$

An important guess as to the form of the universal function $u(f, T)$ was made in 1893 by Wien and had the form

$$u(f, T) = Af^3 e^{-\beta f/T} \qquad (2.8)$$

where A and β are constants. This result was known as Wien's exponential law; it resembles and was loosely based on Maxwell's velocity distribution for gas molecules. Within a year the great German spectroscopist Friedrich Paschen had confirmed Wien's guess by working in the then difficult infrared range of 1 to 4 μm and at temperatures of 400 to 1600 K.[4]

As can be seen in Figure 2.5, Paschen had made most of his measurements in the maximum energy region of a body heated to 1500 K and had found good

---

[4] We should point out the great difficulty in making blackbody radiation measurements and the singular advances made by German spectroscopists in the crucial areas of blackbody sources, sensitive detectors, and techniques for operating far into the infrared region. In fact, it is dubious whether Planck would have found the correct blackbody law as quickly without his close association with the experimentalists at the Physikalisch Technische Reichsanstalt of Berlin (a sort of German National Bureau of Standards)—Otto Lummer, Ernst Pringsheim, Heinrich Rubens, and Ferdinand Kurlbaum.

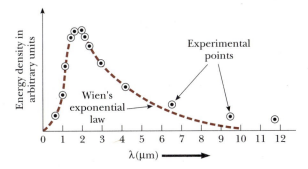

Figure 2.5   Discrepancy between Wien's law and experimental data for a black body at 1500 K.

agreement with Wien's exponential law. In 1900, however, Lummer and Pringsheim extended the measurements to 18 μm, and Rubens and Kurlbaum went even farther—to 60 μm. Both teams concluded that Wien's law failed in this region (see Fig. 2.5). The experimental setup used by Rubens and Kurlbaum is shown in Figure 2.6. It is interesting to note that these historic experiments involved the measurement of blackbody radiation intensity at a fixed wavelength and variable temperature. Typical results measured at $\lambda = 51.2$ μm and over the temperature range of -200 to $+1500°$C are shown in Figure 2.7 from the paper by Rubens and Kurlbaum.

Enter Planck

On a Sunday evening early in October of 1900, Max Planck discovered the famous blackbody formula, which truly ushered in the quantum theory. Planck's proximity to the Reichsanstalt experimentalists was extremely impor-

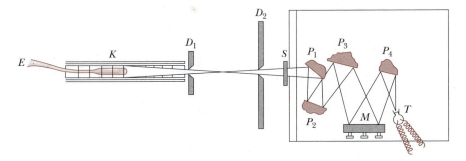

Figure 2.6   Apparatus for measuring blackbody radiation at a single wavelength in the far infrared region. The experimental technique that disproved Wien's law and was so crucial to the discovery of the quantum theory was the method of residual rays (*Restrahlen*). In this technique, one isolates a narrow band of far infrared radiation by causing white light to undergo multiple reflections from alkalide halide crystals (P_1–P_4). Because each alkali halide has a maximum reflection at a characteristic wavelength, quite pure bands of far infrared radiation may be obtained with repeated reflections. These pure bands can then be directed onto a thermopile (T) to measure intensity. E is a thermocouple used to measure the temperature of the blackbody oven, K.

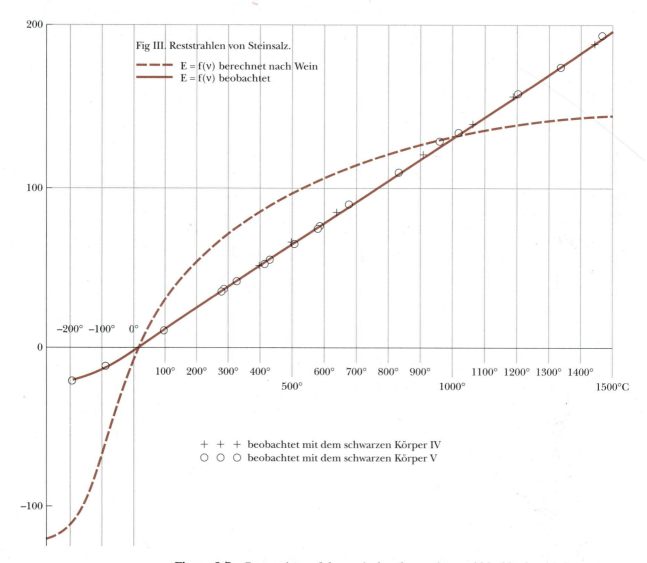

Figure 2.7   Comparison of theoretical and experimental blackbody emission curves at 51.2 μm and over the temperature range of -188 to 1500°C. The title of this modified figure is "Residual Rays from Rocksalt." "Berechnet nach" means "calculated according to," and "beobachtet" means "observed." The vertical axis is emission intensity in arbitrary units. (*From H. Rubens and S. Kurlbaum,* Ann. Physik *4:649, 1901.*)

tant for his discovery—earlier in the day he had heard from Rubens that his latest measurements showed that $u(f, T)$, the spectral energy density, was proportional to T for long wavelengths or low frequency. Planck knew that Wien's law agreed well with the data at high frequency and indeed had been working hard for several years to derive Wien's exponential law from the principles of statistical mechanics and Maxwell's laws. Interpolating between the two limiting forms (Wien's exponential law and an energy density proportional to temperature), he immediately found a general formula which he sent to Rubens, on a postcard, the same evening. His formula was[5]

$$u(f, T) = \frac{8\pi h f^3}{c^3} \left(\frac{1}{e^{hf/k_B T} - 1} \right) \qquad (2.9)$$

where h is Planck's constant $= 6.626 \times 10^{-34}\,\text{J·s}$, and k_B is Boltzmann's constant $= 1.380 \times 10^{-23}\,\text{J/K}$. At high frequencies, where $hf/k_B T \gg 1$,

$$\frac{1}{e^{hf/k_B T} - 1} \approx e^{-hf/k_B T}$$

so that

$$u(f, T) = \frac{8\pi h f^3}{c^3} \left(\frac{1}{e^{hf/k_B T} - 1} \right) \approx \frac{8\pi h f^3}{c^3} e^{-hf/k_B T}$$

and we recover Wien's exponential law, Equation 2.8. At low frequencies, where $hf/k_B T \ll 1$,

$$\frac{1}{e^{hf/k_B T} - 1} = \frac{1}{1 + \dfrac{hf}{k_B T} + \cdots - 1} \approx \frac{k_B T}{hf}$$

and

$$u(f, T) = \frac{8\pi h f^3}{c^3} \left(\frac{1}{e^{hf/k_B T} - 1} \right) \approx \frac{8\pi f^2}{c^3} k_B T$$

This result shows that the spectral energy density is proportional to T in the low-frequency or so-called classical region.

We should emphasize that Planck's work entailed much more than clever mathematical manipulation. For more than six years Planck (Fig. 2.8) labored to find a rigorous derivation of the blackbody distribution curve. He was driven, in his own words, by the fact that the emission problem "represents something absolute, and since I had always regarded the search for the absolute as the loftiest goal of all scientific activity, I eagerly set to work." This work was to occupy most of his life as he strove to give his formula an ever deeper physical interpretation and to reconcile the elementary quantum with classical theory.

The Quantum of Energy

Planck's original theoretical justification of Equation 2.9 is rather abstract because it involves arguments based on entropy, statistical mechanics, and several theorems proved earlier by Planck concerning matter and radiation in equilib-

Figure 2.8 Max Planck (1858–1947). The work leading to the "lucky" blackbody radiation formula was described by Planck in his Nobel prize acceptance speech (1920): "But even if the radiation formula proved to be perfectly correct, it would after all have been only an interpolation formula found by lucky guess-work and thus, would have left us rather unsatisfied. I therefore strived from the day of its discovery, to give it a real physical interpretation and this led me to consider the relations between entropy and probability according to Boltzmann's ideas. After some weeks of the most intense work of my life, light began to appear to me and unexpected views revealed themselves in the distance." (*AIP Niels Bohr Library, W. F. Meggers Collection*)

---

[5] Planck originally published his formula as

$$u(\lambda, T) = \frac{C_1}{\lambda^5} \left(\frac{1}{e^{C_2/\lambda T} - 1} \right)$$

where $C_1 = 8\pi ch$ and $C_2 = hc/k_B$. He then found best fit values to the experimental data for C_1 and C_2 and evaluated $h = 6.55 \times 10^{-34}\,\text{J·s}$ and $k_B = N_A/R = 1.345 \times 10^{-23}\,\text{J/K}$. As R, the universal gas constant, was fairly well known at the time, this technique also resulted in another method for finding N_A, Avogadro's number.

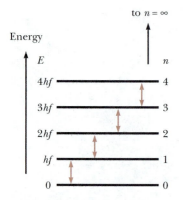

Figure 2.9 Allowed energy levels for an oscillator of natural frequency f. Allowed transitions are indicated by vertical arrows.

rium.[6] We shall give arguments that are easier to visualize physically yet attempt to convey the spirit and revolutionary influence of Planck's original work.

Planck was convinced that blackbody radiation was produced by submicroscopic electric oscillators, which he called resonators. He assumed that the walls of a glowing cavity were composed of literally billions of these resonators (whose exact nature was unknown), all vibrating at different frequencies. Hence, according to Maxwell, each oscillator should emit radiation with a frequency corresponding to its vibration frequency. *According to classical theory, an oscillator of frequency f can have any value of energy and can change its amplitude continuously as it radiates any fraction of its energy.* This is where Planck made his revolutionary proposal. To secure agreement with experiment, *Planck had to assume that the total energy of a resonator with mechanical frequency f could only be an integral multiple of hf* or

$$E_{\text{resonator}} = nhf \qquad n = 1, 2, 3, \ldots \qquad (2.10)$$

where h is a fundamental constant, $h = 6.626 \times 10^{-34}$ J·s, known as Planck's constant. In addition, he concluded that emission of radiation of frequency f occurred when a resonator dropped to the next lowest energy state. *Thus the resonator can change its energy only by the difference ΔE according to*

$$\Delta E = hf \qquad (2.11)$$

That is, it cannot lose just any fraction of its total energy. Figure 2.9 shows the quantized energy levels and allowed transitions proposed by Planck.

[6] M. Planck, *Ann. Physik*, 4:553, 1901.

EXAMPLE 2.3   A Quantum Oscillator versus a Classical Oscillator

Consider the implications of Planck's conjecture that *all* oscillating systems of natural frequency f have discrete energies $E = nhf$ and that the smallest change in energy of the system is given by $\Delta E = hf$.

(a) First compare an atomic oscillator sending out 540-nm light (green) to one sending out 700-nm light (red) by calculating the minimum energy change of each. For the green quantum,

$$\Delta E_{\text{green}} = hf = \frac{hc}{\lambda} = \frac{(6.63 \times 10^{-34}\, \text{J·s})(3.00 \times 10^8\, \text{m/s})}{540 \times 10^{-9}\, \text{m}}$$

$$= 3.68 \times 10^{-19}\, \text{J}$$

Actually, the joule is much too large a unit of energy for describing atomic processes; a more appropriate unit of energy is the electron volt (eV). The electron volt takes the charge on the electron as its unit of charge. By definition, an electron accelerated through a potential difference of one volt has an energy of 1 eV. An electron volt may be converted to joules by noting that

$$E = V \cdot q = 1\, \text{eV} = (1.602 \times 10^{-19}\, \text{C})(1\, \text{J/C})$$

$$= 1.602 \times 10^{-19}\, \text{J}$$

It is also most useful to have expressions for h and hc in terms of electron volts. These are

$$h = 4.136 \times 10^{-15}\, \text{eV·s}$$

$$hc = 1.240 \times 10^{-6}\, \text{eV·m} = 1240\, \text{eV·nm}$$

Returning to our example, we see that the minimum energy change of an atomic oscillator sending out green light is

$$\Delta E_{\text{green}} = \frac{3.68 \times 10^{-19}\, \text{J}}{1.602 \times 10^{-19}\, \text{J/eV}} = 2.30\, \text{eV}$$

For the red quantum the minimum energy change is

$$\Delta E_{\text{red}} = \frac{hc}{\lambda} = \frac{(6.63 \times 10^{-34}\, \text{J·s})(3.00 \times 10^8\, \text{m/s})}{700 \times 10^{-9}\, \text{m}}$$

$$= 2.84 \times 10^{-19}\, \text{J} = 1.77\, \text{eV}$$

Note that the quantum of energy is not uniform under all conditions as is the quantum of charge—the quantum of energy is proportional to the natural frequency of the oscillator. Note, too, that the high frequency of atomic oscillators produces a measurable quantum of energy of several electron volts.

(b) Now consider a pendulum undergoing small oscillations with length $\ell = 1$ m. According to classical theory, if air friction is present, the amplitude and consequently the energy decrease *continuously* with time as shown in Figure 2.10a. Actually, *all* vibrating systems are quantized (according to Equation 2.10) and lose energy in discrete packets or quanta, hf. This would lead to a decrease of the pendulum's energy in a stepwise manner as shown in Figure 2.10b. We shall show that there is no contradiction between quantum theory and the observed behavior of laboratory pendulums and springs.

(a)

(b)

Figure 2.10   (Example 2.3) (a) Observed classical behavior of a pendulum. (b) Predicted quantum behavior of a pendulum.

An energy change of one quantum corresponds to

$$\Delta E = hf$$

where the pendulum frequency f is

$$f = \frac{1}{2\pi}\sqrt{\frac{g}{\ell}} = 0.50 \text{ Hz}$$

Thus

$$\Delta E = (6.63 \times 10^{-34}\,\text{J·s})(0.50\,\text{s}^{-1})$$
$$= 3.3 \times 10^{-34}\,\text{J}$$
$$= 2.1 \times 10^{-15}\,\text{eV}$$

Because the total energy of a pendulum of mass m and length ℓ displaced through an angle θ is

$$E = mg\ell(1 - \cos\theta)$$

we have for a typical pendulum with $m = 100$ g, $\ell = 1.0$ m, and $\theta = 10°$,

$$E = (0.10\,\text{kg})(9.8\,\text{m/s}^2)(1.0\,\text{m})(1 - \cos 10°) = 0.015\,\text{J}$$

Thus the fractional change in energy, $\Delta E/E$, is unobservably small:

$$\frac{\Delta E}{E} = \frac{3.3 \times 10^{-34}\,\text{J}}{1.5 \times 10^{-2}\,\text{J}} = 2.2 \times 10^{-32}$$

Note that the energy quantization of large vibrating systems is unobservable because of their low frequencies. Hence there is no contradiction between Planck's quantum postulate and the behavior of macroscopic oscillators.

Exercise 2   Calculate the quantum number, n, for this pendulum with $E = 1.5 \times 10^{-2}$ J.

Answer 4.6×10^{31}

Exercise 3   An object of mass m on a spring of stiffness k oscillates with an amplitude A about its equilibrium position. Suppose that $m = 300$ g, $k = 10$ N/m, and $A = 10$ cm. (a) Find the total energy. (b) Find the mechanical frequency of vibration of the mass. (c) Calculate the change in amplitude when the system loses one quantum of energy.

Answers   (a) $E_{total} = 0.050$ J (b) $f = 0.92$ Hz

(c) $\Delta E_{quantum} = 6.1 \times 10^{-34}$ J, so

$$\Delta A \approx -\frac{\Delta E}{\sqrt{2Ek}} = -6.1 \times 10^{-34}\,\text{m}$$

Until now we have been concentrating on the remarkable quantum properties of single oscillators of frequency f. Planck explained the continuous spectrum of the black body by assuming that the heated walls contained resonators vibrating at many different frequencies, each emitting light at the same frequency as its vibration frequency. By considering the conditions leading to equilibrium between the wall resonators and the radiation in the blackbody cavity, he was able to show that the spectral energy density $u(f, T)$ could be expressed as the product of the number of oscillators having frequency between f and $f + df$, denoted by $N(f)\ df$, and the average energy emitted per oscillator, $\bar{E}$. Thus we have the important result

$$u(f, T)\ df = \bar{E}\ N(f)\ df \tag{2.12}$$

Furthermore, Planck showed that (see the appendix at the end of this chapter for details)

$$N(f)\ df = \frac{8\pi f^2}{c^3}\ df \tag{2.13}$$

Substituting Equation 2.13 into Equation 2.12 gives

$$u(f, T)\ df = \bar{E}\ \frac{8\pi f^2}{c^3}\ df \tag{2.14}$$

The result shows that the spectral energy density is proportional to the product of the frequency squared and the average oscillator energy. Furthermore, since $u(f, T)$ approaches zero at high frequencies (see Fig. 2.5), $\bar{E}$ must tend to zero at high frequencies faster than $1/f^2$. The fact that the mean oscillator energy must become extremely small when the frequency becomes high guided Planck in the development of his theory. Later we shall see that the failure of $\bar{E}$ to become small at high frequencies in the classical Rayleigh-Jeans theory led to the "ultraviolet catastrophe"—the prediction of an infinite spectral energy density at high frequencies in the ultraviolet region.

O P T I O N A L

2.3   DERIVATION OF THE RAYLEIGH-JEANS LAW AND PLANCK'S LAW

Rayleigh-Jeans Law

Planck's law and the Rayleigh-Jeans law (the classical theory of blackbody radiation) both may be derived using the idea that the blackbody radiation energy per unit volume with frequency between f and $f + df$ can be expressed as the product of the number of oscillators per unit volume in this frequency range with the average energy per oscillator:

$$u(f, T)\ df = \bar{E}\ N(f)\ df \tag{2.12}$$

It is instructive to perform both the Rayleigh-Jeans and Planck calculations to see the effect on $u(f, T)$ of calculating $\bar{E}$ from a continuous distribution of oscillator energies (Rayleigh-Jeans) as opposed to a discrete set of oscillator energies (Planck). We discuss Lord Rayleigh's derivation first because it is a more direct classical calculation.

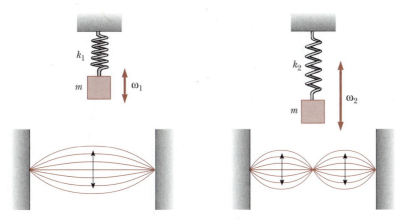

Figure 2.11   A one-dimensional harmonic oscillator is equivalent to a plane-polarized electromagnetic standing wave.

While Planck concentrated on the thermal equilibrium of cavity radiation with oscillating electric charges in the cavity walls, Rayleigh concentrated directly on the electromagnetic waves in the cavity. Rayleigh and Jeans reasoned that the standing electromagnetic waves in the cavity could be considered to have a temperature T, because they constantly exchange energy with the walls and could cause a thermometer to reach the same temperature as the walls. Further, they considered a standing polarized electromagnetic wave to be equivalent to a one-dimensional oscillator (Fig. 2.11). Using the same general idea as Planck, they expressed the energy density as a product of the number of standing waves (oscillators) and the average energy per oscillator. They found the average oscillator energy $\overline{E}$ to be equal to $k_B T$ from Boltzmann's distribution law. According to this fundamental law of statistical mechanics, the probability P of finding an individual system (such as a molecule or an oscillator) with energy E above some minimum energy, E_0, in an ensemble of systems at temperature T is

$$P(E) = P_0 e^{-(E-E_0)/k_B T} \tag{2.15}$$

where P_0 is the probability that a system has the minimum energy. In the case of a *discrete* set of allowed energies, the average energy, $\overline{E}$, is given by

$$\overline{E} = \frac{\Sigma\, E \cdot P(E)}{\Sigma\, P(E)} \tag{2.16}$$

and division by the sum in the denominator serves to normalize the probabilities to one. *In the classical case considered by Rayleigh, an oscillator could have any energy in a continuous range from 0 to ∞.* Thus the sums in Equation 2.16 must be replaced with integrals, and the expression for $\overline{E}$ becomes

$$\overline{E}_{\text{classical}} = \frac{\displaystyle\int_0^\infty E e^{-E/k_B T}\, dE}{\displaystyle\int_0^\infty e^{-E/k_B T}\, dE} = k_B T$$

The calculation of $N(f)$ is a bit more complicated but is of importance here as well as in the free electron model of metals. The appendix at the end of this chapter gives the derivation of the density of modes, $N(f)\, df$. One finds

$$N(f)\, df = \frac{8\pi f^2}{c^3}\, df \tag{2.46}$$

Density of standing waves in a cavity

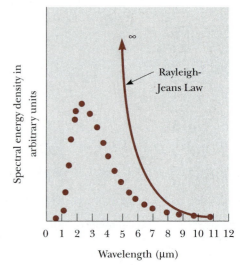

Figure 2.12   The failure of the classical Rayleigh-Jeans law (Equation 2.18) to fit the observed spectrum of a black body heated to 1000 K.

or in terms of wavelength

$$N(\lambda) \; d\lambda = \frac{8\pi}{\lambda^4} \; d\lambda \tag{2.47}$$

The spectral energy density is simply the density of modes multiplied by $k_B T$, or

$$u(f, \; T) \; df = \frac{8\pi f^2}{c^3} \; k_B T \; df \tag{2.17}$$

In terms of wavelength,

$$u(\lambda, \; T) \; d\lambda = \frac{8\pi}{\lambda^4} \; k_B T \; d\lambda \tag{2.18}$$

Rayleigh-Jeans blackbody law

However, as one can see from Figure 2.12, this classical expression, known as the Rayleigh-Jeans law, does not agree with the experimental data. Equation 2.18 diverges as $\lambda \rightarrow 0$, predicting unlimited energy emission in the ultraviolet region, which was dubbed the "ultraviolet catastrophe." One is forced to conclude that rigorous classical theory fails miserably to explain blackbody radiation.

Planck's Law

As mentioned earlier, Planck concentrated on the energy states of resonators in the cavity walls and used the condition that the resonators and cavity radiation were in equilibrium to determine the spectral quality of the radiation. By thermodynamic reasoning (and apparently unaware of Rayleigh's derivation), he arrived at the same expression for $N(f)$ as Rayleigh. However, Planck arrived at a different form for $\overline{E}$ by postulating discrete values of energy for his resonators. Because it is simpler and more direct, we shall continue with Rayleigh's point of view and *quantize the energy of the standing waves or modes in the blackbody cavity to derive Planck's law*. It is interesting, however, that historically Planck quantized only the material resonators in the cavity and maintained the classical wave nature of light in the cavity.

We shall now derive Planck's blackbody formula by quantizing the standing electromagnetic waves in the cavity. Instead of Rayleigh's assumption that a standing wave can have any energy from 0 to ∞, we say that a standing wave of frequency f can have only *discrete energies* $E = nhf$. We again apply the Boltzmann formula to find the probability that an oscillator is in its nth state, where now the probability of occupancy is $P(n) = P_0 e^{-nhf/k_B T}$. Note that in this case the probability is defined at discrete points as shown in Figure 2.13. The expression for the average energy of a standing wave, Equation 2.16, now becomes

$$\overline{E} = \frac{\sum_{n=0}^{\infty} (nhf) \, P_0 e^{-nhf/k_B T}}{\sum_{n=0}^{\infty} P_0 e^{-nhf/k_B T}} \qquad (2.19)$$

Using

$$\frac{1}{1-r} = \sum_{n=0}^{\infty} r^n$$

gives

$$\sum_{n=0}^{\infty} e^{-nhf/k_B T} = \frac{1}{(1 - e^{-hf/k_B T})}$$

Substituting this into Equation 2.19 yields

$$\overline{E} = hf(1 - e^{-hf/k_B T}) \sum_{n=0}^{\infty} n e^{-nhf/k_B T} \qquad (2.20)$$

To evaluate the sum, we let $\alpha = hf/k_B T$, and note that

$$\sum n e^{-n\alpha} = -\frac{d}{d\alpha} \sum_{n=0}^{\infty} e^{-n\alpha}$$

$$= -\frac{d}{d\alpha} \left(\frac{1}{1 - e^{-\alpha}} \right) = \frac{e^{-\alpha}}{(1 - e^{-\alpha})^2}$$

Therefore,

$$\sum_{n=0}^{\infty} n e^{-nhf/k_B T} = \frac{e^{-\alpha}}{(1 - e^{-\alpha})^2}$$

and the expression for $\overline{E}$ becomes

$$\overline{E} = \frac{hf e^{-hf/k_B T}}{1 - e^{-hf/k_B T}} = \frac{hf}{e^{hf/k_B T} - 1} \qquad (2.21)$$

Multiplying $\overline{E}$ by $N(f)$ gives the Planck distribution formula:

$$u(f, T) \, df = \frac{8\pi f^2}{c^3} \left(\frac{hf}{e^{hf/k_B T} - 1} \right) df \qquad (2.9)$$

or, in terms of wavelength, λ

$$u(\lambda, T) \, d\lambda = \frac{8\pi hc \, d\lambda}{\lambda^5 \, (e^{hc/\lambda k_B T} - 1)} \qquad (2.22)$$

This result shows that the ultraviolet catastrophe is avoided because the $\overline{E}$ term dominates the f^2 term at high frequencies. One can qualitatively understand why $\overline{E}$ tends to zero at high frequencies by noting that the first allowed oscillator level (hf) is so large compared to the average thermal energy available ($k_B T$) that there is almost zero probability that the first excited state is occupied.

In summary, Planck arrived at his blackbody formula by making two startling assumptions: (1) the energy of a charged oscillator of frequency f is limited to discrete values nhf and (2) during emission or absorption of light, the change in energy of an oscillator is hf. But Planck was every bit the "unwilling revolutionary." From most of Planck's early correspondence one gains the impression that his concept of quantization was really a desperate calculational device, and moreover a device that applied only in the one case of blackbody radiation. It remained for the great Albert Einstein, the incarnation of physics in the 20th century, to elevate quantization to the level of a universal phenomenon by showing that light itself was quantized.

Figure 2.13 Probability that an oscillator of frequency f at temperature T is in one of its quantized energy states, nhf.

Planck blackbody law

EXAMPLE 2.4   Derivation of Stefan's Law from the Planck Distribution

In this example, we show that the Planck spectral distribution formula leads to the experimentally observed Stefan law for the total radiation emitted by a black body at all wavelengths,

$$e_{total} = 5.67 \times 10^{-8} \, T^4 \, \text{W} \cdot \text{m}^{-2} \cdot \text{K}^{-4}$$

Solution   Considering that Stefan's law is an expression for the total power per unit area radiated at all wavelengths, we must integrate the expression for $u(\lambda, T) \, d\lambda$ given by Equation 2.22 over λ and use Equation 2.7 for the connection between the energy density inside the blackbody cavity and the power emitted per unit area of blackbody surface. So we have

$$e_{total} = \frac{c}{4} \int_{\lambda=0}^{\infty} u(\lambda, T) \, d\lambda = \int_0^{\infty} \frac{2\pi hc^2}{\lambda^5(e^{hc/\lambda k_B T} - 1)} \, d\lambda$$

If we make the change of variable $x = hc/\lambda k_B T$, we find that the integral assumes a form common in tables:

$$e_{total} = \frac{2\pi k_B^4 T^4}{c^2 h^3} \int_0^{\infty} \frac{x^3}{(e^x - 1)} \, dx$$

Using

$$\int_0^{\infty} \frac{x^3}{(e^x - 1)} \, dx = \frac{\pi^4}{15}$$

we find

$$e_{total} = \frac{2\pi^5 k_B^4}{15 c^2 h^3} \, T^4 = \sigma T^4$$

Finally, substituting for k_B, c, and h, we have

$$\sigma = \frac{(2)(3.141)^5 (1.381 \times 10^{-23} \, \text{J/K})^4}{(15)(2.998 \times 10^8 \, \text{m/s})^2 (6.626 \times 10^{-34} \, \text{J} \cdot \text{s})^3}$$
$$= 5.67 \times 10^{-8} \, \text{W} \cdot \text{m}^{-2} \cdot \text{K}^{-4}$$

Exercise 4   Show that

$$\int_{\lambda=0}^{\infty} \frac{2\pi hc^2}{\lambda^5(e^{hc/\lambda k_B T} - 1)} \, d\lambda = \frac{2\pi k_B^4 T^4}{h^3 c^2} \int_{x=0}^{\infty} \frac{x^3}{(e^x - 1)} \, dx$$

2.4   LIGHT QUANTIZATION AND THE PHOTOELECTRIC EFFECT

We now turn to the year 1905, in which the next important development in quantum theory took place. The year 1905 was an incredible one for the "willing revolutionary" Albert Einstein (Fig. 2.14). In this year Einstein produced three immortal papers on three different topics, each revolutionary, and each worthy of a Nobel prize. All three papers contained balanced, symmetric, and unifying new results achieved by spare and clean logic and simple mathematics. The first work, entitled "A Heuristic[7] Point of View About the Generation and Transformation of Light," formulated the theory of light quanta and explained the photoelectric effect.[8] The second paper was entitled "On the Motion of Particles Suspended in Liquids as Required by the Molecular-Kinetic Theory of Heat." It explained Brownian motion and provided strong proof of the reality of atoms.[9] The third paper, which is perhaps his most famous, contained the invention of the theory of special relativity[10] and was entitled "On the Electrodynamics of Moving Bodies." It is interesting to note that when Einstein was awarded the Nobel prize in 1922, the Swedish Academy judged his greatest

Figure 2.14 Albert Einstein shown in a playful mood riding his bicycle. The photograph was taken in 1933 in Santa Barbara. (*From the California Institute of Technology Archives*)

[7] A heuristic argument is one that is plausible and enlightening but not rigorously justified.

[8] A. Einstein, *Ann. Physik*, 17:132, 1905 (March).

[9] A. Einstein, *Ann. Physik*, 17:549, 1905 (May).

[10] A. Einstein, *Ann. Physik*, 17:891, 1905 (June).

contribution to physics to have been the theory of the photoelectric effect. No mention was made at all of his theory of relativity!

Let us turn now to the paper concerning the light quantum,[8] in which Einstein crossed swords with Maxwell and challenged the unqualified successes of the wave theory of light. Einstein recognized an inconsistency between Planck's quantization of oscillators in the walls of the black body and his insistence that the cavity radiation consisted of classical electromagnetic waves. By showing that the change in entropy of blackbody radiation was like the change in entropy of an ideal gas consisting of independent particles, Einstein reached the conclusion that light itself is composed of "grains," or discontinuous quanta, of energy.[11] Furthermore, he asserted that light interacting with matter also consists of quanta, and he worked out the implications for photoelectric and photochemical processes. His explanation of the photoelectric effect offered such convincing proof that light consists of energy packets that we should describe it in more detail. First, however, we need to consider the main experimental features of the photoelectric effect and the failure of classical theory to explain this effect.

As noted earlier, Hertz first established that clean metal surfaces emit charges when exposed to ultraviolet light. In 1888 Hallwachs discovered that the emitted charges were negative, and in 1899 J. J. Thomson showed that the emitted charges were electrons. He did this by measuring the charge-to-mass ratio of the particles produced by ultraviolet light and even succeeded in measuring e separately by a cloud chamber technique (see Chapter 3).

The last crucial discovery before Einstein's explanation was made in 1902 by Philip Lenard, who was studying the photoelectric effect with intense carbon arc light sources. He found that electrons are emitted from the metal with a range of velocities and that the maximum kinetic energy of photoelectrons, K_{max}, *does not* depend on the *intensity* of the exciting light. Although he was unable to establish the precise relationship, Lenard also indicated that K_{max} increases with light *frequency*. A typical apparatus used to measure the maximum kinetic energy of photoelectrons is shown in Figure 2.15. K_{max} is easily measured by applying a retarding voltage and gradually increasing it until the most energetic electrons are stopped and the photocurrent becomes zero. At this point,

$$K_{max} = \tfrac{1}{2}m_e v_{max}^2 = eV_s \tag{2.23}$$

where m_e is the mass of the electron, v_{max} is the maximum electron speed, e is the electronic charge, and V_s is the stopping voltage. A plot of the type found by Lenard is shown in Figure 2.16a; it illustrates that K_{max} or V_s is independent of light intensity I. The increase in current (or number of electrons per second) with increasing light intensity shown in Figure 2.16a was expected and could be explained classically. However, the result that K_{max} *does not* depend on the intensity was completely unexpected.   ·

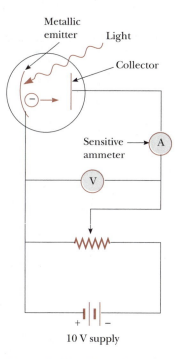

Figure 2.15 Photoelectric effect apparatus.

---

[11] Einstein, as Planck before him, fell back on the unquestionable solidity of thermodynamics and statistical mechanics to derive his revolutionary results. At the time it was well known that the probability, W, for n independent gas atoms to be in a partial volume V of a larger volume V_0 is $(V/V_0)^n$. Einstein showed that light of frequency f and total energy E enclosed in a cavity obeys an identical law, where in this case W is the probability that all the radiation is in the partial volume and $n = E/hf$.

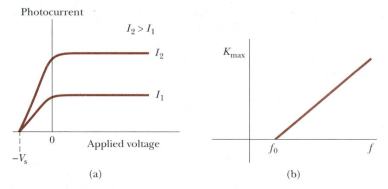

Figure 2.16 (a) A plot of photocurrent vs. applied voltage. The graph shows that K_{max} is independent of light intensity I for light of fixed frequency. (b) A graph showing the dependence of K_{max} on light frequency. Note that the results are independent of I.

Two other experimental results were completely unexpected classically as well. One was *the linear dependence of K_{max} on frequency* shown in Figure 2.16b. Note that Figure 2.16b also shows the existence of a threshold frequency, f_0, below which no photoelectrons are emitted. (Actually a threshold energy called the *work function*, ϕ, is associated with the binding energy of an electron in a metal and is expected classically. That there is an energy barrier holding electrons in a solid is evident from the fact that electrons are not spontaneously emitted from a metal in a vacuum but require high temperatures or incident light as energy sources to promote emission.) The other interesting result impossible to explain classically is that *there is no time lag between the start of illumination and the start of the photocurrent.* Measurements have shown that if there is a time lag, it is less than 10^{-9} s. In summary, as shown in detail in the following example, classical electromagnetic theory has major difficulties explaining the independence of K_{max} and light intensity, the linear dependence of K_{max} on light frequency, and the instantaneous response of the photocurrent.

EXAMPLE 2.5   Maxwell Takes a Licking

For a typical case of photoemission from sodium, show that *classical theory* predicts that (a) K_{max} depends on the incident light intensity, I, (b) K_{max} does not depend on the frequency of the incident light, and (c) there is a long time lag between the start of illumination and the beginning of the photocurrent. The work function for sodium is $\phi = 2.28$ eV and an absorbed power per unit area of 1.00×10^{-7} mW/cm^2 produces a measurable photocurrent in sodium.

Solution (a) According to classical theory, the energy in a light wave is spread out uniformly and continuously over the wavefront. Assuming that all absorption of light occurs in the top atomic layer, that each atom absorbs an equal amount of energy proportional to its cross-sectional area, A, and that each atom funnels this energy into one of its electrons, we find that each electron absorbs an energy K in time t given by

$$K = CIAt$$

where C is a fraction accounting for less than 100% light absorption. Because electrons inside the metal have a range of energies (from zero to several electron volts—see Section 9.4), the most energetic electrons will be emitted with K_{max} once they have absorbed enough energy to overcome the surface energy barrier ϕ. We can express this as

$$K_{max} = CIAt - \phi \qquad (1)$$

Thus, classical theory predicts that for a fixed absorption period, t, at low light intensities no electrons ought to be emitted, thereby contradicting experiment. At higher intensities, above some threshold, electrons should be emitted with higher kinetic energies the higher the light intensity, also contradicting experiment.

(b) According to classical theory, *the intensity of a light wave is proportional to the square of the amplitude of the electric field, E_0^2*, and it is this electric field amplitude that increases with increasing intensity and imparts an increasing acceleration and kinetic energy to an electron. Replacing I with a quantity proportional to E_0^2 in Equation (1) shows that K_{max} should not depend at all on the frequency of the classical light wave, again contradicting the experimental results.

(c) To estimate the time lag between the start of illumination and the emission of electrons, we assume that an electron must accumulate the threshold energy to just overcome the work function. Setting $K_{max} = 0$ in Equation (1) gives

$$0 = CIAt - \phi$$

or

$$t = \frac{\phi}{CIA} = \frac{\phi}{IA}$$

assuming that I is the actual absorbed intensity. Because ϕ and I are given, we need A, the cross-sectional area of an atom, to calculate the time. As an estimate of A we simply use $A = \pi r^2$, where r is a typical atomic radius. Taking $r = 1.0 \times 10^{-8}$ cm, we find $A = \pi \times 10^{-16}$ cm^2. Finally, substituting this value into the expression for t, we obtain

$$t = \frac{2.28 \text{ eV} \times 1.60 \times 10^{-16} \text{ mJ/eV}}{(10^{-7} \text{ mJ/s} \cdot \text{cm}^2)(\pi \times 10^{-16} \text{ cm}^2)}$$
$$= 1.2 \times 10^7 \text{ s} \approx 130 \text{ days}$$

Thus we see that the classical calculation of the time lag for photoemission is in direct opposition to the experimental result, disagreeing by a factor of 10^{16}.

Exercise 5   Why do the I–V curves in Figure 2.16(a) rise gradually between $-V_s$ and 0, that is, why do they not rise abruptly upward at $-V_s$? What statistical information about the conduction electrons inside the metal is contained in the slope of the I–V curve?

Einstein's explanation of the puzzling photoelectric effect was as brilliant for what it focused on as for what it omitted. For example, he stressed that Maxwell's classical theory had been immensely successful in describing the progress of light through space *over long time intervals* but that a different theory might be needed to describe *momentary interactions* of light and matter, as in light emission by oscillators or the transformation of light energy to kinetic energy of the electron in the photoelectric effect. He also focused only on the energy aspect of the light and avoided models or mechanisms concerning the conversion of the quantum of light energy to kinetic energy of the electron. In short, he introduced only those ideas necessary to explain the photoelectric effect. He maintained that the energy of light is not distributed evenly over the classical wave front, but is concentrated in discrete regions (or in "bundles"), called quanta, each containing energy, *hf*. A suggestive image, not to be taken too literally, is shown in Figure 2.17b. Einstein's picture was that a light quan-

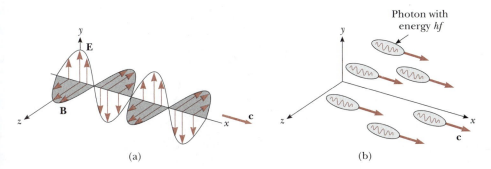

(a)   (b)

Figure 2.17   (a) A classical view of a traveling light wave. (b) Einstein's photon picture of "a traveling light wave."

tum was so localized that it gave all its energy, hf, directly to a single electron in the metal. Thus, according to Einstein, the maximum kinetic energy for emitted electrons is

$$K_{max} = hf - \phi \tag{2.24}$$

Einstein's theory of the photoelectric effect

where ϕ is the work function of the metal, which corresponds to the minimum energy with which an electron is bound in the metal. Table 2.1 lists values of work functions measured for different metals.

Equation 2.24 beautifully explained the puzzling independence of K_{max} and intensity found by Lenard. For a fixed light frequency f, an increase in light intensity means more photons and more photoelectrons per second, although K_{max} remains unchanged according to Equation 2.24. In addition, Equation 2.24 explained the phenomenon of threshold frequency. Light of threshold frequency f_0, which has just enough energy to knock an electron out of the metal surface, causes the electron to be released with zero kinetic energy. Setting $K_{max} = 0$ in Equation 2.24 gives

$$f_0 = \frac{\phi}{h} \tag{2.25}$$

Table 2.1 Work Functions of Selected Metals

Metal	Work Function, ϕ, (in eV)
Na	2.28
Al	4.08
Cu	4.70
Zn	4.31
Ag	4.73
Pt	6.35
Pb	4.14
Fe	4.50

Thus the variation in threshold frequency for different metals is produced by the variation in work function. Note that light with $f < f_0$ has insufficient energy to free an electron. Consequently, the photocurrent is zero for $f < f_0$.

With any theory, one looks not only for explanations of previously observed results but also for new predictions. This was indeed the case here, as Equation 2.24 predicted the result (new in 1905) that K_{max} should vary linearly with f for any material and that the slope of the K_{max} versus f plot should yield the universal constant h (see Fig. 2.18). In 1916, the American physicist Robert

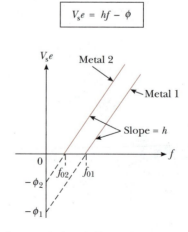

Figure 2.18 Universal characteristics of all metals undergoing the photoelectric effect.

Millikan reported photoelectric measurement data, from which he substantiated the linear relation between K_{max} and f and determined h with a precision of about 0.5%.[12]

EXAMPLE 2.6   The Photoelectric Effect in Zinc

Philip Lenard determined that photoelectrons released from zinc by ultraviolet light were stopped by a voltage of 4.3 V. Find K_{max} and v_{max} for these electrons.

Solution

$$K_{max} = eV_s = (1.6 \times 10^{-19} \text{ C})(4.3 \text{ V}) = 6.9 \times 10^{-19} \text{ J}$$

To find v_{max} we set the work done by the electric field equal to the change in the electron's kinetic energy, to obtain

$$\tfrac{1}{2}m_e v_{max}^2 = eV_s$$

or

$$v_{max} = \sqrt{\frac{2eV_s}{m_e}} = \sqrt{\frac{2(6.9 \times 10^{-19} \text{ J})}{9.11 \times 10^{-31} \text{ kg}}}$$

$$= 1.2 \times 10^6 \text{ m/s}$$

Thus a 4.3-eV electron is rather energetic and moves with a speed of about a million meters per second. Note, however, that this is still only about 0.4% of the speed of light, so relativistic effects are negligible in this case.

EXAMPLE 2.7   The Photoelectric Effect for Iron

Suppose that light of total intensity 1.0 μW/cm^2 falls on a clean iron sample 1.0 cm^2 in area. Assume that the iron sample reflects 96% of the light and that only 3.0% of the absorbed energy lies in the violet region of the spectrum above the threshold frequency.

(a) What intensity is actually available for the photoelectric effect?

Because only 4.0% of the incident energy is absorbed, and only 3.0% of this energy is able to produce photoelectrons, the intensity available is

$$I = (0.030)(0.040)I_0 = (0.030)(0.040)(1.0 \; \mu\text{W/cm}^2)$$

$$= 1.2 \text{ nW/cm}^2$$

(b) Assuming that all the photons in the violet region have an effective wavelength of 250 nm, how many electrons will be emitted per second?

For an efficiency of 100%, one photon of energy hf will produce one electron, so

$$\text{Number of electrons/s} = \frac{1.2 \times 10^{-9} \text{ W}}{hf}$$

$$= \frac{\lambda(1.2 \times 10^{-9})}{hc}$$

$$= \frac{(250 \times 10^{-9} \text{ m})(1.2 \times 10^{-9} \text{ J/s})}{(6.6 \times 10^{-34} \text{ J} \cdot \text{s})(3.0 \times 10^8 \text{ m/s})}$$

$$= 1.5 \times 10^9$$

(c) Calculate the current in the phototube in amperes.

$$i = (1.6 \times 10^{-19} \text{ C})(1.5 \times 10^9 \text{ electrons/s})$$

$$= 2.4 \times 10^{-10} \text{ A}$$

A sensitive electrometer is needed to detect this small current.

(d) If the cutoff frequency is $f_0 = 1.1 \times 10^{15}$ Hz, find the work function, ϕ, for iron.

From Equation 2.25, we have

$$\phi = hf_0 = (4.14 \times 10^{-15} \text{ eV} \cdot \text{s})(1.1 \times 10^{15} \text{ s}^{-1})$$

$$= 4.5 \text{ eV}$$

(e) Find the stopping voltage for iron if photoelectrons are produced by light with $\lambda = 250$ nm.

From the photoelectric equation,

$$eV_s = hf - \phi$$

$$= \frac{hc}{\lambda} - \phi$$

$$= \frac{(4.14 \times 10^{-15} \text{ eV} \cdot \text{s})(3.0 \times 10^8 \text{ m/s})}{250 \times 10^{-9} \text{ m}} - 4.5 \text{ eV}$$

$$= 0.46 \text{ eV}$$

Thus the stopping voltage is 0.46 V.

---

[12] R. A. Millikan, *Phys. Rev.*, 7:355, 1916. Some of the experimental difficulties in the photoelectric effect were the lack of strong monochromatic uv sources, small photocurrents, and large effects of rough and impure metal surfaces on f_0 and k_{max}. Millikan brilliantly circumvented these difficulties by using alkali metal cathodes, which are sensitive in the visible to about 600 nm (thus making it possible to use the strong visible lines of the mercury arc), and machining fresh alkali surfaces while the metal sample was held under high vacuum.

Exercise 6   The following data are found for photoemission from calcium:

λ (nm)	253.6	313.2	365.0	404.7
V_s (V)	1.95	0.98	0.50	0.14

Plot V_s versus f, and from the graph obtain Planck's constant, the threshold frequency, and the work function for calcium.

2.5   THE COMPTON EFFECT AND X-RAYS

Although Einstein introduced the concept that light consists of pointlike quanta of *energy* in 1905, he did not directly treat the *momentum* carried by light until 1906. In that year, in a paper treating a molecular gas in thermal equilibrium with electromagnetic radiation (statistical mechanics again!), Einstein concluded that a light quantum of energy E travels in a single direction (unlike a spherical wave) and carries a momentum directed along its line of motion of E/c, or hf/c. In his own words, "if a bundle of radiation causes a molecule to emit or absorb an energy packet hf, then momentum of quantity hf/c is transferred to the molecule, directed along the line of motion of the bundle for absorption and opposite the bundle for emission."

After developing the first theoretical justification for photon momentum, and treating the photoelectric effect much earlier, it is curious that Einstein bowed out of the picture and carried the treatment of photon momentum no further. The theoretical treatment of photon–particle collisions had to await the insight of Peter Debye and Arthur Compton. In 1923, both men independently realized that the scattering of x-ray photons from electrons could be explained by treating photons as pointlike particles with energy hf and momentum hf/c and by conserving relativistic energy and momentum of the photon–electron pair in a collision.[13,14] This remarkable development completed the particle picture of light by showing that photons, in addition to carrying energy, hf, carry momentum, hf/c, and scatter like particles. Before treating this in detail, a brief introduction to the important topic of x-rays will be given.

X-Rays

X-rays were discovered in 1895 by the German physicist Wilhelm Roentgen. He found that a beam of high-speed electrons striking a metal target produced a new and extremely penetrating type of radiation (Fig. 2.19). Within months of Roentgen's discovery the first medical x-ray pictures were taken, and within several years it became evident that x-rays were electromagnetic vibrations similar to light but with extremely short wavelengths and great penetrating power

---

[13] P. Debye, *Phys. Zeitschr.*, 24:161, 1923. In this paper, Debye acknowledges Einstein's pioneering work on the quantum nature of light.

[14] A. H. Compton, *Phys. Rev.*, 21:484, 1923.

Figure 2.19   X-rays are produced by bombarding a metal target (copper, tungsten, and molybdenum are common) with energetic electrons having energies of 50 to 100 keV.

Figure 2.20   One of the first images made by Roentgen using x-rays (December 22, 1895).

(see Fig. 2.20). Rough estimates obtained from the diffraction of x-rays by a narrow slit showed x-ray wavelengths of about 10^{-10} m, which is of the same order of magnitude as the atomic spacing in crystals. Because the best artificially ruled gratings of the time had spacings of 10^{-7} m, Max von Laue in Germany and William Henry Bragg and William Lawrence Bragg (a father and son team) in England suggested using single crystals such as calcite as natural three-dimensional gratings, the periodic atomic arrangement in the crystals constituting the grating rulings.

A particularly simple method of analyzing the scattering of x-rays from parallel crystal planes was proposed by W. L. Bragg in 1912. Consider two successive planes of atoms as shown in Figure 2.21. Note that adjacent atoms *in a single plane* A will scatter constructively if the angle of incidence, θ_i, equals the angle of reflection, θ_r. Atoms in *successive planes* (A and B) will scatter constructively at an angle θ if the path length difference for rays (1) and (2) is a whole number

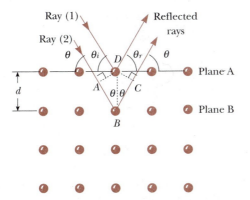

Figure 2.21   Bragg scattering of x-rays from successive planes of atoms. Constructive interference occurs for *ABC* equal to an integral number of wavelengths.

of wavelengths, $n\lambda$. From the diagram, constructive interference will occur when

$$AB + BC = n\lambda \qquad n = 1, 2, 3, \ldots$$

and because $AB = BC = d\sin\theta$, it follows that

Bragg equation

$$n\lambda = 2d\sin\theta \qquad n = 1, 2, 3, \ldots \qquad (2.26)$$

where n is the order of the intensity maximum, λ is the x-ray wavelength, d is the spacing between planes, and θ is the angle of the intensity maximum measured from plane A. Note that there are several maxima for a fixed d and λ corresponding to $n = 1, 2, 3, \ldots$. Equation 2.26 is known as the Bragg equation; it was used with great success by the Braggs to determine atomic positions in crystals. A diagram of a Bragg x-ray spectrometer is shown in Figure 2.22a.

(a)

(b)

Figure 2.22   (a) A Bragg crystal x-ray spectrometer. The crystal is rotated about an axis through P. (b) The x-ray spectrum of a metal target consists of a broad continuous spectrum plus a number of sharp lines, which are due to the characteristic x-rays. Those shown were obtained when 35-keV electrons bombarded a molybdenum target. Note that $1 \text{ pm} = 10^{-12} \text{ m} = 10^{-3} \text{ nm}$.

The crystal is slowly rotated until a strong reflection is observed, which means that Equation 2.26 holds. If λ is known, d can be calculated and, from the series of d values found, the crystal structure may be determined. (See Problem 39.)

The actual x-ray emission spectrum produced by a metal target bombarded by electrons is interesting in itself and is shown in Figure 2.22b. Although the broad, continuous spectrum is well explained by classical electromagnetic theory, a feature of Figure 2.22b, λ_{min}, shows proof of the photon theory. The broad continuous x-ray spectrum shown in Figure 2.22b results from glancing or indirect scattering of electrons from metal atoms. In such collisions only part of the electron's energy is converted to electromagnetic radiation. This radiation is called *bremsstrahlung* (German for braking radiation), which refers to the radiation given off by any charged particle when it is decelerated. The minimum continuous x-ray wavelength λ_{min} is found to be independent of target composition and depends only on the tube voltage, V. It may be explained by attributing it to the case of a head-on electron–atom collision in which all of the incident electron's kinetic energy is converted to electromagnetic energy in the form of a single x-ray photon. For this case we have

$$eV = hf = \frac{hc}{\lambda_{min}}$$

or

$$\lambda_{min} = \frac{hc}{eV} \tag{2.27}$$

where V is the x-ray tube voltage.

Superimposed on the continuous spectrum are sharp x-ray lines labeled K_α and K_β, which are like sharp lines emitted in the visible light spectra. The sharp lines depend on target composition and provide evidence for discrete atomic energy levels separated by thousands of electron volts, as explained in Chapter 8.

The Compton Effect

Let us now turn to the year 1922 and the experimental confirmation by the American Arthur Holly Compton that x-ray photons behave like particles with momentum hf/c. For some time prior to 1922, Compton and his coworkers had been accumulating evidence that showed that classical wave theory failed to explain the scattering of x-rays from free electrons. In particular, classical theory predicted that incident radiation of frequency f_0 should accelerate an electron in the direction of propagation of the incident radiation, and that it should cause forced oscillations of the electron and reradiation at frequency f', where $f' \leq f_0$ (see Fig. 2.23a).[15] Also, according to classical theory, the frequency or wavelength of the scattered radiation should depend on the length of time the electron was exposed to the incident radiation as well as on the intensity of the incident radiation.

---

[15] This decrease in frequency of the reradiated wave is caused by a double Doppler shift, first because the electron is receding from the incident radiation, and second because the electron is a moving radiator as viewed from the fixed lab frame. See *Quantum Theory*, D. Bohm, Prentice-Hall, 1961, p. 35.

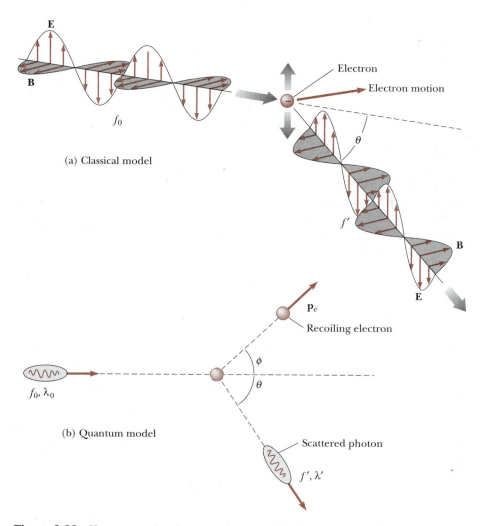

Figure 2.23   X-ray scattering from an electron: (a) the classical model, (b) the quantum model.

Imagine the surprise when Compton showed, experimentally, that the wavelength shift of x-rays scattered at a given angle is absolutely independent of the intensity of radiation and the length of exposure, and depends only on the scattering angle. Figure 2.23b shows the quantum model of the transfer of momentum and energy between an individual x-ray photon and an electron. Note that the quantum model easily explains the lower scattered frequency f', because the incident photon gives some of its energy to the recoiling electron.

A schematic diagram of the apparatus used by Compton is shown in Figure 2.24a. In the original experiment Compton measured the dependence of scattered x-ray intensity on wavelength at three different scattering angles of $45°$, $90°$, and $135°$. The wavelength was measured with a rotating crystal spectrometer, and the intensity was determined by an ionization chamber that generated a current proportional to the x-ray intensity. Monochromatic x-rays of wavelength $\lambda_0 = 0.71$ Å constituted the incident beam. A carbon target with a low

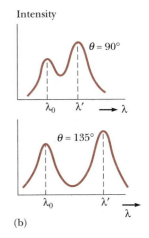

(a)   (b)

Figure 2.24   (a) Schematic diagram of Compton's apparatus. The wavelength was measured with a rotating crystal spectrometer using graphite (carbon) as the target. The
intensity was determined by a movable ionization chamber that generated a current
proportional to the x-ray intensity. (b) Scattered x-ray intensity versus wavelength of
Compton scattering at $\theta = 0°$, $45°$, $90°$, and $135°$.

atomic number, Z, was used because atoms with small Z have a higher percentage of loosely bound electrons. The experimental intensity versus wavelength plots observed by Compton for scattering angles of $0°$, $45°$, $90°$, and $135°$
are shown in Figure 2.24b. They show two peaks, one at λ_0 and a shifted peak
at a longer wavelength λ'. The shifted peak at λ' is caused by the scattering of
x-rays from nearly free electrons. Assuming that x-rays behaved like particles,
λ' was predicted by Compton to depend on scattering angle as

$$\lambda' - \lambda_0 = \frac{h}{m_e c}(1 - \cos\theta) \qquad (2.28)$$ **Compton effect**

where m_e = electron mass; the combination of constants $h/m_e c$ is called the
Compton wavelength of the electron and has a currently accepted value of

$$\frac{h}{m_e c} = 0.0243 \text{ Å} = 0.00243 \text{ nm}$$

Compton's careful measurements confirmed the scattering angle dependence
of λ' completely and determined the Compton wavelength of the electron to
be 0.0242 Å, in excellent agreement with the currently accepted value. It is fair
to say that these results were the first to really convince most American physicists
of the fundamental validity of the quantum theory![16]
    Let us now turn to the derivation of Equation 2.28 by a treatment that as-
sumes that the photon exhibits particle-like behavior and collides elastically

---

[16] The peak at λ_0 is caused by x-ray scattering from tightly bound electrons that have an effective
    mass equal to that of the carbon atom. This unshifted peak at λ_0 is also predicted by Equation
    2.28 if the electron mass is replaced with the mass of carbon, which is about 23,000 times the
    mass of the electron.

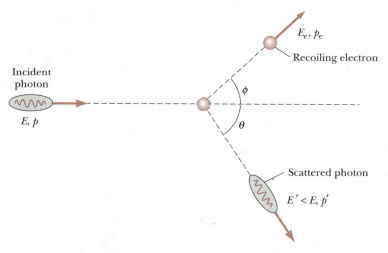

Incident photon

E, p

ϕ

θ

E_e, p_e

Recoiling electron

Scattered photon

$E' < E, p'$

Figure 2.25   Diagram representing Compton scattering of a photon by an electron. The scattered photon has less energy (or longer wavelength) than the incident photon.

with a free electron like a billiard ball. Figure 2.25 shows the photon–electron collision for which energy and momentum are conserved. Because the electron typically recoils at speeds close to the speed of light, we must treat the collision relativistically. The expression for conservation of energy gives

Energy conservation

$$E + m_e c^2 = E' + E_e \qquad (2.29)$$

where E is the energy of the incident photon, E' is the energy of the scattered photon, $m_e c^2$ is the rest energy of the electron, and E_e is the total relativistic energy of the electron after the collision. Likewise, from momentum conservation we have

$$p = p' \cos \theta + p_e \cos \phi \qquad (2.30)$$

$$p' \sin \theta = p_e \sin \phi \qquad (2.31)$$

where p is the momentum of the incident photon, p' is the momentum of the scattered photon, and p_e is the recoil momentum of the electron. Equations 2.30 and 2.31 may be solved simultaneously to eliminate ϕ, the electron scattering angle, which gives the following expression for p_e^2:

$$p_e^2 = (p')^2 + p^2 - 2pp' \cos \theta \qquad (2.32)$$

At this point it is necessary, paradoxically, to invoke the wave nature of light to explain the particle-like behavior of photons. We have already seen that the energy of a photon and the frequency of the associated light wave are related by $E = hf$. If we assume that, in addition, a photon obeys the relativistic expression $E^2 = p^2 c^2 + m^2 c^4$, and that a photon has a mass of zero, we have

$$p_{\text{photon}} = \frac{E}{c} = \frac{hf}{c} = \frac{h}{\lambda} \qquad (2.33)$$

Here again we have a paradoxical situation; a particle property, the photon momentum, is given in terms of a wave property, λ, of an associated light wave.

If the relations $E = hf$ and $p = hf/c$ are substituted into Equations 2.29 and 2.32, these become respectively

$$E_e = hf - hf' + m_e c^2 \qquad (2.34)$$

and

$$p_e{}^2 = \left(\frac{hf'}{c}\right)^2 + \left(\frac{hf}{c}\right)^2 - \frac{2h^2 ff'}{c^2} \cos \theta \qquad (2.35)$$

Because the Compton measurements do not concern the total energy and momentum of the electron, we eliminate E_e and p_e by substituting Equations 2.34 and 2.35 into the expression for the electron's relativistic energy,

$$E_e{}^2 = p_e{}^2 c^2 + m_e{}^2 c^4$$

After some algebra (see Problem 32), one obtains Compton's result for the increase in a photon's wavelength when it is scattered through an angle θ:

$$\lambda' - \lambda_0 = \frac{h}{m_e c} (1 - \cos \theta) \qquad (2.28)$$

EXAMPLE 2.8   The Compton Shift for Carbon

X-rays of wavelength $\lambda = 0.200$ nm are aimed at a block of carbon. The scattered x-rays are observed at an angle of 45.0° to the incident beam. Calculate the wavelength of the scattered x-rays at this angle.

Solution   The shift in wavelength of the scattered x-rays is given by Equation 2.28. Taking $\theta = 45.0°$, we find that

$$\Delta \lambda = \frac{h}{m_e c} (1 - \cos \theta)$$

$$= \frac{6.63 \times 10^{-34} \text{ J} \cdot \text{s}}{(9.11 \times 10^{-31} \text{ kg})(3.00 \times 10^8 \text{ m/s})} (1 - \cos 45.0°)$$

$$= 7.11 \times 10^{-13} \text{ m} = 0.000711 \text{ nm}$$

Hence, the wavelength of the scattered x-ray at this angle is

$$\lambda = \Delta \lambda + \lambda_0 = 0.200711 \text{ nm}$$

Exercise 7   Find the fraction of energy lost by the photon in this collision.

Answer   Fraction $= \Delta E/E = 0.00355$.

EXAMPLE 2.9   X-ray Photons versus Visible Photons

(a) Why are x-ray photons used in the Compton experiment, rather than visible light photons? To answer this question, we shall first calculate the Compton shift for scattering at 90° from graphite for the following cases: (1) very high energy γ-rays from cobalt, $\lambda = 0.0106$ Å; (2) x-rays from molybdenum, $\lambda = 0.712$ Å; and (3) green light from a mercury lamp, $\lambda = 5461$ Å.

Solution   In all cases, the Compton shift formula gives $\Delta \lambda = \lambda' - \lambda_0 = (0.0243 \text{ Å})(1 - \cos 90°) = 0.0243 \text{ Å} = 0.00243$ nm. That is, regardless of the incident wavelength, the same small shift is observed. However, the fractional change in wavelength, $\Delta \lambda / \lambda_0$, is quite different in each case:

γ-rays from cobalt:

$$\frac{\Delta \lambda}{\lambda_0} = \frac{0.0243 \text{ Å}}{0.0106 \text{ Å}} = 2.29$$

X-rays from molybdenum:

$$\frac{\Delta \lambda}{\lambda_0} = \frac{0.0243 \text{ Å}}{0.712 \text{ Å}} = 0.0341$$

Visible light from mercury:

$$\frac{\Delta \lambda}{\lambda_0} = \frac{0.0243 \text{ Å}}{5461 \text{ Å}} = 4.45 \times 10^{-6}$$

Because both incident and scattered wavelengths are simultaneously present in the beam, they can be easily resolved only if $\Delta \lambda / \lambda_0$ is a few percent or if $\lambda_0 \lesssim 1$ °A.

(b) The so-called free electrons in carbon are actually electrons with a binding energy of about 4 eV. Why may this binding energy be ignored for x-rays with $\lambda_0 = 0.712$ Å?

Solution   The energy of a photon with this wavelength is

$$E = hf = \frac{hc}{\lambda} = \frac{12\ 400\ \text{eV}\cdot\text{Å}}{0.712\ \text{Å}} = 17\ 400\ \text{eV}$$

Thus the electron binding energy of 4 eV is negligible in comparison with the incident x-ray energy.

2.6   PARTICLE–WAVE COMPLEMENTARITY

As we have seen, the Compton effect offers ironclad evidence that when light interacts with matter it behaves as if it were composed of particles with energy hf and momentum h/λ. Yet the very success of Compton's theory raises many questions. If the photon is a particle, what can be the meaning of the "frequency" and "wavelength" of the particle, which determine its energy and momentum? Is light in some sense simultaneously a wave and a particle? Although photons have zero mass, is there a simple expression for an effective inertial photon mass that determines a photon's gravitational attraction? What is the spatial extent of a photon, and how does an electron absorb or scatter a photon?

Although answers to some of these questions are possible, it is well to be aware that some demand a view of atomic processes that is too pictorial and literal. Many of these questions issue from the viewpoint of classical mechanics, in which all matter and energy are seen in the context of colliding billiard balls or water waves breaking on a shore. Quantum theory gives light a more flexible nature by implying that different experimental conditions elicit either the wave properties or the particle properties of light. In fact, *both views are necessary and complementary*. Neither model can be used exclusively to describe radiation adequately. A complete understanding is obtained only if the two models are combined in a complementary manner.

The physicist Max Born, an important contributor to the foundations of quantum theory, had this to say about the particle–wave dilemma:

The ultimate origin of the difficulty lies in the fact (or philosophical principle) that we are compelled to use the words of common language when we wish to describe a phenomenon, not by logical or mathematical analysis, but by a picture appealing to the imagination. Common language has grown by everyday experience and can never surpass these limits. Classical physics has restricted itself to the use of concepts of this kind; by analyzing visible motions it has developed two ways of representing them by elementary processes: moving particles and waves. There is no other way of giving a pictorial description of motions—we have to apply it even in the region of atomic processes, where classical physics breaks down.

Every process can be interpreted either in terms of corpuscles or in terms of waves, but on the other hand it is beyond our power to produce proof that it is actually corpuscles or waves with which we are dealing, for we cannot simultaneously determine all the other properties which are distinctive of a corpuscle or of a wave, as the case may be. We can therefore say that the wave and corpuscular descriptions are only to be regarded as complementary ways of viewing one and the same objective process, a process which only in definite limiting cases admits of complete pictorial interpretation . . . [17]

Thus we are left with an uneasy compromise between wave and particle concepts and must accept, at this point, that both are necessary to explain the observed behavior of light. Further considerations of the dual nature of light and indeed of all matter will be taken up again in Chapters 3 and 4.

2.7   DOES GRAVITY AFFECT LIGHT?

O P T I O N A L

It is interesting to speculate on how far the particle model of light may be carried. Encouraged by the successful particle explanation of the photoelectric and Compton effects, one may ask whether the photon possesses an effective gravitational mass, and whether photons will experience an observable change in energy when attracted gravitationally by large masses, such as those of the Sun or Earth.

To investigate these questions, recall that the photon has zero mass, but its effective inertial mass m_i may reasonably be taken to be the mass equivalent of the photon energy, E, or

$$m_i = \frac{E}{c^2} = \frac{hf}{c^2} \qquad (2.36)$$

The same result is obtained if we take the photon momentum and divide by the photon speed c:

$$m_i = \frac{p}{c} = \frac{hf}{c^2}$$

Note that the effective inertial mass determines how the photon responds to an applied force such as that exerted on it during a collision with an electron. The gravitational mass of an object determines the force of gravitational attraction of that object to another, such as the Earth. It is a remarkable fact that the inertial mass of all material bodies is equal to the gravitational mass. (Recent measurements show that these two quantities are equal to within 1 part in 10^{10}.)

Let us assume that the photon, like other objects, also has a gravitational mass equal to its inertial mass. In this case a photon falling from a height H should increase in energy by mgH and thus increase in frequency, although its speed cannot increase and remains at c. In fact, experiments have been carried out that show this increase in frequency and confirm that the photon indeed has an effective gravitational mass of hf/c^2. Figure 2.26 shows a schematic representation of the experiment. An expression for f' in terms of f may be derived by applying conservation of energy to the photon at points A and B.

$$KE_B + PE_B = KE_A + PE_A$$

Because the photon's kinetic energy is $E = pc = hf$ and its potential energy is mgH, where $m = hf/c^2$, we have

$$hf' + 0 = hf + \left(\frac{hf}{c^2}\right)gH$$

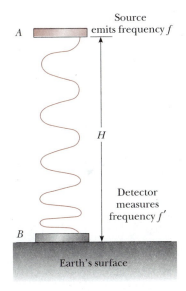

Figure 2.26 Schematic diagram of the falling-photon experiment.

[17] M. Born, *Atomic Physics*, New York, Hafner Publishing Co., fourth edition, 1946, p. 92.

or

$$f' = f\left(1 + \frac{gH}{c^2}\right) \qquad (2.37)$$

The fractional change in frequency, $\Delta f/f$, is given by

$$\frac{\Delta f}{f} = \frac{f' - f}{f} = \frac{gH}{c^2} \qquad (2.38)$$

For $H = 50$ m, we find

$$\frac{\Delta f}{f} = \frac{(9.8 \text{ m/s}^2)(50 \text{ m})}{(3.0 \times 10^8 \text{ m/s})^2} = 5.4 \times 10^{-15}$$

This incredibly small shift has actually been measured (with difficulty)![18] The shift amounts to only about $1/250$ of the line width of the monochromatic γ-ray photons used in the falling photon experiment.

The increase in frequency for a photon falling inward suggests a decrease in frequency for a photon that escapes outward to infinity against the gravitational pull of a star (see Fig. 2.27). This effect, known as a gravitational red shift, would cause a photon emitted in the blue region to be shifted in frequency toward the red end of the spectrum. An expression for the red shift may be derived once again by conserving photon energy:

$$[KE + PE]_{R=\infty} = [KE + PE]_{R=R_s}$$

Using hf for the photon's kinetic energy and $-GMm/R$ for its potential energy, with m equal to hf/c^2 and R_s equal to the star's radius, yields

$$hf' - 0 = hf - \frac{GM}{R_s}\left(\frac{hf}{c^2}\right) \qquad (2.39)$$

or

$$f' = f\left(1 - \frac{GM}{R_s c^2}\right) \qquad (2.40)$$

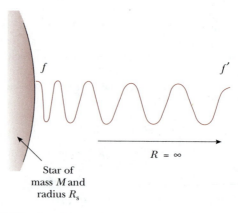

f

f'

$R = \infty$

Star of
mass M and
radius R_s

Figure 2.27   Gravitational red shift from a high-density star.

[18] R. V. Pound and G. A. Rebka, Jr., *Phys. Rev. Lett.*, 4:337, 1960.

EXAMPLE 2.10   The Red Shift for a White Dwarf

White dwarf stars are extremely massive, compact stars that have a mass on the order of the Sun's mass concentrated in a volume similar to that of the Earth. Calculate the gravitational red shift for 300-nm light emitted from such a star.

Solution   We can write Equation 2.40 in the alternate form

$$\frac{f' - f}{f} = \frac{\Delta f}{f} = \frac{GM}{R_s c^2}$$

Using the values

$$M = \text{mass of Sun} = 1.99 \times 10^{30} \text{ kg}$$
$$R_s = \text{radius of Earth} = 6.37 \times 10^6 \text{ m}$$
$$G = 6.67 \times 10^{-11} \text{ N} \cdot \text{m}^2/\text{kg}^2$$

we find

$$\frac{\Delta f}{f} = \frac{(6.67 \times 10^{-11} \text{ N} \cdot \text{m}^2/\text{kg}^2)(1.99 \times 10^{30} \text{ kg})}{(6.37 \times 10^6 \text{ m})(3.00 \times 10^8 \text{ m/s})^2}$$
$$= 2.31 \times 10^{-4}$$

Because $\Delta f/f \approx df/f$, and $df = (c/\lambda^2)\, d\lambda$ (from $f = c/\lambda$), we find $|df/f| = d\lambda/\lambda$. Therefore, the shift in wavelength is

$$\Delta \lambda = (300 \text{ nm})(2.31 \times 10^{-4}) = 0.0695 \text{ nm} \approx 0.7 \text{ Å}$$

Note that this is a red shift, so the observed wavelength would be 300.07 nm.

One more observation about Equation 2.40 is irresistible. Is it possible for a very massive star in the course of its life cycle to become so dense that the term $GM/R_s c^2$ becomes greater than 1? In that case Equation 2.39 suggests that the photon cannot escape from the star, because escape requires more energy than the photon initially possesses. Such a star is called a *black hole* because it emits no light and acts like a celestial vacuum cleaner for all nearby matter and radiation. Even though the black hole itself is not luminous, it may be possible to observe it indirectly in two ways. One way is through the gravitational attraction the black hole would exert on a normal luminous star if the two constituted a binary star system. In this case the normal star would orbit the center of mass of the black hole/normal star pair, and the orbital motion might be detectable. A second indirect technique for "viewing" a black hole would be to search for x-rays produced by inrushing matter attracted to the black hole. Although the black hole itself would not emit x-rays, an x-ray emitting region of roughly stellar diameter should be observable, as shown in Figure 2.28. X-rays are produced by the heating of the infalling matter as it circulates, is compressed, and eventually falls into the black hole. Such an intense nonluminous point source of x-rays has been detected in the constellation of the Swan. This source, designated Cygnus X-1, is believed by most astronomers to be a black hole; it possesses a luminosity, or power output, of 10^{30} W in the 2- to 10-keV x-ray range.

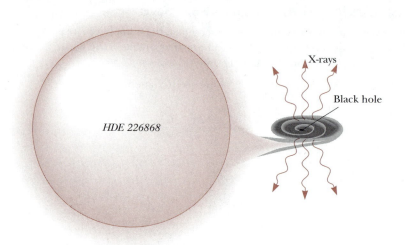

Figure 2.28   The Cygnus X-1 black hole. The stellar wind from HDE 226868 pours matter onto a huge disk around its black hole companion. The infalling gases are heated to enormous temperatures as they spiral toward the black hole. The gases are so hot that they emit vast quantities of x-rays.

SUMMARY

The work of Maxwell and Hertz in the late 1800s conclusively showed that light, heat radiation, and radio waves were all electromagnetic waves differing only in frequency and wavelength. Thus it astonished scientists to find that the spectral distribution of radiation from a heated cavity could not be explained in terms of classical wave theory. Planck was forced to introduce the concept of the quantum of energy in order to derive the correct blackbody formula. According to Planck, the atomic oscillators responsible for blackbody radiation can have only *discrete*, or quantized, *energies* given by

$$E = nhf \tag{2.10}$$

where n is an integer, h is Planck's constant, and f is the oscillator's natural frequency. Using general thermodynamic arguments Planck was able to show that $u(f, T)$, the blackbody radiation energy per unit volume with frequency between f and $f + df$, could be expressed as the product of the number of oscillators per unit volume in this frequency range, $N(f)\ df$, and the average energy emitted per oscillator, $\bar{E}$. That is,

$$u(f, T)\ df = \bar{E}\ N(f)\ df \tag{2.12}$$

If $\bar{E}$ is calculated by allowing a *continuous* distribution of oscillator energies, the incorrect Rayleigh-Jeans law is obtained. If $\bar{E}$ is calculated from a *discrete* set of oscillator energies (following Planck), the correct blackbody radiation formula is obtained:

$$u(f, T)\ df = \frac{8\pi f^2}{c^3}\left(\frac{hf}{e^{hf/k_{\mathrm{B}}T} - 1}\right)df \tag{2.9}$$

Planck quantized the energy of atomic oscillators, but Einstein extended the concept of quantization to light itself. In Einstein's view, light of frequency f consists of a stream of particles, called *photons*, each with energy $E = hf$. The photoelectric effect, a process in which electrons are ejected from a metallic surface when light of sufficiently high frequency is incident on the surface, can be simply explained with the photon theory. According to this theory, the maximum kinetic energy of the ejected photoelectron, K_{max}, is given by

$$K_{max} = hf - \phi \tag{2.23}$$

where ϕ is the work function of the metal.

Although the idea that light consists of a stream of photons with energy hf was put forward in 1905, the idea that these photons also carry momentum was not experimentally confirmed until 1923. In that year it was found that x-rays scattered from free electrons suffer a simple shift in wavelength with scattering angle, known as the Compton shift. When an x-ray of frequency f is viewed as a *particle* with *energy hf* and *momentum hf/c*, x-ray–electron scattering can be correctly analyzed to yield the Compton shift formula:

$$\Delta\lambda = \frac{h}{m_e c}(1 - \cos\theta) \tag{2.28}$$

Here m_e is the mass of the electron, and θ is the x-ray scattering angle.

The striking success of the photon theory in explaining interactions between light and electrons contrasts sharply with the success of classical wave theory in explaining the polarization, reflection, and interference of light. This leaves us with the dilemma of whether light is a wave or a particle. The currently accepted view suggests that light has both wave and particle characteristics, and that these characteristics together constitute a complementary view of light.

Calculation of the Number of Modes for Waves in a Cavity

APPENDIX TO CHAPTER 2

We wish to solve for the number of standing waves that fall within the frequency range from f to $f + df$ for waves confined to a cubical cavity of side L. Because $N(f)$ is known experimentally to be the same for a cavity of any shape and for walls of any material, we can choose the simplest shape—a cube—and the simplest boundary conditions—waves that vanish at the boundaries. Starting with Maxwell's equations, it can be shown that the electric field obeys a wave equation that may be separated into time-dependent and time-independent parts. The time-dependent equation has a simple sinusoidal solution with frequency ω, and each time-independent component of the electric field satisfies an equation of the type

$$\frac{\partial^2 E_x}{\partial x^2} + \frac{\partial^2 E_x}{\partial y^2} + \frac{\partial^2 E_x}{\partial z^2} + k^2 E_x = 0 \tag{2.41}$$

where

$$E_x = E_x(x, y, z)$$

and

$$k = \frac{2\pi}{\lambda} = \frac{2\pi f}{c}$$

Assuming that $E_x = u(x)v(y)w(z)$, we can separate Equation 2.41 into three ordinary differential equations of the type

$$\frac{d^2u}{dx^2} + k_x^2 u = 0 \qquad (2.42)$$

where

$$k^2 = k_x^2 + k_y^2 + k_z^2$$

Equation 2.42 is the simple harmonic oscillator equation and has the solution

$$u(x) = B \cos k_x x + C \sin k_x x$$

Imposing the boundary condition that E_x or u is 0 at $x = 0$ and at $x = L$ leads to $B = 0$ and $k_x L = n_x \pi$, where $n_x = 1, 2, 3, \ldots$. Similar solutions are obtained for $v(y)$ and $w(z)$, giving

$$E_x(x, y, z) = A \sin(k_x x) \sin(k_y y) \sin(k_z z)$$

where

$$k^2 = \frac{\pi^2}{L^2} (n_x^2 + n_y^2 + n_z^2) \qquad (2.43)$$

and n_x, n_y, n_z are positive integers.

To determine the density of modes, we interpret Equation 2.43 as giving the square of the distance from the origin to a point in k-space, or "reciprocal" space. It is called reciprocal space because k has dimensions of $(\text{length})^{-1}$. As shown in Figure 2.29, the axes in k-space are k_x, k_y, and k_z. Because $k_x = n_x \pi/L$, $k_y = n_y \pi/L$, and $k_z = n_z \pi/L$, the points in k-space (or modes) are separated by π/L along each axis, and there is one standing wave in k-space per $(\pi/L)^3$ of volume. The number of standing waves $N(k)$ having wavenumbers between k and $k + dk$ is then simply the volume of k-space between k and $k + dk$ divided by $(\pi/L)^3$. The volume between k and $k + dk$ is $\frac{1}{8}$ the volume of a spherical shell of thickness dk, so that

$$N(k) \, dk = \frac{\frac{1}{2}\pi k^2 \, dk}{(\pi/L)^3} = \frac{Vk^2 \, dk}{2\pi^2} \qquad (2.44)$$

where $V = L^3$ is the cavity volume.

For electromagnetic waves there are two perpendicular polarizations for each mode, so that $N(k)$ in Equation 2.44 must be increased by a factor of 2. Thus, we have for the number of standing waves per unit volume

$$\frac{N(k) \, dk}{V} = \frac{k^2 \, dk}{\pi^2} \qquad (2.45)$$

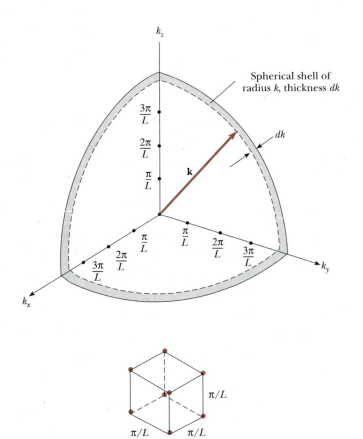

Figure 2.29 A geometrical interpretation of $k^2 = (\pi^2/L^2)(n_x^2 + n_y^2 + n_z^2)$.

To find $N(f)$ we use $k = 2\pi f/c$ in Equation 2.45 to obtain

$$N(f)\,df = \frac{8\pi f^2}{c^3}\,df \qquad (2.46)$$

$N(\lambda)$, the number of modes per unit volume between λ and $\lambda + d\lambda$, may be obtained from Equation 2.46 by using $f = c/\lambda$ to give

$$N(\lambda)\,d\lambda = -\frac{8\pi}{\lambda^4}\,d\lambda \qquad (2.47)$$

QUESTIONS

1. What assumptions were made by Planck in dealing with the problem of blackbody radiation? Discuss the consequences of these assumptions.

2. If the photoelectric effect is observed for one metal, can you conclude that the effect will also be observed for another metal under the same conditions? Explain.

3. Suppose the photoelectric effect occurs in a gaseous target rather than a solid. Will photoelectrons be produced at *all* frequencies of the incident photon? Explain.

4. How does the Compton effect differ from the photoelectric effect?

5. What assumptions were made by Compton in dealing with the scattering of a photon from an electron?

6. An x-ray photon is scattered by an electron. What happens to the frequency of the scattered photon relative to that of the incident photon?

7. Why does the existence of a cutoff frequency in the photoelectric effect favor a particle theory for light rather than a wave theory?

8. All objects radiate energy. Why, then, are we not able to see all objects in a dark room?

9. Which has more energy, a photon of ultraviolet radiation or a photon of yellow light?

10. What effect, if any, would you expect the temperature of a material to have on the ease with which electrons can be ejected from it in the photoelectric effect?

11. Some stars are observed to be reddish, and some are blue. Which stars have the higher surface temperature? Explain.

12. When wood is stacked on a special elevated grate (which is commercially available) in a fireplace, a pocket of burning wood forms beneath the grate whose temperature is higher than that of the burning wood at the top of the stack. Explain how this device provides more heat to the room than a conventional fire does and thus increases the efficiency of the fireplace.

13. What physical process described in this chapter might reasonably be called "the inverse photoelectric effect"? Can you account for this process classically or must it be accounted for by viewing light as a collection of many little particles each with energy hf? Explain.

14. In the photoelectric effect, if the intensity of incident light is very low then the number of photons per second striking the metal surface will be small, and the probability per second of electron emission *per surface atom* will also be small. Account for the observed instantaneous emission of photoelectrons under these conditions.

PROBLEMS

2.1 Light as an Electromagnetic Wave

1. *Classical Zeeman Effect or the Triumph of Maxwell's Equations!* As pointed out in Section 2.1, Maxwell's equations may be used to predict the change in emission frequency when gas atoms are placed in a magnetic field. Consider the situation shown in Figure P2.1. Note that the application of a magnetic field perpendicular to the orbital plane of the electron induces an electric field, which changes the direction of the velocity vector. (a) Using

$$\oint \mathbf{E} \cdot d\mathbf{s} = -\frac{d\Phi_B}{dt}$$

show that the magnitude of the electric field is given by

$$E = \frac{r}{2}\frac{dB}{dt}$$

(b) Using $F\,dt = m\,dv$, calculate the change in speed Δv of the electron. Show that if r remains constant,

$$\Delta v = \frac{erB}{2m_e}$$

(c) Find the change in angular frequency, $\Delta \omega$, of the electron and calculate the numerical value of $\Delta \omega$ for B equal to 1 T. Note that this is also the change in frequency of the light emitted according to Maxwell's equations. Find the fractional change in frequency $\Delta\omega/\omega$ for an ordinary emission line of 500 nm. (d) Actually, the original emission line at ω_0 is split into three components at $\omega_0 - \Delta\omega$, ω_0, and $\omega_0 + \Delta\omega$. The line at $\omega_0 + \Delta\omega$ is produced by atoms with electrons rotating as shown in Figure P2.1, whereas the line at $\omega_0 - \Delta\omega$ is produced by atoms with electrons rotating in the opposite sense. The line at ω_0 is produced by atoms with electronic planes of rotation oriented parallel to **B**. Explain.

2.2 Blackbody Radiation

2. The temperature of your skin is approximately 35°C. What is the wavelength at which the peak occurs in the radiation emitted from your skin?

3. A 2.0-kg mass is attached to a massless spring of force constant $k = 25$ N/m. The spring is stretched 0.40 m from its equilibrium position and released. (a) Find the total energy and frequency of oscillation according to classical calculations. (b) Assume that the energy is quantized and find the quantum number, n, for the system.

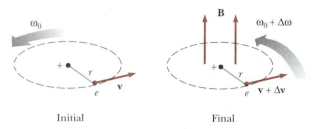

Initial   Final

Figure P2.1

(c) How much energy would be carried away in a one-quantum change?

4. (a) Use Stefan's law to calculate the total power radiated per unit area by a tungsten filament at a temperature of 3000 K. (Assume that the filament is an ideal radiator.) (b) If the tungsten filament of a light bulb is rated at 75 W, what is the surface area of the filament? (Assume that the main energy loss is due to radiation.)

5. Consider the problem of the distribution of blackbody radiation described in Figure 2.3. Note that as T increases, the wavelength λ_{max} at which $u(\lambda, T)$ reaches a maximum shifts toward shorter wavelengths. (a) Show that there is a general relationship between temperature and λ_{max} stating that $T\lambda_{max}$ = constant *(Wien's displacement law).* (b) Obtain a numerical value for this constant. [*Hint:* Start with Planck's radiation law and note that the slope of $u(\lambda, T)$ is zero when $\lambda = \lambda_{max}$.]

2.3 Derivation of the Rayleigh-Jeans Law and Planck's Law (Optional)

6. The essentials of calculating the number of modes of vibration of waves confined to a cavity may be understood by considering a one-dimensional example. (a) Calculate the number of modes (standing waves of different wavelength) with wavelengths between 2.0 cm and 2.1 cm that can exist on a string with fixed ends that is 2 m long. (*Hint:* use $n(\lambda/2) = L$, where $n = 1, 2, 3, 4, 5. \ldots$ Note that a specific value of n defines a specific mode or standing wave with different wavelength.) (b) Calculate, in analogy to our three-dimensional calculation, the number of modes per unit wavelength per unit length, $\dfrac{\Delta n}{L \, \Delta\lambda}$. (c) Show that in general the number of modes per unit wavelength per unit length for a string of length L is given by

$$\frac{1}{L} \left| \frac{dn}{d\lambda} \right| = \frac{2}{\lambda^2}$$

Does this expression yield the same numerical answer as found in (a)? (d) Under what conditions is it justified to replace $\left| \left(\dfrac{\Delta n}{L \, \Delta\lambda} \right) \right|$ with $\left| \left(\dfrac{dn}{L \, d\lambda} \right) \right|$? Is the expression $n = 2L/\lambda$ a continuous function?

2.4 Light Quantization and the Photoelectric Effect

7. Calculate the energy of a photon whose frequency is (a) 5×10^{14} Hz, (b) 10 GHz, (c) 30 MHz. Express your answers in electron volts.

8. Determine the corresponding wavelengths for the photons described in Problem 7.

9. An FM radio transmitter has a power output of 100 kW and operates at a frequency of 94 MHz. How many photons per second does the transmitter emit?

10. The average power generated by the Sun has the value 3.74×10^{26} W. Assuming the average wavelength of the Sun's radiation to be 500 nm, find the number of photons emitted by the Sun in 1 s.

11. A sodium-vapor lamp has a power output of 10 W. Using 589.3 nm as the average wavelength of the source, calculate the number of photons emitted per second.

12. The photocurrent of a photocell is cut off by a retarding potential of 2.92 V for radiation of wavelength 250 nm. Find the work function for the material.

13. The work function for potassium is 2.24 eV. If potassium metal is illuminated with light of wavelength 350 nm, find (a) the maximum kinetic energy of the photoelectrons and (b) the cutoff wavelength.

14. Molybdenum has a work function of 4.2 eV. (a) Find the cutoff wavelength and threshold frequency for the photoelectric effect. (b) Calculate the stopping potential if the incident light has a wavelength of 200 nm.

15. When cesium metal is illuminated with light of wavelength 300 nm, the photoelectrons emitted have a maximum kinetic energy of 2.23 eV. Find (a) the work function of cesium and (b) the stopping potential if the incident light has a wavelength of 400 nm.

16. Consider the metals lithium, beryllium, and mercury, which have work functions of 2.3 eV, 3.9 eV, and 4.5 eV, respectively. If light of wavelength 300 nm is incident on each of these metals, determine (a) which metals exhibit the photoelectric effect and (b) the maximum kinetic energy for the photoelectron in each case.

17. Light of wavelength 500 nm is incident on a metallic surface. If the stopping potential for the photoelectric effect is 0.45 V, find (a) the maximum energy of the emitted electrons, (b) the work function, and (c) the cutoff wavelength.

18. The active material in a photocell has a work function of 2.00 eV. Under reverse-bias conditions (where the polarity of the battery in Figure 2.15 is reversed), the cutoff wavelength is found to be 350 nm. What is the value of the bias voltage?

19. A light source of wavelength λ illuminates a metal and ejects photoelectrons with a maximum kinetic energy of 1.00 eV. A second light source with half the wavelength of the first ejects photoelectrons with a maximum kinetic energy of 4.00 eV. What is the work function of the metal?

20. Figure P2.20 shows the stopping potential versus incident photon frequency for the photoelectric effect for sodium. Use these data points to find (a) the work function, (b) the ratio h/e, and (c) the cutoff wavelength. (d) Find the percent difference between your answer to (b) and the accepted value. (Data taken from R. A. Millikan, *Phys. Rev.* 7:362 [1916].)

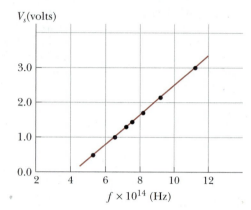

V_s(volts)

$f \times 10^{14}$ (Hz)

Figure P2.20   Some of Millikan's original data for sodium.

21. Photons of wavelength 450 nm are incident on a metal. The most energetic electrons ejected from the metal are bent into a circular arc of radius 20 cm by a magnetic field whose strength is equal to 2.0×10^{-5} T. What is the work function of the metal?

2.5   The Compton Effect and X-Rays

22. Calculate the energy and momentum of a photon of wavelength 500 nm.

23. X-rays of wavelength 0.200 nm are scattered from a block of carbon. If the scattered radiation is detected at 90° to the incident beam, find (a) the Compton shift $\Delta\lambda$ and (b) the kinetic energy imparted to the recoiling electron.

24. X-rays with an energy of 300 keV undergo Compton scattering from a target. If the scattered rays are detected at 30° relative to the incident rays, find (a) the Compton shift at this angle, (b) the energy of the scattered x-ray, and (c) the energy of the recoiling electron.

25. X-rays with a wavelength of 0.040 nm undergo Compton scattering. (a) Find the wavelength of photons scattered at angles of 30°, 60°, 90°, 120°, 150°, 180°, and 210°. (b) Find the energy of the scattered electrons corresponding to these scattered x-rays. (c) Which one of the given scattering angles provides the electron with the greatest energy?

26. Show that a photon cannot transfer all of its energy to a free electron. (*Hint:* Note that energy and momentum must be conserved.)

27. Show that the ratio of the Compton wavelength $\lambda_c = h/m_e c$ to the de Broglie wavelength for a *relativistic* electron is given by

$$\frac{\lambda_c}{\lambda} = \left[\left(\frac{E}{m_e c^2} \right)^2 - 1 \right]^{1/2}$$

where E is the total energy of the electron and m_e is its mass.

28. In the Compton scattering event illustrated in Figure 2.23b, the scattered photon has an energy of 120 keV and the recoiling electron has an energy of 40 keV. Find (a) the wavelength of the incident photon, (b) the angle θ at which the photon is scattered, and (c) the recoil angle ϕ of the electron.

29. Gamma rays (high-energy photons) of energy 1.02 MeV are scattered from electrons that are initially at rest. If the scattering is *symmetric*, that is, if $\theta = \phi$ in Figure 2.25, find (a) the scattering angle θ and (b) the energy of the scattered photons.

30. If the maximum energy given to an electron during Compton scattering is 30 keV, what is the wavelength of the incident photon?

31. A photon of initial energy 0.1 MeV undergoes Compton scattering at an angle of 60°. Find (a) the energy of the scattered photon, (b) the recoil kinetic energy of the electron, and (c) the recoil angle of the electron.

32. Show that the Compton formula

$$\lambda' - \lambda_0 = \frac{h}{m_e c} (1 - \cos\theta)$$

results when expressions for the electron energy (Equation 2.34) and momentum (Equation 2.35) are substituted into the relativistic energy expression,

$$E_e^2 = p_e^2 c^2 + m_e^2 c^4$$

33. An excited iron (Fe) nucleus (mass 57 u) decays to its ground state with the emission of a photon. The energy available from this transition is 14.4 keV. (a) By how much is the photon energy reduced from the full 14.4 keV as a result of having to share energy with the recoiling atom? (b) What is the wavelength of the emitted photon?

34. Find the energy of an x-ray photon that can impart a maximum energy of 50 keV to an electron by Compton collision.

35. Compton used photons of wavelength 0.0711 nm. (a) What is the energy of these photons? (b) What is the wavelength of the photons scattered at an angle of 180° (backscattering case)? (c) What is the energy of the backscattered photons? (d) What is the recoil energy of the electrons in this case?

36. A photon undergoing Compton scattering has an energy after scattering of 80 keV, and the electron recoils with an energy of 25 keV. (a) Find the wavelength of the incident photon. (b) Find the angle at which the photon is scattered. (c) Find the angle at which the electron recoils.

37. In a Compton collision with an electron, a photon of violet light ($\lambda = 4000$ Å) is backscattered through an angle of 180°. (a) How much energy (eV) is transferred to the electron in this collision? (b) Compare your result with the energy this electron would acquire in a photoelectric process with the same photon. (c) Could violet light eject electrons from a metal by Compton collision? Explain.

38. X-radiation from a molybdenum target (0.626 Å) is incident on a crystal with adjacent atomic planes spaced 4.00×10^{-10} m apart. Find the three smallest angles at which intensity maxima occur in the diffracted beam.

39. As a single crystal is rotated in an x-ray spectrometer (Fig. 2.22a), many parallel planes of atoms besides AA and BB produce strong diffracted beams. Two such planes are shown in Figure P2.39. (a) Determine geometrically the interplanar spacings d_1 and d_2 in terms of d_0. (b) Find the angles (with respect to the surface plane AA) of the $n = 1$, 2, and 3 intensity maxima from planes with spacing d_1. Let $\lambda = 0.626$ Å and $d_0 = 4.00$ Å. Note that a given crystal structure (e.g., cubic) has interplanar spacings with characteristic ratios, which produce characteristic diffraction patterns. Thus measurement of the angular position of diffracted x-rays may be used to infer the crystal structure.

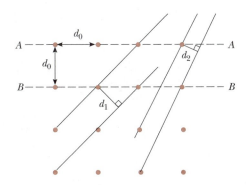

Figure P2.39 Atomic planes in a cubic lattice.

40. *The determination of Avogadro's number with x-rays.* X-rays from a molybdenum target (0.626 Å) are incident on an NaCl crystal, which has the atomic arrangement shown in Figure P2.40. If NaCl has a density of 2.17 g/cm³ and the $n = 1$ diffraction maximum from planes separated by d is found at $\theta = 6.41°$, compute Avogadro's number. (*Hint:* First determine d. Using Figure P2.40, determine the number of NaCl molecules per primitive cell and set the mass per unit volume of the primitive cell equal to the density.)

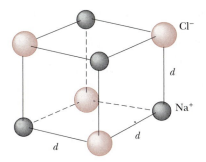

Figure P2.40 The primitive cell of NaCl.

2.7 Does Gravity Affect Light? (Optional)

41. In deriving expressions for the change in frequency of a photon falling or rising in a gravitational field, we have assumed a small change in frequency and a constant photon mass of hf/c^2. Suppose that a star is so dense that Δf is not small. (a) Show that f', the photon frequency at ∞, is related to f, the photon frequency at the star's surface, by

$$f' = fe^{-GM_s/R_s c^2}$$

(b) Show that this expression reduces to Equation 2.40 for small M_s/R_s. (*Hint:* The decrease in photon energy, $h\,df$, as the photon moves dr away from the star is equal to the work done against gravity, $F_G\,dr$.)

42. If the Sun were to contract and become a black hole, (a) what would its approximate radius be and (b) by what factor would its density increase?

Additional Problems

43. Two light sources are used in a photoelectric experiment to determine the work function for a particular metal surface. When green light from a mercury lamp ($\lambda = 546.1$ nm) is used, a retarding potential of 1.70 V reduces the photocurrent to zero. (a) Based on this measurement, what is the work function for this metal? (b) What stopping potential would be observed when using the yellow light from a helium discharge tube ($\lambda = 587.5$ nm)?

44. Ultraviolet light is incident normally on the surface of a certain substance. The work function of the electrons in this substance is 3.44 eV. The incident light has an intensity of 0.055 W/m². The electrons are photoelectrically emitted with a maximum speed of 4.2×10^5 m/s. How fast can electrons be emitted from a square centimeter of the surface? Pretend

that none of the photons are reflected or heat the surface.

45. A 0.500-nm x-ray photon is deflected through 134° in a Compton scattering event. At what angle (with respect to the incident beam) is the recoiling electron found?

46. An electron initially at rest recoils from a head-on collision with a photon. Show that the kinetic energy acquired by the electron is $2hf\alpha/(1 + 2\alpha)$, where α is the ratio of the photon's initial energy to the rest energy of the electron.

3

The Particle Nature of Matter

Chapter Outline

In Chapter 2 we reviewed the evidence for the wave nature of electromagnetic radiation and dealt with major experimental puzzles of the first quarter of the 20th century, which required a particle-like behavior of radiation for their solution. In particular, we discussed Planck's revolutionary idea of energy quantization of oscillators in the walls of a perfect radiator, Einstein's extension of energy quantization to light in the photoelectric effect, and Compton's further confirmation of the existence of the photon as a particle carrying momentum in x-ray scattering experiments.

In this chapter we shall examine the evidence for the particle nature of matter. We only mention briefly the early atomists and concentrate instead on the developments from 1800 onward that dealt with the composition of atoms. In particular, we review the ingenious and fascinating experiments that led to the discoveries of the electron, the proton, the nucleus, and the important Rutherford-Bohr planetary model of the atom.

3.1 THE ATOMIC NATURE OF MATTER

To say that the world is made up of atoms is, today, commonplace. Because the atomic picture of reality is often accepted without question, students can miss out on the rich and fascinating story of how atoms were shown to be real. The discovery and proof of the graininess of the world seem especially fascinating for two reasons. First, because of the size of individual atoms, measurements of atomic properties are usually indirect and necessarily involve clever manipulations of large-scale measurements to infer properties of microscopic particles. Second, the historical evolution of ideas about atomicity shows clearly the real way in which science progresses. This progression is often nonlinear, and it involves an interdependence of physics, chemistry, and mathematics, and the convergence of many different lines of investigation.

There is also an exalted romance in honoring the great atomists who were able to pick out organizing principles from the confusing barrage of market-place ideas of their time: Democritus and Leucippus, who speculated that the unchanging substratum of the world was atoms in motion; the debonair Lavoisier, who established the conservation of matter in many careful chemical experiments; Dalton, who perceived the atomicity of nature in the law of multiple proportions of compounds; Avogadro, who in a most obscure and little appreciated paper, postulated that all pure gases at the same temperature and pressure have the same number of molecules per unit volume; and Maxwell,[1] who showed with his molecular-kinetic theory of gases how macroscopic quantities, such as pressure and temperature, could be derived from averages over distributions of molecular properties. The list could run on and on. We abbreviate it by naming Jean Perrin and the ubiquitous Albert Einstein,[2] who carried on very important theoretical and experimental work concerning Brownian motion, the zigzag movement of small suspended particles caused by molecular impacts. Their work produced additional confirmation of the atomic–molecular hypothesis and resulted in improved values of Avogadro's number as late as the early 1900s.

3.2   THE COMPOSITION OF ATOMS

We now turn our attention to answering the rather dangerous question, "If matter is primarily composed of atoms, what are atoms composed of?" Again, we can point to some primary discoveries that showed that atoms are composed of light, negatively charged particles orbiting a heavy, positively charged nucleus. These were:

- *The discovery of the law of electrolysis in 1833 by Michael Faraday.* Through careful experimental work on electrolysis, Faraday showed that the mass of an element liberated at an electrode is directly proportional to the charge transferred and to the atomic weight of the liberated material but is inversely proportional to the valence of the freed material.
- *The identification of cathode rays as electrons and the measurement of the charge-to-mass ratio (e/m_e) of these particles by Joseph John (J. J.) Thomson in 1897.* Thomson measured the properties of negative particles emitted from different metals and found that the value of e/m_e was always the same.

---

[1] Maxwell was a genius twice over. Either his theory of electricity and magnetism or his kinetic theory of gases would qualify him for that rank.

[2] Much of Einstein's earliest work was concerned with the molecular analysis of solutions and determinations of molecular radii and Avogadro's number. See A. Pais, '*Subtle is the Lord . . . ' The Science and the Life of Albert Einstein*, New York, Oxford University Press, 1982, Chapter 5.

He thus came to the conclusion that the electron is a constituent of all matter!

- *The precise measurement of the electronic charge (e) by Robert Millikan in 1909.* By combining his result for (e) with Thomson's e/m_e value, Millikan showed unequivocally that particles about 1000 times less massive than the hydrogen atom exist.
- *The establishment of the nuclear model of the atom by Ernest Rutherford and co-workers Hans Geiger and Ernest Marsden in 1913.* By scattering fast moving α-particles (charged nuclei of helium atoms emitted spontaneously in radio-active decay processes) from metal foil targets, Rutherford established that atoms consist of a compact positively charged nucleus (diameter ≈ 10^{-14} m) surrounded by a swarm of orbiting electrons (electron cloud diameter ≈ 10^{-10} m).

Let us describe these developments in more detail. We start with a brief example of Faraday's experiments, in particular the electrolysis of molten common salt (NaCl). Faraday found that if 96 500 C of charge (1 faraday) is passed through such a molten solution, 23.0 g of Na will deposit on the cathode and 35.5 g of chlorine gas will bubble off the anode (Fig. 3.1). In this case exactly one gram atomic weight or mole of each element is released because both are monovalent. For divalent and trivalent elements, exactly $\frac{1}{2}$ and $\frac{1}{3}$ of a mole, respectively, would be released. As expected, doubling the quantity of charge passed doubles the mass of the neutral element liberated. Faraday's results may be given in equation form as

$$m = \frac{(q)\,(\text{molar mass})}{(96\ 500\ \text{C})\,(\text{valence})} \tag{3.1}$$

Faraday's law of electrolysis

where m is the mass of the liberated substance in grams, q is the total charge passed in coulombs, the molar mass is in grams, and the valence is dimensionless.

Figure 3.1   Electrolysis of molten NaCl.

EXAMPLE 3.1   The Electrolysis of BaCl$_2$

How many grams of barium and chlorine (cough!) will you get if you pass a current of 10.0 A through molten BaCl$_2$ for 1 h? Barium has a molar weight of 137 g and a valence of 2. Chlorine has a molar weight of 35.5 g and a valence of 1.

Solution   Using Equation 3.1 and $q = It$, where I is the current, we have

$$m_{Ba} = \frac{(q)(\text{molar mass})}{(96\ 500\ \text{C})(\text{valence})}$$

$$= \frac{(10.0\ \text{C/s})(3600\ \text{s})(137\ \text{g})}{(96\ 500\ \text{C})(2)} = 25.6\ \text{g}$$

$$m_{Cl} = \frac{(10.0\ \text{C/s})(3600\ \text{s})(35.5\ \text{g})}{(96\ 500\ \text{C})(1)} = 13.2\ \text{g}$$

Faraday's law of electrolysis is explained in terms of an atomic picture shown in Figure 3.1. Charge passes through the molten solution in the form of ions, which possess an excess or deficiency of one or more electrons. Under the influence of the electric field produced by the battery, these ions move to the anode or cathode, where they respectively lose or gain electrons and are liberated as neutral atoms.

Although it was far from clear in 1833, Faraday's law of electrolysis confirmed three important parts of the atomic picture. First, it offered proof that matter consists of molecules, and that molecules consist of atoms. Second, it showed that charge is quantized, because only integral numbers of charges are transferred at the electrodes. Third, it showed that the subatomic parts of atoms are positive and negative charges, although the mass and the size of the charge of these subatomic particles remained unknown.

The next major step explaining the composition of atoms was taken by Joseph John (J. J.) Thomson. His discovery in 1897[3] that the "rays" seen in low-pressure gas discharges were actually caused by negative particles (electrons) ended a debate dating back nearly 30 years: were cathode rays material particles or waves of some type? Contrary to our rather blasé present acceptance of the electron, many of Thomson's distinguished contemporaries responded with utter disbelief to the idea that electrons were a constituent of all matter. Much of the opposition to Thomson's discovery stemmed from the fact that it required the abandonment of the recently established concept of the atom as an indivisible entity. Thomson's discovery of the electron disturbed this newly established order in atomic theory and provoked startling new developments—Rutherford's nuclear model and the first satisfactory theory of the emission of light by atomic systems, the Bohr model of the atom.

Figure 3.2 shows the original vacuum tube used by Thomson in his e/m_e experiments. Figure 3.3 shows the various parts of the Thomson apparatus for easy reference. Electrons are accelerated from the cathode to the anode, collimated by slits in the anodes, and then allowed to drift into a region of crossed (perpendicular) electric and magnetic fields. The simultaneously applied **E** and **B** fields are first adjusted to produce an undeflected beam. If the **B** field is then turned off, the **E** field alone produces a measurable beam deflection on the phosphorescent screen. From the size of the deflection and the measured values of **E** and **B** the charge-to-mass ratio, e/m_e, may be determined. The truly

---

[3] J. J. Thomson, *Phil. Mag.* 44:269, 1897.

Figure 3.2 The original e/m_e tube used by J. J. Thomson *(after Figure 1.3, p. 7, R. L. Sproull and W. A. Phillips,* Modern Physics, *3rd ed., New York, John Wiley & Sons, 1980).*

ingenious feature of this experiment is the manner in which Thomson measured v_x, the horizontal velocity component of the beam. He did this by balancing the magnetic and electric forces. In effect he created a *velocity selector,* which could select out of the beam those particles having a velocity within a narrow range of values. This device was extensively used in the first quarter of the 20th century in charge-to-mass measurements (q/m) on many particles and in early mass spectrometers.

To gain a clearer picture of the Thomson experiment, let us analyze the electron's motion in his apparatus. Figure 3.5 shows the trajectory of a beam of negative particles entering the **E** and **B** field regions with horizontal velocity v_x. Consider first only an **E** field between the plates. For this case v_x remains constant throughout the motion because there is no force acting in the x di-

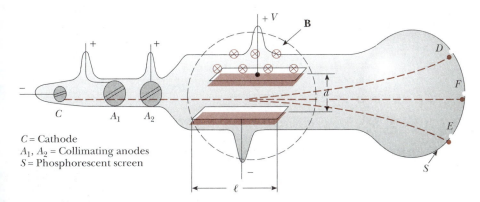

C = Cathode
A_1, A_2 = Collimating anodes
S = Phosphorescent screen

Figure 3.3 A diagram of Thomson's e/m_e tube (patterned after J. J. Thomson, *Philosophical Magazine* (5)44:293, 1897). Electrons subjected to an electric field alone land at D, while those subjected to a magnetic field alone land at E. When both electric and magnetic fields are present and properly adjusted, the electrons experience no net deflection and land at F.

rection. The y component of velocity, v_y, is constant everywhere except between the plates, where the electron experiences a constant upward acceleration due to the electric force and follows a parabolic path. To solve for the deflection angle, θ, we must solve for v_x and v_y. Because v_y initially is zero, the electron leaves the plates with a y component of velocity given by

$$v_y = a_y t \tag{3.2}$$

Because $a_y = F/m_e = Ee/m_e = Ve/m_e d$, and $t = \ell/v_x$, where d and ℓ are the dimensions of the region between the plates and V is the applied potential, we obtain

$$v_y = \frac{V\ell e}{m_e v_x d} \tag{3.3}$$

From Figure 3.5, $\tan \theta = v_y/v_x$, so using Equation 3.3 we obtain

$$\tan \theta = \frac{V\ell}{v_x^2 d} \left(\frac{e}{m_e} \right) \tag{3.4}$$

Assuming small deflections, $\tan \theta \approx \theta$, so we have

$$\theta \approx \frac{V\ell}{v_x^2 d} \left(\frac{e}{m_e} \right) \tag{3.5}$$

Note that θ, the beam deflection, V, the voltage applied to the horizontal deflecting plates, and d and ℓ, the spacing and length, respectively, of the horizontal deflecting plates can all be measured. Hence, one needs only to measure v_x to determine e/m_e. Thomson determined v_x by applying a **B** field and adjusting its magnitude to just balance the deflection of the still present **E** field. Equating the magnitudes of the electric and magnetic forces gives

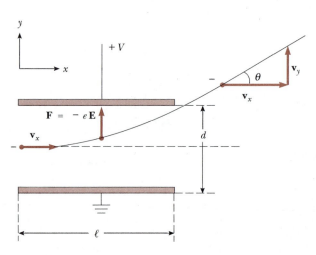

Figure 3.5   Deflection of negative particles by an electric field.

$$qE = qv_x B$$

or

$$v_x = \frac{E}{B} = \frac{V}{Bd} \qquad (3.6)$$

Substituting Equation 3.6 into Equation 3.5 immediately yields a formula for e/m_e entirely in terms of measurable quantities:

$$\frac{e}{m_e} = \frac{V\theta}{B^2 \ell d} \qquad (3.7)$$

The currently accepted value of e/m_e is 1.758803×10^{11} C/kg. Although Thomson's original value was only about 1.0×10^{11} C/kg, prior experiments on the electrolysis of hydrogen ions had given q/m values for hydrogen of about 10^8 C/kg. It was clear almost immediately that Thomson had discovered a particle with a mass about 1000 times smaller than the smallest atom! In his observations, Thomson noted that the e/m_e ratio was independent of the discharge gas and the cathode metal. Furthermore, the particles emitted when electrical discharges were passed through different gases were found to be the same as those observed in the photoelectric effect. Based on these observations, Thomson concluded that these particles must be a universal constituent of all matter. Humanity had achieved its first glimpse into the subatomic world!

EXAMPLE 3.2   Deflection of an Electron Beam by E and B Fields

Using the accepted e/m_e value, calculate the magnetic field required to produce a deflection of 0.20 radians in Thomson's experiment, assuming the values $V = 200$ V, $\ell = 5.0$ cm, and $d = 1.5$ cm (the approximate values used by Thomson). Compare this value of B to the Earth's magnetic field.

Solution   Because $e/m_e = V\theta/B^2\ell d$, solving for B gives

$$B = \sqrt{\frac{V\theta}{\ell d (e/m_e)}}$$

so

$$B = \left[\frac{(200 \text{ V})(0.20 \text{ rad})}{(0.050 \text{ m})(0.015 \text{ m})(1.76 \times 10^{11} \text{ C/kg})} \right]^{1/2}$$

$$= [3.03 \times 10^{-7} \text{ V} \cdot \text{kg/m}^2 \text{ C}]^{1/2}$$

$$= [3.03 \times 10^{-7} \text{ N}^2/(\text{m/s})^2 \text{ C}^2]^{1/2}$$

$$= 5.5 \times 10^{-4} \text{ N}/(\text{m/s}) \cdot \text{C}$$

$$= 5.5 \times 10^{-4} \text{ T}$$

As the Earth's magnetic field has a magnitude of about 0.5×10^{-4} T, we require a field 11 times as strong as the Earth's field.

Exercise 1   Find the horizontal speed v_x for this case.

Answer 2.4×10^7 m/s $= 0.080c$, where c is the speed of light.

Millikan's Value of the Elementary Charge

In 1897, Thomson had been unable to determine e or m_e separately. However, about two years later this great British experimentalist had bracketed the accepted value of e (1.602×10^{-19} C) with values of 2.3×10^{-19} C for charges emitted from zinc illuminated by ultraviolet light and 1.1×10^{-19} C for charges produced by ionizing x-rays and radium emissions. He was also able to conclude that "e is the same in magnitude as the charge carried by the hydrogen atom in the electrolysis of solutions." The technique used by Thomson and his students to measure e is especially interesting because it represents the first use of the cloud chamber technique in physics and also formed the starting point for the famous Millikan oil-drop experiment. Charles Wilson, one of Thomson's students, had discovered that ions act as condensation centers for water droplets when damp air is cooled by a rapid expansion. Thomson used this idea to form charged clouds by using the apparatus shown in Figure 3.6a. Here Q is the measured total charge of the cloud, W is the measured weight of the cloud, and v is the rate of fall or terminal speed. Thomson assumed that the cloud was composed of spherical droplets having a constant mass (no evaporation) and that the magnitude of the drag force D on a single droplet was given by Stokes's law,

$$D = 6\pi a \eta v \tag{3.8}$$

where a is the droplet radius, η is the viscosity of air, and v is the terminal speed of the droplet. The following procedure was used to find a and w, the weight of a single drop. Because v is constant, the droplet is in equilibrium under the combined action of its weight, w, and the drag force, D, as shown in Figure 3.6b. Hence, we require that $w = D$, or

$$w = \tfrac{4}{3}\pi a^3 \rho g = D = 6\pi a \eta v$$

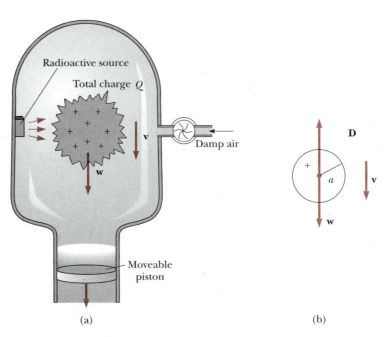

(a)   (b)

Figure 3.6   (a) A diagram of Thomson's apparatus for determining e. (b) A single droplet in the cloud.

Robert Millikan (1868–1953). Although Millikan *(left)* studied Greek as an undergraduate at Oberlin College, he fell in love with physics during his graduate training after teaching school for a few years. He was the first student to receive a PhD in physics from Columbia University in 1895. Following post-doctoral work in Germany under Planck and Nernst, Millikan obtained an academic appointment at the University of Chicago in 1910, where he worked with Michelson. He received the Nobel prize in 1923 for his famous experimental determination of the electronic charge. He is also remembered for his careful experimental work to verify the theory of the photoelectric effect deduced by Einstein. Following World War I, he transferred to the California Institute of Technology, where he worked in atmospheric physics and remained until his retirement. (*Courtesy AIP Emilio Segrè Visual Archives*)

so

$$a = \sqrt{\frac{9\eta v}{2\rho g}} \qquad (3.9)$$

where ρ is the mass density of the droplet and g is the free-fall acceleration. Once w is obtained, the number of drops n (or number of ions) is given by W/w and the electronic charge e is equal to Q/n, assuming that each droplet carries only one electronic charge. Although ingenious, this method is inaccurate because the theory applies only to a single particle and the particles are all assumed to be identical in order to compare the theory to experiments performed on a cloud.

The tremendous advance of Millikan was made possible by his clever idea of making the experiment "fit" the theory. By observing single droplets he eliminated the problems of assuming all particles to be identical and of making uncertain measurements on a cloud. Millikan's basic idea was to measure the rate of fall of a single drop acted on by gravity and drag forces, apply Stokes's law to determine the drop radius and mass, then to measure its upward velocity in an opposing electric field, and hence determine the total charge on an individual drop.[4]

---

[4] Actually the idea of allowing charges to "fall" under a combined gravitational and electric field was first applied to charged clouds of water vapor by H. A. Wilson in 1903. Millikan switched from water to oil to avoid the problems of a changing droplet mass and radius caused by water evaporation.

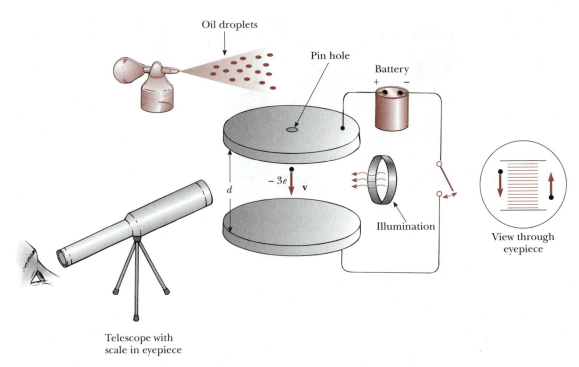

Oil droplets

Pin hole

Battery
+   −

Illumination

View through
eyepiece

Telescope with
scale in eyepiece

$-3e$ $\mathbf{v}$

d

Figure 3.7   A schematic view of the Millikan oil-drop apparatus.

A schematic of the Millikan apparatus is shown in Figure 3.7. Oil droplets charged by an atomizer are allowed to pass through a small hole in the upper plate of a parallel-plate capacitor. If these droplets are illuminated from the side, they appear as brilliant stars against a dark background, and the rate of fall of individual drops may be determined.[5] If an electrostatic field of several thousand volts per meter is applied to the capacitor plates, the drop may move slowly upward, typically at rates of *hundredths* of a centimeter per second. Because the rate of fall is comparable, a single droplet with constant mass and radius may be followed for hours, alternately rising and falling, by simply turning the electric field on and off. The atomicity of charge is shown directly by the observation that after a long series of measurements of constant upward velocities one observes a discontinuous change or jump to a different *upward* velocity (higher or lower). This discontinuous change is caused by the attraction of an ion to the charged droplet and a consequent change in droplet charge. Such changes become more frequent when a source of ionizing radiation is placed between the plates.

Millikan's determination of the electronic charge

The quantitative analysis of the Millikan experiment starts with Newton's second law applied to the oil drop, $\Sigma F_y = ma_y$. Because the drag force D, is large, a constant velocity of fall is quickly achieved, and all measurements are made for the case $a_y = 0$, or $\Sigma F_y = 0$. If we assume that the magnitude of the

[5] Perhaps the reason for the failure of "Millikan's Shining Stars" as a poetic and romantic image has something to do with the generations of physics students who have experienced hallucinations, near blindness, migraine attacks, etc. while repeating his experiment!

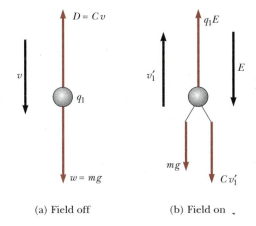

(a) Field off   (b) Field on

Figure 3.8   The forces on a charged oil droplet in the Millikan experiment.

drag force is proportional to the speed ($D = Cv$), and refer to Figure 3.8, we find

$$Cv - mg = 0 \qquad \text{(field off)}$$

$$q_1 E - mg - Cv' = 0 \qquad \text{(field on)}$$

Eliminating C from these expressions gives

$$q_1 = \frac{mg}{E}\left(\frac{v + v_1'}{v}\right) \tag{3.10}$$

When the droplet undergoes a discontinuous change in its *upward speed* from v_1' to v_2' (m, g, E, and v remaining constant), its new charge q_2 is given by

$$q_2 = \frac{mg}{E}\left(\frac{v + v_2'}{v}\right) \tag{3.11}$$

Dividing Equation 3.10 by Equation 3.11 gives

$$\frac{q_1}{q_2} = \frac{v + v_1'}{v + v_2'} \tag{3.12}$$

Equation 3.12 constitutes a remarkably direct and powerful proof of the atomicity of charge, because if successive speed ratios are ratios of whole numbers, successive charges on the drop must be multiples of the same elementary charge! Millikan's experimental results beautifully confirmed this atomicity of charge to within about 1% accuracy.[6]

Up to this point our arguments have been quite general and have assumed only that the drag force on the droplet is proportional to its velocity. To determine the actual value of the electronic charge, e, the mass of the drop must be determined, as can be seen from Equation 3.10. As noted earlier, the droplet radius a may be determined from the application of Stokes's law. This

[6] R. A. Millikan, *Phys. Rev.* 1911, p. 349.

value of a, in turn, can be used to find m from the oil density, ρ. In this procedure, a is

$$a = \sqrt{\frac{9\eta v}{2\rho g}} \qquad (3.9)$$

and the mass of the droplet can be expressed as

$$m = \rho \cdot \text{Volume} = \rho \tfrac{4}{3}\pi a^3 \qquad (3.13)$$

An example of this technique for determining e using Stokes's law is given in Example 3.3.

EXAMPLE 3.3   Experimental Determination of e

In a Millikan experiment the distance of rise or fall of a droplet is 0.600 cm and the average time of fall (field off) is 21.0 s. The observed successive rise times are 46.0, 15.5, 28.1, 12.9, 45.3, and 20.0 s.

(a) Prove that charge is quantized.

Solution   Charge is quantized if q_1/q_2, q_2/q_3, q_3/q_4, and so on are ratios of small whole numbers. Because

$$\frac{q_1}{q_2} = \frac{v + v_1'}{v + v_2'}, \qquad \frac{q_2}{q_3} = \frac{v + v_2'}{v + v_3'}, \qquad \text{etc.,}$$

we must find the speeds. Thus

$$v = \frac{\Delta y}{\Delta t} = \frac{0.600 \text{ cm}}{21.0 \text{ s}} = 0.0286 \text{ cm/s}$$

$$v_1' = \frac{\Delta y}{\Delta t} = \frac{0.600 \text{ cm}}{46.0 \text{ s}} = 0.0130 \text{ cm/s}$$

$$v_2' = 0.600/15.5 = 0.0387 \text{ cm/s}$$

$$v_3' = 0.600/28.1 = 0.0214 \text{ cm/s}$$

$$v_4' = 0.600/12.9 = 0.0465 \text{ cm/s}$$

$$v_5' = 0.600/45.3 = 0.0132 \text{ cm/s}$$

$$v_6' = 0.600/20.0 = 0.0300 \text{ cm/s}$$

so

$$\frac{q_1}{q_2} = \frac{v + v_1'}{v + v_2'} = \frac{0.0286 + 0.0130}{0.0286 + 0.0387} = 0.618 \approx \frac{3}{5}$$

$$\frac{q_2}{q_3} = \frac{v + v_2'}{v + v_3'} = \frac{0.0286 + 0.0387}{0.0286 + 0.0214} = 1.35 \approx \frac{4}{3}$$

$$\frac{q_3}{q_4} = \frac{v + v_3'}{v + v_4'} = \frac{0.0286 + 0.0214}{0.0286 + 0.0465} = 0.666 \approx \frac{2}{3}$$

$$\frac{q_4}{q_5} = \frac{v + v_4'}{v + v_5'} = \frac{0.0286 + 0.0465}{0.0286 + 0.0132} = 1.80 \approx \frac{9}{5}$$

$$\frac{q_5}{q_6} = \frac{v + v_5'}{v + v_6'} = \frac{0.0286 + 0.0132}{0.0286 + 0.0300}$$

$$= 0.713 \approx 8/11 \text{ or } 7/10$$

(b) If the oil density is 858 kg/m³ and the viscosity of air is 1.83×10^{-5} kg/m·s, find the radius, volume, and mass of the drop used in this experiment.

Solution   The radius of the drop is

$$a = \sqrt{\frac{9\eta v}{2\rho g}}$$

$$= \left[\frac{9(1.83 \times 10^{-5} \text{ kg/m·s})(0.0286 \times 10^{-2} \text{ m/s})}{2(858 \text{ kg/m}^3)(9.81 \text{ m/s}^2)} \right]^{1/2}$$

$$= [2.80 \times 10^{-12} \text{ m}^2]^{1/2}$$

$$= 1.67 \times 10^{-6} \text{ m, or } 1.67 \ \mu\text{m}$$

The volume is

$$V = \tfrac{4}{3}\pi a^3 = 1.95 \times 10^{-17} \text{ m}^3$$

The mass is

$$m = \rho V = (858 \text{ kg/m}^3)(1.95 \times 10^{-17} \text{ m}^3)$$

$$= 1.67 \times 10^{-14} \text{ kg}$$

(c) Calculate the successive charges on the drop, and from these results determine the electronic charge. Assume a plate separation of 1.60 cm and a potential difference of 4550 V for the parallel plate capacitor.

Solution   To calculate the charge on each drop using Equation 3.10, we must first calculate the magnitude of the electric field, E. Thus,

$$E = \frac{V}{d} = \frac{4550 \text{ V}}{0.0160 \text{ m}} = 2.84 \times 10^5 \text{ V/m}$$

Now we can find the charges on the drop.

$$q_1 = \left(\frac{mg}{E}\right)\left(\frac{v + v_1'}{v}\right)$$

$$= \frac{(1.67 \times 10^{-14} \text{ kg})(9.81 \text{ m/s}^2)}{(2.84 \times 10^5 \text{ V/m})}\left(\frac{0.0286 + 0.0130}{0.0286}\right)$$

$$= 8.39 \times 10^{-19} \text{ C}$$

Likewise, $q_2 = 13.6 \times 10^{-19}$ C, $q_3 = 10.1 \times 10^{-19}$ C, $q_4 = 15.2 \times 10^{-19}$ C, $q_5 = 8.43 \times 10^{-19}$ C,   and $q_6 = 11.8 \times 10^{-19}$ C.

To find the average value of e, we shall use the fact that at the time of Millikan's work e was known to be between 1.5×10^{-19} C and 2.0×10^{-19} C. Dividing q_1 through q_6 by these values gives the range of integral charges on each drop. Thus, the range of q_1 is $8.39/1.5 = 5.6$ electronic charges to $8.39/2.0 = 4.2$ electronic charges. Similarly, q_2 has 9.1 to 5.8 charges, q_3 has 6.7 to 5.1 charges, q_4 has 10.1 to 7.6 charges, q_5 has 5.6 to 4.2 charges, and q_6 has 7.9 to 5.9 charges. Because there must be an integral number of charges on each drop, we pick an integer in the middle of the allowed range. Therefore, in terms of the electronic charge e, we conclude that $q_1 = 5e$, $q_2 = 8e$, $q_3 = 6e$, $q_4 =$

$9e$, $q_5 = 5e$, and $q_6 = 7e$. Using the values found above, we find

$$e_1 = q/5 = 1.68 \times 10^{-19} \text{ C}$$

$$e_2 = q_1/8 = 1.70 \times 10^{-19} \text{ C}$$

$$e_3 = q_2/6 = 1.68 \times 10^{-19} \text{ C}$$

$$e_4 = q_3/9 = 1.69 \times 10^{-19} \text{ C}$$

$$e_5 = q_4/5 = 1.69 \times 10^{-19} \text{ C}$$

$$e_6 = q_5/7 = 1.69 \times 10^{-19} \text{ C}$$

Taking the average of these values, we find the value of the electronic charge to be $e = 1.688 \times 10^{-19}$ C for this data set.

Stokes's law, as Millikan was aware, is only approximately correct for tiny spheres moving through a gas. The expression $D = 6\pi a \eta v$ holds quite accurately for a 0.1-cm radius ball bearing moving through a liquid or for any case where the moving object radius, a, is large compared with the mean free path, L, of the surrounding molecules. (The mean free path is essentially the average distance between molecules.) In the Millikan experiment, however, a is of the same order of magnitude as the mean free path of air at STP. Consequently Stokes's law overestimates the drag force, because the droplet actually moves for appreciable times through a frictionless "vacuum." Millikan corrected Stokes's law by using a drag force whose magnitude is

$$D = \frac{6\pi a \eta v}{1 + \alpha(L/a)} \tag{3.14}$$

and found that $\alpha = 0.81$ gave the most consistent values of e for drops of different radii. Further corrections to Stokes's law were made by Perrin and Roux, and corrections to Stokes's law and the correct value of e remained a controversial issue for more than 20 years. The currently accepted value of the magnitude of the electronic charge is

$$e = 1.60217733 \times 10^{-19} \text{ C}$$

Electronic charge

Rutherford's Model of the Atom

The early years of the 20th century were generally a period of incredible ferment, change, and activity in physics, including the advent of relativity, quantum theory, and atomic and subatomic physics. In fact, hardly had the reality of pristine, indivisible atoms been established ("They are the only material things which still remain in the precise condition in which they first began to exist" wrote Maxwell in 1872), when Thomson announced their divisibility in 1899: "Electrification essentially involves the splitting of the atom, a part of the mass of the atom getting free and becoming detached from the original atom." Further daring assaults on the indivisibility of the chemical atom came from the experimental work on radioactivity by Marie Curie and by Ernest Rutherford and Frederick Soddy, who explained radioactive transformations of elements in terms of the emission of subatomic particles. The porosity of atoms was also known before 1910 from Lenard's experiments, which showed that electrons are easily transmitted through thin metal and mica foils. All these discoveries, plus the suspicion that the intricate atomic spectral lines (light emitted at a discrete set of frequencies characteristic of each element) must be produced by charge rattling around inside the atom, led to various proposals concerning the internal structure of atoms. The most famous of these early atomic models was the Thomson "plum-pudding" model (1898). This proposal viewed the atom as a homogeneous sphere of uniformly distributed mass and positive charge in which were embedded, like raisins in a plum pudding, negatively charged electrons, which just balanced the positive charge to produce electrically neutral atoms. Although such models did possess electrical stability against collapse or explosion of the atom, they failed to explain the rich line spectra of even the simplest atom, hydrogen.

The key to understanding the mysterious line spectra and the correct model of the atom were both furnished by Ernest Rutherford and his students Hans Geiger and Ernest Marsden through a series of experiments conducted from 1909 to 1914. Noticing that a beam of finely collimated α particles was broadened on passing through a metal foil yet easily penetrated the thin film of metal, they embarked on a series of experiments to probe the distribution of mass within the atom by observing in detail the scattering of α particles from foils. These experiments ultimately led Rutherford to the discovery that most of the atomic mass and all of the positive charge lie in a minute central nucleus of the atom. The accidental chain of events and the clever capitalization on the accidental discoveries leading up to Rutherford's monumental nuclear theory of the atom are nowhere better described than in Rutherford's own essay summarizing the development of the theory of atomic structure:[7]

Rutherford's nuclear model

. . . I would like to use this example to show how you often stumble upon facts by accident. In the early days I had observed the scattering of α-particles, and Dr. Geiger in my laboratory had examined it in detail. He found, in thin pieces of heavy metal, that the scattering was usually small, of the order of one degree. One day Geiger came to me and said, 'Don't you think that young Marsden, whom I am training in radioactive methods, ought to begin a small research?' Now I had thought that, too, so I said, 'Why not let him see if any α-particles can be scattered through a large angle?' I may tell you in confidence that I did not believe that they would be, since we knew that the α-particle was a very fast, massive particle, with a great deal of energy, and you could show that if the scattering was due to the accumulated effect of a number of small scatterings the chance of an α-particle's being scattered backwards was very small. Then I remember two or three days later Geiger coming to me in great excitement and saying, 'We have been able to get some of the α-particles coming backwards . . . ' It was quite the most incredible event that has ever happened to me in my life. It was almost as incredible as if you fired a 15-inch shell at a piece of tissue paper and it came back and hit you. On consideration, I realized that this scattering backwards must be the result of a single collision, and when I made calculations I saw that it was impossible to get anything of that order of magnitude unless

[7] An essay on "The Development of the Theory of Atomic Structure," 1936, Lord Rutherford, published in *Background to Modern Science*, New York, Macmillan Company, 1940.

you took a system in which the greater part of the mass of the atom was concentrated in a minute nucleus. It was then that I had the idea of an atom with a minute massive center carrying a charge. I worked out mathematically what laws the scattering should obey, and I found that the number of particles scattered through a given angle should be proportional to the thickness of the scattering foil, the square of the nuclear charge, and inversely proportional to the fourth power of the velocity. These deductions were later verified by Geiger and Marsden in a series of beautiful experiments.

The essential experimental features of Rutherford's apparatus are shown in Figure 3.9. Alpha-particles emitted with speeds of about 2×10^7 m/s struck a thin gold foil several thousand atomic layers thick. Most of the α's passed straight through the foil along the line DD' (again showing the porosity of the atom), but some were scattered at an angle ϕ. The number of scattered α's at each angle per unit detector area and per unit time was measured by counting the scintillations produced by scattered α's on the ZnS screen. These scintillations were counted with the aid of the microscope. The distance from the point where α particles strike the foil to the zinc sulfide screen is denoted R in Figure 3.9.

Rutherford's basic insight was that because the mass and kinetic energy of the α's are large, even a nearly head-on collision with a particle with the mass of a hydrogen atom would deflect the α particle only slightly and knock the hydrogen atom straight ahead. Multiple scattering of the α particles in the foil accounted for the small broadening (about 1°) originally observed by Rutherford, but it could not account for the occasional large-scale deflections. On the other hand,[8] if all of the positive charge in an atom is assumed to be concentrated at a single central point and not spread out throughout the atom, the electric repulsion experienced by an incident α particle in a head-on collision becomes much greater. Because the charge and mass of the gold atom are

Figure 3.9   A schematic view of Rutherford's α-scattering apparatus.

[8] A dangerous expression that never fails to bring to mind President Truman's infamous request: "If you know of any one-handed economists, bring them to me."

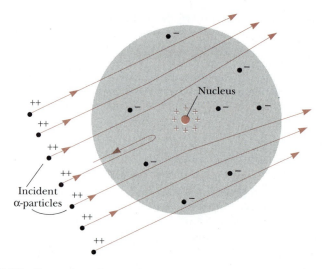

Figure 3.10   Scattering of α particles by a dense, positively charged nucleus.

concentrated at the nucleus, large deflections of the α particle could be experienced in a *single* collision with the massive nucleus. This situation is shown in Figure 3.10.

EXAMPLE 3.4   Collision of an α Particle with a Proton

(a) An α particle of mass m_α and speed v_α strikes a stationary proton with mass m_p. If the collision is elastic and head-on, show that the speed of the proton after the collision, v_p, and the speed of the α particle after the collision, v'_α, are given by

$$v_p = \left(\frac{2m_\alpha}{m_\alpha + m_p} \right) v_\alpha$$

and

$$v'_\alpha = \left(\frac{m_\alpha - m_p}{m_\alpha + m_p} \right) v_\alpha$$

(b) Calculate the percent change in velocity for an α particle colliding with a proton.

Solution   (a) Because the collision is elastic, the total kinetic energy is conserved; therefore

$$\tfrac{1}{2} m_\alpha v_\alpha{}^2 = \tfrac{1}{2} m_\alpha v'_\alpha{}^2 + \tfrac{1}{2} m_p v_p{}^2 \qquad (1)$$

Conservation of momentum for this one-dimensional collision yields

$$m_\alpha v_\alpha = m_p v_p + m_\alpha v'_\alpha \qquad (2)$$

Solving Equation (1) for $(m_\alpha v'_\alpha)^2$ yields

$$(m_\alpha v'_\alpha)^2 = m_\alpha (m_\alpha v_\alpha{}^2 - m_p v_p{}^2) \qquad (3)$$

Solving Equation (2) for $(m_\alpha v'_\alpha)^2$ and equating this to Equation (3) gives

$$(m_\alpha v_\alpha)^2 + (m_p v_p)^2 - 2m_\alpha m_p v_\alpha v_p = (m_\alpha v_\alpha)^2 - m_\alpha m_p v_p{}^2$$

or

$$(m_p v_p)(m_p v_p - 2m_\alpha v_\alpha + m_\alpha v_p) = 0$$

The solutions to this equation are

$$v_p = 0$$

and

$$v_p = \left(\frac{2m_\alpha}{m_\alpha + m_p} \right) v_\alpha \qquad (4)$$

Because the proton must move when struck by the heavy α particle, Equation (4) is the only physically reasonable solution for v_p. The solution for v'_α follows immediately from the substitution of Equation (4) into Equation (2).

(b) Because an α particle consists of two protons and two neutrons, $m_\alpha = 4m_p$. Thus,

$$v_p = \left(\frac{2m_\alpha}{m_\alpha + m_p}\right) v_\alpha = \left(\frac{8m_p}{5m_p}\right) v_\alpha = 1.60\, v_\alpha$$

$$v'_\alpha = \left(\frac{m_\alpha - m_p}{m_\alpha + m_p}\right) v_\alpha = \left(\frac{3m_p}{5m_p}\right) v_\alpha = 0.60\, v_\alpha$$

The percent change in velocity of the α particle is

$$\% \text{ change in } v_\alpha = \left(\frac{v'_\alpha - v_\alpha}{v_\alpha}\right) \times 100\% = -40\%$$

Exercise 2   An α particle with initial velocity v_α undergoes an elastic, head-on collision with an electron initially at rest. Using the fact that an electron's mass is about $1/2000$ of the proton mass, calculate the final velocities of the electron and α particle, and the percent change in velocity of the α particle.

Answers $v_e = 1.998v_\alpha$, $v'_\alpha = 0.9998v_\alpha$, and the percent change in $v_\alpha = -0.02\%$.

In his analysis, Rutherford assumed that large-angle scattering is produced by a single nuclear collision and that the repulsive force between an α particle and a nucleus separated by a distance r is given by Coulomb's law,

$$F = k\frac{(2e)(Ze)}{r^2} \tag{3.15}$$

where $+2e$ is the charge on the α, $+Ze$ is the nuclear charge, and k is the Coulomb constant. With this assumption, Rutherford was able to show that the number of α particles entering the detector per unit time, Δn, at an angle ϕ is given by

$$\Delta n = \frac{k^2 Z^2 e^4 N n A}{4R^2 (\frac{1}{2}m_\alpha v_\alpha{}^2)^2 \sin^4(\phi/2)} \tag{3.16}$$

Here R and ϕ are defined in Figure 3.9, N is the number of nuclei per unit area of the foil (and is thus proportional to the foil thickness), n is the total number of α particles incident on the target per unit time, and A is the area of the detector. The dependence of scattering on foil thickness, α particle speed, and scattering angle was almost immediately confirmed experimentally by Geiger and Marsden.[9] Because the values of atomic number (Z) were uncertain at the time, the scattering dependence on Z could not be directly checked. Turning the argument around, however, and assuming the correctness of Equation 3.16, one could find the value of Z that gave the best fit of this equation to the experimental data points. An illustration of this sensitive technique for determining Z for a silver foil is shown in Figure 3.11. Note that changing Z produces only a vertical shift in the graph and not a change in shape.

Much of the remarkable experimental work of the ingenious Lord Rutherford can be credited to an ability to use his current discoveries to probe even deeper into nature's mysteries. For example, he turned his studies of the transmission of radioactive particles through matter into a sensitive and delicate

---

[9] H. Geiger and E. Marsden, "Deflection of α-Particles through Large Angles," *Phil. Mag.* (6)25:605, 1913.

Figure 3.11   Comparison of theory and experiment for α particle scattering from a silver foil. (*From E. Rutherford, J. Chadwick, and J. Ellis*, Radiations from Radioactive Substances, *Cambridge University Press, Cambridge, 1951.*)

technique for probing the atom. Another example was his clever technique for measuring the size of the nucleus. Realizing that Equation 3.15 would hold only if the α particle did not have enough energy to deform or penetrate the scattering nucleus, he systematically looked for the threshold α energy at which departures from his scattering equation occurred, the idea being that at this threshold energy the α should be just penetrating the nuclear radius at its distance of closest approach. Following Rutherford, we set the kinetic energy of the α at infinity equal to the potential energy of the system (α + target nucleus) at the distance of closest approach, $d_{\min}$, or

$$\tfrac{1}{2}m_\alpha v_\alpha^2 = k \frac{(Ze)(2e)}{d_{\min}} \tag{3.17}$$

Equation 3.17 may then be solved for $d_{\min}$ to determine the distance of closest approach. In the case when the kinetic energy of the α is so high that Equation 3.16 begins to fail, this distance of closest approach is approximately equal to the nuclear radius.

Rutherford was confronted with the experimental dilemma that no failures of Equations 3.15 or 3.16 were found for heavy metal foils with the most energetic naturally occurring α particles available to him (≈ 8 MeV). Showing characteristic economy, instead of embarking on a particle accelerator program, he made use of metals like aluminum with lower Z's and hence lower Coulomb barriers to α penetration.[10] Thus, in 1919, he was able to determine the nuclear radius of aluminum to be about 5×10^{-15} m.

---

[10] Rutherford was famous for the remark to his graduate students, "There is no money for apparatus—we shall have to use our heads" (A. Keller, *Infancy of Atomic Physics: Hercules in His Cradle*, Oxford, Clarendon Press, 1983, p. 215).

EXAMPLE 3.5   Estimate of the Radius of the Aluminum Nucleus

In 1919, Rutherford was able to show a breakdown in Equation 3.16 for 7.7-MeV α particles scattered at large angles from aluminum nuclei ($Z = 13$). Estimate the radius of the aluminum nucleus from these facts.

Solution   Rutherford's scattering formula will no longer hold when α particles begin to penetrate or touch the nucleus. When the α particle is very far from the aluminum nucleus, its kinetic energy is 7.7 MeV. This is also the total energy of the system (α particle plus aluminum nucleus), because the aluminum nucleus is at rest and the potential energy is zero for an infinite separation of particles. When the α particle is at the point of closest approach to the aluminum nucleus, its kinetic energy is zero, and for a head-on collision we take their separation to be the radius of the aluminum nucleus. At this point, the kinetic energy of the system is zero, and the total energy is just the potential energy of the system. Applying conservation of energy gives

$$K_\alpha = \text{Potential energy at closest approach} = \frac{k(Ze)(2e)}{d_{min}}$$

or

$$d_{min} = k\frac{2Ze^2}{K_\alpha}$$
$$= \frac{2(13)(1.60 \times 10^{-19}\ \text{C})^2(8.99 \times 10^9\ \text{N} \cdot \text{m}^2/\text{C}^2)}{(7.7 \times 10^6\ \text{eV})(1.60 \times 10^{-19}\ \text{J/eV})}$$
$$= 4.9 \times 10^{-15}\ \text{m}$$

The overall success of the Rutherford nuclear model was truly striking. Rutherford and his students had shown that the model of the atom with a massive nucleus of diameter 10^{-14} m and combined proton mass equal to approximately half of the nuclear mass agreed in all particulars with experiment. As with all great discoveries, however, the idea of the nuclear atom raised a swarm of questions at the next deeper level: (1) If there are only Z protons in the nucleus, what composes the other half of the nuclear mass? (2) What provides the cohesive force to keep many protons confined in the incredibly small distance of 10^{-14} m? (3) How do the electrons move around the nucleus to form a stable atom, and how does their motion account for the observed spectral lines?

Rutherford had no precise answer to the first question. He speculated that the difference between the mass of Z protons and the total nuclear mass could be accounted for by groupings of neutral particles, each consisting of a bound electron–proton pair. This conjecture seemed especially satisfying because it built the atom out of the most fundamental particles then known to exist.

In answer to the second question, Rutherford cautiously held that electrical forces provided the cement to hold the nucleus together. He wrote, "The nucleus, though of minute dimensions, is in itself a very complex system consisting of positively and negatively charged bodies bound closely together by intense electrical forces." In fact, it was not until 1921 that it was clearly recognized that the Coulomb force did not hold the nucleus together and that a completely new and very strong type of force binds protons together. Interestingly it was James Chadwick, the discoverer of the neutron, who first recognized that a new force field of much more than electric intensity was at work in the nucleus.[11] Perhaps Rutherford's magnificent achievement of explaining α scattering with the Coulomb law blinded him to the possibility that this was not the ultimate law at work within the nucleus.

[11] J. Chadwick and E. S. Biele, *Phil. Mag.* 42:923, 1921.

Figure 3.12   Bohr (on the right) and Rutherford (on the left) were literal as well as intellectual supports for each other. This photograph of Bohr and Rutherford sitting back to back was taken at a rowing regatta in June, 1923 at Cambridge University. (*AIP Niels Bohr Library, and* Physics Today *October 1985, an issue devoted to Bohr.*)

The answer to the third question was not to be given by Rutherford. That was to be the masterwork of Niels Bohr (Fig. 3.12). Even so, with characteristic insight, Rutherford speculated on a planetary model of the atom or, more precisely, that negative charges revolved around the dense positive core as the planets revolved around the Sun.[12]

3.3   THE BOHR ATOM

Bohr's original quantum theory of spectra was one of the most revolutionary, I suppose, that was ever given to science, and I do not know of any theory that has been more successful . . . I consider the work of Bohr one of the greatest triumphs of the human mind.

Lord Rutherford

Then it is one of the greatest discoveries.

Albert Einstein, on hearing of Bohr's theoretical calculation of the Rydberg constants for hydrogen and singly ionized helium

Spectral Series

Before looking in detail at the first successful theory of atomic dynamics, we review the experimental work on line spectra that served as the impetus for and the clear confirmation of the first quantum theory of the atom. As already pointed out in Chapter 2, glowing solids and liquids (and even gases at the high densities found in stars) emit a continuous distribution of wavelengths.

---

[12] Thomson and Nagaoka had worked even earlier with planetary atomic models in 1904.

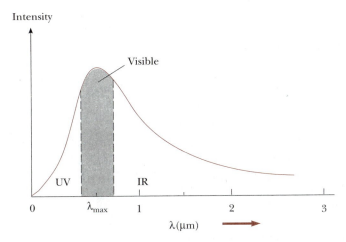

Figure 3.13   Intensity versus wavelength for a body heated to 6000 K.

This distribution exhibits a common shape for the intensity-versus-wavelength curve, and the peak in this curve shifts toward shorter wavelengths with increasing temperature. This universal "blackbody" curve is shown in Figure 3.13.

In sharp contrast to this continuous spectrum is the discrete line spectrum emitted by a low-pressure gas subject to an electric discharge. When the light from such a low-pressure gas discharge is examined with a spectroscope, it is found to consist of a few bright lines of pure color on a dark background. This contrasts sharply with the continuous rainbow of colors seen when a glowing solid is viewed through a spectroscope. Furthermore, as can be seen from Figure 3.14, the wavelengths contained in a given line spectrum are characteristic of the particular element emitting the light. (Also see the inside front cover.) The simplest line spectrum is observed for atomic hydrogen, and we shall de-

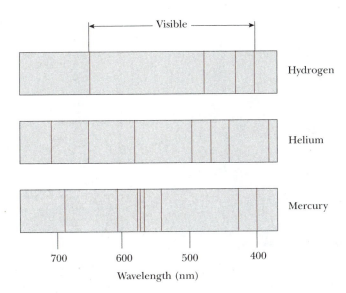

Figure 3.14   Emission line spectra of a few representative elements.

Figure 3.15 Gustav Robert Kirchhoff (1824–1887). Yes, this is the same fellow who brought us the circuit loop theorem and established the connection between the absorption and emission of an object. (See Section 2.2.) (*AIP Niels Bohr Library, W. F. Meggers Collection*)

scribe this spectrum in detail. Other atoms, such as mercury, helium, and neon, give completely different line spectra. Because no two elements emit the same line spectrum, this phenomenon represents a practical and sensitive technique for identifying the elements present in unknown samples. In fact, by 1860 spectroscopy had advanced so far in the hands of Gustav Robert Kirchhoff (Figure 3.15) and Robert Wilhelm von Bunsen (Figure 3.16) at the University of Heidelberg that they were able to discover two new elements, rubidium and cesium, by observing new sequences of spectral lines in mineral samples. Improvements in instruments and techniques resulted in an enormous growth in spectral analysis in Europe from 1860 to 1900. Even the European public imagination was captured by spectroscopy when spectroscopic techniques were used to show that "celestial" meteorites consisted only of known Earth elements after all.

Kirchhoff's immense contribution to spectroscopy is also shown by another advance he made in 1859—the foundation of **absorption spectroscopy** and the explanation of Fraunhofer's dark D-lines in the solar spectrum.[13] In 1814, Joseph Fraunhofer had passed the continuous spectrum from the Sun through a narrow slit and then through a prism. He observed the surprising result of nearly 1000 fine dark lines, or gaps, in the continuous rainbow spectrum of the Sun, and he assigned the letters A, B, C, D . . . to the most prominent dark lines. These lines and many more are shown in Figure 3.17. Kirchhoff correctly deduced that the mysterious dark lines are produced by a cloud of vaporized atoms in the Sun's outer, cooler layers, which absorb at discrete frequencies the intense continuous radiation from the center of the Sun. Further he showed that the Fraunhofer D-lines were produced by vaporized sodium, and that they had the same wavelengths as the strong yellow lines in the emission spectrum of sodium. Kirchhoff also correctly deduced that all of Fraunhofer's dark lines should be attributable to absorption

---

[13] G. Kirchhoff, *Monatsber.* Berlin, 1859, p. 662.

Figure 3.16 Robert Wilhelm von Bunsen (1811–1899). Bunsen is pictured with his most famous invention, the gas laboratory burner named for him. The greatest achievement of this fine chemist, however, was the development, with Kirchhoff, of the powerful analytical method of spectral analysis. (*AIP Niels Bohr Library, W. F. Meggers Collection*)

Figure 3.17 The spectrum of the Sun as viewed with a high-resolution spectrograph, showing many dark Fraunhofer lines. Wavelengths are indicated above the spectrum in Ångstrom units (10^{-8} cm). (*Palomar Observatory photograph.*)

by different elements present in the Sun. In a single stroke he opened the way to determining the elemental composition of stars trillions of miles from the Earth. His ingenious and simple method for demonstrating the presence of sodium vapor in the solar atmosphere is shown in schematic form in Figure 3.18.

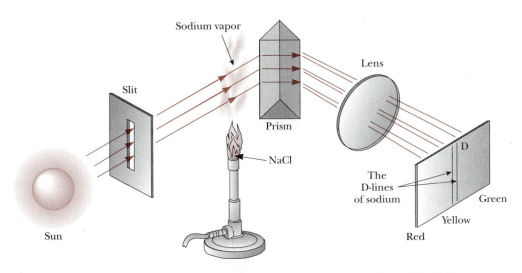

Figure 3.18 Kirchhoff's experiment explaining the Fraunhofer D-lines. The D-lines darken noticeably when sodium vapor is introduced between the slit and the prism.

Today absorption spectroscopy is certainly as important as emission spectroscopy for qualitative and quantitative analyses of elements and molecular groups. In general one obtains an absorption spectrum by passing light from a continuous source [whether in the ultraviolet (uv), visible (vis), or infrared (IR) regions] through a gas of the element being analyzed. The absorption spectrum consists of a series of dark lines superimposed on the otherwise continuous spectrum emitted by the source. It is found that each line in the absorption spectrum of an element coincides with a line in the emission spectrum of that same element; however, not all of the emission lines are present in an absorption spectrum. The differences between emission and absorption spectra are complicated in general and depend on the temperature of the absorbing vapor.

An interesting use of the coincidence of absorption and emission lines is made in the atomic absorption spectrometer, a device routinely used to measure parts per million (ppm) of metals in unknowns. For example, if sodium is to be measured, a sodium lamp emitting a line spectrum is chosen as the light source. The unknown is heated in a hot flame (usually oxyacetylene) to vaporize the sample, to break the chemical bonds of sodium to other elements, and to produce a gas of elemental sodium. The spectrometer is then tuned to a wavelength for which both absorption and emission lines exist (say one of the D-lines at 588.99 or 589.59 nm), and the amount of darkening or decrease in intensity is measured with a sensitive photomultiplier. The decrease in intensity is a measure of the sodium concentration. With proper calibration, concentrations of 0.1 ppm can be measured with this extremely selective technique. Atomic absorption spectroscopy has been a useful technique in analyzing heavy metal contamination of the food chain. For example, the first determinations of high levels of mercury in tuna fish were made with atomic absorption.

From 1860 to 1885 spectroscopic measurements accumulated voluminously, burying frenzied theoreticians under a mountain of data. Accurate measurements of four visible emission lines of hydrogen had recently been made by Anders Ångström, when in 1885 a Swiss school teacher, Johann Jakob Balmer, published a paper with the unpretentious title "Notice Concerning the Spectral Lines of Hydrogen." By trial and error Balmer had found a formula that correctly predicted the wavelengths of Ångström's four visible lines: H_α (red), H_β (green), H_γ (blue), and H_δ (violet). Figure 3.19 shows these and other lines in the emission spectrum of hydrogen. Balmer gave his formula in the form

$$\lambda \text{ (cm)} = C_2 \left(\frac{n^2}{n^2 - 2^2} \right) \qquad n = 3, 4, 5 \ldots \qquad (3.18)$$

where $C_2 = 3645.6 \times 10^{-8}$ cm, a constant called the convergence limit because it gave the wavelength of the line with largest n value ($n = \infty$). Also, note that $n = 3, 4, 5, \ldots$ where H_α has $n = 3$, H_β has $n = 4$, and so forth. Although only four lines were known to Balmer when he started his paper, by the time he had finished, ten more lines in the violet and ultraviolet had been measured. Much to his delight and satisfaction, these lines agreed with his empirical formula to within 0.1%! Encouraged by his success and because he was a bit of a numerologist, Balmer suggested that other hydrogen series might exist of the form

Figure 3.19 The Balmer series of spectral lines for hydrogen (emission spectrum).

Table 3.1   Some Spectral Series for the Hydrogen Atom

Lyman Series (uv)	$n_f = 1$	$n_i = 2, 3, 4 \ldots$
Balmer Series (vis-uv)	$n_f = 2$	$n_i = 3, 4, 5 \ldots$
Paschen Series (IR)	$n_f = 3$	$n_i = 4, 5, 6 \ldots$
Brackett Series IR)	$n_f = 4$	$n_i = 5, 6, 7 \ldots$
Pfund Series (IR)	$n_f = 5$	$n_i = 6, 7, 8 \ldots$

$$\lambda = C_3 \left(\frac{n^2}{n^2 - 3^2} \right) \qquad n = 4, 5, 6, \ldots \tag{3.19}$$

$$\lambda = C_4 \left(\frac{n^2}{n^2 - 4^2} \right) \qquad n = 5, 6, 7, \ldots \tag{3.20}$$

As we now know, his speculations were correct, and these series do indeed exist. In today's notation all of these series are given by a single formula

$$\frac{1}{\lambda} = R \left(\frac{1}{n_f{}^2} - \frac{1}{n_i{}^2} \right) \tag{3.21}$$

where n_f and n_i are *integers*. The Rydberg constant R is the same constant for all series and has the value

$$R = 1.0973732 \times 10^7 \text{ m}^{-1} \tag{3.22}$$

Note that for a given series, n_f has a constant value. Furthermore, for a given series $n_i = n_f + 1, n_f + 2, \ldots$. Table 3.1 lists the name of each series (named after their discoverers) and the integers that define the series.

Bohr's Quantum Model of the Atom

In April of 1913, a young Danish physicist, Niels Bohr (who had recently been working with both Thomson and Rutherford), published a three-part paper that shook the world of physics to its foundations.[14] Not only did this young rebel give the first successful theory of atomic line spectra, but in the process he overthrew some of the most cherished principles of the reigning king of electromagnetism, James Clerk Maxwell.

From our point of view, Bohr's model may seem only a reasonable next step, but it appeared astounding, confounding, and incredibly bold to his contemporaries. As mentioned earlier, both Thomson and Rutherford realized that the electrons must revolve about the nucleus in order to avoid falling directly into it. They, along with Bohr, also realized that according to Maxwell's theory, centripetally accelerated charges revolving with orbital frequency f should radiate light waves of frequency f. Unfortunately, pushed to its logical conclusion, this classical model leads to disaster. As the electron radiates energy, its orbit radius steadily decreases and its frequency of revolution increases. This leads

[14] N. Bohr, "On the Constitution of Atoms and Molecules," *Phil. Mag.* 26:1, 1913. Also, N. Bohr, *Nature* 92:231, 1913.

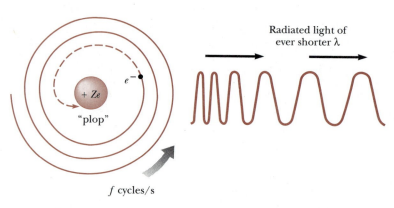

Figure 3.20   The classical model of the nuclear atom.

to an ever increasing frequency of emitted radiation and an ultimate cata-strophic collapse of the atom as the electron plunges into the nucleus (Fig. 3.20).

These deductions of electrons falling into the nucleus and a continuous emission spectrum from elements were boldly circumvented by Bohr. He simply postulated that classical radiation theory, which had been confirmed by Hertz's detection of radio waves using large circuits, did not hold for atomic sized systems. Moreover, he drew on the work of Planck and Einstein as sources of the correct theory of atomic systems. He overcame the problem of a classical electron that continually lost energy by applying Planck's ideas of quantized energy levels to orbiting atomic electrons. Thus he postulated that electrons in atoms are generally confined to certain stable, nonradiating energy levels **Stationary states** and orbits known as **stationary states**. He applied Einstein's concept of the photon to arrive at an expression for the frequency of the light emitted when the electron jumps from one stationary state to another. Thus if ΔE is the separation of two possible electronic stationary states, then $\Delta E = hf$, where h is Planck's constant and f is the frequency of the emitted light regardless of the frequency of the electron's orbital motion. In this way, by combining certain principles of classical mechanics with new quantum principles of light emission, **Assumptions of the Bohr** Bohr arrived at a theory of the atom that agreed remarkably with experiment.
theory   Now that we have looked at the general principles of Bohr's model of hy-drogen and at the detailed experimental spectra already discovered by 1913, let us examine Bohr's quantum theory in detail. The basic ideas of the Bohr theory as it applies to an atom of hydrogen are as follows:

- The electron moves in circular orbits about the proton under the influence of the Coulomb force of attraction, as in Figure 3.21. So far nothing new!
- Only certain orbits are stable. These stable orbits are ones in which the electron does not radiate. Hence the energy is fixed or stationary, and ordinary classical mechanics may be used to describe the electron's motion in these stable orbits.
- Radiation is emitted by the atom when the electron "jumps" from a more energetic initial stationary state to a lower state. This "jump" cannot be visualized or treated classically. In particular, the frequency f of the photon emitted in the jump *is independent of the frequency of the electron's orbital motion.* Instead, the frequency of the light emitted is related to the change in the

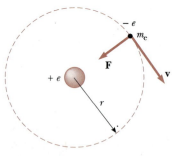

Figure 3.21 Diagram repre-senting Bohr's model of the hy-drogen atom.

atom's energy and is given by the Planck-Einstein formula

$$E_i - E_f = hf \tag{3.23}$$

where E_i is the energy of the initial state, E_f is the energy of the final state, and $E_i > E_f$.

- The size of the allowed electron orbits is determined by an additional quantum condition imposed on the electron's orbital angular momentum. Namely, the allowed orbits are those for which the electron's orbital angular momentum about the nucleus is an integral multiple of $\hbar = h/2\pi$,

$$m_e vr = n\hbar \qquad n = 1, 2, 3, \ldots \tag{3.24}$$

Using these four assumptions, we can now calculate the allowed energy levels and emission wavelengths of the hydrogen atom. Recall that the electrical potential energy of the system shown in Figure 3.21 is given by $U = qV = -ke^2/r$, where k (the Coulomb constant) has the value $1/4\pi\epsilon_0$. Thus, the total energy of the atom, which contains both kinetic and potential energy terms, is given by

$$E = K + U = \tfrac{1}{2}m_e v^2 - k\frac{e^2}{r} \tag{3.25}$$

Applying Newton's second law to this system, we see that the Coulomb attractive force on the electron, ke^2/r^2, must equal the mass times the centripetal acceleration of the electron, or

$$\frac{ke^2}{r^2} = \frac{m_e v^2}{r}$$

From this expression, we immediately find the kinetic energy to be

$$K = \frac{m_e v^2}{2} = \frac{ke^2}{2r} \tag{3.26}$$

Substituting this value of K into Equation 3.25 gives the total energy of the atom as

$$E = -\frac{ke^2}{2r} \tag{3.27}$$

Note that the total energy is negative, indicating a bound electron–proton system. This means that energy in the amount of $ke^2/2r$ must be added to the atom to remove the electron to infinity and leave it motionless. An expression for r, the radius of the electron orbit, may be obtained by solving Equations 3.24 and 3.26 for v and equating the results:

$$r_n = \frac{n^2 \hbar^2}{m_e k e^2} \qquad n = 1, 2, 3, \ldots \tag{3.28}$$

Radii of Bohr orbits in hydrogen

Equation 3.28 shows that only certain orbits are allowed and that these preferred orbits follow from the nonclassical step of requiring the electron's an-

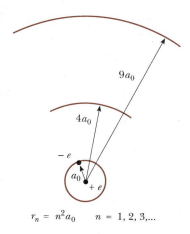

$$r_n = n^2 a_0 \qquad n = 1, 2, 3, \ldots$$

Figure 3.22   The first three Bohr orbits for hydrogen.

gular momentum to be an integral multiple of $\hbar$. The orbit for which $n = 1$ has the smallest radius, is called the **Bohr radius,** a_0. The value is

$$a_0 = \frac{\hbar^2}{m_e ke^2} = 0.529 \text{ Å} = 0.0529 \text{ nm} \tag{3.29}$$

The fact that Bohr's theory gave a value for a_0 in good agreement with the experimental size of hydrogen without any empirical calibration of orbit size was considered a striking triumph for this theory, The first three Bohr orbits are shown to scale in Figure 3.22.

The quantization of the orbit radii immediately leads to energy quantization. This can be seen by substituting $r_n = n^2 a_0$ into Equation 3.27, giving for the allowed energy levels

Energy levels of hydrogen

$$E_n = -\frac{ke^2}{2a_0}\left(\frac{1}{n^2}\right) \qquad n = 1, 2, 3, \ldots \tag{3.30}$$

Inserting numerical values into Equation 3.30 gives

$$E_n = -\frac{13.6}{n^2} \text{ eV} \qquad n = 1, 2, 3, \ldots \tag{3.31}$$

Quantum numbers

The integers n corresponding to the discrete, or quantized, values of the atom's energy have the special name *quantum numbers*. Quantum numbers are central to quantum theory and in general refer to the set of integers that label the discrete values of important atomic quantities, such as energy and angular momentum. The lowest stationary, or nonradiating, state is called the **ground state,** has $n = 1$, and has an energy $E_1 = -13.6$ eV. The next state, or **first excited state,** has $n = 2$, and an energy $E_2 = E_1/2^2 = -3.4$ eV. An energy level diagram showing the energies of these discrete energy states and the corresponding quantum numbers is shown in Figure 3.23. The uppermost level, corresponding to $n = \infty$ (or $r = \infty$) and $E = 0$, represents the state for which the electron is removed from the atom and is motionless. The minimum energy required to ionize the atom (that is, to completely remove an electron in the ground state from the proton's influence) is called the **ionization energy.** As

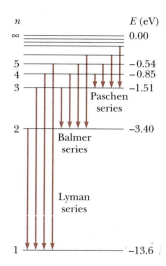

Figure 3.23 An energy level diagram for hydrogen. In such diagrams the discrete allowed energies are plotted on the vertical axis. Nothing is plotted on the horizontal axis, but the horizontal extent of the diagram is made large enough to show allowed transitions. Note that the quantum numbers are given on the left.

can be seen from Figure 3.23, the ionization energy for hydrogen based on Bohr's calculation is 13.6 eV. This constituted another major achievement for the Bohr theory, because the ionization energy for hydrogen had already been measured to be precisely 13.6 eV.

Equation 3.30 together with Bohr's third postulate can be used to calculate the frequency of the photon emitted when the electron jumps from an outer orbit to an inner orbit:

$$f = \frac{E_i - E_f}{h} = \frac{ke^2}{2a_0h}\left(\frac{1}{n_f^2} - \frac{1}{n_i^2}\right) \tag{3.32}$$

Because the quantity actually measured is wavelength, it is convenient to convert frequency to wavelength using $c = f\lambda$ to get

$$\frac{1}{\lambda} = \frac{f}{c} = \frac{ke^2}{2a_0hc}\left(\frac{1}{n_f^2} - \frac{1}{n_i^2}\right) \tag{3.33}$$

Emission wavelengths of hydrogen

The remarkable fact is that the *theoretical* expression, Equation 3.33, is identical to Balmer's *empirical* relation

$$\frac{1}{\lambda} = R\left(\frac{1}{n_f^2} - \frac{1}{n_i^2}\right) \tag{3.34}$$

provided that the combination of constants $ke^2/2a_0hc$ is equal to the experimentally determined Rydberg constant, $R = 1.0973732 \times 10^7$ m^{-1}. When Bohr demonstrated the agreement of these two quantities to a precision of about 1% late in 1913, it was recognized as the crowning achievement of his quantum theory of hydrogen. Furthermore, Bohr showed that all of the observed spectral series for hydrogen mentioned previously in this section have a natural interpretation in his theory. These spectral series are shown as transitions between energy levels in Figure 3.23.

Bohr immediately extended his model for hydrogen to other elements in which all but one electron had been removed. Ionized elements such as He^+, Li^{2+} and Be^{3+} were suspected to exist in hot stellar atmospheres, where frequent atomic collisions occurred with enough energy to completely remove one or more atomic electrons. Bohr showed that several mysterious lines observed in the Sun and stars could not be due to hydrogen, but were correctly predicted by his theory if attributed to singly ionized helium. In general, to describe a single electron orbiting a fixed nucleus of charge $+Ze$, Bohr's theory gives

$$r_n = (n^2)\frac{a_0}{Z} \tag{3.35}$$

and

$$E_n = -\frac{ke^2}{2a_0}\left(\frac{Z^2}{n^2}\right) \qquad n = 1, 2, 3, \ldots \tag{3.36}$$

EXAMPLE 3.6 Spectral Lines from the Star ζ-Puppis

The mysterious lines observed by the American astronomer Edward Charles Pickering in 1896 in the spectrum of the star ζ-Puppis fit the empirical formula

$$\frac{1}{\lambda} = R\left(\frac{1}{(n_f/2)^2} - \frac{1}{(n_i/2)^2}\right)$$

Show that these lines can be explained by the Bohr theory as originating from He^+.

Solution He^+ has $Z = 2$. Thus the allowed energy levels are given by Equation 3.36 as

$$E_n = \frac{ke^2}{2a_0}\left(\frac{4}{n^2}\right)$$

Using $hf = E_i - E_f$ we find

$$f = \frac{E_i - E_f}{h} = \frac{ke^2}{2a_0 h}\left(\frac{4}{n_f^2} - \frac{4}{n_i^2}\right)$$

$$= \frac{ke^2}{2a_0 h}\left(\frac{1}{(n_f/2)^2} - \frac{1}{(n_i/2)^2}\right)$$

or

$$\frac{1}{\lambda} = \frac{f}{c} = \frac{ke^2}{2a_0 hc}\left(\frac{1}{(n_f/2)^2} - \frac{1}{(n_i/2)^2}\right)$$

This is the desired solution, because $R \equiv ke^2/2a_0 hc$.

EXAMPLE 3.7 An Electronic Transition in Hydrogen

The electron in a hydrogen atom at rest makes a transition from the $n = 2$ energy state to the $n = 1$ ground state. (a) Find the wavelength, frequency, and energy (eV) of the emitted photon.

Solution We can use Equation 3.34 directly to obtain λ, with $n_i = 2$ and $n_f = 1$:

$$\frac{1}{\lambda} = R\left(\frac{1}{n_f^2} - \frac{1}{n_i^2}\right)$$

$$= R\left(\frac{1}{1^2} - \frac{1}{2^2}\right) = \frac{3R}{4}$$

$$\lambda = \frac{4}{3R} = \frac{4}{3(1.097 \times 10^7 \text{ m}^{-1})}$$

$$= 1.215 \times 10^{-7} \text{ m} = 121.5 \text{ nm}$$

This wavelength lies in the ultraviolet region.

Because $c = f\lambda$, the frequency of the photon is

$$f = \frac{c}{\lambda} = \frac{3.00 \times 10^8 \text{ m/s}}{1.215 \times 10^{-7} \text{ m}} = 2.47 \times 10^{15} \text{ Hz}$$

The energy of the photon is given by $E = hf$, so

$$E = hf$$

$$= (4.136 \times 10^{-15} \text{ eV·s})(2.47 \times 10^{15} \text{ Hz}) = 10.2 \text{ eV}$$

(b) The calculation of part (a) assumes that *all* of the $n = 2$ to $n = 1$ transition energy is carried off by the photon; however, this is technically incorrect because some of this energy must go into the recoil motion of the atom. By setting the momentum of the system (atom + photon) equal to zero after the emission, and assuming that the recoil energy of the atom is small compared with the $n = 2$ to $n = 1$ energy level separation, find the momentum and energy of the recoiling hydrogen atom.

Solution   Because momentum is conserved, and the total momentum before emission is zero, the total momentum after emission must also be zero. The photon and atom therefore move off in opposite directions, with

$$mv = \frac{E_{photon}}{c}$$

where m and v are the mass and recoil speed of the hydrogen atom, E_{photon} is the actual energy of the photon (less than 10.2 eV), and c is the speed of light. Because the energy difference between the $n = 2$ and $n = 1$ levels, E, is the source of both the photon energy and the recoil kinetic energy of the atom, we can write

$$E = E_{photon} + \tfrac{1}{2}mv^2$$

Because the atom is massive, we can assume that its recoil speed v and kinetic energy are so small that $E \approx E_{photon}$. Substituting $E_{photon} = 10.2$ eV into the expression for mv yields

$$mv = 10.2 \text{ eV}/c$$

The recoil kinetic energy of the hydrogen atom can now be calculated.

$$K = \frac{1}{2}mv^2 = \frac{1}{2}\frac{(mv)^2}{m} = (0.5)\frac{(10.2 \text{ eV})^2}{mc^2}$$

$$= \frac{(0.5)(10.2 \text{ eV})^2}{938.8 \times 10^6 \text{ eV}} = 5.56 \times 10^{-8} \text{ eV}$$

Thus the fraction of the energy difference between the $n = 2$ and $n = 1$ levels that goes into atomic recoil energy is *very* small, approximately 5 parts per billion:

$$\frac{K}{E} = \frac{5.56 \times 10^{-8} \text{ eV}}{10.2 \text{ eV}} = 5.4 \times 10^{-9}$$

Evidently the process of simply equating the photon's energy to the atomic energy level separation yields accurate answers because little energy is needed to conserve momentum.

Exercise 3   Check the approximation that $E \approx E_{photon}$ made in Example 3.7 by directly calculating the recoil kinetic energy of the hydrogen atom, $\tfrac{1}{2}mv^2$. *Hint:* Solve $mv = E_{photon}/c$ and $E = E_{photon} + \tfrac{1}{2}mv^2$ simultaneously to show $v \approx E/mc$, calculate the numerical value of $\tfrac{1}{2}mv^2$, and compare this answer to the result given in Example 3.7.

Exercise 4   What is the wavelength of the photon emitted by hydrogen when the electron makes a transition from the $n = 3$ state to the $n = 1$ state?

Answer $9/8R = 102.6$ nm.

EXAMPLE 3.8   The Balmer Series for Hydrogen

The Balmer series for the hydrogen atom corresponds to electronic transitions that terminate in the state of quantum number $n = 2$, as shown in Figure 3.23. (a) Find the longest wavelength photon emitted and determine its energy.

Solution   The longest wavelength photon in the Balmer series results from the transition from $n = 3$ to $n = 2$. Using Equation 3.34 gives

$$\frac{1}{\lambda} = R\left(\frac{1}{n_f^2} - \frac{1}{n_i^2}\right)$$

$$\frac{1}{\lambda_{max}} = R\left(\frac{1}{2^2} - \frac{1}{3^2}\right) = \frac{5}{36}R$$

$$\lambda_{max} = \frac{36}{5R} = \frac{36}{5(1.097 \times 10^7 \text{ m}^{-1})} = 656.3 \text{ nm}$$

This wavelength is in the red region of the visible spectrum.

The energy of this photon is

$$E_{photon} = hf = \frac{hc}{\lambda_{max}}$$

$$= \frac{(6.626 \times 10^{-34} \text{ J} \cdot \text{s})(3.00 \times 10^8 \text{ m/s})}{656.3 \times 10^{-9} \text{ m}}$$

$$= 3.03 \times 10^{-19} \text{ J} = 1.89 \text{ eV}$$

We could also obtain the energy of the photon by using the expression $hf = E_3 - E_2$, where E_2 and E_3 are the energy levels of the hydrogen atom, which can be calculated from Equation 3.31. Note that this is the lowest energy photon in this series because it involves the smallest energy change.

(b) Find the shortest wavelength photon emitted in the Balmer series.

Solution   The shortest wavelength photon in the Balmer series is emitted when the electron makes a transition from $n = \infty$ to $n = 2$. Therefore

$$\frac{1}{\lambda_{min}} = R\left(\frac{1}{2^2} - \frac{1}{\infty}\right) = \frac{R}{4}$$

$$\lambda_{min} = \frac{4}{R} = \frac{4}{1.097 \times 10^7 \text{ m}^{-1}} = 364.6 \text{ nm}$$

This wavelength is in the ultraviolet region and corresponds to the series limit.

Exercise 5   Find the energy of the shortest wavelength photon emitted in the Balmer series for hydrogen.

Answer   3.40 eV.

Although the theoretical derivation of the line spectrum was a remarkable feat in itself, the scope, breadth, and impact of Bohr's monumental achievement is truly seen only when it is realized what else he treated in his three-part paper of 1913:

- He explained why fewer lines are seen in the absorption spectrum of hydrogen than in the emission spectrum.
- He explained the emission of x-rays from atoms.
- He explained the nuclear origin of β particles.
- He explained the chemical properties of atoms in terms of the electron shell model.
- He explained how atoms associate to form molecules.

Two of these topics, the comparison of absorption and emission in hydrogen and the shell structure of atoms, are of such general importance that they deserve a bit more explanation.

We have already pointed out that a gas will absorb at wavelengths that correspond exactly to some emission lines, but not every line present in emission is seen as a dark absorption line. Bohr explained absorption as the reverse of emission; that is, an electron in a given energy state can absorb only a photon of the exact frequency required to produce a "jump" from a lower energy state to a higher energy state. Ordinarily hydrogen atoms are in the ground state ($n = 1$) and so only the high-energy Lyman series corresponding to transitions from the ground state to higher energy states is seen in absorption. The longer wavelength Balmer series corresponding to transitions originating in the first excited state ($n = 2$) is not seen because the average thermal energy of each atom is insufficient to raise the electron to the first excited state. That is, the number of electrons in the first excited state is insufficient at ordinary temperatures to produce measurable absorption.

EXAMPLE 3.9   Hydrogen in Its First Excited State

Calculate the temperature at which many atoms of monatomic hydrogen will be in the first excited state ($n = 2$). What series should be prominent in absorption at this temperature? (Calculate both from $N_2/N_1 = \exp(-\Delta E/k_B T)$ and from $\frac{3}{2}k_B T =$ average thermal energy.)

Solution   At room temperature almost all hydrogen atoms are in the ground state with an energy of -13.6 eV.

The first excited state ($n = 2$) has an energy equal to $E_2 = -3.4$ eV. Therefore, the hydrogen atoms must gain an energy of 10.2 eV to reach the first excited state. If the atoms are to obtain this energy from heat, we must have

$$\frac{3}{2}k_B T = \frac{\text{average thermal energy}}{\text{atom}} \approx 10.2 \text{ eV}$$

or

$$T = \frac{10.2 \text{ eV}}{(3/2)k_B} = \frac{10.2 \text{ eV}}{(1.5)(8.62 \times 10^{-5} \text{ eV/K})}$$

$$= 79\,000 \text{ K}$$

Let us check this result by using the Boltzmann distribution. In Section 2.3 we saw that the probability of finding an atom with energy E at temperature T is

$$P(E) = P_0 e^{-(E-E_0)/k_B T}$$

where P_0 is the probability of finding the atom in the ground state of energy E_0. From this expression, it follows that the ratio of the number of atoms in two different energy levels in thermal equilibrium at temperature T is

$$\frac{N_2}{N_1} = \frac{P(E_2)}{P(E_1)} = e^{-(E_2-E_1)/k_B T}$$

where N_2 is the number in the upper level, N_1 is the number in the lower level, and ΔE is the energy separation of the two levels. Let us use this equation to determine the temperature at which approximately 10% of the hydrogen atoms are in the $n = 2$ state.

$$\frac{N_2}{N_1} = 0.10 = e^{-(10.2 \text{ eV})/k_B T}$$

or

$$\ln(0.10) = -\frac{10.2 \text{ eV}}{k_B T}$$

Solving for T gives

$$T = -\frac{10.2 \text{ eV}}{k_B \ln(0.10)} = -\frac{10.2 \text{ eV}}{(8.62 \times 10^{-5} \text{ eV/K}) \ln(0.10)}$$

$$= 51\,000 \text{ K}$$

Thus the two estimates agree in order of magnitude and show that the Balmer series will only be seen in absorption if the absorbing gas is quite hot, as in a stellar atmosphere.

The final comment on Bohr's work concerns his development of electronic shell theory to treat multielectron atoms. In part II of his paper he attempted to find stable electronic arrangements subject to the conditions that the total angular momentum of all the electrons is quantized and, simultaneously, that the total energy is a minimum. This is a difficult problem, and one that becomes more difficult as more electrons are introduced into a system. Nevertheless Bohr had considerable success in explaining the chemical activity of multielectron atoms, as he was able to show that neutral hydrogen could easily add another electron to become H^-, and that neutral helium was particularly stable with a closed innermost shell of two electrons and a high ionization potential. He also proposed that lithium ($Z = 3$) had an electronic arrangement consisting of two electrons in one orbit near the nucleus and the third in a large, loosely bound outer orbit. This explains the tendency of lithium atoms to lose an electron and "take a positive charge in chemical combinations with other elements." Although we cannot afford the luxury of looking in detail at all of Bohr's predictions about multielectron atoms, his basic ideas of shell structure are as follows:

- Electrons of elements with higher atomic number form stable concentric rings, with definite numbers of electrons allowed for each ring or shell.
- The number of electrons in the outermost ring determines the valency.[15]

---

[15] G. N. Lewis, an American chemist, contributed much to our understanding of shell structure in 1916, building on Bohr's remarkable foundation.

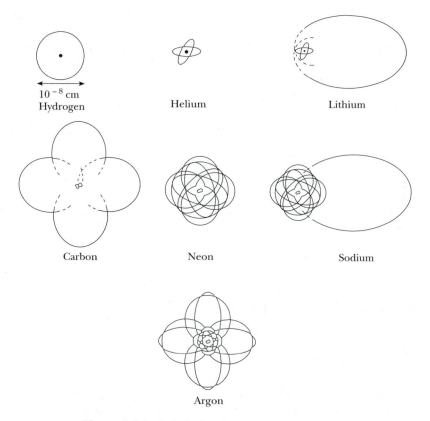

Figure 3.24   Bohr's sketches of electronic orbits.

With almost magical insight, Bohr ended part II of his classic paper with the explanation of the similar chemical properties of the iron group (Fe, Co, Ni) and the rare earths, which have atomic numbers that progressively increase by one and would not normally be expected to be chemically alike. The answer, according to Bohr, is that the configuration of electrons in the outermost ring of these elements is identical and that it is energetically more favorable to add electrons to inner rings. At the risk of encouraging some to take the idea of electronic orbits too seriously, Figure 3.24 shows some sketches of electronic orbits as drawn by Bohr in the early 1900s.

3.4   BOHR'S CORRESPONDENCE PRINCIPLE, OR WHY IS ANGULAR MOMENTUM QUANTIZED?

Correspondence principle

Where others might have left a wild and lawless gap between the revolutionary new laws that apply to atomic systems and those that hold for classical systems, Bohr provided a gentle and refined continuum in the form of the *correspondence principle*. This principle states that *predictions of quantum theory must correspond to the predictions of classical physics in the region of sizes where classical theory is known to hold*. These classical sizes for length, mass, and time are on the order of centimeters, grams, and seconds and typically involve very large quantum numbers, as can be seen by calculating n for a hydrogen atom with a radius of 1 cm. If the quantum number becomes large because of increased size or mass, we may state the correspondence principle symbolically as

$$\lim_{n \to \infty} [\text{quantum physics}] = [\text{classical physics}]$$

where n is a typical quantum number of the system such as the quantum number for hydrogen. In the hands of Bohr, the correspondence principle became a masterful tool to test revolutionary quantum results, and it also became a source of fundamental postulates about atomic systems—in fact, Bohr used reasoning of this type to arrive at the concept of the quantization of the electron's orbital angular momentum. Both Bohr's idea of discrete, nonradiating energy states and the emission postulate for atoms were foreshadowed by Planck's quantization of the energy of blackbody oscillators and by Einstein's treatment of the photoelectric effect. However, the concept of angular momentum quantization seems to have sprung full blown from Bohr's *Gedankenkuche* (thought kitchen), as so aptly expressed by Einstein. Indeed, in some of his later writings Bohr emphasized the point of view that the quantization of angular momentum was a postulate, underivable from any deeper law, and that its validity depended simply on the agreement of his model with experimental spectra.

What is most interesting is that in his 1913 paper Bohr ingeniously showed that the quantization of angular momentum is a consequence of the smooth and gradual emergence of classical results from quantum theory in the limit of large quantum number. In particular Bohr argued that, according to his correspondence principle, the quantum condition for emission ($\Delta E = hf$) and Maxwell's classical radiation theory (electronic charges with orbital frequency f radiate light waves of frequency f) *must simultaneously hold for the case of extremely large electronic orbits.* This case is shown in Figure 3.25. In this figure, r_1 and r_2 are the radii of two large adjacent orbits that are separated in energy by an amount dE, and ω is the orbital angular frequency of the electron, where ω is approximately constant in a transition between large orbits. (The algebra is simpler if we use angular frequency instead of frequency. Recall that the connection is $\omega = 2\pi f$.) Because we want to determine the allowed values of the angular momentum from the known change in the atom's energy when light is emitted, we need the relation between the total energy of the atom, $E = -ke^2/2r$ (Equation 3.27) and the magnitude of the total angular momentum of the atom, $L = m_e vr = m_e \omega r^2$. Using the fact that the electron is kept in orbit by the Coulomb force, it is possible to show that $1/r = m_e ke^2/L^2$ (see Problem 3.30), so Equation 3.27 becomes

$$E = -\frac{1}{2}\frac{m_e k^2 e^4}{L^2} \tag{3.37}$$

Taking a derivative of Equation 3.37 gives the desired connection between the change in energy and the change in angular momentum for the Bohr atom.

$$\frac{dE}{dL} = \frac{m_e k^2 e^4}{L^3} \tag{3.38}$$

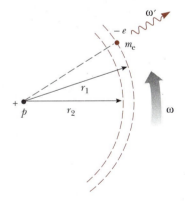

Figure 3.25 The classical limit of the Bohr atom. Note that r_1 and r_2 are the radii of the two adjacent quantum orbits, in which the electron has orbital angular frequencies of ω_1 and ω_2. We assume $r_1 \approx r_2 \approx r$ and $\omega_1 \approx \omega_2 \approx \omega$; ω' is the angular frequency of a photon emitted in a transition from r_1 to r_2.

Finally we obtain dE/dL in terms of ω, the electron orbital angular frequency, by using $L^3 = m_e k^2 e^4/\omega$ (see Problem 3.30). Thus

$$\frac{dE}{dL} = \frac{m_e k^2 e^4}{(m_e k^2 e^4/\omega)} = \omega \tag{3.39}$$

Now consider the emission of a photon of energy $dE = \hbar\omega'$ when the electron makes a transition from r_1 to r_2. Equation 3.39 becomes

$$dE = \omega\, dL$$

or

$$\hbar\omega' = \omega\, dL \tag{3.40}$$

where ω' is the photon angular frequency and ω is the electron orbital angular frequency. Ordinarily, ω' *and* ω are not simply related. Here, however, because we are dealing with large orbits, the correspondence principle tells us that the quantum theory must predict the same frequency for the emitted light as Maxwell's law of radiation. Because Maxwell's classical theory requires the electron to radiate light of the same frequency as its orbital motion frequency, $\omega' = \omega$, and Equation 3.40 becomes

$$\hbar\omega = \omega\, dL$$

or

$$dL = \hbar \tag{3.41}$$

Equation 3.41 shows that the change in electronic angular momentum for a transition between adjacent, large electronic orbits is *always* $\hbar$. This means that the magnitude of total angular momentum of the electron in a specific orbit may be taken to have a value equal to an integral multiple of $\hbar$, or

$$L = m_e v r = n\hbar \tag{3.42}$$

for $n =$ large integers.

Bohr's truly great insight was to realize that although Equation 3.42 was derived for the case of large electron orbits, it was a universal quantum principle applicable to all systems and of wider applicability than Maxwell's law of radiation. Such a bold and far-seeing vision was characteristic of the man so aptly described by Einstein in the following quote: "That this insecure and contradictory foundation [physics from 1910 to 1920] was sufficient to enable a man of Bohr's unique instinct and tact to discover the major laws of the spectral lines and of the electron shells of the atoms together with their significance for chemistry appeared to me like a miracle—and appears to me as a miracle even today. This is the highest form of musicality in the sphere of thought."

3.5   DIRECT CONFIRMATION OF ATOMIC ENERGY LEVELS: THE FRANCK-HERTZ EXPERIMENT

In the preceding sections we have shown the involved trail of reasoning indirectly proving the existence of quantized energy levels in atoms from observations of the optical line spectra emitted by different elements. Now we turn to

Figure 3.26   Franck-Hertz apparatus. A drop of pure mercury is sealed into an evacuated tube. The tube is heated to 185°C during measurements to provide a high enough density of mercury to ensure many electron–atom collisions.

a simpler and more direct experimental proof of the existence of discrete energy levels in atoms involving their excitation by collision with low-energy electrons. The first experiment of this type was performed by James Franck and Gustav Hertz (a nephew of Heinrich Hertz) in 1914 on mercury (Hg) atoms. It provided clear experimental proof of the existence of quantized energy levels in atoms and showed that the levels deduced from electron bombardment agreed with those deduced from optical line spectra. Furthermore it confirmed the universality of energy quantization in atoms, because the quite different physical processes of photon emission and electron bombardment yielded the same energy levels.

Figure 3.26 shows a schematic of a typical college laboratory device similar to the Franck-Hertz apparatus. Electrons emitted by the filament are accelerated over a relatively long region (≈ 1 cm) by the positive potential on the grid, V. The electrons can reach the collector and be registered on the electrometer (a sensitive ammeter) if they have sufficient energy to overcome the retarding potential of about 1.5 V set up over the short distance (≈ 1 mm) between grid and collector. At low electron energies or accelerating voltages, perfectly elastic collisions occur between the electrons and Hg atoms in which the *sum* of the kinetic energies of both electron and atom are conserved. Because the Hg atom is much more massive than the electron, the electron transfers very little kinetic energy to the atom in a collision. (See Problem 3.38.) Even after multiple collisions the electron reaches the grid with a kinetic energy of approximately e times V and will be collected if the accelerating voltage V is greater than 1.5 V. When V is modestly increased, more electrons reach the collector and the current, I, rises.

As the accelerating voltage is increased further, a threshold voltage is reached at which inelastic collisions occur right at the grid where the electrons reach an energy of e times V. In these inelastic collisions, electrons can transfer almost all of their kinetic energy to the atom, raising it to its first excited state. (See Problem 39 and Question 10.) Electrons that have collided inelastically will be unable to overcome the retarding potential and consequently I will

Figure 3.27   Current as a function of voltage in the Franck-Hertz experiment. To obtain these data, the filament voltage was set at 6.0 V and the tube heated to 185°C. (Data taken by Bob Rodick, Utica College, class of 1992.)

decrease for this threshold voltage. Figure 3.27 shows a typical plot of current versus accelerating voltage with the first weak current dip (*A*) occurring at a threshold voltage of slightly more than 7 V. When the voltage is increased once again, the inelastic collision region moves progressively closer to the filament, and the electrons that were stopped by an inelastic collision are reaccelerated, reaching the collector and causing another rise in current (*B*). Another dip (*C*) occurs when *V* has been increased enough for an electron to have two successive inelastic collisions: an electron excites an atom halfway between filament and grid, loses all its energy, and then is reaccelerated to excite another atom at the grid, finally ending up with insufficient energy to be collected. This process takes place periodically with increasing grid voltage giving rise to equally spaced maxima and minima in the *I–V* curve as shown in Figure 3.27.

If the adjacent maxima and minima separations of Figure 3.27 are carefully averaged, one finds an average of 4.9 ± 0.1 V, or a ground to first excited state separation of 4.9 ± 0.1 eV. Note, however, that the first minimum does not occur at 4.9 V but at about 7.1 V. The extra energy (7.1 − 4.9 = 2.2 eV) is required because the filament and collector are made of different metals with different work functions. (Recall that the work function is the energy needed to pull an electron out of a metal—see Chapter 2.) While the filament, like all good emitters, has a low work function, the collector has a high work function, and this work function energy must be supplied to extract an electron from the collector so a current can flow in the circuit.

Not only did Franck and Hertz show with simple ammeter and voltmeter measurements that atoms can only accept discrete amounts of energy from an electron beam, they also showed that the energy levels obtained from electron bombardment agreed with the spectroscopic results. Reasoning that an Hg atom actually excited to an energy level 4.9 eV above its ground state could return to its ground state by emitting a single photon (as Bohr had just postulated), they calculated the wavelength of such a photon to be

$$\Delta E = hf = \frac{hc}{\lambda}$$

or

$$\lambda = \frac{hc}{\Delta E} = \frac{1240 \text{ eV} \cdot \text{nm}}{4.9 \text{ eV}} = 253 \text{ nm} \qquad (3.43)$$

Because glass is not transparent to such ultraviolet radiation, they constructed a quartz apparatus and carefully measured the radiation emitted, finding radiation of wavelength 254 nm to be emitted as soon as the accelerating voltage exceeded 4.9 V. For this careful, direct, and clever experimental confirmation of Bohr's basic ideas of discrete energy levels in atoms and the process of photon emission, Franck and Hertz were awarded the Nobel prize in 1925.

SUMMARY

The determination of the composition of atoms relies heavily on four classic experiments:

- **Faraday's law of electrolysis**, which may be stated as

$$m = \frac{(q)(\text{molar mass})}{(96\,500 \text{ C})(\text{valence})} \qquad (3.1)$$

where m is the mass liberated at an electrode and q is the total charge passed through the solution. Faraday's law shows that atoms are composed of positive and negative charges and that atomic charges always consist of multiples of some unit charge.

- **J. J. Thomson's determination of e/m** and that the electron is a part of all atoms. Thomson measured e/m_e of electrons from a variety of elements by measuring the deflection of an electron beam by an electric field. He then applied a magnetic field to just cancel the electric deflection in order to determine the electron velocity. The charge-to-mass ratio in the Thomson experiment is

$$\frac{e}{m_e} = \frac{V\theta}{B^2 \ell d} \qquad (3.7)$$

where V/d is the magnitude of the applied electric field, ℓ is the length of the vertical deflecting plates, θ is the deflection produced by the electric field, and B is the applied magnetic field.

- **Millikan's determination of the fundamental charge, e.** By balancing the electric and gravitational force on individual oil drops, Millikan was able to determine the fundamental electric charge and to show that charges always occur in multiples of e. The quantum of charge may be determined from the relation

$$ne = \left(\frac{mg}{E}\right)\left(\frac{v + v_1'}{v}\right) \qquad (3.11)$$

where n is an integer, m is the mass of the drop, E is the magnitude of the electric field, v is the terminal speed of the drop with field off (falling), and v_1' is the terminal speed of the drop with field on (rising).

- **Rutherford's scattering of α particles from gold atoms**, which established the nuclear model of the atom. By measuring the rate of scattering of α particles into an angle ϕ, Rutherford was able to establish that most of the mass and all of the positive charge of an atom, $+Ze$, are concentrated in a minute volume of the atom with a radius of about 10^{-14} m.

The explanation of the motion of electrons within the atom and of the rich and elaborate series of spectral lines emitted by the atom was given by Bohr. Bohr's theory was based partly on classical mechanics and partly on some startling new quantum ideas. **Bohr's postulates** were:

- Electrons move about the nucleus in circular orbits determined by Coulomb's and Newton's laws.

- Only certain orbits are stable. The electron does not radiate electromagnetic energy in these special orbits, and because the energy is constant with time these are called *stationary states*.

- A spectral line of frequency f is emitted when an electron jumps from an initial orbit of energy E_i to a final orbit of energy E_f, where

$$hf = E_i - E_f \qquad (3.23)$$

- The sizes of the stable electron orbits are determined by requiring the electron's angular momentum to be an integral multiple of $\hbar$:

$$m_e vr = n\hbar \qquad n = 1, 2, 3, \ldots \qquad (3.24)$$

These postulates lead to quantized orbits and quantized energies for a single electron orbiting a nucleus with charge $+Ze$, given by

$$r_n = \frac{n^2 a_0}{Z} \qquad (3.35)$$

and

$$E_n = -\frac{ke^2}{2a_0}\frac{Z^2}{n^2} = -\frac{13.6\, Z^2}{n^2}\text{ eV} \qquad (3.36)$$

where n is an integer and $a_0 = \hbar^2/m_e ke^2 = 0.529$ Å $= 0.0529$ nm is the Bohr radius.

As a bridge between the familiar domain of classical physics and the more uncertain domain of atomic systems and quantum theory, Bohr provided the *correspondence principle*. This principle states that predictions of quantum theory must correspond to the predictions of classical physics in the region of sizes where classical theory is known to hold.

Direct experimental evidence of the quantized energy of atoms is provided by the Franck-Hertz experiment. This experiment shows that mercury atoms can only accept discrete amounts of energy from a bombarding electron beam.

SUGGESTIONS FOR FURTHER READING

1. G. Holton, *Introduction to Concepts and Theories in Physical Science*, Reading, MA, Addison-Wesley, 1952. This is an accurate, humane, and eminently readable overview of the history and progression of scientific concepts from the Greeks to quantum theory.
2. J. Perrin, *Atoms*, translated by D. L. Hammick, New York, D. Van Nostrand Co., 1923. This is a superb account at first hand of the evidence for atoms at the beginning of the 20th century.
3. G. Thomson, *J. J. Thomson and the Cavendish Laboratory in His Day*, New York, Doubleday and Co., 1965. This book offers a fascinating account by Thomson's son of the *e/m* experiment and other works.
4. R. A. Millikan, *Electrons (+ and −), Protons, Neutrons, Mesotrons, and Cosmic Rays*, Chicago, University of Chicago Press, 1947. This book gives detailed accounts of the determination of *e*.
5. *Physics Today*, October 1985, Special Issue: Niels Bohr Centennial. This magazine contains three articles dealing with Bohr's scientific, social, and cultural contributions to the physics community. Many interesting photos of the key players are included.

QUESTIONS

1. The Bohr theory of the hydrogen atom is based on several assumptions. Discuss these assumptions and their significance. Do any of these assumptions contradict classical physics?
2. Suppose that the electron in the hydrogen atom obeyed classical mechanics rather than quantum mechanics. Why should such a *hypothetical* atom emit a continuous spectrum rather than the observed line spectrum?
3. Can the electron in the ground state of the hydrogen atom absorb a photon of energy (a) *less* than 13.6 eV and (b) *greater* than 13.6 eV?
4. Explain the concept of an atomic stationary state. Why is this idea of central importance in explaining the stability of the Bohr atom?
5. Does Bohr's correspondence principle apply only to quantum theory? Can you give an example of the application of this principle to relativity theory?
6. On the basis of Bohr's ideas, explain why all emission lines are not seen in absorption.
7. The results of classical measurements and calculations are sometimes called classical numbers. Contrast and explain the differences between quantum numbers and classical numbers.
8. What factor causes the finite width of the peaks in the $I-V$ curve of the Franck-Hertz experiment?
9. An electron with a kinetic energy of 4.9 eV (mass = 5.49×10^{-4} u) collides inelastically with a stationary mercury atom (mass = 201 u). Explain qualitatively why almost 100% of the electron's energy can go into raising the atom to its first excited state.
10. Why don't other current dips corresponding to excitation of the mercury atom's second excited state, third excited state, and so forth show up in the Franck-Hertz experiment? (*Hint:* At the high density of mercury vapor used in the experiment, the probability of a 4.9-eV electron experiencing an inelastic collision is approximately 1.)

PROBLEMS

3.2   The Composition of Atoms

1. Using the Faraday (96 500 C) and Avogadro's number, determine the electronic charge. Explain your reasoning.
2. *Weighing a copper atom in an electrolysis experiment.* A standard experiment involves passing a current of several amperes through a copper sulfate solution ($CuSO_4$) for a period of time and determining the mass of copper plated onto the cathode. If it is found that a current of 1.00 A flowing for 3600 s deposits 1.185 g of copper, find (a) the number of copper atoms deposited, (b) the weight of a copper atom, and (c) the molar mass of copper.
3. A mystery particle enters the region between the plates of a Thomson apparatus as shown in Figure 3.5. The deflection angle θ is measured to be 0.20 radians (downwards) for this particle when $V = 2000$ V, $\ell = 10.0$ cm, and $d = 2.00$ cm. If a perpendicular magnetic field of magnitude 4.57×10^{-2} T is applied simultaneously with the electric field, the particle passes through the plates without deflection. (a) Find q/m for this particle. (b) Identify the particle. (c) Find the horizontal speed with which the particle entered the plates. (d) Must we use relativistic mechanics for this particle?
4. Figure P3.4 shows a cathode ray tube for determining e/m_e without applying a magnetic field. In this case v_x

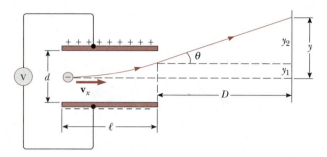

Figure P3.4   Deflection of a charged particle by an electric field.

may be found by measuring the rise in temperature when a known amount of charge is stopped in a target. If V, ℓ, d, D, and y are measured, e/m_e may be found. Show that

$$\frac{e}{m_e} = \frac{y v_x^2 d}{V\ell[(\ell/2) + D]}$$

5. *A Thomson type experiment with relativistic electrons:* One of the earliest experiments to show that $p = \gamma mv$ (rather than $p = mv$) was that of Neumann. [G. Neumann, *Ann. Physik*, 45:529 (1914)]. The apparatus shown in Figure P3.5 is identical to Thomson's except that the source of high-speed electrons is a radioactive radium source, and the magnetic field **B** is arranged to act on the electron over its entire trajectory from source to detector. The combined electric and magnetic fields act as a *velocity selector*, only passing electrons with speed v, where $v = V/Bd$ (Equation 3.6), while in the region where there is only a magnetic field the electron moves in a circle of radius r, with r given by $p = Bre$. This latter region (**E** = 0, **B** = constant) acts as a *momentum selector* be-

cause electrons with larger momenta have paths with larger radii. (a) Show that the radius of the circle described by the electron is given by $r = (l^2 + y^2)/2y$. (b) Typical values for the Neumann experiment were $d = 2.51 \times 10^{-4}$ m, $B = 0.0177$ T, and $l = 0.0247$ m. For $V = 1060$ volts, y, the most critical value, was measured to be 0.0024 ± 0.0005 m. Show that these values disagree with the y value calculated from $p = mv$ but agree with the y value calculated from $p = \gamma mv$ within experimental error. (*Hint:* Find v from Equation 3.6, use $mv = Bre$ or $\gamma mv = Bre$ to find r, and use r to find y.)

6. In a Millikan oil drop experiment, the condenser plates are spaced 2.00 cm apart, the potential across the plates is 4000 V, the rise or fall distance is 4.00 mm, the density of the oil droplets is 0.800 g/cm³, and the viscosity of the air is 1.81×10^{-5} kg·m⁻¹s⁻¹. The average time of fall in the absence of an electric field is 15.9 s. The following different rise times in seconds are observed when the field is turned on: 36.0, 17.3, 24.0, 11.4, 7.54. (a) Find the radius and mass of the drop used in this experiment. (b) Calculate the charge on each drop, and show that charge is quantized by considering both the size of each charge and the amount of charge gained (lost) when the rise time changes. (c) Determine the electronic charge from these data. You may assume that e lies between 1.5 and 2.0×10^{-19} C.

7. Actual data from one of Millikan's early experiments are as follows:

$a = 0.000276$ cm

$\rho = 0.9561$ g/cm³

Average time of fall = 11.894 s

Rise or fall distance = 10.21 mm

Plate separation = 16.00 mm

Average potential difference between plates = 5085 V

Sequential rise times in seconds: 80.708, 22.366, 22.390, 22.368, 140.566, 79.600, 34.748, 34.762, 29.286, 29.236

Find the average value of e by requiring that the difference in charge for drops with different rise times be equal to an *integral* number of elementary charges.

8. A parallel beam of α particles with fixed kinetic energy is normally incident on a piece of gold foil. (a) If 100 α particles per minute are detected at 20°, how many will be counted at 40, 60, 80, and 100°? (b) If the kinetic energy of the incident α particles is doubled, how many scattered α particles will be observed at 20°? (c) If the original α particles were incident on a copper foil of the same thickness, how many scattered α particles would be

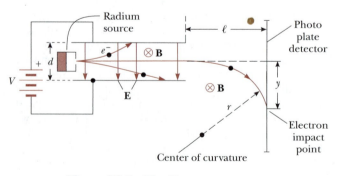

Figure P3.5 .The Neumann apparatus.

detected at 20°? Note that $\rho_{Cu} = 8.9$ g/cm^3 and $\rho_{Au} = 19.3$ g/cm^3.

9. It is observed that α particles with kinetic energies of 13.9 MeV and higher, incident on Cu foils, do not obey Rutherford's $(\sin \phi/2)^{-4}$ law. Estimate the nuclear size of copper from this observation, assuming that the Cu nucleus remains fixed in a head-on collision with an α particle.

10. A typical Rutherford scattering apparatus consists of an evacuated tube containing a polonium-210 α source (5.2-MeV α's), collimators, a gold foil target, and a special alpha-detecting film. The detecting film simultaneously measures all the alphas scattered over a range from $\phi = 2.5$ to 12.5°. (See Fig. P3.10.) The total number of counts measured over a week's time falling in a specific ring (denoted by its average scattering angle) and the corresponding ring area are given in Table 3.2. (a) Find the counts per area at each angle and correct these values for the angle independent background. The background correction may be found from a seven-day count taken with the beam blocked with a metal shutter in which 72 counts were measured evenly distributed over the total detector area of 8.50 cm^2. (b) Show that the corrected counts per unit area are proportional to $\sin^{-4}(\phi/2)$ or, in terms of the Rutherford formula, Equation 3.16,

$$\frac{\Delta n}{A} = \frac{C}{\sin^4(\phi/2)}$$

Notes: If a plot of $(\Delta n/A)$ versus ϕ will not fit on a single sheet of graph paper, try plotting $\log(\Delta n/A)$ versus $\log[1/(\sin \phi/2)^4]$. This plot should yield a straight line with a slope of one and an intercept that gives C. Explain why this technique works.

Table 3.2. Data to be Used in Problem 10

Angle (degrees)	Counts/Ring	Ring Area (cm^2)	Counts/Area
2.5	605	0.257	
3.5	631	0.360	
4.5	520	0.463	
5.5	405	0.566	
6.5	301	0.669	
7.5	201	0.772	
8.5	122	0.875	
9.5	78	0.987	
10.5	65	1.08	
11.5	66	1.18	
12.5	44	1.29	

(a)

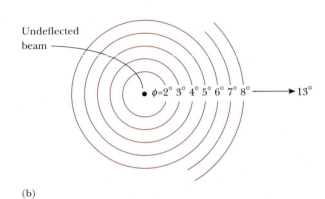

(b)

Figure P3.10   (a) Side view of Rutherford's scattering apparatus: ϕ is the scattering angle. (b) End view of the Rutherford apparatus showing the film detector end cap with grid marking the angle ϕ. The α particles damage the film emulsion and after development show up as dots within the rings.

3.3   The Bohr Atom

11. Calculate the wavelengths of the first three lines in the Balmer series for hydrogen.

12. Calculate the wavelengths of the first three lines in the Lyman series for hydrogen.

13. (a) What value of n is associated with the Lyman series line in hydrogen whose wavelength is 102.6 nm? (b) Could this wavelength be associated with the Paschen or Brackett series?

14. (a) Use Equation 3.35 to calculate the radii of the first, second, and third Bohr orbits of hydrogen. (b) Find the electron's speed in the same three orbits. (c) Is a relativistic correction necessary? Explain.

15. (a) Construct an energy level diagram for the He$^+$ ion, for which $Z = 2$. (b) What is the ionization energy for He$^+$?

16. Construct an energy level diagram for the Li^{2+} ion, for which $Z = 3$.

17. What is the radius of the first Bohr orbit in (a) He$^+$, (b) Li^{2+}, and (c) Be^{3+}?

18. A hydrogen atom initially in its ground state ($n = 1$) absorbs a photon and ends up in the state for which $n = 3$. (a) What is the energy of the absorbed photon?

(b) If the atom returns to the ground state, what photon energies could the atom emit?

19. A photon is emitted from a hydrogen atom that undergoes an electronic transition from the state $n = 3$ to the state $n = 2$. Calculate (a) the energy, (b) the wavelength, and (c) the frequency of the emitted photon.

20. What is the energy of the photon that could cause (a) an electronic transition from the $n = 4$ state to the $n = 5$ state of hydrogen and (b) an electronic transition from the $n = 5$ state to the $n = 6$ state?

21. (a) Calculate the longest and shortest wavelengths for the Paschen series. (b) Determine the photon energies corresponding to these wavelengths.

22. Find the potential energy and kinetic energy of an electron in the ground state of the hydrogen atom.

23. A hydrogen atom is in its ground state ($n = 1$). Using the Bohr theory of the atom, calculate (a) the radius of the orbit, (b) the linear momentum of the electron, (c) the angular momentum of the electron, (d) the kinetic energy, (e) the potential energy, and (f) the total energy.

24. A hydrogen atom initially at rest in the $n = 3$ state decays to the ground state with the emission of a photon. (a) Calculate the wavelength of the emitted photon. (b) Estimate the recoil momentum of the atom, and the kinetic energy of the recoiling atom. Where does this energy come from?

25. Calculate the frequency of the photon emitted by a hydrogen atom making a transition from the $n = 4$ to the $n = 3$ state. Compare your result with the frequency of revolution for the electron in these two Bohr orbits.

26. Calculate the longest and shortest wavelengths in the Lyman series for hydrogen, indicating the underlying electronic transition that gives rise to each. Are any of the Lyman spectral lines in the visible spectrum? Explain.

27. Show that Balmer's formula, $\lambda = C_2 \left(\dfrac{n^2}{n^2 - 2^2} \right)$, reduces to the Rydberg formula, $\dfrac{1}{\lambda} = R \left(\dfrac{1}{2^2} - \dfrac{1}{n^2} \right)$, provided that $(2^2 / C_2) = R$. Check that $(2^2 / C_2)$ has the same numerical value as R.

28. *The Auger process.* An electron in chromium makes a transition from the $n = 2$ state to the $n = 1$ state without emitting a photon. Instead, the excess energy is transferred to an outer electron (in the $n = 4$ state), which is ejected by the atom. (This is called an *Auger process*, and the ejected electron is referred to as an *Auger electron.*) Use the Bohr theory to find the kinetic energy of the Auger electron.

29. An electron initially in the $n = 3$ state of a one-electron atom of mass M at rest undergoes a transition to the $n = 1$ ground state. (a) Show that the recoil speed of the atom from emission of a photon is given approximately by

$$v = \frac{8hR}{9M}$$

(b) Calculate the percent of the $3 \rightarrow 1$ transition energy that is carried off by the recoiling atom if the atom is deuterium.

30. Apply classical mechanics to an electron in a stationary state of hydrogen to show that $L^2 = m_e k e^2 r$ and $L^3 = m_e k^2 e^4 / \omega$. Here k is the Coulomb constant, L is the magnitude of the orbital angular momentum of the electron, and m_e, e, r, and ω are the mass, charge, orbit radius, and orbital angular frequency of the electron, respectively.

31. (a) Find the frequency of the electron's orbital motion, f_e, around a fixed nucleus of charge $+Ze$ by using Equation 3.24 and $f_e = (v / 2\pi r)$ to obtain

$$f_e = \frac{m_e k^2 Z^2 e^4}{2\pi \hbar^3} \left(\frac{1}{n^3} \right)$$

(b) Show that the frequency of the photon emitted when an electron jumps from an outer to inner orbit can be written

$$f_{photon} = \frac{k Z^2 e^2}{2 a_0 h} \left(\frac{1}{n_f^2} - \frac{1}{n_i^2} \right)$$
$$= \frac{m_e k^2 e^4 Z^2}{2\pi \hbar^3} \left(\frac{n_i + n_f}{2 n_i^2 n_f^2} \right) (n_i - n_f)$$

For an electronic transition between adjacent orbits, $n_i - n_f = 1$ and

$$f_{photon} = \frac{m_e k^2 Z^2 e^4}{2\pi \hbar^3} \left(\frac{n_i + n_f}{2 n_i^2 n_f^2} \right)$$

Now examine the factor

$$\left(\frac{n_i + n_f}{2 n_i^2 n_f^2} \right)$$

By substitution of integers show that

$$\frac{1}{n_i^3} < \frac{n_i + n_f}{2 n_i^2 n_f^2} < \frac{1}{n_f^3}$$

(c) What do you conclude about the frequency of emitted radiation compared with the frequencies of orbital revolution in the initial and final states? What happens as $n_i \rightarrow \infty$?

32. Wavelengths of spectral lines depend to some extent on the nuclear mass. This occurs because the nucleus is not an infinitely heavy stationary mass and both the electron and nucleus actually revolve around their common center of mass. It can be shown that a system of this type is entirely equivalent to a single particle of reduced mass μ that revolves around the position of the heavier par-

(a)   (b)

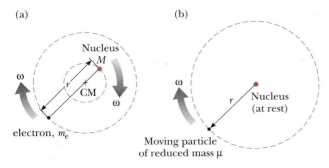

Figure P3.32 (a) Both the electron and the nucleus actually revolve around the center of mass. (b) To calculate the effect of nuclear motion, the nucleus can be considered to be at rest, and m_e is replaced by the reduced mass μ.

ticle at a distance equal to the electron–nucleus separation. See Figure P3.32. Here $\mu = m_e M/(m_e + M)$, where m_e is the electron mass and M is the nuclear mass. To take the moving nucleus into account in the Bohr theory we replace m_e with μ. Thus Equation 3.30 becomes

$$E_n = \frac{-\mu k e^2}{2 m_e a_0}\left(\frac{1}{n^2}\right)$$

and Equation 3.33 becomes

$$\frac{1}{\lambda} = \frac{\mu k e^2}{2 m_e a_0 hc}\left(\frac{1}{n_f^{\,2}} - \frac{1}{n_i^{\,2}}\right) = \left(\frac{\mu}{m_e}\right) R\left(\frac{1}{n_f^{\,2}} - \frac{1}{n_i^{\,2}}\right)$$

Determine the corrected values of wavelength for the first Balmer line ($n = 3$ to $n = 2$ transition) taking nuclear motion into account for (a) hydrogen, ^{1}H, (b) deuterium, ^{2}H, and (c) tritium, ^{3}H. (Deuterium was actually discovered in 1932 by Harold Urey who measured the small wavelength difference between ^{1}H and ^{2}H.)

33. A *muon* is a particle with a charge equal to that of an electron and a mass equal to 207 times the mass of an electron. Muonic lead is formed when ^{208}Pb captures a muon to replace an electron. Assume that the muon moves in such a small orbit that it "sees" a nuclear charge of $Z = 82$. According to the Bohr theory, what are the radius and energy of the ground state of muonic lead? Use the concept of reduced mass introduced in Problem 32.

34. A muon (Problem 33) is captured by a deuteron (an ^{2}H nucleus) to form a muonic atom. (a) Find the energy of the ground state and the first excited state. (b) What is the wavelength of the photon emitted when the atom makes a transition from the first excited state to the

ground state? Use the concept of reduced mass introduced in Problem 32.

35. *Positronium* is a hydrogen-like atom consisting of a positron (a positively charged electron) and an electron revolving around each other. Using the Bohr model, find the allowed radii (relative to the center of mass of the two particles) and the allowed energies of the system. Use the concept of reduced mass introduced in Problem 32.

3.4 The Correspondence Principle

36. (a) Calculate the frequency of revolution and the orbit radius of the electron in the Bohr model of hydrogen for $n = 100$, 1000 and 10 000. (b) Calculate the photon frequency for transitions from the n to $n - 1$ states for the same values of n as in (a) and compare with the revolution frequencies found in (a). (c) Explain how your results verify the correspondence principle.

37. Use Bohr's model of the hydrogen atom to show that when the atom makes a transition from the state n to the state $n - 1$, the frequency of the emitted light is given by

$$f = \frac{2\pi^2 m_e k^2 e^4}{h^3}\left[\frac{2n - 1}{(n - 1)^2 n^2}\right]$$

Show that as $n \to \infty$, the expression above varies as $1/n^3$ and reduces to the classical frequency one would expect the atom to emit. (*Hint:* To calculate the classical frequency, note that the frequency of revolution is $v/2\pi r$, where r is given by Equation 3.28.) This is an example of the correspondence principle, which requires that the classical and quantum models agree for large values of n.

3.5 The Franck-Hertz Experiment

38. An electron with kinetic energy less than 100 eV collides head-on in an elastic collision with a massive mercury atom at rest. (a) If the electron reverses direction in the collision (like a ball hitting a wall) show that the electron loses only a tiny fraction of its initial kinetic energy given by

$$\frac{\Delta K}{K} = \frac{4M}{m_e(1 + M/m_e)^2}$$

where m_e is the electron mass and M is the mercury atom mass. (b) Using the accepted values for m_e and M, show that

$$\frac{\Delta K}{K} \approx \frac{4 m_e}{M}$$

and calculate the numerical value of $\Delta K/K$.

Additional Problems

39. An electron collides inelastically and head on with a mercury atom at rest. (a) If the separation of the first excited state and the ground state of the atom is exactly 4.9 eV, what is the minimum initial electron kinetic energy needed to raise the atom to its first excited state and also conserve momentum? Assume that the collision is completely inelastic. (b) What is the initial speed of the electron in this case? (c) What is the speed of the electron and atom after the collision? (d) What is the kinetic energy (in electron volts) of the electron after collision? Is the approximation that the electron loses all of its kinetic energy in an inelastic collision justified?

40. If the Franck-Hertz experiment could be performed with high-density monatomic hydrogen, at what voltage separations would the current dips appear? Take the separation between ground state and first excited state to be exactly 10.2 eV and be sure to consider momentum as well as energy in arriving at your answer.

41. Liquid oxygen has a bluish color, meaning that it preferentially absorbs light toward the red end of the visible spectrum. Although the oxygen molecule (O_2) does not strongly absorb visible radiation, it does absorb strongly at 1269 nm, which is in the infrared region of the spectrum. Research has shown that it is possible for two colliding O_2 molecules to absorb a single photon, sharing its energy equally. The transition that both molecules undergo is the same transition that results when they absorb 1269-nm radiation. What is the wavelength of the single photon that causes this double transition? What is the color of this radiation?

42. Two hydrogen atoms collide head-on and end up with zero kinetic energy. Each then emits a photon with a wavelength of 121.6 nm ($n = 2$ to $n = 1$ transition). At what speed were the atoms moving before the collision?

43. Four possible transitions for a hydrogen atom are listed below.

$$\text{(A)} \; n_i = 2; \; n_f = 5$$
$$\text{(B)} \; n_i = 5; \; n_f = 3$$
$$\text{(C)} \; n_i = 7; \; n_f = 4$$
$$\text{(D)} \; n_i = 4; \; n_f = 7$$

(a) Which transition emits the photons having the shortest wavelength? (b) For which transition does the atom gain the most energy? (c) For which transition(s) does the atom lose energy?

4

Matter Waves

Chapter Outline

In the previous chapter we discussed some important discoveries and theoretical concepts concerning the *particle nature of matter*. We now point out some of the shortcomings of these theories and introduce the fascinating and bizarre *wave properties of particles*. Especially notable are Count Louis de Broglie's remarkable ideas about how to represent electrons (and other particles) as waves and the experimental confirmation of de Broglie's hypothesis by the electron diffraction experiments of Davisson and Germer. We shall also see how the notion of representing a particle as a localized wave or wave group leads naturally to limitations on simultaneously measuring position and momentum of the particle. Finally, we discuss the passage of electrons through a double slit as a way of understanding the wave–particle duality of matter.

4.1 THE PILOT WAVES OF DE BROGLIE

By the early 1920s scientists recognized that the Bohr theory contained many inadequacies:

- It failed to predict the observed intensities of spectral lines.
- It had only limited success in predicting emission and absorption wavelengths for multielectron atoms.

Figure 4.1 Louis de Broglie was a member of an aristocratic French family that produced marshals, ambassadors, foreign ministers, and at least one duke, his older brother Maurice de Broglie. Louis de Broglie came rather late to theoretical physics as he first studied history. Only after serving as a radio operator in World War I did he follow the lead of his older brother and begin his studies of physics. Maurice de Broglie was an outstanding experimental physicist in his own right and conducted experiments in the palatial family mansion in Paris. (*AIP Meggers Gallery of Nobel Laureates*)

- It failed to provide an equation of motion governing the time development of atomic systems starting from some initial state.
- It overemphasized the particle nature of matter and could not explain the newly discovered wave–particle duality of light.
- It did not supply a general scheme for "quantizing" other systems, especially those without periodic motion.

The first bold step toward a new mechanics of atomic systems was taken by Louis Victor de Broglie in 1923 (Fig. 4.1). In his doctoral dissertation he postulated that *because photons have wave and particle characteristics, perhaps all forms of matter have wave as well as particle properties.* This was a radical idea with no experimental confirmation at that time. According to de Broglie, electrons had a dual particle–wave nature. Accompanying every electron was a wave (not an electromagnetic wave!), which guided, or "piloted," the electron through space. He explained the source of this assertion in his 1929 Nobel Prize acceptance speech:

> On the one hand the quantum theory of light cannot be considered satisfactory since it defines the energy of a light corpuscle by the equation $E = hf$ containing the frequency f. Now a purely corpuscular theory contains nothing that enables us to define a frequency; for this reason alone, therefore, we are compelled, in the case of light, to introduce the idea of a corpuscle and that of periodicity simultaneously. On the other hand, determination of the stable motion of electrons in the atom introduces integers, and up to this point the only phenomena involving integers in physics were those of interference and of normal modes of vibration. This fact suggested to me the idea that electrons too could not be considered simply as corpuscles, but that periodicity must be assigned to them also.

Let us follow de Broglie's reasoning in more detail. He concluded that the wavelength and frequency of a *matter wave* associated with any moving object be given by

$$\lambda = \frac{h}{p} \qquad (4.1)$$

de Broglie wavelength

and

$$f = \frac{E}{h} \qquad (4.2)$$

where h is Planck's constant, p is the relativistic momentum, and E is the total relativistic energy of the object. Recall from Chapter 1 that p and E can be written as

$$p = \gamma m v \qquad (4.3)$$

and

$$E^2 = p^2 c^2 + m^2 c^4 = \gamma^2 m^2 c^4 \qquad (4.4)$$

where $\gamma = (1 - v^2/c^2)^{-1/2}$ and v is the object's speed. Equations 4.1 and 4.2 immediately suggest that it should be easy to calculate the speed of a de Broglie wave from the product λf. However, as we will show later, this is not the speed of the particle. Since the correct calculation is a bit complicated, we postpone it to Section 4.3. Before taking up the question of the velocity of matter waves, we prefer first to give some introductory examples of the use of $\lambda = h/p$ and a brief description of how de Broglie waves provide a physical picture of the Bohr theory of atoms.

De Broglie's Explanation of Quantization in the Bohr Model

Bohr's model of the atom had many shortcomings and problems. For example, as the electrons revolve around the nucleus, how can one understand the fact that only certain electronic energies are allowed? Why do all atoms of a given element have precisely the same physical properties regardless of the infinite variety of starting velocities and positions of the electrons in each atom?

De Broglie's great insight was to recognize that although these are deep problems for particle theories, wave theories of matter handle these problems neatly by means of interference. For example, a plucked guitar string, although initially subjected to a wide range of wavelengths, supports only standing wave patterns that have nodes at each end. Thus only a discrete set of wavelengths is allowed for standing waves, while other wavelengths not included in this discrete set rapidly vanish by destructive interference. This same reasoning can be applied to electron matter waves bent into a circle around the nucleus. Although initially a continuous distribution of wavelengths may be present, corresponding to a distribution of initial electron velocities, most wavelengths and velocities rapidly die off. The residual standing wave patterns thus account for the identical nature of all atoms of a given element and show that atoms are more like vibrating drum heads with discrete modes of vibration than like miniature solar systems. This point of view is emphasized in Figure 4.2, which shows the standing wave pattern of the electron in the hydrogen atom corresponding to the $n = 3$ state of the Bohr theory.

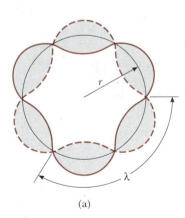

(a)

Figure 4.2 Standing waves fit to a circular Bohr orbit. In this particular diagram, three wavelengths are fit to the orbit, corresponding to the $n = 3$ energy state of the Bohr theory.

Another aspect of the Bohr theory that is also easier to visualize physically by using de Broglie's hypothesis is the quantization of angular momentum. One simply assumes that the allowed Bohr orbits arise because the electron matter waves again interfere constructively when an integral number of wavelengths exactly fits into the circumference of a circular orbit. Thus

$$n\lambda = 2\pi r \tag{4.5}$$

where r is the radius of the orbit. From Equation 4.1, we see that $\lambda = h/m_e v$. Substituting this into Equation 4.5, and solving for $m_e vr$, the angular momentum of the electron, gives

$$m_e vr = n\hbar \tag{4.6}$$

Note that this is precisely the Bohr condition for the quantization of angular momentum.

EXAMPLE 4.1 Why Don't We See the "Fuzzy" Edges of a Baseball?

An object will appear "wavelike" if it exhibits interference or diffraction, both of which require scattering objects or apertures of about the same size as the wavelength. A baseball of mass 140 g traveling at a speed of 60 mph (27 m/s) has a wavelength given by

$$\lambda = \frac{h}{p} = \frac{6.63 \times 10^{-34}\,\text{J} \cdot \text{s}}{(0.14\,\text{kg})(27\,\text{m/s})} = 1.7 \times 10^{-34}\,\text{m}$$

Even a nucleus (whose size is $\approx 10^{-15}$ m) is much too large to diffract this incredibly small wavelength! This explains why all macroscopic objects appear particle-like.

EXAMPLE 4.2 What Size Particles Do Exhibit Diffraction?

A particle of charge q and mass m is accelerated from rest through a small potential difference V. (a) Find its de Broglie wavelength, assuming that the particle is nonrelativistic.

Solution When a charge is accelerated from rest through a potential difference V, its gain in kinetic energy $\frac{1}{2}mv^2$ must equal the loss in potential energy qV. That is,

$$\tfrac{1}{2}mv^2 = qV$$

Because $p = mv$, we can express this in the form

$$\frac{p^2}{2m} = qV \quad \text{or} \quad p = \sqrt{2mqV}$$

Substituting this expression for p into the de Broglie relation $\lambda = h/p$ gives

$$\lambda = \frac{h}{p} = \frac{h}{\sqrt{2mqV}}$$

(b) Calculate λ if the particle is an electron and $V = 50$ V.

Solution The de Broglie wavelength of an electron accelerated through 50 V is

$$\lambda = \frac{h}{\sqrt{2m_e qV}}$$

$$= \frac{6.63 \times 10^{-34}\,\text{J} \cdot \text{s}}{\sqrt{2(9.11 \times 10^{-31}\,\text{kg})(1.6 \times 10^{-19}\,\text{C})(50\,\text{V})}}$$

$$= 1.7 \times 10^{-10}\,\text{m} = 1.7\,\text{Å}$$

This wavelength is of the order of atomic dimensions and the spacing between atoms in a solid. Such low-energy electrons are normally used in electron diffraction experiments.

Exercise 1 (a) Show that the de Broglie wavelength for an electron accelerated from rest through a *large* potential difference, V, is

$$\lambda = \frac{12.27}{V^{1/2}} \left(\frac{Ve}{2m_e c^2} + 1 \right)^{-1/2} \tag{4.7}$$

where λ is in angstroms and V is in volts. (b) Calculate the percent error introduced when $\lambda = 12.27/V^{1/2}$ is used instead of the correct relativistic expression for 10 MeV electrons.

Answer (b) 230%

4.2   THE DAVISSON–GERMER EXPERIMENT

Direct experimental proof that electrons possess a wavelength $\lambda = h/p$ was furnished by the diffraction experiments of Clinton J. Davisson (1881–1958) and Lester H. Germer (1896–1971) at the Bell Laboratories in New York City in 1927 (Fig. 4.3).[1] In fact de Broglie had already suggested in 1924 that a stream of electrons traversing a small aperture should exhibit diffraction phenomena. In 1925, Einstein was led to the necessity of postulating matter waves from an analysis of fluctuations of a molecular gas. In addition, he noted that a molecular beam should show small but measurable diffraction effects. In the same year, Walter Elsasser, much impressed with de Broglie's and Einstein's ideas on matter waves, pointed out that the slow electron scattering experiments of C. J. Davisson and C. H. Kunsman at the Bell Labs could be explained by electron diffraction.

Clear-cut proof of the wave nature of electrons was obtained in 1927 by the work of Davisson and Germer in the United States and George P. Thomson (the son of J. J. Thomson) in England. Both cases are intriguing not only for their physics but also for their human interest. The first case was an accidental discovery, and the second involved the discovery of the particle properties of the electron by the father and the wave properties by the son.

The crucial experiment of Davisson and Germer was an offshoot of an attempt to understand the arrangement of atoms on the surface of a nickel sample by elastically scattering a beam of low-speed electrons from a polycrystalline nickel target. A schematic drawing of their apparatus is shown in Figure 4.4. Note that their device allowed for the variation of three experimental parameters—electron energy; nickel target orientation, α; and scattering angle, ϕ. Before a fortunate accident occurred, the results seemed quite pedestrian. For constant electron energies of about 100 eV, the scattered intensity rapidly decreased as ϕ increased. But then someone dropped a flask of liquid air on the glass vacuum system, rupturing the vacuum and oxidizing the nickel target, which had been at high temperature. To remove the oxide, the sample was

Figure 4.3   Clinton J. Davisson (left) and Lester H. Germer (center) at Bell Laboratories in New York City. (*Bell Laboratories, courtesy AIP Emilio Segrè Visual Archives*)

[1] Clinton J. Davisson and Lester H. Germer, *Phys. Rev.* 30:705, 1927.

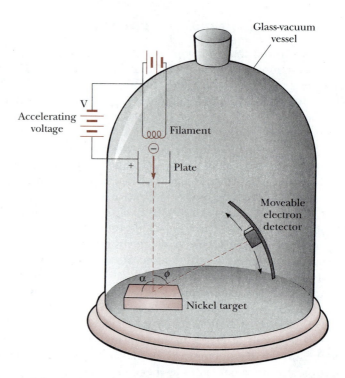

Figure 4.4   A schematic diagram of the Davisson–Germer apparatus.

reduced by heating it cautiously[2] in a flowing stream of hydrogen. When the apparatus was reassembled, quite different results were found: strong variations in the intensity of scattered electrons with angle were observed, as shown in Figure 4.5. The prolonged heating had evidently annealed the nickel sample, causing large single-crystal regions to develop. These crystalline regions fur-

Figure 4.5   A polar plot of scattered intensity versus scattering angle for 54-eV electrons, based on the original work of Davisson and Germer. The scattered intensity is proportional to the distance of the point from the origin.

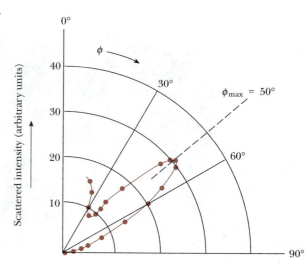

[2] At present this can be done without the slightest fear of "stinks or bangs," because 5% hydrogen–95% argon safety mixtures are commercially available.

nished the extended regular lattice needed to observe electron diffraction. Once Davisson and Germer realized that it was the elastic scattering from *single crystals* that produced such unusual results (1925), they initiated a thorough investigation of elastic scattering from large single crystals with predetermined crystallographic orientation. Even these experiments were not conducted at first as a test of de Broglie's wave theory, however. Following discussions with Richardson, Born, and Franck, the experiments and their analysis finally culminated in 1927 in the proof that electrons experience diffraction where the electron wavelength is given by $\lambda = h/p$.

The idea that electrons behave like waves when interacting with the atoms of a crystal is so striking that Davisson and Germer's proof deserves closer scrutiny. In effect, they calculated the wavelength of electrons from a simple diffraction formula and compared this result with de Broglie's formula $\lambda = h/p$. Although they tested this result over a wide range of target orientations and electron energies, we consider in detail only the simple case shown in Figures 4.4 and 4.5 with $\alpha = 90.0°$, $V = 54.0$ V, and $\phi = 50.0°$, corresponding to the $n = 1$ diffraction maximum. In order to calculate the de Broglie wavelength for this case, we first obtain the velocity of a nonrelativistic electron accelerated through a potential difference V from the energy relation

$$\tfrac{1}{2}m_e v^2 = Ve$$

Substituting $v = \sqrt{2Ve/m_e}$ into the de Broglie relation gives

$$\lambda = \frac{h}{m_e v} = \frac{h}{\sqrt{2Vem_e}} \tag{4.8}$$

Thus the wavelength of 54.0 V electrons is

$$\lambda = \frac{6.63 \times 10^{-34}\,\text{J·s}}{\sqrt{2(54.0\,\text{V})(1.60 \times 10^{-19}\,\text{C})(9.11 \times 10^{-31}\,\text{kg})}}$$
$$= 1.67 \times 10^{-10}\,\text{m} = 1.67\,\text{Å}$$

The experimental wavelength may be obtained by considering the nickel atoms to be a reflection diffraction grating, as shown in Figure 4.6. Only the surface layer of atoms is considered because low-energy electrons, unlike x-rays,

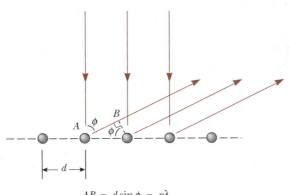

$$AB = d\sin\phi = n\lambda$$

Figure 4.6   Constructive interference from electron matter waves scattered at an angle ϕ.

Figure 4.7 Diffraction of 50-kV electrons from a film of Cu₃Au. The alloy film was 400 Å thick. (*Courtesy of the late Dr. L. H. Germer.*)

do not penetrate deeply into the crystal. Constructive interference occurs when the path length difference between two adjacent rays is an integral number of wavelengths or

$$d \sin \phi = n\lambda \qquad (4.9)$$

As d was known to be 2.15 Å from x-ray diffraction measurements, Davisson and Germer calculated λ to be

$$\lambda = (2.15 \text{ Å})(\sin 50.0°) = 1.65 \text{ Å}$$

which is in excellent agreement with the de Broglie formula.

It is interesting to note that while the diffraction lines from low-energy reflected electrons are quite broad (see Fig. 4.5), the lines from high-energy electrons transmitted through metal foils are quite sharp (see Fig. 4.7). This effect occurs because hundreds of atomic planes are penetrated by high-energy electrons, and consequently Equation 4.9, which treats diffraction from a surface layer, no longer holds. Instead, the Bragg law, $2d \sin \theta = n\lambda$, applies to high-energy electron diffraction. The maxima are extremely sharp in this case because if $2d \sin \theta$ is not exactly equal to $n\lambda$, there will be no diffracted wave. This occurs because there are scattering contributions from so many atomic planes that eventually the path length difference between the wave from the first plane and some deeply buried plane will be *an odd multiple of $\lambda/2$*, resulting in complete cancellation of these waves (see Problem 13).

If de Broglie's postulate is true for all matter, then any object of mass m has wavelike properties and a wavelength $\lambda = h/p$. In the years following Davisson and Germer's discovery, experimentalists tested the universal character of de Broglie's postulate by searching for diffraction of other "particle" beams. In subsequent experiments, diffraction was observed for helium atoms (Estermann and Stern in Germany) and hydrogen atoms (Johnson in the United States). Following the discovery of the neutron in 1932, it was shown that neutron beams of the appropriate energy also exhibit diffraction when incident on a crystalline target (Fig. 4.8).

Figure 4.8 Diffraction pattern of neutrons produced by a single crystal of NaCl. (*Courtesy of Dr. E. O. Wollan.*)

EXAMPLE 4.3   Thermal Neutrons

What kinetic energy (in electron volts) should neutrons have if they are to be diffracted from crystals?

Solution Appreciable diffraction will occur if the de Broglie wavelength of the neutron is of the same order of magnitude as the interatomic distance. Taking $\lambda = 1.00$ Å, we find

$$p = \frac{h}{\lambda} = \frac{6.63 \times 10^{-34}\,\text{J·s}}{1.00 \times 10^{-10}\,\text{m}} = 6.63 \times 10^{-24}\,\text{kg·m/s}$$

The kinetic energy is given by

$$K = \frac{p^2}{2m_n} = \frac{(6.63 \times 10^{-24}\,\text{J·s})^2}{2(1.66 \times 10^{-27}\,\text{kg})}$$

$$= 1.32 \times 10^{-20}\,\text{J} = 0.0825\,\text{eV}$$

Note that these neutrons are nonrelativistic because K is much less than the neutron rest energy of 940 MeV, and so our use of the classical expression $K = p^2/2m_n$ is justified. Because the average thermal energy of a particle in thermal equilibrium is $\frac{1}{2}k_B T$ per degree of freedom, neutrons at room temperature (300 K) possess a kinetic energy of

$$K = \frac{3}{2}k_B T = (1.50)(8.62 \times 10^{-5}\,\text{eV/K})(300\,\text{K})$$

$$= 0.0388\,\text{eV}$$

Thus "thermal neutrons," or neutrons in thermal equilibrium with matter at room temperature, possess energies of the right order of magnitude to diffract appreciably from single crystals. Neutrons produced in a nuclear reactor are far too energetic to produce diffraction from crystals, and must be slowed down in a graphite column as they leave the reactor. In the graphite moderator, repeated collisions with carbon atoms ultimately reduce the average neutron energies to the average thermal energy of the carbon atoms. When this occurs, these so-called "thermalized" neutrons possess a distribution of velocities and a corresponding distribution of de Broglie wavelengths with average wavelengths comparable to crystal spacings.

Exercise 2 *Monochromatic Neutrons.* A beam of neutrons with a single wavelength may be produced by means of a mechanical velocity selector of the type shown in Figure 4.9. (a) Calculate the speed of neutrons with a wavelength of 1.00 Å. (b) What rotational speed (in rpm) should the shaft have in order to pass neutrons with wavelength of 1.00 Å?

Answers   (a) 3.99×10^3 m/s   (b) 13 300 rpm

Neutrons with a range of velocities

Disk A

0.5 m

Disk B

ω

Figure 4.9   A neutron velocity selector. The slot in disk B is offset from the slot in disk A by 10°.

4.3   WAVE GROUPS AND DISPERSION

The matter wave representing a moving particle must reflect the fact that the particle has a large probability of being found in a small region of space only at a specific time. This means that a traveling sinusoidal matter wave of infinite extent and constant amplitude cannot properly represent a localized moving particle. What is needed is a pulse, or "wave group" of limited spatial extent. Such a pulse can be formed by adding sinusoidal waves with different wavelengths. The resulting wave group can then be shown to move with a speed v_g (the group speed) identical to the classical particle speed. This argument is shown schematically in Figure 4.10 and will be treated in detail after the introduction of some general ideas about wave groups.

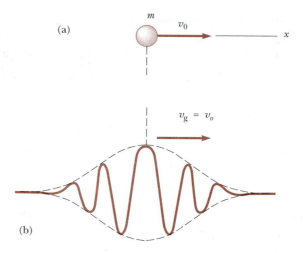

Figure 4.10 Representing a particle with matter waves: (a) particle of mass m and speed v_0. (b) Superposition of many matter waves with a spread of wavelengths centered on $\lambda_0 = h/mv_0$ correctly represents a particle.

Actually, all observed waves are limited to definite regions of space (pulses, wave groups, or wave packets in the case of matter waves). The plane wave with an exact wavelength and infinite extension is an abstraction. Water waves from a stone dropped into a pond, light waves emerging from a briefly opened shutter, a wave generated on a taut rope by a single flip of one end, and a sound wave emitted by a discharging capacitor must all be modeled by wave groups. A wave group consists of a superposition of waves with *different wavelengths*, with the amplitude and phase of each component wave chosen so that the waves interfere constructively over a small region of space. Outside of this region the combination of waves produces a net amplitude that approaches zero rapidly as a result of destructive interference. Perhaps the most familiar physical example in which wave groups arise is the phenomenon of beats. Beating occurs when two sound waves of slightly different wavelength (and hence different frequency) are combined. The resultant sound wave has a frequency equal to the average of the two combining waves and an amplitude that fluctuates, or "beats," at a rate given by the difference of the two original frequencies. This case is illustrated in Figure 4.11.

Figure 4.11 Beats are formed by the combination of two waves of slightly different frequencies traveling in the same direction. (a) The individual waves. (b) The combined wave has an amplitude (broken line) that oscillates in time.

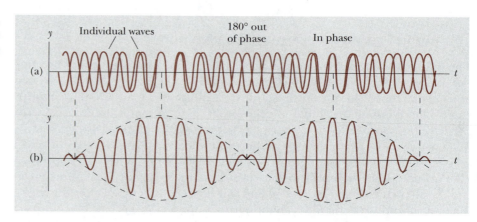

Let us examine this same situation mathematically. Consider a one-dimensional wave propagating in the positive x direction with a *phase speed v_p*. Note that v_p is the speed of a point of constant phase on the wave, such as a wave crest or trough. This traveling wave with wavelength λ, frequency f, and amplitude A may be described by

$$y = A \cos \left(\frac{2\pi x}{\lambda} - 2\pi f t \right) \tag{4.10}$$

where λ and f are related by

$$v_p = \lambda f \tag{4.11}$$

A more compact form for Equation 4.10 results if we take $\omega = 2\pi f$ (where ω is the angular frequency) and $k = 2\pi/\lambda$ (where k is the wavenumber). With these substitutions the infinite wave becomes

$$y = A \cos(kx - \omega t) \tag{4.12}$$

with

$$v_p = \frac{\omega}{k} \tag{4.13}$$

Phase velocity

Let us now form the superposition of two waves of equal amplitude both traveling in the positive x direction but with slightly different wavelengths, frequencies, and phase velocities. The resultant amplitude y is given by

$$y = y_1 + y_2 = A \cos(k_1 x - \omega_1 t) + A \cos(k_2 x - \omega_2 t)$$

Using the trigonometric identity

$$\cos a + \cos b = 2 \cos \tfrac{1}{2}(a - b) \cdot \cos \tfrac{1}{2}(a + b)$$

we find

$$y = 2A \cos \frac{1}{2}\{(k_2 - k_1)x - (\omega_2 - \omega_1)t\} \cdot \cos \tfrac{1}{2}\{(k_1 + k_2)x - (\omega_1 + \omega_2)t\} \tag{4.14}$$

For the case of two waves with slightly different values of k and ω, we see that $\Delta k = k_2 - k_1$ and $\Delta\omega = \omega_2 - \omega_1$ are small, but $(k_1 + k_2)$ and $(\omega_1 + \omega_2)$ are large. Thus, Equation 4.14 may be interpreted as a broad sinusoidal envelope

$$2A \cos \left(\frac{\Delta k}{2} x - \frac{\Delta\omega}{2} t \right)$$

limiting or modulating a high-frequency wave within the envelope

$$\cos \left[\tfrac{1}{2}(k_1 + k_2)x - \tfrac{1}{2}(\omega_1 + \omega_2)t \right]$$

This superposition of two waves is shown in Figure 4.12.

Although our model is primitive and does not really represent a pulse limited to a small region of space, it shows several interesting features common to more complicated models. For example, the envelope and the wave within the envelope actually move at different speeds. The speed of either the high-frequency wave or the envelope is given by dividing the coefficient of the t term by the

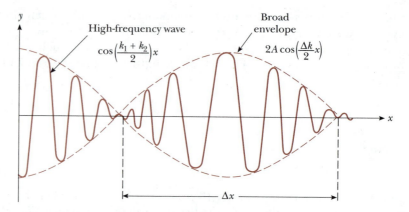

Figure 4.12 Superposition of two waves of slightly different wavelengths resulting in primitive wave groups; t has been set equal to zero in Equation 4.14.

coefficient of the x term as was done in Equations 4.12 and 4.13. *For the wave within the envelope*

$$v_{\text{p}} = \frac{(\omega_1 + \omega_2)/2}{(k_1 + k_2)/2} \approx \frac{\omega_1}{k_1} = v_1$$

Thus, the high-frequency wave moves at the phase velocity v_1 of one of the waves or at v_2 because $v_1 \approx v_2$. The envelope or group described by $2A \cos[(\Delta k/2)x - (\Delta\omega/2)t]$ moves with a different velocity, however, the group velocity given by

$$v_{\text{g}} = \frac{(\omega_2 - \omega_1)/2}{(k_2 - k_1)/2} = \frac{\Delta\omega}{\Delta k} \tag{4.15}$$

Another general characteristic of wave groups for waves of any type is both a limited duration in time, Δt, and a limited extent in space Δx. It is found that the smaller the spatial width of the pulse, Δx, the larger the range of wavelengths or wavenumbers, Δk, needed to form the pulse. This may be stated mathematically as

$$\Delta x \, \Delta k \approx 1 \tag{4.16}$$

Likewise if the time duration, Δt, of the pulse is small, we require a wide spread of frequencies, $\Delta\omega$, to form the group. That is

$$\Delta t \, \Delta\omega \approx 1 \tag{4.17}$$

In pulse electronics, this condition is known as the "response time–bandwidth formula."[3] In this situation Equation 4.17 shows that in order to amplify a voltage pulse of time width Δt without distortion, a pulse amplifier must equally amplify all frequencies in a frequency band of width $\Delta\omega$.

[3] It should be emphasized that Equations 4.16 and 4.17 are true in general and that the quantities Δx, Δk, Δt, and $\Delta\omega$ represent the spread in values present in an *arbitrary* pulse formed from the superposition of two or *more* waves.

Equations 4.16 and 4.17 are important in a general sense because they constitute "uncertainty relations," or "reciprocity relations," for pulses of any kind—electromagnetic, sound, or even matter waves. In particular Equation 4.16 shows that Δx, the uncertainty in spatial extent of a pulse, is inversely proportional to Δk, the range of wavenumbers making up the pulse: *both Δx and Δk cannot become arbitrarily small, but as one decreases the other must increase.*

It is interesting that our simple two-wave model also shows the general principles given by Equations 4.16 and 4.17. If we call (rather artificially) the spatial extent of our group the distance between adjacent minima (labeled Δx in Figure 4.12), we find from the envelope term $2A \cos(\frac{1}{2} \Delta k \, x)$ the condition $\frac{1}{2} \Delta k \, \Delta x = \pi$ or

$$\Delta k \, \Delta x = 2\pi \tag{4.18}$$

Here, $\Delta k = k_2 - k_1$ is the range of wavenumbers present. Likewise, if x is held constant and t is allowed to vary in the envelope portion of Equation 4.14, $2A \cos\{\frac{1}{2}(k_2 - k_1)x - \frac{1}{2}(\omega_2 - \omega_1)t\}$, the result is $\frac{1}{2}(\omega_2 - \omega_1) \, \Delta t = \pi$ or

$$\Delta\omega \, \Delta t = 2\pi \tag{4.19}$$

Thus Equations 4.18 and 4.19 agree with the general principles, respectively, of $\Delta k \, \Delta x \approx 1$ and $\Delta\omega \, \Delta t \approx 1$.

The addition of only two waves with discrete frequencies is instructive but produces an infinite wave instead of a true pulse. In the general case, many waves having a continuous distribution of wavelengths must be added to form a packet that is finite over a limited range and really zero everywhere else. In this case Equation 4.15 becomes

$$v_g = \left. \frac{d\omega}{dk} \right|_{k_0} \tag{4.20}$$

Group velocity

where the derivative is to be evaluated at k_0, the central wavenumber of the many waves present. The connection between the group velocity and the phase velocity of the composite waves is easily obtained. Because $\omega = k v_p$, we find

$$v_g = \left. \frac{d\omega}{dk} \right|_{k_0} = \left. v_p \right|_{k_0} + \left. k \frac{dv_p}{dk} \right|_{k_0} \tag{4.21}$$

where v_p is the phase velocity and is, in general, a function of k or λ. Materials in which the phase velocity varies with wavelength are said to exhibit *dispersion*. An example of a dispersive medium is glass, in which the index of refraction varies with wavelength, and different colors of light travel at different speeds. Media in which the phase velocity does not vary with wavelength (such as vacuum for electromagnetic waves) are termed nondispersive. The term *dispersion* arises from the fact that the individual harmonic waves that form a pulse travel at different phase velocities and cause an originally sharp pulse to change shape and become spread out, or dispersed. As an example, dispersion of a laser pulse after traveling 1 km along an optical fiber is shown in Figure 4.13. In a nondispersive medium where all waves have the same velocity, the group velocity is equal to the phase velocity. In a dispersive medium the group velocity can be less than or greater than the phase velocity, depending on the sign of dv_p/dk as shown by Equation 4.21.

Figure 4.13 Dispersion in a 1-ns laser pulse. A pulse that starts with the width shown by the vertical lines has a time width of approximately 30 ns after traveling 1 km along an optical fiber.

EXAMPLE 4.4   Group Velocity in a Dispersive Medium

In a particular substance the phase velocity of waves doubles when the wavelength is halved. Show that wave groups in this system move at twice the *central* phase velocity.

Solution   From the given information, the dependence of phase velocity on wavelength must be

$$v_p = A'/\lambda = Ak$$

for some constants A' and A. From Equation 4.21 we obtain

$$v_g = v_p \Big|_{k_0} + k \frac{dv_p}{dk} \Big|_{k_0}$$
$$= Ak_0 + Ak_0 = 2Ak_0$$

Thus,

$$v_g = 2v_p \Big|_{k_0}$$

EXAMPLE 4.5   Group Velocity in Deep Water Waves

Newton showed that the phase velocity of deep water waves having wavelength λ is given by

$$v_p = \sqrt{\frac{g\lambda}{2\pi}}$$

where g is the free-fall acceleration, and where the minor contribution of surface tension has been ignored. Show that in this case the velocity of a group of these waves is one half of the phase velocity of the central wavelength.

Solution   Because $k = 2\pi/\lambda$, we can write v_p as

$$v_p = (g/k)^{1/2}$$

Therefore, we find

$$v_g = v_p \Big|_{k_0} + k \frac{dv_p}{dk} \Big|_{k_0} = \left(\frac{g}{k_0}\right)^{1/2} - \frac{1}{2}\left(\frac{g}{k_0}\right)^{1/2}$$
$$= \frac{1}{2}\left(\frac{g}{k_0}\right)^{1/2} \equiv \frac{1}{2} v_p \Big|_{k_0}$$

Matter Wave Packets

We are now in a position to apply our general theory of wave groups to electrons. We shall show both the dispersion of de Broglie waves and the satisfying result that the wave packet and the particle move at the same velocity. According to de Broglie, *individual* matter waves have a frequency f and a wavelength λ given by

$$f = \frac{E}{h} \quad \text{and} \quad \lambda = \frac{h}{p}$$

where E and p are the relativistic energy and momentum of the particle, respectively. The phase speed of these matter waves is given by

$$v_p = f\lambda = \frac{E}{p} \tag{4.22}$$

The phase speed can be expressed as a function of p or k alone by substituting $E = (p^2 c^2 + m^2 c^4)^{1/2}$ into Equation 4.22:

$$v_p = c \sqrt{1 + \left(\frac{mc}{p}\right)^2} \tag{4.23}$$

The dispersion relation for de Broglie waves can be obtained as a function of k by substituting $p = h/\lambda = \hbar k$ into Equation 4.23. This gives

Phase velocity of matter waves

$$v_p = c \sqrt{1 + \left(\frac{mc}{\hbar k}\right)^2} \tag{4.24}$$

Equation 4.24 shows that individual de Broglie waves representing a particle of mass m show dispersion even in *empty space* and always travel at a speed that is *greater than* or at least equal to c. Because these component waves travel at different speeds, the width of the wave packet, Δx, spreads or disperses as time progresses as will be seen in detail in Chapter 5. To obtain the group speed, we use

$$v_g = \left[v_p + k \frac{dv_p}{dk} \right]_{k_0}$$

and Equation 4.24. After some algebra, we find

$$v_g = \frac{c}{\left[1 + \left(\dfrac{mc}{\hbar k_0} \right)^2 \right]^{1/2}} = \frac{c^2}{v_p} \Bigg|_{k_0} \tag{4.25}$$

Solving for the phase speed from Equation 4.22 we find

$$v_p = \frac{E}{p} = \frac{\gamma mc^2}{\gamma mv} = \frac{c^2}{v}$$

where v is the particle's speed. Finally, substituting $v_p = c^2/v$ into Equation 4.25 for v_g shows that the group velocity of the matter wave packet is the same as the particle speed, in agreement with our intuition that the matter wave envelope should move at the same speed as the particle.

Introducing The MP DESKTOP

(Those wishing to bypass the software examples may do so without interrupting the logical progression of material; however, we do not recommend this approach.)

The following example is the first in a series of examples, exercises, and problems that use the simulation software bundled with your *Modern Physics* text. The software—dubbed the MP DESKTOP—brings significant computing power to the study of quantum physics. To use the software you will need an IBM PC (or close compatible) with a VGA display and color monitor; a mouse is *strongly* recommended. (A 486 processor, or better, with a clock speed of at least 50 MHz is desirable for displaying the evolution of wave packets.) Start by copying the program files on the disk to your hard drive. (This *must* be done because the disk files are stored in a compressed format.) The INSTALL utility on the program disk automates this chore: just run INSTALL, and at the prompt, specify the destination drive and directory (the default is C:\MPDESK). The main program file *MP.EXE* is now ready to run.

The MP DESKTOP is a sophisticated interactive learning tool with considerable flexibility and scope. The examples in this and subsequent chapters are designed as a guide to using the DESKTOP while illustrating its potential for clarifying some of the more abstract ideas inherent in wave motion and quantum physics. For each example the goal is clearly stated, and those DESKTOP features pertinent to the study at hand are introduced in a step-by-step fashion. This incremental approach shields the user from the full complexity of the program during the learning period; however, it is only by working through all examples that the student will come to appreciate the full capabilities of the DESKTOP.

Should you require assistance while using the DESKTOP, don't hesitate to consult the *Help* menu. Here you will find details on all aspects of the DESKTOP organized according to task, with cross-references (hyperlinks) to related topics. Also note the QuickHelp feature: clicking the **right** mouse button on any DESKTOP object brings up a summary of that object's function, along with any special usage notes. Further details on installing and using the MP DESKTOP are given in Appendix D at the end of this book.

All examples and end-of-chapter problems requiring use of the MP DESKTOP are clearly identified with the MP logo **MP**.

EXAMPLE 4.6   Phase and Group Velocity: A Simulation

MP In this example, we will use the MP DESKTOP to construct a wave group consisting of a "fast" oscillation modulated by a much slower one. We will confirm that the group velocity (in this case, the speed of the envelope modulating the individual waves) is given by Equation 4.15.

To begin, launch the application by running the main program file *MP.EXE;* this brings up the MP DESKTOP. Select *Task* from the menu bar and *Waves-Classical* from the drop-down task list. A wave group $f(x, t)$ can now be formed by adding harmonic waves with different wavenumbers k, amplitudes $a(k)$, and frequencies $\omega(k)$:

$$f(x, t) = \sum_k a(k) \cos (ks - \omega t)$$

The frequencies and amplitudes for this example are saved in data files furnished with the program. To set the frequencies, click the Frequency button 「w(k)」 located just below the menu bar.

NOTE: The DESKTOP is fully mouse-aware, and is easiest to use with a mouse. If you don't have a mouse or are allergic to rodents, you can still navigate among objects on the DESKTOP with the *Tab* and *Shift-Tab* keys; selections are made by pressing the *Enter* key when the desired object has the focus (indicated by the dashed rectangle).

A dialog box opens, presenting a list of frequency files. Select *EX4-6.W* (for Example 4.6) from this list, then select *OK.* In this example, angular frequencies are related to wavenumbers by the rule $\omega(k) = |k|^\alpha$. (The function $\omega(k)$ is known as a *dispersion relation;* this particular form for $\omega(k)$ was chosen for its simplicity and flexibility—by assigning different values to α, different group velocities are obtained.) After pressing *OK,* you will be presented with another dialog box showing the formula for $\omega(k)$, the active range of wavenumbers ($20 \leq k \leq 25$) and the default parameter value $\alpha = 0.1$. Accept the information as presented by clicking the *OK* button.

The bar graph in the left window of the screen shows the frequencies (y axis) plotted against the wavenumbers

(x axis). Values for ω and k can be found with the MP Analyzer. Select [🔍] to activate the Analyzer. With the mouse (or the direction keys [→], [←]), place the cursor on any one of the bars; the values of k and $\omega(k)$ are displayed (Fig. 4.14). (Keyboard users note that the [Ins] key toggles cursor movement between coarse and fine settings.) Selecting [🔍] again deactivates the Analyzer.

Figure 4.14 (Example 4.6) Bar graph showing angular frequencies $\omega(k)$ for six wavenumbers k in the range $20 \leq k \leq 25$ (arbitrary units).

NOTE: *Classical Waves* are calculated on the spatial domain extending from -1 to $+1$ in whatever length unit the user chooses; the unit for wavenumbers is ALWAYS the reciprocal of this length unit. The frequency unit is specific to the application and would be reflected in the equation for $\omega(k)$: in the present case there is no model system, and the frequency unit is arbitrary; the unit of time is ALWAYS the reciprocal of this frequency unit. For a more exhaustive discussion of units, see the *Help-Contents* entry on the DESKTOP menu bar.

The six bars show that six harmonic waves will be used to construct the wave group. To mix only two waves with wavenumbers $k = 20$ and $k = 25$, we must reset the wavenumber increment to 5 (the separation between successive wavenumbers defaults to unity). Do this by selecting *Options* from the menu bar and *Control Panel* from the drop-down menu. This brings up the Desktop Control Panel dialog box. In the *Fourier* field (bottom left) input area for *Mesh Size*, type in 5 and select *OK*. The bar graph now should show only two bars.

To set the component wave amplitudes, select the Amplitude button a(k) and proceed as before in setting frequencies. We will mix the waves with equal amplitude, so you will want the file *UNIFORM.A* (the extension *.A* is for amplitude). The amplitudes $a(k)$ now overlay the frequencies on the bar plot; in the right window of the screen is plotted the superposition $f(x)$ of the two harmonic waves at time $t = 0$.

Values for $f(x)$ also can be read from the graph with the MP Analyzer. Select [icon] to activate the Analyzer, then click anywhere in the left window to redirect the Analyzer to the waveform plot. The wave disturbance f is read at the current location of the graphics cursor (vertical line). With the mouse (or the direction keys [→], [←]), place the cursor on the wave crest near the center of the window. The crest is found at $x = 0$.

The current time t is recorded in the time box below the menu bar, and is always zero to start. The Advance button [icon] to the immediate right of the box starts the clock. Click on this button. The waveform moves to the right some distance, showing that it is a traveling wave. Click again to stop. The duration between successive "ticks" of the clock can be manipulated "on the fly" by holding down the [Alt] or [Ctrl] keys while the clock is running. [Alt] speeds up evolution tenfold; [Ctrl] speeds up evolution a hundredfold. Try out these keys to see the effect. (To change the fundamental time step, select *Control Panel* from the *Options* menu, enter the appropriate *Time Step* in the *Evolution* field of the Desktop Control Panel and select *OK*.)

We can measure the wave speed directly with the Analyzer. First, reset the clock by clicking on [icon] *while holding down the Shift key*. The clock runs *backward*, stopping at $t = 0$. Next, keep your eye on the central crest and click on [icon] again, stopping the clock only after the crest has moved a measurable distance. Realign the graphics cursor with the crest and record the new location: this is the distance the crest has moved (Fig. 4.15). The phase velocity v_p is this distance divided by the elapsed time. You should find $v_p \approx 0.060$.

Figure 4.15   (Example 4.6) Measuring the phase velocity of a wave group formed as the superposition of two harmonic waves.

For a mix of just two harmonic waves, the group velocity v_g refers to the speed of the envelope modulating the individual oscillations. Monitoring the position of the envelope is easier if we track a node rather than an antinode. Reset the clock to zero as above, and locate a node of the envelope. Start the clock and let it run until this node has moved a measurable distance (Fig. 4.16). Calculate the speed; you should find $v_g \approx 0.006$.

Figure 4.16   (Example 4.6) Measuring the group velocity of a wave group formed from two harmonic waves.

Finally, your direct measurement of group velocity can be compared with the prediction of Equation 4.15. This requires frequencies and wavenumbers for the component waves. Both can be found by activating the left window for analysis. To do this, click on the left window. The MP Analyzer now reads the relevant information for each of the component waves separately. Use these data to compute the group velocity from Equation 4.15. Similarly verify your measurement of phase velocity.

Exercise 3   *Negative Group Velocities.* Demonstrate that the velocity of a wave group may be negative even when the component wave velocities are all positive. (*Hint:* change the sign of α in the dispersion relation $\omega(k)$.) Is there a value for α that makes the group velocity zero?

Exercise 4   *The Beat Phenomenon.* The wave group of Example 4.6 consists of a "fast" oscillation modulated by a much slower one. Use the MP Analyzer to find the wavelengths λ of both oscillations. With each wavelength we associate a frequency f as $f\lambda = v$, where v is the speed of the disturbance. These are the frequencies detected by an observer stationed at a fixed location as the wave passes him by. If this were a sound wave, the lower frequency would be heard as "beats" modulating a higher tone. Determine this beat frequency from your measured values of λ and v. ["Beats" actually are heard at twice the calculated frequency, because the ear cannot distinguish wave crests (maxima) from troughs (minima)].

Answer $f = 0.0024$, giving a beat frequency of 0.0048.

EXAMPLE 4.7   Reciprocity Relations: A Simulation

MP This example uses the MP DESKTOP to verify the reciprocity relation $\Delta x\,\Delta k \approx 1$ (Eq. 4.16) for some simple wave groups.

Begin by readying the DESKTOP to display classical waves, as described in Example 4.6. Fix the component wave frequencies $\omega(k)$ as before by loading the file *EX4-6.W* (click on $\boxed{\omega(k)}$). When the dialog box appears detailing $\omega(k)$, select a new domain of wavenumbers by entering -5 and $+5$ in the *Start* and *Stop* fields, respectively. Accept the new values by selecting *OK*. Next, choose component wave amplitudes by loading the file *EX4-7.A* (click on $\boxed{a(k)}$). In this example the individual waves are mixed with amplitudes that decrease with increasing wavenumber as $a(k) = \exp(-\mu|k|)$. Before proceeding, check to ensure that consecutive wavenumbers differ by unity, the default step size (select *Options* from the menu bar, then *Control Panel*, and inspect the *Mesh Size* entry in the *Fourier* field).

The waveform for the group appears in the right window of the screen. This is the superposition $f(x)$ (at time $t = 0$) of the (11) harmonic wave components whose amplitudes and frequencies are depicted in the left window.

The waveform $f(x)$ can be assigned a width, which we denote by $2\,\Delta x$ (Δx is the half-width). Similarly, the range of wavenumbers that are mixed appreciably to obtain that waveform is prescribed by the width $2\,\Delta k$ of the amplitude function $a(k)$. These widths are readily found with the help of the MP Analyzer.

Select $\boxed{\text{Q}}$ to activate the Analyzer. With the mouse (or the direction keys $[\rightarrow]$, $[\leftarrow]$), place the graphics cursor on the peak of the waveform and note the value of f. Now move the cursor left until $f(x)$ drops to half of its peak value. Press the [.] key to mark this position, and record the value of x. Do the same to the right of the peak. The difference between the marked positions—the full width at half maximum—is a reasonable choice for $2\,\Delta x$ (Fig. 4.17). You should find $\Delta x \approx \frac{1}{2}\,(0.82)$.

$$x = 0.412000067$$
$$f = 3.356507880$$

Figure 4.17   (Example 4.7) Measuring the width of a wavepacket. Vertical lines mark the places where $f(x)$ drops to one-half of its peak value.

Now switch to the left window by clicking on that window, and repeat the above procedure to estimate Δk, the half-width for the amplitude function $a(k)$. You will have to visually interpolate between the discrete values (Fig. 4.18); you should find $\Delta k \approx \frac{1}{2}\,(7)$. For this case, then, the product takes the value $\Delta x\,\Delta k \approx 1.4$.

k = 3.000000000
■w = 1.116123174
□a = 0.548811636

Figure 4.18   (Example 4.7) Assigning a width to the amplitude function $a(k)$.

Next, change the width Δk to see the effect on the waveform. Click on $\boxed{a(k)}$ to edit $a(k)$. In the *Parameters* field of the dialog box change μ from 0.2 (the default) to 0.6. Press *OK* to accept the change and update the display. Determine the new widths Δx and Δk. Notice how the width of the waveform grows as the amplitude width shrinks: the behaviors are reciprocal. Compute the new value for the product; you should find $\Delta x \, \Delta k \approx 0.6$. Experiment with a few more choices for μ to confirm your findings.

4.4 FOURIER INTEGRALS

O P T I O N A L

In this section we show in detail how to construct wave groups, or pulses, that are truly localized in space or time and also show that reciprocity relations of the type $\Delta k \, \Delta x \approx 1$ and $\Delta \omega \, \Delta t \approx 1$ hold for these pulses.

To form a true pulse that is zero everywhere outside of a finite spatial range Δx requires adding together an infinite number of harmonic waves with continuously varying wavelengths and amplitudes. This addition can be done with a Fourier integral, which is defined as follows:

$$f(x) = \frac{1}{\sqrt{2\pi}} \int_{-\infty}^{+\infty} a(k) \, e^{ikx} \, dk \tag{4.26}$$

Here $f(x)$ is a spatially localized wave group, $a(k)$ gives the amount or amplitude of the wave with wavenumber $k = (2\pi/\lambda)$ to be added, and $e^{ikx} = \cos kx + i \sin kx$ is Euler's compact expression for a harmonic wave. The amplitude distribution function $a(k)$ can be obtained if $f(x)$ is known by using the symmetric formula

$$a(k) = \frac{1}{\sqrt{2\pi}} \int_{-\infty}^{+\infty} f(x) \, e^{-ikx} \, dx \tag{4.27}$$

Equations 4.26 and 4.27 apply to the case of a spatial pulse at fixed time, but it is important to note that they are mathematically identical to the case of a time pulse passing a fixed position. This case is common in electrical engineering and involves adding together a continuously varying set of frequencies:

$$V(t) = \frac{1}{\sqrt{2\pi}} \int_{-\infty}^{+\infty} g(\omega) e^{i\omega t} \, d\omega \tag{4.28}$$

$$g(\omega) = \frac{1}{\sqrt{2\pi}} \int_{-\infty}^{+\infty} V(t) e^{-i\omega t} \, dt \tag{4.29}$$

where $V(t)$ is the strength of a signal as a function of time, and $g(\omega)$ is the *spectral content* of the signal and gives the amount of the harmonic wave with frequency ω present.

Let us now consider several examples of how to use Equations 4.26 through 4.29 and how they lead to uncertainty relationships of the type $\Delta\omega \, \Delta t \approx 1$ and $\Delta k \, \Delta x \approx 1$.

EXAMPLE 4.8

This example compares the spectral contents of infinite and truncated sinusoidal waves. A truncated sinusoidal wave is a wave cut off or truncated by a shutter as shown in Figure 4.19. (a) What is the spectral content of an infinite sinusoidal wave $e^{i\omega_0 t}$?

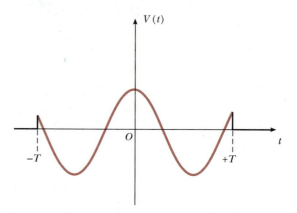

Figure 4.19   (Example 4.8) The real part of a truncated sinusoidal wave.

(b) Find and sketch the spectral content of a truncated sinusoidal wave given by

$$V(t) = e^{i\omega_0 t} \qquad -T < t < +T$$

$$V(t) = 0 \qquad \text{otherwise}$$

(c) Show that for this truncated sinusoid $\Delta t \, \Delta\omega = \pi$, where Δt and $\Delta\omega$ are the half-widths of $v(t)$ and $g(\omega)$, respectively.

Solution   (a) The spectral content consists of a single strong contribution at the frequency ω_0.

(b)
$$g(\omega) = \frac{1}{\sqrt{2\pi}} \int_{-\infty}^{+\infty} f(t) \, e^{-i\omega t} \, dt = \frac{1}{\sqrt{2\pi}} \int_{-T}^{+T} e^{i(\omega_0 - \omega)t} \, dt$$

$$= \frac{1}{\sqrt{2\pi}} \int_{-T}^{+T} [\cos(\omega_0 - \omega)t + i\sin(\omega_0 - \omega)t] \, dt$$

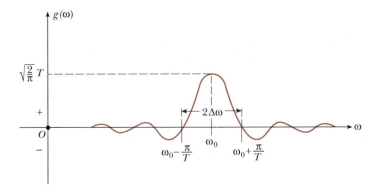

Figure 4.20   (Example 4.8) The Fourier transform of a truncated sinusoidal wave. The curve shows the amount of a given frequency that must be added to produce the truncated wave.

Because the sine term is an odd function and the cosine is even, the integral reduces to

$$g(\omega) = \frac{2}{\sqrt{2\pi}} \int_0^T \cos(\omega_0 - \omega)t \, dt = \sqrt{\frac{2}{\pi}} \frac{\sin(\omega_0 - \omega)T}{(\omega_0 - \omega)} = \sqrt{\frac{2}{\pi}} \, (T) \frac{\sin(\omega_0 - \omega)T}{(\omega_0 - \omega)T}$$

A sketch of $g(\omega)$ (Figure 4.20) shows a typical $\sin Z/Z$ profile centered on ω_0. Note that both positive and negative amounts of different frequencies must be added to produce the truncated sinusoid. Furthermore, the strongest frequency contribution comes from the frequency region near $\omega = \omega_0$ as expected.
(c) Δt clearly equals T and $\Delta\omega$ may be taken to be half the width of the main lobe of $g(\omega)$, $\Delta\omega = \pi/T$. Thus, we get

$$\Delta\omega \, \Delta t = \frac{\pi}{T} \times T = \pi$$

and we see that the product of the spread in frequency, $\Delta\omega$, and the spread in time, Δt, is a constant independent of T.

EXAMPLE 4.9   A Matter Wave Packet

(a) Show that the matter wave packet composed of the continuous set of wavenumbers centered on k_0 and spanning a range Δk all having an amplitude of one (Fig. 4.21) has the form

$$f(x) = \frac{\Delta k}{\sqrt{2\pi}} \frac{\sin(\Delta k \cdot x/2)}{(\Delta k \cdot x/2)} e^{ik_0 x}$$

Solution

$$f(x) = \frac{1}{\sqrt{2\pi}} \int_{-\infty}^{+\infty} a(k) e^{ikx} \, dk = \frac{1}{\sqrt{2\pi}} \int_{k_0 - (\Delta k/2)}^{k_0 + (\Delta k/2)} e^{ikx} \, dk = \frac{1}{\sqrt{2\pi}} \frac{e^{ik_0 x}}{x} 2 \sin(\Delta k \cdot x/2)$$

$$= \frac{\Delta k}{\sqrt{2\pi}} \frac{\sin(\Delta k \cdot x/2)}{(\Delta k \cdot x/2)} e^{ik_0 x}$$

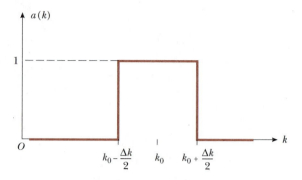

Figure 4.21   (Example 4.9) A simple amplitude distribution specifying a uniform contribution of all wavenumbers from $k_0 - \Delta k/2$ to $k_0 + \Delta k/2$. Although we have used only positive k's here, both positive and negative k values are allowed, in general corresponding to waves traveling to the right ($k > 0$) or left ($k < 0$).

(b) Observe that this wave packet is a complex function. Later in this chapter we shall see how the definition of probability density results in a real function, but for the time being consider only the real part of $f(x)$ and make a sketch of its behavior, showing its envelope and the cosine function within. Determine Δx, and show that an uncertainty relation of the form $\Delta x \, \Delta k \approx 1$ holds.

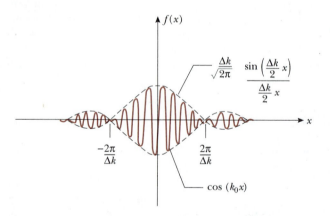

Figure 4.22   (Example 4.9) The real part of the wave packet formed by the uniform amplitude distribution shown in Figure 4.21.

Solution   The real part of the wave packet is shown in Figure 4.22 where the width of the main lobe is $\Delta x = 4\pi/\Delta k$. This immediately gives the uncertainty relation $\Delta x \, \Delta k = 4\pi$. Note that the constant on the right-hand side of the uncertainty relation depends on the shape chosen for $a(k)$ and the precise definition of Δx and Δk.

Exercise 5   Assume that a narrow triangular voltage pulse $V(t)$ arises in some type of radar system (see Fig. 4.23). (a) Find and sketch the spectral content $g(\omega)$. (b) Show that a relation of the type $\Delta \omega \, \Delta t \approx 1$ holds. (c) If the width of the pulse is $2\tau = 10^{-9}$ s, what range of frequencies must this system pass if the pulse is to be undistorted? Take $\Delta t = \tau$ and define $\Delta \omega$ similarly.

Figure 4.23   (Exercise 5).

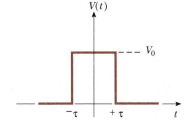

Figure 4.24   (Exercise 6).

Answer   (a) $g(\omega) = \sqrt{\dfrac{2}{\pi}} \left(\dfrac{1}{\omega^2 \tau}\right) (1 - \cos \omega \tau)$ (b) $\Delta \omega \, \Delta t = 2\pi$
(c) $2\Delta f = 4.00 \times 10^9$ Hz

Exercise 6   (a) Find and sketch the spectral content of the rectangular pulse of width 2τ shown in Figure 4.24. (b) Show that a reciprocity relation $\Delta \omega \, \Delta t \approx \pi$ holds in this case. Take $\Delta t = \tau$ and define $\Delta \omega$ similarly. (c) What range of frequencies is required to compose a pulse of width $2\tau = 1 \ \mu s$? A pulse of width $2\tau = 1$ ns?

Answer   (a) $g(\omega) = \sqrt{2/\pi} \ V_0 \{\sin(\omega\tau)\}/\omega$ (c) 2×10^6 Hz; 2×10^9 Hz

Exercise 7   *A Matter Wave Packet.* (a) Find and sketch the real part of the matter wave pulse shape $f(x)$ for a gaussian amplitude distribution $a(k)$, where

$$a(k) = A e^{-\alpha^2 (k - k_0)^2}$$

Note that $a(k)$ is peaked at k_0 and has a width that decreases with increasing α. [*Hint:* In order to put $f(x) = (2\pi)^{-1/2} \int_{-\infty}^{\infty} a(k) e^{ikx} \, dk$ into the standard form $\int_{-\infty}^{\infty} e^{-az^2} \, dz$, complete the square in k.] (b) By comparing the result for the real part of $f(x)$ to the standard form of a gaussian function with width Δx, $f(x) \propto A e^{-(x/2 \, \Delta x)^2}$, show that the width of the matter wave pulse is $\Delta x = \alpha$. (c) Find the width Δk of $a(k)$ by writing $a(k)$ in standard gaussian form and show that $\Delta x \, \Delta k = \frac{1}{2}$, independent of α.

Answer   (a) $f(x) = \dfrac{A}{\alpha\sqrt{2}} e^{-x^2/4\alpha^2} e^{ik_0 x}$ (b) $\mathrm{Re} \, f(x) = \dfrac{A}{\alpha\sqrt{2}} e^{-x^2/4\alpha^2} \cos k_0 x$
(c) $\Delta k = \frac{1}{2}\alpha$

Constructing Moving Wave Packets

Figure 4.22 represents a snapshot of the wave packet at $t = 0$. To construct a moving wave packet representing a moving particle, we replace kx in Equation 4.26 with $(kx - \omega t)$. Thus, the representation of the moving wave packet becomes

$$f(x, t) = \frac{1}{\sqrt{2\pi}} \int_{-\infty}^{+\infty} a(k) e^{i(kx - \omega t)} \, dk \tag{4.30}$$

It is important to realize that here $\omega = \omega(k)$, that is, ω is a function of k and, hence, depends on the type of wave and the medium traversed. In general it is hard to solve this integral analytically. The *QM WaveGroups-1D* task section of the accompanying MP DESKTOP software package effectively solves this integral numerically for the dispersion relation $\omega(k)$ describing nonrelativistic matter waves, given the initial packet shape specified by $f(x, 0)$ or, equivalently, $a(k)$. This application of the MP DESKTOP will be examined at length in Chapter 5.

4.5 THE HEISENBERG UNCERTAINTY PRINCIPLE

In the period 1924–25, Werner Heisenberg, the son of a professor of Greek and Latin at the University of Munich, invented a complete theory of quantum mechanics called matrix mechanics. This theory overcame some of the problems with the Bohr theory of the atom, such as the postulate of "unobservable" electron orbits. Heisenberg's formulation was based primarily on measurable quantities such as the transition probabilities for electronic jumps between quantum states. Because transition probabilities depend on the initial and final states, Heisenberg's mechanics used variables labeled by two subscripts. Although at first Heisenberg presented his theory in the form of noncommuting algebra, Max Born quickly realized that this theory could be more elegantly described by matrices. Consequently, Born, Heisenberg, and Pascual Jordan soon worked out a comprehensive theory of matrix mechanics. Although the matrix formulation was quite elegant, it attracted little attention outside of a small group of gifted physicists because it was difficult to apply in specific cases, involved mathematics unfamiliar to most physicists, and was based on rather vague physical concepts.

Although we will investigate this remarkable form of quantum mechanics no farther, we do wish to discuss another of Heisenberg's brilliant discoveries, the *uncertainty principle* elucidated in a famous paper in 1927. In this paper Heisenberg introduced the notion that it is impossible to determine simultaneously with unlimited precision the position *and* momentum of a particle. In words we may state the uncertainty principle as follows:

Momentum–position uncertainty principle

> If a measurement of position is made with precision Δx and a simultaneous measurement of momentum in the x direction is made with precision Δp_x, then the product of the two uncertainties can never be smaller than $\hbar/2$. That is,
>
> $$\Delta p_x \, \Delta x \geq \frac{\hbar}{2} \tag{4.31}$$

In his paper of 1927, Heisenberg was careful to point out that the inescapable uncertainties Δp_x and Δx do not arise from imperfections in practical measuring instruments. Rather, they arise from the need to use a large range of wave numbers, Δk, to represent a matter wave packet localized in a small region, Δx. The uncertainty principle represents a sharp break with the ideas of classical physics, in which it is assumed that, with enough skill and ingenuity, it is possible to simultaneously measure a particle's position and momentum to any desired degree of precision. As shown in Example 4.10, however, there is no contradiction between the uncertainty principle and classical laws for macroscopic systems because of the small value of $\hbar$.

One can show that $\Delta p_x \, \Delta x \geq \hbar/2$ comes from the uncertainty relation governing any type of wave pulse formed by the superposition of waves with different wavelengths. In Section 4.3 we found that to construct a wave group localized in a small region Δx, we had to add up a large range of wavenumbers Δk, where $\Delta k \, \Delta x \approx 1$ (Eq. 4.16). The precise value of the number on the right-hand side of Equation 4.16 depends on the functional form $f(x)$ of the wave group as well as on the specific definition of Δx and Δk. A different choice of

This photograph of Werner Heisenberg was taken around 1924 (University of Hamburg). Heisenberg obtained his PhD in 1923 at the University of Munich where he studied under Arnold Sommerfeld and became an enthusiastic mountain climber and skier. Later, he worked as an assistant to Max Born at Göttingen and Niels Bohr in Copenhagen. While physicists such as de Broglie and Schrödinger tried to develop pictorialized models of the atom, Heisenberg, with the help of Born and Pascual Jordan, developed an abstract mathematical model called matrix mechanics to explain the wavelengths of spectral lines. The more successful wave mechanics by Schrödinger announced a

BIOGRAPHY

WERNER HEISENBERG
(1901–1976)

few months later was shown to be equivalent to Heisenberg's approach. Heisenberg made many other significant contributions to physics, including his famous uncertainty principle, for which he received the Nobel prize in 1932, the prediction of two forms of molecular hydrogen, and theoretical models of the nucleus. During World War II he was director of the Max Planck Institute at Berlin where he was in charge of German research on atomic weapons. Following the war, he moved to West Germany and became director of the Max Planck Institute for Physics at Göttingen.

(Photo courtesy of the University of Hamburg)

$f(x)$ or a different rule for defining Δx and Δk (or both) will give a slightly different number. With Δx and Δk defined as standard deviations, it can be shown that the smallest number, $\frac{1}{2}$, is obtained for a gaussian wavefunction.[4] In this minimum uncertainty case we have

$$\Delta x \, \Delta k = \tfrac{1}{2}$$

For any other choice of $f(x)$

$$\Delta x \, \Delta k \geqslant \tfrac{1}{2} \tag{4.32}$$

and using $\Delta p_x = \hbar \, \Delta k$, $\Delta x \, \Delta k \geqslant \frac{1}{2}$ immediately becomes

$$\Delta p_x \, \Delta x \geqslant \frac{\hbar}{2} \tag{4.33}$$

The basic meaning of $\Delta p \, \Delta x \geqslant \hbar/2$ is that as one uncertainty increases the other decreases. In the extreme case as one uncertainty approaches ∞, the other must approach zero. This extreme case is illustrated by a plane wave $e^{ik_0 x}$ that has a precise momentum $\hbar k_0$ and an infinite extent—that is, the wavefunction is not concentrated in any segment of the x axis.

Another important uncertainty relation involves the uncertainty in energy of a wave packet ΔE, and the time Δt taken to measure that energy. Starting with $\Delta \omega \, \Delta t \geqslant \frac{1}{2}$ as the minimum form of the time–frequency uncertainty principle, and using the de Broglie relation for the connection between the matter

---

[4] See Section 5.7 for a definition of the standard deviation and Problem 5.35 for a complete mathematical proof of this statement.

wave energy and frequency, $E = \hbar\omega$, we immediately find the *energy–time uncertainty principle*

Energy–time uncertainty principle

$$\Delta E\,\Delta t \geq \frac{\hbar}{2} \qquad (4.34)$$

Equation 4.34 states that the precision with which we can know the energy of some system is limited by the time available for measuring the energy. A common application of the energy–time uncertainty is in calculating the lifetimes of very short-lived subatomic particles whose lifetimes cannot be measured directly, but whose uncertainty in energy or mass can be measured. (See Problem 30.)

A Different View of the Uncertainty Principle

While we have indicated that $\Delta p_x\,\Delta x \geq \hbar/2$ arises from the general theory of forming pulses or wave groups, there is a more concrete physical way to view the origin of the uncertainty principle. We consider certain idealized experiments and show that it is impossible to carry out an experiment that allows the position and momentum of a particle to be simultaneously measured with an accuracy that violates the uncertainty principle. The most famous thought experiment along these lines was introduced by Heisenberg himself and involves the measurement of an electron's position by means of a microscope (Fig. 4.25).

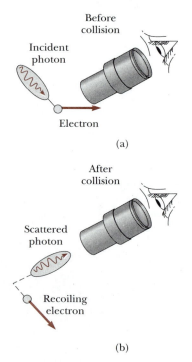

Figure 4.25   A thought experiment for viewing an electron with a powerful microscope. (a) The electron is shown before colliding with the photon. (b) The electron recoils (is disturbed) as a result of the collision with the photon.

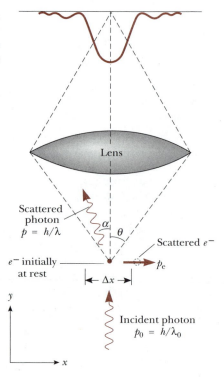

Figure 4.26   The Heisenberg microscope. Conservation of momentum requires $p_{e^-} = (h\sin\alpha)/\lambda$. Because of diffraction by the lens opening, the electron may be anywhere in the region Δx.

Because we know that light can scatter from and perturb the electron, let us minimize this effect by considering the scattering of only one light quantum (Fig. 4.26). To be collected by the lens, the photon may be scattered through any angle ranging from $-\theta$ to $+\theta$, which consequently imparts to the electron an x momentum value ranging from $+(h \sin \theta)/\lambda$ to $-(h \sin \theta)/\lambda$. Thus the uncertainty in the electron's momentum is $\Delta p_x = (2h \sin \theta)/\lambda$. After passing through the lens, the photon lands somewhere on the screen, but the image and consequently the position of the electron is "fuzzy" because the photon is diffracted on passing through the lens aperture. According to physical optics, the smallest distance Δx between two points in an object that will produce separated images in a microscope is given by $\Delta x = \lambda/(2 \sin \theta)$. Here 2θ is the angle subtended by the objective lens, as shown in Figure 4.26.[5] Multiplying the expressions for Δp_x and Δx, we find for the electron

$$\Delta p_x \, \Delta x \approx \left(\frac{2h}{\lambda} \sin \theta \right) \left(\frac{\lambda}{2 \sin \theta} \right) = h$$

in agreement with the uncertainty relation. Note also that this principle is inescapable and relentless! If Δx is reduced by increasing θ, there is an equivalent increase in the uncertainty of the electron's momentum.

Examination of this simple experiment shows several key physical properties that lead to the uncertainty principle:

- The indivisible nature of light particles or quanta. (Nothing less than a single photon can be used!)
- The wave property of light as shown in diffraction.
- The impossibility of predicting or measuring the precise classical path of a single scattered photon and hence of knowing the precise momentum transferred to the electron.[6]

We conclude this section with some examples of the types of calculations that can be done with the uncertainty principle. In the spirit of Fermi or Heisenberg, these "back of the envelope calculations" are surprising for their simplicity and essential description of quantum systems of which the details are unknown.

**EXAMPLE 4.10   The Uncertainty Principle
Changes Nothing for
Macroscopic Objects**

(a) Show that the spread of velocities caused by the uncertainty principle does not have measurable consequences for macroscopic objects (objects that are large compared with atoms) by considering a 100-g racquetball confined to a room 15 m on a side. Assume the ball is moving at 2.0 m/s along the x axis.

Solution

$$\Delta p_x \geq \frac{\hbar}{2 \, \Delta x} = \frac{1.05 \times 10^{-34} \, \text{J} \cdot \text{s}}{2 \times 15 \, \text{m}} = 3.5 \times 10^{-36} \, \text{kg} \cdot \text{m/s}$$

Thus the minimum spread in velocity is given by

$$\Delta v_x = \frac{\Delta p_x}{m} = \frac{3.5 \times 10^{-36} \, \text{kg} \cdot \text{m/s}}{0.100 \, \text{kg}} = 3.5 \times 10^{-35} \, \text{m/s}$$

---

[5] The resolving power of the microscope is treated clearly in F. A. Jenkins and H. E. White, *Fundamentals of Optics*, 4th ed., New York, McGraw-Hill Book Co., 1976, pp. 332–334.

[6] Attempts to measure the photon's position by scattering electrons from it in a Compton process only serve to make its path to the lens more uncertain.

This gives a relative uncertainty of

$$\frac{\Delta v_x}{v_x} = \frac{3.5 \times 10^{-35}}{2.0} = 1.8 \times 10^{-35}$$

which is certainly unmeasurable.

(b) If the ball were to suddenly move along the y axis perpendicular to its well-defined classical trajectory along x, how far would it move in one second? Assume that the ball moves in the y direction with the top speed in the spread Δv_y produced by the uncertainty principle.

Solution   It is important to realize that uncertainty relations hold in the y and z directions as well as in the x direction. This means that $\Delta p_x \Delta x \geq \hbar/2$, $\Delta p_y \Delta y \geq \hbar/2$, and $\Delta p_z \Delta z \geq \hbar/2$ and because all the position uncertainties are equal, all of the velocity spreads are equal. Consequently, we have $\Delta v_y = 3.5 \times 10^{-35}$ m/s and the ball moves 3.5×10^{-35} m in the y direction in one second. This distance is again an immeasurably small quantity, being 10^{-20} times the size of a nucleus!

Exercise 8   How long would it take the ball to move 50 cm in the y direction? (The age of the universe is thought to be 15 billion years, give or take a few billion.)

EXAMPLE 4.11   Do Electrons Exist Within the Nucleus?

Estimate the kinetic energy of an electron confined within a nucleus of size 1.0×10^{-14} m by using the uncertainty principle.

Solution   Taking Δx to be the half width of the confinement length in the equation $\Delta p_x \geq \dfrac{\hbar}{2 \Delta x}$, we have

$$\Delta p_x \geq \frac{6.58 \times 10^{-16} \text{ eV} \cdot \text{s}}{1.0 \times 10^{-14} \text{ m}} \times \frac{3.00 \times 10^8 \text{ m/s}}{c}$$

or

$$\Delta p_x \geq 2.0 \times 10^7 \, \frac{\text{eV}}{c}$$

This means that measurements of the component of momentum of electrons trapped inside a nucleus would range from less than -20 MeV/c to greater than $+20$ MeV/c and that some electrons would have momentum at least as large as 20 MeV/c. Because this appears to be a large momentum, to be safe we calculate the electron's energy relativistically.

$$E^2 = p^2 c^2 + (m_e c^2)^2$$
$$= (20 \text{ MeV}/c)^2 c^2 + (0.511 \text{ MeV})^2 = 400 \text{ (MeV)}^2$$

or

$$E \geq 20 \text{ MeV}$$

Finally, the kinetic energy of an intranuclear electron is

$$K = E - m_e c^2 \geq 19.2 \text{ MeV}$$

Since electrons emitted in radioactive decay of the nucleus (beta decay) have energies much less than 19.5 MeV (about 1 MeV or less) and it is known that no other mechanism could carry off an intranuclear electron's energy during the decay process, we conclude that electrons observed in beta decay do not come from within the nucleus but are actually created at the instant of decay.

EXAMPLE 4.12   The Width of Spectral Lines

Although an excited atom can radiate at any time from $t = 0$ to $t = \infty$, the average time after excitation at which a group of atoms radiates is called the **lifetime**, τ. (a) If $\tau = 1.0 \times 10^{-8}$ s, use the uncertainty principle to compute the line width Δf produced by this finite lifetime.

Solution   We use $\Delta E \Delta t \approx \hbar/2$, where $\Delta E = h \Delta f$, and where $\Delta t = 1.0 \times 10^{-8}$ s is the average time available to measure the excited state. Thus,

$$\Delta f = \frac{1}{4\pi \times 10^{-8} \text{ s}} = 8.0 \times 10^6 \text{ Hz}$$

Note that ΔE is the uncertainty in energy of the excited state. It is also the uncertainty in the energy of the photon emitted by an atom in this state.

(b) If the wavelength of the spectral line involved in this process is 500 nm, find the fractional broadening $\Delta f/f$.

Solution   First, we find the center frequency of this line as follows:

$$f_0 = \frac{c}{\lambda} = \frac{3.0 \times 10^8 \text{ m/s}}{500 \times 10^{-9} \text{ m}} = 6.0 \times 10^{14} \text{ Hz}$$

Hence,

$$\frac{\Delta f}{f_0} = \frac{8.0 \times 10^6 \text{ Hz}}{6.0 \times 10^{14} \text{ Hz}} = 1.3 \times 10^{-8}$$

This narrow natural line width can be seen with a sensitive interferometer. Usually, however, temperature and pressure effects overshadow the natural line width and broaden the line through mechanisms associated with the Doppler effect and collisions.

Exercise 9   Using the nonrelativistic Doppler formula, calculate the Doppler broadening of a 500-nm line emitted by a hydrogen atom at 1000 K. Calculate the broadening by considering the atom to be moving either directly toward or away from an observer with an energy of $\frac{3}{2} k_B T$.

Answer   0.0083 nm, or 0.083 Å.

EXAMPLE 4.13   Measuring the Momentum of an Atom from the Doppler Shift

Instead of concentrating on position measurements for a particle, let us concentrate on measuring the momentum of an atom by using the Doppler shift in order to see if we can get around the uncertainty principle (Fig. 4.27). Assume that a hydrogen atom in an excited state has a lifetime of $\tau = 1.00 \times 10^{-8}$ s and that when the atom is at rest the wavelength radiated is 500 nm. (a) If the Doppler shift in frequency of the light emitted from the moving hydrogen atom is measured to be $\Delta f = 1.00 \times 10^{10}$ Hz, find p and Δp for the atom.

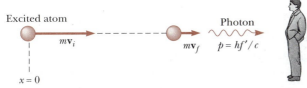

Excited atom

$m\mathbf{v}_i$

$m\mathbf{v}_f$ $p = hf'/c$ Photon

$x = 0$

Figure 4.27   (Example 4.13) Measuring the momentum of an atom from the Doppler shift of emitted radiation.

Solution   The relation between the Doppler shift and speed, v, for $v \ll c$ is

$$\frac{v}{c} = \frac{\Delta f}{f} = \frac{f' - f}{f}$$

where f is the frequency radiated by the atom at rest and f' is the frequency measured when the atom is moving at the speed v. In this case we have

$$\frac{v}{c} = \frac{\Delta f}{f} = \frac{\Delta f}{c/\lambda} = \frac{(1.00 \times 10^{10} \text{ Hz})(500 \times 10^{-9} \text{ m})}{3.00 \times 10^8 \text{ m/s}}$$

or

$$v = 5.00 \times 10^3 \text{ m/s}$$

Hence, the momentum of the hydrogen atom is

$$p = mv = (1.67 \times 10^{-27} \text{ kg})(5.00 \times 10^3 \text{ m/s})$$
$$= 8.35 \times 10^{-24} \text{ kg} \cdot \text{m/s}$$

Note, however, that there is still an uncertainty in v because the measurement of f' is uncertain because the time τ determined the uncertainty in the energy of the excited state, and hence the uncertainty in frequency of the transition. To find the uncertainty in v and hence in p, we differentiate the Doppler shift expression to obtain

$$dv = \frac{c}{f} df'$$

Now, using $\Delta E \, \Delta t \geqslant \hbar/2$ with $\Delta E = h \, \Delta f'$ and $\Delta t = \tau$ gives $\Delta f'$ or $df' \geqslant 1/4\pi\tau$. Substituting this expression for df' into $dv = \frac{c}{f} df'$ gives for the minimum uncertainty in v

$$dv = \frac{c}{4\pi f \tau}$$

Thus, Δp is given by

$$\Delta p = m \, dv = \frac{mc}{4\pi f \tau}$$
$$= \frac{(1.67 \times 10^{-27} \text{ kg})(3.00 \times 10^8 \text{ m/s})}{4\pi(6.00 \times 10^{14} \text{ Hz})(1.00 \times 10^{-8} \text{ s})}$$
$$= 6.65 \times 10^{-27} \text{ kg} \cdot \text{m/s}$$

(b) Estimate the uncertainty in position of the atom as the product of a velocity and a time uncertainty.

Solution   As can be seen in Figure 4.27, the emission of a photon with momentum hf'/c causes an abrupt change Δv in the atom's speed, where $m \, \Delta v = hf'/c$, or

$$\Delta v = \frac{hf'}{mc}$$

Because the precise time of emission of the photon is uncertain, this introduces an uncertainty in the atom's position. Most probably, a photon is emitted between $t = 0$ and $t = 2\tau$ (where τ is the lifetime of the atom), and the atom's position is uncertain by

$$\Delta x = (\Delta v)(\tau) = \frac{hf' \tau}{mc}$$
$$= \frac{(h)(6.00 \times 10^{14} \text{ Hz})(1.00 \times 10^{-8} \text{ s})}{(1.67 \times 10^{-27} \text{ kg})(3.00 \times 10^8 \text{ m/s})}$$
$$= (1.19 \times 10^{25})h = 0.800 \times 10^{-8} \text{ m}$$

(c) Find the value of $\Delta p_x \, \Delta x$ for this example, and verify that the uncertainty relation is satisfied.

Solution

$$\Delta p_x \, \Delta x = (6.65 \times 10^{-27} \text{ kg} \cdot \text{m/s})(0.800 \times 10^{-8} \text{ m})$$
$$= 5.32 \times 10^{-35} \text{ J} \cdot \text{s} \approx \hbar/2.$$

4.6 IF ELECTRONS ARE WAVES, WHAT'S WAVING?

Although we have discussed in some detail the notion of de Broglie matter waves, we have not discussed the precise nature of the field $\Psi(x, y, z, t)$ or wave function that represents the matter waves. We have delayed this discussion because Ψ (Greek letter psi) is rather abstract. Ψ is definitely *not* a measurable disturbance requiring a medium for propagation like a water wave or a sound wave. Instead the stuff that is waving requires no medium. Furthermore, Ψ is in general represented by a complex number and is used to calculate the probability of finding the particle at a given time in a small volume of space. If any of this seems confusing, you should not lose heart, as the nature of the wave function has been confusing people since its invention. It even confused its inventor, Erwin Schrödinger, who incorrectly interpreted $\Psi * \Psi$ as the electric charge density.[7] The great philosopher of the quantum theory, Bohr, immediately objected to this interpretation. Subsequently, Max Born offered the currently accepted statistical view of $\Psi * \Psi$ in late 1926. The confused state of affairs surrounding Ψ at that time was nicely described in a poem by Walter Huckel:

> Erwin with his psi can do
> Calculations quite a few.
> But one thing has not been seen
> Just what does psi really mean?
> (*English translation by Felix Bloch*)

The currently held view is that a particle is described by a function $\Psi(x, y, z, t)$ called the **wavefunction**. The quantity $\Psi * \Psi = |\Psi|^2$ represents the probability per unit volume of finding the particle at a time t in a small volume of space centered on (x, y, z). We will treat methods of finding Ψ in much more detail in Chapter 5, but for now all we require is the idea that *the probability of finding a particle is directly proportional to* $|\Psi|^2$.

4.7 THE WAVE–PARTICLE DUALITY

The Description of Electron Diffraction in Terms of Ψ

In this chapter and previous chapters we have seen evidence for both the wave properties and the particle properties of electrons. Historically, the particle properties were first known and connected with a definite mass, a discrete charge, and detection or localization of the electron in a small region of space. Following these discoveries came the confirmation of the wave nature of electrons in scattering at low energy from metal crystals. In view of these results and because of the everyday experience of seeing the world in terms of *either* grains of sand or diffuse water waves, it is no wonder that we are tempted to simplify the issue and ask, "Well, is the electron a wave or a particle?" The answer is that **electrons are very delicate and rather plastic–they behave like either particles or waves, depending on the kind of experiment**

---

[7] $\Psi*$ represents the complex conjugate of Ψ. Thus, if $\Psi = a + ib$, then $\Psi* = a - ib$. In exponential form, if $\Psi = Ae^{i\theta}$, then $\Psi* = Ae^{-i\theta}$. Note that $\Psi*\Psi = |\Psi|^2$; $a, b, A,$ and θ are all real quantities.

performed on them. In any case, it is impossible to measure both the wave and particle properties simultaneously.[8] The view of Bohr was expressed in an idea known as *complementarity*. As different as they are, both wave and particle views are needed and they complement each other to fully describe the electron. The view of Feynman[9] was that both electrons and photons behave in their own inimitable way. This is like nothing we have seen before, because we do not live at the very tiny scale of atoms, electrons, and photons.

Complementarity

Perhaps the best way to crystallize our ideas about the wave–particle duality is to consider a "simple" double-slit electron diffraction experiment. This experiment highlights much of the mystery of the wave–particle paradox, shows the impossibility of *measuring simultaneously* both wave and particle properties, and illustrates the use of the wavefunction, Ψ, in determining interference effects. A schematic of the experiment with monoenergetic (single wavelength) electrons is shown in Figure 4.28. A parallel beam of electrons falls on a double slit, which has individual openings much smaller than D so that single-slit diffraction effects are negligible. At a distance from the slits much greater than D

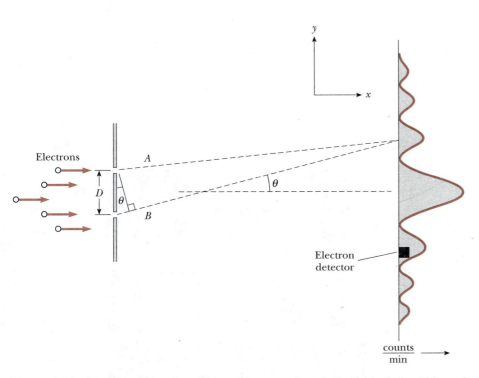

Figure 4.28   Electron diffraction. D is much greater than the individual slit widths and much less than the distance between the slits and the detector.

---

[8] Many feel that the elder Bragg's remark, originally made about light, is a more satisfying answer: electrons behave like waves on Mondays, Wednesdays, and Fridays, like particles on Tuesdays, Thursdays, and Saturdays, and like nothing at all on Sundays.

[9] R. Feynman, *The Character of Physical Law,* Cambridge, Mass., MIT Press, 1982.

is an electron detector capable of detecting individual electrons. It is important to note that the detector always registers discrete particles localized in space and time. In a real experiment this can be achieved if the electron source is weak enough (see Fig. 4.29): *In all cases if the detector collects electrons at different positions for a long enough time, a typical wave interference pattern for the counts per minute or probability of arrival of electrons is found* (see Fig. 4.28). If one imagines a single electron to produce in-phase "wavelets" at the slits, standard wave theory can be used to find the angular separation, θ, of the central probability maximum from its neighboring minimum. The minimum occurs when the path length difference between A and B in Figure 4.28 is half a wavelength, or

$$D \sin \theta = \lambda/2$$

As the electron's wavelength is given by $\lambda = h/p_x$, we see that

$$\sin \theta \approx \theta = \frac{h}{2p_x D} \tag{4.35}$$

for small θ. Thus we can see that the dual nature of the electron is clearly shown in this experiment: *although the electrons are detected as particles at a localized spot at some instant of time, the probability of arrival at that spot is determined by finding the intensity of two interfering matter waves.*

But there is more. What happens if one slit is covered during the experiment? In this case one obtains a symmetric curve peaked around the center of the open slit, much like the pattern formed by bullets shot through a hole in armor plate. Plots of the counts per minute or probability of arrival of electrons

Figure 4.29 Gradual accumulation of interference fringes after the diffraction of increasing number of electrons. (*From P. G. Medi, G. F. Missiroli, and G. Pozzi,* Am J. of Phys. *44:306, 1976*).

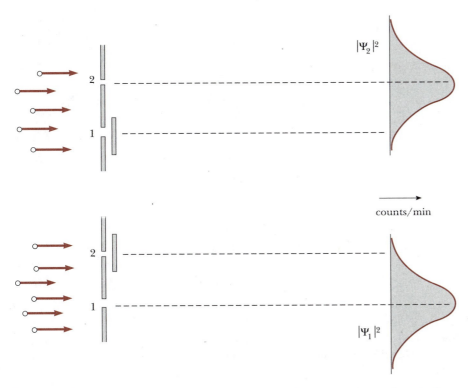

Figure 4.30   The probability of finding electrons at the screen with either the lower or upper slit closed.

with the lower or upper slit closed are shown in Figure 4.30. These are expressed as the appropriate square of the absolute value of some wavefunction, $|\Psi_1|^2 = \Psi_1{}^*\Psi_1$ or $|\Psi_2|^2 = \Psi_2{}^*\Psi_2$, where Ψ_1 and Ψ_2 represent the cases of the electron passing through slit 1 and slit 2, respectively. If an experiment is now performed with slit 2 blocked half of the time and then slit 1 blocked during the remaining time, the accumulated pattern of counts per minute is completely different from the case with both slits open. Note in Figure 4.31 that there is no longer a maximum probability of arrival of an electron at $\theta = 0$. In fact, *the interference pattern has been lost and the accumulated result is simply the sum of the individual results.* The results shown in Figure 4.31 are easier to understand and more reasonable than the interference effects seen with both slits open. When only one slit is open at a time, we know the electron has the same localizability and indivisibility at the slits as we measure at the detector, because the electron clearly goes through slit 1 or slit 2. Thus, the total must be analyzed as the sum of those electrons that come through slit 1, $|\Psi_1|^2$, and those that come through slit 2, $|\Psi_2|^2$. When both slits are open, it is tempting to assume that the electron goes through either slit 1 or slit 2, and that the counts per minute are again given by $|\Psi_1|^2 + |\Psi_2|^2$. We know, however, that the experimental results contradict this. Thus, our assumption that the electron is localized and goes through only one slit when both slits are open must be wrong (a painful conclusion!). Somehow the electron must be simultaneously present at both slits to exhibit interference.

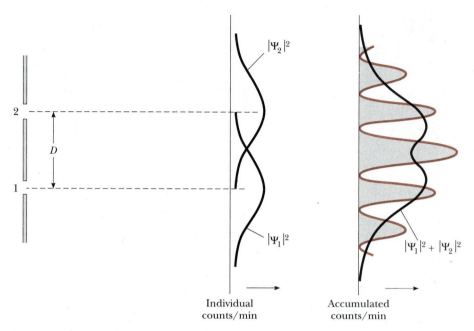

Individual
counts/min

Accumulated
counts/min

Figure 4.31   Accumulated results from the two-slit electron diffraction experiment with each slit closed half the time. For comparison, the results with both slits open are shown in color.

To find the probability of detecting the electron at a particular point on the screen with both slits open, we may say that the electron is in a *superposition state* given by

$$\Psi = \Psi_1 + \Psi_2$$

Thus, the probability of detecting the electron at the screen is equal to the quantity $|\Psi_1 + \Psi_2|^2$ and not $|\Psi_1|^2 + |\Psi_2|^2$. Because matter waves that start out in phase at the slits in general travel different distances to the screen (see Fig. 4.28), Ψ_1 and Ψ_2 will possess a relative phase difference ϕ at the screen. Using a phasor diagram (Fig. 4.32) to find $|\Psi_1 + \Psi_2|^2$ immediately yields

$$|\Psi|^2 = |\Psi_1 + \Psi_2|^2 = |\Psi_1|^2 + |\Psi_2|^2 + 2|\Psi_1||\Psi_2|\cos\phi$$

Figure 4.32   Phasor diagram to represent the addition of two complex wavefunctions, Ψ_1 and Ψ_2, differing in phase by ϕ.

Table 4.1

Case	Wavefunction	Counts/Minute at Screen
Electron is measured to pass through slit 1 or slit 2	Ψ_1 or Ψ_2	$|\Psi_1|^2 + |\Psi_2|^2$
No measurements made on electron at slits	$\Psi_1 + \Psi_2$	$|\Psi_1|^2 + |\Psi_2|^2 + 2|\Psi_1||\Psi_2|\cos\phi$

Note that the term $2|\Psi_1||\Psi_2|\cos\phi$ is an interference term that predicts the interference pattern actually observed in this case. For ease of comparison, a summary of the results found in both cases is given in Table 4.1.

A Thought Experiment: Measuring Through Which Slit the Electron Passes

Another way to view the electron double-slit experiment is to say that the electron passes through the upper or lower slit only when one *measures* the electron to do so. Once one measures unambiguously which slit the electron passes through (yes, you guessed it . . . here comes the uncertainty principle again . . .), the act of measurement disturbs the electron's path enough to destroy the delicate interference pattern.

Let us look again at our two-slit experiment to see in detail how the interference pattern is destroyed.[10] To determine which slit the electron goes

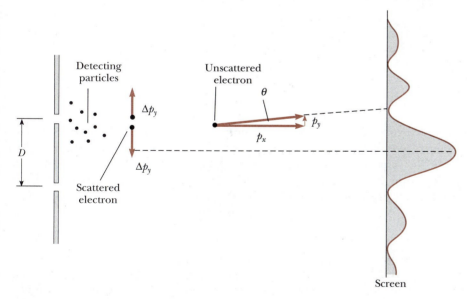

Figure 4.33   A thought experiment to determine through which slit the electron passes.

---

[10] Although we shall use the uncertainty principle in its standard form, it is worth noting that an alternative statement of the uncertainty principle involves this pivotal double-slit experiment: *It is impossible to design any device to determine through which slit the electron passes that will not at the same time disturb the electron and destroy the interference pattern.*

through, imagine that a group of particles is placed right behind the slits as shown in Figure 4.33. If we use the recoil of a small particle to determine which slit the electron goes through, we must have the uncertainty in the detecting particle's position, $\Delta y \ll D$. Also, during the collision the detecting particle suffers a change in momentum Δp_y equal and opposite to the change in momentum experienced by the electron, as shown in Figure 4.33. An undeviated electron landing at the first minimum *and producing an interference pattern* has

$$\tan \theta \approx \theta = \frac{p_y}{p_x} = \frac{h}{2p_x D}$$

from Equation 4.35. Thus, we require that an electron scattered by a detecting particle have

$$\frac{\Delta p_y}{p_x} \ll \theta = \frac{h}{2p_x D}$$

or

$$\Delta p_y \ll \frac{h}{2D}$$

if the interference pattern is not to be distorted. Because the change in momentum of the scattered electron is equal to the change in momentum of the detecting particle, $\Delta p_y \ll h/2D$ also applies to the detecting particle. Thus, we have for the detecting particle

$$\Delta p_y \, \Delta y \ll \frac{h}{2D} \cdot D$$

or

$$\Delta p_y \, \Delta y \ll \frac{h}{2}$$

This is a clear violation of the uncertainty principle. *Hence we get the result that the small uncertainties needed both to see the interference and to know which slit the electron goes through are impossible, because they violate the uncertainty principle.* If Δy is small enough to determine which slit the electron goes through, Δp_y is so large that electrons heading for the first minimum are scattered into adjacent maxima and the interference pattern is destroyed.

Exercise 12   In a real experiment it is likely that some electrons would miss the detecting particles. Thus, we would really have two categories of electrons arriving at the detector: those measured to pass through a definite slit and those not observed, or just missed, at the slits. In this case what kind of pattern of counts per minute would be accumulated by the detector?

Answer   A mixture of an interference pattern $|\Psi_1 + \Psi_2|^2$ (those not measured) and $|\Psi_1|^2 + |\Psi_2|^2$ (those measured) would result.

4.8   A FINAL NOTE

Scientists once viewed the world as being made up of distinct and unchanging parts that interact according to strictly deterministic laws of cause and effect. In the classical limit this is fundamentally correct because classical processes involve so many quanta that deviations from the average are imperceptible. At the atomic level, however, we have seen that a given piece of matter (an electron, say) is not a distinct and unchanging part of the universe obeying completely deterministic laws. Such a particle exhibits wave properties when it interacts with a metal crystal and particle properties a short while later when it registers on a Geiger counter. Thus, rather than viewing the electron as a distinct and separate part of the universe with an intrinsic particle nature, we are led to the view that the electron and indeed all particles are changing entities possessing the potential to cycle endlessly between wave and particle behavior. We also find that it is much more difficult to separate the object measured from the measuring instrument at the atomic level, because the type of measuring instrument determines whether wave properties or particle properties are observed.

SUMMARY

Every lump of matter of mass m and momentum p has wavelike properties with wavelength given by the de Broglie relation

$$\lambda = \frac{h}{p} \tag{4.1}$$

By applying this wave theory of matter to electrons in atoms, de Broglie was able to explain the appearance of integers in certain Bohr orbits as a natural consequence of electron wave interference. In 1927, Davisson and Germer demonstrated directly the wave nature of electrons by showing that low-energy electrons were diffracted by single crystals of nickel. In addition, they confirmed Equation 4.1.

Although the wavelength of matter waves can be experimentally determined, it is important to understand that they are not just like other waves because their frequency and phase velocity cannot be directly measured. In particular, the phase velocity of an individual matter wave is greater than the velocity of light and varies with wavelength or wavenumber

$$v_{\mathrm{p}} = f\lambda = \left(\frac{E}{h}\right)\left(\frac{h}{p}\right) = c\left[1 + \left(\frac{mc}{\hbar k}\right)^2\right]^{1/2} \tag{4.24}$$

To represent a particle properly, a superposition of matter waves with different wavelengths, amplitudes, and phases must be chosen to interfere constructively over a limited region of space. The resulting wave packet or group can then be shown to travel with the same speed as the classical particle. In addition, a wave packet localized in a region Δx contains a range of wavenumbers Δk, where $\Delta x \, \Delta k \geqslant \frac{1}{2}$. Because $p_x = \hbar k$, this implies that there is an uncertainty principle for position and momentum:

$$\Delta p_x \Delta x \geq \frac{\hbar}{2} \tag{4.31}$$

In a similar fashion one can show that an energy–time uncertainty relation exists given by

$$\Delta E \Delta t \geq \frac{\hbar}{2} \tag{4.34}$$

In quantum mechanics matter waves are represented by a wavefunction $\Psi(x, y, z, t)$. The probability of finding a particle represented by Ψ in a small volume centered at (x, y, z) at time t is proportional to $|\Psi|^2$. The wave–particle duality of electrons may be seen by considering the passage of electrons through two narrow slits and their arrival at a viewing screen. We find that although the electrons are detected as particles at a localized spot on the screen, the probability of arrival at that spot is determined by finding the intensity of two interfering matter waves.

Although we have seen the importance of matter waves or wavefunctions in this chapter, we have provided no method of finding Ψ for a given physical system. In the next chapter we introduce the Schrödinger wave equation. The solutions to this important differential equation will provide us with the wavefunctions for a given system.

SUGGESTIONS FOR FURTHER READING

1. David Bohm, *Quantum Theory*, Englewood Cliffs, N.J., Prentice-Hall, 1951. Chapters 3 and 6 in this book give an excellent account of wave packets and the wave–particle duality of matter at a more advanced level.
2. Richard Feynman, *The Character of Physical Law*, Cambridge, Mass., The MIT Press, 1982, Chapter 6. This monograph is an incredibly lively and readable treatment of the double-slit experiment presented in Feynman's inimitable fashion.
3. Banesh Hoffman, *The Strange Story of the Quantum*, New York, Dover Publications, 1959. This short book is a beautifully written nonmathematical discussion of the history of quantum mechanics.

QUESTIONS

1. Is light a wave or a particle? Support your answer by citing specific experimental evidence.
2. Is an electron a particle or a wave? Support your answer by citing some experimental results.
3. An electron and a proton are accelerated from rest through the same potential difference. Which particle has the longer wavelength?
4. If matter has a wave nature, why is this wave-like character not observable in our daily experiences?
5. In what ways does Bohr's model of the hydrogen atom violate the uncertainty principle?
6. Why is it impossible to measure the position and speed of a particle simultaneously with infinite precision?
7. Suppose that a beam of electrons is incident on three or more slits. How would this influence the interference pattern? Would the state of an electron depend on the number of slits? Explain.
8. In describing the passage of electrons through a slit and arriving at a screen, Feynman said that "electrons arrive in lumps, like particles, but the probability of arrival of these lumps is determined as the intensity of the waves would be. It is in this sense that the electron behaves

sometimes like a particle and sometimes like a wave." Elaborate on this point in your own words. (For a further discussion of this point, see R. Feynman, *The Character of Physical Law*, Cambridge, Mass., MIT Press, 1982, Chapter 6.)

9. Do you think that most major experimental discoveries are made by careful planning or by accident? Cite examples.

10. In the case of accidental discoveries, what traits must the experimenter possess to capitalize on the discovery?

11. Are particles even things? An extreme view of the plasticity of electrons and other particles is expressed in this famous quote of Heisenberg: "The invisible elementary particle of modern physics does not have the property of occupying space any more than it has properties like color or solidity. Fundamentally, it is not a material structure in space and time but only a symbol that allows the laws of nature to be expressed in especially simple form."

Are you satisfied with viewing science as a set of predictive rules or do you prefer to see science as a description of an objective world of things, in the case of particle physics, tiny, scaled-down things? What problems are associated with each point of view?

PROBLEMS

4.1 The Pilot Waves of De Broglie

1. Calculate the de Broglie wavelength for a proton moving with a speed of 10^6 m/s.

2. Calculate the de Broglie wavelength for an electron with kinetic energy (a) 50 eV and (b) 50 keV.

3. Calculate the de Broglie wavelength of a 74-kg person who is running at a speed of 5.0 m/s.

4. The "seeing" ability, or resolution, of radiation is determined by its wavelength. If the size of an atom is of the order of 0.1 nm, how fast must an electron travel to have a wavelength small enough to "see" an atom?

5. To "observe" small objects, one measures the diffraction of particles whose de Broglie wavelength is approximately equal to the object's size. Find the kinetic energy (in electron volts) required for electrons to resolve (a) a large organic molecule of size 10 nm, (b) atomic features of size 0.10 nm, and (c) a nucleus of size 10 fm. Repeat these calculations using alpha particles in place of electrons.

6. An electron and a proton each have kinetic energy equal to 50 keV. What are their de Broglie wavelengths?

7. Calculate the de Broglie wavelength of a proton that is accelerated through a potential difference of 10 MV.

8. Show that the de Broglie wavelength of an electron accelerated from rest through a small potential difference V is given by $\lambda = 1.226/\sqrt{V}$, where λ is in nanometers and V is in volts.

9. Find the de Broglie wavelength of a ball of mass 0.20 kg just before it strikes the Earth after being dropped from a building 50 m tall.

10. An electron has a de Broglie wavelength equal to the diameter of the hydrogen atom. What is the kinetic energy of the electron? How does this energy compare with the ground-state energy of the hydrogen atom?

11. For an electron to be confined to a nucleus, its de Broglie wavelength would have to be less than 10^{-14} m. (a) What would be the kinetic energy of an electron confined to this region? (b) On the basis of this result, would you expect to find an electron in a nucleus? Explain.

12. Through what potential difference would an electron have to be accelerated to give it a de Broglie wavelength of 1.00×10^{-10} m?

4.2 The Davisson–Germer Experiment

13. Figure P4.13 shows the top three planes of a crystal with planar spacing d. If $2d \sin \theta = 1.01\lambda$ for the two waves shown, and high-energy electrons of wavelength λ penetrate many planes deep into the crystal, which

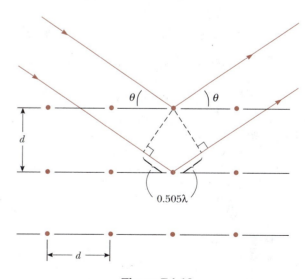

Figure P4.13

atomic plane produces a wave that cancels the surface reflection? This is an example of how extremely narrow maxima in high-energy electron diffraction are formed—that is, there are no diffracted beams unless $2d \sin \theta$ is equal to an integral number of wavelengths.

14. (a) Show that the formula for Low Energy Electron Diffraction (LEED) when electrons are incident perpendicular to a crystal surface may be written as

$$\sin \phi = \frac{nhc}{d(2m_e c^2 K)^{1/2}}$$

where n is the order of the maximum, d is the atomic spacing, m_e is the electron mass, K is the electron's kinetic energy, and ϕ is the angle between the incident and diffracted beams. (b) Calculate the atomic spacing in a crystal that has consecutive diffraction maxima at $\phi = 24.1°$ and $\phi = 54.9°$ for 100-eV electrons.

4.3   Wave Groups and Dispersion

15. Show that the group velocity for a nonrelativistic free electron is also given by $v_g = p/m_e = v_0$, where v_0 is the electron's velocity.

16. When a pebble is tossed into a pond, a circular wave pulse propagates outward from the disturbance. If you are alert (and it's not a sleepy afternoon in late August), you will see a fine structure in the pulse consisting of surface ripples moving inward through the circular disturbance. Explain this effect in terms of group and phase velocity if the phase velocity of ripples is given by $v_p = \sqrt{2\pi S/\lambda \rho}$, where S is the surface tension and ρ is the density of the liquid.

17. The dispersion relation for free relativistic electron waves is

$$\omega(k) = \sqrt{c^2 k^2 + (m_e c^2/\hbar)^2}$$

Obtain expressions for the phase velocity v_p and group velocity v_g of these waves and show that their product is a constant, independent of k. From your result, what can you conclude about v_g if $v_p > c$?

18. **MP** Use the MP DESKTOP to superimpose two harmonic waves of equal amplitude having wavenumbers $k = 20$ and $k = 25$ and dispersion relation $\omega(k) = |k|$. Animate the resulting waveform and describe its evolution. Compare the phase and group velocities for this case. Refer to Example 4.6 for particulars on using the DESKTOP for such applications.

19. **MP** Set up the MP DESKTOP to display a single harmonic wave with wavenumber $k = 10$ and frequency $\omega = 100$. (*Hint:* Load the frequency file *EX 4-6.W*, and edit the interval limits and function ex-

pression as needed.) Measure the (phase) velocity of this wave, and compare with the predicted value $v_p = \omega/k$. Investigate the related case where the sign of the wavenumber is reversed but the frequency is unchanged.

20. **MP** *Standing waves* are formed by mixing pairs of waves with the same frequency and wavenumbers of equal magnitude but opposite sign. Set up the MP DESKTOP to display a standing wave formed from two traveling waves with frequency $\omega = 100$, wavenumbers $k = \pm 10$, and amplitudes $a(k) = 1$. The appropriate frequency and amplitude data can be found in the files *PR4-20.W* and *UNIFORM.A*, respectively. Animate the wave to see how it evolves, and measure its period. (*Hint:* For improved accuracy, reduce the time step to 0.001 in the *Control Panel* dialog box available from the *Options* menu.) Now mix the same two traveling waves with slightly different amplitudes, say 0.90 and 1.10 (take amplitudes from the file *STANDING.A*). How does this new wave evolve? Such "galloping" waves were first noticed using computer simulations (see W. G. Harter, *CIP*, Sept/Oct 1991).

21. **MP** Use the MP DESKTOP to simulate the propagation of deep-water waves, as discussed in Example 4.5. What dispersion relation $\omega(k)$ is appropriate to this application? Fashion a wave packet from the amplitude function $a(k) = 1$ over the range of wavenumbers from $k = 0$ to $k = 10$ rad/m, using 1 rad/m for the wavenumber increment (a mix of 11 waves). Describe the evolution of this packet, and convince yourself that the phase velocity exceeds the group velocity for this case.

4.5   The Heisenberg Uncertainty Principle

22. A ball of mass 50 g moves with a speed of 30 m/s. If its speed is measured to an accuracy of 0.1%, what is the minimum uncertainty in its position?

23. A proton has a kinetic energy of 1.0 MeV. If its momentum is measured with an uncertainty of 5.0%, what is the minimum uncertainty in its position?

24. We wish to measure simultaneously the wavelength and position of a photon. Assume that the wavelength measurement gives $\lambda = 6000$ Å with an accuracy of one part in a million, i.e., $\Delta\lambda/\lambda = 10^{-6}$. What is the minimum uncertainty in the position of the photon?

25. A woman on a ladder drops small pellets toward a spot on the floor. (a) Show that, according to the uncertainty principle, the miss distance must be at least

$$\Delta x = \left(\frac{\hbar}{2m}\right)^{1/2} \left(\frac{H}{2g}\right)^{1/4}$$

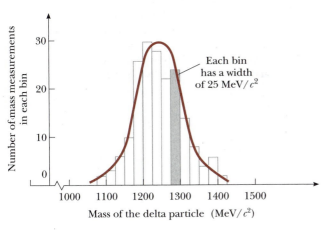

Figure P4.30 Histogram of mass measurements of the delta particle.

where H is the initial height of each pellet above the floor and m is the mass of each pellet. (b) If $H = 2.0$ m and $m = 0.50$ g, what is Δx?

26. A beam of electrons is incident on a slit of variable width. If it is possible to resolve a 1% difference in momentum, what slit width would be necessary to resolve the interference pattern of the electrons if their kinetic energy is (a) 0.010 MeV, (b) 1.0 MeV, and (c) 100 MeV?

27. Suppose Fuzzy, a quantum-mechanical duck, lives in a world in which $h = 2\pi$ J·s. Fuzzy has a mass of 2.0 kg and is initially known to be within a region 1.0 m wide. (a) What is the minimum uncertainty in his speed? (b) Assuming this uncertainty in speed to prevail for 5.0 s, determine the uncertainty in position after this time.

28. An electron of momentum p is at a distance r from a stationary proton. The system has a kinetic energy $K = p^2/2m_e$ and potential energy $U = -ke^2/r$. Its total energy is $E = K + U$. If the electron is bound to the proton to form a hydrogen atom, its average position is at the proton but the uncertainty in its position is approximately equal to the radius r of its orbit. The electron's average momentum will be zero, but the uncertainty in its momentum will be given by the uncertainty principle. Treat the atom as a one-dimensional system in the following: (a) Estimate the uncertainty in the electron's momentum in terms of r. (b) Estimate the electron's kinetic, potential, and total energies in terms of r. (c) The actual value of r is the one that minimizes the total energy, resulting in a stable atom. Find that value of r and the resulting total energy. Compare your answer with the predictions of the Bohr theory.

29. An excited nucleus with a lifetime of 0.100 ns emits a γ-ray of energy 2.00 MeV. Can the energy width (uncertainty in energy, ΔE) of this 2.00-MeV γ emission line be directly measured if the best gamma detectors can measure energies to ±5 eV?

30. Typical measurements of the mass of a subatomic delta particle ($m \approx 1230$ MeV/c^2) are shown in Figure P4.30. Although the lifetime of the delta is much too short to measure directly, it can be calculated from the energy–time uncertainty principle. Estimate the lifetime from the full width at half maximum of the mass measurement distribution shown.

4.7 The Wave–Particle Duality

31. A monoenergetic beam of electrons is incident on a single slit of width 0.50 nm. A diffraction pattern is formed on a screen 20 cm from the slit. If the distance between successive minima of the diffraction pattern is 2.1 cm, what is the energy of the incident electrons?

32. A neutron beam with a selected speed of 0.40 m/s is directed through a double slit with a 1.0-mm separation. An array of detectors is placed 10 m from the slit. (a) What is the de Broglie wavelength of the neutrons? (b) How far off axis is the first zero-intensity point on the detector array? (c) Can we say which slit any particular neutron passed through? Explain.

33. A two-slit electron diffraction experiment is done with slits of *unequal* widths. When only slit 1 is open, the number of electrons reaching the screen per second is 25 times the number of electrons reaching the screen per second when only slit 2 is open. When both slits are open, an interference pattern results in which the destructive interference is not complete. Find the ratio of the probability of an electron arriving at an interference maximum to the probability of an electron arriving at an adjacent interference minimum. (*Hint:* Use the superposition principle.)

Additional Problems

34. Robert Hofstadter won the 1961 Nobel prize in physics for his pioneering work in scattering 20-GeV electrons from nuclei. (a) What is the γ-factor for a 20-GeV electron, where $\gamma = (1 - v^2/c^2)^{-1/2}$? What is the momentum of the electron in kg·m/s? (b) What is the wavelength of a 20-GeV electron and how does it compare with the size of a nucleus?

35. An air rifle is used to shoot 1.0-g particles at 100 m/s through a hole of diameter 2.0 mm. How far from the rifle must an observer be to see the beam spread by 1.0 cm because of the uncertainty principle? Compare this answer with the diameter of the Universe (2×10^{26} m).

36. An atom in an excited state 1.8 eV above the ground state remains in that excited state 2.0 μs before moving to the ground state. Find (a) the frequency of the emitted photon, (b) its wavelength, and (c) its approximate uncertainty in energy.

37. A π^0 meson is an unstable particle produced in high-energy particle collisions. It has a mass–energy equivalent of about 135 MeV, and it exists for an average lifetime of only 8.7×10^{-17} s before decaying into two gamma rays. Using the uncertainty principle, estimate the fractional uncertainty $\Delta m/m$ in its mass determination.

5

Quantum Mechanics in One Dimension

Chapter Outline

We have seen that associated with any particle is a matter wave called the *wavefunction.* How this wavefunction affects our description of a particle and its behavior is the subject of *quantum mechanics,* or *wave mechanics.* This scheme, developed from 1925 to 1926 by Schrödinger, Heisenberg, and others, makes it possible to understand a host of phenomena involving elementary particles, atoms, molecules, and solids. In this and subsequent chapters, we shall describe the basic features of wave mechanics and its application to simple systems. The relevant concepts for particles confined to motion along a line (the x-axis) are developed in the present chapter.

5.1 THE BORN INTERPRETATION

The wavefunction Ψ contains within it all the information that can be known about the particle. That basic premise forms the cornerstone of our investigation: one of our objectives will be to discover how information may be extracted from the wavefunction; the other, to learn how to obtain this wavefunction for a given system.

The currently held view connects the wavefunction Ψ with *probabilities* in the manner first proposed by Max Born in 1925:

The probability that a particle will be found in the infinitesimal interval dx about the point x, denoted by $P(x)\ dx$, is

Born interpretation of Ψ

185

ax Born was a German theoretical physicist who made major contributions in many areas of physics, including relativity, atomic and solid-state physics, matrix mechanics, the quantum mechanical treatment of particle scattering ("Born approximation"), the foundations of quantum mechanics (Born interpretation of Ψ), optics, and the kinetic theory of liquids. Born received the doctorate in physics from the University of Göttingen in 1907, and he acquired an extensive knowledge of mathematics as the private assistant to the great German mathematician David Hilbert. This strong mathematical background proved a great asset when he was quickly able to reformulate Heisenberg's quantum theory in a more consistent way with matrices.

In 1921, Born was offered a post at the University of Göttingen, where he helped build one of the strongest physics centers of the twentieth century. This group consisted, at one time or another, of the mathematicians Hilbert, Courant, Klein, and Runge and the physicists Born, Jordan, Heisenberg, Franck, Pohl, Heitler, Herzberg, Nordheim, and Wig-

B I O G R A P H Y

—

MAX BORN
(1882–1970)

ner, among others. In 1926, shortly after Schrödinger's publication of wave mechanics, Born applied Schrödinger's methods to atomic scattering and developed the Born approximation method for carrying out calculations of the probability of scattering of a particle into a given solid angle. This work furnished the basis for Born's startling (in 1926) interpretation of $|\Psi|^2$ as the probability density. For this so-called statistical interpretation of $|\Psi|^2$ he was awarded the Nobel prize in 1954.

Fired by the Nazis, Born left Germany in 1933 for Cambridge and eventually the University of Edinburgh, where he again became the leader of a large group investigating the statistical mechanics of condensed matter. In his later years Born campaigned against atomic weapons, wrote an autobiography, and translated German humorists into English.

(AIP Emilio Segrè Visual Archives)

$$P(x)\ dx = |\Psi(x, t)|^2\ dx \tag{5.1}$$

Therefore, although it is not possible to specify with certainty the location of a particle, it is possible to assign *probabilities* for observing it at any given position. The quantity $|\Psi|^2$, the square of the absolute value of Ψ, represents the *intensity* of the matter wave and is computed as the product of Ψ with its complex conjugate, that is, $|\Psi|^2 = \Psi^*\Psi$. Notice that Ψ itself is not a measurable quantity; however, $|\Psi|^2$ is measurable and is just the probability per unit length, or **probability density** $P(x)$, for finding the particle at the point x at time t. For example, the intensity distribution in a light diffraction pattern is a measure of the probability that a photon will strike a given point within the pattern. Because of its relation to probabilities, we insist that $\Psi(x, t)$ be a *single-valued and continuous function of x* and t, so that no ambiguities can arise concerning the predictions of the theory. The wavefunction Ψ also should be *smooth,* a condition that will be elaborated later as it is needed.

Because the particle must be somewhere along the x axis, the sum of the probabilities over all values of x must be 1:

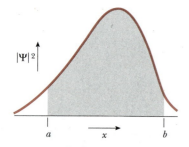

Figure 5.1 The probability for a particle to be in the interval $a \leqslant x \leqslant b$ is the area under the curve from a to b of the probability density function $|\Psi(x, t)|^2$.

$$\int_{-\infty}^{\infty} |\Psi(x, t)|^2\ dx = 1 \tag{5.2}$$

Any wavefunction satisfying Equation 5.2 is said to be *normalized*. Normalization is simply a statement that the particle exists at all times. The probability of finding the particle in any finite interval $a \leqslant x \leqslant b$ is

$$P = \int_a^b |\Psi(x, t)|^2 \, dx \qquad (5.3)$$

That is, the probability is just the area included under the curve of probability density between the points $x = a$ and $x = b$ (Fig. 5.1).

EXAMPLE 5.1   Normalizing the Wavefunction

The initial wavefunction of a particle is given as $\Psi(x, 0) = C \exp(-|x|/x_0)$, where C and x_0 are constants. Sketch this function. Find C in terms of x_0 such that $\Psi(x, 0)$ is normalized.

Solution   The given wavefunction is symmetric, decaying exponentially from the origin in either direction as shown in Figure 5.2. The *decay length* x_0 represents the distance over which the wave amplitude is diminished by the factor $1/e$ from its maximum value $\Psi(0, 0) = C$.

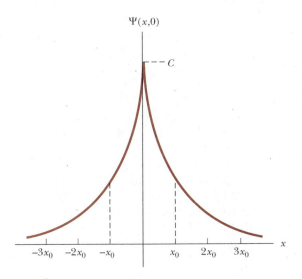

Figure 5.2   (Example 5.1) The symmetric wavefunction $\Psi(x, 0) = C \exp(-|x|/x_0)$. At $x = \pm x_0$ the wave amplitude is down by the factor $1/e$ from its peak value $\Psi(0, 0) = C$. C is a normalizing constant whose proper value is $C = 1/\sqrt{x_0}$.

The normalization requirement is

$$1 = \int_{-\infty}^{\infty} |\Psi(x, 0)|^2 \, dx = C^2 \int_{-\infty}^{\infty} e^{-2|x|/x_0} \, dx$$

Because the integrand is unchanged when x changes sign (it is an even function), we may evaluate the integral over the whole axis as twice that over the half axis $x > 0$, where $|x| = x$. Then

$$1 = 2C^2 \int_0^{\infty} e^{-2x/x_0} \, dx = 2C^2 \left(\frac{x_0}{2} \right) = C^2 x_0$$

Thus, we must take $C = 1/\sqrt{x_0}$ for normalization.

EXAMPLE 5.2   Calculating Probabilities

Calculate the probability that the particle in the preceding example will be found in the interval $-x_0 \leq x \leq x_0$.

Solution   The probability is the area under the curve of $|\Psi(x, 0)|^2$ from $-x_0$ to $+x_0$, and is obtained by integrating the probability density over the specified interval:

$$P = \int_{-x_0}^{x_0} |\Psi(x, 0)|^2 \, dx = 2 \int_0^{x_0} |\Psi(x, 0)|^2 \, dx$$

where the second step follows because the integrand is an even function, as discussed in Example 5.1. Thus,

$$P = 2C^2 \int_0^{x_0} e^{-2x/x_0} \, dx = 2C^2 \, (x_0/2)(1 - e^{-2})$$

Substituting $C = 1/\sqrt{x_0}$ into this expression gives for the probability $P = 1 - e^{-2} = 0.8647$, or about 86.5%, independent of x_0.

The fundamental problem of quantum mechanics is this: given the wavefunction at some initial instant, say $t = 0$, to find the wavefunction at any subsequent time t. The wavefunction $\Psi(x, 0)$ represents the initial information that must be specified; once this is known, however, the wave propagates according to prescribed laws of nature.

Because it describes how a given system evolves, quantum mechanics is a dynamical theory much like Newtonian mechanics. There are, of course, important differences. In Newton's mechanics, the state of a particle at $t = 0$ is specified by giving its initial position $x(0)$ and velocity $v(0)$ —just two numbers; quantum mechanics demands an entire wavefunction $\Psi(x, 0)$ —an infinite set of numbers corresponding to the wavefunction value at every point x. But both theories describe how this state changes with time when the forces acting on the particle are known. In Newton's mechanics $x(t)$ and $v(t)$ are calculated from Newton's second law; in quantum mechanics $\Psi(x, t)$ must be calculated from another law— *The Schrödinger equation.*

5.2 WAVEFUNCTION FOR A FREE PARTICLE

A free particle is one subject to no force. This special case can be studied using prior assumptions without recourse to the Schrödinger equation. The development underscores the role of the initial conditions in quantum physics.

The wavenumber k and frequency ω of free particle matter waves are given by the de Broglie relations

$$k = p/\hbar \qquad \text{and} \qquad \omega = E/\hbar \tag{5.4}$$

For nonrelativistic particles ω is related to k as

$$\omega(k) = \frac{\hbar k^2}{2m} \tag{5.5}$$

which follows from the classical connection $E = p^2/2m$ between the energy E and momentum p for a free particle.[1]

For the wavefunction itself, we should take

Plane wave representation for a free particle

$$\Psi_k(x, t) = Ae^{i(kx - \omega t)} = A\{\cos(kx - \omega t) + i\sin(kx - \omega t)\} \tag{5.6}$$

where $i = \sqrt{-1}$ is the imaginary unit. This is an oscillation with wavenumber k, frequency ω, and amplitude A. Because the variables x and t occur only in the combination $kx - \omega t$, the oscillation is a *traveling wave,* as befits a free particle in motion. Further, the particular combination expressed by Equation 5.6 is that of a plane wave,[2] for which the probability density $|\Psi|^2 (= A^2)$ is *uni-*

---

[1] *The functional* form for $\omega(k)$ was discussed in Section 4.3 for relativistic particles, where $E = \sqrt{(cp)^2 + (mc^2)^2}$. In the nonrelativistic case ($v \ll c$) this reduces to $E = p^2/2m + mc^2$. The rest energy $E_0 = mc^2$ can be disregarded in this case if we agree to make E_0 our energy reference. By measuring all energies from this level, we are in effect setting E_0 equal to zero.

[2] For a plane wave, the wave fronts (points of constant phase) constitute planes perpendicular to the direction of wave propagation. In the present case the constant phase requirement $kx - \omega t = constant$ demands only that x be fixed, so the wave fronts occupy the y-z planes.

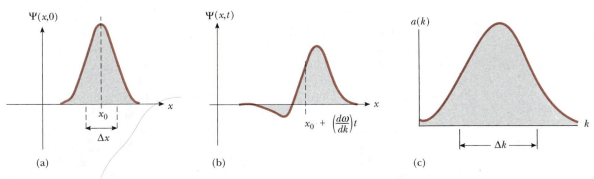

Figure 5.3   (a) A wave packet $\Psi(x, 0)$ formed from a superposition of plane waves. (b) The same wave packet some time t later (real part only). Because $v_p = \omega/k = \hbar k/2m$, the plane waves with smaller wavenumber move at slower speeds, and the packet becomes distorted. The body of the packet propagates with the group speed $d\omega/dk$ of the plane waves. (c) The spectral content $a(k)$ for this packet, indicating the amplitude of each plane wave in the superposition. A narrow wave packet requires a broad spectral content, and vice versa. That is, the widths Δx and Δk are inversely related as $\Delta x\,\Delta k \approx 1$.

form. That is, the probability of finding this particle in any interval of the x axis is the same as that for any other interval of equal length and does not change with time. The plane wave is the simplest traveling waveform with this property—it expresses the reasonable notion that there are no special places for a free particle to be found. The particle's location is completely unknown at $t = 0$ and remains so for all time; however, its momentum and energy are known precisely as $p = \hbar k$ and $E = \hbar\omega$, respectively.

But not *all* free particles are described by Equation 5.6. For instance, we may establish (by measurement) that our particle initially is in some range Δx about x_0. In that case, $\Psi(x, 0)$ must be a *wave packet* concentrated in this interval, as shown in Figure 5.3a. The plane wave description is inappropriate now because the initial wave shape is not given correctly by $\Psi_k(x, 0) = e^{ikx}$. Instead, a *sum* of plane waves with different wavenumbers must be used to represent the packet. Because k is unrestricted, the sum actually is an integral here and we write

$$\Psi(x, 0) = \int_{-\infty}^{\infty} a(k)\, e^{ikx}\, dk \qquad (5.7)$$

Representing a particle with a wave group

The coefficients $a(k)$ specify the amplitude of the plane wave with wavenumber k in the mixture and are chosen to reproduce the initial wave shape. For a given $\Psi(x, 0)$, the required $a(k)$ can be found from the theory of Fourier integrals. We shall not be concerned with the details of this analysis here; the essential point is that it can be done for a packet of any shape (see optional Section 4.4). If each plane wave constituting the packet is assumed to propagate independently of the others according to Equation 5.6, the packet at any time t is

$$\Psi(x, t) = \int_{-\infty}^{\infty} a(k)\, e^{i\{kx - \omega(k)t\}}\, dk \qquad (5.8)$$

Notice that the initial data are used only to establish the amplitudes $a(k)$; subsequently, the packet develops according to the evolution of its plane wave constituents. Because each of these constituents moves with a different velocity $v_p = \omega/k$ (the phase velocity), we have dispersion (see Section 4.3) and *the packet changes its shape as it propagates* (Fig. 5.3b). The speed of propagation is given by the group velocity $d\omega/dk$ of the plane waves forming the packet. It is also clear that Equation 5.8 no longer describes a particle with precise values of momentum and energy. To construct a wave packet (that is, localize the particle), a mixture of wavenumbers (hence, particle momenta) is necessary, as indicated by the different $a(k)$. Together, the $a(k)$ furnish the so-called **spectral content** of the packet, which might look like that sketched in Figure 5.3c. The narrower the desired packet $\Psi(x, 0)$, the broader is the function $a(k)$ needed to represent that packet. If Δx denotes the packet width and Δk the extent of the corresponding $a(k)$, one finds that the product always is a number of order unity, that is, $\Delta x \, \Delta k \approx 1$. Together with $p = \hbar k$ this implies an uncertainty principle:

$$\Delta x \, \Delta p_x \sim \hbar \tag{5.9}$$

In closing, we point out that Equations 5.7 and 5.8 solve the fundamental problem of quantum mechanics for *free* particles subject to *any* initial condition $\Psi(x, 0)$.

EXAMPLE 5.3   Constructing a Wave Packet

Find the wavefunction $\Psi(x, 0)$ that results from taking the function $a(k) = (C\alpha/\sqrt{\pi}) \exp(-\alpha^2 k^2)$, where C and α are constants. Estimate the product $\Delta x \, \Delta k$ for this case.

Solution   The function $\Psi(x, 0)$ is given by the integral of Equation 5.7 or

$$\Psi(x, 0) = \int_{-\infty}^{\infty} a(k) e^{ikx} \, dk = \frac{C\alpha}{\sqrt{\pi}} \int_{-\infty}^{\infty} e^{(ikx - \alpha^2 k^2)} \, dk$$

To evaluate the integral, we first complete the square in the exponent as

$$ikx - \alpha^2 k^2 = -\left(\alpha k - \frac{ix}{2\alpha}\right)^2 - \frac{x^2}{4\alpha^2}$$

The second term on the right is constant for the integration over k; to integrate the first term we change variables with the substitution $z = \alpha k - ix/2\alpha$, obtaining

$$\Psi(x, 0) = \frac{C}{\sqrt{\pi}} e^{-x^2/4\alpha^2} \int_{-\infty}^{\infty} e^{-z^2} \, dz$$

The integral now is a standard one whose value is known to be $\sqrt{\pi}$. Then

(a)

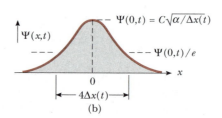

(b)

Figure 5.4   (Example 5.3) (a) The gaussian wavefunction $\Psi(x, 0) = C \exp\{-(x/2\alpha)^2\}$, representing a particle initially localized around $x = 0$. C is the amplitude. At $x = \pm 2\alpha$ the amplitude is down from its maximum value by the factor $1/e$; accordingly, α is identified as the width of the gaussian, $\alpha = \Delta x$. (b) The gaussian wavefunction of Figure 5.4a at time t (apart from a phase factor). The width has increased to $\Delta x(t) = \sqrt{\alpha^2 + (\hbar t/2m\alpha)^2}$ and the amplitude is reduced by the factor $\sqrt{\alpha/\Delta x(t)}$.

$$\Psi(x, 0) = Ce^{-x^2/4\alpha^2} = Ce^{-(x/2\alpha)^2}$$

This function $\Psi(x, 0)$, called a **gaussian function,** has a single maximum at $x = 0$ and decays smoothly to zero on either side of this point (Fig. 5.4a). The width of this gaussian packet becomes larger with increasing α. Accordingly, it is reasonable to identify α with Δx, the initial degree of localization. By the same token, $a(k)$ also is a gaussian function, but with amplitude $C\alpha/\sqrt{\pi}$ and width $1/2\alpha$ (since $\alpha^2 k^2 = k^2/[2(1/2\alpha)]^2$. Thus, $\Delta k = 1/2\alpha$ and $\Delta x \, \Delta k = 1/2$, *independent of α.* The multiplier C is a scale factor chosen to normalize $\Psi(x, 0)$.

Because our gaussian packet is made up of many individual waves all moving with different speeds, the shape of the packet changes over time. In Problem 4 it is shown that the packet *disperses,* its width growing ever larger with the passage of time as

$$\Delta x(t) = \sqrt{\{\Delta x(0)\}^2 + \left[\frac{\hbar t}{2m \, \Delta x(0)}\right]^2}$$

Similarly, the peak amplitude diminishes steadily in order to keep the waveform normalized for all times (Fig. 5.4b). The wave as a whole does not propagate, because for every wavenumber k present in the wave group there is an equal admixture of the plane wave with the opposing wavenumber $-k$.

EXAMPLE 5.4 Dispersion of Matter Waves

An atomic electron initially is localized to a region of space 0.10 nm wide (atomic size). How much time elapses before this localization is destroyed by dispersion? Repeat the calculation for a 1.0-g marble initially localized to 0.10 mm.

Solution Taking for the initial state a gaussian wave shape, we may use the results of the previous example. In particular, the extent of the matter wave after a time t has elapsed is

$$\Delta x(t) = \sqrt{[\Delta x(0)]^2 + \frac{\hbar t}{2m} \Delta x(0)^2}$$

where $\Delta x(0)$ is its initial width. The packet has effectively dispersed when $\Delta x(t)$ becomes appreciable compared to $\Delta x(0)$, say $\Delta x(t) = 10 \, \Delta x(0)$. This happens when $\hbar t/2m = \sqrt{99}[\Delta x(0)]^2$, or $t = \sqrt{99}(2m/\hbar)[\Delta x(0)]^2$.

The electron is initially localized to 0.10 nm ($= 10^{-10}$ m), and its mass is $m_e = 9.11 \times 10^{-31}$ kg. Thus, the electron wave packet disperses after a time

$$t = \sqrt{99} \left\{ \frac{(2)(9.11 \times 10^{-31} \text{ kg})}{1.055 \times 10^{-34} \text{ J} \cdot \text{s}} \right\} (1.00 \times 10^{-10} \text{ m})^2$$
$$= 1.7 \times 10^{-15} \text{ s}$$

The same calculation for a 1.0-g marble localized to $0.10 \text{ mm} = 10^{-4}$ m gives

$$t = \sqrt{99} \left\{ \frac{(2)(10^{-3} \text{ kg})}{1.055 \times 10^{-34} \text{ J} \cdot \text{s}} \right\} (10^{-4} \text{ m})^2$$
$$= 1.9 \times 10^{24} \text{ s}$$

or about 6.0×10^{16} years! This is more than ten million times the currently accepted value for the age of the universe. With its much larger mass, the marble does not show the quantum effects of dispersion on any measurable time scale and will, for all practical purposes, remain localized "forever." By contrast, the localization of an atomic electron is destroyed in a time that is very short, even by atomic standards.

EXAMPLE 5.5 Modeling Dispersion with MP DESKTOP

MP In this example we will use the MP DESKTOP to show dispersion of a gaussian wave packet and confirm some results of Example 5.3.

Launch the MP DESKTOP application by running the main program file *MP.EXE.* Next, select *Task* from the menu bar and *QM WaveGroups-1D* from the drop-down task list. The DESKTOP is now ready to solve the wave equation posed in the task field just below the menu bar:

$$-\frac{\partial^2 \Psi}{\partial x^2} + U(x)\Psi(x, t) = i\frac{\partial \Psi}{\partial t}$$

This is Schrödinger's *time-dependent* equation expressed in dimensionless units of length and time.

NOTE: For *QM WaveGroups* the unit of length (δ) is set implicitly in the file describing the potential energy function $U(x)$. In the present example, the potential energy is zero everywhere, so the length unit can be chosen arbitrarily. The length unit, in turn, sets all other physical units: in particular, the energy unit (ϵ) must satisfy $\epsilon \cdot \delta^2 = \hbar^2/2m$ for the equation in the task field to map into Schrödinger's equation. In describing electrons, atomic units ($\delta = 1$ bohr [0.529 Å] and $\epsilon = 1$ Ry [13.60 eV]) are a convenient choice satisfying this criterion, and are the ones adopted in this example. The time unit becomes $\hbar/\epsilon$, and the unit of momentum is $\hbar/\delta$. For a more exhaustive discussion of units, see the *Help-Contents* entry on the DESKTOP menu bar.

To specify the potential energy, click the Potential button $\boxed{\text{u(x)}}$ located just below the menu bar. A dialog box opens, presenting a list of potential files. Select *EX5-5.U* (for Example 5.5) from this list, then select *OK.*

Figure 5.5 (Example 5.5) Determining the width of the initial gaussian wave packet $\Psi(x, 0)$. Vertical lines mark the positions where $\Psi(x, 0)$ drops to $1/e$ of its peak value. The width of the gaussian is taken as one-quarter the separation distance between these lines, as discussed in Example 5.3.

Another dialog box opens, detailing the potential function $U(x)$ in use. We see that $U(x) = 0$ everywhere, to describe a free particle. The domain of definition extends from $x = -6$ to $x = +6$ in whatever length units we choose for this problem. To be specific, we will take this length unit to be 1 bohr [0.529 Å]. Accept the information as presented by clicking the *OK* button.

Next, click the Amplitude button $\boxed{f(x)}$ and select *GAUSS.PSI* to specify the initial wavefunction $\Psi(x, 0)$. This is a gaussian function, $\Psi(x, 0) = \exp[-\beta(x - \alpha)^2 + i\sigma x]$, centered on $x = \alpha$ with a width controlled by the parameter β. Click the *OK* button to load this selection with the default values $\alpha = 0$, $\beta = 2$, $\sigma = 0$ (the significance of σ will be explored in Exercise 1). The gaussian waveform is computed and is superimposed on the potential.

Figure 5.6 (Example 5.5) The appearance of the gaussian wave packet at $t = 0.400$. Vertical lines mark the positions where $\Psi(x, t)$ drops to $1/e$ of its peak value. The width of the gaussian is taken as one-quarter the separation distance between these lines, as discussed in Example 5.3.

Next, measure the width of $\Psi(x, 0)$. Click the Analyzer button ![Analyzer button] to activate the Analyzer. With the mouse (or the direction keys [→], [←]), place the graphics cursor on the peak of the wavefunction and note the value (it should be 1.00). Now move the cursor left until $\Psi(x, 0)$ drops to $1/e \approx 0.368$ of its peak value. Press the [.] key to mark this position, and record the value of x. Do the same to the right of the peak. Following the discussion of Example 5.3, the difference between the marked positions is identified as $4 \Delta x(0)$, where $\Delta x(0)$ is the initial width of the gaussian wavepacket. You should find $\Delta x(0) \approx 0.353$ bohrs (Fig. 5.5).

The current time is recorded in the time box below the menu bar and is always zero to start. Click the Advance button ![Advance button] to the immediate right of the time box to start the clock. The initial wavepacket immediately begins to spread—clear evidence of *dispersion* for our electron wavefunction. Notice the color changes. These result from *color-for-phase* plotting. The colors represent the phase of the wavefunction at each point, according to the color scheme found on the face of the Color-for Phase button ![Color-for-Phase button]. Phase angle increases counterclockwise from zero (solid black line) to 180° around the top half of the control, and from zero to $-180°$ clockwise around the bottom half of the control. The modulus, or absolute value, of the wavefunction at each point is plotted in the usual way as the height above the baseline. The values for modulus and phase can be read from the graph with the Analyzer. Clicking the Color-for-Phase button toggles the wavefunction display between a modulus–phase plot and a plot showing only the real part of Ψ. For complex-valued wavefunctions, the modulus–phase display is far more informative and is the default plotting style.

Stop the clock at $t = 0.400$ by clicking again on the Advance button ![Advance button]. (If you have gone past this time, run the clock backward by *holding down the Shift key* when selecting Advance, or type the desired value for t in the time box and press *Enter*.) Determine the width $\Delta x(t)$ at the new time by making the appropriate measurements with the Analyzer. You should find $\Delta x(t) \approx 1.185$ bohrs (Fig. 5.6). Finally, we compare the measured width at this time with the prediction of Example 5.3:

$$\Delta x(t) = \sqrt{[\Delta x(0)]^2 + \left[\frac{\hbar t}{2m \, \Delta x(0)}\right]^2}$$

For electrons, the unit of energy (ϵ) consistent with length measurements in bohrs is the rydberg ($\epsilon = 13.60$ eV), so the unit of time is $\hbar/\epsilon = 4.84 \times 10^{-17}$ s. Then, using $m = 9.11 \times 10^{-31}$ kg for the electron mass, and remembering that 1 bohr $= 0.529 \times 10^{-10}$ m, we find

$$\frac{\hbar t}{2m \, \Delta x(0)} = \frac{(1.055 \times 10^{-34} \text{ J} \cdot \text{s})(0.400)(4.84 \times 10^{-17} \text{ s})}{2(9.11 \times 10^{-31} \text{ kg})(0.353)(0.529 \times 10^{-10} \text{ m})}$$

$$= 0.600 \times 10^{-10} \text{ m}$$

or about 1.135 bohrs. Finally, we calculate

$$\Delta x(t) = \sqrt{(0.353)^2 + (1.135)^2} = 1.188 \text{ bohrs}$$

in excellent agreement with the measured value.

Exercise 1   Edit the initial wavefunction file *GAUSS.PSI*, changing σ from zero to $\sigma = 2$. Describe qualitatively how $\Psi(x, 0)$ differs from the original as a result of this change. Also examine $\Psi(x, t)$ to discover how the evolution of the wavefunction is altered with a nonzero value for σ.

5.3   WAVEFUNCTIONS IN THE PRESENCE OF FORCES

For a particle acted on by a force F, $\Psi(x, t)$ must be found from **Schrödinger's equation:**

$$-\frac{\hbar^2}{2m}\frac{\partial^2 \Psi}{\partial x^2} + U(x)\Psi = i\hbar \frac{\partial \Psi}{\partial t} \tag{5.10}$$

The Schrödinger wave equation

Again, we assume knowledge of the initial wavefunction $\Psi(x, 0)$. In this expression, $U(x)$ is the potential energy function for the force F; that is, $F = -dU/dx$. Schrödinger's equation is not derivable from any more basic principle, but is one of the *laws* of quantum physics. As with any law, its "truth" must be gauged ultimately by its ability to make predictions that agree with experiment.

Erwin Schrödinger was an Austrian theoretical physicist best known as the creator of wave mechanics. As a young man he was a good student who liked mathematics and physics, but also Latin and Greek for their logical grammar. He received a doctorate in physics from the University of Vienna. Although his work in physics was interrupted by World War I, by 1920 he had produced important papers on statistical mechanics, color vision, and general relativity, which he at first found quite difficult to understand. Expressing his feelings about a scientific theory in the remarkably open and outspoken way he maintained throughout his life, Schrödinger found general relativity initially "depressing" and "unnecessarily complicated." Other Schrödinger remarks in this vein, with which some readers will enthusiastically agree, are as follows: The Bohr–Sommerfeld quantum theory was "unsatisfactory, even disagreeable." "I . . . feel intimidated, not to say repelled, by what seem to me the very difficult methods [of matrix mechanics] and by the lack of clarity."

Shortly after de Broglie introduced the concept of matter waves in 1924, Schrödinger began to develop a new relativistic atomic theory based on de Broglie's ideas, but his failure to include electron spin led to the failure of this theory for hydrogen. By January of 1926, by treating the electron as a nonrelativistic particle, however, Schrödinger had introduced his famous wave equation and successfully obtained the energy values and wavefunctions for hydrogen. As Schrödinger himself pointed out, an outstanding feature of his approach was that the discrete energy values

B I O G R A P H Y

ERWIN SCHRÖDINGER
(1887–1961)

emerged from his wave equation in a natural way (as in the case of standing waves on a string), and in a way superior to the artificial postulate approach of Bohr. Another outstanding feature of Schrödinger's wave mechanics was that it was easier to apply to physical problems than Heisenberg's matrix mechanics, because it involved a partial differential equation very similar to the classical wave equation. Intrigued by the remarkable differences in conception and mathematical method of wave and matrix mechanics, Schrödinger did much to hasten the universal acceptance of all of quantum theory by demonstrating the mathematical equivalence of the two theories in 1926.

Although Schrödinger's wave theory was generally based on clear physical ideas, one of its major problems

in 1926 was the physical interpretation of the wavefunction Ψ. Schrödinger felt that the electron was ultimately a wave, Ψ was the vibration amplitude of this wave, and $\Psi^*\Psi$ was the electric charge density. As mentioned in Chapter 4, Born, Bohr, Heisenberg, and others pointed out the problems with this interpretation and presented the currently accepted view that $\Psi^*\Psi$ is a probability, and that the electron is ultimately no more a wave than a particle. Schrödinger never accepted this view, but registered his "concern and disappointment" that this "transcendental, almost psychical interpretation" had become "universally accepted dogma."

In 1927 Schrödinger, at the invitation of Max Planck, accepted the chair of theoretical physics at the University of Berlin, where he formed a close friendship with Planck and experienced six stable and productive years. In 1933, disgusted with the Nazis like so many of his colleagues, he left Germany. After several moves reflecting the political instability of Europe, he eventually settled at the Dublin Institute for Advanced Studies. Here he spent 17 happy, creative years working on problems in general relativity, cosmology, and the application of quantum physics to biology. This last effort resulted in a fascinating short book, *What is Life?*, which induced many young physicists to investigate biological processes with chemical and physical methods. In 1956, he returned home to his beloved Tyrolean mountains. He died there in 1961.

(AIP Emilio Segrè Visual Archives)

The Schrödinger equation propagates the initial wave forward in time. To see how this works, suppose $\Psi(x, 0)$ has been given. Then the left-hand side (LHS) of Schrödinger's equation can be evaluated and Equation 5.10 gives $\partial\Psi/\partial t$ at $t = 0$, the initial rate of change of the wavefunction. From this we compute the wavefunction a short time δt later as $\Psi(x, \delta t) = \Psi(x, 0) +$

$[\partial \Psi / \partial t]_0 \delta t$. This allows the LHS to be reevaluated, now at $t = \delta t$. With each such repetition, Ψ is advanced another step δt into the future. Continuing the process generates Ψ at any later time t. Such repetitious calculations are ideally suited to computers, and the method just outlined may be used to solve the Schrödinger equation numerically.

But how can we obtain an explicit mathematical expression for $\Psi(x, t)$? Returning to the free particle case, we see that the plane waves $\Psi_k(x, t)$ of Equation 5.6 serve a dual purpose: on the one hand, they represent particles whose momentum (hence, energy) is known precisely; on the other, they become the building blocks for constructing wavefunctions satisfying any initial condition. The question naturally arises: do analogous functions exist when forces are present? The answer is yes! To obtain them we look for solutions to the Schrödinger equation having the separable form[3]

$$\Psi(x, t) = \psi(x)\phi(t) \tag{5.11}$$

where $\psi(x)$ is a function of x only and $\phi(t)$ is a function of t only. (Note that the plane waves have just this form, with $\psi(x) = e^{ikx}$ and $\phi(t) = e^{-i\omega t}$.) Substituting Equation 5.11 into Equation 5.10 and dividing through by $\psi(x)\phi(t)$ gives

$$-\frac{\hbar^2}{2m}\frac{\psi''(x)}{\psi(x)} + U(x) = i\hbar\frac{\phi'(t)}{\phi(t)}$$

where primes denote differentiation with respect to the arguments. Now the LHS of this equation is a function of x only,[4] and the RHS is a function of t only. Since we can assign any value of x, independently of t, the two sides of the equation can be equal only if each is equal to the same constant, which we call E.[5] This yields two equations determining the unknown functions $\psi(x)$ and $\phi(t)$. The resulting equation for the time-dependent function $\phi(t)$ is

$$i\hbar\frac{d\phi}{dt} = E\phi(t) \tag{5.12}$$

This can be integrated immediately to give $\phi(t) = e^{-i\omega t}$ with $\omega = E/\hbar$: the same result obtained for free particles! The equation for the space function $\psi(x)$ is

$$-\frac{\hbar^2}{2m}\frac{d^2\psi}{dx^2} + U(x)\psi(x) = E\psi(x) \tag{5.13}$$

Wave equation for matter waves in separable form

---

[3] Obtaining solutions to partial differential equations in separable form is called *separation of variables*. On separating variables, a partial differential equation in, say, N variables is reduced to N ordinary differential equations, each involving only a single variable. The technique is a general one which may be applied to many (but not all!) of the partial differential equations encountered in science and engineering applications.

[4] Implicitly we have assumed that the potential energy $U(x)$ is a function of x only. For potentials that also depend on t (e.g., those arising from a time-varying electric field), solutions to the Schrödinger equation in separable form generally do not exist.

[5] More explicitly, changing t cannot affect the LHS because this depends only on x. Since the two sides of the equation are equal, we conclude that changing t cannot affect the RHS either. It follows that the RHS must reduce to a constant. The same argument with x replacing t shows the LHS also must reduce to this same constant.

Equation 5.13 is called the *time-independent Schrödinger equation.* Explicit solutions to this equation cannot be written for an arbitrary potential energy function $U(x)$. But whatever its form, $\psi(x)$ must be well-behaved on account of its connection with probabilities. In particular, $\psi(x)$ must be everywhere finite, single-valued, and continuous. Furthermore, $\psi(x)$ must be "smooth," that is, the slope of the wave $d\psi/dx$ also must be continuous wherever $U(x)$ has a finite value.[6]

For free particles we take $U(x) = 0$ in Equation 5.13 (to give $F = -dU/dx = 0$) and find that $\psi(x) = e^{ikx}$ is a solution with $E = \hbar^2 k^2/2m$. Thus, for free particles the separation constant E becomes the total particle energy; this identification continues to be valid when forces are present. The wavefunction $\psi(x)$ will change, however, with the introduction of forces, because particle momentum (hence k) is no longer constant.

The separable solutions to Schrödinger's equation describe conditions of particular physical interest. One feature shared by all such wavefunctions is especially noteworthy: because $|e^{-i\omega t}|^2 = e^{+i\omega t}e^{-i\omega t} = e^0 = 1$, we have

$$|\Psi(x, t)|^2 = |\psi(x)|^2 \tag{5.14}$$

This equality expresses the time-independence of all probabilities calculated from $\Psi(x, t)$. For this reason solutions in separable form are called **stationary states.** Thus, *for stationary states the probabilities are static* and can be calculated from the time-independent wavefunction $\psi(x)$.

5.4 THE PARTICLE IN A BOX

Of the problems involving forces, the simplest is that of particle confinement. Consider a particle moving along the x axis between the points $x = 0$ and $x = L$, where L is the length of the "box." Inside the box the particle is free; at the edges, however, it experiences strong forces that serve to contain it. A simple example is a ball bouncing elastically between two impenetrable walls (Fig. 5.7). A more sophisticated one is a charged particle moving along the axis of aligned metallic tubes held at different potentials, as shown in Figure 5.8a. The central tube is grounded, so that a test charge inside this tube has zero potential energy and experiences no electric force. When both outer tubes are held at a high electric potential V, there are no electric fields within them, but strong repulsive fields arise in the gaps at 0 and L. The potential energy $U(x)$ for this situation is sketched in Figure 5.8b. As V is increased without limit and the gaps are simultaneously reduced to zero, we approach the idealization known as the *infinite square well,* or "box" potential (Fig. 5.8c).

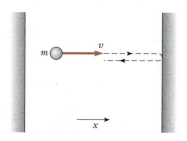

Figure 5.7 A particle of mass m and speed v bouncing elastically between two impenetrable walls.

---

[6] On rearrangement, the Schrödinger equation specifies the second derivative of the wavefunction $d^2\psi/dx^2$ at any point as

$$\frac{d^2\psi}{dx^2} = \frac{2m}{\hbar^2}[U(x) - E]\psi(x)$$

It follows that if $U(x)$ is finite at x, the second derivative also is finite here and the slope $d\psi/dx$ will be continuous.

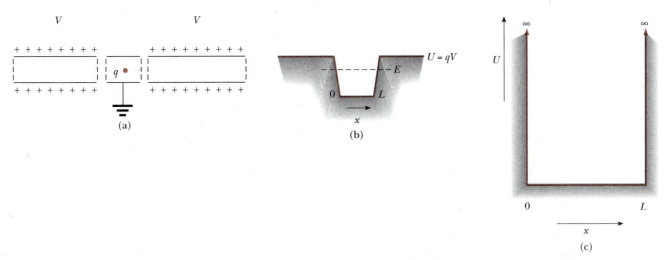

Figure 5.8   (a) Aligned metallic cylinders serve to confine a charged particle. The inner cylinder is grounded, while the outer ones are held at some high electric potential *V*. A charge *q* moves freely within the cylinders, but encounters electric forces in the gaps separating them. (b) The potential energy seen by this charge. A charge whose total energy is less than *qV* is confined to the central cylinder by the strong repulsive forces in the gaps at $x = 0$ and $x = L$. (c) As *V* is increased and the gaps between cylinders are narrowed, the potential energy approaches that of the infinite square well.

From a classical viewpoint, our particle simply bounces back and forth between the confining walls of the box. Its speed remains constant, as does its kinetic energy. Furthermore, classical physics places no restrictions on the values of its momentum and energy. The quantum description is quite different, and leads to the interesting phenomenon of energy quantization.

We are interested in the time-independent wavefunction $\psi(x)$ of our particle. Because it is confined to the box, the particle can *never* be found outside, which requires ψ to be zero in the exterior regions $x < 0$ and $x > L$. Inside the box $U(x) = 0$ and Equation 5.13 for $\psi(x)$ becomes, after rearrangement,

$$\frac{d^2\psi}{dx^2} = -k^2\psi(x) \qquad \text{with} \qquad k^2 = \frac{2mE}{\hbar^2}$$

Independent solutions to this equation are $\sin kx$ and $\cos kx$, indicating that k is the wavenumber of oscillation. The most general solution is a linear combination of these two,

$$\psi(x) = A \sin kx + B \cos kx \qquad \text{for } 0 < x < L \qquad (5.15)$$

This interior wave must match the exterior wave at the walls of the box for $\psi(x)$ to be continuous everywhere.[7] Thus, we require the interior wave to vanish

---

[7] Although $\psi(x)$ must be continuous everywhere, the slope of $d\psi/dx$ is *not* continuous at the walls of the box, where $U(x)$ becomes infinite (cf. footnote 6).

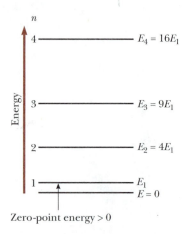

Figure 5.9 Energy level diagram for a particle confined to a one-dimensional box of width L. The lowest allowed energy is E_1 with value $\pi^2\hbar^2/2mL^2$.

at $x = 0$ and $x = L$:

$$\psi(0) = B = 0 \qquad \text{[continuity at } x = 0\text{]}$$

$$\psi(L) = A \sin kL = 0 \qquad \text{[continuity at } x = L\text{]} \qquad (5.16)$$

The last condition requires that $kL = n\pi$, where n is any positive integer.[8] Because $k = 2\pi/\lambda$, this is equivalent to fitting an integral number of half-wavelengths into the box (see Fig. 5.11a). Using $k = n\pi/L$, we find that the particle energies are *quantized*, being restricted to the values

$$E_n = \frac{\hbar^2 k^2}{2m} = \frac{n^2\pi^2\hbar^2}{2mL^2} \qquad n = 1, 2, \dots \qquad (5.17)$$

The lowest allowed energy is given by $n = 1$ and is $E_1 = \pi^2\hbar^2/2mL^2$. This is the **ground state**. Because $E_n = n^2E_1$, the **excited states** for which $n = 2, 3, 4, \dots$ have energies $4E_1, 9E_1, 16E_1, \dots$. An energy level diagram is given in Figure 5.9. Notice that $E = 0$ is not allowed; that is, *the particle can never be at rest*. The least energy the particle can have, E_1, is called the **zero-point energy**. This result clearly contradicts the classical prediction, for which $E = 0$ is an acceptable energy, as are all positive values of E.

EXAMPLE 5.6   Energy Quantization for a Macroscopic Object

A small object of mass 1.00 mg is confined to move between two rigid walls separated by 1.00 cm. (a) Calculate the minimum speed of the object. (b) If the speed of the object is 3.00 cm/s, find the corresponding value of n.

Solution   Treating this as a particle in a box, the energy of the particle can only be one of the values given by Equation 5.17, or

$$E_n = \frac{n^2\pi^2\hbar^2}{2mL^2} = \frac{n^2h^2}{8mL^2}$$

The minimum energy results from taking $n = 1$. For $m = 1.00$ mg and $L = 1.00$ cm we calculate

$$E_1 = \frac{(6.626 \times 10^{-34} \text{ J} \cdot \text{s})^2}{8.00 \times 10^{-10} \text{ kg} \cdot \text{m}^2} = 5.49 \times 10^{-58} \text{ J}$$

Because the energy is all kinetic, $E_1 = mv_1^2/2$ and the *minimum* speed v_1 of the particle is

$$v_1 = \sqrt{2(5.49 \times 10^{-58} \text{ J})/(1.00 \times 10^{-6} \text{ kg})}$$
$$= 3.31 \times 10^{-26} \text{ m/s}$$

This speed is immeasurably small, so that for practical purposes the object can be considered to be at rest. Indeed, the time required for an object with this speed to move the 1.00 cm separating the walls is about 3×10^{23} s, or about one million times the present age of the universe! It is reassuring to verify that quantum mechanics applied to macroscopic objects does not contradict our everyday experiences.

If, instead, the speed of the particle is $v = 3.00$ cm/s, then its energy is

---

[8] For $n = 0$ ($E = 0$) Schrödinger's equation requires $d^2\psi/dx^2 = 0$, whose solution is given by $\psi(x) = Ax + B$ for some choice of constants A and B. For this wavefunction to vanish at $x = 0$ and $x = L$, both A and B must be zero, leaving $\psi(x) = 0$ everywhere. In such a case the particle is nowhere to be found; that is, no description is possible when $E = 0$. Also, the inclusion of negative integers $n < 0$ produces no new states, because changing the sign of n merely changes the sign of the wavefunction, leading to the same probabilities as for positive integers.

$$E = \frac{mv^2}{2} = \frac{(1.00 \times 10^{-6} \text{ kg})(3.00 \times 10^{-2} \text{ m/s})^2}{2}$$

$$= 4.50 \times 10^{-10} \text{ J}$$

This, too, must be one of the special values E_n. To find which one, we solve for the quantum number n, obtaining

$$n = \frac{\sqrt{8mL^2E}}{h} = \frac{\sqrt{(8.00 \times 10^{-10} \text{ kg·m}^2)(4.50 \times 10^{-10} \text{ J})}}{6.626 \times 10^{-34} \text{ J·s}}$$

$$= 9.05 \times 10^{23}$$

Notice that the quantum number representing a typical speed for this (ordinary size) object is enormous. In fact, the value of n is so large that we would never be able to distinguish the quantized nature of the energy levels. That is, the difference in energy between two consecutive states with quantum numbers $n_1 = 9.05 \times 10^{23}$ and $n_2 = 9.05 \times 10^{23} + 1$ is only about 10^{-33} J, much too small to be detected experimentally. This is another example that illustrates the working of Bohr's correspondence principle, which asserts that quantum predictions must agree with classical results for large masses and lengths.

EXAMPLE 5.7 Model of an Atom

An atom can be viewed as a number of electrons moving around a positively charged nucleus, where the electrons are subject mainly to the Coulombic attraction of the nucleus (which actually is partially "screened" by the intervening electrons). The potential well that each electron "sees" is sketched in Figure 5.10. Use the model of a particle in a box to estimate the energy (in eV) required to raise an atomic electron from the state $n = 1$ to the state $n = 2$, assuming the atom has a radius of 0.100 nm.

Solution Taking the length L of the box to be 0.200 nm (the diameter of the atom), $m_e = 511$ keV/c^2, and $\hbar c = 197.3$ eV·nm for the electron, we calculate

$$E_1 = \frac{\pi^2 \hbar^2}{2m_e L^2}$$

$$= \frac{\pi^2 (197.3 \text{ eV·nm}/c)^2}{2(511 \times 10^3 \text{ eV}/c^2)(0.200 \text{ nm})^2}$$

$$= 9.40 \text{ eV}$$

and

$$E_2 = (2)^2 E_1 = 4(9.40 \text{ eV}) = 37.6 \text{ eV}$$

Therefore, the energy that must be supplied to the electron is

$$\Delta E = E_2 - E_1 = 37.6 \text{ eV} - 9.40 \text{ eV} = 28.2 \text{ eV}$$

We could also calculate the wavelength of the photon that would cause this transition by identifying ΔE with the photon energy hc/λ, or

$$\lambda = hc/\Delta E = (1.24 \times 10^3 \text{ eV·nm})/(28.2 \text{ eV}) = 44.0 \text{ nm}$$

This wavelength is in the far ultraviolet region, and it is interesting to note that the result is roughly correct. Although this oversimplified model gives a good estimate for transitions between lowest lying levels of the atom, the estimate gets progressively worse for higher energy transitions.

Exercise 2 Calculate the minimum speed of an atomic electron modeled as a particle in a box with walls that are 0.200 nm apart.

Answer 1.82×10^6 m/s.

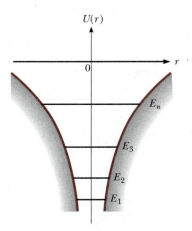

Figure 5.10 (Example 5.7) Model of the potential energy versus r for the one-electron atom.

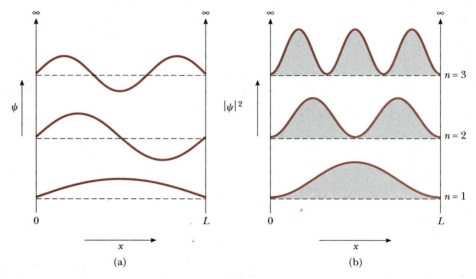

Figure 5.11   The first three allowed stationary states for a particle confined to a one-dimensional box. (a) The wavefunctions for $n = 1$, 2, and 3. (b) The probability distributions for $n = 1$, 2, and 3.

Returning to the wavefunctions, we have from Equation 5.15 (with $k = n\pi/L$ and $B = 0$)

Stationary states for a particle in a box

$$\psi_n(x) = A \sin\left(\frac{n\pi x}{L}\right) \qquad \text{for } 0 < x < L \text{ and } n = 1, 2, \ldots \quad (5.18)$$

For each value of the **quantum number** n there is a specific wavefunction $\psi_n(x)$ describing the state of the particle with energy E_n. Figure 5.11 shows plots of ψ_n versus x and of the probability density $|\psi_n|^2$ versus x for $n = 1$, 2, and 3, corresponding to the three lowest allowed energies for the particle. For $n = 1$, the probability of finding the particle is largest at $x = L/2$—this is the *most probable position* for a particle in this state. For $n = 2$, $|\psi|^2$ is a maximum at $x = L/4$ and again at $x = 3L/4$: both points are equally likely places for a particle in this state to be found.

There are also points within the box where it is impossible to find the particle. Again for $n = 2$, $|\psi|^2$ is zero at the midpoint, $x = L/2$; for $n = 3$, $|\psi|^2$ is zero at $x = L/3$ and at $x = 2L/3$, and so on. But how does our particle get from one place to another when there is no probability for its ever being at points in between? It is as if there were *no path at all*, and not just that the probabilities $|\psi|^2$ express our ignorance about a world somehow hidden from view. Indeed, what is at stake here is the very essence of a *particle* as something that gets from one place to another by occupying all intervening positions. The objects of quantum mechanics are not particles, but more complicated things having both particle *and wave* attributes.

Actual probabilities can be computed only after ψ_n is normalized, that is, we must be sure that all probabilities sum to unity:

$$1 = \int_{-\infty}^{\infty} |\psi_n(x)|^2 \, dx = A^2 \int_0^L \sin^2\left(\frac{n\pi x}{L}\right) dx$$

The integral is evaluated with the help of the trigonometric identity $2\sin^2\theta = 1 - \cos 2\theta$:

$$\int_0^L \sin^2\left(\frac{n\pi x}{L}\right) dx = \frac{1}{2}\int_0^L [1 - \cos(2n\pi x/L)] \, dx$$

Only the first term contributes to the integral, because the cosine integrates to $\sin(2n\pi x/L)$, which vanishes at the limits 0 and L. Thus, normalization requires $1 = A^2 L/2$, or

$$A = \sqrt{\frac{2}{L}} \tag{5.19}$$

EXAMPLE 5.8   Probabilities for a Particle in a Box

A particle is known to be in the ground state of an infinite square well with length L. Calculate the probability that this particle will be found in the middle half of the well, that is, between $x = L/4$ and $x = 3L/4$.

Solution   The probability density is given by $|\psi_1|^2$ with $n = 1$ for the ground state. Thus, the probability is

$$P = \int_{L/4}^{3L/4} |\psi_1|^2 \, dx = \left(\frac{2}{L}\right)\int_{L/4}^{3L/4} \sin^2(\pi x/L) \, dx$$

$$= \left(\frac{1}{L}\right)\int_{L/4}^{3L/4} [1 - \cos(2\pi x/L)] \, dx$$

$$= \left(\frac{1}{L}\right)\left[\frac{L}{2} - \left(\frac{L}{2\pi}\right)\sin(2\pi x/L)\, \Big|_{L/4}^{3L/4}\right]$$

$$= \frac{1}{2} - \left(\frac{1}{2\pi}\right)[-1-1] = 0.818$$

Notice this is considerably larger than $\frac{1}{2}$, which would be expected for a classical particle that spends equal time in all parts of the well.

Exercise 3   Repeat the calculation of Example 5.8 for a particle in the nth state of the infinite square well, and show that the result approaches the classical value $\frac{1}{2}$ in the limit $n \to \infty$.

5.5   THE FINITE SQUARE WELL

O P T I O N A L

The "box" potential is an oversimplification that is never realized in practice. Given sufficient energy, a particle can escape the confines of any well. The potential energy for a more realistic situation—the finite square well—is shown in Figure 5.12, and essentially is that depicted in Figure 5.8b before taking the limit $V \to \infty$. A classical particle with energy E greater than the well height U can penetrate the gaps at $x = 0$ and $x = L$ to enter the outer region. Here it moves freely, but with reduced speed corresponding to a diminished kinetic energy $E - U$.

A classical particle with energy E less than U is permanently bound to the region $0 < x < L$. Quantum mechanics asserts, however, that there is some probability that the particle can be found *outside* this region! That is, the wavefunction generally is nonzero outside the well, and so the probability of finding the particle here also is

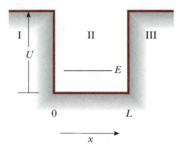

Figure 5.12 Potential energy diagram for a well of finite height U and width L. The energy E of the particle is less than U.

nonzero. For stationary states, the wavefunction $\psi(x)$ is found from the time-independent Schrödinger equation. Outside the well where $U(x) = U$, this is

$$\frac{d^2\psi}{dx^2} = \alpha^2 \psi(x) \qquad x < 0 \text{ and } x > L$$

where $\alpha^2 = 2m(U - E)/\hbar^2$ is a constant. Because $U > E$, α^2 necessarily is *positive* and the independent solutions to this equation are the *real* exponentials $e^{+\alpha x}$ and $e^{-\alpha x}$. The positive exponential must be rejected in region III where $x > L$ to keep $\psi(x)$ finite as $x \rightarrow \infty$; likewise, the negative exponential must be rejected in region I where $x < 0$ to keep $\psi(x)$ finite as $x \rightarrow -\infty$. Thus, the exterior wave takes the form

$$\psi(x) = Ae^{+\alpha x} \qquad \text{for } x < 0$$
$$\psi(x) = Be^{-\alpha x} \qquad \text{for } x > L \qquad (5.20)$$

The coefficients A and B are determined by matching this wave smoothly onto the wavefunction in the well interior. Specifically, we require $\psi(x)$ and its first derivative $d\psi/dx$ to be *continuous* at $x = 0$ and again at $x = L$. This can be done only for certain values of E, corresponding to the allowed energies for the bound particle. For these energies, the matching conditions specify the entire wavefunction except for a multiplicative constant, which then is determined by normalization. Figure 5.13 shows the wavefunctions and probability densities that result for the three lowest allowed particle energies. Note that in each case the waveforms join smoothly at the boundaries of the potential well.

The fact that ψ is nonzero at the walls *increases* the de Broglie wavelength in the well (compared with that in the infinite well), and this in turn lowers the energy and momentum of the particle. This observation can be used to approximate the allowed energies for the bound particle.[9] The wavefunction penetrates the exterior region

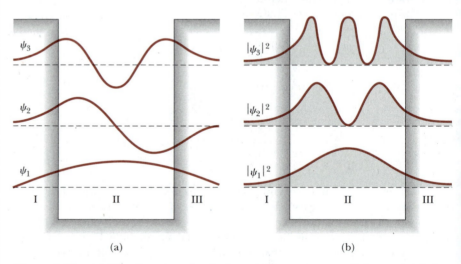

(a)   (b)

Figure 5.13 (a) Wavefunctions for the lowest three energy states for a particle in a potential well of finite height. (b) Probability densities for the lowest three energy states for a particle in a potential well of finite height.

[9] This specific approximation method was reported by S. Garrett in the *Am. J. Phys.*, 47:195–196, 1979.

on a scale of length set by the *penetration depth* δ given by

$$\delta = \frac{1}{\alpha} = \frac{\hbar}{\sqrt{2m(U - E)}} \qquad (5.21)$$

Penetration depth

Specifically, at a distance δ beyond the well edge, the wave amplitude has fallen to $1/e$ of its value at the edge, and approaches zero exponentially in the exterior region. That is, the exterior wave is essentially zero beyond a distance δ on either side of the potential well. If it were truly zero beyond this distance, the allowed energies would be those for an infinite well of length $L + 2\delta$, or (compare Equation 5.17)

$$E_n \approx \frac{n^2 \pi^2 \hbar^2}{2m(L + 2\delta)^2} \qquad n = 1, 2, \ldots \qquad (5.22)$$

Approximate energies for a particle in a well of finite height

The allowed energies for a particle bound to the finite well are given approximately by Equation 5.22 so long as δ is small compared with L. But δ itself is energy-dependent according to Equation 5.21. Thus, Equation 5.22 becomes an *implicit* relation for E that must be solved numerically for a given value of n. The approximation is best for the lowest lying states, and breaks down completely as E approaches U, where δ becomes infinite. From this we infer (correctly) that the number of bound states is limited by the height U of our potential well. Particles with energies E exceeding U are not bound to the well, that is, they may be found with comparable probability in the exterior regions. The case of unbound states will be taken up in the following chapter.

EXAMPLE 5.9   A Bound Electron

Estimate the ground state energy for an electron confined to a potential well of width 0.200 nm and height 100 eV.

Solution   We solve Equations 5.21 and 5.22 together, using an iterative procedure. Because we expect $E \ll U$ ($= 100$ eV), we estimate the decay length δ by first neglecting E to get

$$\delta \approx \frac{\hbar}{\sqrt{2mU}} = \frac{(197.3 \text{ eV} \cdot \text{nm}/c)}{\sqrt{2(511 \times 10^3 \text{ eV}/c^2)(100 \text{ eV})}}$$

$$= 0.0195 \text{ nm}$$

Thus, the effective width of the (infinite) well is $L + 2\delta = 0.239$ nm, for which we calculate the ground state energy

$$E \approx \frac{\pi^2 (197.3 \text{ eV} \cdot \text{nm}/c)^2}{2(511 \times 10^3 \text{ eV}/c^2)(0.239 \text{ nm})^2} = 6.58 \text{ eV}$$

From this E we calculate $U - E = 93.42$ eV and a new decay length

$$\delta \approx \frac{(197.3 \text{ eV} \cdot \text{nm}/c)}{\sqrt{2(511 \times 10^3 \text{ eV}/c^2)(93.42 \text{ eV})}} = 0.0202 \text{ nm}$$

This, in turn, increases the effective well width to 0.240 nm and lowers the ground state energy to $E = 6.53$ eV. The iterative process is repeated until the desired accuracy is achieved. Because our estimate for E has changed by only 0.05 eV, we may be content with this value. Indeed, another iteration gives the same result to the accuracy reported. This is in good agreement with the exact value, about 6.55 eV for this case.

EXAMPLE 5.10 Finding Bound States with MP DESKTOP

MP In this example we revisit Example 5.9—an electron confined to a finite well of width 0.200 nm and height 100 eV—to illustrate the use of the MP DESKTOP in finding bound state wavefunctions and energies.

Launch the MP DESKTOP application by running the main program file *MP.EXE*. Next, select *Task* from the menu bar and *EigenStates-1D* from the drop-down task list. The DESKTOP is now ready to solve the (eigenvalue) problem posed in the task field just below the menu bar:

$$-\frac{d^2\psi}{dx^2} + U(x)\,\psi(x) = E\cdot\psi(x)$$

This is Schrödinger's *time-independent* equation expressed in dimensionless units of length and energy.

NOTE: For *EigenStates-1D* the unit of length (δ) is set implicitly in the file describing the potential energy function $U(x)$. The length unit, in turn, sets all other physical units: in particular, the energy unit (ϵ) must satisfy $\epsilon\cdot\delta^2 = \hbar^2/2m$ to map the equation in the task field to the Schrödinger equation. In describing electrons, atomic units ($\delta = 1$ bohr [0.529 Å] and $\epsilon = 1$ Ry [13.60 eV]) are a convenient choice satisfying this criterion and are the ones adopted in this example. For a more exhaustive discussion of units, see the *Help-Contents* entry on the DESKTOP menu bar.

To specify the potential energy, click the Potential button $\boxed{u(x)}$ located just below the menu bar. A dialog box opens, presenting a list of potential files. Select *EX5-10.U* (for Example 5.10) from this list, then select *OK*. Another dialog box opens, detailing the potential function $U(x)$ in use. We see that $U(x)$ describes a finite square well extending from $x = -1/0.529$ to $x = +1/0.529$: with a length unit of 1 bohr (0.529 Å), the equivalent range is -1.00 Å to $+1.00$ Å, or -0.100 nm to $+0.100$ nm. For length measured in bohrs, energy must be measured in rydbergs (1 Ry = 13.60 eV). Thus, a well height of 100 eV becomes $100/13.60 = 7.353$ Ry. The well height (ϵ) may be edited, but for now simply accept the information as presented by clicking the *OK* button.

The potential $U(x)$ is graphed in the main window of the screen; the current energy E is recorded in the energy box just below the menu bar. In the energy box, input a "guess" for the energy of the ground state. A good starting value would be the energy of an electron in an infinite well of the same width, $\pi^2\hbar^2/2mL^2 = 9.40$ eV, or $9.40/13.60 = 0.691$ Ry. So type in $E = 0.69$ and press *Enter*. The wavefunction for this energy is computed and overlays the potential (Fig. 5.14).

The discontinuous jump near the center of the screen is evidence that this value for E is unacceptable.

NOTE: The DESKTOP integrates the Schrödinger equation inward—first from the left, then from the right—to a common (matching) point. The individual waves are then scaled to make their slopes agree at this point. If E is an eigenvalue, their values also will agree here; otherwise a discontinuity results.

Repeat the procedure with a somewhat lower value, say $E = 0.50$, and observe the consequences. The "jump" is smaller, suggesting we are closer to the true value. Continue with a few more trials until there is no perceptible discontinuity in the wavefunction.

Figure 5.14   (Example 5.10) The trial wavefunction that results for $E = 0.69$ Ry, against the backdrop of the square well potential.

NOTE: The direction keys [↑], [↓] furnish a quick and convenient way to make small changes to the energy. Just press [↑] ([↓]) to increase (decrease) the digit at the cursor by one unit; then press *Enter* to update the display.

(Notice that the sense of the discontinuity reverses if your trial value for the energy is too low.) Next, click the Accuracy button ⌊L⌋ to the right of the energy box: the button label changes from **L**(ow) to **M**(edium), signaling improved precision in the search. Make fine adjustments to E as necessary to generate a smooth wavefunction. Two more precision levels—**H**(igh) and **S**(uper)—are available if desired. Continue the search process until E is known confidently to three decimal places (Fig. 5.15).

You should find $E = 0.483$ Ry, or approximately $(0.483)(13.60) = 6.57$ eV. This is to be compared with the exact ground state energy for this case, 6.55 eV. As this example illustrates, energy values obtained using the DESKTOP are typically accurate to better than 0.5%.

Exercise 4   **Wavefunction in the Forbidden Region.**   Classically, an electron bound to a finite well has insufficient energy to penetrate beyond the limits of the well, yet the quantum wavefunction is nonzero here, decreasing exponentially from the well edge according to Equations 5.20 and 5.21. Make appropriate measurements on the ground state waveform of Example 5.10 to verify this exponential decay into the classically forbidden region, and confirm the result for the decay length given by

Equation 5.21. [$\psi(x)$ can be studied from the DESKTOP by selecting 🔍 to activate

the MP Analyzer. The wavefunction is read at the current location of the graphics cursor, which can be repositioned using the mouse or the direction keys [→], [←]. Keyboard users note that the [Ins] key toggles cursor movement between coarse and fine settings.]

Figure 5.15   (Example 5.10) For $E = 0.483$ Ry, there is no discernible discontinuity in the wavefunction. This is the ground state of the finite square well.

EXAMPLE 5.11   Energy of a Finite Well: Exact Treatment

Impose matching conditions on the interior and exterior wavefunctions and show how these lead to energy quantization for the finite square well.

Solution   The exterior wavefunctions are the decaying exponential functions given by Equation 5.20 with decay constant $\alpha = [2m(U - E)/\hbar^2]^{1/2}$. The interior wave is an oscillation with wavenumber $k = (2mE/\hbar^2)^{1/2}$ having the same form as that for the infinite well, Equation 5.15; here we write it as

$$\psi(x) = C \sin kx + D \cos kx \qquad \text{for } 0 < x < L$$

To join this smoothly onto the exterior wave, we insist that the wavefunction and its slope be continuous at the well edges $x = 0$ and $x = L$. At $x = 0$ the conditions for smooth joining require

$$A = D \qquad [\text{continuity of } \psi]$$

$$\alpha A = kC \qquad \left[\text{continuity of } \frac{d\psi}{dx}\right]$$

Dividing the second equation by the first eliminates A, leaving

$$\frac{C}{D} = \frac{\alpha}{k}$$

In the same way, smooth joining at $x = L$ requires

$$C \sin kL + D \cos kL = Be^{-\alpha L} \qquad [\text{continuity of } \psi]$$

$$kC \cos kL - kD \sin kL = -\alpha Be^{-\alpha L} \qquad \left[\text{continuity of } \frac{d\psi}{dx}\right]$$

Again dividing the second equation by the first eliminates B. Then replacing C/D with α/k gives

$$\frac{(\alpha/k)\cos kL - \sin kL}{(\alpha/k)\sin kL + \cos kL} = -\frac{\alpha}{k}$$

For a specified well height U and width L, this last relation can only be satisfied for special values of E (E is contained in both k and α). For any other energies, the waveform will not match smoothly at the well edges, leaving a wavefunction that is physically inadmissable. (Note that the equation cannot be solved explicitly for E; rather, solutions must be obtained using numerical or graphical methods.)

Exercise 5   Use the result of Example 5.11 to verify that the ground state energy for an electron confined to a square well of width 0.200 nm and height 100 eV is about 6.52 eV.

5.6   THE QUANTUM OSCILLATOR

As a final example of a potential well for which exact results can be obtained, let us examine the problem of a particle subject to a linear restoring force $F = -Kx$. Here x is the displacement of the particle from equilibrium ($x = 0$) and K is the force constant. The corresponding potential energy is given by $U(x) = \frac{1}{2}Kx^2$. The prototype physical system fitting this description is a mass on a spring, but the mathematical description actually applies to any object limited to small excursions about a point of stable equilibrium.

Consider the general potential function sketched in Figure 5.16. The positions a, b, and c all label equilibrium points where the force $F = -dU/dx$ is zero. Further, a and c are examples of *stable* equilibria, but b is *unstable*. The stability of equilibrium is decided by examining the forces in the immediate neighborhood of the equilibrium point. Just to the left of a, for example, $F = -dU/dx$ is positive, that is, the force is directed to the right; conversely, to the right of a the force is directed to the left. Thus, we see that a particle displaced slightly from equilibrium at a encounters a force driving it back to the equilibrium point (restoring force). Similar arguments show that the equilibrium at c also is stable. On the other hand, a particle displaced in either direction from point b experiences a force that drives it further away from equilibrium—an unstable condition. In general, stable and unstable equilibria are marked by potential curves that are concave or convex, respectively, at the equilibrium point. To put it another way, the curvature of $U(x)$ is positive ($d^2U/dx^2 > 0$) at a point of stable equilibrium, and negative ($d^2U/dx^2 < 0$) at a point of unstable equilibrium.

Near a point of stable equilibrium such as a (or c), $U(x)$ can be fit quite well by a parabola:

$$U(x) = U(a) + \frac{1}{2}K(x - a)^2 \qquad (5.23)$$

Of course, the curvature of this parabola ($= K$) must match that of $U(x)$ at the equilibrium point $x = a$:

$$K = \left.\frac{d^2U}{dx^2}\right|_a \qquad (5.24)$$

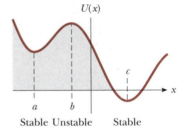

$U(x)$

a ... b ... c ... x

Stable ... Unstable ... Stable

Figure 5.16   A general potential function $U(x)$. The points labeled a and c are positions of stable equilibrium, for which $dU/dx = 0$ and $d^2U/dx^2 > 0$. Point b is a position of unstable equilibrium, for which $dU/dx = 0$ and $d^2U/dx^2 < 0$.

Further, $U(a)$, the potential energy at equilibrium, may be taken as zero if we agree to make this our energy reference, that is, if we subsequently measure all energies from this level. In the same spirit, the coordinate origin may be placed at $x = a$, in effect allowing us to set $a = 0$. With $U(a) = 0$ and $a = 0$, Equation 5.23 becomes the spring potential once again; in other words, *a particle limited to small excursions about any stable equilibrium point behaves as if it were attached to a spring with a force constant K prescribed by the curvature of the true potential at equilibrium.* In this way the oscillator becomes a first approximation to the vibrations occurring in many real systems.

Harmonic approximation to vibrations occurring in real systems

The motion of a classical oscillator with mass m is simple harmonic vibration at the angular frequency $\omega = \sqrt{K/m}$. If the particle is removed from equilibrium a distance A and released, it will oscillate between the points $x = -A$ and $x = +A$ (A is the amplitude of vibration), with total energy $E = \frac{1}{2}KA^2$. By changing the initial point of release A, the classical particle can in principle be given any (nonnegative) energy whatsoever, including zero.

The quantum oscillator is described by the potential energy $U(x) = \frac{1}{2}Kx^2 = \frac{1}{2}m\omega^2x^2$ in the Schrödinger equation. After a little rearrangement we get

$$\frac{d^2\psi}{dx^2} = \frac{2m}{\hbar^2}\left(\frac{1}{2}m\omega^2x^2 - E\right)\psi(x) \tag{5.25}$$

as the equation for the stationary states of the oscillator. The mathematical technique for solving this equation is beyond the level of this text. (The exponential and trigonometric forms for ψ employed before will not work here because of the presence of x^2 in the potential.) It is instructive, however, to make some intelligent guesses and verify their accuracy by direct substitution. The *ground state* wavefunction should possess the following attributes:

1. ψ should be *symmetric* about the midpoint of the potential well $x = 0$.
2. ψ should be *nodeless*, but approaching zero for $|x|$ large.

Both expectations are derived from our previous experience with the lowest energy states of the infinite and finite square wells, which you might want to review at this time. The symmetry condition (1) requires ψ to be some function of x^2; further, the function must have no zeros (other than at infinity) to meet the nodeless requirement (2). The simplest choice fulfilling both demands is the gaussian form

$$\psi(x) = C_0 e^{-\alpha x^2} \tag{5.26}$$

for some as-yet-unknown constants C_0 and α. Taking the second derivative of $\psi(x)$ in Equation 5.26 gives (as you should verify)

$$\frac{d^2\psi}{dx^2} = \{4\alpha^2x^2 - 2\alpha\}C_0\,e^{-\alpha x^2} = \{4\alpha^2x^2 - 2\alpha\}\psi(x)$$

which has the same structure as Equation 5.25. Comparing like terms between them, we see that we have a solution provided that both

$$4\alpha^2 = \frac{2m}{\hbar^2}\frac{1}{2}m\omega^2 \quad \text{or} \quad \alpha = \frac{m\omega}{2\hbar} \tag{5.27}$$

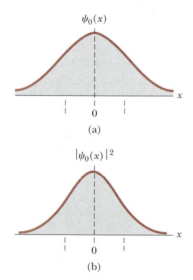

$\psi_0(x)$

(a)

$|\psi_0(x)|^2$

(b)

Figure 5.17 (a) Wavefunction for the ground state of a particle in the oscillator potential well. (b) The probability density for the ground state of a particle in the oscillator potential well. The dashed vertical lines mark the limits of vibration for a classical particle with the same energy, $x = \pm A = \pm\sqrt{\hbar/m\omega}$.

and

$$\frac{2mE}{\hbar^2} = 2\alpha = \frac{m\omega}{\hbar} \qquad \text{or} \qquad E = \tfrac{1}{2}\hbar\omega \qquad (5.28)$$

In this way we discover that the oscillator ground state is described by the wavefunction $\psi_0(x) = C_0 \exp(-m\omega x^2/2\hbar)$, and that the energy of this state is $E_0 = \tfrac{1}{2}\hbar\omega$. The multiplier C_0 is reserved for normalization (see Example 5.12 below). The ground state wave ψ_0 and associated probability density $|\psi_0|^2$ are illustrated in Figure 5.17. The dashed vertical lines mark the limits of vibration for a classical oscillator with the same energy. Note the considerable penetration of the wave into the classically forbidden regions $x > A$ and $x < -A$. A detailed analysis shows that the particle actually can be found in these nonclassical regions about 16% of the time (see Example 5.14).

EXAMPLE 5.12   Normalizing the Oscillator Ground State Wavefunction

Normalize the oscillator ground state wavefunction found in the preceding paragraph.

Solution With $\psi_0(x) = C_0 e^{-m\omega x^2/2\hbar}$, the integrated probability is

$$\int_{-\infty}^{\infty} |\psi_0(x)|^2 \, dx = C_0^2 \int_{-\infty}^{\infty} e^{-m\omega x^2/\hbar} \, dx$$

Evaluation of the integral requires advanced techniques. We shall be content here simply to quote the formula

$$\int_{-\infty}^{\infty} e^{-ax^2} \, dx = \sqrt{\frac{\pi}{a}} \qquad a > 0$$

In our case we identify a with $m\omega/\hbar$ and obtain

$$\int_{-\infty}^{\infty} |\psi_0(x)|^2 \, dx = C_0^2 \sqrt{\frac{\pi\hbar}{m\omega}}$$

Normalization requires this integrated probability to be 1, leading to

$$C_0 = \left(\frac{m\omega}{\pi\hbar}\right)^{1/4}$$

EXAMPLE 5.13   Limits of Vibration for a Classical Oscillator

Obtain the limits of vibration for a classical oscillator having the same total energy as the quantum oscillator in its ground state.

Solution The ground state energy of the quantum oscillator is $E_0 = \tfrac{1}{2}\hbar\omega$. At its limits of vibration $x = \pm A$, the classical oscillator has transformed all this energy into elas-

tic potential energy of the spring, given by $\tfrac{1}{2}KA^2 = \tfrac{1}{2}m\omega^2A^2$. Therefore

$$\tfrac{1}{2}\hbar\omega = \tfrac{1}{2}m\omega^2A^2 \qquad \text{or} \qquad A = \sqrt{\hbar/m\omega}$$

The classical oscillator vibrates in the interval given by $-A \le x \le A$, having insufficient energy to exceed these limits.

EXAMPLE 5.14   The Quantum Oscillator in the Nonclassical Region

Calculate the probability that a quantum oscillator in its ground state will be found outside the range permitted for a classical oscillator with the same energy.

Solution Because the classical oscillator is confined to the interval $-A \le x \le A$, where A is its amplitude of vibration, the question is one of finding the quantum oscillator outside this interval. From the previous example we have $A = \sqrt{\hbar/m\omega}$ for a classical oscillator with energy $\tfrac{1}{2}\hbar\omega$. The quantum oscillator with this energy is described by the wavefunction $\psi_0(x) = C_0 \exp(-m\omega x^2/2\hbar)$, with $C_0 = (m\omega/\pi\hbar)^{1/4}$ from Example 5.12. The probability in question is found by integrating the probability density $|\psi_0|^2$ in the region beyond the classical limits of vibration, or

$$P = \int_{-\infty}^{-A} |\psi_0|^2 \, dx + \int_{A}^{\infty} |\psi_0|^2 \, dx$$

From the symmetry of ψ_0, the two integrals contribute equally to P, so

$$P = 2\left(\frac{m\omega}{\pi\hbar}\right)^{1/2} \int_{A}^{\infty} e^{-m\omega x^2/\hbar} \, dx$$

Changing variables from x to $z = \sqrt{m\omega/\hbar}\; x$ and using $A = \sqrt{\hbar/m\omega}$ (corresponding to $z = 1$) leads to

$$P = \frac{2}{\sqrt{\pi}} \int_1^\infty e^{-z^2}\, dz$$

Expressions of this sort are encountered frequently in probability studies. With the lower limit of integration changed to a variable, say y, the result for P defines the *complementary error function* erfc(y). Values of the error function may be found in tables. In this way we obtain $P = \text{erfc}(1) = 0.157$, or about 16%.

To obtain excited states of the oscillator, a procedure can be followed similar to that for the ground state. The first excited state should be antisymmetric about the midpoint of the oscillator well ($x = 0$) and display exactly one node. By virtue of the antisymmetry, this node must occur at the origin, so that a suitable trial solution would be $\psi(x) = x\exp(-\alpha x^2)$. Substituting this form into Equation 5.25 yields the same α as before, along with the first excited state energy $E_1 = \frac{3}{2}\hbar\omega$.

Continuing in this manner, we could generate ever-higher-lying oscillator states with their respective energies, but the procedure rapidly becomes too laborious to be practical. What is needed is a systematic approach, such as that provided by the method of *power series expansion*.[10] Pursuing this method would take us too far afield, but the result for the allowed oscillator energies is quite simple, and sufficiently important that it be included here:

Energy levels for the harmonic oscillator

$$E_n = (n + \tfrac{1}{2})\hbar\omega \qquad n = 0, 1, 2, \ldots \qquad (5.29)$$

The energy level diagram following from Equation 5.29 is given in Figure 5.18. Note the uniform spacing of levels, widely recognized as the hallmark of the harmonic oscillator spectrum. The energy difference between adjacent levels is just $\Delta E = \hbar\omega$. In these results we find, for the first time, the quantum justification for Planck's revolutionary hypothesis concerning his *cavity resonators* (see Section 2.2). In deriving his blackbody radiation formula, Planck assumed that these resonators (oscillators), which made up the cavity walls, could possess only those energies that were multiples of $hf = \hbar\omega$. Although Planck could not have foreseen the zero-point energy $\hbar\omega/2$, it would make no difference: his resonators still would emit or absorb light energy in the bundles $\Delta E = hf$ necessary to the blackbody spectrum.

The probability densities for some of the oscillator states are plotted in Figure 5.19. The dashed lines, representing the classical probability densities for the same energy, are provided for comparison (see Problem 27 for the calculation of classical probabilities). Note that as n increases, agreement between the classical and quantum probabilities improves, as expected from the correspondence principle.

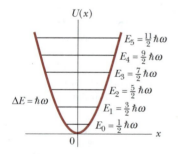

Figure 5.18 Energy level diagram for the quantum oscillator. Note that the levels are equally spaced, with a separation equal to $\hbar\omega$. The ground state energy is E_0.

$U(x)$

$E_5 = \frac{11}{2}\hbar\omega$
$E_4 = \frac{9}{2}\hbar\omega$
$E_3 = \frac{7}{2}\hbar\omega$
$E_2 = \frac{5}{2}\hbar\omega$
$E_1 = \frac{3}{2}\hbar\omega$
$E_0 = \frac{1}{2}\hbar\omega$
$\Delta E = \hbar\omega$

[10] The method of power series expansion as applied to the problem of the quantum oscillator is developed in any more advanced quantum mechanics text. See, for example, E. E. Anderson, *Modern Physics and Quantum Mechanics*, Philadelphia, W. B. Saunders Company, 1971.

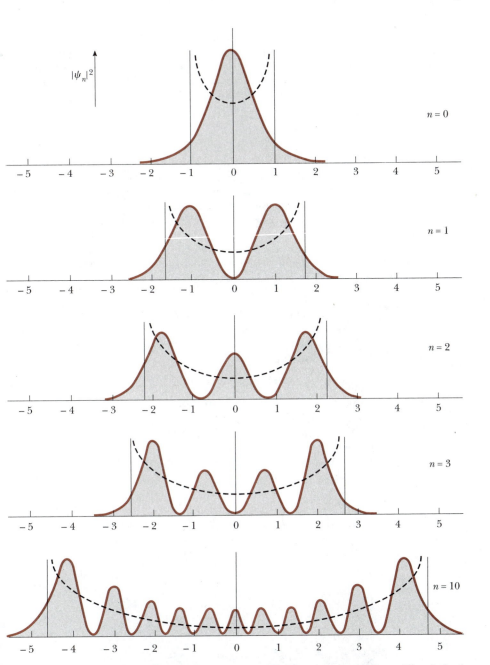

Figure 5.19   Probability densities for a few states of the quantum oscillator. The dashed curves represent the classical probabilities corresponding to the same energies.

EXAMPLE 5.15   Quantization of Vibrational Energy

The energy of a quantum oscillator is restricted to be one of the values $(n + \frac{1}{2})\hbar\omega$. How can this quantization apply to the motion of a mass on a spring, which seemingly can vibrate with any amplitude (energy) whatever?

Solution   The discrete values for the allowed energies of the oscillator would go unnoticed if the spacing between adjacent levels were too small to be detected. At the macroscopic level, a laboratory mass m of, say, 0.0100 kg on a spring having force constant $K = 0.100$ N/m (a typical value) would oscillate with angular frequency $\omega = \sqrt{K/m} = 3.16$ rad/s. The corresponding period of vibration is $T = 2\pi/\omega = 1.99$ s. In this case the quantum level spacing is only

$$\Delta E = \hbar\omega = (6.582 \times 10^{-16} \text{ eV·s})(3.16 \text{ rad/s})$$
$$= 2.08 \times 10^{-15} \text{ eV}$$

Such small energies are far below present limits of detection.

At the atomic level, however, much higher frequencies are commonplace. Consider the vibrational frequency of the hydrogen molecule. This behaves as an oscillator with $K = 510.5$ N/m and reduced mass $\mu = 8.37 \times 10^{-28}$ kg. The angular frequency of oscillation is therefore

$$\omega = \sqrt{\frac{K}{\mu}} = \sqrt{\frac{510.5 \text{ N/m}}{8.37 \times 10^{-28} \text{ kg}}} = 7.81 \times 10^{14} \text{ rad/s}$$

At such frequencies, the quantum of energy $\hbar\omega$ is 0.513 eV, which can be measured easily!

EXAMPLE 5.16   Oscillator Wavefunctions and Energies: A Simulation

MP This example uses the MP DESKTOP to exhibit wavefunctions for several of the lowest energy states of the quantum oscillator.

Begin by readying the DESKTOP to display quantum wavefunctions, as described in Example 5.10. Choose the harmonic oscillator potential $U(x)$ by loading the file

EX5-16.U (click on U(x)). Note the simple form $U(x) = x^2$; this results from measuring distances and energies in units of $(\hbar/m\omega)^{1/2}$ and $\frac{1}{2}\hbar\omega$, respectively, where $\omega = \sqrt{K/m}$ is the classical frequency of oscillation. $U(x)$ is defined on the interval extending from $x = -5$ to $x = +5$ (in units of $(\hbar/m\omega)^{1/2}$). Accept this potential by selecting *OK*.

NOTE: The unit of length (δ) and the unit of energy (ϵ) must satisfy $\epsilon \cdot \delta^2 = \hbar^2/2m$ in order to map the equation in the task field to the Schrödinger equation. In this example we choose $\epsilon = \frac{1}{2}\hbar\omega$ for convenience; then $\delta^2 = \hbar/m\omega$ as given. Notice that ϵ is just the ground state energy of the quantum oscillator, whereas δ is the vibration amplitude of a *classical* oscillator with the same energy ϵ. For a more exhaustive discussion of units, see the *Help-Contents* entry on the DESKTOP menu bar.

The allowed energies of the quantum oscillator are $(n + \frac{1}{2})\hbar\omega$, or $E = 2n + 1$ in program units. In the energy box just below the menu bar, type in $E = 1$ and press *Enter* to display the ground state wavefunction ($n = 0$).

Click ▨ to activate the MP Analyzer. With the mouse (or the direction keys [→], [←]), position the graphics cursor at those locations x where $U(x) = E$, and mark them by pressing the [.] key. These are the classical turning points (Fig. 5.20). The region beyond, where $U(x) > E$, is energetically forbidden to a classical

Figure 5.20   (Example 5.16) Ground state wavefunction of the quantum oscillator, against the backdrop of the oscillator well. The vertical lines mark the classical turning points, where $U(x) = E$.

object, because its kinetic energy would have to be negative there. Notice how much of the ground state wavefunction "spills over" into these forbidden regions (Example 5.14 shows how this nonclassical effect can be quantified.)

Explore several other oscillator wavefunctions by entering appropriate values for E in the energy box. Notice how the symmetry of the wavefunctions alternates between even and odd as the energy increases. Also count the nodes of the wavefunction to verify the rule that the nth excited state has exactly n nodes (Fig. 5.21).

NOTE: In exploring these wavefunctions you may find it necessary to scroll the image up, down, or sideways. Do this by clicking on the vertical or horizontal scroll bars bordering the graphics field. Also, the image may be rescaled horizontally or vertically by clicking the Horizontal Scale button ◀▮▶ or the Vertical Scale button ▲▼, respectively. Finally, you can zoom in or out by clicking the Zoom button.

Figure 5.21   (Example 5.16) The third excited state ($n = 3$) of the quantum oscillator. The wavefunction possesses odd symmetry about the center of the oscillator well, and exhibits three nodes.

5.7   EXPECTATION VALUES

It should be evident by now that two distinct types of measurable quantities are associated with a given wavefunction $\Psi(x, t)$. One type—like the energy E for the stationary states—is fixed by the quantum number labeling the wave. Therefore, every measurement of this quantity performed on the system described by Ψ yields the *same* value. Quantities such as E we call *sharp* to distinguish them from others—like the position x—for which the wavefunction Ψ furnishes only probabilities. We say x is an example of a dynamic quantity that is *fuzzy*. In the following paragraphs we discuss what more can be learned about these "fuzzy" quantities.

Sharp and fuzzy variables

A particle described by the wavefunction Ψ may occupy various places x with probability given by the wave intensity there $|\Psi(x)|^2$. Predictions made this way from Ψ can be tested by making repeated measurements of the particle position. Table 5.1 shows results that might be obtained in a hypothetical experiment of this sort. The table consists of 18 entries, each one representing the actual position of the particle recorded in that particular measurement. We see that the entry 5.4 occurs most often (in 3 of the 18 trials); it represents the most probable position based on the data available. The probability associated with this position, again based on the available data, is $3/18 = 0.167$. Of course, these numbers will fluctuate as additional measurements are taken, but they should approach limiting values. The theoretical predictions refer to these limiting values. Obviously, a good test of the theory would require much more data than we have shown in this illustration.

The information in Table 5.1 also can be used to find the average position of the particle:

$$\bar{x} = \frac{(2.5 + 3.7 + 1.4 + \ldots + 5.4 + 5.3)}{18} = 5.46$$

This same number can be found in a different way. First, order the table entries by value, starting with the smallest: 1.4, 2.5, 3.3, . . . , 5.4, 6.2, . . . , 8.0, 8.8. Now take each value, multiply by its frequency of occurrence, and sum the results:

$$1.4 \left(\frac{1}{18}\right) + 2.5 \left(\frac{1}{18}\right) + \ldots + 5.4 \left(\frac{3}{18}\right)$$
$$+ 6.2 \left(\frac{2}{18}\right) + \ldots + 8.8 \left(\frac{1}{18}\right) = 5.46$$

The two procedures are equivalent, but the latter involves a sum over *ordered values* rather than individual table entries. We may generalize this last expression to include other values for the position of the particle, provided we weight each one by its observed frequency of occurrence (in this case zero). This allows

Table 5.1 Hypothetical Data Set for Position of a Particle as Recorded in Repeated Trials

Trial	Position [arbitrary units]	Trial	Position [arbitrary units]	Trial	Position [arbitrary units]
1	$x_1 = 2.5$	7	$x_7 = 8.0$	13	$x_{13} = 4.2$
2	$x_2 = 3.7$	8	$x_8 = 6.4$	14	$x_{14} = 8.8$
3	$x_3 = 1.4$	9	$x_9 = 4.1$	15	$x_{15} = 6.2$
4	$x_4 = 7.9$	10	$x_{10} = 5.4$	16	$x_{16} = 7.1$
5	$x_5 = 6.2$	11	$x_{11} = 7.0$	17	$x_{17} = 5.4$
6	$x_6 = 5.4$	12	$x_{12} = 3.3$	18	$x_{18} = 5.3$

us to write a general prescription to calculate the average particle position from any data set:

$$\bar{x} = \sum xP_x \qquad (5.30)$$

The sum now includes all values of x, each weighted by its frequency or probability of occurrence P_x. Because the possible values of x are distributed continuously over the entire range of real numbers, the sum in Equation 5.30 really should be an integral, and P_x should refer to the probability of finding the particle in the infinitesimal interval dx about the point x; that is, the probability $P_x \rightarrow P(x)\ dx$, where $P(x)$ is the probability density. In quantum mechanics, $P(x) = |\Psi|^2$ and the average value of x, written $\langle x \rangle$, is called the **expectation value.** Then

$$\langle x \rangle = \int_{-\infty}^{\infty} x|\Psi(x, t)|^2\ dx \qquad (5.31)$$

Average position of a particle

Notice that $\langle x \rangle$ may be a function of time. For a stationary state, however, $|\Psi|^2$ is static and, as a consequence, $\langle x \rangle$ is independent of t.

In similar fashion we find that the average or expectation value for *any* function of x, say $f(x)$, is

$$\langle f \rangle = \int_{-\infty}^{\infty} f(x)|\Psi|^2\ dx \qquad (5.32)$$

With $f(x) = U(x)$, Equation 5.32 becomes $\langle U \rangle$, the average potential energy of the particle. With $f(x) = x^2$ the quantum uncertainty in particle position may be found. To see how this is done, we return to Table 5.1 and notice that the entries scatter about the average value. The amount of scatter is measured by the standard deviation σ of the data, defined as

$$\sigma = \sqrt{\frac{\Sigma(x_i - \bar{x})^2}{N}} \qquad (5.33)$$

where N is the number of data points, in this case 18. Writing out the square under the radical gives

$$\frac{\Sigma(x_i)^2}{N} - 2(\bar{x})\frac{\Sigma(x_i)}{N} + (\bar{x})^2 \sum \left(\frac{1}{N}\right) = \overline{(x^2)} - 2(\bar{x})(\bar{x}) + (\bar{x})^2$$
$$= \overline{(x^2)} - (\bar{x})^2$$

and so

$$\sigma = \sqrt{\overline{(x^2)} - (\bar{x})^2}$$

From Equation 5.33 we see that if the standard deviation were zero, all data entries would be identical and equal to the average. In that case the distribution is *sharp;* otherwise the data exhibit some spread (as in Table 5.1), and the standard deviation is greater than zero. In quantum mechanics the standard deviation, written Δx, is often called the *uncertainty* in position. The preceding development implies that the quantum uncertainty in position can be calculated from expectation values as

$$\Delta x = \sqrt{\langle x^2 \rangle - \langle x \rangle^2} \tag{5.34}$$

The degree to which particle position is fuzzy is given by the magnitude of Δx: Note that the position is *sharp* only if $\Delta x = 0$.

EXAMPLE 5.17   Standard Deviation from Averages

Compute $\overline{(x^2)}$ and the standard deviation for the data given in Table 5.1.

Solution   Squaring the data entries of Table 5.1 and adding the results gives $\Sigma(x_i)^2 = 603.91$. Dividing this by the number of data points, $N = 18$, we find $\overline{(x^2)} = 603.91/18 = 33.55$. Then

$$\sigma = \sqrt{33.55 - (5.46)^2} = 1.93$$

for this case.

EXAMPLE 5.18   Location of a Particle in a Box

Compute the average position $\langle x \rangle$ and the quantum uncertainty in this value, Δx, for the particle in a box, assuming it is in the ground state.

Solution   The possible particle positions within the box are weighted according to the probability density given by $|\Psi|^2 = (2/L) \sin^2(n\pi x/L)$, with $n = 1$ for the ground state. The average position is calculated as

$$\langle x \rangle = \int_{-\infty}^{\infty} x |\Psi|^2 \, dx = (2/L) \int_0^L x \sin^2 \left(\frac{\pi x}{L} \right) dx$$

Making the change of variable $\theta = \pi x/L$ (so that $d\theta = \pi \, dx/L$) gives

$$\langle x \rangle = \frac{2L}{\pi^2} \int_0^\pi \theta \sin^2 \theta \, d\theta$$

The integral is evaluated with the help of the trigonometric identity $2 \sin^2 \theta = 1 - \cos 2\theta$, giving

$$\langle x \rangle = \frac{L}{\pi^2} \left(\int_0^\pi \theta \, d\theta - \int_0^\pi \theta \cos 2\theta \, d\theta \right)$$

An integration by parts shows that the second integral vanishes, whereas the first integrates to $\pi^2/2$. Thus, $\langle x \rangle =$

$L/2$ as expected, because there is equal probability of finding the particle in the left half or the right half of the box.

$\langle x^2 \rangle$ is computed in much the same way, but with an extra factor of x in the integrand. After changing variables to $\theta = \pi x/L$, we get

$$\langle x^2 \rangle = \frac{L^2}{\pi^3} \left(\int_0^\pi \theta^2 \, d\theta - \int_0^\pi \theta^2 \cos 2\theta \, d\theta \right)$$

The first integral evaluates to $\pi^3/3$; the second may be integrated twice by parts to get

$$\int_0^\pi \theta^2 \cos 2\theta \, d\theta = - \int_0^\pi \theta \sin 2\theta \, d\theta$$

$$= \frac{1}{2} \theta \cos 2\theta \, \Big|_0^\pi = \pi/2$$

Then

$$\langle x^2 \rangle = \frac{L^2}{\pi^3} \left(\frac{\pi^3}{3} - \frac{\pi}{2} \right) = \frac{L^2}{3} - \frac{L^2}{2\pi^2}$$

Finally, the uncertainty in position for this particle is

$$\Delta x = \sqrt{\langle x^2 \rangle - \langle x \rangle^2} = L \sqrt{\frac{1}{3} - \frac{1}{2\pi^2} - \frac{1}{4}} = 0.181L$$

This is an appreciable figure, amounting to nearly one fifth the size of the box. Consequently, the whereabouts of such a particle are largely unknown. With some confidence, we may assert only that the particle is likely to be in the range $L/2 \pm 0.181L$.

Finally, notice that none of these results depend on the time, because in a stationary state t enters only through the exponential factor $e^{-i\omega t}$, which cancels when Ψ is combined with Ψ^* in the calculation of averages. Thus, it is generally true that, *in a stationary state, all averages, as well as probabilities, are time-independent.*

We have learned how to predict the average position of a particle $\langle x \rangle$, the uncertainty in this position Δx, the average potential energy of the particle $\langle U \rangle$, and so on. But what about the average momentum $\langle p \rangle$ of the particle, or its average kinetic energy $\langle K \rangle$? These could be calculated if $p(x)$, the momentum as a function of x, were known. In classical mechanics, $p(x)$ may be obtained from the equation for the classical path $x(t)$. Differentiating this function once gives the velocity $v(t)$. Then inverting $x(t)$ to get t as a function of x, and substituting this result into $v(t)$, gives $v(x)$ and the desired relation $p(x) = mv(x)$. In quantum mechanics, however, *x and t are independent variables—there is no path*, nor any function connecting p with x! If there were, then p could be found from x using $p(x)$, and both x and p would be known precisely, in violation of the uncertainty principle.

To obtain $\langle p \rangle$ we must use a different approach: we identify the time derivative of the average particle position with the average velocity of the particle. After multiplication by m, this gives the average momentum $\langle p \rangle$:

$$\langle p \rangle = m \frac{d\langle x \rangle}{dt} \tag{5.35}$$

Average momentum of a particle

Equation 5.35 cannot be derived from anything we have said previously. When applied to macroscopic objects where the quantum uncertainties in position and momentum are small, the averages $\langle x \rangle$ and $\langle p \rangle$ become indistinguishable from "the" position and "the" momentum of the object, and Equation 5.35 reduces to the classical definition of momentum.

Another expression for $\langle p \rangle$ follows from Equation 5.35 by substituting $\langle x \rangle$ from Equation 5.31 and differentiating under the integral sign. Using Schrödinger's equation to eliminate time derivatives of Ψ and its conjugate Ψ^* gives (after much manipulation!)

$$\langle p \rangle = \int_{-\infty}^{\infty} \Psi^* \left(\frac{\hbar}{i} \right) \frac{\partial \Psi}{\partial x} \, dx \tag{5.36}$$

Exercise 6   Show that $\langle p \rangle = 0$ for *any* state of a particle in a box.

5.8   OBSERVABLES AND OPERATORS

An **observable** *is any particle property that can be measured.* The position and momentum of a particle are observables, as are its kinetic and potential energies.[11] Quantum mechanics associates an *operator* with each of these observables. Using this operator, one can calculate the average value of the corresponding observable. An operator here refers to an operation to be performed on whatever function follows the operator. The quantity operated on is called the *operand*. In this language a constant c becomes an operator, whose meaning is

---

[11] By contrast, the wavefunction Ψ, although clearly indispensable to the quantum description, is not directly measurable and so is *not* an observable.

understood by supplying any function $f(x)$ to obtain $cf(x)$. Here the operator c means "multiplication by the constant c." A more complicated operator is d/dx which, after supplying an operand $f(x)$, clearly means "take the derivative of $f(x)$ with respect to x." Still another example is $(d/dx)^2 = (d/dx)(d/dx)$. Supplying the operand $f(x)$ gives $(d/dx)^2 f(x) = (d/dx)(df/dx) = d^2f/dx^2$. Hence, $(d/dx)^2$ means "take the second derivative with respect to x, that is, take the indicated derivative twice."

The operator concept is useful in quantum mechanics because all expectation values we have encountered so far can be written in the same general form, namely,

$$\langle Q \rangle = \int_{-\infty}^{\infty} \Psi^*[Q]\Psi \, dx \tag{5.37}$$

Operators in quantum mechanics

In this expression, Q is the observable and $[Q]$ is the associated operator. The *order* of terms in Equation 5.37 *is* important; it indicates that the operand for $[Q]$ always is Ψ. Comparing the general form with that for $\langle p \rangle$ in Equation 5.36 shows that the momentum operator is $[p] = (\hbar/i)(\partial/\partial x)$. Similarly, writing $x|\Psi|^2 = \Psi^* x \Psi$ in Equation 5.31 implies that the operator for position is $[x] = x$. From $[x]$ and $[p]$ the operator for any other observable can be found. For instance, the operator for x^2 is just $[x^2] = [x]^2 = x^2$. For that matter, the operator for potential energy is simply $[U] = U([x]) = U(x)$, meaning that average potential energy is computed as

$$\langle U \rangle = \int_{-\infty}^{\infty} \Psi^*[U]\Psi \, dx = \int_{-\infty}^{\infty} \Psi^* U(x)\Psi \, dx$$

Still another example is the kinetic energy K. Classically, K is a function of p: $K = p^2/2m$. Then the kinetic energy operator is $[K] = ([p])^2/2m = (-\hbar^2/2m)\partial^2/\partial x^2$, and average kinetic energy is found from

$$\langle K \rangle = \int_{-\infty}^{\infty} \Psi^*[K]\Psi \, dx = \int_{-\infty}^{\infty} \Psi^*\left(-\frac{\hbar^2}{2m}\frac{\partial^2\Psi}{\partial x^2}\right) dx$$

To find the average total energy for a particle, we sum the average kinetic and potential energies to get

$$\langle E \rangle = \langle K \rangle + \langle U \rangle = \int_{-\infty}^{\infty} \Psi^*\left\{-\frac{\hbar^2}{2m}\frac{\partial^2}{\partial x^2} + U(x)\right\}\Psi \, dx \tag{5.38}$$

The form of this result suggests that the term in the braces is the operator for total energy. This operator is called the **hamiltonian,** symbolized by $[H]$:

$$[H] = -\frac{\hbar^2}{2m}\frac{\partial^2}{\partial x^2} + U(x) \tag{5.39}$$

The designation $[E]$ is reserved for another operator, which arises as follows: Inspection of Schrödinger's equation (Equation 5.10) shows that it can be written neatly as $[H]\Psi = i\hbar\partial\Psi/\partial t$. Using this in Equation 5.38 gives an equivalent expression for $\langle E \rangle$, and leads to the identification of the *energy operator*

$$[E] = i\hbar\frac{\partial}{\partial t} \tag{5.40}$$

Table 5.2   Common Observables and Associated Operators

Observable	Symbol	Associated Operator
position	x	x
momentum	p	$\dfrac{\hbar}{i}\dfrac{\partial}{\partial x}$
potential energy	U	$U(x)$
kinetic energy	K	$-\dfrac{\hbar^2}{2m}\dfrac{\partial^2}{\partial x^2}$
hamiltonian	H	$-\dfrac{\hbar^2}{2m}\dfrac{\partial^2}{\partial x^2} + U(x)$
total energy	E	$i\hbar\dfrac{\partial}{\partial t}$

Notice that $[H]$ is an operation involving only the spatial coordinate x, whereas $[E]$ depends only on the time t. That is, $[H]$ and $[E]$ really are two different operators, but they produce identical results when applied to any solution of Schrödinger's equation. This is because the LHS of Schrödinger's equation is simply $[H]\Psi$, while the RHS is none other than $[E]\Psi$ (cf. Equation 5.10)! Table 5.2 summarizes the observables we have discussed and their associated operators.

Quantum Uncertainty and the Eigenvalue Property

In Section 5.7 we showed how Δx, the quantum uncertainty in position, could be found from the expectation values $\langle x^2 \rangle$ and $\langle x \rangle$. But the argument given there applies to *any* observable, that is, the quantum uncertainty ΔQ for any observable Q is calculated as

$$\Delta Q = \sqrt{\langle Q^2 \rangle - \langle Q \rangle^2} \qquad (5.41)$$

Again, if $\Delta Q = 0$, Q is said to be a *sharp* observable, and all measurements of Q yield the *same value*. More often, however, $\Delta Q > 0$ and repeated measurements reveal a distribution of values—as in Table 5.1 for the observable x. In such cases, we say the observable is *fuzzy*, suggesting that, prior to actual measurement, the particle cannot be said to possess a unique value of Q.

In classical physics all observables are sharp. The extent to which sharp observables can be specified in quantum physics is limited by uncertainty principles, such as

$$\Delta x \, \Delta p_x \geqslant \frac{1}{2}\hbar \qquad (5.42)$$

The uncertainties here are to be calculated from Equation 5.41. Equation 5.42 says that, no matter what the state of the particle, the spread in distributions obtained in measurements of x and of p_x will be inversely related; when one is small, the other

O P T I O N A L

Quantum uncertainty for any observable Q

will be large. Alternatively, if the position of the particle is quite "fuzzy," its momentum can be relatively "sharp," and vice versa. The degree to which both may be simultaneously sharp is limited by the size of $\hbar$. The incredibly small value of $\hbar$ in SI units is an indication that quantum ideas are unnecessary at the macroscopic level.

Despite restrictions imposed by uncertainty principles, some observables in quantum physics may still be sharp. The energy E of all stationary states is one example. In the free particle plane waves of Section 5.2 we have another: the plane wave with wavenumber k

$$\Psi_k(x,\ t) = e^{i(kx - \omega t)}$$

describes a particle with momentum $p = \hbar k$. Evidently, momentum is a sharp observable for this wavefunction. We find that the action of the momentum operator in this instance is especially simple:

$$[p]\Psi_k(x,\ t) = \left(\frac{\hbar}{i}\frac{\partial}{\partial x}\right)e^{i(kx-\omega t)} = \hbar k \Psi_k(x,\ t)$$

that is, *the operation [p] returns the original function multiplied by a constant*. This is an example of an *eigenvalue problem* for the operator $[p]$.[12] The wavefunction Ψ_k is the **eigenfunction** and the constant, in this case $\hbar k$, is the **eigenvalue**. Notice that the eigenvalue is just the sharp value of particle momentum for this wave. This connection between sharp observables and eigenvalues is a general one: *for an observable Q to be sharp, the wavefunction must be an eigenfunction of the operator for Q. Further, the sharp value for Q in this state is the eigenvalue*. In this way the eigenvalue property can serve as a simple *test* for sharp observables, as the following examples illustrate.

Eigenfunctions and eigenvalues

EXAMPLE 5.19   Plane Waves and Sharp Observables

Use the eigenvalue test to show that the plane wave $\Psi_k(x,\ t) = e^{i(kx-\omega t)}$ is one for which total energy is a sharp observable. What value does the energy take in this case?

Solution   To decide the issue we examine the action of the energy operator $[E]$ on the candidate function $e^{i(kx-\omega t)}$. Since taking a derivative with respect to t of this function is equivalent to multiplying the function by $-i\omega$, we have

$$[E]e^{i(kx-\omega t)} = \left(i\hbar\frac{\partial}{\partial t}\right)e^{i(kx-\omega t)} = \hbar\omega\, e^{i(kx-\omega t)}$$

showing that $e^{i(kx-\omega t)}$ is an eigenfunction of the energy operator $[E]$ and the eigenvalue is $\hbar\omega$. Thus, energy is a sharp observable and has the value $\hbar\omega$ in this state.

It is instructive to compare this result with the outcome found by using the other energy operator $[H]$. The hamiltonian for a free particle is simply the kinetic energy operator $[K]$, because the potential energy is zero in this case. Then

$$[H]e^{i(kx-\omega t)} = \left(-\frac{\hbar^2}{2m}\frac{\partial^2}{\partial x^2}\right)e^{i(kx-\omega t)}$$

$$= \left(-\frac{\hbar^2}{2m}\right)(ik)^2 e^{i(kx-\omega t)}$$

---

[12] The eigenvalue problem for any operator $[Q]$ is $[Q]\psi = q\psi$; that is, the result of the operation $[Q]$ on some function ψ is simply to return a multiple q of the same function. This is possible only for certain special functions ψ, the *eigenfunctions*, and then only for certain special values of q, the *eigenvalues*. Generally, $[Q]$ is known; the eigenfunctions and eigenvalues are found by imposing the eigenvalue condition.

Again, the operation returns the original function with a multiplier, so that $e^{i(kx - \omega t)}$ also is an eigenfunction of $[H]$. The eigenvalue in this case is $\hbar^2 k^2 / 2m$, which also must be the sharp value of particle energy. The equivalence with $\hbar \omega$ follows from the dispersion relation for free particles (cf. Footnote 1).

Exercise 7   Show that total energy is a sharp observable for *any* stationary state.

EXAMPLE 5.20   Sharp Observables for a Particle in a Box

Are the stationary states of the infinite square well eigenfunctions of $[p]$? of $[p]^2$? If so, what are the eigenvalues? Discuss the implications of these results.

Solution   The candidate function in this case is any one of the square well wave-functions $\Psi(x, t) = \sqrt{2/L} \sin(n\pi x / L) \, e^{-iE_n t / \hbar}$. Because the first derivative gives $(d/dx) \sin(n\pi x / L) = (n\pi / L) \cos(n\pi x / L)$, we see at once that the operator $[p]$ will *not* return the original function Ψ, and so these are not eigenfunctions of the momentum operator. They *are*, however, eigenfunctions of $[p]^2$. In particular, we have $(d^2 / dx^2) \sin(n\pi x / L) = -(n\pi / L)^2 \sin(n\pi x / L)$, so that

$$[p]^2 \Psi(x, t) = -(\hbar / i)^2 \left(\frac{n\pi}{L} \right)^2 \Psi(x, t)$$

$$= \left(\frac{n\pi \hbar}{L} \right)^2 \Psi(x, t)$$

The eigenvalue is the multiplier $(n\pi \hbar / L)^2$. Thus, the squared-momentum (or magnitude of momentum) is sharp for such states, and repeated measurements of p^2 (or $|p|$) for the state labeled by n will give identical results equal to $(n\pi \hbar / L)^2$ (or $n\pi \hbar / L$). By contrast, the momentum itself is not sharp, meaning that different values for p will be obtained in successive measurements. In particular, it is the sign or direction of momentum that is fuzzy, consistent with the classical notion of a particle bouncing back and forth between the walls of the "box."

SUMMARY

In quantum mechanics, matter waves (or de Broglie waves) are represented by a wavefunction $\Psi(x, t)$. The probability that a particle constrained to move along the x axis will be found in an interval dx at time t is given by $|\Psi|^2 \, dx$. These probabilities summed over all values of x must total 1 (certainty). That is,

$$\int_{-\infty}^{\infty} |\Psi|^2 \, dx = 1 \tag{5.2}$$

This is called the **normalization condition.** Furthermore, the probability that the particle will be found in any interval $a \leqslant x \leqslant b$ is obtained by integrating the **probability density** $|\Psi|^2$ over this interval.

Aside from furnishing probabilities, the wavefunction can be used to find the average, or **expectation value,** of any dynamical quantity. The average position of a particle at any time t is

$$\langle x \rangle = \int_{-\infty}^{\infty} \Psi^* x \Psi \, dx \tag{5.31}$$

In general, the average value of any observable Q at time t is

$$\langle Q \rangle = \int_{-\infty}^{\infty} \Psi^*[Q]\Psi \, dx \qquad (5.37)$$

where $[Q]$ is the associated operator. The operator for position is just $[x] = x$, and that for particle momentum is $[p] = (\hbar/i)\partial/\partial x$.

The wavefunction Ψ must satisfy the **Schrödinger equation**

$$-\frac{\hbar^2}{2m}\frac{\partial^2\Psi}{\partial x^2} + U(x)\Psi(x, t) = i\hbar\frac{\partial\Psi}{\partial t} \qquad (5.10)$$

Separable solutions to this equation, called **stationary states**, are $\Psi(x, t) = \psi(x)e^{-i\omega t}$, with $\psi(x)$ a time-independent wavefunction satisfying the **time-independent Schrödinger equation**

$$-\frac{\hbar^2}{2m}\frac{d^2\psi}{dx^2} + U(x)\psi(x) = E\psi(x) \qquad (5.13)$$

The approach of quantum mechanics is to solve Equation 5.13 for ψ and E, given the potential energy $U(x)$ for the system. In doing so, we must require

- that $\psi(x)$ be continuous,
- that $\psi(x)$ be finite for all x including $x = \pm\infty$,
- that $\psi(x)$ be single-valued, and
- that $d\psi/dx$ be continuous wherever $U(x)$ is finite.

Explicit solutions to Schrödinger's equation can be found for several potentials of special importance. For a free particle the stationary states are the **plane waves** $\psi(x) = e^{ikx}$ of wavenumber k and energy $E = \hbar^2 k^2/2m$. The particle momentum in such states is $p = \hbar k$, but the location of the particle is completely unknown. A free particle known to be in some range Δx is described not by a plane wave, but by a **wave packet,** or **group,** formed from a superposition of plane waves. The momentum of such a particle is not known precisely, but only to some accuracy Δp that is related to Δx by the **uncertainty principle**

$$\Delta x \, \Delta p_x \geq \tfrac{1}{2}\hbar \qquad (5.42)$$

For a particle confined to a one-dimensional box of length L, the stationary state waves are those for which an integral number of half-wavelengths can be fit inside, that is $L = n\lambda/2$. In this case the energies are **quantized** as

$$E_n = \frac{n^2\pi^2\hbar^2}{2mL^2} \qquad n = 1, 2, 3, \ldots \qquad (5.17)$$

and the wavefunctions within the box are given by

$$\psi_n(x) = \sqrt{\frac{2}{L}} \sin\left(\frac{n\pi x}{L}\right) \qquad n = 1, 2, 3, \ldots \qquad (5.18)$$

For the **harmonic oscillator** the potential energy function is $U(x) = \frac{1}{2}m\omega^2 x^2$, and the total particle energy is quantized according to the relation

$$E_n = \left(n + \frac{1}{2}\right)\hbar\omega \qquad n = 0, 1, 2, \ldots \qquad (5.29)$$

The lowest energy is $E_0 = \frac{1}{2}\hbar\omega$; the separation between adjacent energy levels is uniform and equal to $\hbar\omega$. The wavefunction for the oscillator ground state is

$$\psi_0(x) = C_0 e^{-\alpha x^2} \tag{5.26}$$

where $\alpha = m\omega/2\hbar$ and C_0 is a normalizing constant. The oscillator results apply to any system executing small-amplitude vibrations about a point of stable equilibrium. The effective spring constant in the general case is

$$K = m\omega^2 = \left.\frac{d^2U}{dx^2}\right|_a \tag{5.24}$$

with the derivative of the potential evaluated at the equilibrium point a.

The stationary state waves for any potential share the following attributes:

- their time dependence is $e^{-i\omega t}$,
- they yield probabilities that are time-independent,
- all average values obtained from stationary states are time-independent, and
- the energy in any stationary state is a **sharp observable,** that is, repeated measurements of particle energy performed on identical systems always yield the same result, $E = \hbar\omega$.

For other observables, such as position, repeated measurements usually yield different results. We say these observables are **fuzzy.** Their inherent "fuzziness" is reflected by the spread in results about the average value, as measured by the standard deviation, or **uncertainty.** The uncertainty in any observable Q can be calculated from expectation values as

$$\Delta Q = \sqrt{\langle Q^2 \rangle - \langle Q \rangle^2} \tag{5.41}$$

SUGGESTIONS FOR FURTHER READING

1. L. de Broglie, *New Perspectives in Physics*, New York, Basic Books Inc., 1962. A series of essays by Louis de Broglie on various aspects of theoretical physics and on the history and philosophy of science. A large part of this work is devoted to de Broglie's developing attitudes toward the interpretation of wave mechanics and wave–particle duality.
2. For an in-depth look at the problems of interpretation and measurement surrounding the formalism of quantum theory see M. Jammer, *The Philosophy of Quantum Mechanics*, New York, John Wiley and Sons, Inc., 1974.
3. A concise, solid introduction to the basic principles of quantum physics with applications may be found in Chapter 41 of R. Serway's *Physics for Scientists and Engineers with Modern Physics* (4th ed.), Philadelphia, Saunders College Publishing, 1996.
4. A novel but delightfully refreshing exposition of quantum theory is presented by R. Feynman, R. Leighton, and M. Sands in *The Feynman Lectures on Physics Vol. III, Modern Physics*, Reading, Mass., Addison-Wesley Publishing Co., 1965. This work is more advanced and sophisticated than other introductory texts in the field, though still somewhat below the intermediate level.

QUESTIONS

1. The probability density at certain points for a particle in a box is zero, as seen in Figure 5.9. Does this imply that the particle cannot move across these points? Explain.
2. Discuss the relation between the zero-point energy and the uncertainty principle.
3. Consider a square well with one finite wall and one infinite wall. Compare the energy and momentum of a particle trapped in this well to the energy and momentum of an identical particle trapped in an infinite well with the same width.

4. Explain why a wave packet moves with the group velocity rather than with the phase velocity.

5. According to Section 5.2, a free particle can be represented by any number of waveforms, depending on the values chosen for the coefficients $a(k)$. What is the source of this ambiguity, and how is it resolved?

6. Because the Schrödinger equation can be formulated in terms of operators as $[H]\Psi = [E]\Psi$, is it incorrect to conclude from this the operator equivalence $[H] = [E]$?

7. For a particle in a box, the squared-momentum p^2 is a sharp observable, but the momentum itself is fuzzy. Explain how this can be so, and how it relates to the classical motion of such a particle.

PROBLEMS

5.1   The Born Interpretation

1. Of the functions graphed in Figure P5.1, which are candidates for the Schrödinger wavefunction of an actual physical system? For those that are not, state why they fail to qualify.

(a)

(b)

(c)

(d)

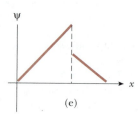

(e)

Figure P5.1

2. A particle is described by the wavefunction

$$\psi(x) = \begin{cases} A \cos\left(\dfrac{2\pi x}{L}\right) & \text{for } -\dfrac{L}{4} \le x \le \dfrac{L}{4} \\ 0 & \text{otherwise} \end{cases}$$

(a) Determine the normalization constant A. (b) What is the probability that the particle will be found between $x = 0$ and $x = L/8$ if a measurement of its position is made?

5.2   Wavefunction for a Free Particle

3. A free electron has a wavefunction

$$\psi(x) = A \sin(5 \times 10^{10}\, x)$$

where x is measured in meters. Find (a) the electron's de Broglie wavelength, (b) the electron's momentum, and (c) the electron's energy in electron volts.

4. *Spreading of a Gaussian Wave Packet.* The gaussian wave packet $\Psi(x, 0)$ of Example 5.3 is built out of plane waves according to the spectral function $a(k) = (C\alpha/\sqrt{\pi})\exp(-\alpha^2 k^2)$. Calculate $\Psi(x, t)$ for this packet and describe its evolution.

5.3   Wavefunctions in the Presence of Forces

5. In a region of space, a particle with zero energy has a wavefunction

$$\psi(x) = Axe^{-x^2/L^2}$$

(a) Find the potential energy U as a function of x.
(b) Make a sketch of $U(x)$ versus x.

6. The wavefunction of a particle is given by

$$\psi(x) = A \cos(kx) + B \sin(kx)$$

where A, B, and k are constants. Show that ψ is a solution of the Schrödinger equation (Eq. 5.13), assuming the particle is free ($U = 0$), and find the corresponding energy E of the particle.

5.4   The Particle in a Box

7. Show that allowing the state $n = 0$ for a particle in a one-dimensional box violates the uncertainty principle, $\Delta p\, \Delta x \ge \hbar/2$.

8. A bead of mass 5.00 g slides freely on a wire 20.0 cm long. Treating this system as a particle in a one-dimensional box, calculate the value of n corresponding to the state of the bead if it is moving at a speed of 0.100 nm per year (that is, apparently at rest).

9. The nuclear potential that binds protons and neutrons in the nucleus of an atom is often approximated by a square well. Imagine a proton confined in an infinite square well of length 10^{-5} nm, a typical nuclear diameter. Calculate the wavelength and energy associated

with the photon that is emitted when the proton undergoes a transition from the first excited state ($n = 2$) to the ground state ($n = 1$). In what region of the electromagnetic spectrum does this wavelength belong?

10. An electron is contained in a one-dimensional box of width 0.100 nm. (a) Draw an energy level diagram for the electron for levels up to $n = 4$. (b) Find the wavelengths of *all* photons that can be emitted by the electron in making transitions that would eventually get it from the $n = 4$ state to the $n = 1$ state.

11. Consider a particle moving in a one-dimensional box with walls at $x = -L/2$ and $x = L/2$. (a) Write the wavefunctions and probability densities for the states $n = 1$, $n = 2$, and $n = 3$. (b) Sketch the wavefunctions and probability densities. (*Hint:* Make an analogy to the case of a particle in a box with walls at $x = 0$ and $x = L$.)

12. A ruby laser emits light of wavelength 694.3 nm. If this light is due to transitions from the $n = 2$ state to the $n = 1$ state of an electron in a box, find the width of the box.

13. A proton is confined to moving in a one-dimensional box of width 0.200 nm. (a) Find the lowest possible energy of the proton. (b) What is the lowest possible energy of an electron confined to the same box? (c) How do you account for the large difference in your results for (a) and (b)?

14. A particle of mass m_0 is placed in a one-dimensional box of length L. The box is so small that the particle's motion is *relativistic*, so that $E = p^2/2m_0$ is *not valid*. (a) Derive an expression for the energy levels of the particle. (b) If the particle is an electron in a box of length $L = 1.00 \times 10^{-12}$ m, find its lowest possible kinetic energy. By what percent is the nonrelativistic formula for the energy in error?

15. Consider a "crystal" consisting of two nuclei and two electrons as shown in Figure P5.15. (a) Taking into account all the pairs of interactions, find the potential energy of the system as a function of d. (b) Assuming the electrons to be restricted to a one-dimensional box of length $3d$, find the minimum kinetic energy of the two electrons. (c) Find the value of d for which the total energy is a *minimum*. (d) Compare this value of d with the spacing of atoms in lithium, which has a density of 0.53 g/cm^3 and an atomic weight of 7. (This type of calculation can be used to estimate the densities of crystals and certain stars.)

16. An electron is trapped in an infinitely deep potential well 0.300 nm in width. (a) If the electron is in its ground state, what is the probability of finding it within 0.100 nm of the left-hand wall? (b) Repeat (a) for an electron in the 99th energy state above the ground state. (c) Are your answers consistent with the correspondence principle?

17. An electron is trapped at a defect in a crystal. The defect may be modeled as a one-dimensional, rigid-walled box of width 1.00 nm. (a) Sketch the wavefunctions and probability densities for the $n = 1$ and $n = 2$ states. (b) For the $n = 1$ state, find the probability of finding the electron between $x_1 = 0.15$ nm and $x_2 = 0.35$ nm, where $x = 0$ is the left side of the box. (c) Repeat (b) for the $n = 2$ state. (d) Calculate the energies in electron volts of the $n = 1$ and $n = 2$ states.

18. Find the points of maximum and minimum probability density for the nth state of a particle in a one-dimensional box. Check your result for the $n = 2$ state.

19. A 1.00-g marble is constrained to roll inside a tube of length $L = 1.00$ cm. The tube is capped at both ends. Modeling this as a one-dimensional infinite square well, find the value of the quantum number n if the marble is initially given an energy of 1.00 mJ. Calculate the *excitation energy* required to promote the marble to the next available energy state.

5.5 The Finite Square Well

20. Consider a particle with energy E bound to a *finite square well* of height U and width $2L$ situated on $-L \le x \le +L$. Because the potential energy is symmetric about the midpoint of the well, the stationary state waves will be either symmetric or antisymmetric about this point. (a) Show that for $E < U$, the conditions for smooth joining of the interior and exterior waves lead to the following equation for the allowed energies of the symmetric waves:

$$k \tan kL = \alpha \qquad \text{(symmetric case)}$$

where $\alpha = \sqrt{(2m/\hbar^2)(U - E)}$, and $k = \sqrt{2mE/\hbar^2}$ is the wavenumber of oscillation in the interior. (b) Show that the energy condition found in (a) can be rewritten as

$$k \sec kL = \frac{\sqrt{2mU}}{\hbar}$$

Apply the result in this form to an electron trapped at a defect site in a crystal, modeling the defect as a square well of height 5 eV and width 0.2 nm. Write a simple computer program to find the ground state energy for the electron. Give your answer accurate to ± 0.001 eV.

21. Sketch the wavefunction $\psi(x)$ and the probability density $|\psi(x)|^2$ for the $n = 4$ state of a particle in a *finite* potential well.

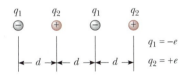

q_1 q_2 q_1 q_2

$q_1 = -e$
$q_2 = +e$

← d →← d →← d →

Figure P5.15

22. **MP** Apply the MP DESKTOP to a proton confined to an atomic nucleus, modeling the nucleus as a square well of width 1.00×10^{-5} nm and height 26.0 MeV. Determine the energies of the ground state and first excited state for this case. Calculate the wavelength of the photon emitted when the proton undergoes a transition from the first excited state to the ground state, and compare your result with the infinite-well model of Problem 9. Refer to Example 5.10 for particulars on using the DESKTOP for such applications.

23. Consider a square well having an infinite wall at $x = 0$ and a wall of height U at $x = L$ (Fig. P5.23). For the case $E < U$, obtain solutions to the Schrödinger equation inside the well $(0 \leqslant x \leqslant L)$ and in the region beyond $(x > L)$ that satisfy the appropriate boundary conditions at $x = 0$ and $x = \infty$. Enforce the proper matching conditions at $x = L$ to find an equation for the allowed energies of this system. Are there conditions for which no solution is possible? Explain.

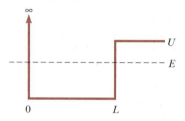

Figure P5.23

5.6 The Quantum Oscillator

24. **MP** Set up the MP DESKTOP to display the first excited state of the harmonic oscillator (see Example 5.16). Mark the classical turning points and estimate the depth of penetration of the wave into the nonclassical region. In what units does your result appear?

25. The wavefunction

$$\psi(x) = Cx\, e^{-\alpha x^2}$$

also describes a state of the quantum oscillator, provided the constant α is chosen appropriately. (a) Using Schrödinger's equation, obtain an expression for α in terms of the oscillator mass m and the classical frequency of vibration ω. What is the energy of this state? (b) Normalize this wave. (*Hint:* See the integral of Problem 33.)

26. Show that the oscillator energies in Equation 5.29 correspond to the classical amplitudes

$$A_n = \sqrt{\frac{(2n+1)\hbar}{m\omega}}$$

27. Obtain an expression for the probability density $P_c(x)$ of a classical oscillator with mass m, frequency ω, and amplitude A. (*Hint:* See Problem 29 for the calculation of classical probabilities.)

28. **MP** *Coherent States.* Set up the MP DESKTOP to display *QM WaveGroups*, following the procedure of Example 5.5, but this time load the harmonic oscillator potential $U(x)$ found in the file *EX5-16.U*. For the initial wavefunction, again take the gaussian waveform from the file *GAUSS.PSI*, but use for the width parameter $\beta = 0.5$. Describe the time evolution of the resulting wavefunction. Is it what you expected? Now displace the gaussian leftward by changing α from zero to $\alpha = -2$. Again describe the time evolution of the resulting wavefunction. What is remarkable about this case? Such wavefunctions, called *coherent states*, are important in the quantum theory of radiation.

5.7 Expectation Values

29. *Classical Probabilities.* (a) Show that the classical probability density describing a particle in an infinite square well of dimension L is $P_c(x) = 1/L$. (*Hint:* The classical probability for finding a particle in dx—$P_c(x)\, dx$—is proportional to the *time* the particle spends in this interval.) (b) Using $P_c(x)$, determine the *classical* averages $\langle x \rangle$ and $\langle x^2 \rangle$ for a particle confined to the well, and compare with the quantum results found in Example 5.18. Discuss your findings in light of the correspondence principle.

30. An electron is described by the wavefunction

$$\psi(x) = \begin{cases} 0 & \text{for } x < 0 \\ Ce^{-x}(1 - e^{-x}) & \text{for } x > 0 \end{cases}$$

where x is in nanometers and C is a constant. (a) Find the value of C that normalizes ψ. (b) Where is the electron most likely to be found; that is, for what value of x is the probability for finding the electron largest? (c) Calculate $\langle x \rangle$ for this electron and compare your result with its most likely position. Comment on any differences you find.

31. For any eigenfunction ψ_n of the infinite square well, show that $\langle x \rangle = L/2$ and that

$$\langle x^2 \rangle = \frac{L^2}{3} - \frac{L^2}{2(n\pi)^2}$$

where L is the well dimension.

32. An electron has a wavefunction

$$\psi(x) = Ce^{-|x|/x_0}$$

where x_0 is a constant and $C = 1/\sqrt{x_0}$ for normalization (see Example 5.1). For this case, obtain expressions for $\langle x \rangle$ and Δx in terms of x_0. Also calculate the probability that the electron will be found within a standard deviation of its average position, that is, in the range

$\langle x \rangle - \Delta x$ to $\langle x \rangle + \Delta x$, and show that this is independent of x_0.

33. Calculate $\langle x \rangle$, $\langle x^2 \rangle$, and Δx for a quantum oscillator in its ground state. *Hint:* Use the integral formula

$$\int_0^\infty x^2 e^{-ax^2}\, dx = \frac{1}{4a}\sqrt{\frac{\pi}{a}} \qquad a > 0$$

34. (a) What value do you expect for $\langle p_x \rangle$ for the quantum oscillator? Support your answer with a symmetry argument rather than a calculation. (b) Energy principles for the quantum oscillator can be used to relate $\langle p_x^2 \rangle$ to $\langle x^2 \rangle$. Use this relation, along with the value of $\langle x^2 \rangle$ from Problem 33, to find $\langle p_x^2 \rangle$ for the oscillator ground state. (c) Evaluate Δp_x, using the results of (a) and (b).

35. From the results of Problems 33 and 34, evaluate $\Delta x\, \Delta p_x$ for the quantum oscillator in its ground state. Is the result consistent with the uncertainty principle? (Note that your computation verifies the minimum uncertainty product; furthermore, the harmonic oscillator ground state is the *only* quantum state for which this minimum uncertainty is realized.)

5.8   Observables and Operators

36. Which of the following functions are eigenfunctions of the momentum operator $[p_x]$? What are the eigenvalues, where appropriate?

(a) $A \sin(kx)$

(b) $A \sin(kx) - A \cos(kx)$

(c) $A \cos(kx) + iA \sin(kx)$

(d) $A\, e^{ik(x-a)}$

Additional Problems

37. **MP**   *Bound States of a Triangular Well.* Use the MP DESKTOP to find the two lowest stationary states for a particle confined to a triangular well (click on $U(x)$ and load the file *PR5-37.U*). How many nodes does each state have? What symmetry (if any) do these wavefunctions possess? (The triangular well mimics the potential seen by an electron in the vicinity of a sheet of positive charge.)

38. **MP**   *The Double Oscillator.* Systems with a pair of stable equilibrium points can often be modeled as a double oscillator with potential energy

$$U(x) = \tfrac{1}{2} m\omega^2 (|x| - a)^2$$

Here ω is the classical frequency of vibration for a mass m around either of the equilibrium points at $x = \pm a$. Use the MP DESKTOP to determine the wavelength of a photon emitted in the transition from the first excited state to the ground state of such an oscillator in the case where $\omega = 1.79 \times 10^{14}$ rad/s, $a = 0.38$ Å, and $m = 2.47$ u. (These values model the nitrogen atom in the ammonia molecule.) (The double oscillator potential—in dimensionless form—can be found in the file *PR5-38.U*).

39. *Nonstationary States.* Consider a particle in an infinite square well described initially by a wave that is a superposition of the ground and first excited states of the well:

$$\Psi(x, 0) = C[\psi_1(x) + \psi_2(x)]$$

(a) Show that the value $C = 1/\sqrt{2}$ normalizes this wave, assuming ψ_1 and ψ_2 are themselves normalized. (b) Find $\Psi(x, t)$ at any later time t. (c) Show that the superposition is *not* a stationary state, but that the *average* energy in this state is the arithmetic mean $(E_1 + E_2)/2$ of the ground and first excited state energies E_1 and E_2.

40. For the nonstationary state of Problem 39, show that the average particle position $\langle x \rangle$ oscillates with time as

$$\langle x \rangle = x_0 + A \cos(\Omega t)$$

where

$$x_0 = \frac{1}{2}\left(\int x|\psi_1|^2\, dx + \int x|\psi_2|^2\, dx \right)$$

$$A = \int x\psi_1^*\psi_2\, dx$$

and $\Omega = (E_2 - E_1)/\hbar$. Evaluate your results for the mean position x_0 and amplitude of oscillation A for an electron in a well 1 nm wide. Calculate the time for the electron to shuttle back and forth in the well once. Calculate the same time classically for an electron with energy equal to the average, $(E_1 + E_2)/2$.

6

Tunneling Phenomena

Chapter Outline

In this chapter the principles of wave mechanics are applied to particles striking a potential *barrier*. Unlike potential wells that attract and *trap* particles, barriers *repel* them. Because barriers have no bound states, the emphasis shifts to determining whether a particle incident on a barrier is reflected or transmitted.

In the course of this study we shall encounter a peculiar phenomenon called *tunneling*. A purely wave-mechanical effect, tunneling nevertheless is essential to the operation of many modern-day devices, and shapes our world on a scale from atomic all the way up to galactic proportions. The chapter includes a discussion of the role played by tunneling in several phenomena of practical interest, such as field emission, radioactive decay, and the operation of the ammonia maser. Finally, the chapter is followed by an essay on the scanning tunneling microscope, or STM, a remarkable device that uses tunneling to make images of surfaces with resolution comparable to the size of a single atom.

6.1 THE SQUARE BARRIER

The square barrier is represented by a potential energy function $U(x)$ that is constant at U in the barrier region, say between $x = 0$ and $x = L$, and zero outside this region. A realization of the square barrier potential using charged hollow cylinders is shown in Figure 6.1a. The outer cylinders are grounded while the central one is held at some positive potential V. For a particle with charge q, the barrier potential energy is $U = qV$. The charge experiences no force except in the gaps separating the cylinders. The force in the gaps is repulsive, tending to expel a positive charge q from the central cylinder. The

Figure 6.1   (a) Aligned metallic cylinders serve as a potential barrier to charged particles. The central cylinder is held at some positive electric potential V, and the outer cylinders are grounded. A charge q whose total energy is less than qV is unable to penetrate the central cylinder classically, but can do so quantum mechanically by a process called *tunneling*. (b) The potential energy seen by this charge in the limit where the gaps between the cylinders have shrunk to zero size. The result is the square barrier potential of height U.

potential energy for the idealized case in which the gaps have shrunk to zero size is the *square barrier*, sketched in Figure 6.1b.

A classical particle incident on the barrier, say from the left, experiences a retarding force on arriving at $x = 0$. Particles with energies E greater than U are able to overcome this force, but suffer a reduction in speed to a value commensurate with their diminished kinetic energy $(=E - U)$ in the barrier region. Such particles continue moving to the right with reduced speed until they reach $x = L$, where they receive a "kick" accelerating them back up to their original speed. Thus, particles having energy $E > U$ are able to cross the barrier with their speed restored to its initial value. By contrast, particles with energy $E < U$ are turned back (reflected) by the barrier, having insufficient energy to cross or even penetrate it. In this way the barrier divides the space into classically *allowed* and *forbidden* regions determined by the particle energy: if $E > U$, the whole space is accessible to the particle; for $E < U$ only the interval to the side of the barrier in which the particle originates is accessible—the barrier region itself is forbidden, and this precludes particle motion on the far side as well.

According to quantum mechanics, however, *there is no region inaccessible to our particle, regardless of its energy*, since the matter wave associated with the particle is nonzero everywhere. A typical wavefunction for this case, illustrated in Figure 6.2a, clearly shows the penetration of the wave into the barrier and beyond. This barrier penetration is in complete disagreement with classical physics. The attendant possibility of finding the particle on the far side of the barrier is called *tunneling*: we say the particle has *tunneled* through the barrier.

The mathematical expression for Ψ on either side of the barrier is easily found. To the left of the barrier the particle is free, so the wavefunction here must be the free particle plane waves introduced in Chapter 5:

$$\Psi(x, t) = Ae^{i(kx - \omega t)} + Be^{i(-kx - \omega t)} \tag{6.1}$$

This wavefunction $\Psi(x, t)$ is actually the *sum* of *two* plane waves. Both have frequency ω and energy $E = \hbar\omega = \hbar^2 k^2/2m$, but the first moves from left to right (wavenumber k), the second from right to left (wavenumber $-k$). Thus, that part of Ψ proportional to A is interpreted as a wave *incident* on the barrier

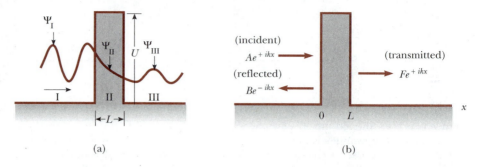

Figure 6.2   (a) A typical stationary state wave for a particle in the presence of a square barrier. The energy E of the particle is less than the barrier height U. Since the wave amplitude is nonzero in the barrier, there is some probability of finding the particle there. (b) Decomposition of the stationary wave into incident, reflected, and transmitted waves.

from the left; that proportional to B as a wave *reflected* from the barrier and moving *to the left* (Fig. 6.2b). The *reflection coefficient R* for the barrier is calculated as the ratio of the reflected probability density to the incident probability density:

Reflection coefficient for a barrier

$$R = \frac{(\Psi^*\Psi)_{\text{reflected}}}{(\Psi^*\Psi)_{\text{incident}}} = \frac{B^*B}{A^*A} = \frac{|B|^2}{|A|^2} \qquad (6.2)$$

In wave terminology R is the fraction of wave intensity in the reflected beam; in particle language, R becomes the likelihood (probability) that a particle incident on the barrier from the left is reflected by it.

Similar arguments apply to the right of the barrier where, again, the particle is free:

$$\Psi(x,\ t) = Fe^{i(kx-\omega t)} + Ge^{i(-kx-\omega t)} \qquad (6.3)$$

This form for $\Psi(x,\ t)$ is valid in the range $x > L$, with the term proportional to F describing a wave traveling to the right, and that proportional to G a wave traveling to the left in this region. The latter has *no* physical interpretation for waves incident on the barrier from the *left*, and so is discarded by requiring $G = 0$. The former is that part of the incident wave that is *transmitted* through the barrier. The relative intensity of this transmitted wave is the *transmission coefficient* for the barrier T:

Transmission coefficient for a barrier

$$T = \frac{(\Psi^*\Psi)_{\text{transmitted}}}{(\Psi^*\Psi)_{\text{incident}}} = \frac{F^*F}{A^*A} = \frac{|F|^2}{|A|^2} \qquad (6.4)$$

The transmission coefficient measures the likelihood (probability) that a particle incident on the barrier from the left penetrates to emerge on the other side. Since a particle incident on the barrier is either reflected or transmitted, the probabilities for these events must sum to unity:

$$R + T = 1 \qquad (6.5)$$

Equation 6.5 expresses a kind of *sum rule* obeyed by the barrier coefficients. Further, the degree of transmission or reflection will depend on particle energy. In the classical case $T = 0$ (and $R = 1$) for $E < U$, but $T = 1$ (and $R = 0$) for $E > U$. The wave-mechanical predictions for the functions $T(E)$ and $R(E)$ are more complicated; to obtain them we must examine the matter wave within the barrier.

To find Ψ in the barrier, we must solve Schrödinger's equation. Let us consider stationary states $\psi(x)e^{-i\omega t}$ whose energy $E = \hbar\omega$ is below the top of the barrier. This is the case $E < U$ for which no barrier penetration is permitted classically. In the region of the barrier $(0 < x < L)$, $U(x) = U$ and the time-independent Schrödinger equation for $\psi(x)$ can be rearranged as

$$\frac{d^2\psi}{dx^2} = \left\{\frac{2m(U-E)}{\hbar^2}\right\}\psi(x)$$

With $E < U$, the term in braces is a positive constant, and solutions to this equation are the *real exponential forms* $e^{\pm\alpha x}$. Since $(d^2/dx^2)e^{\pm\alpha x} = (\alpha)^2 e^{\pm\alpha x}$, we must take the term in braces $= (\alpha)^2$ or

$$\alpha = \frac{\sqrt{2m(U-E)}}{\hbar} \tag{6.6}$$

For wide barriers, the probability of finding the particle should decrease steadily into the barrier; in such cases only the decaying exponential is important, and it is convenient to define a barrier *penetration depth* $\delta = 1/\alpha$. At a distance δ into the barrier, the wavefunction has fallen to $1/e$ of its value at the barrier edge; thus, the probability of finding the particle is appreciable only within about δ of the barrier edge.

The complete wavefunction in the barrier is then

$$\Psi(x, t) = \psi(x)e^{-i\omega t} = Ce^{-\alpha x - i\omega t} + De^{+\alpha x - i\omega t} \quad \text{for } 0 < x < L \tag{6.7}$$

The coefficients C and D are fixed by requiring smooth joining of the wavefunction across the barrier edges; that is, both Ψ and $\partial\Psi/\partial x$ must be continuous at $x = 0$ and $x = L$. Writing out the joining conditions using Equations 6.1, 6.3, and 6.7 for Ψ in the regions to the left, to the right, and within the barrier, respectively, gives

$$A + B = C + D \qquad [\text{continuity of } \Psi \text{ at } x = 0]$$

$$ikA - ikB = \alpha D - \alpha C \qquad \left[\text{continuity of } \frac{\partial\Psi}{\partial x} \text{ at } x = 0\right] \tag{6.8}$$

$$Ce^{-\alpha L} + De^{+\alpha L} = Fe^{ikL} \qquad [\text{continuity of } \Psi \text{ at } x = L]$$

$$(\alpha D)e^{+\alpha L} - (\alpha C)e^{-\alpha L} = ikFe^{ikL} \qquad \left[\text{continuity of } \frac{\partial\Psi}{\partial x} \text{ at } x = L\right]$$

Joining conditions at a square barrier

In keeping with our previous remarks, we have set $G = 0$. Still, there are fewer equations than there are unknowns. This is as it should be, since the amplitude of the incident wave serves only to set the scale for the other amplitudes. That is, doubling the incident wave amplitude simply doubles the amplitudes of the reflected and transmitted waves. Dividing Equations 6.8 through by A furnishes four equations for the ratios B/A, C/A, D/A, and F/A: these equations may be solved by repeated substitution to find B/A and so on in terms of the barrier height U, the barrier width L, and the particle energy E. The result for the transmission coefficient T is (see Problem 7)

$$T(E) = \left\{1 + \frac{1}{4}\left[\frac{U^2}{E(U-E)}\right]\sinh^2\alpha L\right\}^{-1} \tag{6.9}$$

where sinh denotes the hyperbolic sine function: $\sinh x = (e^x - e^{-x})/2$.

Figure 6.3   A sketch of the transmission coefficient $T(E)$ for a square barrier. Oscillation in $T(E)$ with E, and the transmission resonances at E_1, E_2, and E_3, are further evidence for the wave nature of matter.

A sketch of $T(E)$ for the square barrier is shown in Figure 6.3. Equation 6.9 holds only for energies E below the barrier height U. For $E > U$, α becomes imaginary and $\sinh(\alpha L)$ turns oscillatory (see Example 6.3). This leads to fluctuations in $T(E)$ and *transmission resonances* where $T(E) = 1$ (perfect transmission). Such resonances arise from wave interference and constitute further evidence for the wave nature of matter.

EXAMPLE 6.1   Transmission Coefficient for an Oxide Layer

Two copper conducting wires are separated by an insulating oxide layer (CuO). Modeling the oxide layer as a square barrier of height 10.0 eV, estimate the transmission coefficient for penetration by 7.00-eV electrons (a) if the layer thickness is 5.00 nm, and (b) if the layer thickness is 1.00 nm.

Solution   From Equation 6.6 we calculate α for this case, using $\hbar = 1.973$ keV·Å/c and $m_e = 511$ keV/c^2 for electrons to get

$$\alpha = \frac{\sqrt{2m_e(U - E)}}{\hbar}$$

$$= \frac{\sqrt{2(511 \text{ keV}/c^2)(3.00 \times 10^{-3} \text{ keV})}}{1.973 \text{ keV·Å}/c} = 0.8875 \text{ Å}^{-1}$$

The transmission coefficient from Equation 6.9 is then

$$T = \left\{ 1 + \frac{1}{4}\left[\frac{10^2}{7(3)}\right] \sinh^2(0.8875 \text{ Å}^{-1})L \right\}^{-1}$$

Substituting $L = 50.0$ Å (5.00 nm) gives

$$T = 0.963 \times 10^{-38}$$

a fantastically small number! With $L = 10.0$ Å (1.00 nm), however, we find

$$T = 0.657 \times 10^{-7}$$

Thus, reducing the layer thickness by a factor of 5 enhances the likelihood of penetration by nearly 31 orders of magnitude!

EXAMPLE 6.2   Tunneling Current Through an Oxide Layer

A 1.00-mA current of electrons in one of the wires of Example 6.1 is incident on the oxide layer. How much of this current passes through the layer to the adjacent wire if the electron energy is 7.00 eV and the layer thickness is 1.00 nm? What becomes of the remaining current?

Solution   Since each electron carries a charge equal to $e = 1.60 \times 10^{-19}$ C, an electron current of 1.00 mA represents $10^{-3}/(1.60 \times 10^{-19}) = 6.25 \times 10^{15}$ electrons per second impinging on the barrier. Of these, only the fraction T is transmitted, where $T = 0.657 \times 10^{-7}$ from Example 6.1. Thus, the number of electrons per second continuing on to the adjacent wire is

$$(6.25 \times 10^{15})(0.657 \times 10^{-7}) = 4.11 \times 10^8 \text{ electrons/s}$$

This number represents a transmitted current of

$$(4.11 \times 10^8/\text{s})(1.60 \times 10^{-19} \text{ C}) = 6.57 \times 10^{-11} \text{ A}$$

$$= 65.7 \text{ pA (picoamperes)}$$

(Notice that the same transmitted current would be obtained had we simply multiplied the incident current by the transmission coefficient.) The remaining 1.00 mA– 65.7 pA is reflected at the layer. It is important to note that the *measured* conduction current in the wire on the side of incidence is the *net* of the incident and reflected currents, or again 65.7 pA.

EXAMPLE 6.3   Transmission Resonances

Consider a particle incident from the left on a square barrier of width L in the case where the particle energy E *exceeds* the barrier height U. Write the necessary wavefunctions and impose the proper joining conditions to obtain a formula for the transmission coefficient for this case. Show that perfect transmission (resonance) results for special values of particle energy, and explain this phenomenon in terms of the interference of de Broglie waves.

Solution   To the left and right of the barrier the wavefunctions are the free particle waves given by Equations

6.1 and 6.3 (again with $G = 0$ as befits a purely transmitted wave on the far side of the barrier):

$$\Psi(x, t) = Ae^{i(kx-\omega t)} + Be^{i(-kx-\omega t)} \qquad x < 0$$

$$\Psi(x, t) = Fe^{i(kx-\omega t)} \qquad x > L$$

The wavenumber k and frequency ω of these oscillations derive from the particle energy E in the manner characteristic of (nonrelativistic) de Broglie waves: that is, $E = (\hbar k)^2/2m = \hbar\omega$. Within the barrier, the wavefunction also is oscillatory. In effect, the decay constant α of Equation 6.6 has become imaginary, since $E > U$. Introducing a new wavenumber k' as $\alpha = ik'$, the barrier wavefunction becomes

$$\Psi(x, t) = Ce^{i(-k'x-\omega t)} + De^{i(k'x-\omega t)} \qquad 0 < x < L$$

with $k' = [2m(E - U)/\hbar^2]^{1/2}$ a real number.

The barrier wavefunction will join smoothly to the exterior wavefunctions if the wavefunction and its slope are continuous at the barrier edges $x = 0$ and $x = L$. These continuity requirements are identical to Equations 6.8 with the replacement $\alpha = ik'$ everywhere. In particular, we now have

$$A + B = C + D \qquad \text{[continuity of } \Psi \text{ at } x = 0]$$

$$kA - kB = k'D - k'C \qquad \left[\text{continuity of } \frac{\partial \Psi}{\partial x} \text{ at } x = 0\right]$$

$$Ce^{-ik'L} + De^{ik'L} = Fe^{ikL} \qquad \text{[continuity of } \Psi \text{ at } x = L]$$

$$k'De^{ik'L} - k'Ce^{-ik'L} = kFe^{ikL} \qquad \left[\text{continuity of } \frac{\partial \Psi}{\partial x} \text{ at } x = L\right]$$

To isolate the transmission amplitude F/A, we must eliminate from these relations the unwanted coefficients B, C, and D. Dividing the second line by k and adding to the first eliminates B, leaving A in terms of C and D. In the same way, dividing the fourth line by k' and adding the result to the third line gives D (in terms of F), while subtracting the result from the third line gives C (in terms of F). Combining the previous results finally yields A in terms of F:

$$A = \frac{1}{4} Fe^{ikL} \left\{ \left[2 - \left(\frac{k'}{k} + \frac{k}{k'}\right) \right] e^{ik'L} \right.$$
$$\left. + \left[2 + \left(\frac{k'}{k} + \frac{k}{k'}\right) \right] e^{-ik'L} \right\}$$

The transmission probability is $T = |F/A|^2$. Writing $e^{\pm ik'L} = \cos k'L \pm i \sin k'L$, and simplifying, we obtain the final result

$$\frac{1}{T} = \left|\frac{A}{F}\right|^2 = \frac{1}{4} \left| 2 \cos k'L - i \left(\frac{k'}{k} + \frac{k}{k'}\right) \sin k'L \right|^2$$

$$= 1 + \frac{1}{4} \left[\frac{U^2}{E(E - U)}\right] \sin^2 k'L$$

We see that *transmission resonances* occur whenever $k'L$ is a multiple of π. Using $k' = [2m(E - U)/\hbar^2]^{1/2}$, we can express the resonance condition in terms of the particle energy E as

$$E = U + n^2 \frac{\pi^2\hbar^2}{2mL^2} \qquad n = 1, 2, \ldots$$

Particles with these energies are transmitted perfectly ($T = 1$), with no chance of reflection ($R = 0$).

Resonances arise from the interference of the matter wave accompanying a particle. The wave reflected from the barrier can be regarded as the superposition of matter waves reflected from the leading and trailing edges of the barrier at $x = 0$ and $x = L$, respectively. If these reflected waves arrive phase-shifted by odd multiples of 180° or π radians, they will interfere destructively, leaving no reflected wave ($R = 0$) and thus perfect transmission. Now the wave reflected from the rear of the barrier at $x = L$ must travel the extra distance $2L$ before recombining with the wave reflected at the front, leading to a phase difference of $2k'L$. But this wave also suffers an intrinsic phase shift of π radians, having been reflected from a medium with higher optical density.[1] Thus, the condition for destructive interference becomes $2k'L + \pi = (2n + 1)\pi$, or simply $k'L = n\pi$, where $n = 1, 2, \ldots$

Perfect transmission also arises when particles are scattered by a potential *well*, a phenomenon known as the *Ramsauer–Townsend effect* (see Problem 11).

Exercise 1 Verify that for $E \gg U$, the transmission coefficient of Example 6.3 approaches unity. Why is this result expected? What happens to T in the limit as E approaches U?

EXAMPLE 6.4 Scattering Coefficients from the MP DESKTOP

MP In this example we use the MP DESKTOP to obtain transmission and reflection coefficients for an electron with energy 7.00 eV incident on a square barrier 0.100 nm (1.00 Å) wide and 10.0 eV high.

---

[1] This is familiar from the propagation of classical waves: a traveling wave arriving at the interface separating two media is partially transmitted and partially reflected. The reflected portion is phase-shifted 180° only in the case where the wave speed is lower in the medium being penetrated. For matter waves, $p = h/\lambda$, and the wavelength (hence wave speed) is largest in regions where the kinetic energy is smallest. Thus, the matter wave reflected from the front of a barrier suffers no change in phase, but that reflected from the rear is phase-shifted 180°.

Figure 6.4   (Example 6.4) The electron wavefunction against the backdrop of the square barrier. The ratio of electron energy to barrier height is 0.70.

Launch the MP DESKTOP application by running the main program file *MP.EXE*. Next, select *Task* from the menu bar and *EigenStates-1D* from the drop-down task list. The DESKTOP is now ready to solve the (eigenvalue) problem posed in the task field just below the menu bar:

$$-\frac{d^2\psi}{dx^2} + U(x)\,\psi(x) = E\cdot\psi(x)$$

This is Schrödinger's time-independent equation expressed in dimensionless units of length and energy.

NOTE: Recall that the length unit (δ) and the energy unit (ϵ) must satisfy $\epsilon\cdot\delta^2 = \hbar^2/2m$ to map the equation in the task field to the Schrödinger equation. In describing electrons, atomic units ($\delta = 1$ bohr [0.529 Å] and $\epsilon = 1$ Ry [13.60 eV]) are a convenient choice satisfying this criterion and are the ones adopted in this example. For a more exhaustive discussion of units, see the *Help-Contents* entry on the DESKTOP menu bar.

To specify the potential energy, click the Potential button ⌷U(x)⌷ located just below the menu bar. A dialog box opens, presenting a list of potential files. Select *EX6-4.U* (for Example 6.4) from this list, then select *OK*. Another dialog box opens, showing that $U(x)$ is defined over the domain $x = -10$ to $x = +10$ (bohrs) and describes a square barrier extending from $x = 0$ to $x = +1/0.529 =$

1.89 bohrs; that is, the barrier width of 1.00 Å becomes 1.89 bohrs. For length measured in bohrs, energy must be measured in rydbergs (1 Ry = 13.60 eV). Thus, a well height of 10.0 eV becomes (10.0/13.60) Ry. The well height (ϵ) may be edited, but for now simply accept the information as presented by clicking the *OK* button.

The potential $U(x)$ is portrayed as a solid figure in the main window of the screen. In the energy box located just below the menu bar, type in the energy of the incident electron, 7.00/13.60 = 0.514706 Ry, and press *Enter*. The wavefunction for this energy is computed and is superimposed on the potential. Click the Vertical Scroll Bar to shift the graphics image upward until the whole wavefunction is visible.

The wavefunction shown in Figure 6.4 should be compared with that sketched in Figure 6.2(a). Notice the penetration of the wave into and beyond the barrier.

The DESKTOP can resolve this wavefunction into its incident, reflected, and transmitted components. Click the Deconvolute button ⌷→⌷ located on the Graphics ToolBar just to the right of the main window; the original wavefunction is replaced with its constituents, as shown in Figure 6.5. Next, activate the MP Analyzer by clicking 🔍 and position the graphics cursor as necessary to measure the amplitude of the incident, reflected, and transmitted waves. (The cursor is controlled with the mouse or the

Figure 6.5 (Example 6.4) The electron wavefunction of Figure 6.4, showing its incident (i), reflected (r), and transmitted components. The graphics cursor (vertical line) is positioned to record the amplitude of the reflected wave.

direction keys [→], [←]. Keyboard users note that the [Ins] key toggles cursor movement between coarse and fine settings.)

You should find, respectively, $A = 657.814$ for the incident wave amplitude, $B = 486.909$ for the reflected wave amplitude, and $F = 442.311$ for the transmitted wave amplitude. From these we can determine the scattering coefficients at this energy; $R(E) = |B/A|^2 = 0.547886$ and $T(E) = |F/A|^2 = 0.452114$. The simulation results compare favorably with the predictions of Equation 6.9 for this case: $T(E) = 0.451997$; furthermore, we see that the sum rule Equation 6.5 is verified to six-digit accuracy.

EXAMPLE 6.5   Scattering by a Potential Step

The potential step shown in Figure 6.6 may be regarded as a square barrier in the special case where the barrier width L is infinite. Apply the ideas of this section to discuss the quantum scattering of particles incident from the left on a potential step, in the case where the step height U exceeds the total particle energy E.

Solution   The wavefunction everywhere to the right of the origin is the barrier wavefunction given by Equation 6.7. To keep Ψ from diverging for large x we must take $D = 0$, leaving only the decaying wave

$$\Psi(x, t) = Ce^{-\alpha x - i\omega t} \qquad x > 0$$

This must be joined smoothly to the wavefunction on the left of the origin, given by Equation 6.1:

$$\Psi(x, t) = Ae^{ikx - i\omega t} + Be^{-ikx - i\omega t} \qquad x < 0$$

The conditions for smooth joining at $x = 0$ yield

$$A + B = C \qquad \text{[continuity of } \Psi \text{]}$$

$$ikA - ikB = -\alpha C \qquad \left[\text{continuity of } \frac{\partial \Psi}{\partial x} \right]$$

Solving the second equation for C and substituting into the first (with $\delta = 1/\alpha$) gives $A + B = -ik\delta A + ik\delta B$ or

$$\frac{B}{A} = -\frac{(1 + ik\delta)}{(1 - ik\delta)}$$

Figure 6.6   (Example 6.5) The potential step of height U may be thought of as a square barrier of the same height in the limit where the barrier width L becomes infinite. All particles incident on the barrier with energy $E < U$ are reflected.

The reflection coefficient is $R = |B/A|^2 = (B/A)(B/A)^*$, or

$$R = \left(\frac{1 + ik\delta}{1 - ik\delta}\right)\left(\frac{1 - ik\delta}{1 + ik\delta}\right) = 1$$

Thus, an *infinitely* wide barrier reflects *all* incoming particles with energies below the barrier height, in agreement with the classical prediction. Nevertheless, there is a nonzero wave in the step region since

$$\frac{C}{A} = 1 + \frac{B}{A} = \frac{-2ik\delta}{1 - ik\delta} \neq 0$$

But the wavefunction for $x > 0$, $\Psi(x, t) = Ce^{-\alpha x - i\omega t}$, is not a *propagating* wave at all, that is, there is no *net transmission* of particles to the right of the step. However, there will be quantum transmission through a barrier of *finite* width, no matter how wide (cf. Eq. 6.9).

Figure 6.7 (a) Total internal reflection of light waves at a glass–air boundary. An *evanescent* wave penetrates into the space beyond the reflecting surface. (b) Frustrated total internal reflection. The evanescent wave is "picked up" by a neighboring surface, resulting in transmission across the gap. Notice that the light beam *does not* appear in the gap.

The existence of a barrier wave without propagation (as in Example 6.5) is familiar from the optical phenomenon of *total internal reflection* exploited in the construction of beam splitters (Figure 6.7): light entering a right-angle prism is completely reflected at the hypotenuse face, even though an electromagnetic wave, the *evanescent wave*, penetrates into the space beyond. A second prism brought into near contact with the first can "pick up" this evanescent wave, thereby transmitting and redirecting the original beam (Fig. 6.7b). This phenomenon, known as *frustrated total internal reflection*, is the optical analog of tunneling: in effect, photons have tunneled across the gap separating the two prisms.

6.2   BARRIER PENETRATION: SOME APPLICATIONS

In actuality, few barriers can be modeled accurately using the square barrier discussed in the preceding section. Indeed, the extreme sensitivity to barrier constants found there suggests that barrier shape will be important in making reliable predictions of tunneling probabilities. The transmission coefficient for a barrier of arbitrary shape, as specified by some potential energy function $U(x)$, can be found from Schrödinger's equation. For high, wide barriers, where the likelihood of penetration is small, a lengthy treatment yields

$$T(E) \approx \exp\left(-\frac{2}{\hbar}\sqrt{2m}\int\sqrt{U(x) - E}\,dx\right) \tag{6.10}$$

Approximate transmission coefficient of a barrier with arbitrary shape

The integral in Equation 6.10 is to be taken over the *classically forbidden region* where $E < U(x)$. A simple argument leading to this form follows by representing an arbitrary barrier as a succession of square barriers, all of which scatter independently, so that the transmitted wave intensity of one becomes the incident wave intensity for the next, and so forth (see Problem 14).

The use of Equation 6.10 is amply illustrated in the remainder of this section where it is applied to several classic problems in contemporary physics.

Field Emission

In field emission, electrons bound to a metal are literally torn from the surface by the application of a strong electric field. In this way the metal becomes a source that may be conveniently tapped to furnish electrons for many applications. In the past such *cold cathode emission*, as it was known, was a popular way of generating electrons in vacuum tube circuits, producing less electrical "noise" than hot filament sources, where electrons were "boiled off" by heating the metal to a high temperature. Modern applications include the *field emission microscope* (Fig. 6.8) and a related device, the *scanning tunneling microscope* (see the Essay at the end of this chapter), both of which use the escaping electrons to form an image of structural details at the emitting surface.

Field emission is a tunneling phenomenon. Figure 6.9a shows schematically how field emission can be obtained by placing a positively charged plate near the source metal to form, effectively, a parallel plate capacitor. In the gap between the "plates" there is some electric field $\mathcal{E}$, but the electric field inside the metal remains zero due to the shielding by the mobile metal electrons induced at the surface. Thus, an electron in the bulk is virtually free, yet still bound to the metal by a potential well of depth U. The total electron energy E, which includes kinetic energy, is negative to indicate a bound electron; indeed $|E|$ represents the energy needed to free this electron, a value at least equal to the work function of the metal.

Once beyond the surface ($x > 0$), our electron is attracted by the electric force in the gap $F = e\mathcal{E}$, represented by the potential energy $U(x) = -e\mathcal{E}x$. The potential energy diagram is shown in Figure 6.9b, together with the classically allowed and forbidden regions for an electron of energy E. The intersections of E with $U(x)$ at x_1 ($=0$) and x_2 ($=-E/e\mathcal{E}$) mark the *classical turning points*, where a classical particle with this energy would be turned around to keep it from entering the forbidden zone. Thus, from a classical viewpoint, an electron initially confined to the metal has insufficient energy to surmount the potential barrier at the surface, and would remain in the bulk forever! It is only by virtue of its wave character that the electron can tunnel through this barrier to emerge on the other side. The probability of such an occurrence is measured by the transmission coefficient for the (triangular) barrier depicted in Figure 6.9b.

Cathode tip

Fluorescent screen

Figure 6.8 Schematic diagram of a field emission microscope. The intense electric field at the tip of the needle-shaped specimen allows electrons to tunnel through the work function barrier at the surface. Since the tunneling probability is sensitive to the exact details of the surface where the electron passes, the number of escaping electrons varies from point to point with the surface condition, thus providing a picture of the surface under study.

Tunneling model for field emission

(a)

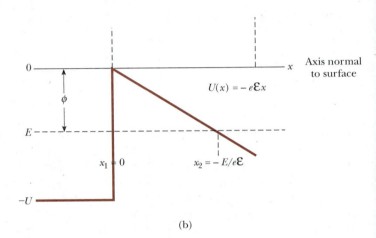

(b)

Figure 6.9 (a) Field emission from a metal surface. (b) The potential energy seen by an electron of the metal. The electric field produces the triangular potential barrier shown, through which electrons can tunnel to escape the metal. Turning points at $x_1 = 0$ and $x_2 = -E/e\mathcal{E}$ delineate the classically forbidden region. (Note that x_2 is positive since E is negative.) Tunneling is greatest for the most energetic electrons, for which $-E$ is equal to the work function ϕ of the metal.

To calculate $T(E)$ we must evaluate the integral in Equation 6.10 over the classically forbidden region from x_1 to x_2. Since $U(x) = -e\mathcal{E}x$ in this region and $E = -e\mathcal{E}x_2$, we have

$$\int \sqrt{U(x) - E}\, dx = \sqrt{e\mathcal{E}} \int_0^{x_2} \sqrt{x_2 - x}\, dx = -\frac{2}{3}\sqrt{e\mathcal{E}}\, \{x_2 - x\}^{3/2}\, \Big|_0^{x_2}$$

$$= \frac{2}{3}\sqrt{e\mathcal{E}}\left(\frac{|E|}{e\mathcal{E}}\right)^{3/2}$$

Using this result in Equation 6.10 gives the transmission coefficient for field emission as

Transmission coefficient for field emission

$$T(E) \approx \exp\left(\left\{-\frac{4\sqrt{2m}\,|E|^{3/2}}{3e\hbar}\right\}\frac{1}{\mathcal{E}}\right) \qquad (6.11)$$

The strong dependence of T on electron energy E in the bulk is evident from this expression. The escape probability is largest for the most energetic electrons; these are the ones most loosely bound and for which $|E| = \phi$, the work function of the metal. It is also apparent that the quantity in curly brackets

Figure 6.10 Field-emission microscope image of the surface of a crystalline platinum alloy with a magnification of 3 000 000×. Individual atoms can be seen on surface layers using this technique. *(Manfred Kage/Peter Arnold, Inc.)*

must have the dimensions of electric field, and represents a characteristic field strength, say $\mathcal{E}_c$, for field emission:

$$\mathcal{E}_c = \frac{4\sqrt{2m}\,|E|^{3/2}}{3e\hbar} \tag{6.12}$$

For $|E| = \phi = 4.0$ eV, a typical value for many metals, we calculate the field strength to be $\mathcal{E}_c = 5.5 \times 10^{10}$ V/m, a strong field by laboratory standards. Measurable emission occurs even with much smaller fields, however, since the emission *rate* grows with the number of electrons per second that collide with the barrier. This *collision frequency* is quite high for a bulk sample containing something like 10^{22} electrons per cubic centimeter, and values in excess of 10^{30} collisions per second per square centimeter are not uncommon (see Problem 17)! In this way field emission rates on the order of 10^{10} electrons per second (currents of about 1 nA) can be realized with applied fields as small as $\mathcal{E}_c/50$, or about 10^9 V/m.

EXAMPLE 6.6 Tunneling in a Parallel Plate Capacitor

Estimate the leakage current due to tunneling that passes across a parallel plate capacitor charged to a potential difference of 10 kV. Take the plate separation to be $d = 0.010$ mm and the plate area to be $A = 1.0$ cm^2.

Solution The number of electrons per second impinging on the plate surface from the bulk is the collision frequency f, about 1.0×10^{30} per second per square centimeter for most metals. Of these, only the fraction given by the transmission coefficient T can tunnel through the potential barrier in the gap to register as a current through the device. Thus, the electron emission rate for a plate of area 1 cm^2 is

$$\lambda = fT(E) = 1.0 \times 10^{30} \exp(-\mathcal{E}_c/\mathcal{E})$$

The electric field $\mathcal{E}$ in the gap is 10 kV/0.010 mm = 1.0×10^9 V/m. Using this and $\mathcal{E}_c = 5.5 \times 10^{10}$ V/m gives for the exponential $\exp\{-55\} = 1.30 \times 10^{-24}$, and an emission rate of $\lambda = 1.30 \times 10^6$ electrons per second. Since each electron carries a charge $e = 1.60 \times 10^{-19}$ C, the tunneling current is

$$I = 2.1 \times 10^{-13} \text{ A} = 0.21 \text{ pA}$$

Alpha Decay

The decay of radioactive elements with the emission of α particles (helium nuclei composed of two protons and two neutrons) was among the longstanding puzzles to which the fledgling wave mechanics was first applied shortly after its inception in 1926. That alpha particles are a disintegration product of such species as radium, thorium, and uranium was well documented as early as 1900, but certain features of this decay remained a mystery, finally unraveled in 1928 in the now-classic works of George Gamow, and R. W. Gurney and E. U. Condon. Their contribution was to recognize that the newly discovered tunnel effect lay behind the two most puzzling aspects of alpha decay:

- All alpha particles emitted from any one source have nearly the same energy, and, for all known emitters, emerge with kinetic energies in the same narrow range, from about 4 to 9 MeV.
- In contrast to the uniformity of energies, the half-life of the emitter (time taken for half of the emitting substance to decay) varies over an enormous range—more than twenty orders of magnitude!—according to the emitting element (Table 6.1).

For instance, alphas emerge from the element thorium with kinetic energy equal to 4.05 MeV, only a little less than half as much as the alphas emitted from polonium (8.95 MeV). Yet the half-life of thorium is 1.4×10^{10} years, compared with only 3.0×10^{-7} *seconds* for the half-life of polonium!

Gamow attributed this striking behavior to a preformed alpha particle rattling around within the nucleus of the radioactive (parent) element, eventually tunneling through the potential barrier to escape as a detectable decay product (Fig. 6.11a). While inside the parent nucleus, the alpha is virtually free, but nonetheless confined to the nuclear potential well by the nuclear force. Once outside the nucleus, the alpha particle experiences only the Coulomb repulsion of the emitting (daughter) nucleus. (The nuclear force on the alpha outside the nucleus is insignificant due to its extremely short range, $\approx 10^{-15}$ m.) Figure 6.11b shows the potential energy diagram for the alpha particle as a function of distance r from the emitting nucleus. The nuclear radius R is about 10^{-14} m, or 10 fm [note that 1 fm (fermi) $= 10^{-15}$ m] for heavy nuclei[2]; beyond this there is only the energy of Coulomb repulsion

Table 6.1   Characteristics of Some Common Alpha Emitters

Element	α-Energy	Half-Life*
$^{212}_{84}\text{Po}$	8.95 MeV	2.98×10^{-7} s
$^{240}_{96}\text{Cm}$	6.40 MeV	27 days
$^{226}_{88}\text{Ra}$	4.90 MeV	1.60×10^3 y
$^{232}_{90}\text{Th}$	4.05 MeV	1.41×10^{10} y

* Note that half-lives range over 24 orders of magnitude when alpha energy changes by a factor of 2.

---

[2] The fermi (fm) is a unit of distance commonly used in nuclear physics.

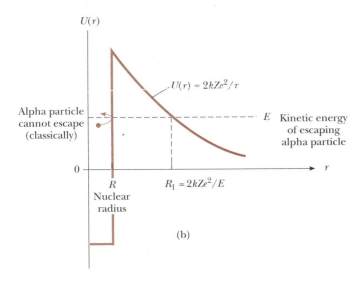

Figure 6.11   (a) Alpha decay of a radioactive nucleus. (b) The potential energy seen by an alpha particle emitted with energy E. R is the nuclear radius, about 10^{-14} m, or 10 fm. Alpha particles tunneling through the potential barrier between R and R_1 escape the nucleus to be detected as radioactive decay products.

$U(r) = kq_1q_2/r$ between the doubly charged alpha ($q_1 = +2e$) and a daughter nucleus with atomic number Z ($q_2 = +Ze$). Classically, even a 9-MeV alpha particle initially bound to the nucleus would have insufficient energy to overcome the Coulomb barrier (≈ 30 MeV high) and escape. But the alpha particle, with its wave attributes, may tunnel *through* the barrier to appear on the outside. The total alpha particle energy E inside the nucleus becomes the observed kinetic energy of the emerging alpha once it has escaped. *It is the sensitivity of the tunneling rate to small changes in particle energy that accounts for the wide range of half-lives observed for alpha emitters.*

   The tunneling probability and associated decay rate are calculated in much the same way as for field emission, apart from the fact that the barrier shape now is Coulombic, rather than triangular. The details of this calculation are given in Example 13.9 (Chapter 13), with the result

Tunneling through the Coulomb barrier

$$T(E) = \exp\left\{ -4\pi Z \sqrt{\frac{E_0}{E}} + 8\sqrt{\frac{ZR}{r_0}} \right\} \qquad (6.13)$$

Transmission coefficient for alpha particles of an unstable nucleus

In this expression, $r_0 = \hbar^2/m_\alpha ke^2$ is a kind of "Bohr" radius for the alpha particle. The mass of the alpha particle is $m_\alpha = 7295 m_e$, so r_0 has the value

$\alpha_0/7295 = 7.25 \times 10^{-5}$ Å, or 7.25 fm. The length r_0, in turn, defines a convenient energy unit E_0 analogous to the rydberg in atomic physics:

$$E_0 = \frac{ke^2}{2r_0} = \left(\frac{ke^2}{2a_0}\right)\left(\frac{a_0}{r_0}\right) = (13.6 \text{ eV})(7295) = 0.0993 \text{ MeV}$$

To obtain decay rates, $T(E)$ must be multiplied by the number of collisions per second that an alpha particle makes with the nuclear barrier. This collision frequency f is the reciprocal of the transit time for the alpha particle crossing the nucleus, or $f = v/2R$ where v is the speed of the alpha particle inside the nucleus. In most cases f is about 10^{21} collisions per second (see Problem 16). The decay rate λ (the probability of alpha emission per unit time) is then

$$\lambda = fT(E) \approx 10^{21} \exp\{-4\pi Z\sqrt{(E_0/E)} + 8\sqrt{Z(R/r_0)}\}$$

The reciprocal of λ has dimensions of time, and is related to the half-life of the emitter $t_{1/2}$ as

$$t_{1/2} = \frac{\ln 2}{\lambda} = \frac{0.693}{\lambda} \tag{6.14}$$

EXAMPLE 6.7 Estimating the Half-lives of Thorium and Polonium

Using the tunneling model developed earlier, estimate the half-lives for alpha decay of the radioactive elements thorium and polonium. The energy of the ejected alphas is 4.05 MeV and 8.95 MeV, respectively, and the nuclear size is about 9.00 fm in both cases.

Solution For thorium ($Z = 90$,) the daughter nucleus has atomic number $Z = 88$, corresponding to the element radium. Using $E = 4.05$ MeV and $R = 9.00$ fm we find for the transmission factor $T(E)$ in Equation 6.13

$$\exp\{-4\pi(88)\sqrt{(0.0993/4.05)} + 8\sqrt{88(9.00/7.25)}\}$$
$$= \exp\{-89.542\} = 1.3 \times 10^{-39}$$

Taking $f = 10^{21}$ Hz, we obtain for the decay rate the value $\lambda = 1.29 \times 10^{-18}$ alphas per second. The associated half-life is, from Equation 6.14,

$$t_{1/2} = \frac{0.693}{1.3 \times 10^{-18}} = 5.4 \times 10^{17} \text{ s}$$
$$= 1.7 \times 10^{10} \text{ y}$$

which compares favorably with the actual value for thorium, 1.4×10^{10} y

If polonium ($Z = 84$) is the radioactive species, the daughter element is lead, with $Z = 82$. Using for the disintegration energy $E = 8.95$ MeV, we obtain for the transmission factor

$$\exp\{-4\pi(82)\sqrt{(0.0993/8.95)} + 8\sqrt{82(9.00/7.25)}\}$$
$$= \exp\{-27.825\} = 8.2 \times 10^{-13}$$

Assuming f is unchanged at 10^{21} collisions per second, we get for this case $\lambda = 8.2 \times 10^8$ alphas per second and a half-life

$$t_{1/2} = \frac{0.693}{8.2 \times 10^8} = 8.4 \times 10^{-10} \text{ s}$$

The measured half-life of polonium is 3.0×10^{-7} s.

Given the crudeness of our method, both estimates should be considered satisfactory. Further, the calculations show clearly how a factor of only 2 in disintegration energy leads to half-lives differing by more than 26 orders of magnitude!

EXAMPLE 6.8 The Leaky Well

MP The potential seen by an alpha particle in an atomic nucleus constitutes a "leaky" well, inasmuch as the alpha can escape the well by tunneling. In this example we explore the stationary states of such a well using the MP DESKTOP and show how they lead to an alternative description of the decay process.

Begin by readying the DESKTOP to find stationary states of Schrödinger's equation: select *Task* from the menu bar and *EigenStates-1D* from the drop-down task list. Click the Potential button $\boxed{U(x)}$ located just below the menu bar, and select *EX6-8.U* (for Example 6.8) from the list of potential files. This is the potential $U(x)$ for a leaky square well, that is, a square well bounded on one side by a barrier of finite width. A particle initially confined to the well eventually escapes by tunneling through this barrier. The leaky well is our simplified model for the potential seen by an alpha particle in a radioactive nucleus. The well itself extends from $x = -1$ to $x = +1$. For this to describe a nucleus with radius 5.00 fm means x must be measured in units of 5.00 fm; this, in turn, requires that energy be measured in units of

$$\frac{\hbar^2}{2M(5.00 \text{ fm})^2} = \frac{(197.3 \text{ MeV} \cdot \text{fm}/c)^2}{2(4.0026)(931.50 \text{ MeV}/c^2)(5.00 \text{ fm})^2}$$

$$= 0.2088 \text{ MeV}$$

NOTE: Recall that the length unit (δ) and the energy unit (ϵ) must satisfy $\epsilon \cdot \delta^2 = \hbar^2/2m$ to map the equation in the task field to the Schrödinger equation. For an exhaustive discussion of units, see the *Help-Contents* entry on the DESKTOP menu bar.

Notice we have used $M = 4.0026$ u (1 u = 931.50 MeV/c^2) here to represent an alpha particle in the nucleus. Thus, a typical nuclear well height (≈ 30 MeV) becomes $30/0.2088 \approx 144$ in program units. The width β of the barrier on the right controls the leakiness of this well; the default $\beta = 0.2$ translates to 1.00 fm in the present case, a reasonable figure for alpha decay. Click

the *OK* button to load the potential with the default values.

The leaky well is portrayed as a solid figure in the main window of the screen. In the energy box just below the menu bar, type in the energy E of an alpha particle confined to an infinite well of the same width, $\pi^2 \hbar^2 / 2M(10.0 \text{ fm})^2$, or $\pi^2/4 = 2.467$ in program units, and press *Enter*. The wavefunction for this energy is computed and is superimposed on the potential.

The wavefunction is everywhere smooth—this is an acceptable wavefunction and energy for the leaky well. Try a few other energies: you will find that *all* positive values of E lead to acceptable wavefunctions! (This phenomenon is known as *weak quantization.*) But notice that, invariably, the wave amplitude is much larger outside the well than inside, that is, these waves describe states for which the alpha is *unbound*, spending most of its time outside the confines of the nucleus. There are, however, special energies for which this "inside–outside" behavior reverses. Examine the wavefunction that results with $E = 2.099$. This is about the closest we can come to a bound state of the leaky well; such states are called *quasibound* (Fig. 6.12). (In fact, the energies of the quasibound states are essentially those of the finite square well obtained by allowing the leaky well barrier to extend all the way to infinity.)

Figure 6.12 (Example 6.8) A quasibound state for an alpha particle in a leaky nuclear well. With negligible wave amplitude outside, the alpha particle in this state will be found with overwhelming probability inside the well (nucleus).

Figure 6.13   (Example 6.8) An unbound state for an alpha particle in the leaky well. The energy of this state, $E = 2.069$, differs by little more than 1% from the value $E = 2.099$ of the quasibound state shown in Figure 6.12.

Notice the extreme sensitivity to energy; with even a small change to E, the state reverts to an unbound one (Fig. 6.13).

The decay of a radioactive nucleus by alpha emission can be understood in terms of these quasibound states and their sensitivity to energy. An alpha particle initially confined to the nucleus must be described by a state that is predominantly quasibound, with the inevitable small admixture of unbound states whose energy is only slightly removed from the quasibound value. The resulting *nonstationary* state evolves so as to produce decay, with the alpha eventually occupying one or more unbound states. We can verify this behavior with the MP DESKTOP.

Open the *Task* menu and select *QM WaveGroups-1D*. The initial wavefunction $\Psi(x, 0)$ will be taken as a sum of stationary states with energies centered on the quasibound value found in Figure 6.12. To do this, click the *Control Panel* entry on the *Options* menu. From the *Control Panel* dialog box, select *Enable Mode* in the *Fourier* field, then click the *OK* button to activate this feature.

NOTE: The DESKTOP allows the initial wavefunction $\Psi(x, 0)$ to be specified directly as a function of x (as defined in a data file with the extension Ψ) or constructed

as a sum of stationary state wavefunctions $\psi_k(x)$ for the potential $U(x)$ in use. The latter method, referred to generally as Fourier mode, is activated as described above and gives $\Psi(x, 0)$ as

$$\Psi(x, 0) = \sum a(k)\psi_k(x)$$

In Fourier mode, the display changes to a split screen and the Frequency and Amplitude buttons appear below the menu bar to allow inputs for the frequency $\omega(k)$ (or energy) and the weighting factor $a(k)$ for each stationary state in the sum.

Now click the Frequency button $\boxed{\omega(k)}$ to load the frequency file *EX6-8.W* specifying the frequencies $\omega(k)$—or equivalently, the energies $E(k)$—of the stationary states in the mixture. Altogether we are mixing nine stationary states, equally spaced in energy around the quasibound value of Figure 6.12. Finally, click the Amplitude button $\boxed{a(k)}$ to load the amplitude file *UNIFORM.A*, which mixes these states with equal weights [$a(k) = 1$]. Figure 6.14a shows the initial wavefunction that results (to show the entire wavefunction the vertical scale has been reduced by clicking $\boxed{\updownarrow}$ several times while holding down the *Shift* key).

NOTE: The colors represent the phase of the wavefunction at each point, according to the color scheme found on the face of the Color-for-Phase button. Phase angle increases counterclockwise from zero (solid black line) to 180° around the top half of the control, and from zero to −180° clockwise around the bottom half of the control. The modulus, or absolute value, of the wavefunction at each point is plotted in the usual way as the height above the baseline.

Since $\Psi(x, 0)$ has comparatively little amplitude outside the well, the alpha particle is almost certainly in the nucleus to start.

   Click the Advance button ▨ to the immediate right of the time box to start the clock. The wavefunction gradually "leaks" out of the well, showing that it is truly a nonstationary state. Click again to stop the clock when $t =$ 5.25. [*Note:* The time unit is $\hbar$ divided by the energy unit, $(6.582 \times 10^{-22} \text{ MeV})/0.2088 \text{ MeV} = 3.152 \times 10^{-21}$ s.

So the time shown is $(5.25)(3.152 \times 10^{-21} \text{ s}) = 1.65 \times 10^{-20}$ s.] The display should look like Figure 6.14b. Compared with Figure 6.14a, the increased wave amplitude outside the well shows the alpha particle now is found with higher probability in the space around the nucleus, signaling nuclear decay. Notice that the alpha particle outside the nucleus will be found with the same energy it had inside, which is just the average energy of this nonstationary state, or about $(2.099)(0.2088 \text{ MeV}) = 0.438 \text{ MeV}$ in this case. Unfortunately, the decay time, or half-life, cannot be obtained reliably from this simulation because our model grossly exaggerates the possibility of reverse tunneling; that is, once outside, the alpha particle can tunnel *back* into the nucleus! In actuality, the probability of reverse tunneling is nearly zero: the space surrounding the nucleus is effectively a probability "sink," a feature not reproduced by the simulation. Despite its complexity, this method of describing nuclear decay with quasibound states is the only *fully quantum-mechanical* treatment available.

(a)   (b)

Figure 6.14   (Example 6.8) (a) The wavefunction $\Psi(x, 0)$ representing an alpha particle initially confined to a nucleus, modeled as a leaky square well. (b) The alpha particle wavefunction $\Psi(x, t)$ at time $t = 1.65 \times 10^{-20}$ s. The comparatively larger wave amplitude outside the well signals nuclear decay.

Ammonia Inversion

The "inversion" of the ammonia molecule is another example of tunneling, this time for an entire atom. The equilibrium configuration of the ammonia (NH_3) molecule is shown in Figure 6.15a: the nitrogen atom is situated at the

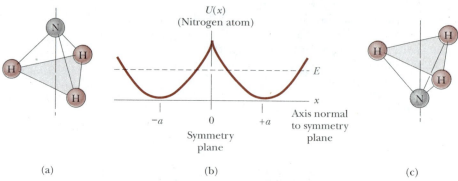

Figure 6.15 (a) The ammonia molecule NH_3. At equilibrium, the nitrogen atom is situated at the apex of a pyramid whose base is the equilateral triangle formed by the three hydrogen atoms. By symmetry, a second equilibrium configuration exists for the nitrogen atom on the opposite side of the plane formed by the hydrogen atoms. (b) The potential energy seen by the nitrogen atom along a line perpendicular to the symmetry plane. The two equilibrium points at $-a$ and $+a$ give rise to the double oscillator potential shown. A nitrogen atom with energy E can tunnel back and forth through the barrier from one equilibrium point to the other, with the result that the molecule alternates between the normal configuration in (a) and the inverted configuration shown in (c).

apex of a pyramid whose base is the equilateral triangle formed by the three hydrogen atoms. But this equilibrium is not truly stable; indeed, there is a *second* equilibrium position for the nitrogen atom on the opposite side of the plane formed by the hydrogen atoms. With its two equilibrium locations, the nitrogen atom of the ammonia molecule constitutes a *double oscillator*, which can be modeled by using the double oscillator potential shown in Figure 6.15b. A nitrogen atom initially located on one side of the symmetry plane will not remain there indefinitely, since there is some probability that it can tunnel through the oscillator barrier to emerge on the other side. When this occurs, the molecule becomes *inverted* (Fig. 6.15c). But the process does not stop there; the nitrogen atom, now on the opposite side of the symmetry plane, has a probability of tunneling *back* through the barrier to take up its original position! The molecule does not just undergo one inversion, but flip-flops repeatedly, alternating between the two classical equilibrium configurations. The "flopping" frequency is fixed by the tunneling rate and turns out to be quite high, approximately 2.4×10^{10} Hz (microwave range of the electromagnetic spectrum)!

Double oscillator representation of the ammonia molecule

We can estimate the tunneling probability for inversion using Equation 6.10. The double oscillator potential of Figure 6.15b is described by the potential energy function

$$U(x) = \tfrac{1}{2}M\omega^2 \,(|x| - a)^2 \tag{6.15}$$

with ω the classical frequency of vibration for the nitrogen atom around either of the equilibrium points $x = \pm a$. We suppose the nitrogen atom possesses the minimum energy of vibration, $E = \tfrac{1}{2}\hbar\omega$. There are four classical turning points for which $U(x) = E$ (see Figure 6.15b); the limits for the tunneling integral are the pair closest to $x = 0$, and these are given by $x = \pm(a - A)$, where A

is the vibration amplitude for the nitrogen atom in a single oscillator well. The vibration amplitude A is found from energy conservation by recognizing that here all the energy is in potential form: $\frac{1}{2}\hbar\omega = \frac{1}{2}M\omega^2 A^2$, or

$$A = \sqrt{\frac{\hbar}{M\omega}} \qquad (6.16)$$

Since the tunneling integral is symmetric about $x = 0$, we may write it as

$$\frac{2}{\hbar}\sqrt{2M}\int_{-(a-A)}^{a-A}\sqrt{U(x) - E}\,dx = \frac{4}{A^2}\int_0^{a-A}\sqrt{(x-a)^2 - A^2}\,dx$$

where $U(x)$ and E have been expressed in terms of A, using Equation 6.16. The integral on the right can be evaluated in terms of hyperbolic functions,[3] with the result

$$\frac{4}{A^2}\int_0^{a-A}\sqrt{(x-a)^2 - A^2}\,dx = \sinh(2y_0) - 2y_0$$

where y_0 is defined by the relation $\cosh(y_0) = a/A$. The transmission coefficient is then

$$T = e^{-[\sinh(2y_0) - 2y_0]} \qquad (6.17)$$

To get the tunneling rate λ (and its reciprocal—the tunneling time) we must multiply T by the frequency with which the nitrogen atom collides with the potential barrier. For an atom vibrating about its equilibrium position, this is the vibration frequency $f = \omega/2\pi$. From Equation 6.16, we see that f is related to the vibration amplitude as $f = \hbar/2\pi MA^2$.

The tunneling rate depends sensitively on the values chosen for a and A. For the equilibrium distance from the symmetry plane, we take $a = 0.38$ Å, an experimental value obtained from x-ray diffraction measurements.[4] The vibration amplitude A is not directly observable, but its value can be inferred from $U(0)$, the height of the potential barrier at $x = 0$, which is known to be 0.25690 eV. Using Equations 6.15 and 6.16, and taking $M = 14$ u for the mass of the nitrogen atom, we find

$$A = \left(\frac{\hbar^2 a^2}{2MU(0)}\right)^{1/4}$$

$$= \left(\frac{(1.973 \text{ keV}\cdot\text{Å}/c)^2(0.38 \text{ Å})^2}{2(14)(931.50 \times 10^3 \text{ keV}/c^2)(0.2569 \times 10^{-3} \text{ keV})}\right)^{1/4}$$

$$= 0.096 \text{ Å}$$

This underestimates the true value for A because we have (incorrectly) identified M as the mass of the nitrogen atom. In fact, M should be the reduced mass of the nitrogen-hydrogen group, about 2.47 u. With this correction, we find $A = 0.148$ Å and a tunneling rate of $\lambda = fT \approx 2.4 \times 10^{12}$ Hz. The observed

---

[3] Introduce a new integration variable y with the substitution $x - a = -A\cosh(y)$, and use the properties of the hyperbolic cosine and sine functions, $\cosh(y) = \frac{1}{2}(e^y + e^{-y})$, $\sinh(y) = \frac{1}{2}(e^y - e^{-y})$, to obtain the final form.

[4] B. H. Bransden and C. J. Joachain, *Physics of Atoms and Molecules*, New York, John Wiley and Sons, Inc., 1983, p. 456.

tunneling rate, 2.4×10^{10} Hz, suggests a somewhat smaller value for A. By trial and error, we find the actual tunneling rate frequency is reproduced with $A = 0.125$ Å, still a reasonable figure for the vibration amplitude of the nitrogen atom in the ammonia molecule.

Notice that because of tunneling, the nitrogen atom to one side of the symmetry plane or the other does *not* constitute a stationary state of the ammonia molecule, since the probability for finding it there changes over time. In fact, the flopping behavior stems from a simple combination of *two* stationary states, as the following example illustrates.

EXAMPLE 6.9   Ammonia Inversion: A Simulation

MP In this example we use the MP DESKTOP to explore the stationary states of the nitrogen atom in the ammonia molecule and to show how these may be combined to describe the time-dependent (nonstationary) flopping state characteristic of the inversion process. (See Problems 5.39 and 5.40 for more on nonstationary states.)

Begin by readying the DESKTOP to find stationary states of Schrödinger's equation: select *Task* from the menu bar and *EigenStates-1D* from the drop-down task list. Click the Potential button $\boxed{\text{U(x)}}$ located just below the menu bar, and select *EX6-9.U* from the list of potential files. This is the double oscillator potential $U(x)$ in dimensionless form

$$U(x) = (|x| - \alpha)^2$$

where distance has been scaled by $A = (\hbar/M\omega)^{1/2}$ and energy by $\frac{1}{2}\hbar\omega$. For the vibrational amplitude of the nitrogen atom in ammonia, we take $A = 0.125$ Å. With M the reduced mass of the nitrogen-hydrogen group (2.47 u), the energy unit is

$$\frac{1}{2}\hbar\omega = \frac{\hbar^2}{2MA^2}$$

$$= \frac{(1.973 \text{ keV}\cdot\text{Å}/c)^2}{2(2.47)(931.50 \times 10^3 \text{ keV}/c^2)(0.125 \text{ Å})^2}$$

$$= 0.05414 \text{ eV}$$

In the ammonia molecule, the equilibrium points for the nitrogen atom lie a distance $a = 0.38$ Å from the basal plane, making $\alpha = 0.38/0.125 \approx 3$. Click the *OK* button to load the potential with the default value $\alpha = 3$.

The double oscillator potential appears as the solid figure in the main window of the screen. The particle energy E is recorded in the energy box just below the menu bar. Start by typing in the energy box $E = 1$ for the ground state of a single oscillator in these units, and press *Enter* to display the waveform. Continue with different values for

E, using the trial-and-error method described in Example 5.10 to obtain the ground state for the double oscillator to the highest precision possible. Do the same for the first excited state. You should find $E = 0.99955$ and $E = 1.0004$ for the ground state and first excited state energies, respectively. The corresponding wavefunctions are shown in Figures 6.16 and 6.17.

Notice how the stationary state wavefunctions are spread evenly throughout both potential wells: in a truly stationary state, there must be equal probability for tunneling through the barrier from either direction. Notice, too, how the symmetry of the states alternates between even and odd, confirming the general remarks made in Chapter 5 for symmetric potentials. The two lowest states are separated in energy by $\Delta E = (1.0004 - 0.99955)(0.05414 \text{ eV}) = 4.60 \times 10^{-5}$ eV, corresponding to a photon of frequency $\Delta E/h = 1.11 \times 10^{10}$ Hz (microwave range).

Inversion of the ammonia molecule can be modeled by superposing these stationary states to form a *nonstationary* (time-dependent) state describing the nitrogen atom initially localized to one side of the basal plane. This nonstationary state evolves in a flopping fashion, as can be seen from the MP DESKTOP.

Open the *Task* menu and select *QM WaveGroups-1D*. The initial wavefunction $\Psi(x, 0)$ will be taken as a sum of the ground and first excited states found in Figures 6.16 and 6.17. To do this, click the *Control Panel* entry on the *Options* menu. From the *Control Panel* dialog box, select *Enable Mode* in the *Fourier* field. Also, in the *Evolution* field choose the *Sum Eigenstates* option and, in the *Time Step* input box enter a value of 10. Finally, click the *OK* button to activate these features.

NOTE: In Fourier mode, the DESKTOP allows the initial wavefunction $\Psi(x, 0)$ to be constructed as a sum of stationary state wavefunctions $\Psi_k(x)$ for the potential $U(x)$ in use:

$$\Psi(x, 0) = \sum a(k)\psi_k(x)$$

Figure 6.16   (Example 6.9) The ground state wavefunction for the double oscillator. The wavefunction is nodeless and symmetric (even) about the point $x = 0$ midway between the oscillator wells.

Figure 6.17   (Example 6.9) The first excited state wavefunction for the double oscillator potential. The first excited state has one node and is antisymmetric (odd) about the point midway between the oscillator wells.

The *Sum Eigenstates* option instructs the DESKTOP to compute $\Psi(x, t)$ as

$$\Psi(x, t) = \sum a(k)\,\psi_k(x)\,\exp[-i\omega(k)\,t]$$

rather than integrating the (time-dependent) Schrödinger equation. When just a few states are being summed, as is the case here, summing eigenstates is the faster, more accurate method. The larger time step also presents no problem in the summation method and is needed to handle the large inversion time encountered in this example.

In Fourier mode, the display changes to a split screen and the Frequency and Amplitude buttons appear below the menu bar to allow inputs for the frequency $\omega(k)$ (or energy) and the weighting factor $a(k)$ for each stationary state in the sum.

Now click the Frequency button $\boxed{\omega(k)}$ to load the frequency file *EX6-9.W* specifying the frequencies $\omega(k)$—or equivalently, the energies $E(k)$—for the two lowest lying states of the ammonia molecule. By convention, bound states are assigned integer quantum numbers k, beginning with $k = 0$ for the ground state. Click the *OK* button to load this selection: the energies appear on a bar graph in the left window of the screen. Finally, click the Amplitude button $\boxed{a(k)}$ and choose *UNIFORM.A* to mix the

ground and first excited states with equal amplitude. Click the *OK* button to load this selection. The amplitudes are superimposed on the energies on the bar plot in the left window of the screen; in the right window is plotted the superposition $\Psi(x, 0)$ of these two stationary states at time $t = 0$ (Fig. 6.18a). By mixing these stationary states, we have produced a wavefunction predominantly in the right-side compartment of the double well: this represents the nitrogen atom initially localized to one side of the basal plane in the ammonia molecule.

The current time is recorded in the time box below the menu bar. Click the Advance button $\boxed{\triangleright}$ to the immediate right of the time box to start the clock. The wavefunction leaks through the intervening barrier, eventually reassembling in the left-side compartment. Click again to stop. Estimate the inversion time for the molecule by allowing the clock to run until the waveform is localized in the left-side compartment (Fig. 6.18b). From the inversion time deduce the flopping frequency, and compare your result with the frequency $\Delta E/h$ separating the two stationary states. [*Note:* The time unit is $\hbar$ divided by the energy unit, $(6.582 \times 10^{-16}\ \text{eV·s})/(5.414 \times 10^{-2}\ \text{eV}) = 1.216 \times 10^{-14}\ \text{s}$.]

(a)   (b)

Figure 6.18 (a) The nonstationary state formed from the superposition of the ground and first excited states represents the nitrogen atom initially localized in the right-side compartment of the double oscillator well. (b) Some time later, the nonstationary state of Figure 6.18a reassembles in the left-side compartment of the double oscillator well, indicating the nitrogen atom is now on the opposite side of the basal plane: the molecule has undergone inversion.

Since the flopping frequency is in the microwave range, the ammonia molecule can serve as an amplifier for microwave radiation. The *ammonia maser* operates on this principle. Because of the small energy difference between the ground and first excited states of the ammonia molecule, ammonia vapor at room temperature has roughly equal numbers of molecules in both states. Having opposite electric dipole moments, these states are easily separated by passing the vapor through a nonuniform electric field. In this way, ammonia vapor can be produced with the unusually large concentrations of excited molecules needed to create the *population inversion* necessary for maser operation. A spontaneous deexcitation to the ground state of one molecule releases a (microwave) photon, which, in turn, induces other molecules to deexcite. The result—much like a chain reaction—produces a photon cascade: an intense burst of coherent microwave radiation. The operation of masers and lasers is discussed in more detail in Chapter 11.

Decay of Black Holes

Once inside the event horizon, nothing—not even light—can escape the gravitational pull of a black hole.[5] That was the view held until 1974, when the brilliant British astrophysicist Stephen Hawking proposed that black holes are indeed radiant objects, emitting a variety of particles by a mechanism involving tunneling through the (gravitational) potential barrier surrounding the black hole. The thickness of this barrier is proportional to the size of the black hole,

---

[5] A brief introduction to black holes is found in Clifford Will's essay "The Renaissance of General Relativity", p. 50.

Stephen W. Hawking (1942–), a British astrophysicist, has made major contributions to problems in cosmology, including the theory of black holes. (*MMP/Manni Mason's Pictures*)

so that the likelihood of a tunneling event initially may be extremely small. As the black hole emits particles, however, its mass and size steadily decrease, making it easier for more particles to tunnel out. In this way emission continues at an ever increasing rate, until eventually the black hole radiates itself out of existence in an explosive climax! Thus, Hawking's scenario leads inexorably to the decay and eventual demise of any black hole. Calculations indicate that a black hole with the mass of our Sun would survive against decay by tunneling for about 10^{66} years. On the other hand, a black hole with the mass of only a billion tons and roughly the size of a proton (such mini black holes are believed to have been formed just after the Big Bang origin of the Universe) should have almost completely evaporated in the 10 billion years that have elapsed since the time of creation.

SUMMARY

For potentials representing barriers there are no bound states, only unbound states describing particles scattering from the barrier. When a particle of energy E meets a barrier of height U, part of the incident matter wave is reflected by the barrier, and part is transmitted through the barrier. Particles incident on the barrier are reflected or transmitted with probability given by the **reflection coefficient R** and **transmission coefficient T,** respectively. These coefficients obey the **sum rule**

$$R(E) + T(E) = 1 \tag{6.5}$$

for any particle energy E. Even when $E < U$, the particle has some nonzero probability of penetrating the barrier. This process, called **tunneling,** is the basic mechanism underlying the phenomena of **field emission, alpha decay** of radioactive nuclei, **ammonia inversion,** particle radiation from **black holes,** and many more. The likelihood of tunneling through a barrier of arbitrary shape $U(x)$ is given approximately as

$$T(E) \approx \exp\left(-\frac{2}{\hbar} \sqrt{2m} \int \sqrt{U(x) - E}\, dx \right) \tag{6.10}$$

The integral in Equation 6.10 is taken over the **classically forbidden** region where $U(x) > E$.

SUGGESTIONS FOR FURTHER READING

1. Many of the topics in this chapter also are treated at about the same level by A. P. French and Edwin F. Taylor in *An Introduction to Quantum Physics*, New York, W. W. Norton and Company, Inc., 1978.
2. Experimental aspects of field emission microscopy and related analytical surface techniques are discussed by Philip F. Kane in Chapter 6 of *Characterization of Solid Surfaces*, New York, Plenum Press, 1974, pp. 133–146.
3. For readable accounts of tunneling from black holes, see S. W. Hawking, "The Quantum Mechanics of Black Holes," *Sci. Am.*, January 1977, pp. 34–40; and J. D. Beckenstein, "Black-hole Thermodynamics," *Phys. Today*, January 1980, pp. 24–31. The second title is somewhat more advanced, and contains additional references.

QUESTIONS

1. Consider a particle with energy E scattered from a potential barrier of height $U > E$. How does the amplitude of the reflected wave change as the barrier height is reduced? How does the amplitude of the incident wave change?
2. An electron and a proton of identical energy encounter the same potential barrier. For which is the probability of transmission greatest, and why?
3. In classical physics, only *differences* in energy have physical significance. Discuss how this carries over to the quantum scattering of particles by examining how adding a constant potential U_0 everywhere affects barrier reflection and transmission coefficients.

4. Suppose a particle with energy E is located within a barrier whose height U is greater than E. Will this particle be found to have a negative kinetic energy? Explain.
5. Explain how a barrier whose width is L might be considered wide for penetration by protons, yet at the same time narrow for penetration by electrons.
6. Discuss the suitability of portraying wavefunctions and potential barriers on the same graph, as in Figure 6.2a. Can a wave whose crest falls below the top of a square barrier ever penetrate the barrier? Explain.

PROBLEMS

6.1 The Square Barrier

1. A particle incident on the potential step of Example 6.5 with a certain energy $E < U$ is described by the wave

$$\psi(x) = \tfrac{1}{2}\{(1 + i)\,e^{ikx} + (1 - i)\,e^{-ikx}\} \qquad \text{for } x \le 0$$

$$\psi(x) = e^{-kx} \qquad \text{for } x \ge 0$$

   (a) Verify by direct calculation that the reflection coefficient is unity in this case. (b) How must k be related to E in order for $\psi(x)$ to solve Schrödinger's equation in the region to the left of the step $(x \le 0)$? to the right of the step $(x > 0)$? What does this say about the ratio E/U? (c) Evaluate the penetration depth $\delta = 1/k$ for 10 MeV protons incident on this step.
2. Consider the step potential of Example 6.5 in the case where $E > U$. (a) Examine the Schrödinger equation to the left of the step to find the form of the solution in the range $x < 0$. Do the same to the right of the step to obtain the solution form for $x > 0$. Complete the solution by enforcing whatever boundary and matching conditions may be necessary. (b) Obtain an expression for the reflection coefficient R in this case, and show that it can be written in the form

$$R = \frac{(k_1 - k_2)^2}{(k_1 + k_2)^2}$$

   where k_1 and k_2 are wavenumbers for the incident and transmitted waves, respectively. Also write an expression for the transmission factor T using the sum rule obeyed by these coefficients.
   (c) Evaluate R and T in the limiting cases of $E \to U$ and $E \to \infty$. Are the results sensible? Explain. (This situation is analogous to the partial reflection and transmission of light striking an interface separating two different media.)

3. Use the results of the preceding problem to calculate the fraction of 25-MeV protons reflected and the fraction transmitted by a 20-MeV step. How do your answers change if the protons are replaced by electrons?
4. A 0.100-mA electron beam with kinetic energy 54.0 eV enters a sharply defined region of *lower* potential where the kinetic energy of the electrons is increased by 10.0 eV. What current is reflected at the boundary? (This simulates electron scattering at normal incidence from a metal surface, as in the Davisson–Germer experiment.)
5. (a) Tunneling of particles through barriers that are high or wide (or both) is very unlikely. Show that for a square barrier with

$$\frac{2mUL^2}{\hbar^2} \gg 1$$

   and particle energies well below the top of the barrier $(E \ll U)$, the probability for transmission is approximately

$$P \approx 16\frac{E}{U}\,e^{-2[\sqrt{2m(U-E)}/\hbar]L}$$

   (The combination UL^2 is sometimes referred to as the barrier *strength*.) (b) Give numerical estimates for the exponential factor in P for each of the following cases: (1) an electron with $U - E = 0.01$ eV and $L = 0.1$ nm; (2) an electron with $U - E = 1$ eV and $L = 0.1$ nm; (3) an alpha particle ($m = 6.7 \times 10^{-27}$ kg) with $U - E = 10^6$ eV and $L = 10^{-15}$ m; and (4) a bowling ball ($m = 8$ kg) with $U - E = 1$ J and $L = 2$ cm (this corresponds to the ball's getting past a barrier 2 cm wide and too high for the ball to slide over).
6. A beam of electrons is incident on a barrier 5 eV high and 1 nm wide. Write a simple computer program to

find what energy the electrons should have if 0.1% of them are to get through the barrier.

7. Starting from the joining conditions, Equations 6.8, obtain the result for the transmission coefficient of a square barrier given in Equation 6.9 (valid when the particle has insufficient energy to penetrate the barrier classically: $E < U$).

8. **MP** Use the MP DESKTOP to investigate the scattering of electrons from a square barrier 1.00 Å thick and 10.0 eV high, in the case where the electron energy is equal to the barrier height, $E = U$ (see Example 6.4). What is the functional form of the wave in the barrier region? Determine the transmission coefficient at this energy, and compare your result with the prediction of Equation 6.9. What does classical physics predict for the probability of transmission in this case?

9. **MP** Set up the MP DESKTOP as in Example 6.4 to investigate the scattering of electrons from a square barrier 1.00 Å thick and 10.0 eV high. Determine the transmission and reflection coefficients for 5.00-eV electrons incident on the barrier. Verify the sum rule, Equation 6.5, and compare your result for $T(E)$ with the prediction of Equation 6.9. What must the barrier thickness be to transmit 5.00-eV *protons* with the same probability?

10. **MP** *Scattering Resonances.* Set up the MP DESKTOP to investigate the scattering of electrons from a square barrier of width 1.00 Å and height 10.0 eV (Example 6.4). Use trial-and-error methods to locate the two lowest energies E giving rise to perfect transmission (look for zero-amplitude reflections). For each energy use the MP Analyzer (click 🔍) to estimate the electron wavelength λ in the barrier (click ◀▮▶ several times for a closeup view of the barrier waveform). Compare λ with the barrier width L and discuss your findings in terms of the interference of waves reflected from the leading and trailing edges of the barrier (see Example 6.3).

11. *The Ramsauer–Townsend Effect.* Consider the scattering of particles from the potential well shown in Figure P6.11. (a) Explain why the waves reflected from the well edges

$x = 0$ and $x = L$ will cancel completely if $2L = \lambda_2$, where λ_2 is the de Broglie wavelength of the particle in region 2. (b) Write expressions for the wavefunctions in regions 1, 2, and 3. Impose the necessary continuity restrictions on Ψ and $\partial \Psi / \partial x$ to show explicitly that $2L = \lambda_2$ leads to no reflected wave in region 1. [This is a crude model for the Ramsauer–Townsend effect observed in the collisions of slow electrons with noble gas atoms like argon, krypton, and xenon. Electrons with just the right energy are diffracted around these atoms as if there were no obstacle in their path (perfect transmission). The effect is peculiar to the noble gases because their closed-shell configurations produce atoms with abrupt outer boundaries.]

12. A model potential of interest for its simplicity is the *delta well*. The delta well may be thought of as a square well of width L and depth S/L in the limit $L \to 0$ (Fig. P6.12). The limit is such that S, the product of the well depth with its width, remains fixed at a finite value known as the *well strength*. The effect of a delta well is to introduce a discontinuity in the slope of the wavefunction at the well site, although the wave itself remains continuous here. In particular, it can be shown that

$$\left. \frac{d\psi}{dx} \right|_{0+} - \left. \frac{d\psi}{dx} \right|_{0-} = -\frac{2mS}{\hbar^2} \psi(0)$$

for a delta well of strength S situated at $x = 0$. (a) Solve Schrödinger's equation on both sides of the well ($x < 0$ and $x > 0$) for the case where particles are incident from the left with energy $E > 0$. Note that in these regions the particles are free, so that $U(x) = 0$. (b) Enforce the continuity of ψ and the slope condition at $x = 0$. Solve the resulting equations to obtain the transmission coefficient T as a function of particle energy E. Sketch $T(E)$ for $E \geq 0$. (c) If we allow E to be negative, we find that $T(E)$ diverges for some particular energy E_0. Find this value E_0. (As it happens, E_0 is the energy of a *bound state* in the delta well. The calculation illustrates a general technique, in which bound states are sought among the singularities of the scattering coefficients for a potential well of arbitrary shape.) (d) What fraction of the particles incident on the well

Figure P6.11

Figure P6.12

with energy $E = |E_0|$ is transmitted and what fraction is reflected?

13. Obtain directly an expression for the reflection coefficient $R(E)$ for the delta well of Problem 12, and verify the sum rule

$$R(E) + T(E) = 1$$

for all particle energies $E > 0$.

6.2 Barrier Penetration

14. A barrier of arbitrary shape can be approximated as a succession of square barriers, as shown in Figure P6.14. Write the transmission coefficient for this barrier using the result of Equation 6.9 for each of the individual barriers, assuming the transmitted wave intensity for one becomes the incident wave intensity for the barrier immediately following it in the series. Show that the form of Equation 6.10 is recovered in the case where $E < U$ and $\alpha L \gg 1$.

(a)

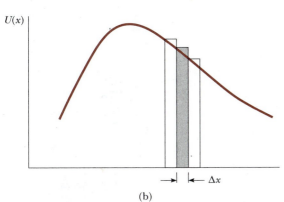

(b)

Figure P6.14

15. Consider an alpha particle confined to a thorium nucleus. Model the nuclear potential as a semiinfinite square well with an infinitely high wall at $r = 0$ and a

wall of height 30.0 MeV at the nuclear radius $R = 9.00$ fm. Use the iterative method of Example 5.9 to estimate the smallest values of energy and velocity permitted for the alpha particle. What conclusion can you draw from the fact that the *ejected* alpha is observed to have a kinetic energy of 4.05 MeV?

16. The attempt frequency of an alpha particle to escape the nucleus is the number of times per second it collides with the nuclear barrier. Estimate this collision frequency in the tunneling model for the alpha decay of thorium, assuming the alpha behaves like a true particle inside the nucleus with *total* energy equal to the observed kinetic energy of decay.

The daughter nucleus for this case (radium) has $Z = 88$ and a radius of 9.00 fm. Take for the overall nuclear barrier 30.0 MeV, measured from the bottom of the nuclear well to the top of the Coulomb barrier (see Fig. 6.10).

17. Verify the claim of Section 6.2 that the electrons of a metal collide with the surface at a rate of about 10^{30} per second per square centimeter. Do this by estimating the collision frequency of electrons in a 1.00-cm cube of copper metal with one face of the cube surface. Assume that each copper atom contributes one conduction electron to the metal (the chemical valence of copper is $+1$), and that these conduction electrons move freely with kinetic energy equal to 7.00 eV. In fact, not all the electrons have this energy; see Chapter 9.

18. **MP** *Resonant Tunneling.* Heterostructures formed from layered semiconductors have characteristics important to many modern electronic devices. Here, we use the MP DESKTOP to study tunneling in a three-layer gallium arsenide/gallium aluminum arsenide (GaAs-Ga$_{1-x}$Al$_x$As) sandwich. The GaAs layer constitutes a potential well between two confining barriers formed by the Ga$_{1-x}$Al$_x$As layers. Unusually large transmission (resonant tunneling) through the double barrier occurs when the energy of the incident electron coincides with that of a bound state (actually a quasibound state—see Example 6.8) in the well. Set up the MP DESKTOP to model this device. (The double barrier potential can be found in the file *PR6-18.U.*) Use 0.25 eV for the barrier height and a layer thickness of 50.0 Å. Note that electrons in these materials behave like free electrons with an effective mass $m^* = 0.067m_e$, only a fraction of the free electron value. Following the method of Example 6.8, find the energy of a quasibound state in the well, and obtain the transmission coefficient $T(E)$ of the device at this energy. Make small adjustments to E to find the energy for peak transmission. Investigate the width of the resonance by varying the electron energy until $T(E)$ falls to half of its peak value. (In reality, the incident electron energy is fixed; however, the device can be "tuned" to resonance by applying a suitable bias voltage, which alters the transparency of the barriers.)

THE SCANNING TUNNELING MICROSCOPE

Roger A. Freedman

The basic idea of quantum mechanics, that particles have properties of waves and vice versa, is among the strangest found anywhere in science. Because of this strangeness, and because quantum mechanics mostly deals with the very small, it might seem to have little practical application. As we will show in this essay, however, one of the basic phenomena of quantum mechanics—the tunneling of particles—is at the heart of a very practical device that is one of the most powerful microscopes ever built. This device, the *scanning tunneling microscope* or *STM*, enables physicists to make highly detailed images of surfaces with resolution comparable to the size of a *single atom*. Such images promise to revolutionize our understanding of structures and processes on the atomic scale.

Before discussing how the STM works, we first look at a sample of what the STM can do. An image made by a scanning tunneling microscope of the surface of a piece of gold is shown in Figure 1. You can easily see that the surface is not uniformly flat, but is a

Paul K. Hansma

*Department of Physics,
University of California,
Santa Barbara*

Figure 1 Scanning tunneling microscope image of the surface of crystalline gold. Successive scans are approximately 1.5 Å apart. The figure is from G. Binnig, H. Rohrer, Ch. Gerber, and E. Stoll, *Surface Sci.*, 144:321, 1984.

1 cm

Figure 2   One design for a scanning tunneling microscope (STM). The sample to be studied is mounted on a plate in the cylindrical dish. The probe extends beneath the left tripod. The micrometer attached to the spring is used to position the sample.

series of terraces separated by steps that are only one atom high. Gentle corrugations can be seen in the terraces, caused by subtle rearrangements of the gold atoms.

What makes the STM so remarkable is the fineness of the detail that can be seen in images such as Figure 1. The *resolution* in this image—that is, the size of the smallest detail that can be discerned—is about 2 Å (2×10^{-10} m). For an ordinary microscope, the resolution is limited by the wavelength of the waves used to make the image. Thus an optical microscope has a resolution of no better than 2000 Å, about half the wavelength of visible light, and so could never show the detail displayed in Figure 1. Electron microscopes can have a resolution of 2 Å by using electron waves of wavelength 4 Å or shorter. From the de Broglie formula $\lambda = h/p$, the electron momentum p required to give this wavelength is 3100 eV/c, corresponding to an electron speed $v = p/m_e = 1.8 \times 10^6$ m/s. Electrons traveling at this high speed would penetrate into the interior of the piece of gold in Figure 1, and so would give no information about individual surface atoms.

The image in Figure 1 was made by Gerd Binnig, Heinrich Rohrer, and collaborators at the IBM Research Laboratory in Zurich, Switzerland. Binnig and Rohrer invented the STM and shared the 1986 Nobel prize in physics for their work. Such is the importance of this device that unlike most Nobel prizes, which come decades after the original work, Binnig and Rohrer received their Nobel prize just six years after their first experiments with an STM.

One design for an STM is shown in Figure 2. The basic idea behind its operation is very simple, as shown in Figure 3. A conducting probe with a very sharp tip is brought near the surface to be studied. Because it is attracted to the positive ions in the surface,

(a)

Material   Empty space

ψ

(b)

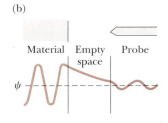

Material   Empty space   Probe

ψ

Figure 3   (a) The wavefunction of an electron in the surface of the material to be studied. The wavefunction extends beyond the surface into the empty region. (b) The sharp tip of a conducting probe is brought close to the surface. The wavefunction of a surface electron penetrates into the tip, so that the electron can "tunnel" from surface to tip. Compare this figure to Figure 6.2a.

an electron in the surface has a lower total energy than would an electron in the empty space between surface and tip. The same thing is true for an electron in the tip. In classical Newtonian mechanics, electrons could not move between the surface and tip because they would lack the energy to escape either material. But because the electrons obey quantum mechanics, they can "tunnel" across the barrier of empty space between the surface and the tip. Let us explore the operation of the STM in terms of the discussion of tunneling in Section 6.1.

For an electron in the apparatus of Figures 2 and 3, a plot of the energy as a function of position would look like Figure 6.2b. The horizontal coordinate in this figure represents electron position. Now L is to be interpreted as the distance between the surface and the tip, so that coordinates less than 0 refer to positions inside the surface material and coordinates greater than L refer to positions inside the tip. The barrier height $U = q\phi$ is the potential energy difference between an electron outside the material and an electron in the material. That is, an electron in the surface or tip has potential energy $-U$ compared with one in vacuum. (We are assuming for the moment that the surface and tip are made of the same material. We will comment on this assumption shortly.) The kinetic energy of an electron in the surface is E so that an amount of energy equal to $(U - E)$ must be given to an electron to remove it from the surface. Thus $(U - E)$ is the work function of an electron in the surface.

For the potential energy curve of Figure 6.2b, one could expect as much tunneling from the surface into the tip as in the opposite direction. In an STM, the direction in which electrons tend to cross the barrier is controlled by applying a voltage between the surface and the tip. With preferential tunneling from the surface into the tip, the tip samples the distribution of electrons in and above the surface. Because of this "bias" voltage, the work functions of surface and tip are different, giving a preferred direction of tunneling. This is also automatically the case if the surface and tip are made of different materials. In addition, the top of the barrier in Figure 6.2b will not be flat but will be tilted to reflect the electric field between the surface and the tip. If the barrier energy U is large compared with the difference between the surface and tip work functions, however, and if the bias voltage is small compared with $\phi = U/q$, we can ignore these complications in our calculations. Then all the results for a square barrier given in Section 6.1 may be applied to an STM. Detailed discussions of the effects that result when these complications are included can be found in the article by P. K. Hansma and J. Tersoff and in the Nobel prize lecture of Binnig and Rohrer (see Suggestions for Further Reading).

The characteristic scale of length for tunneling is set by the work function $(U - E)$. For a typical value $(U - E) = 4.0$ eV, this scale of length is

$$\delta = \frac{\hbar}{\sqrt{2m_e(U - E)}} = \frac{\hbar c}{\sqrt{2m_e c^2 (U - E)}}$$

$$= \frac{1.973 \text{ keV} \cdot \text{Å}}{\sqrt{2(511 \text{ keV})(4.0 \times 10^{-3} \text{ keV})}} = 0.98 \text{ Å} \approx 1.0 \text{ Å}$$

The probability that a given electron will tunnel across the barrier is just the transmission coefficient T (Equation 6.9). If the separation L between surface and tip is not small compared with δ, then the barrier is "wide" and we can use the approximate result of Problem 5 for T. The current of electrons tunneling across the barrier is simply proportional to T. The tunneling current density can be shown to be

$$j = \frac{e^2 V}{4\pi^2 L\delta\hbar} e^{-2L/\delta}$$

In this expression e is the charge of the electron and V is the bias voltage between surface and tip.

We can see from this expression that the STM is very sensitive to the separation L between tip and surface. This is because of the exponential dependence of the tunneling current on L (this is much more important than the $1/L$ dependence). As we saw earlier, typically $\delta \approx 1.0$ Å. Hence increasing the distance L by just 0.01 Å causes the tunneling current to be multiplied by a factor $e^{-2(0.01 \text{ Å})/(1.0 \text{ Å})} \approx 0.98$; that is, the current decreases by 2%—a change that is measurable. For distance L greater than 10 Å (that is, beyond a few atomic diameters), essentially no tunneling takes place. This sensitivity to L is the basis of the operation of the STM: monitoring the tunneling current as the tip is scanned over the surface gives a sensitive measure of the topography of the surface. In this way the STM can measure the height of surface features to within 0.01 Å, or approximately one one-hundredth of an atomic diameter.

The STM also has excellent lateral resolution, that is, resolution of features in the plane of the surface. This is because the tips used are *very* sharp indeed, typically only an atom or two wide at their extreme end. Thus the tip samples the surface electrons only in a very tiny region approximately 2 Å wide, and so can "see" very fine detail. You might think that making such tips would be extremely difficult, but in fact it's relatively easy: sometimes just sharpening the tip on a fine grinding stone (or even with fine sandpaper) is enough to cause the tip atoms to rearrange by themselves into an atomically sharp configuration. (If you find this surprising, you're not alone. Binnig and Rohrer were no less surprised when they discovered this.)

There are two modes of operation for the STM, shown in Figure 4. In the *constant current mode* (Fig. 4a), a convenient operating voltage (typically between 2 mV and 2 V) is first established between surface and tip. The tip is then brought close enough to the surface to obtain measurable tunneling current. The tip is then scanned over the surface

(a) Constant current mode

(b) Constant height mode

Schematic view

One scan

Multiple scans

Figure 4 Scanning tunneling microscopes can be operated in either (a) the constant current mode or (b) the constant height mode. The images of the surface of graphite were made by Richard Sonnenfeld at the University of California at Santa Barbara. The constant height mode was first used by A. Bryant, D. P. E. Smith, and C. F. Quate, *Appl. Phys. Lett.* 48:832, 1986.

while the tunneling current I is measured. A feedback network changes the vertical position of the tip, z, to keep the tunneling current constant, thereby keeping the separation L between surface and tip constant. An image of the surface is made by plotting z versus lateral position (x, y). The simplest scheme for plotting the image is shown in the graph below the schematic view. The height z is plotted versus the scan position x. An image consists of multiple scans displaced laterally from each other in the y direction.

The constant current mode was historically the first to be used and has the advantage that it can be used to track surfaces that are not atomically flat (as in Fig. 1). The feedback network, however, requires that the scanning be done relatively slowly. As a result, the sample being scanned must be held fixed in place for relatively long times to prevent image distortion.

Alternatively, in the *constant height mode* (Fig. 4b), the tip is scanned across the surface at constant voltage and nearly constant height while the current is monitored. In this case the feedback network responds only rapidly enough to keep the average current constant, which means that the tip maintains the same average separation from the surface. The image is then a plot of current I versus lateral position (x, y), as shown in the graph below the schematic. Again, multiple scans along x are displayed laterally displaced in the y direction. The image shows the substantial variation of tunneling current as the tip passes over surface features such as individual atoms.

The constant height mode allows much faster scanning of atomically flat surfaces (100 times faster than the constant current mode), since the tip does not have to be moved up and down over the surface "terrain." This fast scanning means that making an image of a surface requires only a short "exposure time." By making a sequence of such images, researchers may be able to study in real-time processes in which the surfaces rearrange themselves—in effect making an STM "movie."

Individual atoms have been imaged on a variety of surfaces, including those of so-called *layered materials* in which atoms are naturally arranged into two-dimensional layers. Figure 5 shows an example of atoms on one of these layered materials. In this image it is fascinating not only to see individual atoms, but also to note that some atoms are missing. Specifically, there are three atoms missing from Figure 5. Can you find the places where they belong?

$\longmapsto$ 1 nm $\longrightarrow$

Figure 5 Image of atoms on a surface of tantalum disulfide (TaS_2) immersed in liquid nitrogen. 1 nm $= 10^{-9}$ meter $= 10$ Å. The figure is from C. G. Slough, W. W. McNairy, R. V. Coleman, B. Drake, and P. K. Hansma, *Phys. Rev.* B34:994, 1986.

Figure 6   Image of a graphite electrode in an electrolyte used for silver plating. The figure is from R. Sonnenfeld and B. Schardt, *Appl. Phys. Lett.* 49:1172, 1986.

Another remarkable aspect of the STM image in Figure 5 is that it was obtained with the surface and tip immersed in liquid nitrogen. We assumed earlier in this essay that the space between the surface and tip must be empty, but in fact electron tunneling can take place not just through vacuum but also through gases and liquids—even water. This seems very surprising since we think of water, especially water with salts dissolved in it, as a conductor. But water is only an *ionic* conductor. For electrons, water behaves as an insulator just as vacuum behaves as an insulator. Thus electrons can flow through water only by tunneling, which makes scanning tunneling microscopy possible "under water."

As an example, Figure 6 shows individual carbon atoms on a graphite surface. It was obtained for a surface immersed in a silver-plating solution, which is highly conductive for ions but behaves as an insulator for electrons. (The sides of the conducting probe were sheathed with a nonconductor, so that the predominant current into the probe comes from electrons tunneling into the exposed tip. The design of STM used to make this particular image is the one shown in Fig. 2.) Sonnenfeld and Schardt observed atoms on this graphite surface before plating it with silver, after "islands" of silver atoms were plated onto the surface, and after the silver was electrochemically stripped from the surface. Their work illustrates the promise of the scanning tunneling microscope for seeing processes that take place on an atomic scale.

While the original STMs were one-of-a-kind laboratory devices, commercial STMs have recently become available. Figure 7 is an image of a graphite surface in air made with such a commercial STM. Note the high quality of this image and the recognizable rings of carbon atoms. You may be able to see that three of the six carbon atoms in each ring *appear* lower than the other three. All six atoms are in fact at the same level, but the three that appear lower are bonded to carbon atoms lying directly beneath them in the underlying atomic layer. The atoms in the surface layer that appear higher do not lie directly over subsurface atoms, and hence are not bonded to carbon atoms beneath them. For the higher-appearing atoms, some of the electron density that would have been involved in bonding to atoms beneath the surface instead extends into the space above the surface. This extra electron density makes these atoms appear higher in Figure 7, since what the STM maps is the topography of the electron distribution.

Figure 7   Image of a graphite surface in air, obtained with a commercial STM: the Nanoscope II from Digital Instruments in Goleta, California.

The availability of commercial instruments should speed the use of scanning tunneling microscopy in a variety of applications. These include characterizing electrodes for electrochemistry (while the electrode is still in the electrolyte), characterizing the roughness of surfaces, measuring the quality of optical gratings, and even imaging replicas of biological structures.

Perhaps the most remarkable thing about the scanning tunneling microscope is that its operation is based on a quantum mechanical phenomenon—tunneling—that was well understood in the 1920s, yet the STM itself wasn't built until the 1980s. What other applications of quantum mechanics may yet be waiting to be discovered?

Suggestions for Further Reading

G. Binnig, H. Rohrer, Ch. Gerber, and E. Weibel, *Phys. Rev. Lett.* 49:57, 1982. The first description of the operation of a scanning tunneling microscope.

G. Binnig and H. Rohrer, *Sci. Am.*, August 1985, p. 50. A popular description of the STM and its applications.

C. F. Quate, *Phys. Today*, August 1986, p. 26. An overview of the field of scanning tunneling microscopy, including insights into how it came to be developed.

P. K. Hansma and J. Tersoff, *J. Appl. Phys.* 61:R1, 1987. A comprehensive review of the "state of the art" in scanning tunneling microscopy.

G. Binnig and H. Rohrer, *Rev. Mod. Phys.* 59:615, 1987. The text of the lecture given on the occasion of the presentation of the 1986 Nobel prize in physics.

7

Quantum Mechanics in Three Dimensions

Chapter Outline

So far we have shown how quantum mechanics can be used to describe motion in one dimension. Although the one-dimensional case illustrates such basic features of systems as the quantization of energy, we need a full three-dimensional treatment for the applications to atomic, solid-state, and nuclear physics that we will meet in later chapters. In this chapter we extend the concepts of quantum mechanics from one to three dimensions and explore the predictions of the theory for the simplest of real systems—the hydrogen atom.

With the introduction of new degrees of freedom (and the additional co-ordinates needed to describe them) comes a disproportionate increase in the level of mathematical difficulty. To guide our inquiry, we shall rely on classical insights in helping us identify the observables that are likely candidates for quantization. These must come from the ranks of the so-called *sharp observables* and, with few exceptions, are the same as the constants of the classical motion.

7.1 PARTICLE IN A THREE-DIMENSIONAL BOX

Let us explore the workings of wave mechanics in three dimensions through the example of a particle confined to a cubic "box." The box has edge length L and occupies the region $0 < x, y, z < L$, as shown in Figure 7.1. We assume

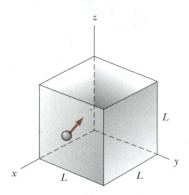

Figure 7.1 A particle confined to move in a cubic box of sides L. Inside the box $U = 0$. The potential energy is infinite at the walls and outside the box.

Schrödinger equation in three dimensions

Figure 7.2 Change in particle velocity (or momentum) during collision with a wall of the box. For elastic collision with a smooth wall, the component normal to the wall is reversed, but the tangential components are unaffected.

the walls of the box are *smooth,* so they exert forces only perpendicular to the surface, and that collisions with the walls are elastic. A classical particle would rattle around inside such a box, colliding with the walls. At each collision, the component of particle momentum normal to the wall is reversed (changes sign), while the other two components of momentum are unaffected (Fig. 7.2). Thus, the collisions preserve the *magnitude* of each momentum component, in addition to the total particle energy. These four quantities—$|p_x|$, $|p_y|$, $|p_z|$, and E—then, are constants of the classical motion, and it should be possible to find quantum states for which all of them are sharp.[1]

The wavefunction Ψ in three dimensions is a function of $\mathbf{r}$ and t. Again, only the magnitude of Ψ is meaningful, and we define the probability density $P(\mathbf{r},t) = |\Psi(\mathbf{r}, t)|^2$, which is now a *probability per unit volume*. Multiplication by the volume element $dV(= dx\,dy\,dz)$ gives the probability of finding the particle within the volume element dV at the point $\mathbf{r}$ at time t.

Since our particle is confined to the box, the wavefunction Ψ must be zero outside and at the walls. The wavefunction inside the box is found from Schrödinger's equation,

$$-\frac{\hbar^2}{2m}\nabla^2\Psi + U(\mathbf{r})\Psi = i\hbar\frac{\partial\Psi}{\partial t} \tag{7.1}$$

We see that $\partial^2/\partial x^2$ in the one-dimensional case is replaced in three dimensions by the **Laplacian,**

$$\nabla^2 = \frac{\partial^2}{\partial x^2} + \frac{\partial^2}{\partial y^2} + \frac{\partial^2}{\partial z^2} \tag{7.2}$$

where $U(\mathbf{r})$ is still the potential energy, but now a function of all the space coordinates: $U(\mathbf{r}) = U(x, y, z)$. Indeed, the Laplacian together with its multiplying constant is just the kinetic energy operator of Table 5.2 extended to include the contributions to the kinetic energy from motion in every direction:

$$-\frac{\hbar^2}{2m}\nabla^2 = \left(-\frac{\hbar^2}{2m}\frac{\partial^2}{\partial x^2}\right) + \left(-\frac{\hbar^2}{2m}\frac{\partial^2}{\partial y^2}\right) + \left(-\frac{\hbar^2}{2m}\frac{\partial^2}{\partial z^2}\right)$$
$$= [K_x] + [K_y] + [K_z] \tag{7.3}$$

This form is consistent with our belief that the cartesian axes identify independent but logically equivalent directions in space. With this identification of the kinetic energy operator in three dimensions, the left-hand side of Equation 7.1 is again the hamiltonian operator $[H]$ applied to Ψ, and the right-hand side is the energy operator $[E]$ applied to Ψ (see Section 5.8). As it does in one dimension, Schrödinger's equation asserts the equivalence of these two operators when applied to the wavefunction of any physical system.

---

[1] Recall from Section 5.7 that sharp observables are those for which there is no statistical distribution of measured values. Indeed, quantum wavefunctions typically are labeled by the sharp observables for that state. (For example, the oscillator states of Section 5.6 are indexed by the quantum number n which, in turn, specifies the sharp value of particle energy E. In this case, the sharp values of energy also are *quantized*, that is limited to the discrete values $(n + \frac{1}{2})\hbar\omega$. It follows that any sharp observable is constant over time (unless the corresponding operator explicitly involves time). The converse—that quantum states exist for which all constants of the classical motion are sharp—is not always true but occurs frequently enough that it can serve as a useful rule of thumb.

Stationary states are those for which all probabilities are constant in time and are given by solutions to Schrödinger's equation in the separable form,

$$\Psi(\mathbf{r},\, t) = \psi(\mathbf{r})\, e^{-i\omega t} \tag{7.4}$$

With this time dependence, the right-hand side of Equation 7.1 reduces to $\hbar\omega\Psi$, leaving $\psi(\mathbf{r})$ to satisfy the **time-independent Schrödinger equation** for a particle whose energy is $E = \hbar\omega$:

$$-\frac{\hbar^2}{2m}\nabla^2\psi(\mathbf{r}) + U(\mathbf{r})\psi(\mathbf{r}) = E\psi(\mathbf{r}) \tag{7.5}$$

The time-independent Schrödinger equation

Since our particle is free inside the box, we take the potential energy $U(\mathbf{r}) = 0$ for $0 < x,\, y,\, z < L$. In this case the spatial wavefunction also is separable; that is, solutions to Equation 7.5 with $U(\mathbf{r}) = 0$ can be found in product form:

$$\psi(\mathbf{r}) = \psi(x,\, y,\, z) = \psi_1(x)\psi_2(y)\psi_3(z) \tag{7.6}$$

Substituting Equation 7.6 into Equation 7.5 and dividing every term by the function $\psi(x,\, y,\, z)$ gives (for $U(\mathbf{r}) = 0$)

$$-\frac{\hbar^2}{2m\psi_1}\frac{d^2\psi_1}{dx^2} - \frac{\hbar^2}{2m\psi_2}\frac{d^2\psi_2}{dy^2} - \frac{\hbar^2}{2m\psi_3}\frac{d^2\psi_3}{dz^2} = E$$

In this form the independent variables are isolated: the first term on the left depends only on x, the second only on y, and the third only on z. To satisfy the equation *everywhere* inside the cube, each of these terms must reduce to a constant:

$$-\frac{\hbar^2}{2m\psi_1}\frac{d^2\psi_1}{dx^2} = E_1$$

$$-\frac{\hbar^2}{2m\psi_2}\frac{d^2\psi_2}{dy^2} = E_2 \tag{7.7}$$

$$-\frac{\hbar^2}{2m\psi_3}\frac{d^2\psi_3}{dz^2} = E_3$$

The stationary states for a particle confined to a cube are obtained from these three separate equations. The energies E_1, E_2, and E_3 are **separation constants** and represent the energy of motion along the three cartesian axes x, y, and z. Consistent with this identification, the Schrödinger equation requires that $E_1 + E_2 + E_3 = E$.

The first of Equations 7.7 is the same as that for the infinite square well in one dimension. Independent solutions to this equation are $\sin k_1 x$ and $\cos k_1 x$, where $k_1 = \sqrt{2mE_1/\hbar^2}$ is the wavenumber of oscillation. Only $\sin k_1 x$ satisfies the condition that the wavefunction must vanish at the wall $x = 0$, however. Requiring the wavefunction to vanish also at the opposite wall $x = L$ implies $k_1 L = n_1\pi$, where n_1 is any positive integer. In other words, we must be able to fit an integral number of half-wavelengths into our box along the direction marked by x. It follows that the magnitude of particle momentum along this direction must be one of the special values

$$|p_x| = \hbar k_1 = n_1 \frac{\pi \hbar}{L} \qquad n_1 = 1, 2, \ldots$$

Identical considerations applied to the remaining two equations show that the magnitudes of particle momentum in all three directions are similarly quantized:

Allowed values of momentum
components for a particle in
a box

$$|p_x| = \hbar k_1 = n_1 \frac{\pi \hbar}{L} \qquad n_1 = 1, 2, \ldots$$

$$|p_y| = \hbar k_2 = n_2 \frac{\pi \hbar}{L} \qquad n_2 = 1, 2, \ldots \tag{7.8}$$

$$|p_z| = \hbar k_3 = n_3 \frac{\pi \hbar}{L} \qquad n_3 = 1, 2, \ldots$$

Notice that $n_i = 0$ is *not* allowed, since that choice leads to ψ_i that is also zero and a wavefunction $\psi(\mathbf{r})$ that vanishes everywhere. Since the momenta are restricted this way, the particle energy (all kinetic) is limited to the following discrete values:

Discrete energies allowed for
a particle in a box

$$E = \frac{1}{2m}(|p_x|^2 + |p_y|^2 + |p_z|^2) = \frac{\pi^2 \hbar^2}{2mL^2}\{n_1^2 + n_2^2 + n_3^2\} \tag{7.9}$$

Thus, confining the particle to the cube serves to quantize its momentum and energy according to Equations 7.8 and 7.9. Note that *three quantum numbers are needed to specify the quantum condition, corresponding to the three independent degrees of freedom for a particle in space.* These quantum numbers specify the values taken by the sharp observables for this system.

Collecting the previous results, we see that the stationary states for this particle are

$$\Psi(x, y, z, t) = A \sin(k_1 x)\sin(k_2 y)\sin(k_3 z)\, e^{-i\omega t} \qquad \text{for } 0 < x, y, z < L \tag{7.10}$$
$$= 0 \qquad \text{otherwise}$$

The multiplier A is chosen to satisfy the normalization requirement. Example 7.1 shows that $A = (2/L)^{3/2}$ for the ground state, and this result continues to hold for the excited states as well.

EXAMPLE 7.1   Normalizing the Box Wavefunctions

Find the value of the multiplier A that normalizes the wavefunction of Equation 7.10 having the lowest energy.

Solution The state of lowest energy is described by $n_1 = n_2 = n_3 = 1$, or $k_1 = k_2 = k_3 = \pi/L$. Since Ψ is nonzero only for $0 < x, y, z < L$, the probability density integrated over the volume of this cube must be unity:

$$1 = \int_0^L dx \int_0^L dy \int_0^L dz\, |\Psi(x, y, z, t)|^2$$

$$= A^2 \left\{ \int_0^L \sin^2(\pi x/L)\, dx \right\}\left\{ \int_0^L \sin^2(\pi y/L)\, dy \right\}$$

$$\times \left\{ \int_0^L \sin^2(\pi z/L)\, dz \right\}$$

Using $2 \sin^2 \theta = 1 - \cos 2\theta$ gives

$$\int_0^L \sin^2(\pi x/L) \, dx = \frac{L}{2} - \frac{L}{4\pi} \sin(2\pi x/L) \, \Big|_0^L = \frac{L}{2}$$

The same result is obtained for the integrations over y and z. Thus, normalization requires

$$1 = A^2 \left(\frac{L}{2}\right)^3$$

or

$$A = \left(\frac{2}{L}\right)^{3/2}$$

Exercise 1   With what probability will the particle described by the wavefunction of Example 7.1 be found in the volume $0 < x, y, z < L/4$?

Answer   0.00075, or about 0.075%

Exercise 2   Modeling a defect trap in a crystal as a three-dimensional box with edge length 5.00 Å, find the values of momentum and energy for an electron bound to the defect site, assuming the electron is in the ground state.

Answer $|p_x| = |p_y| = |p_z| = 1.24 \text{ keV}/c$; $E = 4.51$ eV

The ground state, for which $n_1 = n_2 = n_3 = 1$, has energy

$$E_{1,1,1} = \frac{3\pi^2 \hbar^2}{2mL^2}$$

There are *three* first excited states, corresponding to the three different combinations of n_1, n_2, and n_3 whose squares sum to 6. That is, we obtain the *same* energy for the three combinations $n_1 = 2$, $n_2 = 1$, $n_3 = 1$, or $n_1 = 1$, $n_2 = 2$, $n_3 = 1$, or $n_1 = 1$, $n_2 = 1$, $n_3 = 2$. The first excited state has energy

$$E_{2,1,1} = E_{1,2,1} = E_{1,1,2} = \frac{6\pi^2 \hbar^2}{2mL^2}$$

Note that each of the first excited states is characterized by a different wavefunction: $\psi_{2,1,1}$ has wavelength L along the x axis and wavelength $2L$ along the y and z axes, but for $\psi_{1,2,1}$ and $\psi_{1,1,2}$ the shortest wavelength is along the y axis and the z axis, respectively.

Whenever different states have the same energy, this energy level is said to be **degenerate.** In the example just described, the first excited level is threefold (or triply) degenerate. This system has degenerate levels because of the high degree of symmetry associated with the cubic shape of the box. The degeneracy would be removed, or *lifted*, if the sides of the box were of unequal lengths (see Example 7.3). In fact, the extent of splitting of the levels increases with the degree of asymmetry.

Figure 7.3 is an energy level diagram showing the first five levels of a particle in a cubic box; Table 7.1 lists the quantum numbers and degeneracies of the various levels.

Computer-generated plots of the probability density $|\psi(x, y, z)|^2$ for the ground state and first excited states of a particle in a box are shown in Figure 7.4. Notice that the probabilities for the (degenerate) first excited states differ only in their orientation with respect to the coordinate axes, again a reflection of the cubic symmetry imposed by the box potential.

Figure 7.3   An energy level diagram for a particle confined to a cubic box. The ground state energy is $E_0 = 3\pi^2 \hbar^2/2mL^2$, and $n^2 = n_1^2 + n_2^2 + n_3^2$. Note that most of the levels are degenerate.

Table 7.1 Quantum Numbers and Degeneracies of the Energy Levels for a Particle Confined to a Cubic Box†

n_1	n_2	n_3	n^2	Degeneracy
1	1	1	3	None
1	1	2	6	
1	2	1	6	Threefold
2	1	1	6	
1	2	2	9	
2	1	2	9	Threefold
2	2	1	9	
1	1	3	11	
1	3	1	11	Threefold
3	1	1	11	
2	2	2	12	None

† Note: $n^2 = n_1{}^2 + n_2{}^2 + n_3{}^2$

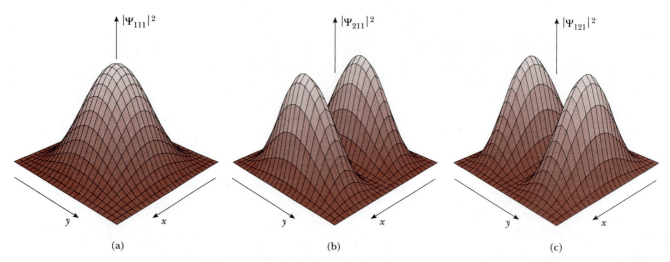

Figure 7.4 (Unnormalized) probability density for a particle in a box: (a) ground state $|\Psi_{111}|^2$; (b) and (c) first excited states, $|\Psi_{211}|^2$ and $|\Psi_{121}|^2$. Plots are for $|\Psi|^2$ in the plane $z = \frac{1}{2}L$. In this plane, $|\Psi_{112}|^2$ (not shown) is indistinguishable from $|\Psi_{111}|^2$.

EXAMPLE 7.2 The Second Excited State

Determine the wavefunctions and energy for the second excited level of a particle in a cubic box of edge L. What is the degeneracy of this level?

Solution The second excited level corresponds to the three combinations of quantum numbers $n_1 = 2$, $n_2 = 2$, $n_3 = 1$ or $n_1 = 2$, $n_2 = 1$, $n_3 = 2$ or $n_1 = 1$, $n_2 = 2$, $n_3 = 2$. The corresponding wavefunctions inside the box are

$$\Psi_{2,2,1} = A \sin\left(\frac{2\pi x}{L}\right) \sin\left(\frac{2\pi y}{L}\right) \sin\left(\frac{\pi z}{L}\right) e^{-iE_{2,2,1}t/\hbar}$$

$$\Psi_{2,1,2} = A \sin\left(\frac{2\pi x}{L}\right) \sin\left(\frac{\pi y}{L}\right) \sin\left(\frac{2\pi z}{L}\right) e^{-iE_{2,1,2}t/\hbar}$$

$$\Psi_{1,2,2} = A \sin\left(\frac{\pi x}{L}\right) \sin\left(\frac{2\pi y}{L}\right) \sin\left(\frac{2\pi z}{L}\right) e^{-iE_{1,2,2}t/\hbar}$$

The level is threefold degenerate, since each of these wave-functions has the same energy,

$$E_{2,2,1} = E_{2,1,2} = E_{1,2,2} = \frac{9\pi^2\hbar^2}{2mL^2}$$

EXAMPLE 7.3   Quantization in a Rectangular Box

Obtain a formula for the allowed energies of a particle confined to a rectangular box with edge lengths L_1, L_2, and L_3. What is the degeneracy of the first excited state?

Solution   For a box having edge length L_1 in the x direction, ψ will be zero at the walls if L_1 is an integral number of half-wavelengths. Thus, the magnitude of particle momentum in this direction is quantized as

$$|p_x| = \hbar k_1 = n_1 \frac{\pi\hbar}{L_1} \qquad n_1 = 1, 2, \ldots$$

Likewise, for the other two directions, we have

$$|p_y| = \hbar k_2 = n_2 \frac{\pi\hbar}{L_2} \qquad n_2 = 1, 2, \ldots$$

$$|p_z| = \hbar k_3 = n_3 \frac{\pi\hbar}{L_3} \qquad n_3 = 1, 2, \ldots$$

The allowed energies are

$$E = (|p_x|^2 + |p_y|^2 + |p_z|^2)/2m$$

$$= \frac{\pi^2\hbar^2}{2m} \left\{ \left(\frac{n_1}{L_1}\right)^2 + \left(\frac{n_2}{L_2}\right)^2 + \left(\frac{n_3}{L_3}\right)^2 \right\}$$

The lowest energy occurs again for $n_1 = n_2 = n_3 = 1$. Increasing one of the integers by 1 gives the next lowest, or first excited level. If L_1 is the largest dimension, then $n_1 = 2$, $n_2 = 1$, $n_3 = 1$ produces the smallest energy increment, and describes the first excited state. Further, so long as both L_2 and L_3 are not equal to L_1, the first excited level is nondegenerate, that is, there is no other state with this energy. If L_2 or L_3 equals L_1, the level is doubly degenerate; if all three are equal, the level will be triply degenerate. Thus, the higher the symmetry, the more degeneracy we find.

7.2   CENTRAL FORCES AND ANGULAR MOMENTUM

The formulation of quantum mechanics in cartesian coordinates is the natural way to generalize from one to higher dimensions, but often it is not the best suited to a given application. For instance, an atomic electron is attracted to the nucleus of the atom by the Coulomb force between opposite charges. This is an example of a *central force*, that is, one directed toward a fixed *point*. The nucleus is the center of force, and the coordinates of choice here are spherical coordinates r, θ, ϕ centered on the nucleus (Fig. 7.5).

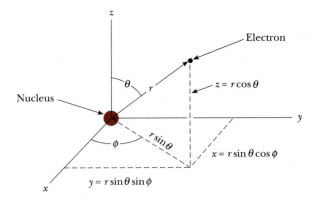

Figure 7.5   The central force on an atomic electron is one directed toward a fixed point, the nucleus. The coordinates of choice here are the spherical coordinates r, θ, ϕ centered on the nucleus.

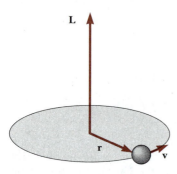

Figure 7.6 The angular momentum **L** of an orbiting particle is perpendicular to the plane of the orbit. If the direction of **L** were known precisely, both the coordinate and momentum in the direction perpendicular to the orbit would be known, in violation of the uncertainty principle.

If the central force is conservative, the particle energy (kinetic plus potential) stays constant and E becomes a candidate for quantization. In that case, the quantum states are stationary waves $\psi(\mathbf{r})e^{-i\omega t}$, with $\omega = E/\hbar$, where E is the sharp value of particle energy.

But for central forces, *angular momentum* **L** about the force center also is constant, and we should expect that wavefunctions ψ can be found for which all three angular momentum components take sharp values. This imposes severe constraints on the form of the wavefunction. In fact, these constraints are so severe that it is impossible to find a wavefunction satisfying all of them at once; that is, *not all components of angular momentum can be known simultaneously!*

The dilemma is reminiscent of our inability to specify simultaneously the position and momentum of a particle. Indeed, if the direction of **L** were known precisely, the particle would be confined to the orbital plane (the plane perpendicular to the vector **L**), and its coordinate in the direction normal to this plane would be accurately known (and unchanging) (Fig. 7.6). In that case, however, the particle could have no momentum out of the orbital plane, so that its linear momentum perpendicular to this plane also would be known (to be zero), in violation of the uncertainty principle. The argument just given may be refined to establish an **uncertainty principle for angular momentum:** *it is impossible to specify simultaneously any two components of angular momentum.* Alternatively, if one component of **L** is sharp, then the remaining two must be "fuzzy." [2]

Uncertainty principle for angular momentum

Along with E and one component of **L**, then, what else might be quantized, or sharp, for central forces? The answer is contained in the following observation: with only one component of **L** sharp, there is no redundancy in having the magnitude |**L**| sharp also. In this way, E, |**L**|, and one component of **L**, say L_z, become the sharp observables subject to quantization in the central force problem.

Wavefunctions for which |**L**| and L_z are both sharp are the *spherical harmonics*. These functions follow from separating the variables in Schrödinger's equation for a central force. We take the time-independent state wavefunction in spherical coordinates r, θ, ϕ to be

Separation of variables for the stationary state wavefunction

$$\psi(\mathbf{r}) = \psi(r, \theta, \phi) = R(r)\,\Theta(\theta)\,\Phi(\phi) \tag{7.11}$$

and write Schrödinger's time-independent equation (Equation 7.5) in these coordinates using the spherical coordinate form for the Laplacian:[3]

$$\nabla^2 = \frac{\partial^2}{\partial r^2} + \left(\frac{2}{r}\right)\frac{\partial}{\partial r} + \frac{1}{r^2}\left[\frac{\partial^2}{\partial \theta^2} + \cot\theta\,\frac{\partial}{\partial \theta} + \csc^2\theta\,\frac{\partial^2}{\partial \phi^2}\right]$$

---

[2] Angular momentum is a notable exception to the argument that constants of the classical motion correspond to sharp observables in quantum mechanics. The failure is rooted in another maxim—the uncertainty principle—which takes precedence. The only instance in which two or more angular momentum components may be known exactly is the trivial case $L_x = L_y = L_z = 0$.

[3] The Laplacian in spherical coordinates is given in any more advanced scientific text or may be found from the cartesian form by following the arguments of Section 7.4.

Separating variables here is complicated because of the way the spherical coordinates are mixed in the Laplacian. To disentangle them, notice that the operator in square brackets [. . .] will involve only θ if $\partial^2/\partial\phi^2$ applied to $\psi(\mathbf{r})$ returns a multiple of $\psi(\mathbf{r})$. This can be achieved by requiring

$$\frac{d^2\Phi}{d\phi^2} = -m_\ell^2\,\Phi(\phi) \qquad (7.12)$$

where m_ℓ is one of two *separation constants*, and is termed the **magnetic quantum number.**[4] The function $\Phi(\phi) = \exp(im_\ell\phi)$ is a solution to Equation 7.12, and is periodic with period 2π if m_ℓ is restricted to integer values. Periodicity is necessary here because all physical properties that derive from the wavefunction should be unaffected by the replacement $\phi \to \phi + 2\pi$, both of which describe the same point in space (see Fig. 7.5).

With the help of Equation 7.12, we can now write

$$\left[\frac{\partial^2}{\partial\theta^2} + \cot\theta\,\frac{\partial}{\partial\theta} + \csc^2\theta\,\frac{\partial^2}{\partial\phi^2}\right]\psi(\mathbf{r}) = \left[\frac{\partial^2}{\partial\theta^2} + \cot\theta\,\frac{\partial}{\partial\theta} - m_\ell^2\,\csc^2\theta\right]\psi(\mathbf{r})$$

The separation of variables is completed by requiring the right-hand side of this equation to return yet another multiple of $\psi(\mathbf{r})$ and is accomplished by demanding

$$\frac{d^2\Theta}{d\theta^2} + \cot\theta\,\frac{d\Theta}{d\theta} - m_\ell^2\,\csc^2\theta\,\Theta(\theta) = -\ell(\ell+1)\Theta(\theta) \qquad (7.13)$$

Equation 7.13 introduces a second separation constant, $\ell(\ell+1)$, involving the **orbital quantum number** ℓ. For $\psi(\mathbf{r})$ to be an acceptable wavefunction, $\Theta(\theta)$ must be bounded and single-valued, conditions which are met only if ℓ is a nonnegative integer, and then only if m_ℓ is limited to absolute values not larger than ℓ. The resulting solutions $\Theta(\theta)$ are polynomials in $\cos\theta$ known as *associated Legendre polynomials*. A few of these polynomials are listed in Table 7.2 for easy reference. The products $\Theta(\theta)\Phi(\phi)$ specify the full angular dependence of the central force wavefunction and are known as *spherical harmonics*, denoted by $Y_\ell^{m_\ell}(\theta, \phi)$. Some spherical harmonics are given in Table 7.3. The constant prefactors are chosen to normalize these functions.[5]

In keeping with our earlier remarks, the separation constants $\ell(\ell+1)$ and m_ℓ should relate to the sharp observables $|\mathbf{L}|$, L_z, and E for central forces. The connection is established by the more detailed arguments of Section 7.4, with the result

$$|\mathbf{L}| = \sqrt{\ell(\ell+1)}\,\hbar \qquad \ell = 0, 1, 2, \ldots \qquad (7.14)$$

$$L_z = m_\ell\hbar \qquad m_\ell = 0, \pm 1, \pm 2, \ldots, \pm\ell$$

Table 7.2 Some Associated Legendre Polynomials $P_\ell{}^{m_\ell}(\cos\theta)$

$P_0{}^0 = 1$

$P_1{}^0 = 2\cos\theta$

$P_1{}^1 = \sin\theta$

$P_2{}^0 = 4(3\cos^2\theta - 1)$

$P_2{}^1 = 4\sin\theta\cos\theta$

$P_2{}^2 = \sin^2\theta$

$P_3{}^0 = 24(5\cos^3\theta - 3\cos\theta)$

$P_3{}^1 = 6\sin\theta(5\cos^2\theta - 1)$

$P_3{}^2 = 6\sin^2\theta\cos\theta$

$P_3{}^3 = \sin^3\theta$

Angular momentum and its z-component are quantized

---

[4] This seemingly peculiar way of writing the constant multiplier is based on the physical significance of the function $\Phi(\phi)$ and is discussed at length in (optional) Section 7.4. The student is referred there for a concise treatment of the quantum central force problem based on the operator methods of Section 5.8.

[5] The normalization is such that the integral of $|Y_\ell^{m_\ell}|^2$ over the surface of a sphere with unit radius is 1.

Table 7.3   The Spherical Harmonics
$$Y_\ell{}^{m_\ell}(\theta,\ \phi)$$

$$Y_0{}^0 = \frac{1}{2\sqrt{\pi}}$$

$$Y_1{}^0 = \frac{1}{2}\sqrt{\frac{3}{\pi}}\cdot\cos\theta$$

$$Y_1{}^{\pm 1} = \mp\frac{1}{2}\sqrt{\frac{3}{2\pi}}\cdot\sin\theta\cdot e^{\pm i\phi}$$

$$Y_2{}^0 = \frac{1}{4}\sqrt{\frac{5}{\pi}}\cdot(3\cos^2\theta - 1)$$

$$Y_2{}^{\pm 1} = \mp\frac{1}{2}\sqrt{\frac{15}{2\pi}}\cdot\sin\theta\cdot\cos\theta\cdot e^{\pm i\phi}$$

$$Y_2{}^{\pm 2} = \frac{1}{4}\sqrt{\frac{15}{2\pi}}\cdot\sin^2\theta\cdot e^{\pm 2i\phi}$$

$$Y_3{}^0 = \frac{1}{4}\sqrt{\frac{7}{\pi}}\cdot(5\cos^3\theta - 3\cos\theta)$$

$$Y_3{}^{\pm 1} = \mp\frac{1}{8}\sqrt{\frac{21}{\pi}}\cdot\sin\theta\cdot(5\cos^2\theta - 1)\cdot e^{\pm i\phi}$$

$$Y_3{}^{\pm 2} = \frac{1}{4}\sqrt{\frac{105}{2\pi}}\cdot\sin^2\theta\cdot\cos\theta\cdot e^{\pm 2i\phi}$$

$$Y_3{}^{\pm 3} = \mp\frac{1}{8}\sqrt{\frac{35}{\pi}}\cdot\sin^3\theta\cdot e^{\pm 3i\phi}$$

We see the limitation on the magnetic quantum number m_ℓ to values between $-\ell$ and $+\ell$, obtained on purely mathematical grounds from separating variables, has an obvious physical interpretation: the z-component of angular momentum, L_z, must never exceed the magnitude of the vector, $|\mathbf{L}|$! Notice, too, that ℓ and m_ℓ are quantum numbers for angular momentum only; their connection with particle energy E must depend on the potential energy function $U(r)$, and is prescribed along with the *radial wavefunction* $R(r)$ in the final stage of the separation procedure.

To obtain $R(r)$, and the complete wavefunction $\psi(\mathbf{r}) = R(r)Y_\ell{}^{m_\ell}(\theta,\ \phi)$, we must combine the results of Equations 7.11–13 to get

$$\nabla^2[R(r)Y_\ell{}^{m_\ell}(\theta,\ \phi)] = \left[\frac{\partial^2}{\partial r^2} + \left(\frac{2}{r}\right)\frac{\partial}{\partial r} - \frac{\ell(\ell+1)}{r^2}\right]R(r)Y_\ell{}^{m_\ell}(\theta,\ \phi) \quad (7.15)$$

Using this result in the Schrödinger equation (Equation 7.5), and noting that $U(\mathbf{r})$ becomes simply $U(r)$ for a central force seated at the origin, permits cancellation of the spherical harmonic in every term, leaving

Radial wave equation
$$-\frac{\hbar^2}{2m}\left[\frac{d^2R}{dr^2} + \left(\frac{2}{r}\right)\frac{dR}{dr}\right] + \frac{\ell(\ell+1)\hbar^2}{2mr^2}R(r) + U(r)R(r) = ER(r) \quad (7.16)$$

This is the **radial wave equation;** it determines the radial part of the wavefunction ψ and the allowed particle energies E. The reduction from Schrödinger's equation to the radial wave equation represents enormous progress and is valid for any central force. Still, the task of solving this equation for a specified potential $U(r)$ is a difficult one, often requiring methods of considerable sophistication.

EXAMPLE 7.4   Orbital Quantum Number for a Stone

A stone with mass 1.00 kg is whirled in a horizontal circle of radius 1.00 m with a period of revolution equal to 1.00 s. What value of orbital quantum number ℓ describes this motion?

Solution   The speed of the stone in its orbit is

$$v = \frac{2\pi R}{T} = \frac{2\pi(1.00 \text{ m})}{1.00 \text{ s}} = 6.28 \text{ m/s}$$

The corresponding angular momentum has magnitude

$$|\mathbf{L}| = mvR = (1.00 \text{ kg})(6.28 \text{ m/s})(1.00 \text{ m})$$
$$= 6.28 \text{ kg} \cdot \text{m}^2/\text{s}$$

But angular momentum is quantized as $\sqrt{\ell(\ell+1)}\hbar$, which is approximately $\ell\hbar$ when ℓ is large. Then

$$\ell \approx \frac{|\mathbf{L}|}{\hbar} = \frac{6.28 \text{ kg} \cdot \text{m}^2/\text{s}}{1.055 \times 10^{-34} \text{ kg} \cdot \text{m}^2/\text{s}} = 5.96 \times 10^{34}$$

Again, we see that macroscopic objects are described by enormous quantum numbers, so that quantization is not evident on this scale.

EXAMPLE 7.5   The Bohr Atom Revisited

Discuss angular momentum quantization in the Bohr model. What orbital quantum number describes the electron in the first Bohr orbit of hydrogen?

Solution   Angular momentum in the Bohr atom is quantized according to the Bohr postulate

$$|\mathbf{L}| = mvr = n\hbar$$

with $n = 1$ for the first Bohr orbit. Comparing this with the quantum mechanical result, Equation 7.14, we see that the two rules are incompatible! The magnitude $|\mathbf{L}|$ can never be an integral multiple of $\hbar$—the smallest nonzero value consistent with Equation 7.14 is $|\mathbf{L}| = \sqrt{2}\hbar$ for $\ell = 1$.

In fairness to Bohr, we should point out that the Bohr model makes no distinction between the quantization of $\mathbf{L}$ and quantization of its components along the coordinate axes. From the classical viewpoint, the coordinate system can always be oriented to align one of the axes, say the z axis, along the direction of $\mathbf{L}$. In that case $\mathbf{L}$ may be identified with $|L_z|$. The Bohr postulate in this form agrees with the quantization of L_z in Equation 7.14, and indicates $\ell = 1$ for the first Bohr orbit! This conflicting result derives from a false assertion, namely, that we may orient a coordinate axis along the direction of $\mathbf{L}$. The freedom to do so must be abandoned if the quantization rules of Equation 7.14 are correct! This stunning conclusion is one of the great mysteries of quantum physics, and is implicit in the notion of *space quantization.*

7.3   SPACE QUANTIZATION

For wavefunctions satisfying Equations 7.12 and 7.13, the orbital angular momentum magnitude $|\mathbf{L}|$, and L_z, the projection of $\mathbf{L}$ along the z axis, are both sharp and quantized according to the restrictions imposed by the orbital and magnetic quantum numbers, respectively. Together, ℓ and m_ℓ specify the orientation of the angular momentum vector $\mathbf{L}$. The fact that the direction of $\mathbf{L}$ is quantized with respect to an arbitrary axis (the z axis) is referred to as *space quantization.*

Let us look at the possible orientations of $\mathbf{L}$ for a given value of orbital quantum number ℓ. Recall that m_ℓ can have values ranging from $-\ell$ to $+\ell$. If $\ell = 0$, then $m_\ell = 0$ and $L_z = 0$. In fact, for this case $|\mathbf{L}| = 0$, so that all com-

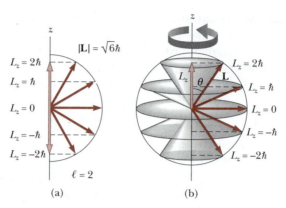

Figure 7.7   (a) The allowed projections of the orbital angular momentum for the case $\ell = 2$. (b) From a three-dimensional perspective, the orbital angular momentum vector **L** lies on the surface of a cone. The fuzzy character of L_x and L_y is depicted by allowing **L** to precess about the z axis, so that L_x and L_y change continually, but L_z maintains the fixed value $m_\ell \hbar$.

ponents of **L** are zero. If $\ell = 1$, then the possible values for m_ℓ are -1, 0, and $+1$, so that L_z may be $-\hbar$, 0, or $+\hbar$. If $\ell = 2$, m_ℓ can be -2, -1, 0, $+1$, or $+2$, corresponding to L_z values of $-2\hbar$, $-\hbar$, 0, $+\hbar$, or $+2\hbar$, and so on. A classical visualization of the algebra describing space quantization for the case $\ell = 2$ is shown in Figure 7.7a. Note that **L** can never be aligned with the z axis, since L_z must be smaller than the total angular momentum L. From a three-dimensional perspective, **L** must lie on the surface of a cone that makes an angle θ with the z axis, as shown in Figure 7.7b. From the figure, we see that θ also is quantized and that its values are specified by the relation

The orientations of L are restricted (quantized)

$$\cos \theta = \frac{L_z}{|\mathbf{L}|} = \frac{m_\ell}{\sqrt{\ell(\ell + 1)}} \tag{7.17}$$

Classically θ can take any value; that is, the angular momentum vector **L** can point in any direction whatsoever. According to quantum mechanics, the possible orientations for **L** are those consistent with Equation 7.17. Furthermore, these special directions have nothing to do with the forces acting on the particle, provided only that these forces are central. Thus, *the rule of Equation 7.17 does not originate with the law of force but derives from the structure of space itself*, hence the name space quantization.

Figure 7.7 is misleading in showing **L** with a specific direction, for which all three components L_x, L_y, and L_z are known exactly. As we have mentioned, there is no quantum state for which this condition is true. If L_z is sharp, then L_x and L_y must be fuzzy. Accordingly, it is more proper to envision the vector **L** of Figure 7.7b as precessing around the z axis so as to trace out a cone in space. This allows the components L_x and L_y to change continually, while L_z maintains the fixed value $m_\ell \hbar$.[6]

---

[6] This precession of the classical vector to portray the inherent fuzziness in L_x and L_y is meant to be suggestive only. In effect, we have identified the quantum averages $\langle L_x \rangle$ and $\langle L_y \rangle$ with their averages over *time* in a classical picture, but the two kinds of averaging in fact are quite distinct.

**EXAMPLE 7.6   Space Quantization
for an Atomic Electron**

Consider an atomic electron in the $\ell = 3$ state. Calculate the magnitude $|\mathbf{L}|$ of the total angular momentum and the allowed values of L_z and θ.

Solution   With $\ell = 3$, Equation 7.14 gives

$$|\mathbf{L}| = \sqrt{3(3 + 1)}\hbar = 2\sqrt{3}\,\hbar$$

The allowed values of L_z are $m_\ell \hbar$, with $m_\ell = 0, \pm 1, \pm 2,$ and ± 3. This gives

$$L_z = -3\hbar, -2\hbar, -\hbar, 0, \hbar, 2\hbar, 3\hbar$$

Finally, we obtain the allowed values of θ from

$$\cos \theta = \frac{L_z}{|\mathbf{L}|} = \frac{m_\ell}{2\sqrt{3}}$$

Substituting the values for m_ℓ gives

$$\cos \theta = \pm 0.866, \quad \pm 0.577, \quad \pm 0.289, \quad \text{and} \quad 0$$

or

$$\theta = \pm 30°, \quad \pm 54.8°, \quad \pm 73.2°, \quad \text{and} \quad 90°$$

Exercise 3   Compare the minimum angles between $\mathbf{L}$ and the z axis for the electron of Example 7.6 and for the 1.00-kg stone of Example 7.4.

Answer   30° for the electron vs. 2.3×10^{-16} degrees for the stone.

7.4 QUANTIZATION OF ANGULAR MOMENTUM AND ENERGY

O P T I O N A L

We saw in Section 7.2 that angular momentum plays an essential role in the quantization of systems with spherical symmetry. Here we develop further the properties of angular momentum from the viewpoint of quantum mechanics and show in more detail how angular momentum considerations facilitate the solution to Schrödinger's equation for central forces.

Our treatment is based on the operator methods introduced in Section 5.8, which the reader should review at this time. In particular, the eigenvalue condition

$$[Q]\Psi = q\Psi \tag{7.18}$$

used there as a *test* for sharp observables becomes here a tool for discovering the form of the *unknown* wavefunction Ψ. (Recall that $[Q]$ denotes the operator for some observable Q, and q is the sharp value of that observable when measured for a system described by the wavefunction Ψ.) We look on Equation 7.18 now as a *constraint* imposed on the wavefunction Ψ that guarantees that the observable Q will take the sharp value q in that state. The more sharp observables we can identify for some system, the more we can learn in advance about the wavefunction describing that system. *But with few exceptions, the sharp observables are just those that are constants of the classical motion.* (See footnote 1.)

Consider the particle energy. If total energy is constant, we should be able to find wavefunctions Ψ for which E is a sharp observable. Otherwise, repeated measurements made on identical systems would reveal a statistical distribution of values for the particle energy, inconsistent with the idea of a quantity not changing over time. Thus, energy conservation suggests $\Delta E = 0$ which, in turn, requires Ψ to be an eigenfunction of the energy operator

$$[E]\Psi = E\Psi$$

Since $[E] = i\hbar \partial/\partial t$, this eigenvalue condition is met by the stationary waves $\psi(\mathbf{r})\,e^{-i\omega t}$, with $E = \hbar\omega$ the sharp value of particle energy.

The argument for energy applies equally well to other constants of the classical motion.[7] If the *only* forces are central, *angular momentum* about the force center is a constant of the motion. This is a vector quantity $\mathbf{L} = \mathbf{r} \times \mathbf{p}$, whose rectangular components are expressed in terms of position and momentum components as

$$L_z = (\mathbf{r} \times \mathbf{p})_z = xp_y - yp_x$$

and so on. In the same way, the *operators* for angular momentum are found from the coordinate and momentum *operators*. From Chapter 5, the operator for x is just the coordinate itself, and the operator for momentum in this direction is $(\hbar/i)\partial/\partial x$. Similar relations should apply to the directions labeled y and z,[8] so that the operator for L_z becomes

$$[L_z] = [x][p_y] - [y][p_x] = \frac{\hbar}{i} \left(x \frac{\partial}{\partial y} - y \frac{\partial}{\partial x} \right) \tag{7.19}$$

Angular momentum finds its natural expression in the spherical coordinates of Figure 7.5. A little geometry applied to Figure 7.5 shows that the spherical-to-cartesian coordinate transformation equations are

$$z = r \cos \theta$$
$$x = r \sin \theta \cos \phi \tag{7.20}$$
$$y = r \sin \theta \sin \phi$$

The inverse transformations are

$$r = \{x^2 + y^2 + z^2\}^{1/2}$$
$$\cos \theta = \frac{z}{r} = z\{x^2 + y^2 + z^2\}^{-1/2} \tag{7.21}$$
$$\tan \phi = \frac{y}{x}$$

The procedure for transcribing operators such as $[L_z]$ from cartesian to spherical form is straightforward, but tedious. For instance, an application of the chain rule gives

$$\frac{\partial f}{\partial z} = \frac{\partial r}{\partial z} \frac{\partial f}{\partial r} + \frac{\partial \theta}{\partial z} \frac{\partial f}{\partial \theta} + \frac{\partial \phi}{\partial z} \frac{\partial f}{\partial \phi}$$

for any function f. On the left we think of f expressed as a function of x, y, z, but on the right the same function is expressed in terms of r, θ, ϕ. The partial derivatives $\partial r/\partial z$, and so on, are to be taken with the aid of the transformation equations, holding x and y fixed. The simplest of these is $\partial \phi/\partial z$. From the inverse transformations in Equations 7.21, we see that ϕ is independent of z, so that $\partial \phi/\partial z = 0$. From the same equations we calculate

---

[7] Exceptions to this rule do exist. For instance, an atomic electron cannot have all three angular momentum components sharp at once, even though in this case all components are constant classically. For such *incompatible* observables, quantum mechanics adopts the broader interpretation of a conserved quantity as one whose *average value does not change over time*, no matter what may be the initial state of the system. With this definition, all components of angular momentum for an atomic electron remain constant, but no more than one can be sharp in a given state.

[8] For the momentum operator along each axis we use the one-dimensional form from Chapter 5, $([p_x] = (\hbar/i)\partial/\partial x$, etc.), consistent with our belief that the cartesian axes identify independent but logically equivalent directions in space.

$$\frac{\partial r}{\partial z} = \frac{1}{2} \{x^2 + y^2 + z^2\}^{-1/2}(2z) = \frac{z}{r} = \cos \theta$$

Now, $\partial \theta / \partial z$ may be obtained from the second of Equations 7.21 by implicit differentiation:

$$-\sin \theta \frac{\partial \theta}{\partial z} = \frac{\partial \{\cos \theta\}}{\partial z} = z(-\tfrac{1}{2})\{x^2 + y^2 + z^2\}^{-3/2}(2z) + \{x^2 + y^2 + z^2\}^{-1/2}$$

Converting the right-hand side back to spherical coordinates gives

$$-\sin \theta \frac{\partial \theta}{\partial z} = -\frac{(r\cos \theta)^2}{r^3} + \frac{1}{r} = \frac{1 - \cos^2 \theta}{r} = \frac{\sin^2 \theta}{r}$$

or $\partial \theta / \partial z = -\sin \theta / r$. Collecting the previous results, we have

$$\frac{\partial}{\partial z} = \frac{\partial r}{\partial z}\frac{\partial}{\partial r} + \frac{\partial \theta}{\partial z}\frac{\partial}{\partial \theta} = \cos \theta \frac{\partial}{\partial r} - \frac{\sin \theta}{r}\frac{\partial}{\partial \theta}$$

In like fashion we can obtain spherical representations for $\partial / \partial x$, $\partial / \partial y$, and ultimately for the angular momentum operators themselves:

$$[L_x] = i\hbar \left\{ \sin \phi \frac{\partial}{\partial \theta} + \cot \theta \cos \phi \frac{\partial}{\partial \phi} \right\}$$

$$[L_y] = -i\hbar \left\{ \cos \phi \frac{\partial}{\partial \theta} - \cot \theta \sin \phi \frac{\partial}{\partial \phi} \right\} \tag{7.22}$$

$$[L_z] = -i\hbar \frac{\partial}{\partial \phi}$$

Notice that the operators for angular momentum in spherical form make no reference to the radial coordinate r—our reward for selecting the right system of coordinates! Now, by insisting that angular momentum be sharp in the quantum state Ψ, and using the differential forms (Equations 7.22) for the angular momentum operators, we can discover how the wavefunction depends on the spherical angles θ and ϕ, *without knowing the details of the central force.*

L_z Is Sharp: The Magnetic Quantum Number

For L_z to be sharp, our wavefunction $\Psi = \psi(r)e^{-i\omega t}$ must be an eigenfunction of $[L_z]$, or

$$-i\hbar \frac{\partial \psi}{\partial \phi} = L_z \psi \tag{7.23}$$

with L_z the eigenvalue (a number). Equation 7.23 prescribes the functional dependence of ψ on the azimuthal angle ϕ to be

$$\psi(\mathbf{r}) = Ce^{i\,L_z\phi/\hbar} \tag{7.24}$$

with C still any function of r and θ. The values taken by L_z must be restricted, however. Since increasing ϕ by 2π radians returns us to the same place in space (see Fig. 7.5), the wavefunction also must return to its initial value, that is, the solutions represented by Equation 7.24 must be periodic in ϕ with period 2π. This will be true if $L_z/\hbar$ is any integer, say m_ℓ, or $L_z = m_\ell \hbar$. The magnetic quantum number m_ℓ indicates the (sharp) value for the z-component of angular momentum in the state described by ψ.

Since all components of **L** are constant for central forces, we should continue by requiring the wavefunction of Equation 7.24 also to be an eigenfunction of the operator for a second angular momentum component, say $[L_x]$, so that L_z and L_x might both be sharp in the state ψ. For ψ to be an eigenfunction of $[L_x]$ with eigenvalue L_x requires

$$i\hbar \left\{ \sin \phi \frac{\partial \psi}{\partial \theta} + \cot \theta \cos \phi \frac{\partial \psi}{\partial \phi} \right\} = L_x \psi$$

This relation must hold for all values of θ and ϕ. In particular, at $\phi = 0$ the requirement is

$$i\hbar \cot \theta \frac{\partial \psi}{\partial \phi} = L_x \psi$$

Since ψ varies exponentially with ϕ according to Equation 7.24, the indicated derivative may be taken and evaluated at $\phi = 0$ to get

$$-L_z \cot \theta \, \psi = L_x \psi$$

which can be satisfied for *all* θ only if $L_x = L_z = 0$, or if ψ vanishes identically. A similar difficulty arises if we attempt to make L_y sharp together with L_z. Thus, *unless the angular momentum is exactly zero (all components), there is no wavefunction for which two or more components are simultaneously sharp!*

|L| Is Sharp: The Orbital Quantum Number

To make further progress, we must look to other constants of the classical motion. In addition to each component of angular momentum, the magnitude |**L**| of the vector also is constant and becomes a candidate for an observable that can be made sharp together with L_z. Consider simply the squared-magnitude $L^2 = \mathbf{L} \cdot \mathbf{L}$. The operator for L^2 is

$$[L]^2 = [L_x]^2 + [L_y]^2 + [L_z]^2$$

Using Equations 7.22, we find the spherical coordinate form for $[L^2]$:

$$[L^2] = -\hbar^2 \left\{ \frac{\partial^2}{\partial \theta^2} + \cot \theta \frac{\partial}{\partial \theta} + \csc^2 \theta \frac{\partial^2}{\partial \phi^2} \right\} \tag{7.25}$$

For L^2 to be sharp requires $[L^2]\psi = |\mathbf{L}|^2 \psi$ or

$$-\hbar^2 \left\{ \frac{\partial^2 \psi}{\partial \theta^2} + \cot \theta \frac{\partial \psi}{\partial \theta} + \csc^2 \theta \frac{\partial^2 \psi}{\partial \phi^2} \right\} = |\mathbf{L}|^2 \psi$$

But from Equation 7.23, we see that differentiating ψ with respect to ϕ is equivalent to multiplication by $iL_z/\hbar = im_\ell$, so that the last equation reduces to

$$-\hbar^2 \left\{ \frac{\partial^2 \psi}{\partial \theta^2} + \cot \theta \frac{\partial \psi}{\partial \theta} - m_\ell^2 \csc^2 \theta \, \psi \right\} = |\mathbf{L}|^2 \psi \tag{7.26}$$

For wavefunctions satisfying Equation 7.26, the magnitude of angular momentum will be sharp at the value |**L**|. The equation prescribes the polar angle, or θ, dependence of ψ. The solutions are not elementary functions but can be investigated with the help of more advanced techniques. The results of such an investigation are reported in Section 7.2 and repeated here for completeness: physically acceptable solutions to Equation 7.26 can be found provided $|\mathbf{L}| = \sqrt{\ell(\ell+1)}\,\hbar$, where ℓ, the orbital quantum number, must be a non-negative integer. Furthermore, the magnetic quantum number, also appearing in Equation 7.26, must be limited as $|m_\ell| \leq \ell$. With these restrictions, the solutions to Equation 7.26 are the associated Legendre polynomials in $\cos\theta$, denoted $P_\ell^{m_\ell}(\cos\theta)$. Several of these polynomials are listed in Table 7.2 for easy reference; you may verify by direct substitution that they satisfy Equation 7.26 for the appropriate values of ℓ and m_ℓ.

The associated Legendre polynomials may be multiplied by $\exp(im_\ell\phi)$ and still satisfy the orbital equation Equation 7.26. Indeed, the products $P_\ell^{m_\ell}(\cos\theta)\exp(im_\ell\phi)$ satisfy both Equations 7.23 and 7.26; thus, they represent waves for which $|\mathbf{L}|$ and L_z are simultaneously sharp. Except for a multiplicative constant, these are just the spherical harmonics $Y_\ell^{m_\ell}(\theta, \phi)$ introduced in Section 7.2.

E Is Sharp: The Radial Wave Equation

For energy to be constant, E should be a sharp observable. The stationary state form $\Psi(\mathbf{r}, t) = \psi(\mathbf{r})e^{-i\omega t}$ results from imposing the eigenvalue condition for the energy operator $[E]$, but what of the other energy operator $[H] = [K] + [U]$ (the hamiltonian)? In fact, requiring $[H]\Psi = E\Psi$ is equivalent to writing the time-independent Schrödinger equation for $\psi(\mathbf{r})$

$$-\frac{\hbar^2}{2m}\nabla^2\psi + U(\mathbf{r})\psi = E\psi \tag{7.27}$$

since $[U] = U(\mathbf{r})$ and the kinetic energy operator $[K]$ in three dimensions is none other than

$$[K] = \frac{[p_x]^2 + [p_y]^2 + [p_z]^2}{2m} = \frac{(\hbar/i)^2\{(\partial/\partial x)^2 + (\partial/\partial y)^2 + (\partial/\partial z)^2\}}{2m} = -\frac{\hbar^2}{2m}\nabla^2$$

The spherical form of the Laplacian given in Section 7.2 may be used to write $[K]$ in spherical coordinates:

$$[K] = -\frac{\hbar^2}{2m}\left\{\frac{\partial^2}{\partial r^2} + \frac{2}{r}\frac{\partial}{\partial r} + \frac{1}{r^2}\left[\frac{\partial^2}{\partial\theta^2} + \cot\theta\frac{\partial}{\partial\theta} + \csc^2\theta\frac{\partial^2}{\partial\phi^2}\right]\right\}$$

Comparing this with the spherical representation for $[L^2]$ from Equation 7.25 shows that $[K]$ is the sum of two terms

$$[K] = [K_{rad}] + [K_{orb}] = [K_{rad}] + \frac{1}{2mr^2}[L^2] \tag{7.28}$$

representing the separate contributions to the kinetic energy from the orbital and radial components of motion. The comparison also furnishes the spherical form for $[K_{rad}]$:

$$[K_{rad}] = -\frac{\hbar^2}{2m}\left\{\frac{\partial^2}{\partial r^2} + \frac{2}{r}\frac{\partial}{\partial r}\right\} \tag{7.29}$$

These expressions for $[K]$ are completely general. When applied to waves for which $|\mathbf{L}|$ is sharp, however, the *operator* $[L^2]$ may be replaced with the *number* $|\mathbf{L}|^2 = \ell(\ell + 1)\hbar^2$. Thus, the Schrödinger equation (Eq. 7.27) becomes

$$[K_{rad}]\psi(\mathbf{r}) + \frac{\ell(\ell + 1)\hbar^2}{2mr^2}\psi(\mathbf{r}) + U(r)\psi(\mathbf{r}) = E\psi(\mathbf{r}) \tag{7.30}$$

For central forces all terms in this equation, *including U(r)*, involve only the spherical coordinate r: the angle variables θ and ϕ have been eliminated from further consideration by the requirement that ψ be an eigenfunction of $[L^2]$! It follows that the solutions to Equation 7.30 take the form of a radial wave $R(r)$ multiplied by a spherical harmonic:

$$\psi(\mathbf{r}) = R(r) Y_\ell^{m_\ell}(\theta, \phi) \qquad (7.31)$$

with the radial wavefunction $R(r)$ also satisfying Equation 7.30.

Equation 7.30 for $R(r)$ is just the *radial wave equation* of Section 7.2; it determines the radial wavefunction $R(r)$ and the allowed particle energies E, once the potential energy function $U(r)$ is specified. As the equation suggests, each angular momentum state will have a different radial wave and a different energy. Indeed, the "effective potential"

$$U_{\text{eff}}(r) = U(r) + \frac{\ell(\ell + 1)\hbar^2}{2mr^2} \qquad (7.32)$$

is different for each orbital quantum number ℓ. By contrast, the magnetic quantum number m_ℓ appears nowhere in these equations. Thus, the radial wave and particle energy remain the same for different m_ℓ values consistent with a given value of ℓ. In particular, for a fixed ℓ *the particle energy E is independent of m_ℓ, and so is at least $2\ell + 1$-fold degenerate.* Such degeneracy—a property of all central forces—stems from the spherical symmetry of the potential and illustrates once again the deep-seated connection between symmetry and the degeneracy of quantum states.

7.5   ATOMIC HYDROGEN AND HYDROGEN-LIKE IONS

In this section we study the hydrogen atom from the viewpoint of wave mechanics. Its simplicity makes atomic hydrogen the ideal testing ground for comparing theory with experiment. Furthermore, the hydrogen atom is the prototype system for the many complex ions and atoms of the heavier elements. Indeed, our study of the hydrogen atom ultimately will enable us to understand the periodic table of the elements, one of the greatest triumphs of quantum physics.

The object of interest is the orbiting electron, with its mass m and charge $-e$, bound by the force of electrostatic attraction to the nucleus, with its much larger mass M and charge $+Ze$. Z is the atomic number; the choice $Z = 1$ describes the hydrogen atom, while for singly ionized helium (He^+) and doubly ionized lithium (Li^{2+}), we take $Z = 2$ and $Z = 3$, respectively, and so on. Ions with only one electron, like He^+ and Li^{2+}, are called *hydrogen-like*. Since $M \gg m$, we will assume that the nucleus does not move but simply exerts the attractive force that binds the electron. This force is the Coulomb force, with its associated potential energy

$$U(r) = \frac{k(+Ze)(-e)}{r} = -\frac{kZe^2}{r} \qquad (7.33)$$

where k is the Coulomb constant.

The hydrogen atom constitutes a central force problem; according to Section 7.2, the stationary states for any central force are

$$\Psi(r, \theta, \phi, t) = R(r) Y_\ell^{m_\ell}(\theta, \phi) e^{-i\omega t} \qquad (7.34)$$

where $Y_\ell^{m_\ell}(\theta, \phi)$ is a spherical harmonic from Table 7.3. The *radial wave function $R(r)$* is found from the *radial wave equation* of Section 7.2,

$$-\frac{\hbar^2}{2m}\left[\frac{d^2R}{dr^2} + \left(\frac{2}{r}\right)\frac{dR}{dr}\right] + \frac{\ell(\ell+1)\hbar^2}{2mr^2}R(r) + U(r)R(r) = ER(r) \quad (7.16)$$

$U(r)$ is the potential energy of the electron in the field of the nucleus. The remaining terms on the left of Equation 7.16 are associated with the kinetic energy of the electron. The term proportional to $\ell(\ell+1)\hbar^2(=|\mathbf{L}^2|)$ is the orbital contribution to kinetic energy, K_{orb}. To see this, consider a particle in circular orbit, in which all the kinetic energy is in orbital form (since the distance from the center remains fixed). For such a particle $K_{orb} = \frac{1}{2}mv^2$ and $|\mathbf{L}| = mvr$. Eliminating the particle speed v, we get

$$K_{orb} = \frac{m}{2}\left(\frac{|\mathbf{L}|}{mr}\right)^2 = \frac{|\mathbf{L}|^2}{2mr^2} \quad (7.35)$$

Although it was derived for circular orbits, this result correctly represents the orbital contribution to kinetic energy of a mass m in general motion with angular momentum $\mathbf{L}$.[9]

The derivative terms in Equation 7.16 are the radial contribution to the kinetic energy, that is, they represent the contribution from electron motion toward or away from the nucleus. The leftmost term is just what we should write for the kinetic energy of a matter wave $\psi = R(r)$ associated with motion along the coordinate line marked by r. The appearance of dR/dr is mysterious, however; in fact, the existence of this term is evidence that *the effective one-dimensional matter wave is not $R(r)$, but $rR(r)$*. In support of this claim we note that

$$\frac{d^2(rR)}{dr^2} = r\left[\frac{d^2R}{dr^2} + \left(\frac{2}{r}\right)\frac{dR}{dr}\right]$$

Then the radial wave equation written for the *pseudo-wavefunction* $g(r) = rR(r)$ takes the same form as Schrödinger's equation in one dimension,

$$\left(-\frac{\hbar^2}{2m}\right)\frac{d^2g}{dr^2} + U_{eff}(r)g(r) = Eg(r) \quad (7.36)$$

but with an *effective potential energy*

$$U_{eff} = \frac{|\mathbf{L}|^2}{2mr^2} + U(r) = \frac{\ell(\ell+1)\hbar^2}{2mr^2} + U(r) \quad (7.37)$$

The magnitude of this pseudo-wavefunction $g(r)$ also furnishes probabilities, as described later in this section.

The solution of Equation 7.36 with $U(r) = -kZe^2/r$ (for one-electron atoms or ions) requires methods beyond the scope of this text. Here, as before, we shall present the results and leave their verification to the interested reader. Acceptable wavefunctions can be found only if the energy E is restricted to be one of the following special values:

$$E_n = -\frac{ke^2}{2a_0}\left\{\frac{Z^2}{n^2}\right\} \qquad n = 1, 2, 3, \ldots \quad (7.38)$$

Allowed energies for hydrogen-like atoms

[9] For any planar orbit we may write $|\mathbf{L}| = rp_\perp$, where $p_\perp$ is the component of momentum in the orbital plane that is normal to the radius vector $\mathbf{r}$. The orbital part of the kinetic energy is then $K_{orb} = p_\perp^2/2m = |\mathbf{L}|^2/2mr^2$.

Table 7.4 The Radial Wavefunctions $R_{n,\ell}(r)$ of Hydrogen-like Atoms for $n = 1$, 2, and 3

n	ℓ	$R(r)$
1	0	$\left(\dfrac{Z}{a_0}\right)^{3/2} 2e^{-Zr/a_0}$
2	0	$\left(\dfrac{Z}{2a_0}\right)^{3/2} \left(2 - \dfrac{Zr}{a_0}\right) e^{-Zr/2a_0}$
2	1	$\left(\dfrac{Z}{2a_0}\right)^{3/2} \dfrac{Zr}{\sqrt{3}\, a_0} e^{-Zr/2a_0}$
3	0	$\left(\dfrac{Z}{3a_0}\right)^{3/2} 2\left[1 - \dfrac{2Zr}{3a_0} + \dfrac{2}{27}\left(\dfrac{Zr}{a_0}\right)^2\right] e^{-Zr/3a_0}$
3	1	$\left(\dfrac{Z}{3a_0}\right)^{3/2} \cdot \dfrac{4\sqrt{2}}{3} \dfrac{Zr}{a_0}\left(1 - \dfrac{Zr}{6a_0}\right) e^{-Zr/3a_0}$
3	2	$\left(\dfrac{Z}{3a_0}\right)^{3/2} \cdot \dfrac{2\sqrt{2}}{27\sqrt{5}}\left(\dfrac{Zr}{a_0}\right)^2 e^{-Zr/3a_0}$

This result is in exact agreement with that found from the simple Bohr theory (see Chapter 3): $a_0 = \hbar^2/m_e ke^2$ is the Bohr radius, 0.529 Å, and $ke^2/2a_0$ is the Rydberg energy, 13.6 eV. The integer n is called the **principal quantum number.** Although n can be any positive integer, the orbital quantum number now is limited to values less than n; that is, ℓ can have only the values

Allowed values of ℓ

$$\ell = 0, 1, 2, \ldots, (n - 1) \tag{7.39}$$

Principal quantum number n

$$n = 1, 2, 3, \ldots$$

orbital quantum number ℓ

$$\ell = 0, 1, 2, \ldots, (n - 1)$$

magnetic quantum number m_ℓ

$$m_\ell = 0, \pm 1, \pm 2, \ldots, \pm \ell$$

The cutoff at $(n - 1)$ is in keeping with the physical significance of these quantum numbers: the magnitude of orbital angular momentum (fixed by ℓ) cannot become arbitrarily large for a given energy (fixed by n). A semiclassical argument expressing this idea leads to $\ell_{max} = n - 1$ (see Problem 21). In the same spirit, the restriction on the magnetic quantum number m_ℓ to values between $-\ell$ and $+\ell$ guarantees that the z-component of angular momentum never exceeds the magnitude of the vector.

The radial waves $R(r)$ for hydrogen-like atoms are products of exponentials with polynomials in r/a_0. These radial wavefunctions are tabulated as $R_{n,\ell}(r)$ in Table 7.4 for principal quantum numbers up to and including $n = 3$.

For hydrogen-like atoms, then, the quantum numbers are n, ℓ, and m_ℓ, associated with the sharp observables E, $|\mathbf{L}|$, and L_z, respectively. Notice that the electron energy E depends on n, but not at all on ℓ or m_ℓ. The energy is independent of m_ℓ because of the spherical symmetry of the atom, and this will be true for any central force that varies only with distance r. The fact that E also is independent of ℓ is a special consequence of the Coulomb force, however, and is not to be expected in heavier atoms, say, where the force on any one electron includes the electrostatic repulsion of the remaining electrons in the atom, as well as the Coulombic attraction of the nucleus.

Table 7.5   Spectroscopic Notation for Atomic Shells and Subshells

n	Shell Symbol	ℓ	Subshell Symbol
1	K	0	s
2	L	1	p
3	M	2	d
4	N	3	f
5	O	4	g
6	P	5	h
. . .		. . .	

For historical reasons, all states with the same principal quantum number n are said to form a **shell.** These shells are identified by the letters K, L, M, . . . , which designate the states for which $n = 1, 2, 3, \ldots$. Likewise, states having the same value of both n and ℓ are said to form a **subshell.** The letters s, p, d, f, . . . are used to designate the states for which $\ell = 0, 1, 2, 3, \ldots$. This *spectroscopic notation* is summarized in Table 7.5.

The shell (and subshell) energies for several of the lowest lying states of the hydrogen atom are illustrated in the energy level diagram of Figure 7.8. The figure also portrays a few of the many electronic transitions possible within the atom. Each such transition represents a change of energy for the atom and must be compensated for by emission (or absorption) of energy in some other form. For *optical transitions*, photons carry off the surplus energy, but not all energy-conserving optical transitions may occur. As it happens, photons also carry angular momentum. To conserve *total* angular momentum (atom +

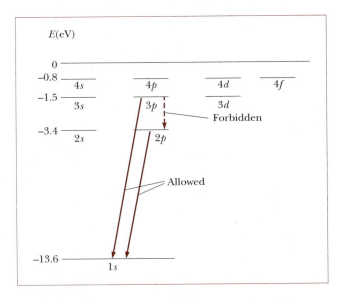

Figure 7.8   Energy level diagram of atomic hydrogen. Allowed photon transitions are those obeying the *selection rule* $\Delta\ell = \pm 1$. The $3p \rightarrow 2p$ transition ($\Delta\ell = 0$) is said to be *forbidden*, though it may still occur (but only rarely).

photon) in optical transitions, the angular momentum of the electron in the initial and final states must differ by exactly one unit, that is

$$|\ell_f - \ell_i| = 1 \qquad \text{or} \qquad \Delta\ell = \pm 1 \qquad (7.40)$$

Equation 7.40 expresses a *selection rule* that must be obeyed in optical transitions.[10] As Figure 7.8 indicates, the transitions $3p \rightarrow 1s$ and $2p \rightarrow 1s$ are allowed by the rule ($\Delta\ell = -1$), but the $3p \rightarrow 2p$ transition is said to be *forbidden* ($\Delta\ell = 0$). (Such transitions can occur, but with negligible probability compared with that of allowed transitions.) Clearly, selection rules play a vital role in the interpretation of atomic spectra.

EXAMPLE 7.7   The $n = 2$ Level of Hydrogen

Enumerate all states of the hydrogen atom corresponding to the principal quantum number $n = 2$, giving the spectroscopic designation for each. Calculate the energies of these states.

Solution   When $n = 2$, ℓ can have the values 0 and 1. If $\ell = 0$, m_ℓ can only be 0. If $\ell = 1$, m_ℓ can be $-1, 0$, or $+1$. Hence, we have a $2s$ state with quantum numbers

$$n = 2, \qquad \ell = 0, \qquad m_\ell = 0$$

and three $2p$ states for which the quantum numbers are

$$n = 2, \qquad \ell = 1, \qquad m_\ell = -1$$

$$n = 2, \qquad \ell = 1, \qquad m_\ell = 0$$
$$n = 2, \qquad \ell = 1, \qquad m_\ell = +1$$

Because all of these states have the same principal quantum number, $n = 2$, they also have the same energy, which can be calculated from Equation 7.38. For $Z = 1$ and $n = 2$, this gives

$$E_2 = -(13.6 \text{ eV})\{1^2/2^2\} = -3.4 \text{ eV}$$

Exercise 4   How many possible states are there for the $n = 3$ level of hydrogen? For the $n = 4$ level?

Answers   Nine states for $n = 3$, and 16 states for $n = 4$.

The Ground State of Hydrogen-like Atoms

The ground state of a one-electron atom or ion with atomic number Z, for which $n = 1$, $\ell = 0$, and $m_\ell = 0$, has energy

$$E_1 = -(13.6 \text{ eV})Z^2 \qquad (7.41)$$

The wavefunction for this state is

$$\psi_{1,0,0} = R_{1,0}(r)Y_0^0 = \pi^{-1/2}(Z/a_0)^{3/2}e^{-Zr/a_0} \qquad (7.42)$$

The constants are such that ψ is normalized.

Notice that $\psi_{1,0,0}$ does not depend on angle, since it is the product of a radial wave with $Y_0^0 = 1/\sqrt{4\pi}$. In fact, all the $\ell = 0$ waves share this feature; that is, *all s-state waves are spherically symmetric.*

The electron described by the wavefunction of Equation 7.42 is found with a probability per unit volume given by

$$|\psi_{1,0,0}|^2 = \frac{Z^3}{\pi a_0^3} e^{-2Zr/a_0}$$

---

[10] $\Delta\ell = 0$ also is allowed by angular momentum considerations but forbidden by parity conservation. Further, since the angular momentum of a photon is just $\pm\hbar$, the z-component of the atom's angular momentum cannot change by more than $\pm\hbar$, giving rise to a *second selection rule*, $\Delta m_\ell = 0, \pm 1$.

(Three-dimensional renditions of the probability per unit volume $|\psi(\mathbf{r})|^2$—often called electron "clouds"—are constructed by making the shading at every point proportional to $|\psi(\mathbf{r})|^2$.)

The probability distribution also is spherically symmetric, as it would be for any s-state wave; that is, the likelihood for finding the electron in the atom is the same at all points equidistant from the center (nucleus). Thus, it is convenient to define another probability function called the **radial probability distribution** with its associated density $P(r)$, such that $P(r)\,dr$ is the probability of finding the electron anywhere in the spherical shell of radius r and thickness dr (Fig. 7.9). The shell volume is its surface area, $4\pi r^2$, multiplied by the shell thickness, dr. Since the probability density $|\psi_{1,0,0}|^2$ is the same everywhere in the shell, we have

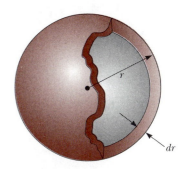

Figure 7.9 $P(r)\,dr$ is the probability that the electron will be found in the volume of a spherical shell with radius r and thickness dr. The shell volume is just $4\pi r^2\,dr$.

$$P(r)\,dr = |\psi|^2 4\pi r^2\,dr$$

or, for the hydrogen-like $1s$ state,

$$P_{1s}(r) = \frac{4Z^3}{a_0^3}\,r^2 e^{-2Zr/a_0} \qquad (7.43)$$

The same result is obtained for $P_{1s}(r)$ from the intensity of the pseudo-matter wave $g(r) = rR(r)$:

$$P(r) = |g(r)|^2 = r^2|R(r)|^2 \qquad (7.44)$$

The radial probability density for any state

In fact, *Equation 7.44 gives the correct radial probability density for any state;* for the spherically symmetric s-states this is the same as $4\pi r^2|\psi|^2$, since then $\psi(r) = (1/\sqrt{4\pi})R(r)$.[11]

A plot of the function $P_{1s}(r)$ is presented in Figure 7.10a; Figure 7.10b shows the $1s$ electron "cloud" or probability per unit volume $|\psi_{1,0,0}|^2$ from which $P_{1s}(r)$ derives. $P(r)$ may be loosely interpreted as the probability of finding the elec-

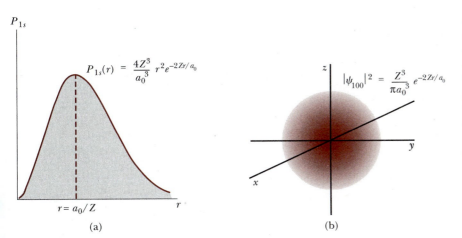

$$P_{1s}(r) = \frac{4Z^3}{a_0^3}\,r^2 e^{-2Zr/a_0}$$

$r = a_0/Z$

(a)

$$|\psi_{100}|^2 = \frac{Z^3}{\pi a_0^3}\,e^{-2Zr/a_0}$$

(b)

Figure 7.10   (a) The curve $P_{1s}(r)$, representing the probability of finding the electron as a function of distance from the nucleus in a $1s$ hydrogen-like state. Note that the probability takes its maximum value when r equals a_0/Z. (b) The spherical electron "cloud" for a hydrogen-like $1s$ state. The shading at every point is proportional to the probability density $|\psi_{1s}(\mathbf{r})|^2$.

---

[11] From its definition, $P(r)\,dr$ always may be found by integrating $|\psi|^2 = |R(r)|^2|Y_\ell{}^{m_\ell}|^2$ over the volume of a spherical shell having radius r and thickness dr. Since the volume element in spherical coordinates is $dV = r^2 dr \sin\theta\,d\theta\,d\phi$, and the integral of $|Y_\ell{}^{m_\ell}|^2$ over angle is unity (see footnote 5), this leaves $P(r) = r^2|R(r)|^2$.

tron at distance r from the nucleus, irrespective of its angular position. Thus, the peak of the curve in Figure 7.10a represents the most probable distance of the $1s$ electron from the nucleus. Furthermore, the normalization condition becomes

$$1 = \int_0^\infty P(r) \; dr \tag{7.45}$$

where the integral is taken over all *possible* values of r. The average distance of the electron from the nucleus is found by weighting each possible distance with the probability that the electron will be found at that distance:

The average distance of an electron from the nucleus

$$\langle r \rangle = \int_0^\infty rP(r) \; dr \tag{7.46}$$

In fact, the average value of *any* function of distance $f(r)$ is obtained by weighting the function value at every distance with the probability at that distance:

$$\langle f \rangle = \int_0^\infty f(r)P(r) \; dr \tag{7.47}$$

EXAMPLE 7.8   Probabilities for the Electron in Hydrogen

Calculate the probability that the electron in the ground state of hydrogen will be found outside the first Bohr radius.

Solution   The probability is found by integrating the radial probability density for this state, $P_{1s}(r)$, from the Bohr radius a_0 to ∞. Using Equation 7.43 with $Z = 1$ for hydrogen gives

$$P = \frac{4}{a_0^3} \int_{a_0}^\infty r^2 e^{-2r/a_0} \; dr$$

We can put the integral in dimensionless form by changing variables from r to $z = 2r/a_0$. Noting that $z = 2$ when $r = a_0$, and that $dr = (a_0/2) \; dz$, we get

$$P = \frac{1}{2} \int_2^\infty z^2 e^{-z} \; dz = -\frac{1}{2} \{z^2 + 2z + 2\}e^{-z} \Big|_2^\infty = 5e^{-2}$$

This is about 0.677, or 67.7%.

EXAMPLE 7.9   The Electron–Proton Separation in Hydrogen

Calculate the most probable distance of the electron from the nucleus in the ground state of hydrogen, and compare this with the average distance.

Solution   The most probable distance is the value of r that makes the radial probability $P(r)$ a maximum. The slope here is zero, so the most probable value of r is obtained by setting $dP/dr = 0$ and solving for r. Using Equation 7.43 with $Z = 1$ for the $1s$, or ground state, of hydrogen, we get

$$0 = \left(\frac{4}{a_0^3}\right)\frac{d}{dr}\{r^2 e^{-2r/a_0}\} = \left(\frac{4}{a_0^3}\right) e^{-2r/a_0} \left\{-\frac{2r^2}{a_0} + 2r\right\}$$

The right-hand side is zero for $r = 0$ and for $r = a_0$. Since $P(0) = 0$, $r = 0$ is a minimum of $P(r)$, not a maximum. Thus, the most probable distance is

$$r = a_0$$

The average distance is obtained from Equation 7.46, which in this case becomes

$$\langle r \rangle = \frac{4}{a_0^3} \int_0^\infty r^3 e^{-2r/a_0} \; dr$$

Again introducing $z = 2r/a_0$, we obtain

$$\langle r \rangle = \frac{a_0}{4} \int_0^\infty z^3 e^{-z} \; dz$$

The definite integral on the right is one of a broader class

$$\int_0^\infty z^n e^{-z}\, dz = n!$$

whose value $n! = n(n-1) \ldots (1)$ is established by repeated integration by parts. Then

$$\langle r \rangle = \frac{a_0}{4}(3!) = \frac{3}{2}a_0$$

The average distance and the most probable distance are not the same, because the probability curve $P(r)$ is not symmetric about the peak distance a_0. Indeed, values of r greater than a_0 are weighted more heavily in Equation 7.46 than values smaller than a_0, so that the average $\langle r \rangle$ actually exceeds a_0 for this probability distribution.

Excited States of Hydrogen-like Atoms

There are four first excited states for hydrogen-like atoms: $\psi_{2,0,0}$, $\psi_{2,1,0}$, $\psi_{2,1,1}$, and $\psi_{2,1,-1}$. All have the same principal quantum number $n = 2$, hence the same total energy

$$E_2 = -(13.6\text{ eV})\frac{Z^2}{4}$$

Accordingly, the first excited level E_2 is fourfold degenerate.

The $2s$ state, $\psi_{2,0,0}$, is again spherically symmetric. Plots of the radial probability density for this and several other hydrogen-like states are shown in Figure 7.11. Note that the plot for the $2s$ state has two peaks. In this case, the most probable distance ($\sim 5a_0/Z$) is marked by the highest peak. An electron in the $2s$ state would be much farther from the nucleus (on the average) than an electron in the $1s$ state. Likewise, the most probable distances are even greater for an electron in any of the $n = 3$ states ($3s$, $3p$, or $3d$). Observations such as these continue to support the old idea of a shell structure for the atom, even in the face of the uncertainties inherent in the wave nature of matter.

The remaining three first excited states, $\psi_{2,1,1}$, $\psi_{2,1,0}$, and $\psi_{2,1,-1}$, have $\ell = 1$ and make up the $2p$ subshell. These states are not spherically symmetric. All of them have the same radial wavefunction $R_{2,1}(r)$, but they are multiplied by different spherical harmonics and thus depend differently on the angles θ and ϕ. For example, the wavefunction $\psi_{2,1,1}$ is

$$\psi_{2,1,1} = R_{2,1}(r)\,Y_1{}^1 = \pi^{-1/2}\left(\frac{Z}{a_0}\right)^{3/2}\left(\frac{Zr}{8a_0}\right)e^{-Zr/2a_0}\sin\theta\, e^{i\phi} \quad (7.48)$$

Notice, however, that the probability density $|\psi_{2,1,1}|^2$ is independent of ϕ and therefore is symmetric about the z axis (Fig. 7.12a). Since $|e^{im_\ell\phi}|^2 = 1$, the same is true for all the hydrogen-like states ψ_{n,ℓ,m_ℓ}, as suggested by the remaining illustrations in Figure 7.12.

The $\psi_{2,1,0}$ state

$$\psi_{2,1,0} = R_{2,1}(r)\,Y_1{}^0 = \pi^{-1/2}\left(\frac{Z}{2a_0}\right)^{3/2}\left(\frac{Zr}{2a_0}\right)e^{-Zr/2a_0}\cos\theta \quad (7.49)$$

has distinct directional characteristics, as shown in Figure 7.13a, and is sometimes designated $[\psi_{2p}]_z$ to indicate the preference for the electron in this state to be found along the z axis. Other highly directional states can be formed by

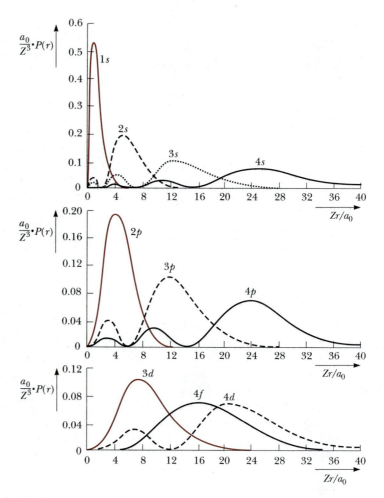

Figure 7.11 The radial probability density function for several states of hydrogen-like atoms.

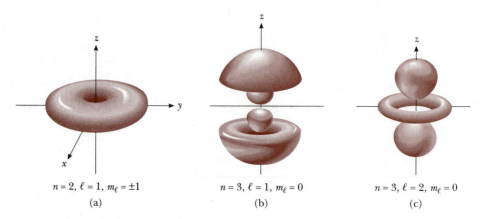

$n = 2, \ell = 1, m_\ell = \pm 1$

(a)

$n = 3, \ell = 1, m_\ell = 0$

(b)

$n = 3, \ell = 2, m_\ell = 0$

(c)

Figure 7.12 (a) The probability density $|\psi_{2,1,1}|^2$ for a hydrogen-like $2p$ state. Note the axial symmetry about the z axis. (b) and (c) The probability densities $|\psi(\mathbf{r})|^2$ for several other hydrogen-like states. The electron "cloud" is axially symmetric about the z axis for all the hydrogen-like states $\psi_{n,\ell m_\ell}(\mathbf{r})$.

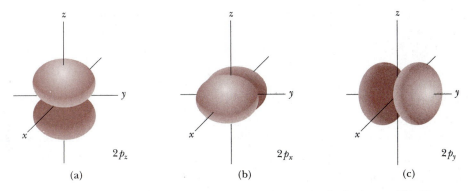

$2p_z$ $2p_x$ $2p_y$

(a)   (b)   (c)

Figure 7.13   (a) Probability distribution for an electron in the hydrogen-like $2p_z$ state, described by the quantum numbers $n = 2$, $\ell = 1$, $m_\ell = 0$. (b) and (c) Probability distributions for the $2p_x$ and $2p_y$ states. The three distributions $2p_x$, $2p_y$, and $2p_z$ have the same structure, but differ in their spatial orientation.

combining the waves with $m_\ell = +1$ and $m_\ell = -1$. Thus, the wavefunctions

$$[\psi_{2p}]_x = \frac{1}{\sqrt{2}}\{\psi_{2,1,1} + \psi_{2,1,-1}\}$$

and

$$[\psi_{2p}]_y = \frac{1}{\sqrt{2}}\{\psi_{2,1,1} - \psi_{2,1,-1}\} \tag{7.50}$$

have large probability densities along the x and y axes, respectively, as shown in Figures 7.13b and 7.13c. The wavefunctions of Equations 7.50, formed as superpositions of first excited state waves with *identical* energies E_2, are themselves stationary states with this same energy; indeed, the three wavefunctions $[\psi_{2p}]_z$, $[\psi_{2p}]_x$, and $[\psi_{2p}]_y$ together constitute an equally good description of the $2p$ states for hydrogen-like atoms. Wavefunctions with a highly directional character, such as these, play an important role in chemical bonding, the formation of molecules, and chemical properties.

The next excited level, E_3, is ninefold degenerate; that is, there are nine different wavefunctions with this energy, corresponding to all possible choices for ℓ and m_ℓ consistent with $n = 3$. Together, these nine states constitute the third shell, with subshells $3s$, $3p$, and $3d$ composed of one, three, and five states, respectively. The wavefunctions for these states may be constructed from the entries in Tables 7.3 and 7.4. Generally, they are more complicated than their second shell counterparts because of the increasing number of nodes in the radial part of the wavefunction.

The progression to higher lying states leads to still more degeneracy and wavefunctions of ever-increasing complexity. The nth shell has energy

$$E_n = -(13.6 \text{ eV})\frac{Z^2}{n^2}$$

and contains exactly n subshells, corresponding to $\ell = 0, 1, \ldots, n - 1$. Within the ℓth subshell there are $2\ell + 1$ orbitals. Altogether, the nth shell contains a total of n^2 states, all with the same energy E_n; that is, *the energy level E_n is n^2 degenerate.* Equivalently, we say the nth shell can hold as many as n^2

electrons, assuming no more than one electron can occupy a given orbital. This argument underestimates the shell capacity by a factor of 2, owing to the existence of electron spin, as discussed in the next chapter.

EXAMPLE 7.10   Studying Degeneracy with the MP DESKTOP

MP This example uses the MP DESKTOP to verify the degeneracy of the first excited state of atomic hydrogen and further explore the degenerate wavefunctions.

Launch the MP DESKTOP application by running the main program file *MP.EXE*. Next, select *Task* from the menu bar and *EigenStates-3D* from the drop-down task list. The DESKTOP is now ready to solve the (eigenvalue) problem posed in the task field just below the menu bar:

$$-\frac{d^2 g}{dr^2} + \left[U(r) + \frac{\ell(\ell + 1)}{r^2} \right] g(r) = E \cdot g(r)$$

This is the *radial wave equation* for the pseudo-wavefunction $g(r) = rR(r)$, expressed in dimensionless units of length and energy.

NOTE: For *EigenStates-3D* the unit of length (δ) and the unit of energy (ϵ) must satisfy $\epsilon \cdot \delta^2 = \hbar^2/2m$ in order to map the equation in the task field to the radial wave equation. In describing electrons, atomic units ($\delta = 1$ bohr [0.529 Å] and $\epsilon = 1$ Ry [13.60 eV]) are a

convenient choice satisfying this criterion and are the ones adopted in this example. For a more exhaustive discussion of units, see the *Help-Contents* entry on the DESKTOP menu bar.

Select the hydrogen atom potential by clicking the Potential button located just below the menu bar and choosing *EX7-10.U3* (for Example 7.10) from the list of potential files. A second dialog box opens, showing the potential function $U(r) = -2\Gamma/r$. This is the Coulomb potential of an electron in the field of Γ protons with r scaled by a_0 (1 bohr) and U scaled by $ke^2/2a_0$ (1 Ry). The default $\Gamma = 1$ describes atomic hydrogen. Click the *OK* button to accept all defaults and load this potential.

The effective potential $U_{\text{eff}}(r) = U(r) + \ell(\ell + 1)/r^2$ is plotted in the main window of the screen; since $\ell = 0$ to start, this is just the Coulomb potential $U(r) = -2/r$. The 2s state of hydrogen has energy -3.4 eV, or $-1/4$ Ry, so type $E = -0.25$ in the energy box just below the menu bar and press *Enter*. The pseudo-wavefunction $g_{2s}(r)$ for this energy is computed and is superimposed on the potential (Fig. 7.14). The smooth appearance of the

Figure 7.14 The pseudo-wavefunction $g_{2s}(r)$ for the 2s state of hydrogen, against the backdrop of the Coulomb potential. The graphics cursor marks the most probable distance from the nucleus for the 2s electron.

Figure 7.15   The hydrogen $2s$ state. Left window: the $2s$ electron "cloud" $|\psi_{2s}(\mathbf{r})|^2$. Right window: the pseudo-wavefunction $g_{2s}(r)$. The cursor marks the most probable distance from the nucleus for the $2s$ electron.

waveform is evidence that -0.25 Ry is one of the allowed energies for the electron in atomic hydrogen. Notice that the $2s$ radial wave has one node, implying there is another wavefunction with lower energy (which one?). The $2s$ pseudo-wavefunction has both a maximum and a minimum: of these, the one with larger absolute value (the maximum) marks the most probable distance of the $2s$ electron from the nucleus. To locate the maximum, activate the MP Analyzer by clicking . The wavefunction is read at the current location of the graphics cursor. Position the cursor on the peak using the mouse or the direction keys [→], [←]. (Keyboard users note that the [Ins] key toggles cursor movement between coarse and fine settings.) You should find the peak at $r \approx 5.23\ a_0$.

NOTE: With the Analyzer activated, you may notice that the distance scale for the graphics display is not linear. In fact, for *Eigenstates-3D* the default scaling for the coordinate r is logarithmic, to accommodate the long-range character of many physically interesting potentials (such as the Coulomb potential).

Now click the Split Screen button to split the

graphics viewport. In the left window is painted the electron "cloud" for this state, actually a two-dimensional rendition in the x-z plane of the probability density function $|\psi_{2s}(\mathbf{r})|^2$. (The z-axis is vertically upward.) The display should look like that shown in Figure 7.15.

NOTE: Having the angular dependence of spherical harmonics, the electron "clouds" all exhibit axial symmetry about the z-axis (vertical). It follows that the x-z slice of the cloud shown by the DESKTOP can be rotated about the vertical axis passing through the origin to reconstruct the entire three-dimensional figure.

Activate the left window for analysis by clicking the [blank] bar just beneath it. The intensity of color at each point is a measure of the probability per unit volume there, in this case $|\psi_{2s}(\mathbf{r})|^2$. Unnormalized values for $|\psi_{2s}(\mathbf{r})|^2$ are read at the intersection of the crosshairs, which can be repositioned using the mouse or keyboard in the usual way. The distribution of probability exhibits clear spherical symmetry, with $|\psi_{2s}(\mathbf{r})|^2$ largest at the site of the nucleus, $\mathbf{r} = 0$. This is the most likely place to find the $2s$ electron in hydrogen and is quite distinct from its most probable distance from the nucleus found above (why?).

The $2p$ state of hydrogen can be similarly studied. Set $\ell = 1$ by clicking the Orbital button [l= 0] located on the Graphics Toolbar to the right of the graphics viewport. This updates the display to show the effective potential for $\ell = 1$ and the $2p$ wavefunction with its associated electron "cloud." Unlike the $2s$ wave, the $2p$ wave has a single peak and no nodes (it is the p state with lowest energy). Return to the right window by clicking the [blank] bar underneath, and locate the peak to find the most probable distance from the nucleus for the $2p$ electron in hydrogen. This time the peak occurs at $r \approx 4a_0$.

In the left window the $2p$ electron cloud is painted to the screen. The spherical symmetry of the $2s$ state is absent, having been replaced with the angular dependence of the spherical harmonic $Y_1{}^{m_\ell}$. Actually this is the probability distribution for $\ell = 1$ and $m_\ell = 0$ (Fig. 7.16).

Figure 7.16   The electron cloud for the $n = 2$, $\ell = 1$, $m_\ell = 0$ state of hydrogen.

For a p electron, we can also have $m_\ell = \pm 1$. To change the magnetic quantum number click the Magnetic button [m= 0], located on the Graphics Toolbar just below the Orbital button, and reactivate the left window by clicking the [blank] bar underneath. This paints to the screen the distribution for $m_\ell = 1$ (Fig. 7.17), and updates the buttonface to reflect the new value for m_ℓ. You may return to the state $m_\ell = 0$ by clicking [m= 0] again, this time *holding down the Shift key*. Toggling between the two states makes it evident that the electron distribution depends markedly on m_ℓ, even though the electron energy is the same for both. (The distributions for $m_\ell = -1$ and $m_\ell = +1$ are identical, however. Why?)

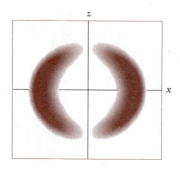

Figure 7.17   The electron cloud for the $n = 2$, $\ell = 1$, $m_\ell = \pm 1$ state of hydrogen.

EXAMPLE 7.11   Directional Characteristics of Hydrogen Wavefunctions

MP This example uses the MP DESKTOP to explore the angular variation of wavefunctions having s, p, d, and f character by examining the $n = 4$ states of atomic hydrogen.

Prepare the DESKTOP as described in Example 7.10, once again selecting the hydrogen-like potential file *EX7-10.U3*. This is the Coulomb potential $U(r) = -2\Gamma/r$ of an electron in the field of Γ protons for the case where r is measured in bohrs (a_0) and U in rydbergs (1 Ry $= -13.6$ eV $= ke^2/2a_0$). The default $\Gamma = 1$ describes atomic hydrogen. Click on the *OK* button to accept all defaults and load this potential.

If necessary, reset $\ell = 0$ by clicking the Orbital button [l= 0] *while holding down the Shift key*. The $n = 4$ wavefunctions of hydrogen ($4s$, $4p$, $4d$, and $4f$) all have energy -0.85 eV or $-1/16$ Ry, so type $E = -0.0625$ in the energy box just below the menu bar and press *Enter*. The pseudo-wavefunction $g_{4s}(r)$ for this energy overlays the potential. The smooth appearance of the waveform is evidence that -0.0625 Ry is one of the allowed energies for the electron in atomic hydrogen.

Click the Split Screen button [⊞] to split the graphics viewport. In the left window is painted the electron cloud for this state, actually a two-dimensional rendition in the x-z plane of the probability density function $|\psi_{4s}(\mathbf{r})|^2$. (The z axis is vertically upward.) The $4s$ cloud exhibits clear spherical symmetry, with the probability for finding the electron largest at the site of the nucleus, $\mathbf{r} = 0$.

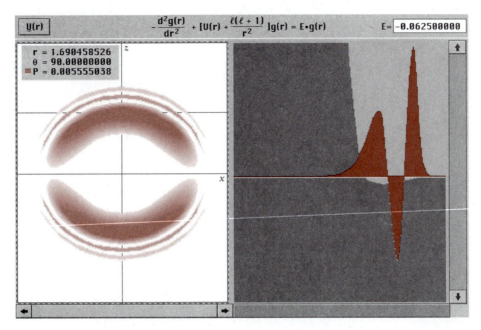

Figure 7.18 The $4p$ state of hydrogen. Left window: the $4p$ electron cloud $|\psi_{4p}(\mathbf{r})|^2$. The crosshairs are centered at $r = 1.69$ [bohrs], $\theta = 90°$, one of (two) locations where $|\psi_{4p}(\mathbf{r})|^2$ is a maximum. Right window: the $4p$ pseudo-wavefunction $g_{4p}(r)$ against the backdrop of the effective potential for a hydrogen p state.

NOTE: The electron clouds all exhibit axial symmetry about the z axis (vertical). It follows that the x-z slice of the cloud shown by the DESKTOP can be rotated about the vertical axis passing through the origin to reconstruct the entire three-dimensional figure.

To examine the $4p$ state of hydrogen, set $\ell = 1$ by clicking the Orbital button $\boxed{\underline{l}= 0}$. The pseudo-wavefunction $g_{4p}(r)$ appears in the right window of the viewport, and the corresponding electron cloud $|\psi_{4p}(\mathbf{r})|^2$ is painted in the left window. The two-lobed structure of this cloud is typical of the symmetry displayed by p states (Fig. 7.18).

Activate the MP Analyzer by clicking $\boxed{}$, and switch to the left window by clicking the [blank] bar underneath. Position the crosshairs on a point where $|\psi_{4p}(\mathbf{r})|^2$ is largest. You should find $r \approx 1.69\ a_0$, and $\theta = \pm 90°$. These are the most likely places for a $4p$ electron to be.

Repeat the previous procedure for the $4d$ state ($\ell = 2$). Observe the distinctly different symmetry of the d wave; also notice that a d electron is found with great-

est probability at a larger distance from the nucleus (Fig. 7.19). Overall, the $4d$ state is more *extended* than the $4p$, that is, a $4d$ electron can be found with appreciable probability further away from the nucleus. These trends continue with the $4f$ state, for which $\ell = 3$. As Figure 7.20 indicates, the symmetry of f waves is different from that of s, p, or d waves, and the corresponding angular variation is even more complex.

Figure 7.19 The $4d$ state of hydrogen, for which $n = 4$, $\ell = 2$, and $m_\ell = 0$.

Figure 7.20 The 4f state of hydrogen, for which $n = 4$, $\ell = 3$, and $m_\ell = 0$.

Exercise 5 *Symmetry and the Magnetic Quantum Number.* Explore the effect of the magnetic quantum number m_ℓ on the overall appearance and properties of the 4p, 4d, and 4f wavefunctions of hydrogen. (Click <u>m= 0</u> to change m_ℓ.) Can you identify any trends?

SUMMARY

In three dimensions, the matter wave intensity $|\Psi(\mathbf{r}, t)|^2$ represents the probability per unit volume for finding the particle at $\mathbf{r}$ at time t. Probabilities are found by integrating this probability density over the volume of interest.

The wavefunction itself must satisfy the Schrödinger equation

$$-\frac{\hbar^2}{2m}\nabla^2\Psi + U(r)\Psi = i\hbar\frac{\partial\Psi}{\partial t} \tag{7.1}$$

Stationary states are solutions to this equation in separable form: $\Psi(\mathbf{r}, t) = \psi(\mathbf{r})e^{-i\omega t}$ with $\psi(\mathbf{r})$ satisfying

$$-\frac{\hbar^2}{2m}\nabla^2\psi(\mathbf{r}) + U(r)\psi(\mathbf{r}) = E\psi(\mathbf{r}) \tag{7.5}$$

This is the *time-independent Schrödinger equation*, from which we obtain the time-independent wavefunction $\psi(\mathbf{r})$ and the allowed values of particle energy E.

For a particle confined to a cubic box whose sides are L, the magnitudes of the components of particle momentum normal to the walls of the box can be made sharp, as can the particle energy. The sharp momentum values are quantized as

$$|p_x| = n_1\frac{\pi\hbar}{L}$$

$$|p_y| = n_2\frac{\pi\hbar}{L} \tag{7.8}$$

$$|p_z| = n_3\frac{\pi\hbar}{L}$$

and the allowed energies are found to be

$$E = \frac{\pi^2\hbar^2}{2mL^2}\{n_1^2 + n_2^2 + n_3^2\} \tag{7.9}$$

The three *quantum numbers* n_1, n_2, and n_3 are all positive integers, one for each degree of freedom of the particle. Many levels of this system are *degenerate*, that is, there is more than one wavefunction with the same energy.

For particles acted on by a *central force*, the *angular momentum* **L** is a constant of the classical motion and is quantized along with particle energy. Wavefunctions for which the z component L_z and magnitude $|\mathbf{L}|$ of angular momentum are simultaneously sharp are the *spherical harmonics* $Y_\ell{}^{m_\ell}(\theta, \phi)$ in the spherical coordinate angles θ and ϕ. For any central force, angular momentum is quantized by the rules

$$|\mathbf{L}| = \sqrt{\ell(\ell + 1)}\,\hbar$$

and

$$L_z = m_\ell \hbar \qquad (7.14)$$

The *orbital quantum number* ℓ must be a nonnegative integer. For a fixed value of ℓ, the *magnetic quantum number* m_ℓ is restricted to integer values lying between $-\ell$ and $+\ell$. Since $|\mathbf{L}|$ and L_z are quantized differently, the classical freedom to orient the z axis in the direction of **L** must be abandoned. This stunning conclusion is the essence of *space quantization*. Furthermore, no two components of **L**, such as L_z and L_y, can be sharp simultaneously. This implies a lower limit to the uncertainty product $\Delta L_z \Delta L_y$ and gives rise to an uncertainty principle for the components of angular momentum.

A central force of considerable importance is the force on the electron in a one-electron atom or ion. This is the *Coulomb* force, described by the potential energy $U(r) = -kZe^2/r$, where Z is the atomic number of the nucleus. The allowed energies for this case are given by

$$E_n = -\frac{ke^2}{2a_0}\left\{\frac{Z^2}{n^2}\right\} \qquad n = 1, 2, \ldots \qquad (7.38)$$

This coincides exactly with the results obtained from the Bohr theory. The energy depends only on the *principal quantum number n*. For a fixed value of n, the orbital and magnetic quantum numbers are restricted as

$$\ell = 0, 1, 2, \ldots, n - 1$$

$$m_\ell = 0, \pm 1, \pm 2, \ldots, \pm \ell \qquad (7.39)$$

All states with the same principal quantum number n form a *shell*, identified by the letters $K, L, M, \ldots$ (corresponding to $n = 1, 2, 3, \ldots$). All states with the same values of n and ℓ form a *subshell*, designated by the letters $s, p, d, f, \ldots$ (corresponding to $\ell = 0, 1, 2, \ldots$).

The wavefunctions for an electron in hydrogen or a hydrogen-like ion

$$\psi(r, \theta, \phi) = R_{n,\ell}(r)\,Y_\ell{}^{m_\ell}(\theta, \phi)$$

depend on the three quantum numbers n, ℓ, and m_ℓ, and are products of spherical harmonics multiplied by radial wavefunctions $R_{n,\ell}(r)$. The *pseudo-wavefunction* $g(r) = rR_{n,\ell}(r)$ is analogous to the wavefunction $\psi(x)$ in one dimension; the intensity of $g(r)$ gives the *radial probability density*

$$P(r) = |g(r)|^2 = r^2|R_{n,\ell}(r)|^2 \qquad (7.44)$$

$P(r)$ dr is the probability that the electron will be found at a distance between r and $r + dr$ from the nucleus. The *most probable distance* is the one that maximizes $P(r)$ and generally differs from the *average distance* $\langle r \rangle$ calculated as

$$\langle r \rangle = \int_0^\infty rP(r) \; dr \tag{7.46}$$

The most probable values are found to coincide with the radii of the allowed orbits in the Bohr theory.

SUGGESTIONS FOR FURTHER READING

1. For more on the use of angular momentum to simplify the three-dimensional Schrödinger equation in central force applications, see Chapter 11 of *An Introduction to Quantum Physics*, by A. P. French and Edwin F. Taylor, New York: W. W. Norton and Company, Inc., 1978.
2. A discussion of the radial probability density and its use for the lowest states of hydrogen is contained in Chapter 7 of *Modern Physics*, 2nd ed. by Kenneth Krane, New York: John Wiley and Sons, Inc., 1996.
3. Illustrations in perspective of the electron "cloud" $|\psi(\mathbf{r})|^2$ for various hydrogen-like states may be found in *Quantum Physics of Atoms, Molecules, Solids, Nuclei, and Particles*, 2nd ed. by R. Eisberg and R. Resnick, New York: John Wiley and Sons, Inc., 1985.

QUESTIONS

1. Why are three quantum numbers needed to describe the state of a one-electron atom?
2. Compare the Bohr theory with the Schrödinger treatment of the hydrogen atom. Comment specifically on the total energy and orbital angular momentum.
3. How do we know whether a given $2p$ electron in an atom has $m_\ell = 0, +1$, or -1? What value of m_ℓ characterizes a directed orbital such as $[\psi_{2p}]_x$ of Equation 7.50?
4. For atomic s states, the probability density $|\psi|^2$ is largest at the origin, yet the probability for finding the electron a distance r from the nucleus, given by $P(r)$, goes to zero with r. Explain.
5. For the electron in the ground state of hydrogen—as with many other quantum systems—the kinetic and potential energies are fuzzy observables, but their sum, the total energy, is sharp. Explain how this can be so.
6. Discuss the relationship between space quantization and Schrödinger's equation. If the latter were invalid, would space quantization still hold?

PROBLEMS

7.1 Particle in a Three-Dimensional Box

1. A particle of mass m moves in a three-dimensional box with edge lengths L_1, L_2, and L_3. Find the energies of the six lowest states if $L_1 = L$, $L_2 = 2L$, and $L_3 = 2L$. Which of these are degenerate?
2. An electron moves in a cube whose sides have a length of 0.2 nm. Find values for the energy of (a) the ground state and (b) the first excited state of the electron.
3. A particle of mass m moves in a three-dimensional box with sides L. If the particle is in the third excited level, corresponding to $n^2 = 11$, find (a) the energy of the particle, (b) the combinations of n_1, n_2, and n_3 that would give this energy, and (c) the wavefunctions for these different states.

4. A particle of mass m moves in a two-dimensional box of sides L. (a) Write expressions for the wavefunctions and energies as a function of the quantum numbers n_1 and n_2 (assuming the box is in the xy plane). (b) Find the energies of the ground state and first excited state. Is either of these states degenerate? Explain.
5. Assume that the nucleus of an atom can be regarded as a three-dimensional box of width 2×10^{-14} m. If a proton moves as a particle in this box, find (a) the ground state energy of the proton in MeV and (b) the energies of the first and second excited states. (c) What are the degeneracies of these states?
6. Obtain the stationary states for a *free* particle in three dimensions by separating the variables in Schrödinger's

equation. Do this by substituting the separable form $\Psi(\mathbf{r}, t) = \psi_1(x)\psi_2(y)\psi_3(z)\phi(t)$ into the time-dependent Schrödinger equation and dividing each term by $\Psi(\mathbf{r}, t)$. Isolate all terms depending only on x from those depending only on y, and so on, and argue that four separate equations must result, one for each of the unknown functions ψ_1, ψ_2, ψ_3, and ϕ. Solve the resulting equations. What dynamical quantities are sharp for the states you have found?

7. For a particle confined to a cubic box of dimension L, show that the normalizing factor is $A = (2/L)^{3/2}$, the *same* value for *all* the stationary states. How is this result changed if the box has edge lengths L_1, L_2, and L_3, all of which are different?

8. Consider a particle of mass m confined to a three-dimensional cube of length L so small that the particle motion is *relativistic*. Obtain an expression for the allowed particle energies in this case. Compute the ground state energy for an electron if $L = 10$ fm $(10^{-5}$ nm$)$, a typical nuclear dimension.

7.2 Central Forces and Angular Momentum

9. If an electron has an orbital angular momentum of 4.714×10^{-34} J·s, what is the orbital quantum number for this state of the electron?

10. Consider an electron for which $n = 4$, $\ell = 3$, and $m_\ell = 3$. Calculate the numerical value of (a) the orbital angular momentum and (b) the z component of the orbital angular momentum.

11. The orbital angular momentum of the Earth in its motion about the Sun is 4.83×10^{31} kg·m^2/s. Assuming it is quantized according to Equation 7.14, find (a) the value of ℓ corresponding to this angular momentum and (b) the fractional change in $|\mathbf{L}|$ as ℓ changes from ℓ to $\ell + 1$.

7.5 Atomic Hydrogen and Hydrogen-Like Ions

12. The normalized ground state wavefunction for the electron in the hydrogen atom is

$$\psi(r, \theta, \phi) = \frac{1}{\sqrt{\pi}}\left(\frac{1}{a_0}\right)^{3/2} e^{-r/a_0}$$

where r is the radial coordinate of the electron and a_0 is the Bohr radius. (a) Sketch the wavefunction versus r. (b) Show that the probability of finding the electron between r and $r + dr$ is given by $|\psi(r)|^2 4\pi r^2\ dr$. (c) Sketch the probability versus r and from your sketch find the radius at which the electron is most likely to be found. (d) Show that the wavefunction as given is normalized. (e) Find the probability of locating the electron between $r_1 = a_0/2$ and $r_2 = 3a_0/2$.

13. (a) Determine the quantum numbers ℓ and m_ℓ for the He$^+$ ion in the state corresponding to $n = 3$. (b) What is the energy of this state?

14. (a) Determine the quantum numbers ℓ and m_ℓ for the Li^{2+} ion in the states for which $n = 1$ and $n = 2$. (b) What are the energies of these states?

15. **MP** *Singly Ionized Helium.* Set up the MP DESKTOP to display the states of the He$^+$ ion $(Z = 2)$. Explore the $n = 4$ shell wavefunctions of He$^+$ along the lines of Example 7.11 for H, and compare the two sets of results to see the effect of increased atomic number.

16. In obtaining the results for hydrogen-like atoms in Section 7.5, the atomic nucleus was assumed to be immobile due to its much larger mass compared with that of the electron. If this assumption is relaxed, the results remain valid if the electron mass m_e is replaced everywhere by the *reduced mass* μ of the electron–nucleus combination:

$$\mu = \frac{m_e M}{m_e + M}$$

Here M is the nuclear mass. (a) Making this replacement in Equation 7.38, show that a more general expression for the allowed energies of a one-electron atom with atomic number Z is

$$E_n = -\frac{\mu k^2 e^4}{2\hbar^2}\left\{\frac{Z^2}{n^2}\right\}$$

(b) The wavelength for the $n = 3$ to $n = 2$ transition of the hydrogen atom is 656.3 nm (visible red light). What is the wavelength of this same transition in singly ionized helium? In positronium? (Note: Positronium is an "atom" consisting of a bound positron–electron pair. A positron is a positively charged electron.)

17. Calculate the possible values of the z component of angular momentum for an electron in a d subshell.

18. Calculate the angular momentum for an electron in (a) the $4d$ state and (b) the $6f$ state of hydrogen.

19. A hydrogen atom is in the $6g$ state. (a) What is the principal quantum number? (b) What is the energy of the atom? (c) What are the values for the orbital quantum number and the magnitude of the electron's orbital angular momentum? (d) What are the possible values for the magnetic quantum number? For each value, find the corresponding z component of the electron's orbital angular momentum and the angle that the orbital angular momentum vector makes with the z axis.

20. Prove that the nth energy level of an atom has degeneracy equal to n^2.

21. For fixed electron energy, the orbital quantum number ℓ is limited to $n - 1$. Obtain this result from a semiclassical argument using the fact that the largest angular momentum describes circular orbits, where all the kinetic energy is in orbital form. For hydrogen-like atoms, $U(r) = -Zke^2/r$, and the energy in circular orbits becomes

$$E = \frac{|\mathbf{L}|^2}{2mr^2} - \frac{Zke^2}{r}$$

Quantize this relation using the rules of Equations 7.14 and 7.38, together with the Bohr result for the allowed values of r, to show that the largest integer value of ℓ consistent with the total energy is $\ell_{max} = n - 1$.

22. Suppose that a hydrogen atom is in the 2s state. Taking $r = a_0$, calculate values for (a) $\psi_{2s}(a_0)$, (b) $|\psi_{2s}(a_0)|^2$, and (c) $P_{2s}(a_0)$.

23. The radial part of the wavefunction for the hydrogen atom in the 2p state is given by

$$R_{2p}(r) = Are^{-r/2a_0}$$

where A is a constant and a_0 is the Bohr radius. Using this expression, calculate the average value of r for an electron in this state.

24. A dimensionless number that often appears in atomic physics is the *fine-structure constant* α, given by

$$\alpha = \frac{ke^2}{\hbar c}$$

where k is the Coulomb constant. (a) Obtain a numerical value for $1/\alpha$. (b) In scattering experiments, the "size" of the electron is the *classical electron radius*, $r_0 = ke^2/m_e c^2$. In terms of α, what is the ratio of the Compton wavelength, $\lambda = h/m_e c$, to the classical electron radius? (c) In terms of α, what is the ratio of the Bohr radius, a_0, to the Compton wavelength? (d) In terms of α, what is the ratio of the *Rydberg wavelength*, $1/R$, to the Bohr radius?

25. Calculate the average potential and kinetic energies for the electron in the ground state of hydrogen.

26. Compare the most probable distances of the electron from the proton in the hydrogen 2s and 2p states with the radius of the second Bohr orbit in hydrogen, $4a_0$.

27. Compute the probability that a 2s electron of hydrogen will be found inside the Bohr radius for this state, $4a_0$. Compare this with the probability of finding a 2p electron in the same region.

28. **MP** Set up the MP DESKTOP to display the wavefunctions for the $n = 3$ level of atomic hydrogen. Locate the most probable distance from the nucleus for an electron in the 3s state. Do the same for an electron in the 3p and 3d states. What does the simple Bohr theory predict for this case?

29. As shown in Example 7.9, the average distance of the electron from the proton in the hydrogen ground state is 1.5 bohrs. For this case calculate Δr, the uncertainty in distance about the average value, and compare it with the average itself. Comment on the significance of your result.

30. Calculate the uncertainty product $\Delta r \Delta p$ for the 1s electron of a hydrogen-like atom with atomic number Z. (*Hint:* Use $\langle p \rangle = 0$ by symmetry and deduce $\langle p^2 \rangle$ from the average kinetic energy, calculated as in Problem 25.)

Additional Problems

31. An electron outside a dielectric is attracted to the surface by a force $F = -A/x^2$, where x is the perpendicular distance from the electron to the surface, and A is a constant. Electrons are prevented from crossing the surface, since there are no quantum states in the dielectric for them to occupy. Assume that the surface is infinite in extent, so that the problem is effectively one-dimensional. Write the Schrödinger equation for an electron outside the surface $x > 0$. What is the appropriate boundary condition at $x = 0$? Obtain a formula for the allowed energy levels in this case. (*Hint:* compare the equation for $\psi(x)$ with that satisfied by the pseudo-wavefunction $g(r) = rR(r)$ for hydrogen-like atoms.)

32. **MP** *The Spherical Well.* The spherical well is the simplest particle trap with spherical symmetry and is the three-dimensional analog of the square well in one dimension. It is defined by a potential $U(r)$, which is zero everywhere inside a sphere, and takes a large (possibly infinite) positive value outside this sphere. Use the MP DESKTOP to find the ground state energy for an electron bound to a spherical well of radius 2.00 bohrs and height 10.0 Ry ($= 136$ eV). Is the ground state an s state? Why? Also report the most probable distance from the center of the well for this electron. Refer to Examples 5.10 and 7.10 for particulars on using the DESKTOP for such applications. (The spherical well potential—in atomic units—can be found in the file PR7-32.U3).

33. **MP** Use the MP DESKTOP to find the first excited state(s) of the spherical well described in the preceding problem. What orbital quantum numbers ℓ describe these waveforms? Display the electron clouds for the degenerate wavefunctions, and compare your findings with those of Example 7.10.

8

Atomic Structure

Chapter Outline

Much of what we have learned about the hydrogen atom with its single electron can be used directly to describe such single-electron ions as He^+ and Li^{2+}, which are hydrogen-like in their electronic structure. Multielectron atoms, however, such as neutral helium and lithium, introduce extra complications that stem from the interactions among the atomic electrons. Thus, the study of the atom inevitably involves us in the complexities of systems consisting of many interacting electrons. In this chapter we will learn some of the basic principles needed to treat such systems effectively and apply these principles to describe the physics of electrons in atoms.

Being of like charge and confined to a small space, the electrons of an atom repel one another strongly through the Coulomb force. In addition, we shall discover that the atomic electrons behave like tiny bar magnets, interacting magnetically with one another as well as with any external magnetic field applied to the atom. These magnetic properties derive in part from a new concept—*electron spin*—which will be explored at some length in this chapter.

Another new physical idea, known as the *exclusion principle*, is also presented in this chapter. This principle is extremely important in understanding the properties of multielectron atoms and the periodic table. In fact, the implications of the exclusion principle are almost as far-reaching as those of the Schrödinger equation itself.

8.1 ORBITAL MAGNETISM AND THE NORMAL ZEEMAN EFFECT

An electron orbiting the nucleus of an atom should give rise to magnetic effects, much like those arising from an electric current circulating in a wire loop. In particular, the motion of charge generates a magnetic field within the

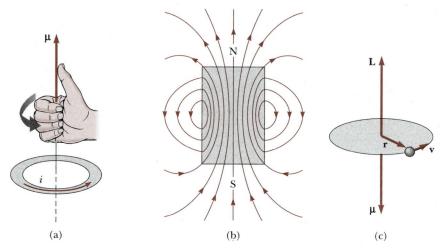

Figure 8.1   (a) The magnetic field in the space surrounding a current-carrying wire loop is that of a magnetic dipole with moment $\boldsymbol{\mu}$ perpendicular to the plane of the loop. The vector $\boldsymbol{\mu}$ points in the direction of the thumb if the fingers of the right hand are curled in the sense of the current i (right-hand rule). (b) The magnetic field in the space surrounding a bar magnet is also that of a magnetic dipole. The dipole moment vector $\boldsymbol{\mu}$ points from the south to the north pole of the magnet. (c) The magnetic moment $\boldsymbol{\mu}$ of an orbiting electron with angular momentum $\mathbf{L}$. Since the electron is negatively charged, $\boldsymbol{\mu}$ and $\mathbf{L}$ point in opposite directions.

atom, and the atom as a whole will be subject to forces and torques when it is placed in an external magnetic field. These magnetic interactions can all be described in terms of a single property of the atom—the magnetic dipole moment.

To calculate the magnetic moment of an orbiting charge, we reason by analogy with a current-carrying loop of wire. The moment $\boldsymbol{\mu}$ of such a loop has magnitude $|\boldsymbol{\mu}| = iA$, where i is the current and A is the area bounded by the loop. The direction of this moment is perpendicular to the plane of the loop, and its sense is given by a right-hand rule, as shown in Figure 8.1a. This characterization of a current loop as a magnetic dipole implies that its magnetic behavior is similar to that of a bar magnet with its N-S axis directed along $\boldsymbol{\mu}$ (Fig. 8.1b).

For a circulating charge q, the (time-averaged) current is simply q/T where T is the orbital period. Furthermore, A/T is just the area swept out per unit time and equals the magnitude of the angular momentum $|\mathbf{L}|$ of the orbiting charge divided by twice the particle mass m.[1] This relation is easily verified for circular orbits, where $|\mathbf{L}| = mvr$, $v = 2\pi r/T$, and $A = \pi r^2$, so that

$$|\mathbf{L}| = m\left(\frac{2\pi r}{T}\right) r = 2m\left(\frac{\pi r^2}{T}\right) = 2m\left(\frac{A}{T}\right) \qquad \text{or} \qquad \frac{A}{T} = \frac{|\mathbf{L}|}{2m}$$

---

[1] This is one of *Kepler's laws* of planetary motion, later shown by Newton to be a consequence of any central force.

The same result holds for orbital motion of any kind (see Problem 2), so that $|\boldsymbol{\mu}| = iA$ becomes

$$\boldsymbol{\mu} = \frac{q}{2m}\mathbf{L}$$

(8.1)

Magnetic moment of an orbiting charge

for the magnetic moment of an orbiting charge q. Since $\mathbf{L}$ is perpendicular to the orbital plane, so too is $\boldsymbol{\mu}$. You may verify that the sense of the vector described by Equation 8.1 is consistent with that expected from the right-hand rule. Thus, the magnetic moment vector is directed along the angular momentum vector, and its magnitude is fixed by the proportionality constant $q/2m$, called the **gyromagnetic ratio.** For electrons, $q = -e$ so $\boldsymbol{\mu}$ actually is opposite the direction of $\mathbf{L}$ (Fig. 8.1c).

On the atomic scale, the elemental unit of angular momentum is $\hbar$. It follows that the natural unit for atomic moments is the quantity $e\hbar/2m_e$, called the **Bohr magneton** and designated by the symbol μ_B. Its value in SI units (joules/tesla) is

$$\mu_B = \frac{e\hbar}{2m_e} = 9.274 \times 10^{-24}\,\text{J/T}$$

(8.2)

Bohr magneton

Since $\boldsymbol{\mu}$ is proportional to $\mathbf{L}$, the orbital magnetic moment is subject to space quantization, as illustrated in Figure 8.2. In particular, the z component of the orbital magnetic moment is fixed by the value of the magnetic quantum number m_ℓ as

$$\mu_z = -\frac{e}{2m_e}L_z = -\frac{e\hbar}{2m_e}m_\ell = -\mu_B m_\ell$$

(8.3)

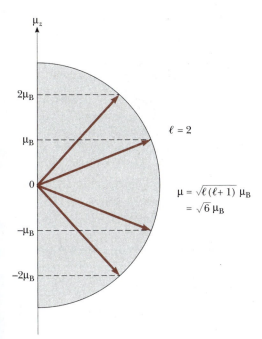

Figure 8.2 The orientations in space and z components of the orbital magnetic moment for the case $\ell = 2$. There are $2\ell + 1 = 5$ different possible orientations.

$\ell = 2$

$\mu = \sqrt{\ell(\ell+1)}\,\mu_B$
$\quad = \sqrt{6}\,\mu_B$

Just as with angular momentum, the magnetic moment vector can be visualized as precessing about the z axis to preserve this sharp value of μ_z while depicting the remaining components μ_x and μ_y as fuzzy.

The interaction of an atom with an applied magnetic field depends on the size and orientation of the atom's magnetic moment. Suppose an external field **B** is imposed in the direction of the z axis. According to classical electromagnetism, the atom experiences a torque

$$\boldsymbol{\tau} = \boldsymbol{\mu} \times \mathbf{B} \tag{8.4}$$

that tends to align its moment with the applied field. Instead of aligning itself with **B**, however, the moment actually precesses around the field direction! This unexpected precession arises because $\boldsymbol{\mu}$ is proportional to the angular momentum **L**. The motion is analogous to that of a spinning top precessing in the Earth's gravitational field: the gravitational torque acting to tip it over instead results in precession because of the angular momentum possessed by the spinning top. Returning to the atomic case, since $\boldsymbol{\tau} = d\mathbf{L}/dt$, we see from the torque equation 8.4 that the *change* in angular momentum $d\mathbf{L}$ is always perpendicular to both **L** and **B**. Figure 8.3 depicts the motion (precession) that results. For atoms in a magnetic field this is known as **Larmor precession.**

From the geometry of Figure 8.3, we see that in a time dt the precession angle increases by $d\phi$, where

$$L \sin \theta \cdot d\phi = |d\mathbf{L}|$$

But Equations 8.1 and 8.4 can be combined to give

$$|d\mathbf{L}| = |\boldsymbol{\tau}|\, dt = \left| \frac{q}{2m_e} LB \sin \theta \right| dt$$

For electrons we take $q = -e$ and the frequency of precession, or **Larmor frequency** ω_L, becomes

$$\omega_L = \frac{d\phi}{dt} = \frac{1}{L \sin \theta} \frac{|d\mathbf{L}|}{dt} = \frac{e}{2m_e} B \tag{8.5}$$

It is useful to introduce the quantum of energy $\hbar \omega_L$ associated with the Larmor frequency ω_L. This energy is related to the work required to reorient the atomic moment against the torque of the applied field. Remembering that the work of a torque $\boldsymbol{\tau}$ to produce an angular displacement $d\theta$ is $dW = \tau\, d\theta$, we have from Equation 8.4

$$dW = -\mu B \sin \theta\, d\theta = d(\mu B \cos \theta) = d(\boldsymbol{\mu} \cdot \mathbf{B})$$

The minus sign signifies that the external torque must *oppose* that produced by the magnetic field **B**. The work done is stored as orientational potential energy of the dipole in the field. Writing $dW = -dU$, we identify the **magnetic potential energy** U as

$$U = -\boldsymbol{\mu} \cdot \mathbf{B} \tag{8.6}$$

Equation 8.6 expresses the fact that the energy of a magnetic dipole in an external magnetic field **B** depends on its orientation in this field. The magnetic

A magnetic moment precesses in a magnetic field

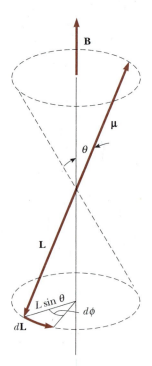

Figure 8.3 Larmor precession of the orbital moment $\boldsymbol{\mu}$ in an applied magnetic field **B**. Since $\boldsymbol{\mu}$ is proportional to **L**, the torque of the applied field causes the moment vector $\boldsymbol{\mu}$ to precess around the direction of **B** with frequency $\omega_L = eB/2m_e$.

The energy of a magnetic moment depends on its orientation in a magnetic field

energy is minimal when $\boldsymbol{\mu}$ and $\mathbf{B}$ are aligned; therefore, this alignment is the preferred orientation. Since the possible orientations for $\boldsymbol{\mu}$ are restricted by space quantization, the magnetic energy is quantized accordingly. Taking the z axis along $\mathbf{B}$, and combining Equations 8.1, 8.3, and 8.6, we find

$$U = \frac{e}{2m_e} \mathbf{L} \cdot \mathbf{B} = \frac{eB}{2m_e} L_z = \hbar\omega_L m_\ell \qquad (8.7)$$

From Equation 8.7 we see that the magnetic energy of an atomic electron depends on the magnetic quantum number m_ℓ and, hence, is quantized. The total energy of this electron is the sum of its magnetic energy U plus whatever energy it had in the absence of an applied field, say E_0. Thus,

$$E = E_0 + \hbar\omega_L m_\ell \qquad (8.8)$$

For atomic hydrogen, E_0 depends only on the principal quantum number n; in more complex atoms the atomic energy also varies according to the subshell label ℓ, as discussed further in Section 8.5.

Unlike energies, the wavefunctions of atomic electrons are unaffected by the application of a magnetic field. This somewhat surprising result can be partly understood by recognizing that, according to classical physics, the only effect of the field is to cause (Larmor) precession around the direction of $\mathbf{B}$. For atomic electrons this translates into precession of $\mathbf{L}$ about the z axis. However, such a precession is already implicit in our semiclassical picture of electron orbits in the absence of external fields, as required by the sharpness of L_z while L_x and L_y remain fuzzy. From this viewpoint, the introduction of an applied magnetic field merely transforms this *virtual precession*[2] into a real one at the Larmor frequency!

EXAMPLE 8.1   Magnetic Energy of the Electron in Hydrogen

Calculate the magnetic energy and Larmor frequency for an electron in the $n = 2$ state of hydrogen, assuming the atom is in a magnetic field of strength $B = 1.00$ T.

Solution   Taking the z axis along $\mathbf{B}$, we calculate the magnetic energy from Equation 8.7 as

$$U = \frac{eB}{2m_e} L_z = \frac{e\hbar}{2m_e} Bm_\ell = \hbar\omega_L m_\ell$$

For a 1.00 T field, the energy quantum $\hbar\omega_L$ has the value

$$\hbar\omega_L = \frac{e\hbar}{2m_e} B = \mu_B B = (9.27 \times 10^{-24} \text{ J/T})(1.00 \text{ T})$$

$$= 9.27 \times 10^{-24} \text{ J} = 5.79 \times 10^{-5} \text{ eV}$$

With $n = 2$, ℓ can be 0 or 1, and m_ℓ is 0 (twice) and ± 1. Thus, the magnetic energy U can be 0, $+\hbar\omega_L$, or $-\hbar\omega_L$. In such applications, the energy quantum $\hbar\omega_L$ is called the **Zeeman energy.** The Zeeman energy divided by $\hbar$ is the Larmor frequency:

$$\omega_L = \frac{5.79 \times 10^{-5} \text{ eV}}{6.58 \times 10^{-16} \text{ eV} \cdot \text{s}} = 8.80 \times 10^{10} \text{ rad/s}$$

---

[2] In zero magnetic field, the precession of the classical vector may be termed virtual (not real) since, even though the same value may not be obtained for L_x (or L_y) in successive measurements, the average value $\langle L_x \rangle$ (or $\langle L_y \rangle$) does not change over time. With $\mathbf{B}$ nonzero, however, it can be shown that $(d^2/dt^2)\langle L_x \rangle = -\omega_L^2 \langle L_x \rangle$ (and similarly for $\langle L_y \rangle$), indicating that $\langle L_x \rangle$ (and $\langle L_y \rangle$) oscillates at the Larmor frequency ω_L.

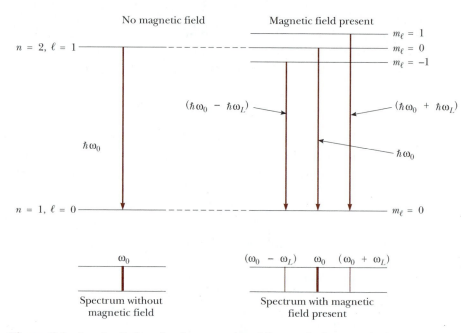

Figure 8.4   Level splittings for the ground and first excited states of a hydrogen atom immersed in a magnetic field **B**. An electron in one of the excited states decays to the ground state with the emission of a photon, giving rise to emission lines at ω_0, $\omega_0 + \omega_L$, and $\omega_0 - \omega_L$. This is the normal Zeeman effect. When **B** = 0, only the line at ω_0 is observed.

Evidence for the existence of atomic moments is the appearance of extra lines in the spectrum of an atom that is placed in a magnetic field. Consider a hydrogen atom in its first excited ($n = 2$) state. For $n = 2$, ℓ can have values 0 and 1. The magnetic field has no effect on the state for which $\ell = 0$, since then $m_\ell = 0$. For $\ell = 1$, however, m_ℓ can take values of 1, 0, and -1, and the first excited level is split into three levels by the magnetic field (Figure 8.4). The original (Lyman) emission line is replaced by the three lines depicted in Figure 8.4. The central line appears at the same frequency ω_0 as it would without a magnetic field. This is flanked on both sides by new lines at frequencies $\omega_0 \pm \omega_L$. Thus, the magnetic field splits the original emission line into three lines. Since ω_L is proportional to B, the amount of splitting increases linearly with the strength of the applied field. This effect of spectral line splitting by a magnetic field is known as the **Zeeman effect** after its discoverer, Pieter Zeeman.

The normal Zeeman effect

Zeeman spectra of atoms excited to higher states should be more complex, since many more level splittings are involved. For electrons excited to the $n = 3$ state of hydrogen, the expected Zeeman lines and the atomic transitions that give rise to them are shown in Figure 8.5. Accompanying each hydrogen line are anywhere from two to six *satellites* at frequencies removed from the original by multiples of the Larmor frequency. But the observed Zeeman spectrum is not this complicated, owing to **selection rules** that limit the transitions to those for which ℓ changes by 1 and m_ℓ changes by 0, +1, or -1. The result is that satellites appear at the Larmor frequency *only* and not at multiples of

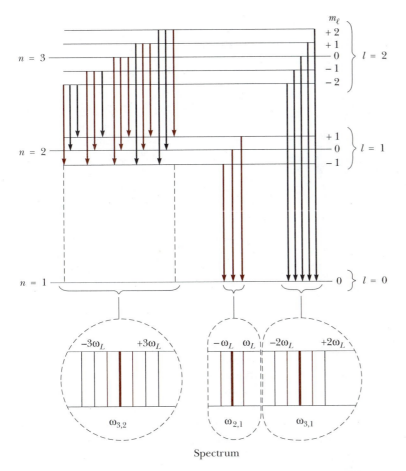

Figure 8.5   Zeeman spectral lines and the underlying atomic transitions that give rise to them for an electron excited to the $n = 3$ state of hydrogen. Because of *selection rules*, only the transitions drawn in color actually occur.

this frequency. The selection rules express conservation of angular momentum for the *system*, taking into account the angular momentum of the emitted photon. (See Section 7.5.)

Finally, even the splitting of an emission line into a triplet of equally spaced lines as predicted here, called the *normal Zeeman effect*, frequently is not observed. More commonly, splittings into four, six, or even more unequally spaced lines are seen. This is the *anomalous Zeeman effect*, which has its roots in the existence of electron spin.

8.2   THE SPINNING ELECTRON

The anomalous Zeeman splittings are only one of several phenomena not explained by the magnetic interactions discussed thus far. Another is the observed doubling of many spectral lines referred to as **fine structure.** Both effects are

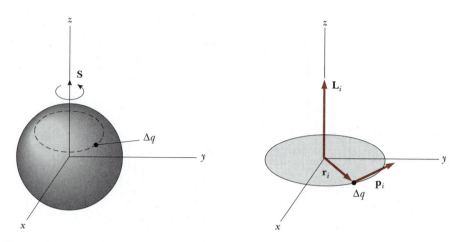

Figure 8.6 A spinning charge q may be viewed as a collection of charge elements Δq orbiting a fixed line, the axis of rotation. The magnetic moments accompanying these orbiting charge elements are summed to give the total magnetic moment of rotation, or spin moment, of the charge q.

attributed to the existence of a new magnetic moment—the spin moment—that arises from the electron spinning on its axis.

We have seen that the orbital motion of charge gives rise to magnetic effects that can be described in terms of the orbital moment $\boldsymbol{\mu}$ of Equation 8.1. Similarly, a charged object in rotation produces magnetic effects related to the spin moment $\boldsymbol{\mu}_s$. The spin moment is found by noting that a rotating body of charge can be viewed as a collection of charge elements Δq with mass Δm all rotating in circular orbits about a fixed line, the axis of rotation (Fig. 8.6). To each of these we should apply Equation 8.1 with $\mathbf{L}$ replaced by $\mathbf{L}_i$, the orbital angular momentum of the ith charge element. If the charge-to-mass ratio is uniform throughout the body, then $\Delta q/\Delta m$ is the ratio of total charge q to total mass m_e and we get for the spin moment

A rotating charge gives rise to a spin magnetic moment

$$\boldsymbol{\mu}_s = \frac{q}{2m_e} \sum \mathbf{L}_i = \frac{q}{2m_e} \mathbf{S} \tag{8.9}$$

where $\mathbf{S}$, the *spin* angular momentum, is the total angular momentum of rotation. The spin angular momentum $\mathbf{S}$ points along the axis of rotation according to a right-hand rule, as shown in Figure 8.6; its magnitude depends on the size and shape of the object, as well as its speed of rotation. If the charge-to-mass ratio is not uniform, the gyromagnetic ratio in Equation 8.9, $q/2m_e$, must be multiplied by a dimensionless constant, the **g-factor,** whose value reflects the detailed charge-to-mass distribution within the body. Note that g-factors different from unity are not expected classically since a given quantity of charge normally is associated with a proportionate amount of material; that is, the charge-to-mass ratio should be constant.[3]

---

[3] It is only fair to caution the reader at this point not to take the classical view of an electron as a tiny charged ball spinning on its axis too literally. Although such a picture is useful in first introducing and visualizing electron spin, it is not technically correct. Several shortcomings of the classical picture are discussed in detail on pp. 309 and 310.

Otto Stern was one of the finest experimental physicists of the twentieth century. Born and educated in Germany (PhD in physical chemistry in 1912), he at first worked with Einstein on theoretical issues in molecular theory, in particular applying the new quantum ideas to theories of the specific heat of solids. From about 1920, Stern devoted himself to his real life's work, the development of the molecular beam method, which enabled him to investigate the properties of free or isolated atoms and culminated in a Nobel prize in 1945. In this method, a thin stream of atoms is introduced into a high-vacuum chamber where the atoms are free, and hence the properties of individual atoms may be investigated by applying external fields or by some other technique.

Stern first used this method to confirm that silver atoms obey the Maxwell speed distribution. Shortly after, in a series of elegant and difficult experiments with Walter Gerlach, Stern showed that silver atoms obey space quantization and succeeded in measuring the magnetic moment of the silver atom. In the period from 1923 to 1933, Stern directed a remarkably productive molecular beam laboratory at the University of Hamburg. With his students and co-workers he directly demonstrated the wave nature of he-

BIOGRAPHY

OTTO STERN
(1888–1969)

lium atoms, and measured the magnetic moments of many atoms. Finally, with a great deal of effort, he succeeded in measuring the very small magnetic moments of the proton and deuteron. For these last important fundamental measurements he was awarded the Nobel prize. In connection with the measurement of the proton's magnetic moment, an interesting story is told by Victor Weisskopf, which should gladden the hearts of experimentalists everywhere:

"There was a seminar held by the theoretical group in Göttingen, and Stern came down and gave a talk on the measurements he was about to

finish of the magnetic moment of the proton. He explained his apparatus, but he did not tell us the result. He took a piece of paper and went to each of us saying, 'What is your prediction of the magnetic moment of the proton?' Every theoretician from Max Born down to Victor Weisskopf said, 'Well, of course, the great thing about the Dirac equation is that it predicts a magnetic moment of one Bohr magneton for a particle of spin one-half!' Then he asked us to write down the prediction; everybody wrote 'one magneton.' Then, two months later, he came to give again a talk about the finished experiment, which showed that the value was 2.8 magnetons. He then projected the paper with our predictions on the screen. It was a sobering experience."*

In protest over Nazi dismissals of some of his closest co-workers, Stern resigned his post in Hamburg and came to the Carnegie Institute of Technology in the United States in 1933. Here he worked on molecular beam research until his retirement in 1946.

*From Victor F. Weisskopf, *Physics in the Twentieth Century; Selected Essays: My Life as a Physicist,* Cambridge, Mass., The MIT Press, 1972.

(Photo courtesy of University of Hamburg, Germany)

The existence of a spin moment for the electron was first demonstrated in 1921 in a classic experiment performed by Otto Stern and Walter Gerlach. Electron spin was unknown at that time; the Stern–Gerlach experiment was originally conceived to demonstrate the space quantization associated with orbiting electrons in atoms. In their experiment, a beam of neutral silver atoms was passed through a *nonuniform* magnetic field created in the gap between the pole faces of a large magnet. The beam was then detected by being deposited on a glass collector plate (Fig. 8.7). A nonuniform field exerts a force on any magnetic moment, so that each atom is deflected in the gap by an amount governed by the orientation of its moment with respect to the direction of

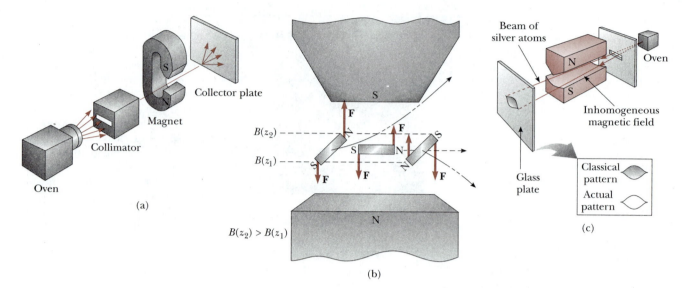

Figure 8.7 The Stern–Gerlach experiment to detect space quantization. (a) A beam of silver atoms is passed through a nonuniform magnetic field and detected on a collector plate. (b) The atoms, with their magnetic moment, are equivalent to tiny bar magnets. In a nonuniform field, each atomic magnet experiences a net force that depends on its orientation. (c) If any moment orientation were possible, a continuous fanning of the beam would be seen at the collector. For space quantization, the fanning is replaced by a set of discrete lines, one for each distinct moment orientation present in the beam.

inhomogeneity (the z axis), as illustrated in Figure 8.7b. If the moment directions are restricted by space quantization as in Figure 8.2, so too are the deflections. Thus, the atomic beam should split into a number of discrete components, one for each distinct moment orientation present in the beam. This is contrary to the classical expectation that any moment orientation (and hence any beam deflection) is possible, and all would combine to produce a continuous fanning of the atomic beam (Fig. 8.7c).

The Stern–Gerlach experiment produced a staggering result: the silver atomic beam was clearly split—but into only *two* components, not the odd number $(2\ell + 1)$ expected from the space quantization of orbital moments! This is all the more remarkable when we realize that silver atoms in their ground state have no orbital angular momentum ($\ell = 0$), since the outermost electron in silver normally would be in an s state. The result was so surprising that the experiment was repeated in 1927 by T. E. Phipps and J. B. Taylor with a beam of hydrogen atoms replacing silver, thereby eliminating any uncertainties arising from the use of the more complex silver atoms. The results, however, were unchanged. From these experiments, we are forced to conclude that there is some contribution to the atomic magnetic moment other than the orbital motion of electrons, and that this moment is subject to space quantization.

Our present understanding of the situation dates to the 1925 paper of Samuel Goudsmit and George Uhlenbeck, then graduate students at the University of Leiden. Goudsmit and Uhlenbeck believed that the unknown moment had its origin in the spinning motion of atomic electrons, with the spin angular momentum obeying the same quantization rules as orbital angular momentum.

The magnetic moment seen in the Stern–Gerlach experiment is attributed to the spin of the outermost electron in silver. Since all allowed orientations of the spin moment should be represented in the atomic beam, the observed splitting presents a dramatic confirmation of space quantization as applied to electron spin, with the number of components $(2s + 1)$ indicating the value of the **spin quantum number** s.

The spin moment suggests that the electron can be viewed as a charge in rotation, although the classical picture of a spinning body of charge must be adjusted to accommodate the wave properties of matter. The resulting *semi-classical model* of electron spin can be summarized as follows:

- The spin quantum number s for the electron is $\frac{1}{2}$! This value is dictated by the observation that an atomic beam passing through the Stern–Gerlach magnet is split into just two components $(= 2s + 1)$. Accordingly, there are exactly two orientations possible for the spin axis, described as the up and down spin states of the electron. This is space quantization again, according to the quantization rules for angular momentum[4] as applied to a spin of $\frac{1}{2}$:

$$S_z = m_s \hbar \qquad \text{where } m_s = \tfrac{1}{2} \text{ or } -\tfrac{1}{2} \qquad (8.10)$$

The two values $\pm \hbar/2$ for S_z correspond to the two possible orientations for **S** shown in Figure 8.8. The value $m_s = +\frac{1}{2}$ refers to the up spin case, some-

Properties of electron spin

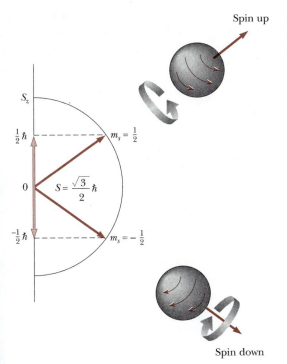

Spin up

Spin down

Figure 8.8   The spin angular momentum also exhibits space quantization. This figure shows the two allowed orientations of the spin vector **S** for a spin $\frac{1}{2}$ particle, such as the electron.

[4] For integer angular momentum quantum numbers, the z-component is quantized as $m_s = 0$, $\pm 1, \ldots \pm s$, which can also be written as $m_s = s, s - 1, \ldots -s$. For $s = \frac{1}{2}$, the latter implies $m_s = \frac{1}{2}$ or $-\frac{1}{2}$.

times designated with an up arrow (↑) or simply a plus sign (+). Likewise, $m_s = -\frac{1}{2}$ is the down spin case (↓) or (−). The fact that s has a nonintegral value suggests that spin is not merely another manifestation of orbital motion, as the classical picture implies.

- The magnitude of the spin angular momentum is

**The spin angular momentum
of an electron**

$$|\mathbf{S}| = \sqrt{s(s+1)}\hbar = \frac{\sqrt{3}}{2}\hbar \qquad (8.11)$$

and *never changes!* This angular momentum of rotation cannot be changed in any way, but is an intrinsic property of the electron, like its mass or charge. The notion that $|\mathbf{S}|$ is fixed contradicts classical laws, where a rotating charge would be slowed down by the application of a magnetic field owing to the Faraday emf that accompanies the changing field (the diamagnetic effect). Furthermore, if the electron were viewed simply as a spinning ball with angular momentum $\hbar\sqrt{3}/2$ subject to classical laws, parts of the ball near its surface would be rotating with velocities in excess of the speed of light![5] All of this is taken to mean that the classical picture of the electron as a charge in rotation must not be pressed too far; ultimately *the spinning electron is a quantum entity defying any simple classical description.*

- The spin moment is given by Equation 8.9 with a *g-factor of 2;* that is, the moment is twice as large as would be expected for a body with spin angular momentum given by Equation 8.10. The value $g = 2$ is required by the amount of beam deflection produced by the Stern–Gerlach magnet; the larger the magnetic moment, the greater will be the deflection of the atomic beam. As already mentioned, any g-factor other than unity implies a nonuniform charge-to-mass ratio in the classical picture and is not to be expected. Thus, the g-factor signals yet another limitation of the classical model (but see Problem 8). The correct g-factor of 2 is predicted by the relativistic quantum theory of the electron put forth by Paul Dirac in 1929.[6]

With the recognition of electron spin we see that an additional quantum number, m_s, is needed to specify the internal, or spin, state of an electron. Thus, the state of an electron in hydrogen must be described by the four quantum numbers n, ℓ, m_ℓ, and m_s. Furthermore, the total magnetic moment now has orbital and spin contributions:

**The total magnetic moment
of an electron**

$$\boldsymbol{\mu} = \boldsymbol{\mu}_0 + \boldsymbol{\mu}_s = \frac{-e}{2m_e}\{\mathbf{L} + g\mathbf{S}\} \qquad (8.12)$$

---

[5] This follows from the extremely small size of the electron. The exact size of the electron is unknown, but an upper limit of 10^{-6} Å is deduced from experiments in which electrons are scattered from other electrons. According to some current theories, the electron may be a true point object, that is, a particle with zero size!

[6] The g-factor for the electron is not exactly 2. The best value to date is $g = 2.00232$. The discrepancy between Dirac's predicted value and the observed value is attributed to the electron interacting with the "vacuum." Such effects are the subject of quantum electrodynamics, developed by Richard Feynman in the early 1950s.

Because of the electron g-factor, the total moment $\boldsymbol{\mu}$ is no longer in the same direction as the total (orbital plus spin) angular momentum $\mathbf{J} = \mathbf{L} + \mathbf{S}$. The component of the total moment $\boldsymbol{\mu}$ along $\mathbf{J}$ is sometimes referred to as the *effective moment*. When the magnetic field $\mathbf{B}$ applied to an atom is weak, it is the effective moment that determines the magnetic energy of atomic electrons according to Equation 8.6. As we shall discover in Section 8.3, the number of possible orientations for $\mathbf{J}$ and, hence, for the effective moment is even, leading to the even number of spectral lines seen in the anomalous Zeeman effect.

EXAMPLE 8.2 Semiclassical Model for Electron Spin

Calculate the angles between the z axis and the spin angular momentum $\mathbf{S}$ of the electron in the up and down spin states. How should we portray the fuzziness inherent in the x and y components of the spin angular momentum?

Solution For the electron, the magnitude of the spin angular momentum is $|\mathbf{S}| = \hbar\sqrt{3}/2$, and the z component of spin is $S_z = \pm\hbar/2$. Thus, the spin vector $\mathbf{S}$ is inclined from the z axis at angles given by

$$\cos\theta = \frac{S_z}{|\mathbf{S}|} = \pm\frac{1}{\sqrt{3}}$$

For the up spin state, we take the plus sign and get $\cos\theta = 0.577$ or $\theta = 54.7°$. The down spin orientation is described by the minus sign and gives $\cos\theta = -0.577$, or $\theta = 125.3°$. Since the axis of rotation coincides with the direction of the spin vector, these are the angles the rotation axis makes with the z axis.

While S_z is sharp in either the up or down spin orientation, both S_x and S_y are fuzzy. This fuzziness may be depicted by allowing the spin vector to precess about the z axis, as we did for the orbital angular momentum in Chapter 7.

Exercise 1 As it happens, the photon is a spin one particle, that is, $s = 1$ for the photon. Calculate the possible angles between the z axis and the spin vector of the photon.

Answer 45°, 90°, and 135°

EXAMPLE 8.3 Zeeman Spectrum of Hydrogen Including Spin

Examine the Zeeman spectrum produced by hydrogen atoms initially in the $n = 2$ state when electron spin is taken into account, assuming the atoms to be in a magnetic field of strength $B = 1.00$ T.

Solution The electron energies now have a magnetic contribution from both the orbital and spin motions. Orienting the z axis along the direction of $\mathbf{B}$, we calculate the magnetic energy from Equations 8.6 and 8.12:

$$U = -\boldsymbol{\mu}\cdot\mathbf{B} = \frac{e}{2m_e}B\{L_z + gS_z\} = \frac{e\hbar}{2m_e}B(m_\ell + gm_s)$$

The energy $(e\hbar/2m_e)B$ is the Zeeman energy $\mu_B B$ or $\hbar\omega_L$; its value in this example is

$$\mu_B B = (9.27 \times 10^{-24}\text{ J/T})(1.00\text{ T}) = 9.27 \times 10^{-24}\text{ J}$$
$$= 5.79 \times 10^{-5}\text{ eV}$$

For the $n = 2$ state of hydrogen, the shell energy is $E_2 = -(13.6\text{ eV})/2^2 = -3.40$ eV. Since m_ℓ takes the values 0 (twice) and ±1, there is an orbital contribution to the magnetic energy $U_0 = m_\ell\hbar\omega_L$ that introduces new levels at $E_2 \pm \hbar\omega_L$, as discussed in Example 8.1. The presence of electron spin splits each of these into a pair of levels, the additional (spin) contribution to the energy being $U_s = (gm_s)\hbar\omega_L$ (Fig. 8.9). Since $g = 2$ and m_s is $\pm\frac{1}{2}$ for the electron, the spin energy in the field $|U_s|$ is again the Zeeman energy $\hbar\omega_L$. Thus, an electron in this shell can have any one of the energies

$$E_2, \quad E_2 \pm \hbar\omega_L, \quad E_2 \pm 2\hbar\omega_L$$

In making a downward transition to the $n = 1$ shell with energy $E_1 = -13.6$ eV, the final state of the electron may have energy $E_1 + \hbar\omega_L$ or $E_1 - \hbar\omega_L$, depending on the orientation of its spin in the applied field. Thus, the energy of transition may be any one of the following possibilities:

$$\Delta E_{2,1}, \quad \Delta E_{2,1} \pm \hbar\omega_L, \quad \Delta E_{2,1} \pm 2\hbar\omega_L, \quad \Delta E_{2,1} \pm 3\hbar\omega_L$$

Photons emitted with these energies have frequencies

$$\omega_{2,1}, \quad \omega_{2,1} \pm \omega_L, \quad \omega_{2,1} \pm 2\omega_L, \quad \omega_{2,1} \pm 3\omega_L$$

Thus, the spectrum would be expected to consist of the original line at $\omega_{2,1}$ flanked on both sides by satellite lines separated from the original by the Larmor frequency, twice the Larmor frequency, and three times this frequency. Notice that the lines at $\omega_{2,1} \pm 2\omega_L$ and $\omega_{2,1} \pm 3\omega_L$ appear solely because of electron spin.

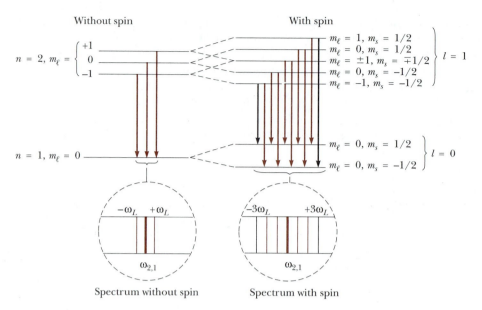

Figure 8.9   (Example 8.3) Predicted Zeeman pattern and underlying atomic transitions for an electron excited to the $n = 2$ state of hydrogen, when electron spin is taken into account. Again, selection rules prohibit all but the colored transitions. Because of the neglect of the spin–orbit interaction, the effect shown here (called the Paschen–Back effect) is observed only in very intense applied magnetic fields.

Again, however, the observed pattern is not the predicted one. Selection rules inhibit transitions unless $m_\ell + m_s$ changes by 0, +1, or −1. This has the effect of eliminating the satellites at $\omega_{2,1} \pm 3\omega_L$. Furthermore, the spin moment and the orbital moment of the electron interact with *each other*, a circumstance not recognized in our calculation. Only when this spin–orbit interaction energy is small compared with the Zeeman energy $\hbar\omega_L$ do we observe the spectral lines predicted here. This is the case for the *Paschen–Back effect*, in which the magnetic field applied to the atom is intense enough to make $\hbar\omega_L$ the dominant energy.

8.3   THE SPIN–ORBIT INTERACTION AND OTHER MAGNETIC EFFECTS

The existence of both spin and orbital moments for the electron inevitably leads to their mutual interaction. This so-called **spin–orbit interaction** is best understood from the vantage point of the orbiting electron, which "sees" the atomic nucleus circling it (Fig. 8.10). The apparent orbital motion of the nucleus generates a magnetic field at the electron site, and the electron spin moment acquires magnetic energy in this field according to Equation 8.6. This can be thought of as an internal Zeeman effect, with **B** arising from the orbital motion of the electron itself. The electron has a higher energy when its spin is up, or aligned with **B**, than when its spin is down, or aligned opposite to **B** (Fig. 8.10b).

The energy difference between the two spin orientations is responsible for the *fine structure doubling* of many atomic spectral lines. For example, the

$2p \rightarrow 1s$ transition in hydrogen is split into two lines because the $2p$ level is actually a spin doublet with a level spacing of about 5×10^{-5} eV (Fig. 8.11), while the $1s$ level remains unsplit (there is no orbital field in a state with zero orbital angular momentum). Similarly, the spin–orbit doubling of the sodium $3p$ level gives rise to the well-known sodium D lines to be discussed in Example 8.4.

The coupling of spin and orbital moments implies that *neither orbital angular momentum nor spin angular momentum is conserved separately*. But total angular momentum $\mathbf{J} = \mathbf{L} + \mathbf{S}$ is conserved, so long as no external torques are present. Consequently, quantum states exist for which $|\mathbf{J}|$ and J_z are sharp observables quantized in the manner we have come to expect for angular momentum:

$$|\mathbf{J}| = \sqrt{j(j+1)}\hbar \qquad (8.13)$$

$$J_z = m_j\hbar \qquad \text{with } m_j = j, j-1, \ldots, -j$$

Permissible values for the total angular momentum quantum number j are

$$j = \ell + s, \quad \ell + s - 1, \ldots, |\ell - s| \qquad (8.14)$$

in terms of the orbital (ℓ) and spin (s) quantum numbers. For an atomic electron $s = \frac{1}{2}$ and $\ell = 0, 1, 2, \ldots$, so $j = \frac{1}{2}$ (for $\ell = 0$) and $j = \ell \pm \frac{1}{2}$ (for $\ell > 0$). These results can be deduced from the vector addition model shown in Figure 8.12a. With $j = \frac{1}{2}$, there are only two possibilities for m_j, namely $m_j = \pm\frac{1}{2}$. For $j = \ell \pm \frac{1}{2}$, the number of possibilities $(2j + 1)$ for m_j becomes either 2ℓ or $2\ell + 2$. Notice that the number of m_j values is always *even* for a single electron, leading to an even number of orientations in the semiclassical

Figure 8.10 (a) *Left:* An electron with angular momentum **L** orbiting the nucleus of an atom. In the up-spin orientation shown here, the spin angular momentum **S** of the electron is essentially aligned with **L**. *Right:* From the viewpoint of the orbiting electron, the nucleus circulates as shown. (b) The apparently circulating nuclear charge is represented by the current i, and causes a magnetic field **B** at the site of the electron. In the presence of **B** the electron spin moment $\boldsymbol{\mu}_s$ acquires magnetic energy $U = -\boldsymbol{\mu}_s \cdot \mathbf{B}$. The spin moment $\boldsymbol{\mu}_s$ is opposite the spin vector **S** for the negatively charged electron. The direction of **B** is given by a right-hand rule: with the thumb pointing in the direction of the current i, the fingers give the sense in which the **B** field circulates about the wire. The magnetic energy is highest for the case shown, where **S** and **L** are "aligned."

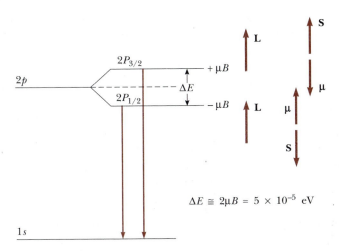

$$\Delta E \cong 2\mu B = 5 \times 10^{-5} \text{ eV}$$

Figure 8.11 The $2p$ level of hydrogen is split by the spin–orbit effect into a doublet separated by the spin–orbit energy $\Delta E = 5 \times 10^{-5}$ eV. The higher energy state is the one for which the spin angular momentum of the electron is "aligned" with its orbital angular momentum. The $1s$ level is unaffected, since no magnetic field arises for orbital motion with zero angular momentum.

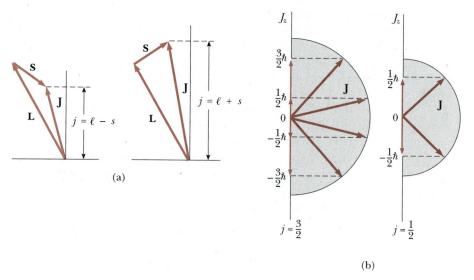

Figure 8.12 (a) A vector model for determining the total angular momentum $\mathbf{J} = \mathbf{L} + \mathbf{S}$ of a single electron. (b) The allowed orientations of the total angular momentum $\mathbf{J}$ for the states $j = \frac{3}{2}$ and $j = \frac{1}{2}$. Notice that there are now an even number of orientations possible, not the odd number familiar from the space quantization of $\mathbf{L}$ alone.

model for $\mathbf{J}$ (Fig. 8.12b), rather than the odd number predicted for the orbital angular momentum $\mathbf{L}$ alone.

A common spectroscopic notation is to use a subscript after a letter to designate the total angular momentum of an atomic electron, where the letter itself (now upper case) describes its orbital angular momentum. For example, the notation $1S_{1/2}$ describes the ground state of hydrogen, where the 1 indicates $n = 1$, the S tells us that $\ell = 0$, and the subscript $\frac{1}{2}$ denotes $j = \frac{1}{2}$. Likewise, the spectroscopic notations for the $n = 2$ states of hydrogen are $2S_{1/2}(\ell = 0, j = \frac{1}{2})$, $2P_{3/2}(\ell = 1, j = \frac{3}{2})$, and $2P_{1/2}(\ell = 1, j = \frac{1}{2})$. Again, the spin–orbit interaction splits the latter two states in energy by about 5×10^{-5} eV.

Spectroscopic notation extended to include spin

EXAMPLE 8.4 The Sodium Doublet

The famed sodium doublet arises from the spin–orbit splitting of the sodium $3p$ level, and consists of the closely spaced pair of spectral lines at wavelengths of 588.995 nm and 589.592 nm. Show on an energy level diagram the electronic transitions giving rise to these lines, labeling the participating atomic states with their proper spectroscopic designations. From the doublet spacing, deduce the magnitude of the spin–orbit energy.

Solution The outer electron in sodium is the first electron to occupy the $n = 3$ shell, and it would go into the lowest energy subshell, the $3s$ or $3S_{1/2}$ level. The next highest levels belong to the $3p$ subshell. The $2(2\ell + 1) = 6$ states of this subshell are grouped into the $3P_{1/2}$ level with

two states, and the $3P_{3/2}$ level with four states. The spin–orbit effect splits these levels by the spin–orbit energy. The outer electron, once it is excited to either of these levels by some means (such as an electric discharge in the sodium vapor lamp), returns to the $3S_{1/2}$ level with the emission of a photon. The two possible transitions $3P_{3/2} \rightarrow 3S_{1/2}$ and $3P_{1/2} \rightarrow 3S_{1/2}$ are shown in Figure 8.13. The emitted photons have nearly the same energy, but differ by the small amount ΔE representing the spin–orbit splitting of the initial levels. Since $E = hc/\lambda$ for photons, ΔE is found as

$$\Delta E = \frac{hc}{\lambda_1} - \frac{hc}{\lambda_2} = \frac{hc(\lambda_2 - \lambda_1)}{\lambda_1 \lambda_2}$$

Figure 8.13   (Example 8.4). The transitions $3P_{3/2} \rightarrow 3S_{1/2}$ and $3P_{1/2} \rightarrow 3S_{1/2}$ that give rise to the sodium doublet. The $3p$ level of sodium is split by the spin–orbit effect, but the $3s$ level is unaffected. In the sodium vapor lamp, electrons normally in the $3s$ level are excited to the $3p$ levels by an electric discharge.

For the sodium doublet, the observed wavelength difference is

$$\lambda_2 - \lambda_1 = 589.592 \text{ nm} - 588.995 \text{ nm} = 0.597 \text{ nm}$$

Using this with $hc = 1240$ eV·nm gives

$$\Delta E = \frac{(1240 \text{ eV·nm})(0.597 \text{ nm})}{(589.592 \text{ nm})(588.995 \text{ nm})} = 2.13 \times 10^{-3} \text{ eV}$$

Exercise 2   Using the spin–orbit interaction energy calculated in Example 8.4, deduce a value for the magnetic field strength at the site of the orbiting $3p$ electron in sodium.

Answer $B = 18.38$ T, a large field by laboratory standards.

8.4   EXCHANGE SYMMETRY AND THE EXCLUSION PRINCIPLE

As mentioned earlier, the existence of spin requires that the state of an atomic electron be specified with four quantum numbers. In the absence of spin–orbit effects these could be n, ℓ, m_ℓ, and m_s; if the spin–orbit interaction is taken into account, m_ℓ and m_s are replaced by j and m_j. In either case four quantum numbers are required, one for each of the four degrees of freedom possessed by a single electron.

In those systems where two or more electrons are present, we might expect to describe each electronic state by giving the appropriate set of four quantum numbers. In this connection an interesting question arises, namely, "How many electrons in an atom can have the same four quantum numbers, that is, be in the same state?" This important question was answered by Wolfgang Pauli in 1925 in a powerful statement known as the **exclusion principle.** The exclusion principle states that *no two electrons in an atom can have the same set of quantum numbers.* We should point out that if this principle were not valid, every electron would occupy the $1s$ atomic state (this being the state of lowest energy), the chemical behavior of the elements would be drastically different, and nature as we know it would not exist!

The exclusion principle follows from our belief that electrons are identical particles— that it is impossible to distinguish one electron from another, even in principle. This seemingly innocuous statement takes on added importance in view of the wave nature of matter, and has far-reaching consequences. To gain an appreciation for this point, let us consider a collision between two electrons, as shown in Figure 8.14. Figures 8.14a and 8.14b depict two distinct events, the scattering effect being much stronger in the latter where the electrons are turned through a larger angle. Each event, however, arises from the same initial condition and leads to the same outcome—both electrons are scattered and emerge at angles

B I O G R A P H Y

WOLFGANG PAULI

(1900–1958)

Wolfgang Pauli was an extremely talented Austrian theoretical physicist who made important contributions in many areas of modern physics. At the age of 21, Pauli gained public recognition with a masterful review article on relativity, which is still considered to be one of the finest and most comprehensive introductions to the subject. Other major contributions were the discovery of the exclusion principle, the explanation of the connection between particle spin and statistics, theories of relativistic quantum electrodynamics, the neutrino hypothesis, and the hypothesis of nuclear spin. An article entitled "The Fundamental Principles of Quantum Mechanics," written by Pauli in 1933 for the *Handbuch der Physik*, is widely acknowledged to be one of the best treatments of quantum physics ever written. Pauli was a forceful and colorful character, well known for his witty and often caustic remarks directed at those who presented new theories in a less than perfectly clear manner. Pauli exerted great influence on his students and colleagues by forcing them with his sharp criticism to a deeper and clearer understanding. Victor Weisskopf, one of

Pauli's famous students, has aptly described him as "the conscience of theoretical physics." Pauli's sharp sense of humor was also nicely captured by Weisskopf in the following anecdote:

"In a few weeks, Pauli asked me to come to Zurich. I came to the big door of his office, I knocked, and no answer. I knocked again and no answer. After about five minutes he said, rather roughly, "Who is it? Come in!" I opened the door, and here was Pauli—it was a very big office—at the other side of the room, at his desk, writing and writing. He said, "Who is this? First I must finish calculating." Again he let me wait for about five minutes and then: "Who is that?" "I am Weisskopf." "Uhh, Weisskopf, ja, you are my new assistant." Then he looked at me and said, "Now, you see I wanted to take Bethe, but Bethe works now on the solid state. Solid state I don't like, although I started it. This is why I took you." Then I said, "What can I do for you, sir?" and he said "I shall give you right away a problem." He gave me a problem, some calculation, and then he said, "Go and work." So I went, and after 10 days or so, he came and said, "Well, show me what you have done." And I showed him. He looked at it and exclaimed: "I should have taken Bethe!"*

*From Victor F. Weisskopf, *Physics in the Twentieth Century: Selected Essays: My Life as a Physicist*. Cambridge, Mass., The MIT Press, 1972, p. 10.

(*Photo taken by S. A. Goudsmit, AIP Niels Bohr Library*)

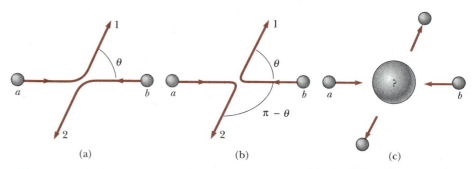

Figure 8.14 The scattering of two electrons as a result of their mutual repulsion. The events depicted in (a) and (b) produce the same outcome for identical electrons, but are nonetheless distinguishable classically because the path taken by each electron is different in the two cases. In this way, the electrons retain their separate identities during collision. (c) According to quantum mechanics, the paths taken by the electrons are blurred by the wave properties of matter. In consequence, once they have interacted, the electrons cannot be told apart in any way!

θ relative to the axis of incidence. Had we not followed their paths, we could not decide which of the two collisions actually occurred, and the separate identities of the electrons would have been lost in the process of collision.

But paths are classical concepts, blurred by the wave properties of matter according to the uncertainty principle. That is, there is an inherent fuzziness to these paths, which blends them inextricably in the collision region, where the electrons may be separated by only a few de Broglie wavelengths. The quantum viewpoint is better portrayed in Figure 8.14c, where the two distinct possibilities (from a classical standpoint) merge into a single quantum event—the scattering of two electrons through an angle θ. The argument makes clear why indistinguishability plays no role in classical physics: all particles, even identical ones, are in fact distinguishable classically through their *paths!* With our acceptance of matter waves, we are forced to conclude that *identical particles cannot be told apart in any way—they are truly indistinguishable*. Incorporating this remarkable fact into quantum theory leads to the exclusion principle discovered by Pauli.

Electrons are truly indistinguishable

Let us see how indistinguishability affects our mathematical description of a two-electron system, say the helium atom. Each electron has kinetic energy as well as electrostatic potential energy in the presence of the doubly charged helium nucleus. These contributions are represented in Schrödinger's equation by terms

$$-\frac{\hbar^2}{2m_e}\nabla_1^2\psi + \frac{k(2e)(-e)}{r_1}\psi$$

where ∇_1^2 is the Laplacian in this electron's coordinate, $\mathbf{r}_1$. For brevity, let us write the sum of both terms simply as $h(1)\psi$, with the label 1 referring to $\mathbf{r}_1$. For the second electron, we write the same expression, except that $\mathbf{r}_1$ must be replaced everywhere by $\mathbf{r}_2$, the coordinate of the second electron. The stationary states for our two-electron system satisfy Schrödinger's time-independent equation,

$$h(1)\psi + h(2)\psi = E\psi \tag{8.15}$$

The fact that $h(1)$ and $h(2)$ are the *same* but for their arguments reflects the indistinguishability of the two electrons.

Equation 8.15 accounts for the electrons' kinetic and potential energies in the field of the helium nucleus but ignores their mutual energy of repulsion. In fact, each electron repels the other through the Coulomb force, leading to an *interaction energy* that must be added to the left-hand side of Equation 8.15. For simplicity, we shall ignore this interaction and treat the electrons as independent objects, each unaffected by the other's presence. In Section 8.5 we show how this *independent particle approximation* can be improved to give a better description of reality.

The two-electron wavefunction will depend on the coordinates of both particles, $\psi = \psi(\mathbf{r}_1, \mathbf{r}_2)$, with $|\psi(\mathbf{r}_1, \mathbf{r}_2)|^2$ representing the probability density for finding one electron at $\mathbf{r}_1$ and the other at $\mathbf{r}_2$. Now, the indistinguishability of electrons requires that a formal interchange of particles produce no observable effects. In particular, all probabilities are unaffected by the interchange, so the wavefunction ψ must be one for which

$$|\psi(\mathbf{r}_1, \mathbf{r}_2)| = |\psi(\mathbf{r}_2, \mathbf{r}_1)|$$

We say that such a wavefunction exhibits **exchange symmetry.** The wavefunction itself may be either even or odd under particle exchange. The former is characterized by the property

$$\psi(\mathbf{r}_1, \mathbf{r}_2) = \psi(\mathbf{r}_2, \mathbf{r}_1) \tag{8.16}$$

and describes a class of particles called *bosons*. Photons belong to this class, as do some more exotic particles such as pions. Electrons, as well as protons and neutrons, are examples of *fermions*, for which

$$\psi(\mathbf{r}_1, \mathbf{r}_2) = -\psi(\mathbf{r}_2, \mathbf{r}_1) \tag{8.17}$$

Thus, our two-electron helium wavefunction must obey Equation 8.17 to account for the indistinguishability of electrons.[7]

To recover the Pauli principle, we must examine the wavefunction more closely. For independent electrons, solutions to Equation 8.15 are easily found. Since each electron "sees" only the helium nucleus, the wavefunction in each coordinate must be an atomic function of the type discussed in Chapter 7. We denote these atomic functions by ψ_a, where a is a collective label for the *four* quantum numbers n, ℓ, m_ℓ, and m_s (or n, ℓ, j, and m_j if spin–orbit effects are included). The products $\psi_a(\mathbf{r}_1)\psi_b(\mathbf{r}_2)$ satisfy our equation, since

$$h(1)\psi_a(\mathbf{r}_1)\psi_b(\mathbf{r}_2) = E_a\psi_a(\mathbf{r}_1)\psi_b(\mathbf{r}_2)$$
$$h(2)\psi_a(\mathbf{r}_1)\psi_b(\mathbf{r}_2) = E_b\psi_a(\mathbf{r}_1)\psi_b(\mathbf{r}_2)$$

E_a and E_b are hydrogen-like energies for the states labeled a and b (see Equation 7.38). Thus,

$$[h(1) + h(2)]\psi_a(\mathbf{r}_1)\psi_b(\mathbf{r}_2) = (E_a + E_b)\psi_a(\mathbf{r}_1)\psi_b(\mathbf{r}_2) \tag{8.18}$$

and $E = E_a + E_b$ is the total energy of this two-electron state.

Notice that the one-electron energies are simply additive, as we might have anticipated for independent particles. Furthermore, the solution $\psi_a(\mathbf{r}_1)\psi_b(\mathbf{r}_2)$ describes one electron occupying the atomic state labeled a and the other the state labeled b. But this product is *not* odd under particle exchange, as required for identical fermions. However, you can verify that $\psi_a(\mathbf{r}_2)\psi_b(\mathbf{r}_1)$ also is a solution to Equation 8.15 with energy $E = E_a + E_b$, corresponding to our two electrons having exchanged states. The antisymmetric combination of these two

$$\psi_{ab}(\mathbf{r}_1, \mathbf{r}_2) = \psi_a(\mathbf{r}_1)\psi_b(\mathbf{r}_2) - \psi_a(\mathbf{r}_2)\psi_b(\mathbf{r}_1) \tag{8.19}$$

does display the correct exchange symmetry, that is,

$$\psi_{ab}(\mathbf{r}_2, \mathbf{r}_1) = \psi_a(\mathbf{r}_2)\psi_b(\mathbf{r}_1) - \psi_a(\mathbf{r}_1)\psi_b(\mathbf{r}_2)$$
$$= -\psi_{ab}(\mathbf{r}_1, \mathbf{r}_2)$$

[7] It is an experimental fact that integer spin particles are bosons, but half-integer spin particles are fermions. This connection between spin and symmetry under particle exchange can be shown to have a theoretical basis when the quantum theory is formulated so as to conform to the requirements of special relativity.

Thus, Equation 8.19 furnishes an acceptable description of the system. Notice, however, that it is now impossible to decide which electron occupies which state—as it should be for identical electrons! Finally, we see that when a and b label the same state ($a = b$), ψ_{ab} is, identically zero—the theory allows no solution (description) in this case, in agreement with the familiar statement of the exclusion principle.

EXAMPLE 8.5   Ground State of the Helium Atom

Construct explicitly the two-electron ground state wavefunction for the helium atom in the independent particle approximation, using the prescription of Equation 8.19. Compare the predicted energy of this state with the measured value, and account in a qualitative way for any discrepancy.

Solution   In the independent-particle approximation, each helium electron "sees" only the doubly charged helium nucleus. Accordingly, the ground state wave is constructed from the lowest energy hydrogen-like wavefunctions, with atomic number $Z = 2$ for helium. These are states for which $n = 1$, $\ell = 0$, and $m_\ell = 0$. Referring to Equation 7.42 of Chapter 7, we find (with $Z = 2$)

$$\psi_{1,0,0}(\mathbf{r}) = \pi^{-1/2}(2/a_0)^{3/2}e^{-2r/a_0}$$

To this orbital function we must attach a spin label ($\pm$) indicating the direction of electron spin. Thus, the one-electron state labels a and b in this example are given by $a = (1, 0, 0, +)$, $b = (1, 0, 0, -)$. Since there is no orbital field to interact with the electron spin, the energies of these two states are identical and are just the hydrogen-like levels of Equation 7.38 with $n = 1$ and $Z = 2$:

$$E_a = E_b = -(2^2/1^2)(13.6\text{ eV}) = -54.4\text{ eV}$$

The antisymmetric two-electron wavefunction for the ground state of helium is then

$$\psi(\mathbf{r}_1, \mathbf{r}_2) = \psi_{1,0,0,+}(\mathbf{r}_1)\psi_{1,0,0,-}(\mathbf{r}_2)$$
$$- \psi_{1,0,0,-}(\mathbf{r}_1)\psi_{1,0,0,+}(\mathbf{r}_2)$$

Both terms have the same spatial dependence but differ as to their spin. The first term of the antisymmetric wave describes electron 1 as having spin up, and electron 2 as having spin down. These spin directions are reversed in the second term. If we introduce the notation $|+, -\rangle$ to describe the two electron spins in the first term, then the second term becomes $|-, +\rangle$, and the total two-electron wavefunction for the helium ground state can be written

$$\psi(\mathbf{r}_1, \mathbf{r}_2) = \pi^{-1}(2/a_0)^3 e^{-2(r_1 + r_2)/a_0}\{|+, -\rangle - |-, +\rangle\}$$

The equal admixture of the spin states $|+, -\rangle$ and $|-, +\rangle$ means the spin of any one of the helium electrons is just as likely to be up as it is to be down. Notice, however, that the spin of the remaining electron is *always* opposite the first. Such spin–spin *correlations* are a direct consequence of the exclusion principle. (The valence electrons in different orbitals of many higher-Z atoms tend to align their spins. This tendency—known as *Hund's rule*—is another example of spin–spin correlations induced by the exclusion principle.)

The total electronic energy of the helium atom in this approximation is the sum of the one-electron energies E_a and E_b:

$$E = E_a + E_b = -54.4\text{ eV} - 54.4\text{ eV} = -108.8\text{ eV}$$

The magnitude of this number, 108.8 eV, represents the energy (work) required to remove both electrons from the helium atom in the independent particle model. The measured value is substantially lower, about 79.0 eV, owing to the mutual repulsion of the two electrons. Specifically, it costs less energy—only about 24.6 eV—to remove the first electron from the atom, since the electron left behind *screens* the nuclear charge, making it appear less positive than a bare helium nucleus.

8.5   ELECTRON INTERACTIONS AND SCREENING EFFECTS

O P T I O N A L

The preceding discussion of the helium atom exposes a difficulty that arises whenever we treat a system with two or more electrons, namely, handling the effects of electron–electron repulsion. Electrons confined to the small space of an atom are expected to exert strong forces on one another. To ignore these altogether, as in the independent-particle model, is simply too crude; to include them exactly is unmanageable, since precise descriptions even for the classical motion in this case are unknown, except through numerical integration. Accordingly, some workable approximation scheme is needed. A most fruitful approach to this problem begins with the notion of an *effective field*.

Any one atomic electron is subject to the Coulomb attraction of the nucleus as well as the Coulomb repulsion of all other electrons in the atom. These influences largely cancel each other, leaving a net effective field with its potential energy $U_{eff}(\mathbf{r})$. U_{eff} may not be Coulombic—or even spherically symmetric—and may be different for each atomic electron. The success of this approach hinges on how simply and accurately we can model the effective potential. A few of the more obvious possibilities are outlined here.

The outermost, or valence, electrons of an atom "see" not the bare nucleus, but one shielded, or *screened*, by the intervening electrons. The attraction is more like that arising from a nucleus with an effective atomic number Z_{eff} somewhat less than the actual number Z, and would be described by

$$U_{eff}(\mathbf{r}) = \frac{k(Z_{eff}e)(-e)}{r} \qquad (8.20)$$

For a Z-electron atom, $Z_{eff} = Z$ would represent no screening whatever; at the opposite extreme is perfect screening by the $Z - 1$ other electrons, giving $Z_{eff} = Z - (Z - 1) = 1$. The best choice for Z_{eff} need not even be integral, and useful values may be deduced from *measurements* of atomic ionization potentials (see Example 8.6). Furthermore, the degree of screening depends on how much time an electron spends near the nucleus, and we should expect Z_{eff} to vary with the shell and subshell labels of the electron in question. In particular, a $4s$ electron is screened more effectively than a $3s$ electron, since its average distance from the nucleus is greater. Similarly, a $3d$ electron is better screened than a $3s$, or even a $3p$ electron (lower angular momentum implies more eccentric classical orbits, with greater penetration into the nuclear region). The use of a Z_{eff} for valence electrons is appropriate whenever a clear distinction exists between these and inner (core) electrons of the atom, as in the alkali metals.

EXAMPLE 8.6 Z_{eff} for the 3s Electron in Sodium

The outer electron of the sodium atom occupies the $3s$ atomic level. The observed value for the ionization energy of this electron is 5.14 eV. From this information, deduce a value of Z_{eff} for the $3s$ electron in sodium.

Solution   Since the ionization energy, 5.14 eV, represents the amount of energy that must be expended to remove the $3s$ electron from the atom, we infer that the energy of the $3s$ electron in sodium is $E = -5.14$ eV. This should be compared with the energy of a $3s$ electron in a hydrogen-like atom with atomic number Z_{eff}, or

$$E = -\frac{Z_{eff}^2}{3^2}(13.6 \text{ eV})$$

Equating this to -5.14 eV and solving for Z_{eff} gives

$$Z_{eff} = 3\sqrt{\frac{5.14}{13.6}} = 1.84$$

In principle, nuclear shielding can be better described by allowing Z_{eff} to vary continuously throughout the atom in a way that mimics the tighter binding accompanying electron penetration into the core. Two functional forms commonly are used for this purpose. For **Thomas–Fermi screening** we write

$$Z_{eff}(r) = Z\, e^{-r/a_{TF}} \qquad (8.21)$$

The Thomas–Fermi atom

where a_{TF} is the *Thomas–Fermi screening length*. According to Equation 8.21, Z_{eff} is very nearly Z close to the nucleus ($r \approx 0$) but drops off quickly in the outer region, becoming essentially zero for $r \gg a_{TF}$. In this way a_{TF} becomes an indicator of atomic size. The Thomas–Fermi model prescribes a_{TF} proportional to $Z^{-1/3}$; the weak variation with Z suggests that all atoms are essentially the same size, regardless of how many electrons they may have. Since the Thomas–Fermi potential is not Coulombic, the one-electron energies that result from the use of Equation 8.21 vary within a given shell; that is, they depend on the principal (n) *and* orbital (ℓ) quantum numbers. The study of these energies and their associated wavefunctions requires numerical methods, or further approximation. The Thomas–Fermi approximation improves with larger values of Z, and so is especially well suited to describe the outer electronic structure of the heavier elements.

In another approach, nuclear shielding is described by

$$Z_{eff}(r) = 1 + \frac{b}{r} \qquad (8.22)$$

where b is again a kind of screening length. This form is appropriate to the alkali metals, where a lone outer electron is responsible for the chemical properties of the atom. From Equation 8.22, this electron "sees" $Z_{eff} \approx 1$ for $r \gg b$, and larger values in the core. The special virtue of Equation 8.22 is that it leads to one-electron energies and wavefunctions that can be found without further approximation. In particular, the energy levels that follow from Equation 8.22 can be shown to be

$$E_n = -\frac{ke^2}{2a_0}\{n - D(\ell)\}^{-2} \qquad (8.23)$$

Quantum defects

where $D(\ell)$ is termed the *quantum defect*, since it indicates the departure from the simple hydrogen-atom level structure. As the notation suggests, the quantum defect for an s electron differs from that for a p or d electron, but all s electrons have the same quantum defect, regardless of their shell label. Table 8.1 lists some quantum defects deduced experimentally for the sodium atom. Of course, taking $b = 0$ in Equation 8.22 causes all quantum defects to vanish, returning us to the hydrogen-like level structure discussed in Chapter 7.

The use of a simple Z_{eff}, or the more complicated forms of the Thomas–Fermi or quantum defect method, still results in a U_{eff} with spherical symmetry; that is, the electrons are affected by a *central field*. The **Hartree theory** discards even this feature in order to achieve more accurate results. According to Hartree, the electron "cloud" in the atom should be treated as a classical body of charge distributed

Table 8.1   Some Quantum Defects for the Sodium Atom

Subshell	s	p	d	f
$D(\ell)$	1.35	0.86	0.01	~ 0

with some volume charge density $\rho(\mathbf{r})$. The potential energy of any one atomic electron is then

$$U_{\text{eff}}(\mathbf{r}) = -\frac{kZe^2}{r} - \int ke\left[\frac{\rho(\mathbf{r}')}{|\mathbf{r} - \mathbf{r}'|}\right] dV' \tag{8.24}$$

The first term is the attractive energy of the nucleus, and the second term is the repulsive energy of all other atomic electrons. This U_{eff} gives rise to a one-electron Schrödinger equation for the energies E_i and wavefunctions ψ_i of this, say the ith, atomic electron.

But the Hartree theory is *self-consistent*, in that the charge density ρ due to the other atomic electrons is itself calculated from the electron waves as

**Hartree's self-consistent
fields**

$$\rho(\mathbf{r}) = -e\sum |\psi_j(\mathbf{r})|^2 \tag{8.25}$$

The sum in Equation 8.25 includes all occupied electron states ψ_j except the ith. In this way the mathematical problem posed by U_{eff} is turned back on itself: we must solve not one Schrödinger equation, but N of them in a single stroke, one for each of the N electrons in the atom! To do this, numerical methods are employed in an iterative solution scheme. An educated guess is made initially for each of the N ground state electron waves. With this, the ρ and U_{eff} for every electron can be computed, and all N Schrödinger equations solved. The resulting wavefunctions are compared with the initial guesses; if discrepancies appear, the calculation is repeated with the new set of electron waves replacing the old. After several such iterations, agreement is attained between the starting and calculated waves. The resulting N electron wavefunctions are said to be fully self-consistent. Implementation of the Hartree method is laborious and demands considerable skill, but the results for atomic electrons are among the best available. Indeed, the Hartree and closely related Hartree–Fock methods are the ones frequently called on today when accurate atomic energy levels and wavefunctions are required.

EXAMPLE 8.7   Quantum Defects for Alkali Atoms

MP This example uses the MP DESKTOP to validate the quantum defect level structure for the alkali atoms, where screening of the nuclear charge is adequately described by Equation 8.22.

Launch the MP DESKTOP application by running the main program file *MP.EXE*. Next, select *Task* from the menu bar and *EigenStates-3D* from the drop-down task list. The DESKTOP is now ready to solve the (eigenvalue) problem posed in the task field just below the menu bar:

$$-\frac{d^2g}{dr^2} + \left[U(r) + \frac{\ell(\ell + 1)}{r^2}\right] g(r) = E \cdot g(r)$$

This is the *radial wave equation* for the pseudo-wavefunction $g(r) = rR(r)$, expressed in dimensionless units of length and energy.

NOTE: Recall that for *EigenStates-3D* the unit of length (δ) and the unit of energy (ϵ) must satisfy $\epsilon \cdot \delta^2 = \hbar^2/2m$ in order to map the equation in the task field to the radial wave equation. In describing electrons, atomic units ($\delta = 1$ bohr [0.529 Å] and $\epsilon = 1$ Ry [13.60 eV]) are a convenient choice satisfying this criterion and are the ones adopted in this example.

Select the potential energy $U(r)$ by clicking the Potential button $\boxed{\text{U(r)}}$ located just below the menu bar and choosing *EX8-7.U3* (for Example 8.7) from the list of potential files. A second dialog box opens for the function $U(r) = -2(1 + \beta/r)/r$. This is the [screened] Coulomb potential implied by Equation 8.22, with r scaled by a_0 (1 bohr) and U scaled by $ke^2/2a_0$ (1 Ry). The default screening length is $\beta = 0.100$ (bohr). Click the *OK* button to accept the default and load this potential.

The effective potential $U_{\text{eff}}(r) = U(r) + \ell(\ell + 1)/r^2$ is plotted in the main window of the screen; since $\ell = 0$ to start, this is the same as the true potential $U(r)$. The 1s state should have energy intermediate between that of atomic hydrogen (-13.6 eV, or -1 Ry) and singly ionized helium (-4 Ry), so type $E = -2$ in the energy box just below the menu bar and press *Enter*. The wavefunction for this energy is computed and is superimposed on the potential. The discontinuity is evidence that this guess for E is incorrect. Continue with a few more trials until there is no perceptible discontinuity in the wavefunction.

NOTE: The direction keys [↑],[↓] furnish a quick and convenient way to make small changes to the energy. Just press [↑] ([↓]) to increase (decrease) the digit at the cursor by one unit; then press *Enter* to update the display. The digit at the cursor can also be changed by clicking the mouse at the location of the cursor.

Next, click the Accuracy button ⌐L⌐ to the right of the energy box: the button label changes from **L**(ow) to **M**(edium), signaling improved precision in the search. Make fine adjustments to E as necessary to generate a smooth waveform. Two more precision levels—**H**(igh) and **S**(uper)—are available if desired. Continue the search process until E is known confidently to three decimal places. You should find $E = -1.873$ Ry ($= -25.47$ eV). The pseudo-wavefunction $g_{1s}(r)$ for this energy should look like that shown in Figure 8.15.

Figure 8.15 (Example 8.7) The pseudo-wavefunction $g_{1s}(r)$ against the backdrop of the (screened) Coulomb potential applicable to the alkali atoms. The screening length is 0.100 bohr.

Now Equation 8.23 prescribes the energy of the $1s$ state (in rydbergs) as

$$E_{1s} = -\frac{1}{[1 - D(0)]^2}$$

Comparing this with the result of the simulation gives $D(0) = 0.269$ for the s-wave quantum defect when the screening length is 0.100 bohr.

NOTE: This value for the s-wave quantum defect is smaller than that observed for any of the alkali elements, due to the small value we have chosen for the screening length ($\beta = 0.100$ bohr). Larger defects can be had by increasing β, but too large a value for β introduces singularities in the wavefunction at $r = 0$ that the DESKTOP is not equipped to handle. For s-waves, β must not exceed 0.125, and even values too close to this limit will produce results that are abnormally sensitive to the choice of display range. More generally, the limitation imposed by the DESKTOP is $\beta \leq \frac{1}{2}(\ell + \frac{1}{2})^2$. Modeling p-wave quantum defects with the DESKTOP is more realistic, as outlined in Problem 20.

If the notion of quantum defects is valid, the energy of the $2s$ state should be

$$E_{2s} = -\frac{1}{[2 - D(0)]^2} = -0.3339$$

To verify this expectation, type this value for E into the energy box and press *Enter*. The resulting wavefunction is smooth everywhere (Fig. 8.16), indicating that $E = -0.3339$ Ry is an acceptable energy. Since the wavefunction has one node, this must be the $2s$ state.

Figure 8.16   (Example 8.7) The first excited state pseudo-wavefunction $g_{2s}(r)$ for the quantum defect potential applicable to the alkali atoms. The screening length is 0.100 bohr. Note the single node characteristic of a $2s$ state.

Knowing the *s*-wave quantum defect, we can display all *s* states of the alkali atom. Try a few higher states to further confirm the validity of Equation 8.23.

Exercise 3   *More Quantum Defects.* Use the method just described to determine the *p*-wave and *d*-wave quantum defects for the alkali potential of Example 8.7. (Click the Orbital button $\boxed{\underline{l}=\;0}$ to change ℓ.)

8.6   THE PERIODIC TABLE

In principle, it is possible to predict the properties of all the elements by applying the procedures of wave mechanics to each one. Because of the large number of interactions possible in multielectron atoms, however, approximations must be used for all atoms except hydrogen. Nevertheless, the electronic structure of even the most complex atoms can be viewed as a succession of filled levels increasing in energy, with the outermost electrons primarily responsible for the chemical properties of the element.

In the central field approximation, the atomic levels can be labeled by the quantum numbers n and ℓ. From the exclusion principle, the maximum number of electrons in one such *subshell* level is $2(2\ell + 1)$. The energy of an electron in this level depends primarily on the quantum number n, and to a lesser extent on ℓ. The levels can be grouped conveniently according to the value of n (the shell label), and all those within a group have energies that increase with increasing ℓ. The order of filling the subshell levels with electrons is as follows: once a subshell is filled, the next electron goes into the vacant level that is lowest in energy. This *minimum energy principle* can be understood by noting that if the electron were to occupy a higher level, it would spontaneously decay to a lower one with the emission of energy.

The chemical properties of atoms are determined predominantly by the least tightly bound, or *valence*, electrons, which are in the subshell of highest energy. The most important factors are the occupancy of this subshell and the energy separation between this and the next higher (empty) subshell. For example, an atom tends to be chemically inert if its highest subshell is full and there is an appreciable energy gap to the next higher subshell, since then electrons are not readily shared with other atoms to form a molecule. The quasi-periodic recurrence of similar highest shell structures as Z increases is responsible for the *periodic system* of the chemical elements.

The specification of n and ℓ for each atomic electron is called the **electron configuration** of that atom. We are now in a position to describe the electron configuration of any atom in its ground state:

Hydrogen has only one electron, which, in its ground state, is described by the quantum numbers $n = 1$, $\ell = 0$. Hence, its electron configuration is designated as $1s^1$.

Helium, with its two electrons, has a ground state electron configuration of $1s^2$. That is, both electrons are in the same (lowest energy) subshell, the $1s$. Since two is the maximum occupancy for an s subshell, the subshell (and in this case also the shell) is said to be *closed*, and helium is inert.

Lithium has three electrons. Two of these are assigned to the $1s$ subshell, and the third must be assigned to the $2s$ subshell, since this is slightly lower in energy than the $2p$ subshell. Hence, the electron configuration of lithium is $1s^2 2s^1$.

With the addition of another electron to make *beryllium*, the $2s$ subshell is closed. The electron configuration of beryllium, with four electrons altogether, is $1s^2 2s^2$. (Beryllium is not inert, however, since the energy gap separating the $2s$ from the next available level—the $2p$—is not very large.)

Boron has a configuration of $1s^2 2s^2 2p^1$. (With spin–orbit doubling, the $2p$ electron in boron actually occupies the $2P_{1/2}$ sublevel, corresponding to $n = 2$, $\ell = 1$, and $j = \frac{1}{2}$.)

Carbon has six electrons, and a question arises of how to assign the two $2p$ electrons. Do they go into the same orbital with paired spins (↑ ↓), or do they occupy different orbitals with unpaired spins (↑ ↑)? Experiments show that the energetically preferred configuration is the latter, in which the spins are aligned. This is an illustration of **Hund's rule,** which states that electrons usually fill different orbitals with unpaired spins, rather than the same orbital with paired spins. Hund's rule can be partly understood by noting that electrons in the same orbital tend to be closer together, where their mutual repulsion contributes to a higher energy than if they were separated in different orbitals. Some exceptions to this rule do occur in those elements with subshells that are nearly filled or half-filled. The progressive filling of the $2p$ subshell illustrating Hund's rule is shown schematically in Figure 8.17. With *neon*, the $2p$ subshell also becomes closed. The neon atom has ten electrons in the configuration $1s^2 2s^2 2p^6$. Since the energy gap separating the $2p$ subshell from the next available level—the $3s$—is quite large, the neon configuration is exceptionally stable and the atom is chemically inert.

A complete list of electron configurations for all the known stable elements is given in Table 8.2. Note that, beginning with *potassium* ($Z = 19$), the $4s$ subshell starts to fill while the $3d$ level remains empty. Only after the $4s$ subshell is closed to form *calcium* does the $3d$ subshell begin to fill. We infer that the $3d$ subshell has a higher energy than the $4s$, even though it belongs to a lower-indexed shell. This should come as no surprise, because the energy separating consecutive shells becomes smaller with increasing n (see the hydrogen-like spectrum), but the energy separating subshells is more nearly constant owing to the screening effect discussed in Section 8.5. (In fact, the $3d$ and $4s$ subshells must be very close in energy, as shown by the electron configuration of *chromium*, in which the $3d$ subshell temporarily regains an electron from the $4s$.) The same phenomenon occurs again with *rubidium* ($Z = 37$), in which the $5s$ subshell begins to fill at the expense of the $4d$ and $4f$ subshells. Energetically, the electron configurations shown in the table imply the following ordering of subshells with respect to energy:

Ordering of subshells by energy

$$1s < 2s < 2p < 3s < 3p < 4s \sim 3d < 4p < 5s < 4d < 5p < 6s < 4f \sim 5d$$
$$< 6p < 7s < 6d \sim 5f \ldots$$

The elements from *scandium* ($Z = 21$) to *zinc* ($Z = 30$) form the first *transition series*. These transition elements are characterized by a progressively filling $3d$ subshell, while the outer electron configuration is unchanged at $4s^2$ (except in the case of copper). Consequently, all the transition elements exhibit similar chemical properties. The belated occupancy of inner d subshells is encountered

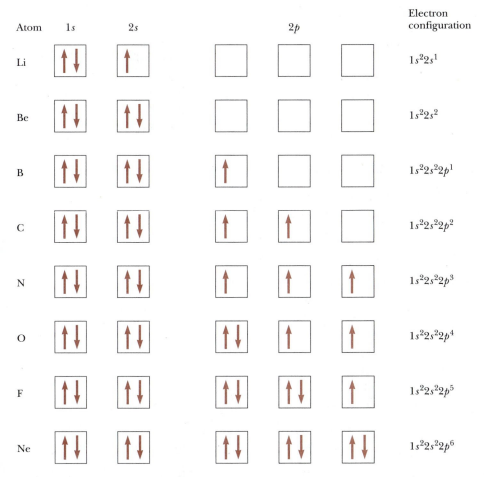

Figure 8.17 Electronic configurations of successive elements from lithium to neon. The filling of electronic states must obey the Pauli exclusion principle and Hund's rule.

again in the second and third transition series, marked by the progressive filling of the $4d$ and $5d$ subshells, respectively. The second transition series includes the elements *yttrium* ($Z = 39$) to *cadmium* ($Z = 48$); the third contains the elements *lutetium* ($Z = 71$) to *mercury* ($Z = 80$).

Related behavior is seen also with the belated filling of the $4f$ and $5f$ subshells. The *lanthanide series*, stretching from *lanthanum* ($Z = 57$) to *ytterbium* ($Z = 70$), is marked by a common $6s^2$ valence configuration, with the added electrons filling out the $4f$ subshell (the energetically close $5d$ levels also are occupied in some instances). The lanthanide elements, or *lanthanides*, also are known as the *rare earths* because of their low natural abundance. *Cerium* ($Z = 58$), which forms 0.00031% by weight of the Earth's crust, is the most abundant of the lanthanides.

In the *actinide series* from *actinium* ($Z = 89$) to *nobelium* ($Z = 102$), the va-

Table 8.2 Electronic Configurations of the Elements

Z	Symbol	Ground Configuration	Ionization Energy (eV)	Z	Symbol	Ground Configuration	Ionization Energy (eV)
1	H	$1s^1$	13.595	26	Fe	$3d^64s^2$	7.87
2	He	$1s^2$	24.581	27	Co	$3d^74s^2$	7.86
				28	Ni	$3d^84s^2$	7.633
3	Li	[He] $2s^1$	5.390	29	Cu	$3d^{10}4s^1$	7.724
4	Be	$2s^2$	9.320	30	Zn	$3d^{10}4s^2$	9.391
5	B	$2s^22p^1$	8.296	31	Ga	$3d^{10}4s^24p^1$	6.00
6	C	$2s^22p^2$	11.256	32	Ge	$3d^{10}4s^24p^2$	7.88
7	N	$2s^22p^3$	14.545	33	As	$3d^{10}4s^24p^3$	9.81
8	O	$2s^22p^4$	13.614	34	Se	$3d^{10}4s^24p^4$	9.75
9	F	$2s^22p^5$	17.418	35	Br	$3d^{10}4s^24p^5$	11.84
10	Ne	$2s^22p^6$	21.559	36	Kr	$3d^{10}4s^24p^6$	13.996
11	Na	[Ne] $3s^1$	5.138	37	Rb	[Kr] $5s^1$	4.176
12	Mg	$3s^2$	7.644	38	Sr	$5s^2$	5.692
13	Al	$3s^23p^1$	5.984	39	Y	$4d5s^2$	6.377
14	Si	$3s^23p^2$	8.149	40	Zr	$4d^25s^2$	6.835
15	P	$3s^23p^3$	10.484	41	Nb	$4d^45s^1$	6.881
16	S	$3s^23p^4$	10.357	42	Mo	$4d^55s^1$	7.10
17	Cl	$3s^23p^5$	13.01	43	Tc	$4d^55s^2$	7.228
18	Ar	$3s^23p^6$	15.755	44	Ru	$4d^75s^1$	7.365
				45	Rh	$4d^85s^1$	7.461
19	K	[Ar] $4s^1$	4.339	46	Pd	$4d^{10}$	8.33
20	Ca	$4s^2$	6.111	47	Ag	$4d^{10}5s^1$	7.574
21	Sc	$3d4s^2$	6.54	48	Cd	$4d^{10}5s^2$	8.991
22	Ti	$3d^24s^2$	6.83	49	In	$4d^{10}5s^25p^1$	5.785
23	V	$3d^34s^2$	6.74	50	Sn	$4d^{10}5s^25p^2$	7.342
24	Cr	$3d^54s$	6.76	51	Sb	$4d^{10}5s^25p^3$	8.639
25	Mn	$3d^54s^2$	7.432	52	Te	$4d^{10}5s^25p^4$	9.01

lence configuration remains $7s^2$, as the $5f$ subshell progressively fills (along with occasional occupancy of the energetically close $6d$).

Table 8.2 also lists the ionization energies of the elements. The ionization energy for each element is plotted against its atomic number Z in Figure 8.18a. We see that the ionization energy tends to increase within a shell, then drops dramatically as the filling of a new shell begins. The behavior repeats, and it is from this recurring pattern that the periodic table gets its name. A similar repetitive pattern is observed in a plot of the atomic volume per atom versus atomic number (see Fig. 8.18b).

The primary features of these plots can be understood from simple arguments. First, the larger nuclear charge that accompanies higher values of Z

Table 8.2 Electronic Configurations of the Elements

Z	Symbol	Ground Configuration	Ionization Energy (eV)	Z	Symbol	Ground Configuration	Ionization Energy (eV)
53	I	$4d^{10}5s^25p^5$	10.454	79	Au	$[\text{Xe}, 4f^{14}5d^{10}]\,6s^1$	9.22
54	Xe	$4d^{10}5s^25p^6$	12.127	80	Hg	$6s^2$	10.434
				81	Tl	$6s^26p^1$	6.106
55	Cs	$[\text{Xe}]\,6s^1$	3.893	82	Pb	$6s^26p^2$	7.415
56	Ba	$6s^2$	5.210	83	Bi	$6s^26p^3$	7.287
57	La	$5d6s^2$	5.61	84	Po	$6s^26p^4$	8.43
58	Ce	$4f5d6s^2$	6.54	85	At	$6s^26p^5$	
59	Pr	$4f^36s^2$	5.48	86	Rn	$6s^26p^6$	10.745
60	Nd	$4f^46s^2$	5.51				
61	Pm	$4f^56s^2$		87	Fr	$[\text{Rn}]\,7s^1$	
62	Fm	$4f^66s^2$	5.6	88	Ra	$7s^2$	5.277
63	Eu	$4f^76s^2$	5.67	89	Ac	$6d7s^2$	6.9
64	Gd	$4f^75d6s^2$	6.16	90	Th	$6d^27s^2$	
65	Tb	$4f^96s^2$	6.74	91	Pa	$5f^26d7s^2$	
66	Dy	$4f^{10}6s^2$	6.82	92	U	$5f^36d7s^2$	4.0
67	Ho	$4f^{11}6s^2$		93	Np	$5f^46d7s^2$	
68	Er	$4f^{12}6s^2$		94	Pu	$5f^67s^2$	
69	Tm	$4f^{13}6s^2$		95	Am	$5f^77s^2$	
70	Yb	$4f^{14}6s^2$	6.22	96	Cm	$5f^76d7s^2$	
71	Lu	$4f^{14}5d6s^2$	6.15	97	Bk	$5f^86d7s^2$	
72	Hf	$4f^{14}5d^26s^2$	7.0	98	Cf	$5f^{10}7s^2$	
73	Ta	$4f^{14}5d^36s^2$	7.88	99	Es	$5f^{11}7s^2$	
74	W	$4f^{14}5d^46s^2$	7.98	100	Fm	$5f^{12}7s^1$	
75	Re	$4f^{14}5d^56s^2$	7.87	101	Mv	$5f^{13}7s^2$	
76	Os	$4f^{14}5d^66s^2$	8.7	102	No	$5f^{14}7s^2$	
77	Ir	$4f^{14}5d^76s^2$	9.2	103	Lw	$5f^{14}6d7s^2$	
78	Pt	$4f^{14}5d^86s^2$	8.88	104	Ku	$5f^{14}6d^27s^2$	

Note: The bracket notation is used as a shorthand method to avoid repetition in indicating inner shell electrons. Thus, [He] represents $1s^2$, [Ne] represents $1s^22s^22p^6$, [Ar] represents $1s^22s^22p^63s^23p^6$, and so on.

tends to pull the electrons closer to the nucleus and binds them more tightly. Were this the only effect, the ionization energy simply would increase and the atomic volume would decrease steadily with increasing Z. But the innermost, or core, electrons screen the nuclear charge, making it less effective in binding the outer electrons. The screening effect varies in a complicated way from one element to the next, but it is most pronounced for a lone electron outside a closed shell, as in the alkali metals (Li, Na, K, Rb, Cs, and Fr). For these configurations the ionization energy dips sharply, only to rise again as the nuclear charge intensifies at higher Z. The variation in ionization energy is mirrored by the behavior of atomic volume, which peaks at the alkali configurations and becomes smaller as the screening effect subsides.

(a)

(b)

Figure 8.18 (a) Ionization energy of the elements versus atomic number Z. (b) Atomic volume of the elements versus atomic number Z. The recurring pattern with increasing atomic number exemplifies the behavior from which the periodic table gets its name.

8.7   X-RAY SPECTRA AND MOSELEY'S LAW

Electronic transitions within the inner shells of heavier atoms are accompanied by large energy transfers. If the excess energy is carried off by a photon, there results x-ray emission at specific wavelengths peculiar to the emitting atom. This is the explanation for the discrete x-ray line spectra produced when energetic electrons slam into a metal target, as discussed earlier in Section 2.5.

The inner electrons of high Z elements are bound tightly to the atom, since they see a nuclear charge essentially unscreened by the remaining electrons. Consider the case of molybdenum (Mo), with atomic number $Z = 42$ (see Table 8.2). The innermost, or K shell, electrons have $n = 1$ and energy (from Equation 7.38)

$$E_1 = -\frac{ke^2}{2a_0}\left\{\frac{Z^2}{1^2}\right\} = -(13.6\text{ eV})(42)^2 = -23990.4\text{ eV}$$

Thus, 23.9904 keV must be supplied to dislodge a K shell electron from the Mo atom.[8] Energies of this magnitude are routinely delivered via electron im-

Figure 8.19   Origin of x-ray spectra. The K series (K_α, K_β, K_γ, . . .) originates with electrons in higher-lying shells making a downward transition to fill a vacancy in the K shell. In the same way, the filling of vacancies created in higher shells produces the L series, the M series, and so on.

[8] In contrast, only 7.10 eV (the ionization energy from Table 8.2) is required to free the outermost, or 5s, electron.

pact: electrons accelerated to kilovolt energies collide with atoms of a molybdenum target, giving up most or all of their energy to one atom in a single collision. If large enough, the collision energy may excite one of the K shell electrons to a higher vacant level or free it from the atom altogether. (Recall there are two electrons in a filled K shell.) In either case, a vacancy, or *hole*, is left in the K shell. This hole is quickly filled by another, higher-lying atomic electron, with the energy of transition released in the form of a photon. The exact energy (and wavelength) of the escaping photon depends on the energy of the electron filling the vacancy, giving rise to an entire K *series* of emission lines, denoted in order of increasing energy (decreasing wavelength) by $K_\alpha, K_\beta, K_\gamma, \ldots$. In the same way, the filling of vacancies left in higher shells produces the L *series*, the M *series*, and so on, as illustrated in Figure 8.19.

The energy of the longest wavelength photons in a series can be estimated from simple screening arguments. For the K_α line, the K shell hole is filled by an electron from the L shell ($n = 2$). But an electron in the L shell is partially screened from the nucleus by the remaining K shell electron, and so sees a nuclear charge of only $Z - 1$. Thus, the energy of the K_α photon can be approximated as an $n = 2$ to $n = 1$ transition in a one-electron atom with an

Henry G. J. Moseley (1887–1915) discovered a direct way to measure Z, the atomic number, from the characteristic x-ray wavelength emitted by an element. Moseley's work not only established the correct sequence of elements in the periodic table but also provided another confirmation of the Bohr model of the atom, in this case at x-ray energies. One wonders what other major discoveries Moseley would have made if he had not been killed in action at the age of 27 in Turkey in the first world war. *(University of Oxford, Museum of the History of Science/Courtesy AIP Niels Bohr Library)*

Figure 8.20   The original data of Moseley showing the relationship between atomic number Z and the characteristic x-ray frequencies. The gaps at $Z = 43$, 61, and 75 represent elements unknown at the time of Moseley's work. (There are also several errors in the atomic number designations for the elements.) (© *From H. G. J. Moseley, Philos. Mag. (6), 27:703, 1914)*

effective nuclear charge of $Z - 1$:

$$E[K_\alpha] = -\frac{ke^2}{2a_0}\frac{(Z-1)^2}{2^2} + \frac{ke^2}{2a_0}\frac{(Z-1)^2}{1^2} = \frac{ke^2}{2a_0}\frac{3(Z-1)^2}{4} \quad (8.26)$$

For molybdenum ($Z = 42$), this is $E[K_\alpha] = 17.146$ keV, corresponding to a wavelength

$$\lambda[K_\alpha] = \frac{hc}{E[K_\alpha]} = \frac{12.4 \text{ keV·Å}}{17.146 \text{ keV}} = 0.723 \text{ Å}$$

For comparison, the observed K_α line of molybdenum has wavelength 0.7095 Å, in reasonable agreement with our calculation.

In a series of careful experiments conducted from 1913 to 1914, the British physicist H. G. J. Moseley measured the wavelength of K_α lines for numerous elements, and confirmed the validity of Equation 8.26, known as **Moseley's law.** According to Moseley's law, a plot of the square root of photon frequency $(E/h)^{1/2}$ versus atomic number Z should yield a straight line. Such a *Moseley plot*, as it is called, is reproduced here as Figure 8.20. Before Moseley's work, atomic numbers were mere placeholders for the elements appearing in the periodic table, the elements being ordered according to their mass. By measuring their K_α lines, Moseley was able to establish the correct sequence of elements in the periodic table, a sequence properly based on atomic number rather than atomic mass. The gaps in Moseley's data at $Z = 43, 61$, and 75 represent elements unknown at the time of his work.

SUMMARY

The magnetic behavior of atoms is characterized by their *magnetic moment*. The orbital moment of an atomic electron is proportional to its orbital angular momentum:

$$\boldsymbol{\mu} = \frac{-e}{2m_e}\mathbf{L} \quad (8.1)$$

The constant of proportionality, $-e/2m_e$, is called the *gyromagnetic ratio*. Since $\mathbf{L}$ is subject to space quantization, so too is the atomic moment $\boldsymbol{\mu}$. Atomic moments are measured in Bohr magnetons, $\mu_B = e\hbar/2m_e$; the SI value of μ_B is 9.27×10^{-24} J/T.

An atom subjected to an external magnetic field $\mathbf{B}$ experiences a magnetic torque, which results in precession of the moment vector $\boldsymbol{\mu}$ about the field vector $\mathbf{B}$. The frequency of precession is the *Larmor frequency* ω_L given by

$$\omega_L = \frac{eB}{2m_e} \quad (8.5)$$

Associated with the Larmor frequency is the energy quantum $\hbar\omega_L = \mu_B B$. The ℓth subshell level of an atom placed in a magnetic field is split by the field into $2\ell + 1$ sublevels separated by the Larmor energy $\hbar\omega_L$. This is the *Zeeman effect*, and $\hbar\omega_L$ is also known as the Zeeman energy. The magnetic contribution to the energy of the atom is

$$U = \hbar\omega_L m_\ell \tag{8.7}$$

where m_ℓ is the same *magnetic* quantum number discussed in Chapter 7. Equation 8.7 is a special case of the more general result

$$U = -\boldsymbol{\mu} \cdot \mathbf{B} \tag{8.6}$$

for the magnetic potential energy U of any magnetic moment $\boldsymbol{\mu}$ in an applied field $\mathbf{B}$.

In addition to any orbital magnetic moment, the electron possesses an intrinsic magnetic moment called the *spin* moment $\boldsymbol{\mu}_s$. In a classical picture, the spin moment arises from the rotation of the electron on its axis and is proportional to the angular momentum of rotation, or spin, $\mathbf{S}$. The magnitude of the spin angular momentum is

$$|\mathbf{S}| = \frac{\sqrt{3}}{2}\hbar \tag{8.11}$$

corresponding to a *spin* quantum number $s = \frac{1}{2}$, analogous to the orbital quantum number ℓ. The z component of $\mathbf{S}$ is quantized as

$$S_z = m_s \hbar \tag{8.10}$$

where the *spin magnetic* quantum number m_s is the analog of the orbital magnetic quantum number m_ℓ. For the electron, m_s can be either $+\frac{1}{2}$ or $-\frac{1}{2}$, which describes the up and down spin states, respectively. Equations 8.10 and 8.11 imply that electron spin also is subject to space quantization. This was confirmed experimentally by Stern and Gerlach, who observed that a beam of neutral silver atoms passed through a nonuniform magnetic field was split into two distinct components. The same experiment shows that the spin moment is related to the spin angular momentum by

$$\boldsymbol{\mu}_s = -\frac{e}{m_e}\mathbf{S}$$

This is *twice* as large as the orbital moment for the same angular momentum. The anomalous factor of 2 is called the *g-factor* of the electron. With the recognition of spin, four quantum numbers—n, ℓ, m_ℓ, and m_s—are needed to specify the state of an atomic electron.

The spin moment of the electron interacts with the magnetic field arising from its orbital motion. The energy difference between the two spin orientations in this orbital field is responsible for the *fine structure doubling* of atomic spectral lines. With the spin–orbit interaction, atomic states are labeled by a quantum number j for the *total* angular momentum $\mathbf{J} = \mathbf{L} + \mathbf{S}$. The value of j is included as a subscript in the spectroscopic notation of atomic states. For example, $3P_{1/2}$ specifies a state for which $n = 3$, $\ell = 1$, and $j = \frac{1}{2}$. For each value of j, there are $2j + 1$ possibilities for the total magnetic quantum number m_j.

The *exclusion principle* states that no two electrons can be in the same quantum state; that is, no two electrons can have the same four quantum numbers.

The exclusion principle derives from the notion that electrons are identical particles called *fermions*. Fermions are described by wavefunctions that are *antisymmetric* in the electron coordinates. Wavefunctions that are symmetric in the particle coordinates describe another class of objects called *bosons*, to which no exclusion principle applies. All known particles are either fermions or bosons. An example of a boson is the photon.

Using the exclusion principle and the principle of minimum energy, one can determine the electronic configurations of the elements. This serves as a basis for understanding atomic structure and the physical and chemical properties of the elements.

One can understand the discrete x-ray line spectra emitted by different metals in terms of electronic transitions within inner shells. When electron bombardment creates a vacancy in an inner K or L shell, a higher-lying atomic electron quickly fills the vacancy, giving up its excess energy as an x-ray photon. According to *Moseley's law*, the square root of this photon frequency should be proportional to the atomic number of the emitting atom.

SUGGESTIONS FOR FURTHER READING

1. A classic work on the physics of atoms is H. E. White, *Introduction to Atomic Spectra*, New York, McGraw-Hill, 1934.

The following sources contain more extensive discussions of the topics found in this chapter:

2. A. P. French and E. F. Taylor, *An Introduction to Quantum Physics*, New York, W. W. Norton and Company, Inc., 1978.

3. R. Eisberg and R. Resnick, *Quantum Physics of Atoms, Molecules, Solids, Nuclei, and Particles*, 2nd ed., New York, John Wiley and Sons, Inc., 1985.

4. B. H. Bransden and C. J. Joachain, *Physics of Atoms and Molecules*, New York, John Wiley and Sons, Inc., 1990.

QUESTIONS

1. Why is the direction of the orbital angular momentum of an electron opposite that of its magnetic moment?

2. Why is an inhomogeneous magnetic field used in the Stern–Gerlach experiment?

3. Could the Stern–Gerlach experiment be performed with ions rather than neutral atoms? Explain.

4. Describe some experiments that would support the conclusion that the spin quantum number for electrons can have only the values $\pm\frac{1}{2}$.

5. Discuss some of the consequences of the exclusion principle.

6. Why do lithium, potassium, and sodium exhibit similar chemical properties?

7. From Table 8.2, we find that the ionization energies for Li, Na, K, Rb, and Cs are 5.390, 5.138, 4.339, 4.176, and 3.893 eV, respectively. Explain why these values are to be expected in terms of the atomic structures.

8. Although electrons, protons, and neutrons obey the exclusion principle, some particles that have integral spin, such as photons (spin = 1), do not. Explain.

9. How do we know that a photon has a spin of 1?

10. An energy of about 21 eV is required to excite an electron in a helium atom from the 1s state to the 2s state. The same transition for the He$^+$ ion requires about twice as much energy. Explain why this is so.

11. Discuss degeneracy as it applies to a multielectron atom. Can a one-electron atom have degeneracy? Explain.

12. The absorption or emission spectrum of a gas consists of lines that broaden as the density of gas molecules increases. Why do you suppose this occurs?

13. For a one-electron atom or ion, spin–orbit coupling splits all states except s states into doublets. Why are s states exceptions to this rule?

14. Why is it approximately correct to neglect the screening effect of outer shell electrons (e.g., electrons in the M and N shells) on an electron in the L shell?

PROBLEMS

8.1 Orbital Magnetism and the Normal Zeeman Effect

1. In the technique known as electron spin resonance (ESR), a sample containing unpaired electrons is placed in a magnetic field. Consider the simplest situation, that in which there is only one electron and therefore only two possible energy states, corresponding to $m_s = \pm\frac{1}{2}$. In ESR, the electron's spin magnetic moment is "flipped" from a lower energy state to a higher energy state by the absorption of a photon. (The lower energy state corresponds to the case in which the magnetic moment $\boldsymbol{\mu}_s$ is aligned with the magnetic field, and the higher energy state corresponds to the case where $\boldsymbol{\mu}_s$ is aligned against the field.) What is the photon frequency required to excite an ESR transition in a magnetic field of 0.35 T?

2. Show that for a mass m in orbit with angular momentum $\mathbf{L}$ the rate at which area is swept out by the orbiting particle is

$$\frac{dA}{dt} = \frac{|\mathbf{L}|}{2m}$$

(*Hint:* First show that in its displacement $d\mathbf{r}$ *along the path*, the particle sweeps out an area $dA = \frac{1}{2}|\mathbf{r} \times d\mathbf{r}|$, where $\mathbf{r}$ is the position vector of the particle drawn from some origin.)

8.2 The Spinning Electron

3. How many different sets of quantum numbers are possible for an electron for which (a) $n = 1$, (b) $n = 2$, (c) $n = 3$, (d) $n = 4$, and (e) $n = 5$? Check your results to show that they agree with the general rule that the number of different sets of quantum numbers is equal to $2n^2$.

4. List the possible sets of quantum numbers for an electron in (a) the $3d$ subshell and (b) the $3p$ subshell.

5. The force on a magnetic moment $\boldsymbol{\mu}_z$ in a nonuniform magnetic field B_z is given by

$$F_z = \mu_z \frac{dB_z}{dz}$$

If a beam of silver atoms travels a horizontal distance of 1 m through such a field and each atom has a speed of 100 m/s, how strong must the field gradient dB_z / dz be in order to deflect the beam 1 mm?

6. Consider the original Stern–Gerlach experiment employing an atomic beam of silver, for which the magnetic moment is due entirely to the spin of the single valence electron of the silver atom. Assuming the magnetic field B has magnitude 0.500 T, compute the energy difference in electron volts of the silver atoms in the two exiting beams.

7. When the idea of electron spin was introduced, the electron was thought to be a tiny charged sphere (today it is considered a point object with no extension in space). Find the equatorial speed under the assumption that the electron is a uniform sphere of radius 3×10^{-6} nm, as early theorists believed, and compare your result to the speed of light, c.

8. Consider a right circular cylinder of radius R, with mass M uniformly distributed throughout the cylinder volume. The cylinder is set into rotation with angular speed ω about its longitudinal axis. (a) Obtain an expression for the angular momentum $\mathbf{L}$ of the rotating cylinder. (b) If charge Q is distributed uniformly over the *curved* surface only, find the magnetic moment $\boldsymbol{\mu}$ of the rotating cylinder. Compare your expressions for $\boldsymbol{\mu}$ and $\mathbf{L}$ to deduce the g-factor for this object.

9. An exotic elementary particle called the omega minus (symbol Ω^-) has spin $\frac{3}{2}$. Calculate the magnitude of the spin angular momentum for this particle and the possible angles the spin angular momentum vector makes with the z axis. Does the Ω^- obey the Pauli exclusion principle? Explain.

8.3 The Spin–Orbit Interaction and Other Magnetic Effects

10. Consider a single-electron atom in the $n = 2$ state. Find all possible values for j and m_j for this state.

11. Find all possible values of j and m_j for a d electron.

12. Give the spectroscopic notation for the following states: (a) $n = 7$, $\ell = 4$, $j = \frac{9}{2}$; (b) all the possible states of an electron with $n = 6$ and $\ell = 5$.

13. An electron in an atom is in the $4F_{5/2}$ state. (a) Find the values of the quantum numbers n, ℓ, and j. (b) What is the magnitude of the electron's total angular momentum? (c) What are the possible values for the z component of the electron's total angular momentum?

14. (a) Starting with the expression $\mathbf{J} = \mathbf{L} + \mathbf{S}$ for the total angular momentum of an electron, derive an expression for the scalar product $\mathbf{L} \cdot \mathbf{S}$ in terms of the quantum numbers j, ℓ, and s. (b) Using $\mathbf{L} \cdot \mathbf{S} = |\mathbf{L}| |\mathbf{S}| \cos \theta$, where θ is the angle between $\mathbf{L}$ and $\mathbf{S}$, find the angle between the electron's orbital angular momentum and spin angular momentum for the following states: (1) $P_{1/2}$, $P_{3/2}$ and (2) $H_{9/2}$, $H_{11/2}$.

15. *Spin–Orbit Energy in an Atom.* Estimate the magnitude of the spin–orbit energy for an atomic electron in the hydrogen $2p$ state. (*Hint:* From the vantage point of the moving electron, the nucleus circles it in an orbit with radius equal to the Bohr radius for this state. Treat the

orbiting nucleus as a current in a circular wire loop and use the result from classical electromagnetism,

$$B = \frac{2k_m\mu}{r^3}$$

for the B field at the center of loop with radius r and magnetic moment μ. Here $k_m = 10^{-7}$ N/A^2 is the magnetic constant in SI units.)

8.4   Exchange Symmetry and the Exclusion Principle

16. Show that the symmetric combination of two single-particle wavefunctions

$$\psi_{ab}(\mathbf{r}_1, \mathbf{r}_2) = \psi_a(\mathbf{r}_1)\psi_b(\mathbf{r}_2) + \psi_a(\mathbf{r}_2)\psi_b(\mathbf{r}_1)$$

displays the exchange symmetry characteristic of bosons, Equation 8.16. Is it possible for two bosons to occupy the same quantum state? Explain.

17. Eight identical, noninteracting particles are placed in a cubical box of sides $L = 0.200$ nm. Find the lowest energy of the system (in electron volts) and list the quantum numbers of all occupied states if (a) the particles are electrons and (b) the particles have the same mass as the electron but do not obey the exclusion principle.

8.5   Electron Interactions and Screening Effects (Optional)

18. The claim is made in Section 8.5 that a d electron is screened more effectively from the nuclear charge in an atom than is a p electron or an s electron. Give a classical argument based on the definition of angular momentum $\mathbf{L} = \mathbf{r} \times \mathbf{p}$ that indicates that smaller values of angular momentum are associated with orbits of larger eccentricity. Verify this quantum mechanically by calculating the probability that a $2p$ electron of hydrogen will be found inside the $n = 1$ atomic shell and comparing this with the probability of finding a hydrogen $2s$ electron in this same region. For which is the probability largest, and what effect does this have on the degree of screening? The relevant wavefunctions may be found in Table 7.4 of Chapter 7.

19. **MP** *Multielectron Atoms.* For atoms containing many electrons, the potential seen by the outer, or valence, electrons is often described by the Thomas–Fermi form

$$U(r) = -\frac{Zke^2}{r} e^{-r/a_{TF}}$$

where Z is the atomic number and a_{TF} is the Thomas–Fermi screening length. Use the MP DESKTOP to find the lowest valence energy and wavefunction for gold (Au), taking $Z = 79$ and $a_{TF} = 0.39$ bohr. (According to Table 8.2, gold has a valence electron configuration of $6s^1$.) How many nodes does this wavefunction exhibit?

Use the results of this study to estimate the ionization energy of gold, and compare with the experimental value given in Table 8.2. Also report the most probable distance from the nucleus for the $6s$ electron in gold, according to your findings. What size would you assign to the gold atom? How does this size compare with that of the hydrogen atom? (The Thomas–Fermi potential may be found in the file *PR8-19.U3*. The screening length a_{TF} is symbolized in the file by α.)

20. **MP** According to Table 8.1, the p-wave quantum defect for sodium is 0.86. What is the energy of the $2p$ level in sodium? the $3p$ level? Use the MP DESKTOP to determine the screening length b in Equation 8.22 that reproduces the observed p-state energies for the sodium atom. Based on your findings, locate the most probable distance from the nucleus for the $2p$ electrons in sodium. (*Note:* See Example 8.7 for using the DESKTOP to study quantum defects. Also be aware that the screening length b is symbolized by β in the potential file *EX8-7.U3*.)

8.6   The Periodic Table

21. (a) Write out the electronic configuration for oxygen ($Z = 8$). (b) Write out the values for the set of quantum numbers n, ℓ, m_ℓ, and m_s for each of the electrons in oxygen.

22. Which electronic configuration has a lower energy: [Ar]$3d^44s^2$ or [Ar]$3d^54s^1$? Identify this element and discuss Hund's rule in this case. (*Note:* The notation [Ar] represents the filled configuration for Ar.)

23. Which electronic configuration has the lesser energy and the greater number of unpaired spins: [Kr]$4d^95s^1$ or [Kr]$4d^{10}$? Identify this element and discuss Hund's rule in this case. (*Note:* The notation [Kr] represents the filled configuration for Kr.)

24. Devise a table similar to that shown in Figure 8.17 for atoms with 11 through 19 electrons. Use Hund's rule and educated guesswork.

8.7   X-ray Spectra and Moseley's Law

25. (a) Show that Moseley's law for K$_\alpha$ radiation may be expressed as

$$\sqrt{f} = \sqrt{\frac{3}{4}\left(\frac{13.6 \text{ eV}}{h}\right)}\,(Z - 1)$$

where f is the x-ray frequency and Z is the atomic number. (b) Check the agreement of the original 1914 data shown in Figure 8.20 with Moseley's law. Do this by comparing the least-squares slope and intercept of the K$_\alpha$ line in Figure 8.20 to the theoretical slope and intercept predicted by Moseley's law. (c) Is the screened charge seen by the L shell electron equal to $Z - 1$?

26. (a) Derive an equation similar to that in Problem 25, but for L_α x-rays. Assume, as in the case of K_α x-rays, that electrons in the shell of origin (in this case M) produce no screening, and that all screening is caused by electrons in the inner shells (in this case L and K). (b) Test your equation by comparing its slope and intercept with that of the experimental L_α line in Figure 8.20. (c) From the intercept of the experimental L_α line, calculate the average screened charge seen by the M shell electron.

9

Statistical Physics

Chapter Outline

Thermodynamics is based on macroscopic or bulk properties, such as temperature and pressure of a gas. In this chapter we explain thermodynamic properties by means of microscopic properties, such as the motion of individual atoms. The goal of this microscopic approach, known as *statistical physics*, or *statistical mechanics*, is to explain the relationships between thermodynamic bulk properties using a more fundamental atomic picture. It is possible in principle to calculate the detailed motion of individual atoms from Newton's laws or the Schrödinger equation. The number of atoms in the average size sample ($\approx 10^{22}$ atoms/cm^3), however, makes such calculations impractical, and we must rely on a statistical physics approach.

In this chapter we introduce the laws of statistical physics and discuss systems of particles that obey either classical or quantum mechanics. We will show how a certain amount of energy may be shared or distributed among a large number of particles in thermal equilibrium at the absolute temperature, T. We investigate this energy distribution by calculating the average number of particles with a specific energy or, what is essentially the same thing, by finding the probability that a single particle has a certain energy.

9.1   THE MAXWELL-BOLTZMANN DISTRIBUTION

The satisfying explanation of thermodynamics in terms of averages over atomic properties was given in the second half of the 1800s by three physicists: James Clerk Maxwell, Ludwig Boltzmann, and Josiah Willard Gibbs. Maxwell, a Scottish professor at Cambridge, was extremely impressed by the work of Rudolf Clausius in explaining the apparent contradiction between the high speed of gas molecules at room temperature (about 400 m/s) and the slow diffusion rate of a gas. Clausius had explained this riddle by reasoning that gas molecules do not all travel at a single high speed, but that there is a well-defined distribution of molecular speeds in a gas that depends on the gas temperature; furthermore the gas molecules collide and hence follow long zigzag paths from one spot to another. Building on this idea, Maxwell was able to derive the functional form of the equilibrium speed distribution, which is the number of gas molecules per unit volume having speeds between v and $v + dv$ at a specific temperature. Applying the theory of statistics to this distribution, Maxwell was able to calculate the temperature dependence of quantities such as the average molecular speed, the most probable speed, and the dispersion, or width, of the speed distribution.

In 1872, Boltzmann, an Austrian professor at the University of Vienna, profoundly impressed with Darwin's ideas on evolution, took Maxwell's work a step further. He not only wanted to establish the properties of the equilibrium or most probable distribution, but he also wished to describe the evolution in time of the gas toward the Maxwellian distribution—the so-called approach-to-equilibrium problem. With the use of a time-dependent speed distribution function and his kinetic equation, Boltzmann was able to show that a system of particles that starts off with a non-Maxwellian speed distribution steadily approaches and eventually achieves an equilibrium Maxwellian speed distribution.

James Clerk Maxwell (1831–1879) is generally regarded as the greatest theoretical physicist of the 19th century. *(Courtesy AIP Emilio Segrè Visual Archives)*

Is the universe a gambling casino? *(Courtesy of Tropicana Casino And Resort)*

Ludwig Boltzmann (1844–1908), an Austrian theoretical physicist. *(Courtesy AIP Niels Bohr Library, Lande Collection)*

Assumptions of the Maxwell-Boltzmann distribution

Boltzmann, a staunch advocate of the reality of molecules, was subjected to personal attacks at the hands of some who rejected the molecular theory of matter in the late 1800s. Depressed over the lack of universal acceptance of his theories, he committed suicide in 1908.

Gibbs, in contrast to Boltzmann, led a rather sheltered and secluded life as a professor at Yale. The son of a Yale professor, he lived his adult life in the same house in New Haven in which he had grown up, quietly establishing statistical mechanics and the kinetic theory of gases on a rigorous mathematical basis. Gibbs published his work in the obscure *Transactions of the Connecticut Academy of Arts and Sciences,* and his work remained relatively unknown during his lifetime.

Having briefly discussed the contributions of Maxwell, Boltzmann, and Gibbs to statistical mechanics, let us examine the underlying assumptions and explicit form of the Maxwell-Boltzmann distribution. The basic assumptions are:

- The particles are identical in terms of physical properties but distinguishable in terms of position. It will be demonstrated later in this chapter that this assumption is equivalent to the statement that the particle size is small compared with the average distance between particles.
- *The equilibrium distribution is the most probable way of distributing the particles among various allowed energy states* subject to the constraints of a fixed number of particles and fixed total energy.
- There is no theoretical limit on the fraction of the total number of particles in a given energy state, but the density of particles is sufficiently low and the temperature sufficiently high that no more than one particle is likely to be in a given state at the same time.

To make these assumptions more concrete, let us consider the analysis of a simple system of distinguishable particles. In particular, we wish to consider the distribution of a total energy of $8E$ among six particles. To work with a diagram of reasonable size, Figure 9.1a enumerates the 20 possible ways of sharing an energy of $8E$ among six *indistinguishable* particles. Since we are actually interested in *distinguishable* particles, each of the 20 arrangements can be decomposed into many distinguishable substates, or *microstates*, as shown explicitly for one arrangement in Figure 9.1b. The number of microstates for each of the 20 arrangements is given in parentheses in Figure 9.1a and may be computed from the relation

$$N_{\text{MB}} = \frac{N!}{n_1! \, n_2! \, n_3! \, \ldots} \tag{9.1}$$

where N is the number of particles and n_1, n_2, n_3, . . . are the numbers of particles in occupied states of a certain energy. This result may be understood by arguing that the first energy level may be assigned in N ways, the second in $N - 1$ ways, and so on, giving $N!$ in the numerator. The factor in the denominator of Equation 9.1 corrects for indistinguishable order arrangements when more than one particle occupies the same energy level. As an example of the use of Equation 9.1 consider the energy distribution of six particles, with two having energy $1E$, one having energy $6E$, and three having energy 0. This energy distribution is shown in the fourth diagram from the left in the top line of Figure 9.1a. In this case, $N = 6$, $n_1(0E) = 3$ (that is, the number of particles in the 0 energy state is 3), $n_2(1E) = 2$, $n_3(2E) = 0$, $n_4(3E) = 0$,

Josiah Willard Gibbs (1839–1903), an American theoretical physicist who made significant contributions to statistical mechanics. *(Burndy Library/Courtesy AIP Emilio Segrè Visual Archives)*

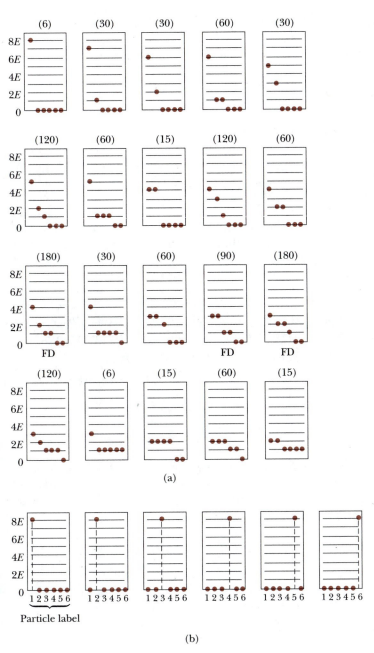

Figure 9.1   (a) The 20 arrangements of six *indistinguishable* particles with a total energy of 8E. (b) The decomposition of the upper left-hand arrangement of part (a) into six distinguishable states for *distinguishable* particles.

$n_5(4E) = 0$, $n_6(5E) = 0$, $n_7(6E) = 1$, $n_8(7E) = 0$, and $n_9(8E) = 0$. Since only the numbers of particles in occupied levels appear in the denominator of Equation 9.1, we find that the number of distinguishable microstates for this case is

$$N_{\mathrm{MB}} = \frac{6!}{3!2!1!} = 60$$

in agreement with the number in parentheses in the fourth diagram from the left in the top line of Figure 9.1a.

To find the *average number of particles with a particular value of energy*, we use the formula

$$\overline{n_j} = n_{j_1}p_1 + n_{j_2}p_2 + \ldots \tag{9.2}$$

Here $\overline{n_j}$ is the average number of particles in the jth energy level, n_{j_1} is the number of particles found in the jth level in arrangement 1, n_{j_2} is the number of particles found in the jth level in arrangement 2, p_1 is the probability of observing arrangement 1, p_2 is the probability of arrangement 2, and so on. Using the *basic postulate of statistical mechanics*, that *any individual microstate is as likely as any other individual microstate*, we may go on to calculate the various p's and $\overline{n_j}$'s. For example, since there are 1287 microstates (the sum of all the numbers in the parentheses), and 6 distinguishable ways of obtaining arrangement 1, we see that $p_1 = 6/1287$. Using these ideas and Equation 9.2, we calculate the average number of particles with energy 0 as follows:

$$
\begin{aligned}
\overline{n_0} = &\ (5)(6/1287) + (4)(30/1287) + (4)(30/1287) + (3)(60/1287) \\
&+ (4)(30/1287) + (3)(120/1287) + (2)(60/1287) + (4)(15/1287) \\
&+ (3)(120/1287) + (3)(60/1287) + (2)(180/1287) \\
&+ (1)(30/1287) + (3)(60/1287) + (2)(90/1287) + (2)(180/1287) \\
&+ (1)(120/1287) + (0)(6/1287) + (2)(15/1287) \\
&+ (1)(60/1287) + (0)(15/1287) \\
= &\ 2.307
\end{aligned}
$$

Now notice that *it is easy to calculate the probability of finding a particle with energy 0 if we imagine reaching randomly into a box containing the six particles with total energy* $8E$. This probability, $p(0)$, is simply the average number of particles with energy 0 divided by the total number of particles:

$$p(0) = \frac{\overline{n_0}}{6} = \frac{2.307}{6} = 0.385$$

It is left as a problem (Problem 1) to show that the probabilities of finding a particle with energies from $1E$ through $8E$ are as follows:

$$p(1E) = 0.256$$
$$p(2E) = 0.167$$
$$p(3E) = 0.0978$$
$$p(4E) = 0.0543$$
$$p(5E) = 0.0272$$
$$p(6E) = 0.0117$$
$$p(7E) = 0.00388$$
$$p(8E) = 0.000777$$

These results, which are plotted in Figure 9.2, show for this simple system, with only six distinguishable particles and a fixed total energy $8E$, an approximately exponential decrease in probability with energy. (See Problem 5.) The rapid decrease in probability with increasing energy shown in Figure 9.2 shows that we are more likely to find the energy uniformly distributed among many particles of the system rather than concentrated in a few particles.

One may rigorously derive the Maxwell-Boltzmann distribution for a system at the absolute temperature T containing a large number of particles by using calculus to maximize the expression for the number of ways of distributing the particles among the allowed energy states subject to two constraints. These constraints are (1) the total number of particles is constant at any temperature and (2) the total system energy is fixed at a given temperature (see, for example, supplemental reading 1). One finds an exponential form

$$f_{MB} = Ae^{-E_i/k_B T} \qquad (9.3)$$

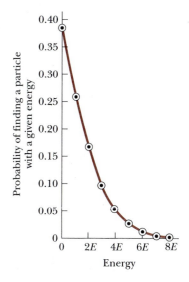

Figure 9.2 The distribution function for an assembly of six distinguishable particles with a total energy of $8E$.

Maxwell-Boltzmann distribution

where f_{MB} is the probability of finding a particle with energy E_i, or in the language of statistical mechanics, the probability that a state with energy E_i is occupied at the absolute temperature T. If the number of states with the same energy E_i is denoted by g_i (g_i is called the *degeneracy* or *statistical weight*), then the number of particles, n_i, with energy E_i is equal to the product of the statistical weight with the probability that the state E_i is occupied, or

$$n_i = g_i f_{MB} \qquad (9.4)$$

The parameter A in Equation 9.3 is a normalization coefficient, which is similar to the normalization constant in quantum physics. A is determined by requiring the number of particles in the system to be constant or

$$\sum n_i = N \qquad (9.5)$$

where N is the total number of particles in the system.

When the allowed energy states are numerous and closely spaced, the discrete quantities are replaced by continuous functions as follows:

$$g_i \longrightarrow g(E)$$

$$f_{MB} \longrightarrow Ae^{-E/k_B T}$$

where $g(E)$ is the *density of states* or the number of energy states per unit volume in the interval dE. In a similar manner, Equations 9.4 and 9.5 may be replaced as follows:

$$n_i = g_i f_{MB} \longrightarrow n(E)\, dE = g(E) f_{MB}(E)\, dE \qquad (9.6)$$

$$\sum n_i = N \longrightarrow \frac{N}{V} = \int_0^\infty n(E)\, dE = \int_0^\infty g(E) f_{MB}(E)\, dE \qquad (9.7)$$

Number of particles per unit volume with energy between E and $E + dE$

where $n(E)\, dE$ is the number of particles per unit volume with energies between E and $E + dE$. Note that Equations 9.6 and 9.7 may also be used for a system of quantum particles, provided that $g(E)$ and $f_{MB}(E)$ are replaced with the appropriate density of states and quantum distribution functions.

EXAMPLE 9.1   Emission Lines from Stellar Hydrogen

(a) Find the populations of the first and second excited states *relative* to the ground state for atomic hydrogen at room temperature, assuming that hydrogen obeys Maxwell-Boltzmann statistics.

Solution   For a gas at ordinary pressures, the atoms maintain the discrete quantum levels of isolated atoms. Recall that the discrete energy levels of atomic hydrogen are given by $E_n = (-13.6/n^2)$ eV and the degeneracy by $g_n = 2n^2$. Thus we have

Ground state: $\quad\quad\quad\quad E_1 = -13.6$ eV $\quad g_1 = 2$

First excited state: $\quad\quad E_2 = -3.40$ eV $\quad g_2 = 8$

Second excited state: $\quad E_3 = -1.51$ eV $\quad g_3 = 18$

Using Equation 9.4 gives

$$\frac{n_2}{n_1} = \frac{g_2 A e^{-E_2/k_B T}}{g_1 A e^{-E_1/k_B T}} = \frac{g_2}{g_1} e^{(E_1 - E_2)/k_B T}$$

$$= \frac{8}{2} \exp\{(-10.2 \text{ eV})/(8.617 \times 10^{-5} \text{ eV/K})(300 \text{ K})\}$$

$$= 4e^{-395} \approx 0$$

The ratio of n_3/n_1 will be even smaller. Thus, essentially all atoms are in the ground state at 300 K.

(b) Find the populations of the first and second excited states relative to the ground state for hydrogen heated to 20 000 K in a star.

Solution   When a gas is at very high temperatures (as in a flame, under electric discharge, or in a star), detectable numbers of atoms are in excited states. In this case, $T = 20\,000$ K, $k_B T = 1.72$ eV, and we find

$$\frac{n_2}{n_1} = \frac{g_2}{g_1} e^{(E_1 - E_2)/k_B T} = 4e^{-10.2/1.72} = 0.0107$$

$$\frac{n_3}{n_1} = \frac{g_3}{g_1} e^{(E_1 - E_3)/k_B T} = 9e^{-12.1/1.72} = 0.00807$$

(c) Find the emission strengths of the spectral lines corresponding to the transitions $E_3 \rightarrow E_1$ and $E_3 \rightarrow E_2$ relative to $E_2 \rightarrow E_1$ at 20 000 K.

Solution   The strength of an emission or absorption line is proportional to the number of atomic transitions per unit time. For particles obeying Maxwell-Boltzmann statistics, the number of transitions per unit time from some initial state (i) to some final state (f) equals the product of the population of the initial state and the probability

for the transition $i \rightarrow f$. Note that *the transition rate for particles obeying MB statistics depends only on the initial population, since there are no restrictions on the number of particles in the final state.* Returning to the calculation of the emission strength, S, we find the relative values:

$$\frac{S(3 \rightarrow 1)}{S(2 \rightarrow 1)} = \frac{n_3 P(3 \rightarrow 1)}{n_2 P(2 \rightarrow 1)}$$

$$\frac{S(3 \rightarrow 2)}{S(2 \rightarrow 1)} = \frac{n_3 P(3 \rightarrow 2)}{n_2 P(2 \rightarrow 1)}$$

In reality, the transition probabilities depend on the wavefunctions of the states involved, but to simplify matters we assume equal probabilities of transition; that is, $P(2 \rightarrow 1) = P(3 \rightarrow 1) = P(3 \rightarrow 2)$. This yields

$$\frac{S(3 \rightarrow 1)}{S(2 \rightarrow 1)} = \frac{n_3}{n_2} = \frac{g_3}{g_2} e^{(E_2 - E_3)/k_B T} = \frac{18}{8} e^{-1.89/1.73}$$
$$= 0.75$$

$$\frac{S(3 \rightarrow 2)}{S(2 \rightarrow 1)} = 0.75$$

If the emission lines are narrow, the measured heights of the $3 \rightarrow 1$ and $3 \rightarrow 2$ lines will be 75% of the height of the $2 \rightarrow 1$ line, as shown in Figure 9.3. For broader lines, the area under the peaks must be used as the experimental measure of emission strength.

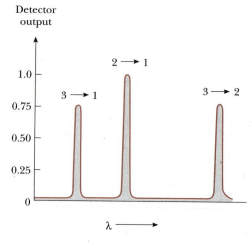

Figure 9.3   The predicted emission spectrum for the $2 \rightarrow 1$, $3 \rightarrow 1$, and $3 \rightarrow 2$ transitions for atomic hydrogen at 20 000 K.

The Maxwell Speed Distribution for Gas Molecules in Thermal Equilibrium at Temperature T

Maxwell's important formula for the equilibrium speed distribution, or the number of molecules with *speeds* between v and $v + dv$ in a gas at temperature T, may be found by using the Maxwell-Boltzmann distribution in its continuous form (Equations 9.6 and 9.7). In particular, we wish to show that

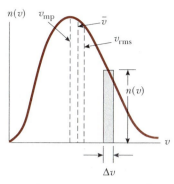

$$n(v)\ dv = \frac{4\pi N}{V}\left(\frac{m}{2\pi k_B T}\right)^{3/2} v^2 e^{-mv^2/2k_B T}\ dv \qquad (9.8)$$

where $n(v)\ dv$ is the number of gas molecules per unit volume with speeds between v and $v + dv$, N/V is the total number of molecules per unit volume, m is the mass of a gas molecule, k_B is Boltzmann's constant, and T is the absolute temperature. This speed distribution function is sketched in Figure 9.4. The v^2 term determines the behavior of the distribution as $v \to 0$, and the exponential term determines what happens as $v \to \infty$.

Assuming an ideal gas of point particles such that the energy of each molecule consists only of translational kinetic energy, we have

$$E = \tfrac{1}{2}mv^2$$

for each molecule. Since the gas molecules have translational velocities that are continuously distributed from 0 to ∞, the energy distribution of molecules is also continuous and we may write the number of molecules per unit volume with energy between E and $E + dE$ as

$$n(E)\ dE = g(E)f_{MB}(E)\ dE = g(E)Ae^{-mv^2/2k_B T}\ dE$$

To find the density of states, $g(E)$, we introduce the concept of **velocity space**. According to this idea, the velocity of each molecule may be represented by a velocity vector with components v_x, v_y, and v_z or by a point in a velocity space with coordinate axes v_x, v_y, and v_z (Fig. 9.5). From Figure 9.5 we note that the number of states $f(v)\ dv$ with speeds between v and $v + dv$ is proportional to the volume of the spherical shell between v and $v + dv$:

$$f(v)\ dv = C4\pi v^2\ dv \qquad (9.9)$$

where C is some constant. Because $E = \tfrac{1}{2}mv^2$, each speed v corresponds to a single energy E, and the number of energy states, $g(E)\ dE$, with energies between E and $E + dE$ is the same as the number of states with speeds between v and $v + dv$. Thus,

$$g(E)\ dE = f(v)\ dv = C4\pi v^2\ dv$$

Substituting this expression for $g(E)\ dE$ into our expression for $n(E)\ dE$, we obtain

$$n(E)\ dE = A4\pi v^2 e^{-mv^2/2k_B T}\ dv$$

where the constant C has been absorbed into the normalization coefficient A. Since the number of molecules with energy between E and $E + dE$ equals the number of molecules with speed between v and $v + dv$, we may write

$$n(E)\ dE = n(v)\ dv = A4\pi v^2 e^{-mv^2/2k_B T}\ dv \qquad (9.10)$$

Figure 9.4 The speed distribution of gas molecules at some temperature. The number of molecules in the range Δv is equal to the area of the shaded rectangle, $n(v)\Delta v$. The most probable speed, v_{mp}, the average speed, $\bar{v}$, and the root-mean-square speed, v_{rms}, are indicated.

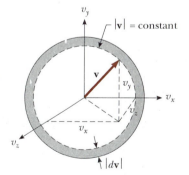

Figure 9.5 Velocity space. The number of states with speeds between v and $v + dv$ is proportional to the volume of a spherical shell with radius v and thickness dv.

To find A we use the fact that the total number of particles per unit volume is N/V:

$$\frac{N}{V} = \int_0^\infty n(v)\ dv = \int_0^\infty 4\pi A v^2 e^{-mv^2/2k_B T}\ dv \qquad (9.11)$$

Because

$$\int z^{2j} e^{-az^2}\ dz = \frac{1 \cdot 3 \cdot 5 \cdots (2j-1)}{2^{j+1} a^j} \sqrt{\frac{\pi}{a}} \qquad j = 1, 2, 3, \ldots$$

we find with $j = 1$ and $a = m/2k_B T$

$$\frac{N}{V} = \frac{(4\pi A)}{2^2(m/2k_B T)} \sqrt{\frac{\pi 2 k_B T}{m}} = A \left(\frac{2\pi k_B T}{m}\right)^{3/2}$$

or

$$A = \frac{N}{V} \left(\frac{m}{2\pi k_B T}\right)^{3/2}$$

Thus, the normalization coefficient A depends on the number of particles per unit volume, the particle mass, and the temperature. Substituting this value for A into Equation 9.10, we finally obtain Maxwell's famous 1859 result:

$$n(v)\ dv = \frac{4\pi N}{V} \left(\frac{m}{2\pi k_B T}\right)^{3/2} v^2 e^{-mv^2/2k_B T}\ dv \qquad (9.8)$$

To find the average speed, $\bar{v}$, indicated in Figure 9.4, we multiply $n(v)\ dv$ by v, integrate over all speeds from 0 to ∞, and divide the result by the total number of molecules per unit volume:

$$\bar{v} = \frac{\displaystyle\int_0^\infty v n(v)\ dv}{N/V} = \frac{4\pi(N/V)(m/2\pi k_B T)^{3/2} \displaystyle\int_0^\infty v^3 e^{-mv^2/2k_B T}\ dv}{N/V}$$

Using the definite integral formula

$$\int_0^\infty z^3 e^{-az^2}\ dz = \frac{1}{2a^2}$$

we have

$$\bar{v} = 4\pi \left(\frac{m}{2\pi k_B T}\right)^{3/2} \left(\frac{1}{2}\right) \left(\frac{2k_B T}{m}\right)^2 = \sqrt{\frac{8k_B T}{\pi m}} \qquad (9.12)$$

This famous result, first proved by Maxwell, shows that the average speed of the molecules in a gas is proportional to the square root of the temperature and inversely proportional to the square root of the molecular mass. The root mean square speed may be found by finding the average of v^2, denoted $\overline{v^2}$, and then taking its square root. Consequently, we have

$$\overline{v^2} = \frac{\displaystyle\int_0^\infty v^2 n(v)\ dv}{N/V} = 4\pi \left(\frac{m}{2\pi k_B T}\right)^{3/2} \int_0^\infty v^4 e^{-mv^2/2k_B T}\ dv$$

Using the definite integral formula

$$\int_0^\infty z^4 e^{-az^2}\, dz = \frac{3}{8a^2}\sqrt{\frac{\pi}{a}}$$

gives

$$\overline{v^2} = 4\pi \left(\frac{m}{2\pi k_B T}\right)^{3/2}\left[\frac{3}{8(m/2k_B T)^2}\right]\left(\frac{2\pi k_B T}{m}\right)^{1/2} = \frac{3k_B T}{m} \qquad (9.13)$$

Since the root mean square speed, v_{rms}, is defined as $v_{rms} = \sqrt{\overline{v^2}}$, we have

$$v_{rms} = \sqrt{\frac{3k_B T}{m}} \qquad (9.14)$$

Note that v_{rms} is *not* the same as the average speed, $\bar{v}$, but is about 10% greater, as indicated in Figure 9.4. The derivation of the most probable speed, v_{mp}, is left to Problem 2.

As a final note to this section, we observe that Equation 9.13 may be rewritten as

$$\tfrac{1}{2}m\overline{v^2} = \overline{K} = \tfrac{3}{2}k_B T$$

In this form, Equation 9.13 is consistent with the result known as the *equipartition of energy*, or *the equipartition theorem. According to this theorem, a molecule at temperature T has an average energy of $k_B T/2$ for each independent mode of motion or so-called degree of freedom.* In this case there are 3 degrees of freedom corresponding to translational motion of the molecule along the independent x, y, and z directions in space; hence the average kinetic energy in each independent direction is $k_B T/2$:

$$\tfrac{1}{2}m\overline{v_x^2} = \tfrac{1}{2}m\overline{v_y^2} = \tfrac{1}{2}m\overline{v_z^2} = \tfrac{1}{2}k_B T$$

The total average kinetic energy consequently equals three times $k_B T/2$ in agreement with Equation 9.13:

Equipartition of Energy

$$\tfrac{1}{2}m\overline{v^2} = \tfrac{1}{2}m\overline{v_x^2} + \tfrac{1}{2}m\overline{v_y^2} + \tfrac{1}{2}m\overline{v_z^2} = \tfrac{3}{2}k_B T$$

Note that degrees of freedom are not only associated with translational velocities. A degree of freedom is also associated with each rotational velocity as well. In fact, each *variable* that occurs squared in the formula for the energy of a particular system represents a degree of freedom subject to the equipartition of energy. For example, a one-dimensional harmonic oscillator with $E = \tfrac{1}{2}mv_x^2 + \tfrac{1}{2}kx^2$ has 2 degrees of freedom, one associated with its kinetic energy and the other with its potential energy. Thus, each oscillator in a group in thermal equilibrium at T has $\overline{K} = \tfrac{1}{2}m\overline{v_x^2} = \tfrac{1}{2}k_B T$ and $\overline{U} = \tfrac{1}{2}\overline{kx^2} = \tfrac{1}{2}k_B T$. The average total energy of each oscillator is then $\overline{E}_{total} = \overline{K} + \overline{U} = k_B T/2 + k_B T/2 = k_B T$. This result will be of use to us shortly when we model the atoms of a solid as a system of vibrating harmonic oscillators.

Exercise 1   (a) Show that the formula for the number of molecules with energies between E and $E + dE$ in an ideal gas with absolute temperature T is given *explicitly* in terms of E by

$$n(E)\ dE = \frac{2\pi(N/V)}{(\pi k_B T)^{3/2}} E^{1/2} e^{-E/k_B T}\ dE$$

(b) Use this result to show that the total energy per unit volume of the gas is given by

$$E_{\text{total}} = \frac{3}{2} \frac{NkT}{V}$$

in agreement with the equipartition theorem.

Exercise 2   Confirm Maxwell's 1859 result that the "spread-outness" of the speed distribution increases as $\sqrt{T}$. Do this by finding the standard deviation of the molecular speeds.

Answer

$$\sigma_v = \sqrt{3 - \frac{8}{\pi}} \cdot \sqrt{\frac{k_B T}{m}}$$

9.2   QUANTUM STATISTICS, INDISTINGUISHABILITY, AND THE PAULI EXCLUSION PRINCIPLE

If we reexamine the assumptions that lead to the Maxwell-Boltzmann distribution, keeping the quantum mechanical wave nature of particles in mind, we immediately find a problem with the assumption of distinguishability. Since particles exhibit wave-like behavior, they are necessarily fuzzy and are not distinguishable when they are close together because their wavefunctions overlap. (See Section 8.4, Exchange Symmetry and the Exclusion Principle, for a review of this issue.) If trading molecule A for molecule B no longer counts as a different configuration, then the number of ways a given energy distribution can be realized changes, as does the equilibrium or most probable distribution. Thus the classical Maxwell-Boltzmann distribution must be replaced by a quantum distribution when there is wavefunction overlap or when the particle concentration is high. The MB distribution is a valid approximation to the correct quantum distribution, however, in the common case of gases at ordinary conditions. Quantum statistics are required only for cases involving high particle concentrations, such as electrons in a metal[1] or photons in a blackbody cavity.

It is useful to develop a criterion to determine when the classical distribution is valid. We may say that the *Maxwell-Boltzmann distribution is valid when the average distance between particles, d, is large compared with the quantum uncertainty in particle position*, Δx, or

$$\Delta x \ll d \qquad\qquad (9.15)$$

---

[1] The density of conduction electrons in a metal is several thousand times the density of molecules in a gas at standard temperature and pressure.

To find Δx we use the uncertainty principle and evaluate Δp_x for a particle of mass m. For such a particle that is part of a system of particles in thermal equilibrium at temperature T, $\overline{p_x} = 0$ and $\overline{p_x^2}/2m = k_B T/2$ from the equipartition theorem. Thus

$$\Delta p_x = \sqrt{\overline{p_x^2} - (\overline{p_x})^2} = \sqrt{mk_B T} \qquad (9.16)$$

Substituting this expression for Δp_x into $\Delta p_x \, \Delta x \geq \hbar/2$, we find

$$\Delta x \geq \frac{\hbar}{2\sqrt{mk_B T}} \qquad (9.17)$$

As mentioned before, the uncertainty in particle position, Δx, must be much less than the average distance, d, between particles if the particles are to be distinguishable and the Maxwell-Boltzmann distribution is to hold. Substituting $d = (V/N)^{1/3}$ and $\Delta x = \hbar/2\sqrt{mk_B T}$ into the relation $\Delta x \ll d$ gives

$$\frac{\hbar}{2\sqrt{mk_B T}} \ll \left(\frac{V}{N}\right)^{1/3}$$

or cubing both sides,

$$\left(\frac{N}{V}\right)\frac{\hbar^3}{8(mk_B T)^{3/2}} \ll 1 \qquad (9.18)$$

Criterion for the validity of Maxwell-Boltzmann statistics

Equation 9.18 shows that the Maxwell-Boltzmann distribution holds for low particle concentration and for high particle mass and temperature.

EXAMPLE 9.2   When Can We Use Maxwell-Boltzmann Statistics?

(a) Is Maxwell-Boltzmann statistics valid for hydrogen gas at standard temperature and pressure?

Solution   Under standard conditions, 1 mol of H_2 gas (corresponding to 6.02×10^{23} molecules) occupies a volume of 22.4×10^{-3} m³. Using $m_{H_2} = 3.34 \times 10^{-27}$ kg, and $k_B T = 3.77 \times 10^{-21}$ J, we find

$$\left(\frac{N}{V}\right)\frac{\hbar^3}{8(mk_B T)^{3/2}} = \left(\frac{6.02 \times 10^{23}}{22.4 \times 10^{-3} \text{ m}^3}\right)$$

$$\times \frac{(1.055 \times 10^{-34})^3 (\text{J}\cdot\text{s})^3}{8[(3.34 \times 10^{-27} \text{ kg})(3.77 \times 10^{-21} \text{ J})]^{3/2}}$$

$$= 8.83 \times 10^{-8}$$

This is much less than one, and from the condition given by Equation 9.18, we conclude that even hydrogen, the lightest gas, is described by Maxwell-Boltzmann statistics.

(b) Is Maxwell-Boltzmann statistics valid for conduction electrons in silver at 300 K?

Solution   Silver has a density of 10.5 g/cm³ and a molar weight of 107.9 g. Assuming one free electron per silver atom, the density of free electrons in silver is found to be

$$\frac{10.5 \text{ g/cm}^3}{107.9 \text{ g/mol}}(6.02 \times 10^{23} \text{ electrons/mol})$$

$$= 5.86 \times 10^{22} \text{ electrons/cm}^3$$

$$= 5.86 \times 10^{28} \text{ electrons/m}^3$$

Note that the density of free electrons in silver is about 2000 times greater than the density of hydrogen gas molecules at STP; that is

$$\frac{(N/V)_{\text{electrons in Ag}}}{(N/V)_{H_2 \text{ at STP}}} = \frac{5.86 \times 10^{28} \text{ m}^{-3}}{2.69 \times 10^{25} \text{ m}^{-3}} = 2180$$

Using $\hbar^3 = 1.174 \times 10^{-102}$ (J·s)³, $m_e = 9.109 \times 10^{-31}$ kg, and $k_B T = 4.14 \times 10^{-21}$ J (at $T = 300$ K), we find

$$\left(\frac{N}{V}\right)\frac{\hbar^3}{8(m_e k_B T)^{3/2}} = 4.64$$

Comparing this result to the condition given by Equation 9.18, we conclude that the Maxwell-Boltzmann distribution *does not hold* for electrons in silver because of the small mass of the electron and because of the high free-electron density. We shall see that the correct quantum distribution for electrons is the Fermi-Dirac distribution.

There are two distributions valid for quantum physics, the **Bose-Einstein distribution** and the **Fermi-Dirac distribution.** For each of these, we make the assumption that the particles are identical and indistinguishable, and the assumption of maximum probability. To obtain the Bose-Einstein distribution, we retain the assumption of no theoretical limit on the number of particles per state. Particles that obey the Bose-Einstein distribution are called **bosons** and are observed to have *integral spin*. Some examples of bosons are the alpha particle ($S = 0$), the photon ($S = 1$), and the deuteron ($S = 1$). To obtain the Fermi-Dirac distribution, we again retain the first two assumptions but now stipulate that only one particle can occupy a given quantum state. (See Section 8.4 on the Pauli exclusion principle.) Particles that obey the Fermi-Dirac distribution are called **fermions** and are observed to have *half integral spin*. Some important examples of fermions are the electron, the proton, and the neutron, all with spin $\frac{1}{2}$.

To see the essential changes in the distribution function introduced by quantum statistics, let us return to our simple system of six particles with a total energy of $8E$. *First we consider the Bose-Einstein case; the particles are indistinguishable and there is no limit on the number of particles in a particular energy state.* Figure 9.1a was drawn to represent this situation. Since the particles are indistinguishable, each of the 20 arrangements shown in Figure 9.1a is equally likely, so the probability of each arrangement is $1/20$. The average number of particles in a particular energy level may be calculated by again using Equation 9.2. The average number of particles in the zero energy level is found to be

$$\overline{n_0} = (5)(1/20) + (4)(1/20) + (4)(1/20) + (3)(1/20) + (4)(1/20)$$
$$+ (3)(1/20) + (2)(1/20) + (4)(1/20) + (3)(1/20) + (3)(1/20)$$
$$+ (2)(1/20) + (1)(1/20) + (3)(1/20) + (2)(1/20) + (2)(1/20)$$
$$+ (1)(1/20) + (0)(1/20) + (2)(1/20) + (1)(1/20) + (0)(1/20)$$
$$= 49/20 = 2.45$$

Similarly, we find $\overline{n_1} = 31/20 = 1.55$, $\overline{n_2} = 18/20 = 0.90$, $\overline{n_3} = 9/20 = 0.45$, $\overline{n_4} = 6/20 = 0.30$, $\overline{n_5} = 3/20 = 0.15$, $\overline{n_6} = 2/20 = 0.10$, $\overline{n_7} = 1/20 = 0.05$, and $\overline{n_8} = 0.05$. Once again using the idea that the probability of finding a particle with a given energy, $p(E)$, is simply the average number of particles with that energy divided by the total number of particles we find

$$p(0) = \frac{\overline{n_0}}{6} = \frac{2.45}{6} = 0.408$$

In like manner we find: $p(1E) = 0.258$, $p(2E) = 0.150$, $p(3E) = 0.0750$, $p(4E) = 0.0500$, $p(5E) = 0.0250$, $p(6E) = 0.0167$, $p(7E) = 0.00833$, and $p(8E) = 0.00833$. A plot of these values in Figure 9.6 shows that the Bose-Einstein distribution gives results similar, but not identical, to the Maxwell-Boltzmann distribution. In general, the Bose-Einstein distribution tends to have more particles in the lowest energy levels. At higher energies, the curves come together and both exhibit a rapid decrease in probability with increasing energy.

To illustrate the distinctive shape of the Fermi-Dirac distribution, again consider our simple example of six indistinguishable particles with energy $8E$. *Since the particles are fermions, we impose the constraint that no more than two particles can*

Figure 9.6 The distribution function for six indistinguishable particles with total energy $8E$ (bosons).

be assigned to a given energy state (corresponding to electrons with spin up and down). There are only three states (denoted by FD) out of the 20 arrangements shown in Figure 9.1a that meet this additional constraint imposed by the Pauli exclusion principle. Since each of these arrangements is equally likely, each has a probability of occurrence of 1/3, and we may again use Equation 9.2 to calculate the average number of fermions in the zero energy level as follows:

$$\overline{n_0} = (2)(1/3) + (2)(1/3) + (2)(1/3) = 2.00$$

Similarly, we find for the average number of fermions with energies of $1E$ through $8E$ the following:

$$\overline{n_1} = 5/3 = 1.67, \overline{n_2} = 3/3 = 1, \overline{n_3} = 3/3 = 1, \overline{n_4}$$
$$= 1/3 = 0.33, \overline{n_5} = \overline{n_6} = \overline{n_7} = \overline{n_8} = 0$$

Finally we find for the probabilities of finding a fermion with energies 0 through $8E$:

$$p(0) = 2.00/6 = 0.333, p(1E) = 0.278, p(2E) = 0.167, p(3E)$$
$$= 0.167, p(4E) = 0.0550, \text{ and } p(5E) = p(6E) = p(7E) = p(8E) = 0$$

When this distribution is plotted, we see that a distinctly different distribution from the Maxwell-Boltzmann or Bose-Einstein curves results. Although it is not entirely clear that the points plotted in Figure 9.7 conform to the smooth curve drawn, consideration of systems with more than six particles proves this to be the case (see Problem 10).

When large numbers of quantum particles are considered, *continuous* distribution functions may be rigorously obtained for both the Bose-Einstein and Fermi-Dirac cases. By maximizing the number of ways of distributing the indistinguishable quantum particles among the allowed energy states, again subject to the two constraints of a fixed number of particles and a fixed total energy, we find the distribution functions to have the explicit forms:

$$f_{BE}(E) = \frac{1}{Be^{E/k_B T} - 1} \tag{9.19}$$

$$f_{FD}(E) = \frac{1}{He^{E/k_B T} + 1} \tag{9.20}$$

where $f(E)$ is the probability of finding a particle in a particular state of energy E at a given absolute temperature T. As noted earlier, the *number of particles per unit volume* with energy between E and $E + dE$ is given by

$$n(E)\ dE = g(E)f_{BE}(E)\ dE \tag{9.21}$$

or

$$n(E)\ dE = g(E)f_{FD}(E)\ dE \tag{9.22}$$

Thus the parameters B and H in Equations 9.19 and 9.20 may be determined from the total number of particles, N, since integrating Equations 9.21 and 9.22 yields

$$\left(\frac{N}{V}\right)_{bosons} = \int_0^\infty \frac{g(E)}{Be^{E/k_B T} - 1}\ dE \tag{9.23}$$

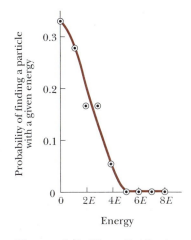

Figure 9.7 The distribution function for six indistinguishable particles with total energy $8E$ constrained so that no more than two particles occupy the same energy state (fermions).

and

$$\left(\frac{N}{V}\right)_{\text{fermions}} = \int_0^\infty \frac{g(E)}{He^{E/k_B T} + 1}\, dE \tag{9.24}$$

In general we find that B and H depend on the system temperature and particle density as shown by Equations 9.23 and 9.24. For a system of bosons that are not fixed in number with temperature, Equation 9.23 no longer serves to determine B. By maximizing the ways of distributing the bosons among allowed states subject to the single constraint of fixed energy, it can be shown that the coefficient B in Equation 9.19 is equal to one. This is a particularly important case since both photons in a blackbody cavity and phonons in a solid are bosons whose numbers per unit volume increase with increasing temperature. (We define phonons shortly.) Thus,

$$f(E) = \frac{1}{e^{E/k_B T} - 1} \qquad \text{(for photons or phonons)}$$

For the case of the Fermi-Dirac distribution, H depends strongly on temperature and is often written in an explicitly temperature-dependent form as $H = e^{-E_F/k_B T}$, where E_F is called the Fermi energy.[2] With this substitution, Equation 9.20 changes to the more common form

Fermi-Dirac distribution

$$f_{\text{FD}}(E) = \frac{1}{e^{(E - E_F)/k_B T} + 1} \tag{9.25}$$

This expression shows the meaning of the Fermi energy: the probability of finding an electron with an energy equal to the Fermi energy is exactly 1/2 at any temperature.

A plot comparing the Maxwell-Boltzmann, Bose-Einstein, and Fermi-Dirac distributions as functions of energy at a common temperature of 5000 K is shown in Figure 9.8. Note that for large E, all occupation probabilities decrease to zero as $e^{-E/k_B T}$. For small values of E, the FD probability saturates at 1, the

Figure 9.8 A comparison of Maxwell-Boltzmann, Bose-Einstein, and Fermi-Dirac distribution functions at 5000 K.

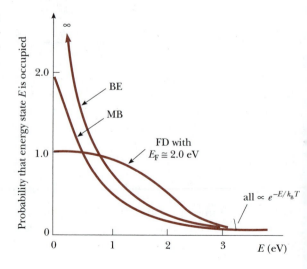

[2] If we force the functional form of H to be $H = e^{-E_F/k_B T}$, E_F will have a weak dependence itself on T. Fortunately, this dependence of E_F on T is so weak that we can ignore it here.

MB probability constantly increases but remains finite, and the BE probability tends to infinity. This very high probability for bosons to have low energies means that at low temperatures most of the particles drop into the ground state. When this happens, a new phase of matter with different physical properties can occur. This change in phase for a system of bosons is called a **Bose-Einstein condensation,** and it occurs in liquid helium at a temperature of 2.18 K. Below 2.18 K liquid helium becomes a mixture of the normal liquid and a phase with all molecules in the ground state. The ground state phase, called liquid helium II, exhibits many remarkable properties, one being zero viscosity. For more on Bose condensation, see the essay by Steven Chu at the end of this chapter.

Enrico Fermi (1901–1954)
(National Accelerator Laboratory)

Electrons and other fermions cannot all occupy the same energy state, so no condensation is possible for fermions. For small values of E such that $E \ll E_F$, the factor $e^{(E-E_F)/k_B T}$ is small and $f_{FD} \approx 1$. Thus, every energy state at low energy is filled but with only two electrons, as required by the exclusion principle.

The history of the discovery of the quantum distributions is interesting. The first quantum distribution to be discovered was the Bose-Einstein function introduced in 1924 by Satyendra Nath Bose, an Indian physicist working in isolation. His paper, which contained a new proof of the Planck formula for blackbody radiation, was sent to Einstein. In this paper, Bose applied the normal methods of statistical mechanics to light quanta but treated the quanta as absolutely indistinguishable. Einstein was impressed by Bose's work and proceeded to translate the paper into German for publication in the *Zeitschrift für Physik*.[3] To obtain the quantum theory of the ideal gas, Einstein extended the method to molecules in several papers published in 1924 and 1925.

In 1925, Wolfgang Pauli, after an exhaustive study of the quantum numbers assigned to atomic levels split by the Zeeman effect, announced his new and fundamental principle of quantum theory, the exclusion principle: **Two electrons in an atom cannot have the same set of quantum numbers**. In 1926, Enrico Fermi obtained the second type of quantum statistics that occurs in

The exclusion principle

Satyendra Nath Bose (1894–1974) appears on the far right in the first row. *(Max Planck Institute/Courtesy AIP Emilio Segrè Visual Archives)*

[3] S. N. Bose, *Z. Phys.* 26:178, 1924.

Paul Adrien Maurice Dirac (1902–1984). *(Courtesy AIP Emilio Segrè Visual Archives)*

nature by combining Pauli's exclusion principle with the requirement of indistinguishability. Paul Dirac is also credited for this work, since he performed a more rigorous quantum mechanical treatment of these statistics in 1926. The empirical observation that particles with integral spin obey BE statistics and that particles with half-integral spin obey FD statistics was explained much later (in 1940) by Pauli, using relativity and causality arguments.

9.3 APPLICATIONS OF BOSE-EINSTEIN STATISTICS

Blackbody Radiation

In this section we shall apply BE statistics to the problem of determining the energy density (energy per unit volume) of electromagnetic radiation in an enclosure heated to temperature T, now treating the radiation as a gas of photons. (In Chapter 2 we discussed the importance of this blackbody problem for quantum physics.) Since photons have spin 1, they are bosons and follow Bose-Einstein statistics and the number of photons per unit volume with energy between E and $E + dE$ is $n(E)\ dE = g(E)f_{BE}(E)\ dE$. The energy density of photons in the range from E to $E + dE$ is

$$u(E)\ dE = En(E)\ dE = \frac{g(E)E\ dE}{e^{E/k_B T} - 1} \tag{9.26}$$

To complete our calculation, we need the factor $g(E)$, the density of states for photons in an enclosure. This important calculation, given in the appendix to Chapter 2, shows that the determination of allowed modes for electromagnetic waves or photons confined to an enclosure depends on the electric field vanishing at the boundaries. From Equation 2.46 we have for the number of photon states per unit volume with frequencies between f and $f + df$

$$N(f)\ df = \frac{8\pi f^2\ df}{c^3} = \frac{8\pi (hf)^2 d(hf)}{(hc)^3} = \frac{8\pi E^2\ dE}{(hc)^3}$$

using $E = hf$ for photons. Since the number of photon states per unit volume with frequencies between f and $f + df$ is equal to the number of photon states with energies between E and $E + dE$, we have

$$N(f)\ df = \frac{8\pi E^2\ dE}{(hc)^3} = g(E)\ dE$$

Thus we find that the density of states for photons is

$$g(E) = \frac{8\pi E^2}{(hc)^3} \tag{9.27}$$

Substituting Equation 9.27 into Equation 9.26 gives an expression for the energy density:

$$u(E)\ dE = \frac{8\pi}{(hc)^3} \frac{E^3\ dE}{e^{E/k_B T} - 1} \tag{9.28}$$

Converting from photon energy to frequency using $E = hf$ in Equation 9.28, we immediately retrieve the Planck blackbody formula:

$$u(f, T) = \frac{8\pi h}{c^3} \frac{f^3}{e^{hf/k_B T} - 1} \tag{2.9}$$

Thus the Planck formula for a black body follows directly and simply from Bose-Einstein statistics.

EXAMPLE 9.3   Photons in a Box

(a) Find an expression for the number of photons per unit volume in a cavity at temperature T with energies between E and $E + dE$.

Solution

$$n(E) \; dE = g(E)f(E) \; dE = \frac{8\pi E^2 dE}{(hc)^3(e^{E/k_B T} - 1)}$$

(b) Find an expression for the total number of photons per unit volume (all energies).

Solution

$$\frac{N}{V} = \int_0^\infty n(E) \; dE = \frac{8\pi (k_B T)^3}{(hc)^3} \int_0^\infty \frac{(E/k_B T)^2 (dE/k_B T)}{e^{E/k_B T} - 1}$$

or

$$\frac{N}{V} = 8\pi (k_B T/hc)^3 \int_0^\infty \frac{z^2 \, dz}{e^z - 1}$$

(c) Calculate the number of photons/cm^3 inside a cavity whose walls are heated to 3000 K. Compare this with a cavity whose walls are at 3.00 K.

Solution   From standard tables,

$$\int_0^\infty \frac{z^2 \, dz}{e^z - 1} \approx 2.40$$

Therefore,

$$\frac{N}{V} \text{ (at 3000 K)}$$

$$= (8\pi) \left[\frac{(8.62 \times 10^{-5} \text{ eV/K})(3000 \text{ K})}{1.24 \times 10^{-4} \text{ eV} \cdot \text{cm}} \right]^3 \cdot (2.40)$$

$$= 5.47 \times 10^{11} \text{ photons/cm}^3$$

Likewise, N/V (at 3.00 K) $= 5.47 \times 10^2$ photons/cm^3. Thus, the photon density decreases by a factor of 10^9 when the temperature drops from 3000 K to 3.00 K.

Einstein's Theory of Specific Heat: Quantized Lattice Waves, or Phonons

Recall that the molar specific heat of a substance, C, is the ratio of the differential thermal energy dU added to a mole of substance divided by the resulting differential increase in temperature, dT, or

$$C = \frac{dU}{dT} \tag{9.29}$$

Thus C has units of calories per mole per kelvin (cal/mol·K). To develop a theoretical expression for comparison to the experimental curves of C versus T measured for different elemental solids, we need an expression for U, the internal thermal energy of the solid, as a function of the solid's temperature, T. Differentiation of this expression will then yield the specific heat as a function of temperature.

To find an expression for U, let us model the solid as a collection of atoms vibrating *independently* on springs with equal force constants in the x, y, and z directions, each atom being represented by three equivalent one-dimensional harmonic oscillators. The internal energy of each atom may then be calculated from the classical equipartition theorem. For a one-dimensional harmonic oscillator we have 2 degrees of freedom: one for its kinetic energy and one for its potential energy. (This may be understood by realizing that thermal energy added to a system may go into energy of motion or into work done to stretch the springs holding the atoms in place.) Since the equipartition theorem states that the average thermal energy per degree of freedom should be $k_B T/2$, the internal energy per atom of a solid should be ($k_B T/2$ per degree of freedom) $\times$ (2 degrees of freedom per one-dimensional oscillator) $\times$ (three oscillators per atom) $= 3\ k_B T$. As a mole contains Avogadro's number of atoms, N_A, the total internal energy per mole, U, is predicted to be

$$U = 3N_A k_B T = 3RT \tag{9.30}$$

where R is the universal gas constant given by $R = N_A k_B = 8.31 \text{ J/mol} \cdot \text{K} = 1.99 \text{ cal/mol} \cdot \text{K}$. Using $C = dU/dT$ we immediately see that C should be constant with temperature:

$$C = \frac{d}{dT}(3RT) = 3R = 5.97 \text{ cal/mol} \cdot \text{K} \tag{9.31}$$

The specific heat of many solids is indeed constant with temperature, especially at higher temperatures, as can be seen in Figure 9.9, showing good agreement with the classical idea that the average thermal energy is $k_B T/2$ per degree of freedom. As can also be seen in Figure 9.9, however, the specific heat of all solids drops sharply at some temperature and approaches zero as the temperature approaches zero kelvin.

The explanation of why classical physics failed to give the correct value of specific heat at all temperatures was given by Einstein in 1907. He assumed that sound, or lattice, waves in the solid had quantized energies and could be considered to be particles which we now call *phonons*, just as he had found that electromagnetic waves had quantized energies and could be analyzed as particles called *photons*. Making Einstein's assumptions that each one-dimensional atomic oscillator in the solid has the same frequency, ω, and that each atomic oscillator can be modeled as a quantum harmonic oscillator with a set of energy levels of the form $\hbar\omega(n + 1/2)$, we find that varying numbers of phonons with energy $\hbar\omega$ are constantly emitted and absorbed by the quantum oscillators. Figure 9.10 shows the allowed energy levels of the atomic oscillators in the solid and how quantized acoustic or mechanical waves with the energy $\hbar\omega$ are emitted

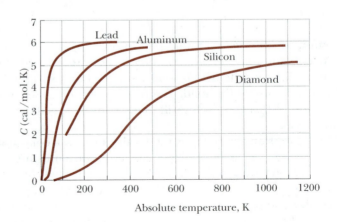

Figure 9.9 The dependence of specific heat on temperature for several solid elements.

Figure 9.10 Quantum oscillator energy levels. Dots indicate the number of oscillators occupying each level. Process (1) is an upward transition produced by the absorption of a phonon with energy $\hbar\omega$. Process (2) is a downward transition accompanied by the emission of a phonon with energy $\hbar\omega$.

and absorbed. Since oscillator transitions are only allowed between adjacent levels, all phonons have the same energy, $\hbar\omega$. In addition, phonons are bosons, and since the number of phonons is not fixed but increases rapidly with temperature, the proper distribution function for phonons is the Bose-Einstein distribution with $B = 1$. Thus *the probability that a phonon with quantized energy $\hbar\omega$ exists in a solid at temperature T is*

$$f_{\text{phonon}}(\hbar\omega) = \frac{1}{e^{\hbar\omega/k_{\text{B}}T} - 1} \tag{9.32}$$

Now note that the internal thermal energy per mole of solid, U, is primarily the energy carried by the phonons contained in a mole of solid. (We show in the next section that the contribution of free electrons is small.) Since this phonon energy per mole of solid can be expressed as the product of f_{phonon}, $e^{-\hbar\omega/k_{\text{B}}T}$, the energy per phonon, $\hbar\omega$, and the statistical weight per mole of solid, $g = 3N_{\text{A}}$, we find

$$U = 3N_{\text{A}} \frac{\hbar\omega}{e^{\hbar\omega/k_{\text{B}}T} - 1} \tag{9.33}$$

Finally we obtain the molar specific heat:

$$C = \frac{dU}{dT} = 3R \left(\frac{\hbar\omega}{k_{\text{B}}T}\right)^2 \frac{e^{\hbar\omega/k_{\text{B}}T}}{(e^{\hbar\omega/k_{\text{B}}T} - 1)^2} \tag{9.34}$$

It is left as an exercise to show that Equation 9.34 predicts that C approaches zero for small T as $e^{-\hbar\omega/k_{\text{B}}T}$ and that C approaches $3R$ for large T.

To understand Equation 9.34 qualitatively, it is useful to define a quantity, $\bar{u}$, the average energy carried by a phonon at temperature, T. $\bar{u}$ may be defined as energy $\hbar\omega$ times the probability that the phonon with energy $\hbar\omega$ exists or

$$\bar{u} = \frac{\hbar\omega}{e^{\hbar\omega/k_{\text{B}}T} - 1} \tag{9.35}$$

Note that the vibrating atoms of the solid have quantized energy levels spaced $\hbar\omega$ apart. For high temperatures such that $\hbar\omega \ll k_{\text{B}}T$, the energy level spacing $\hbar\omega$ is small relative to the average thermal energy carried per phonon, $\bar{u} \approx$

$k_B T$, and thus the atomic energies appear to be continuous, and the classical result $C = 3R$ holds. For low temperatures such that $\hbar\omega \gg k_B T$, the average thermal energy of a phonon is less than the spacing between adjacent atomic energy levels, $\hbar\omega$, and there is insufficient thermal energy in a phonon to raise an atom to its next higher energy level. In this case atoms are unable to absorb energy from the surroundings, and the specific heat tends to zero. (See Problem 14.)

A final point to note is that Equation 9.34 has only one adjustable parameter, ω, the phonon vibration frequency. Frequently ω is given in terms of an equivalent temperature T_E, called the *Einstein temperature*, where

$$\hbar\omega = k_B T_E \tag{9.36}$$

In his pioneering 1907 paper, Einstein found excellent agreement between his formula and Heinrich Weber's data on diamond with $T_E = 1300$ K. This agreement is shown in Figure 9.11.

The too rapid falloff of the Einstein formula at low temperatures hinted at in Figure 9.11 was confirmed in 1911 by Hermann Nernst. Although it was generally felt that the problem with Einstein's result was in the assumption of a single vibration frequency for all atoms and hence for all phonons, no one really knew how to treat a band or spread of phonon frequencies corresponding to groups of neighboring atoms moving together. In 1912, however, Peter Debye obtained the experimentally observed temperature dependence of $C \propto T^3$ for low temperatures by assuming that there was a distribution of allowed phonon frequencies and that both transverse and longitudinal phonon vibrations existed in solids.

Exercise 4   Show that Equation 9.34 predicts that C approaches zero for small T as $e^{-\hbar\omega/k_B T}$ and that C approaches $3R$ for large T.

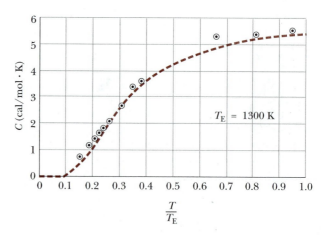

Figure 9.11   Einstein's specific heat formula fitted to Weber's experimental data for diamond. This figure is adapted from A. Einstein, *Ann. Physik.* 4(22):180, 1907.

EXAMPLE 9.4   What Are Phonons Anyway?

As we have seen, a solid at temperature T can be viewed as a system of quantized harmonic oscillators in equilibrium with a "gas" of phonons. The oscillators can only absorb a phonon and rise in energy, however, if the average phonon energy $\bar{u}$ is approximately equal to the oscillator energy level spacing, $\hbar\omega$. Thus, for $\bar{u} \approx \hbar\omega$ the phonons are coupled to the atomic oscillators, and the oscillators can exchange energy freely with the phonon gas, but for $\bar{u} \ll \hbar\omega$ the phonons are effectively isolated from the atoms and cannot add energy to the atoms. In the following example we show that the phonons in diamond are effectively decoupled from the atomic oscillators at room temperature but can provide energy at a temperature of 1500 K.

(a) Calculate the vibration frequency of the carbon atoms in diamond if the Einstein temperature is 1300 K. Also find the energy level spacing for the carbon atoms.

Solution   Since $\hbar\omega = k_B T_E$ the frequency of vibration of carbon atoms in diamond is

$$\omega = \frac{k_B T_E}{\hbar} = \frac{(8.62 \times 10^{-5} \text{ eV/K})(1300 \text{ K})}{6.58 \times 10^{-16} \text{ eV·s}}$$

$$= 1.70 \times 10^{14} \text{ Hz}$$

The spacing between adjacent oscillator energy levels in carbon is

$$\hbar\omega = (6.58 \times 10^{-16} \text{ eV·s})(1.70 \times 10^{14} \text{ Hz}) = 0.112 \text{ eV}$$

(b) Calculate the average phonon energy $\bar{u}$ at room temperature and at 1500 K and compare with the carbon energy level spacing $\hbar\omega$. Do the phonons have sufficient energy on average to excite carbon atoms at 300 K? at 1500 K?

Solution   The average phonon energy $\bar{u}$ at room temperature (300 K) is

$$\bar{u} = \frac{\hbar\omega}{e^{\hbar\omega/k_B T} - 1}$$

$$= \frac{0.112 \text{ eV}}{e^{0.112 \text{ eV}/(8.62 \times 10^{-5} \text{ eV/K})(300\text{K})} - 1}$$

$$= 0.00149 \text{ eV}$$

while the average phonon energy at 1500 K is

$$\bar{u} = \frac{0.112 \text{ eV}}{e^{0.112 \text{ eV}/(8.62 \times 10^{-5} \text{ eV/K})(1500 \text{ K})} - 1} = 0.0813 \text{ eV}$$

Comparing, we see that $\bar{u}$ at 300 K is about 0.01 $\hbar\omega$, and $\bar{u}$ at 1500 K is approximately equal to $\hbar\omega$. This means that at 300 K a phonon, on average, lacks sufficient energy to excite a carbon atom to the next higher state. At 1500 K, however, the average phonon possesses enough energy to excite carbon, and the phonon bath is in thermal equilibrium with the carbon atoms; that is, the phonons freely exchange energy with the atoms as the phonons are both emitted and absorbed by the carbon atoms.

Exercise 5   (a) Calculate the vibration frequency of lead atoms and their energy level spacing if the Einstein temperature of lead is 70 K. (b) Explain the low Einstein temperature of lead relative to that for diamond in terms of the physical properties of lead. (c) Calculate the average phonon energy in lead at room temperature. Do the phonons have enough energy to excite lead atoms at 300 K?

9.4   AN APPLICATION OF FERMI-DIRAC STATISTICS: THE FREE-ELECTRON GAS THEORY OF METALS

Since the outer electrons are weakly bound to individual atoms in a metal, we can treat these outer conduction electrons as a gas of fermions trapped within a cavity formed by the metallic surface. Once again, many interesting physical quantities, such as the average energy, Fermi energy, specific heat, and thermionic emission rate, may be derived from the expression for the concentration of electrons with energies between E and $E + dE$:

$$n(E) \; dE = g(E) f_{FD}(E) \; dE \tag{9.22}$$

Recall that the probability of finding an electron in a particular energy state E is given by the Fermi-Dirac distribution function,

$$f_{FD}(E) = \frac{1}{e^{(E-E_F)/k_B T} + 1}$$

Plots of this function versus energy are shown in Figure 9.12 for the cases $T = 0$ K and $T > 0$. Note that at $T = 0$ K, we have $f_{FD} = 1$ for $E < E_F$ and $f_{FD} = 0$ for $E > E_F$. Thus all states with energies less than E_F are completely filled and all states with energies greater than E_F are empty. This is in sharp contrast to the predictions of MB and BE statistics, in which all particles condense to a state of zero energy at absolute zero. In fact, far from having zero speed, a conduction electron in a metal with the cut-off energy E_F has a speed v_F given by

$$\tfrac{1}{2}m_e v_F{}^2 = E_F \tag{9.37}$$

where v_F is called the *Fermi speed*. Substituting a typical value of 5 eV for the Fermi energy yields the remarkable result that electrons at the Fermi level possess speeds of the order of 10^6 m/s at 0 K!

Figure 9.12b shows that as T increases, the distribution rounds off slightly, with states between E and $E - k_B T$ losing population and states between E and $E + k_B T$ gaining population. In general, E_F also depends on temperature, but the dependence is weak in metals, and we may say that $E_F(T) \approx E_F(0)$ up to several thousand kelvin.

Let us now turn to the calculation of the density of states, $g(E)$, for conduction electrons in a metal. Since the electrons may be viewed as a system of matter waves whose wavefunctions vanish at the boundaries of the metal, we obtain the same result for electrons as for electromagnetic waves confined to a cavity. In the latter case, we found (see the appendix to Chapter 2) that the number of states per unit volume with wavenumber between k and $k + dk$ is

$$g(k)\ dk = \frac{k^2\ dk}{2\pi^2} \tag{2.44}$$

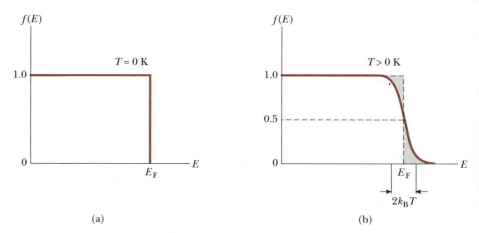

(a)

(b)

Figure 9.12   A comparison of the Fermi-Dirac distribution functions at (a) absolute zero and (b) finite temperature.

To apply this expression to electrons in a metal, we must multiply it by a factor of 2 to account for the two allowed spin states of an electron with a given momentum or energy:

$$g(k) \; dk = \frac{k^2 \; dk}{\pi^2} \tag{9.38}$$

To obtain $g(E)$ from $g(k)$, we assume nonrelativistic free electrons. Thus

$$E = \frac{p^2}{2m_e} = \frac{\hbar^2 k^2}{2m_e}$$

or

$$k = \left(\frac{2m_e E}{\hbar^2}\right)^{1/2} \tag{9.39}$$

and

$$dk = \frac{1}{2}\left(\frac{2m_e}{\hbar^2}\right)^{1/2} E^{-1/2} \; dE \tag{9.40}$$

Substituting Equations 9.39 and 9.40 into Equation 9.38 yields

$$g(E) \; dE = DE^{1/2} \; dE \tag{9.41}$$

Density of states for conduction electrons

where

$$D = \frac{8\sqrt{2}\,\pi m_e^{3/2}}{h^3} \tag{9.42}$$

Thus the key expression for the number of electrons per unit volume with energy between E and $E + dE$ becomes

$$n(E) \; dE = \frac{DE^{1/2} \; dE}{e^{(E-E_F)/k_B T} + 1} \tag{9.43}$$

Number of electrons per unit volume with energy between E and dE

Figure 9.13 is a plot of $n(E)$ versus E, showing the product of an increasing density of states and the decreasing FD distribution. Since

$$\frac{N}{V} = \int_0^\infty n(E) \; dE = D\int_0^\infty \frac{E^{1/2} \; dE}{e^{(E-E_F)/k_B T} + 1} \tag{9.44}$$

we can determine the Fermi energy as a function of the electron concentration, N/V. For arbitrary T, Equation 9.44 must be integrated numerically. At $T = 0$ K, the integration is simple since $f_{FD}(E) = 1$ for $E < E_F$ and is zero for $E > E_F$. Therefore at $T = 0$ K, Equation 9.44 becomes

$$\frac{N}{V} = D\int_0^{E_F} E^{1/2} \; dE = \tfrac{2}{3}DE_F^{3/2} \tag{9.45}$$

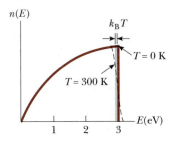

Figure 9.13 The number of electrons per unit volume with energy between E and $E + dE$. Note that $n(E) = g(E)f_{FD}(E)$.

Table 9.1 Calculated Values of Various Parameters for Metals Based on the Free-Electron Theory

Metal	Electron Concentration (m^{-3})	Fermi Energy (eV)	Fermi Speed (m/s)	Fermi Temperature (K)
Li	4.70×10^{28}	4.72	1.29×10^6	5.48×10^4
Na	2.65×10^{28}	3.23	1.07×10^6	3.75×10^4
K	1.40×10^{28}	2.12	0.86×10^6	2.46×10^4
Cu	8.49×10^{28}	7.05	1.57×10^6	8.12×10^4
Ag	5.85×10^{28}	5.48	1.39×10^6	6.36×10^4
Au	5.90×10^{28}	5.53	1.39×10^6	6.41×10^4

Substituting the value of D from Equation 9.42 into Equation 9.45 gives for the Fermi energy at 0 K, $E_F(0)$,

$$E_F(0) = \frac{h^2}{2m_e} \left(\frac{3N}{8\pi V} \right)^{2/3} \tag{9.46}$$

Equation 9.46 shows a gradual increase in $E_F(0)$ with increasing electron concentration. This is expected, because the electrons fill the available energy states, two electrons per state, in accordance with the Pauli exclusion principle up to a maximum energy E_F. Representative values of $E_F(0)$ for various metals calculated from Equation 9.46 are given in Table 9.1. This table also lists values of the Fermi velocity and the **Fermi temperature,** T_F, defined by

$$T_F \equiv E_F/k_B \tag{9.47}$$

As a final note, it is interesting that a long-standing puzzle concerning the anomalously small contribution of the conduction electron "gas" to the heat capacity of a solid has a qualitative solution in terms of the Fermi-Dirac distribution. If conduction electrons behaved classically, warming a gas of N electrons from 0 to 300 K should result in an average energy increase of $3k_B T/2$ for *each* particle, or a total thermal energy per mole, U, given by

$$U = N_A(\tfrac{3}{2}k_B T) = \tfrac{3}{2}RT$$

Thus the electronic heat capacity per mole should be given by

$$C_{el} = \frac{dU}{dT} = \tfrac{3}{2}R$$

assuming one free electron per atom. An examination of Figure 9.13, however, shows that on heating from 0 K, very few electrons become excited and gain an energy $k_B T$. Only a small fraction f within $k_B T$ of E_F can be excited thermally.

The fraction f may be approximated by the ratio of the area of a thin rectangle of width $k_B T$ and height $n(E_F)$ to the total area under $n(E)$. Thus

$$f \approx \frac{\text{Area of shaded rectangle in Figure 9.13}}{\text{Total area under } n(E)}$$

$$= \frac{(k_B T) g(E_F)}{D \int_0^{E_F} E^{1/2} \, dE} = \frac{(k_B T) D(E_F)^{1/2}}{\frac{2}{3} D E_F^{3/2}} = \frac{3}{2} \frac{k_B T}{E_F} = \frac{3}{2} \frac{T}{T_F}$$

Since only $f \cdot N$ of the electrons gain an energy of the order of $k_B T$, the actual total thermal energy gained per mole is

$$U = \left(\frac{3}{2} \frac{T}{T_F} \right) (N_A k_B T) = \frac{3}{2} \frac{R T^2}{T_F}$$

From this result, we find that the electronic heat capacity is

$$C_{el} = \frac{dU}{dT} \approx 3R \frac{T}{T_F}$$

Substituting $T = 300$ K and $T_F = 5 \times 10^4$ K, we find a very small value for the electronic heat capacity at ordinary temperatures:

$$C_{el} \approx 3R \left(\frac{300 \text{ K}}{50\,000 \text{ K}} \right) = 0.018R$$

Thus, the electrons contribute only $0.018R/1.5R$, or about 1% of the classically expected amount, to the heat capacity.

EXAMPLE 9.5   The Fermi Energy of Gold

(a) Calculate the Fermi energy for gold at 0 K.

Solution   The density of gold is 19.32 g/cm³, and its molar weight is 197 g/mole. Assuming each gold atom contributes one free electron to the Fermi gas, we can calculate the electron concentration as follows:

$$\frac{N}{V} = (19.32 \text{ g/cm}^3) \left(\frac{1}{197 \text{ g/mol}} \right)$$
$$\times (6.02 \times 10^{23} \text{ electrons/mol})$$
$$= 5.90 \times 10^{22} \text{ electrons/cm}^3$$
$$= 5.90 \times 10^{28} \text{ electrons/m}^3$$

Using Equation 9.46, we find

$$E_F(0) = \frac{h^2}{2m_e} \left(\frac{3N}{8\pi V} \right)^{2/3}$$

$$= \frac{(6.625 \times 10^{-34} \text{ J} \cdot \text{s})^2}{2(9.11 \times 10^{-31} \text{ kg})} \left(\frac{3 \times 5.90 \times 10^{28} \text{ m}^{-3}}{8\pi} \right)^{2/3}$$

$$= 8.85 \times 10^{-19} \text{ J} = 5.53 \text{ eV}$$

(b) Calculate the Fermi speed for gold at 0 K.

Solution   Since $\frac{1}{2} m_e v_F^2 = E_F$,

$$v_F = \left(\frac{2E_F}{m_e} \right)^{1/2} = \left(\frac{2 \times 5.85 \times 10^{-19} \text{ J}}{9.11 \times 10^{-31} \text{ kg}} \right)^{1/2}$$
$$= 1.39 \times 10^6 \text{ m/s}$$

(c) Calculate the Fermi temperature for gold at 0 K.

Solution   The Fermi temperature is given by

$$T_F = \frac{E_F}{k_B} = \frac{5.53 \text{ eV}}{8.62 \times 10^{-5} \text{ eV/K}} = 64\,000 \text{ K}$$

Thus a gas of classical particles would have to be heated to about 64 000 K to have an average energy per particle equal to the Fermi energy at 0 K!

SUMMARY

Statistical physics deals with the distribution of a fixed amount of energy among a number of particles that are identical and indistinguishable in any way (quantum particles) or identical particles which are distinguishable in the classical limit of narrow particle wavepackets and low particle density. In most situations, one is not interested in the energies of all the particles at a given instant, but rather in the time average of the number of particles in a particular energy level. The average number of particles in a given energy level is of special interest in spectroscopy because the intensity of radiation emitted or absorbed is proportional to the number of particles in a particular energy state.

For a system described by a continuous distribution of energy levels, the number of particles per unit volume with energy between E and $E + dE$ is given by

$$n(E)\ dE = g(E)f(E)\ dE \tag{9.6}$$

where $g(E)$ is the density of states or the number of energy states per unit volume in the interval dE, and $f(E)$ is the probability that a particle is in the energy state E. The function $f(E)$ is called the *distribution function*.

Three distinct distribution functions are used, depending on whether the particles are distinguishable, and whether there is a restriction on the number of particles in a given energy state:

- **Maxwell-Boltzmann Distribution (Classical).** The particles are distinguishable, and there is no limit on the number of particles in a given energy state.

$$f_{MB}(E) = Ae^{-E/k_B T} \tag{9.3}$$

- **Bose-Einstein Distribution (Quantum).** The particles are indistinguishable, and there is no limit on the number of particles in a given energy state.

$$f_{BE}(E) = \frac{1}{Be^{E/k_B T} - 1} \tag{9.19}$$

- **Fermi-Dirac Distribution (Quantum).** The particles are indistinguishable, and there can be no more than one particle per quantum state.

$$f_{FD}(E) = \frac{1}{e^{(E - E_F)/k_B T} + 1} \tag{9.25}$$

where E_F is the *Fermi energy*. At $T = 0$ K, all levels below E_F are filled, and all levels above E_F are empty.

At low particle concentrations and high temperature, most systems are best described by Maxwell-Boltzmann statistics. The criterion that determines when the classical Maxwell-Boltzmann distribution is valid is written

$$\left(\frac{N}{V}\right) \frac{h^3}{(8mk_B T)^{3/2}} \ll 1 \tag{9.18}$$

where N/V is the particle concentration, m is the particle mass, and T is the absolute temperature. For high particle concentration, low particle mass, and modest temperature, there is considerable overlap between the particles' wave-functions, and quantum distributions must be used to describe these systems of indistinguishable particles.

A system of photons in thermal equilibrium at temperature T is described by the Bose-Einstein distribution with $B = 1$ and a density of states given by

$$g(E) = \frac{8\pi E^2}{(hc)^3} \tag{9.27}$$

Thus, the concentration of photons with energies between E and $E + dE$ is

$$n(E) \; dE = \frac{8\pi E^2}{(hc)^3}\left(\frac{1}{e^{E/k_B T} - 1}\right) dE$$

Phonons, which are quantized lattice vibrations of a solid, are also described by the Bose-Einstein distribution with $B = 1$.

Free (conduction) electrons in metals obey the Pauli exclusion principle, and we must use the Fermi-Dirac distribution to treat such a system. The density of states for electrons in a metal is

$$g(E) = \frac{8\sqrt{2}\pi m_e^{3/2}}{h^3} E^{1/2} \tag{9.41}$$

hence the number of electrons per unit volume with energy between E and $E + dE$ is

$$n(E) \; dE = \frac{8\sqrt{2}\pi m_e^{3/2} E^{1/2}}{h^3(e^{(E-E_F)/k_B T} + 1)} dE \tag{9.43}$$

An expression for the Fermi energy as a function of electron concentration may be obtained by integrating Equation 9.43. One finds

$$E_F(0) = \frac{h^2}{2m_e}\left(\frac{3N}{8\pi V}\right)^{2/3} \tag{9.46}$$

where $E_F(0)$ is the Fermi energy at 0 K. The small electronic contribution to the heat capacity of a metal can be explained by noting that only a small fraction of the electrons near E_F gain $k_B T$ in thermal energy when the metal is heated from 0 K to T K.

SUGGESTIONS FOR FURTHER READING

1. A. Beiser, *Concepts of Modern Physics*, 5th edition, New York, McGraw-Hill Book Co., 1995.

More advanced treatments of statistical physics may be found in the following books:

2. P. M. Morse, *Thermal Physics*, New York, Benjamin, 1965.
3. C. Kittel, *Thermal Physics*, New York, Wiley, 1969.
4. D. Griffiths, *Introduction to Quantum Mechanics*, Englewood Cliffs, New Jersey, Prentice-Hall, 1994.

QUESTIONS

1. Discuss the basic assumptions of Maxwell-Boltzmann, Fermi-Dirac, and Bose-Einstein statistics. How do they differ, and what are their similarities?

2. Explain the role of the Pauli exclusion principle in describing the electrical properties of metals.

PROBLEMS

9.1   The Maxwell-Boltzmann Distribution

1. Verify that for a system of six distinguishable particles with total energy $8E$ the probabilities of finding a particle with energies $1E$ through $8E$ are: 0.256, 0.167, 0.0978, 0.0543, 0.0272, 0.0117, 0.00388, 0.000777

2. Show that the most probable speed of a gas molecule is

$$v_{mp} = \sqrt{\frac{2k_B T}{m}}$$

Note that the most probable speed corresponds to the point where the Maxwellian speed distribution curve, $n(v)$, has a maximum.

3. Figure P9.3 shows an apparatus similar to that used by Otto Stern in 1920 to verify the Maxwell speed distribution. A collimated beam of gas molecules from an oven O is allowed to enter a rapidly rotating cylinder when slit S is coincident with the beam. The pulse of molecules created by the rapid rotation of S then strikes and adheres to a glass plate detector, D. The velocity of a molecule may be determined from its position on the glass plate (fastest molecules to the right). The number of molecules arriving with a given velocity may be determined by measuring the density of molecules deposited on D at a given position. Suppose that the oven contains a gas of bismuth molecules (Bi_2) at 850 K, and that the cylinder is 10 cm in diameter and rotates at 6250 rpm. (a) Find the distance from A of the impact points of molecules traveling at $\bar{v}$, v_{rms}, and v_{mp}. (b) Why do you suppose that measurements were originally made with Bi_2 instead of O_2 or N_2?

4. Energy levels known as tunneling levels have been observed from CN^- ions incorporated into KCl crystals. These levels arise from the rotational motion of CN^- ions as they tunnel, at low temperatures, through barriers separating crystalline potential minima. According to one model, the tunneling levels should consist of four equally spaced levels in the far IR with spacing of 12.41×10^{-5} eV (1 cm^{-1}). Assuming that the CN^- ions obey Maxwell-Boltzmann statistics and using Figure P9.4, calculate and sketch the expected appearance of the absorption spectra (strength of absorption vs. energy) at 4 K and 1 K. Assume equal transition probabilities for all allowed transitions, and make the strength of absorption proportional to the peak height. Use $k_B = 8.62 \times 10^{-5}$ eV/K.

$$\Delta = 1 \text{ cm}^{-1} = 12.41 \times 10^{-5} \text{ eV}$$

Figure P9.4   Tunneling energy levels. Allowed transitions are indicated by vertical arrows. The degeneracy of each level is indicated in parentheses.

5. Fit an exponential curve $n = Ae^{-BE}$ to Figure 9.2 to see how closely the system of six distinguishable particles comes to an exponential distribution. Use the values at energies of 0 and $1E$ to determine A and B.

6. The energy difference between the first excited state of mercury and the ground state is 4.86 eV. (a) If a sample

Figure P9.3   A schematic drawing of an apparatus used to verify the Maxwell speed distribution.

of mercury vaporized in a flame contains 10^{20} atoms in thermal equilibrium at 1600 K, calculate the number of atoms in the $n = 1$ (ground) and $n = 2$ (first excited) states. (Assume the Maxwell-Boltzmann distribution applies and that the $n = 1$ and $n = 2$ states have equal statistical weights.) (b) If the mean lifetime of the $n = 2$ state is τ seconds, the transition probability is $1/\tau$ and the number of photons emitted per second by the $n = 2$ state is n_2/τ, where n_2 is the number of atoms in state 2. If the lifetime of the $n = 2$ state is 100 ns, calculate the power (in watts) emitted by the hot atoms.

7. Consider a molecule with a permanent electric dipole moment **p** placed in an electric field **ℰ**, with **p** aligned either parallel or antiparallel to **ℰ**. (a) Recall that the energy of a dipole in an electric field is given by $E = -\mathbf{p} \cdot \mathbf{ℰ}$, and show that this system has two allowed energy states separated by $2p\mathcal{E}$. (b) For simplicity assume a ground state energy of 0 and an excited state energy of $2p\mathcal{E}$ and degeneracies in the ratio $g(2p\mathcal{E})/g(0) = 2/1$. For a collection of N molecules obeying Maxwell-Boltzmann statistics, calculate the ratio of the number of molecules in the excited state to the number in the ground state at temperature T. (c) For high T such that $k_B T \gg 2p\mathcal{E}$, the ratio of the number of molecules in the upper state to the number in the lower is 2 to 1. Taking reasonable estimates of $p = 1.0 \times 10^{-30}$ C · m and $\mathcal{E} = 1.0 \times 10^6$ V/m, find the temperature at which the ratio has fallen by a measurable 10% to 1.9 to 1. (d) Calculate the average energy $\bar{E}$ at T and show that $\bar{E} \rightarrow 0$ as $T \rightarrow 0$ and $\bar{E} \rightarrow 4p\mathcal{E}/3$ as $T \rightarrow \infty$. (e) Find E_{total} from $\bar{E}$, and show that the heat capacity for this two-level system is

$$C = \left(\frac{Nk_B}{2}\right)\left(\frac{2p\mathcal{E}}{k_B T}\right)^2 \left(\frac{e^{2p\mathcal{E}/k_B T}}{(1 + \frac{1}{2}e^{2p\mathcal{E}/k_B T})^2}\right)$$

(f) Sketch C as a function of $2p\mathcal{E}/k_B T$. Find the value of $2p\mathcal{E}/k_B T$ at which C is a maximum, and explain, in physical terms, the dependence of C on T.

8. Use the distribution function given in Exercise 9.1,

$$n(E)\, dE = \frac{2\pi(N/V)}{(\pi k_B T)^{3/2}} E^{1/2} e^{-E/k_B T}\, dE$$

to find (a) the most probable kinetic energy of gas molecules at temperature T, (b) the mean kinetic energy at T, and (c) the root-mean-square kinetic energy at T.

9. The light from a heated atomic gas is shifted in frequency because of the random thermal motion of light-emitting atoms toward or away from an observer. Estimate the fractional Doppler shift ($\Delta f/f_0$) assuming that light of frequency f_0 is emitted in the rest frame of each atom, that the light-emitting atoms are iron atoms in a star at temperature 6000 K, and that the atoms are moving relative to an observer with the mean speed

$$\bar{v} = \pm\sqrt{\frac{8k_B T}{\pi m}}$$

Must we use the relativistic Doppler shift formulas

$$f = f_0 \frac{\sqrt{1 \pm v/c}}{\sqrt{1 \mp v/c}}$$

for this calculation? Such thermal Doppler shifts are measurable and are used to determine stellar surface temperatures.

9.2 Quantum Statistics

10. To obtain a more clearly defined picture of the Fermi-Dirac distribution, consider a system of 20 Fermi-Dirac particles sharing 94 units of energy. By drawing diagrams like Figure P9.10, show that there are nine different microstates. Using Equation 9.2, calculate and plot the average number of particles in each energy level from 0 to $14E$. Locate the Fermi energy at 0 K on your plot from the fact that electrons at 0 K fill all the levels consecutively up to the Fermi energy. (At 0 K the system no

15E

10E

5E

0

1 Microstate.......8 others?

Figure P9.10 One of the nine equally probable microstates for 20 FD particles with a total energy of 94E.

longer has 94 units of energy, but has the *minimum* amount of 90E.)

11. Helium atoms have spin 0 and are therefore bosons. (a) Must we use the Bose-Einstein distribution at standard temperature and pressure to describe helium gas, or will the Maxwell-Boltzmann distribution suffice? (b) Helium becomes a liquid with a density of 0.145 g/cm³ at 4.2 K and atmospheric pressure. Must the Bose-Einstein distribution be used in this case? Explain.

9.3   Applications of Bose-Einstein Statistics

12. (a) Find the average energy per photon for photons in thermal equilibrium with a cavity at temperature T. (b) Calculate the average photon energy in electron volts at $T = 6000$ K. *Hint:* Two useful integrals are

$$\int_0^\infty \frac{z^2 dz}{e^z - 1} \approx 2.41 \quad \text{and} \quad \int_0^\infty \frac{z^3 dz}{e^z - 1} = \frac{\pi^4}{15}$$

13. (a) Show that the specific heat of any substance in the Einstein model equals 5.48 cal/mol·K at $T = T_E$. (b) Using this result, estimate the Einstein temperatures of lead, aluminum, and silicon from Figure 9.9. (c) Using the temperatures found in part (b), calculate the Einstein specific heats of each element at 50-K intervals and compare with the experimental results shown in Figure 9.9. You may wish to photocopy Figure 9.9 and plot your calculated values on this figure for easy comparison.

14. Show that the average energy per phonon, $\bar{u}$ (Eq. 9.35), is approximately equal to $k_B T$ for $k_B T \gg \hbar\omega$ and that $\bar{u} \cong 0$ for $k_B T \ll \hbar\omega$. How do these results explain the variation of the Einstein specific heat with temperature?

9.4   Fermi-Dirac Statistics

15. The Fermi energy of copper at 0 K is 7.05 eV. (a) What is the average energy of a conduction electron in copper at 0 K? (b) At what temperature would the average energy of a molecule of an ideal gas equal the energy obtained in (a)? (See Problem 17)

16. The Fermi energy of aluminum is 11.63 eV. (a) Assuming that the free electron model applies to aluminum, calculate the number of free electrons per unit volume at low temperatures. (b) Determine the valence of aluminum by dividing the answer found in part (a) by the number of aluminum atoms per unit volume as calculated from the density and the atomic weight. Note that aluminum has a density of 2.70 g/cm³.

17. Show that the average kinetic energy of a conduction electron in a metal at 0 K is given by $\bar{E} = 3E_F/5$. By way

of contrast, note that *all* of the molecules in an ideal gas at 0 K have *zero energy!* *Hint:* Use the standard definition of an average given by

$$\bar{E} = \frac{\int_0^\infty E g(E) f_{FD}(E) \; dE}{N/V}$$

18. Although we usually apply Fermi-Dirac statistics to free electrons in a conductor, Fermi-Dirac statistics apply to any system of spin $\frac{1}{2}$ particles, including protons and neutrons in a nucleus. Since protons are distinguishable from neutrons, assume that each set of nucleons independently obeys the Fermi-Dirac distribution and that the number of protons equals the number of neutrons. Using these ideas, estimate E_F and $\bar{E}$ for the nucleons in Zn. (Zn has 30 protons, 34 neutrons, and a radius of 4.8×10^{-15} m.) Are your answers reasonable? Explain.

19. Show that Equation 9.46 can be expressed as

$$E_F = (3.65 \times 10^{-19}) n^{2/3} \text{ eV}$$

where E_F is in electron volts when n is in electrons per cubic meter.

20. Calculate the probability that a conduction electron in copper at 300 K has an energy equal to 99% of the Fermi energy.

21. Find the probability that a conduction electron in a metal has an energy equal to the Fermi energy at the temperature 300 K.

22. Sodium is a monovalent metal having a density of 0.971 g/cm³ and a molar mass of 23.0 g/mol. Use this information to calculate (a) the density of charge carriers, (b) the Fermi energy, and (c) the Fermi speed for sodium.

23. Calculate the energy of a conduction electron in silver at 800 K if the probability of finding the electron in that state is 0.95. Assume that the Fermi energy for silver is 5.48 eV at this temperature.

24. Consider a cube of gold 1 mm on an edge. Calculate the approximate number of conduction electrons in this cube whose energies lie in the range from 4.000 to 4.025 eV at 300 K. Assume $E_F(300 \text{ K}) = E_F(0)$.

25. (a) Consider a system of electrons confined to a three-dimensional box. Calculate the ratio of the number of allowed energy levels at 8.5 eV to the number of allowed energy levels at 7.0 eV. (b) Copper has a Fermi energy of 7.0 eV at 300 K. Calculate the ratio of the number of occupied levels at an energy of 8.5 eV to the number of occupied levels at the Fermi energy. Compare your answer with that obtained in part (a).

CALCULATOR/COMPUTER PROBLEMS

26. Consider a system of 10^4 oxygen molecules per cubic centimeter at a temperature T. Write a program that will enable you to calculate and plot a graph of the Maxwell distribution $n(v)$ as a function of v and T. Use your program to evaluate $n(v)$ for speeds ranging from $v = 0$ to $v = 2000$ m/s (in intervals of 100 m/s) at temperatures of (a) 300 K and (b) 1000 K. (c) Make graphs of $n(v)$ versus v and use the graph at $T = 1000$ K to estimate the number of molecules per cubic centimeter having speeds between 800 m/s and 1000 m/s at $T = 1000$ K. (d) Calculate and indicate on each graph the root-mean-square speed, the average speed, and the most probable speed (see Problem 2).

27. Write a program that will enable you to calculate the Fermi-Dirac distribution function, Equation 9.25, versus energy. Use your program to plot $f(E)$ versus E for (a) $T = 0.2T_F$, and (b) $T = 0.5T_F$, where T_F is the Fermi temperature, defined by Equation 9.47.

28. Copper has a Fermi energy of 7.05 eV at 300 K and a conduction electron concentration of 8.49×10^{28} m^{-3}. Write programs that will enable you to plot (a) the density of states, $g(E)$, versus E, (b) the particle distribution function, $n(E)$, as a function of energy at $T = 0$ K, and (c) the particle distribution function versus E at $T = 1000$ K. Your energy scales should range from $E = 0$ to $E = 10$ eV.

Ludwig Boltzmann spent the summer of 1905 teaching at the University of California at Berkeley. Some of his fellow Viennese, including the cartoonist who drew this image, were amused at the thought of Boltzmann lecturing on theoretical physics in the "Wild West." You can read more about Boltzmann's experience in L. Boltzmann, "A German Professor's Trip to El Dorado," *Physics Today,* January 1992, p. 44. (*University of Vienna/Courtesy AIP Emilio Segrè Visual Archives*)

Essay

LASER MANIPULATION OF ATOMS

by Steven Chu

Stanford University

The ability to control charged particles at a distance with electric and magnetic fields has led to an enormous number of advances in science and technology. Applications include particle accelerators, x-ray sources, television, and vacuum tubes. Until recently, our ability to control the motion of neutral particles has been much more limited. Usually, the device we use to hold an object such as the tips of your fingers or the wall of a bottle do not exert any force on an object until the device is brought within a few atomic diameters of the object. Once the atoms at the surface of your fingers are atomically close to an object, the electrons on your finger are repelled by the electrons of the object. The electric forces generated by this repulsion allow us to grasp neutral objects.

In the last two decades, scientists have developed the ability to manipulate neutral particles, such as atoms, molecules, and micron-sized particles. This new ability has quickly led to a wide number of applications, including the study of atom–atom interactions and light–matter interactions in an entirely new regime, the production of new quantum states of matter and the construction of exquisitely accurate atomic clocks and accelerometers.

LASER COOLING

The revolution in the control of neutral atoms is based on the ability to laser-cool atoms to extremely low temperatures. How cold? The surface of the Sun (5000 K) is 18 times hotter than the freezing point of water and 1250 times hotter than the temperature at which helium liquefies (4 K). By comparison, atoms have been cooled by lasers to less than 10^{-9} K, a billion times colder than liquid helium temperatures.

The first methods of laser cooling were simple. Since photons have momentum, light can exert forces on atoms. For a photon of frequency f, its momentum is $p_\gamma = E/c = hf/c$. As an example, a sodium atom absorbing one quantum of yellow light at 589 nm will recoil with a velocity change $\Delta v = p_\gamma / m_{Na} = 3$ cm/s. Although typical atomic speeds are on the order of 10^5 cm/s, laser light directed against a beam of atoms can be used to slow the atoms down since a laser can induce over 10^7 photon absorptions per second. Typically, atoms in a thermal beam can be stopped in a few milliseconds.

Atoms absorb light only if the laser is tuned to a resonance frequency. An atom moving toward a laser beam with velocity v will experience an upward Doppler fractional frequency shift given by $\Delta f/f = v/c$. Thus, an atom moving toward a laser beam at 10^5 cm/s will be in resonance with the laser beam only if the light is tuned 1.7 GHz below the atomic resonance. As the atom slows down, the laser frequency has to be continuously increased in order to keep it in resonance.

Once the atoms are fairly cold, further laser cooling is accomplished by surrounding the atom with counterpropagating laser beams tuned slightly to the low-frequency side of an atomic resonance as in Figure 1. The atoms absorb more photons from the beam opposing its motion because of the Doppler shift, independent of the direction of motion. By surrounding the atom with six laser beams, cooling in all three dimensions is accomplished. This method of laser cooling was first proposed in 1975 by Theodore Hänsch and Arthur Schawlow at Stanford University and first demonstrated by Steve Chu and his colleagues at AT&T Bell Labs in 1985. At that time, sodium atoms were cooled from over 500 K to 240 μK. The sea of photons surrounding the atoms was dubbed "optical molasses" because the light field serves as a viscous damping medium that slows the motion of the atoms.

Figure 1 An atom irradiated by two counterpropagating laser beams tuned to frequency f_L will see frequencies $f_L \pm \Delta f_D$. If f_L is tuned below the atomic resonance, the atom will receive more photon momentum kicks from the beam opposing the motion.

As the atoms cool to temperatures at which the average Doppler shift is a small fraction of the resonance linewidth, the differential absorption, and hence the cooling force, decreases. An equilibrium temperature is reached when the cooling rate of optical molasses balances the heating rate due to the random absorption and reemission of photons. For a simple atom consisting of one ground and one excited state separated by $\hbar\omega$, the minimum equilibrium temperature was calculated to be $k_B T = \hbar\omega/2$. In 1988, a NIST* group at Gaithersburg, Maryland, led by William Phillips discovered that atoms could be cooled in optical molasses to temperatures far below this theoretical minimum temperature.

Jean Dalibard and Claude Cohen-Tannoudji at the École Normale Supérieure, and Steve Chu, now at Stanford, realized that the low temperatures were due to the interplay between several physical effects. (1) Atoms with nonzero angular momentum have several Zeeman sublevels in the ground and excited states. (2) These Zeeman energy levels shift with the presence of light by different amounts, depending on the strength of the coupling of the atom to the light field. If the laser is tuned below the resonance frequency, the states are shifted to lower energy. (3) For a given polarization of the light field, the atom is preferentially driven into the quantum states with the lowest energy. (The population of specific Zeeman states due to polarized light is known as "optical pumping." Alfred Kastler was awarded a Nobel prize in 1966 for his studies of this effect.)

Two counterpropagating laser beams with orthogonal linear polarization generate a laser field having spatially dependent polarization as in Figure 2. An atom in a region in space where the light has positive helicity (σ^+) will optically pump into a low-energy ground state $m_F = +F$. If it then moves into a region of space where the light has negative helicity (σ^-), the $+F$ state becomes the highest-energy ground state. The kinetic energy of the atom is thus converted into internal energy. This internal energy is dissipated by the optical pumping effect, which drives the atom into the $m_F = -F$ state, the low-energy state for σ^- light. In effect, the atom slows down as it rolls up a potential hill created by the light shift of the atomic energy states. Near the top of the hill, the atom optically pumps into the new low-energy state; that is, it finds itself at the bottom of a new hill. Optical molasses with polarization gradients is sometimes referred to as Sisyphus cooling, after the mythical condemned man who faced the perpetual torment of rolling a rock up a hill only to have it roll back down again. This cooling process has cooled sodium atoms to 35 μK and cesium atoms to 3 μK.

Figure 2 (a) Polarization gradients that result from the superposition of orthogonal linearly polarized light. The local polarization of the light changes from positive helicity (σ^+) to linear to negative helicity polarization (σ^-) in a distance $\lambda/4$. (b) The energy levels of an atom with an $F = 1/2$ ground state and an $F = 3/2$ excited state irradiated with positive-helicity light. The dashed lines show the allowed spontaneous emission paths, and the dotted lines denote the energy levels in the absence of the light shifts. Because the excitation light is always trying to add $+\hbar$ unit of angular momentum to the atom, repeated excitation and spontaneous emission will optically pump the atom into the $+1/2$ ground state. The light shifts are greatly exaggerated with respect to the separation of the energy levels. (From S. Chu, *Science* 253: 861–866, 1991).

* National Institute of Standards and Technology

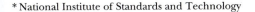

Other laser cooling techniques that use a number of quantum coherence tricks and that are given exotic names, such as "coherent population trapping" and "stimulated Raman cooling" have cooled atoms to temperatures in the nanokelvin range at which the average speed is less than the speed change an atom experiences if it recoils from a single photon. An explanation of these cooling methods can be found in the articles listed at the end of this essay.

Once atoms are cooled to low temperatures corresponding to speeds on the order of 1 cm/s, they can be easily manipulated with light, as well as static magnetic or electric fields. For example, atoms cooled to microkelvin energies have been tossed upward with light to form an atomic fountain. As they follow the ballistic trajectory dictated by gravity, they are in a nearly perturbation-free environment, and very accurate measurements of atomic energy level splittings are possible.

A fundamental limitation to the accuracy of any measurement is the Heisenberg uncertainty principle, which states that the quantum measurement time Δt and the uncertainty in the energy measurement ΔE must be greater than $\hbar/2$: $\Delta E\,\Delta t \geqslant \hbar/2$. For an atom in an atomic beam the measurement time is limited by the transit time of the atom through the measuring device. With an atomic fountain, the measurement time has been increased from a few milliseconds to roughly one second. This thousandfold increase in the measurement time will undoubtedly lead to more accurate atomic clocks. The current time standard, defined by the energy level splitting between two ground states of the cesium atom, is accurate to one part in 10^{14}.

Atomic fountains have also been used to construct atom interferometers. Similar to an optical interferometer, the atom interferometer splits the atom into a superposition of two coherent states that separate spatially and recombine to form interference fringes. Atom interferometers make extremely sensitive inertial sensors because of the long transit time of the atoms through the device. The Stanford group has measured g, the acceleration of an atom due to gravity, with a resolution of one part in 10^{8} with an atom interferometer, and it is likely that the uncertainty will be reduced to less than one part in 10^{11}. A portable version of this device could replace the mechanical "g" meters now used in oil exploration. A low value of g could signify the presence of porous, oil-laden rock, which has a lower density than solid rock.

ATOM TRAPPING

Atoms cooled below a millikelvin can also be held in space with either static magnetic fields, laser fields, or a combination of a weak magnetic field and circularly polarized light. Atom traps have been used to accumulate a large number of laser-cooled atoms, confine them for further cooling, and use them for studies of atomic collisions at very low temperatures.

A magnetic trap exerts forces on an atom via the atom's magnetic moment. The potential energy of a magnetic dipole in a magnetic field **B** is given by $-\boldsymbol{\mu} \cdot$ B, where $\boldsymbol{\mu}$ is the magnetic moment, typically on the order of 1 μ_B. Atoms were first magnetically trapped in 1985 by the NIST group with a "spherical quadrupole field" as shown in Figure 3. At the center of the trapping coils, the **B**-field is 0 and increases linearly as one moves radially outward. An atom with its magnetic dipole aligned antiparallel to the **B**-field minimizes its energy by seeking regions where the field magnitude is smallest; that is, the atom experiences a force that drives it to the center of the trap. If the atom moves slowly in the trap, it can remain aligned antiparallel to the magnetic field, even if the field changes direction. Classically, if the magnetic moment precesses rapidly around a slowly changing magnetic field, it continues to spin around a changing field axis. Quantum mechanically, the state of the atom "adiabatically" follows external field conditions that define that quantum state.

Atoms can also be held with a laser beam by using intense light tuned far from an atomic resonance to polarize the atom. The induced dipole moment **p** on the atom

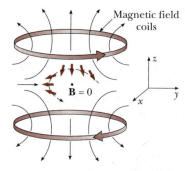

Figure 3 A spherical quadrupole trap showing some of the magnetic field lines and the direction of the currents. The heavy dashed line indicates an atom moving in the trap with its magnetic moment (bright colored arrows) remaining antiparallel to the field lines.

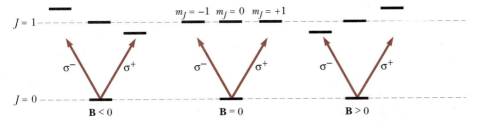

Figure 4   In a spherical quadrupole magnetic field, the energy levels of an atom with a $J = 0$ ground state and a $J = 1$ excited state will be shifted in energy in a spatially dependent way. For example, along the z axis of the field given in Figure 3, the **B** field is positive and increases linearly along the $+z$ axis. It reverses direction along the $-z$ axis. An atom irradiated by two counterpropagating beams tuned below the transition frequency as in optical molasses will cool. If the light is circularly polarized as shown, the momentum exchange will also trap the atom by tending to force it to the **B** = 0 region. Because σ^- light excites the transition $m_J = 0 \rightarrow m_J = -1$, and σ^+ light excites $m_J = 0 \rightarrow m_J = +1$, the σ^- beam is slightly more in resonance with an atom located in the positive z direction than the σ^+ light. Consequently, more σ^- photons than σ^+ photons are scattered by an atom in the $+z$ position.

points in the same direction as the driving electric field as long as the frequency of the light is well below the atomic resonance frequency. Thus, the potential energy of the atom in the light field, $-\mathbf{p} \cdot \boldsymbol{\varepsilon}$, causes the atom to seek regions of highest laser intensity. The first optical trap, demonstrated by the Bell Labs group in 1986, was based on a single focused laser beam. Since the focal point of a laser beam can be easily moved with mirrors or lenses, this trap was dubbed "optical tweezers." If the light is tuned above the atomic resonance, an atom is repelled by the light. A box of light formed from focused sheets of light has also been used to trap atoms by the Stanford group. By using an optical microscope to focus the laser light and simultaneously observe samples, this type of trap has also been used to manipulate particles, such as micron-sized spheres, bacteria, and viruses and individual molecules of DNA in aqueous solution.

The most widely used atom trap, first demonstrated at Bell Labs in 1987, uses a combination of a weak spherical magnetic quadrupole field and circularly polarized light. The magnetic field shifts the Zeeman energy levels of the atom as in Figure 4. When illuminated with circularly polarized light, an atom to the right of the trap center will predominately scatter σ^+ light, and atoms left of center will scatter more σ^- light. Thus, the atoms are pushed to the trap center by the same scattering force used in the first laser cooling experiments. The advantage of this trap is that it combines laser trapping with laser cooling and needs only low-intensity laser light and weak magnetic fields.

Another advantage is that atoms can be loaded directly inside a low-pressure vapor cell without the initial slowing from an atomic beam. Over 10^{10} atoms can be collected in a magneto-optic/cell trap in a few seconds.

BOSE CONDENSATION

Laser-cooled atoms in traps allow one to study atomic collisions in the new domain of ultracold temperatures. At very low temperatures, the atoms will no longer exhibit a Maxwell-Boltzmann distribution of energies. Depending on whether a gas of atoms has total angular momentum $0\hbar$, $1\hbar$, $2\hbar$, . . . or $\hbar/2$, $3\hbar/2$, $5\hbar/2$, . . . the atoms obey either Bose-Einstein or Fermi-Dirac statistics. One of the most dramatic quantum gas effects predicted by Einstein should occur when the de Broglie wavelength $\lambda = h/p$ of Bose atoms is comparable to the interatomic spacing. At the critical phase space density of $n\lambda^3 = 2.612$, where n is the density of atoms in the gas, a sizable fraction of the atoms should begin to condense into a single quantum state.

In 1995, a group at the University of Colorado at Boulder led by Eric Cornell and Carl Wieman verified this prediction for a gas of rubidium atoms by combining many of the cooling and trapping techniques developed in the previous ten years. They started by cooling and trapping atoms in a magneto-optic trap. Further cooling was accomplished using optical molasses and Sisyphus cooling in the absence of a magnetic field. At this point, the phase space density of the atoms was increased above that of an intense thermal atomic beam by 13 orders of magnitude. The atoms were then optically pumped into a particular atomic Zeeman state and loaded into a magnetic trap.

The spherical quadrupole magnetic trap they used was modified to correct a known flaw of the NIST trap. If an atom in the trap ventures close to the zero magnetic field point, the precession frequency of the atom goes to zero and its spin can no longer follow the direction of the field line. The spin could then flip to become parallel to the field and hence antitrapped: the trap has a "hole" at its center. The Boulder group solved the problem by rotating the zero point of the magnetic field with additional magnetic field coils. If the rotation frequency is made much faster than the atomic motion in the trap, the atoms experience a time-averaged potential centered on the point of rotation. Since the hole rotates outside of the ensemble of atoms, the atoms never see the hole.

The final cooling was accomplished by "evaporization" from the magnetic trap, a cooling technique first demonstrated by Harold Hess, Tom Greytak, Dan Kleppner, and colleagues at MIT in 1989. Evaporization allows the very hottest atoms to leave the sample, thereby reducing the average energy of the remaining atoms. Collisions rethermalize the sample to produce more hot atoms, and as the evaporization proceeds, the evaporization barrier is lowered in order to continue the cooling process. Using this technique, they were able to increase the phase space density by another five orders of magnitude.

Figure 5 shows the velocity distribution of atoms cooled in the vicinity of the Bose condensation threshold. Just above the required phase space density, the atomic energy is distributed in all directions equally among the many quantum states, in accord with the equipartition theorem of statistical physics. As the threshold condition is crossed, a central peak at $v = 0$ begins to form, signifying the onset of a condensation. With further cooling, most of the atoms condense into the ground quantum state of the system. Since the trap potential is not spherically symmetrical, the ground quantum state is also asymmetrical.

Once the atoms are in a Bose condensate, their energy is defined in terms of the localization energy $\Delta x \, \Delta p \geq \hbar/2$. Since the atoms remain in the Bose condensate as the confining forces of the trap are relaxed, the effective temperature $k_B T/2 = 3\Delta p^2/2m$ can be brought into the picokelvin range.

The activity in laser cooling and atom trapping has exploded into many areas of physics, chemistry, biology, and medicine. One of the joys of science is that many of

Figure 5   Computer-generated images of the velocity distribution of a cloud of rubidium atoms. Before the Bose-Einstein threshold is reached, the velocity distribution is uniform. The leftmost image was obtained just before the appearance of the Bose-Einstein condensate. Immediately after the appearance of the condensate (middle image), an asymmetrical velocity spike centered at $v = 0$ appears above the broader uniform distribution. At still higher phase space densities (rightmost image), most of the remaining atoms have condensed into a ground quantum state. The field of view of each frame is $200 \times 270 \ \mu$m, corresponding to the distance the atoms have moved in about 1/20 of a second. (Images courtesy of Eric Cornell and Mike Matthews, JILA Research Team, University of Colorado at Boulder.)

these applications were not foreseen by the inventors of the field in the early days. Similarly, the applications of a Bose condensate of atoms cannot be predicted. Since the condensate is analogous to photons in a laser cavity, one can, in principle, construct an exquisitely intense atomic beam. Perhaps most of the uses of an intensely bright beam of atoms in a single quantum state will be invented by the next generation of scientists. This is an exciting time for physics.

Suggestions for Further Reading

For general reviews of cooling and trapping, see S. Chu, *Sci. Am.*/February, 1992; and S. Chu, *Science* 253:861, 1991.

For a description of newer methods of laser cooling, see C. Cohen-Tannoudji and W. D. Phillips, *Phys. Today* 43:33, 1990. Raman cooling is described in M. Kasevich and S. Chu, *Phys. Rev. Lett.* 69:1741, 1992.

The first Bose-Einstein condensation of a gas is described in M. H. Anderson, J. R. Ensher, M. R. Matthews, C. E. Wieman, and E. A. Cornell, *Science* 269:198, 1995.

10

Molecular Structure

Chapter Outline

Except for the inert gases, elements generally combine to form chemical compounds whose basic unit is the *molecule*, an aggregate of individual atoms joined by chemical bonds. The physical and chemical properties of molecules derive from their constituent atoms—their arrangement, the manner and degree to which they interact, and their individual electronic structures.

The properties of molecules can be studied experimentally by examining their spectra. As with atoms, a molecule can emit or absorb photons, with accompanying electronic transitions among the allowed energy levels of the molecule. The resulting emission or absorption spectrum is different for each molecule, and acts as a sort of fingerprint of its electronic structure.

But molecules also emit or absorb energy in ways not found in atoms. Molecules can rotate, storing energy in the form of kinetic energy of rotation; and they can vibrate, and so possess energy of vibration. As we shall discover, both the rotational and vibrational energies are quantized, and so give rise to their own unique spectra. It follows that molecular spectra are vastly more complicated than atomic spectra, but also carry a good deal more information. In particular, the vibration–rotation spectrum tells us how the individual atoms that form the molecule are arranged, and the strength of their interaction.

In this chapter, we shall describe the bonding mechanisms in molecules, the various modes of molecular excitation, and the radiation emitted or absorbed by molecules. In the course of our study, we shall encounter the quantum origins of the chemical bond, and discover why some atoms bond to form a molecule and others do not. Central to this inquiry are the roles played by the exclusion principle and tunneling, both nonclassical ideas that underscore the importance of wave mechanics to the study of molecular structure.

10.1 BONDING MECHANISMS: A SURVEY

Two atoms combine to form a molecule because of a net attractive force between them. Furthermore, the total energy of the stable bound molecule is *less* than the total energy of the separated atoms; the energy difference, or *binding energy*, represents the energy that must be supplied to break apart the molecule into its constituent atoms.

Fundamentally, the bonding mechanisms in a molecule are primarily due to electrostatic forces between atoms (or ions). When two atoms are separated by an infinite distance, the force between them is zero, as is their electrostatic energy. As the atoms are brought closer together, both attractive and repulsive forces come into play. At very large separations, the dominant forces are attractive in nature. For small separations, repulsive forces between like charges begin to dominate. The potential energy of the pair can be positive or negative, depending on the separation between the atoms.

The total potential energy of two atoms often is approximated by the expression

$$U = -\frac{A}{r^n} + \frac{B}{r^m}$$

where r is the internuclear separation, A and B are constants associated with the attractive and repulsive forces, and n and m are small integers. Figure 10.1 presents a sketch of the total potential energy versus internuclear separation. Note that the potential energy for large separations is negative, corresponding to a net attractive force. At the equilibrium separation, the attractive and repulsive forces just balance and the potential energy has its minimum value.

A complete description of the binding mechanisms in molecules is a highly complex problem because it involves the mutual interactions of many particles. In this section, we shall discuss some simplified models in the following order of decreasing bond strength: the ionic bond, the covalent bond, the hydrogen bond, and the van der Waals bond.

Figure 10.1 The total particle energy as a function of the internuclear separation for a system of two atoms.

Ionic Bonds

Ionic bonds are fundamentally due to the Coulomb attraction between oppositely charged ions. A familiar example of an ionically bonded molecule is sodium chloride, NaCl, or common table salt. Sodium, which has an electronic configuration $1s^2 2s^2 2p^6 3s$, gives up its $3s$ valence electron to form a Na^+ ion. The energy required to ionize the atom to form Na^+ is 5.1 eV. Chlorine, which has an electronic configuration $1s^2 2s^2 2p^5$, is one electron short of the closed-shell structure of argon. Because closed-shell configurations are energetically more favorable, the Cl^- ion is more stable than the neutral Cl atom. The energy released when an atom takes on an electron is called the **electron affinity.** For chlorine, the electron affinity is 3.7 eV. Therefore, the energy required to form Na^+ and Cl^- from isolated atoms is $5.1 - 3.7 = 1.4$ eV. As the ions are brought closer together, their mutual energy decreases due to electrostatic attraction. At sufficiently small separations, the energy of formation becomes negative, indicating that the ion pair now is energetically preferred over neutral Na and Cl atoms.

Figure 10.2   Total energy versus the internuclear separation for Na^+ and Cl^- ions. The energy required to separate the NaCl molecule into neutral atoms of Na and Cl is 4.2 eV.

The total energy versus the internuclear separation for Na^+ and Cl^- ions is shown in Figure 10.2. Note that the total energy of the molecule has a minimum value of -4.2 eV at the equilibrium separation of about 0.24 nm. The energy required to separate the NaCl molecule into neutral sodium and chlorine atoms, called the **dissociation energy,** is equal to 4.2 eV.

When the two ions are brought closer than 0.24 nm, the electrons in closed shells begin to overlap, which results in a repulsion between the closed shells. This repulsion is partly electrostatic in origin, and partly a result of the identity of electrons. Because they must obey the exclusion principle (Chapter 8), some electrons in overlapping shells are forced into higher energy states, and the system energy increases, as if there were a repulsive force between the ions.

Covalent Bonds

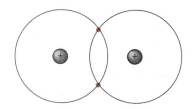

Figure 10.3   Classical orbit model for the covalent bond formed by the two $1s$ electrons of the H_2 molecule.

A covalent bond between two atoms can be visualized as the sharing of electrons supplied by one or both atoms that form the molecule. Many diatomic molecules, such as H_2, F_2, and CO, owe their stability to covalent bonds. In the case of the H_2 molecule, the two electrons are equally shared between the nuclei, and form a so-called molecular orbital. The two electrons are more likely to be found between the two nuclei, hence the electron density is large in this region. The formation of the molecular orbital from the s orbitals of the two hydrogen atoms is represented in Figure 10.3. Because of the exclusion principle, the two electrons in the ground state of H_2 must have antiparallel spins. It is interesting to note that if a third H atom is brought near the H_2 molecule, the third electron would have to occupy a higher-energy quantum state because of the exclusion principle, which is an energetically unfavorable situation. Hence, the H_3 molecule is not stable and does not form.

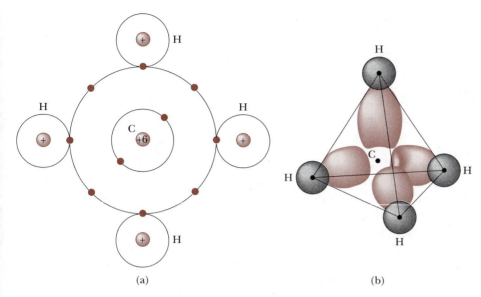

(a)   (b)

Figure 10.4   (a) Classical orbit model for the four covalent bonds in the CH_4 molecule. (b) Quantum-mechanical picture of the spatial arrangement of the four covalent bonds of the CH_4 molecule. The carbon atom is at the center of a tetrahedron with hydrogen atoms at its corners. The orbitals supplied by carbon are actually sp^3 hybrid orbitals as explained at the end of Section 10.5. These orbitals have two lobes, and only the larger lobes, which are shown here greatly narrowed for ease of depiction, participate in the bonding. Each C—H bond consists of an overlapping $1s$ orbital from hydrogen and an sp^3 hybrid orbital from carbon. (*Adapted from D. Ebbing, General Chemistry, 5th ed., Boston, Houghton Mifflin Co., 1996.*)

More complex stable molecules, such as H_2O, CO_2, and CH_4, are also formed by covalent bonds. Consider methane, CH_4, a typical organic molecule shown schematically in the electron-sharing diagram of Figure 10.4a. Note that covalent bonds are formed between the carbon atom and each of the four hydrogen atoms. The spatial electron distribution of the four covalent bonds is shown in Figure 10.4b. The four hydrogen nuclei are at the corners of a regular tetrahedron, with the carbon nucleus at the center.

The Hydrogen Bond

Because hydrogen has only one electron, it is expected to form a covalent bond with only one other atom. In some molecules, however, hydrogen forms a different type of bond between two atoms or ions, called a **hydrogen bond.** One example of a hydrogen bond shown in Figure 10.5 is the hydrogen difluoride ion, $(HF_2)^-$. The two negative fluorine ions are bound by the positively charged proton between them. This is a relatively weak chemical bond, with a binding energy of only about 0.1 eV. Although the hydrogen bond is weak, it is the mechanism responsible for linking giant biological molecules and polymers. For example, in the case of the famous DNA molecule that has a double helix structure, hydrogen bonds link the turns of the helix.

Figure 10.5 Hydrogen bonding in the $(HF_2)^-$ molecular ion. The two negative fluorine ions are bound by the positively charged proton between them.

Van der Waals Bonds

Atoms that do not form ionic or covalent bonds still are attracted to each other by weak electrostatic forces. For this reason, at sufficiently low temperatures at which thermal excitations are negligible, substances will condense into a liquid and then solidify (with the exception of helium, which does not solidify at atmospheric pressure). The weak electrostatic attractions between atoms or molecules are called **van der Waals forces.**

There are actually three types of van der Waals forces, which we shall briefly describe. The first type, called the *dipole–dipole force*, occurs between two molecules each having a permanent electric dipole moment. For example, polar molecules such as HCl and H_2O have permanent electric dipole moments and attract other polar molecules. In effect, one molecule interacts with the electric field produced by another molecule.

The second type of van der Waals force is a *dipole-induced force* in which a polar molecule having a permanent electric dipole moment induces a dipole moment in a nonpolar molecule.

The third type of van der Waals force is called the *dispersion force*, or the *London* force. The dispersion force is an attractive force that occurs between two nonpolar atoms or molecules. In this case, the interaction results from the fact that although the average dipole moment of a nonpolar molecule is zero, the average of the square of the dipole moment is nonzero because of charge fluctuations. Consequently, two nonpolar molecules near each other tend to be correlated so as to produce an attractive force, which is the London force.

Since all three types of van der Waals forces are dipolar in origin, they all fall off with separation distance r as $1/r^7$; however, the proportionality constant is different for each type.

10.2   MOLECULAR ROTATION AND VIBRATION

The various means by which a molecule can store energy internally can be divided into several categories: (1) electronic, accompanied by electron rearrangement among the allowed energy levels of the molecule; (2) rotational, corresponding to rotation of the molecule about its center of mass; and (3) vibrational, represented by the vibration of the molecule's constituent atoms.[1]

The electronic contribution to the energy of a molecule is difficult to determine because each electron interacts with all the others and with the atomic nuclei forming the molecule. The allowed electronic levels of a molecule will be taken up in Section 10.4. The remaining two contributions to molecular energy—rotation and vibration—are far easier to describe and are discussed in this section.

---

[1] To these *internal* modes we should add for completeness the translational mode, describing the motion of the molecule's center of mass through space. Since the quantization of translational energy is unrelated to internal structure, however, the translational mode is unimportant for the interpretation of molecular spectra.

Molecular Rotation

Let us consider the rotation of a molecule about a fixed axis. We shall confine our discussion to a diatomic molecule, although the same ideas can be extended to polyatomic molecules. As shown in Figure 10.6a, the diatomic molecule has only two rotational degrees of freedom, corresponding to rotations about the y and z axes, that is, the axes perpendicular to the molecular axis.[2]

The energy of a rotating molecule is all kinetic. Let m_1 and m_2 denote the atomic masses, with speeds v_1 and v_2. For a molecule in rotation, the speeds v_1 and v_2 are interrelated. In terms of the angular velocity of rotation ω, we have

$$v_1 = \omega r_1 \quad \text{and} \quad v_2 = \omega r_2$$

where r_1 is the distance of m_1 from the axis of rotation, and similarly for r_2 (see Fig. 10.6a). The angular momentum of rotation about the z axis is

$$L = m_1 v_1 r_1 + m_2 v_2 r_2 = \{m_1 r_1{}^2 + m_2 r_2{}^2\}\omega = I\omega$$

$$\boxed{E_1 = \frac{\hbar^2}{2I}}$$

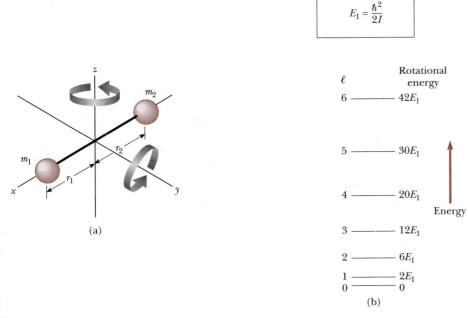

(a)

(b)

Figure 10.6   (a) A diatomic molecule oriented along the x axis has two rotational degrees of freedom, corresponding to rotation about the y and z axes. (b) Allowed rotational energies of a diatomic molecule as calculated using Equation 10.5.

---

[2] The excitation energy for rotations about the molecular axis is so large that such modes are not observable. This follows because nearly all the molecular mass is concentrated within nuclear dimensions of the rotation axis, giving a negligibly small moment of inertia about the internuclear line (see Problem 11).

The quantity in curly brackets $\{ \ . \ . \ . \ \}$ is the *moment of inertia*, denoted I. With this identification, the energy of rotation is

$$E_{\text{rot}} = \tfrac{1}{2}m_1 v_1{}^2 + \tfrac{1}{2}m_2 v_2{}^2 = \tfrac{1}{2}I\omega^2$$

Eliminating ω from the preceding two equations gives the simple result

$$E_{\text{rot}} = \frac{L^2}{2I} \tag{10.1}$$

Comparing Equation 10.1 with the kinetic energy of a particle in translation, $p^2/2m$, we see that the moment of inertia I of the molecule measures its resistance to changes in rotation in the same way that the mass m of a single particle measures its resistance to changes in translation. The value for I depends on the rotation axis, however. For the important case where the axis of rotation passes through the center of mass,[3] we have $m_1 r_1 = m_2 r_2$, and r_1 and r_2 can be written in terms of the atomic separation $R_0 = r_1 + r_2$ as

$$R_0 = \left(\frac{m_2}{m_1} + 1 \right) r_2 = \left(1 + \frac{m_1}{m_2} \right) r_1$$

Then the moment of inertia about the center of mass, I_{CM}, becomes

$$I_{\text{CM}} = \left(\frac{m_1 m_2}{m_1 + m_2} \right) R_0{}^2 = \mu R_0{}^2 \tag{10.2}$$

where μ is the **reduced mass** of the molecule,

$$\mu = \frac{m_1 m_2}{m_1 + m_2} \tag{10.3}$$

Unlike moment of inertia, which is a *property* of the molecule, angular momentum L is a *dynamical variable;* in the transition to quantum mechanics, L^2 becomes quantized as discussed in Chapter 7. The correct quantization rule is

$$L^2 = \ell(\ell + 1)\hbar^2 \qquad \ell = 0, 1, 2, \ . \ . \ . \tag{10.4}$$

which, in turn, restricts the energy of rotation to be one of the values

$$E_{\text{rot}} = \frac{\hbar^2}{2I_{\text{CM}}} \ell(\ell + 1) \qquad \ell = 0, 1, 2, \ . \ . \ . \tag{10.5}$$

In the context of molecular rotation, the integer ℓ is called the **rotational quantum number.** Thus, we see that *the rotational energy of the molecule is quantized and depends on the moment of inertia of the molecule.* The allowed rotational energies of a diatomic molecule are sketched in Figure 10.6b. These results apply also to polyatomic molecules provided the appropriate generalization of I_{CM} is used.

---

[3] Since there is no axis about which the molecule is constrained to rotate, the correct axis for calculating rotational energy is one passing through the center of mass. Energy associated with rotation about any other axis would include some energy of translation, as well as energy of rotation.

Table 10.1 Microwave Absorption Lines for Several Rotational Transitions of the CO Molecule

Rotational Transition	Wavelength of Absorption Line (m)	Frequency of Absorption Line (Hz)
$\ell = 0 \rightarrow \ell = 1$	2.60×10^{-3}	1.15×10^{11}
$\ell = 1 \rightarrow \ell = 2$	1.30×10^{-3}	2.30×10^{11}
$\ell = 2 \rightarrow \ell = 3$	8.77×10^{-4}	3.46×10^{11}
$\ell = 3 \rightarrow \ell = 4$	6.50×10^{-4}	4.61×10^{11}

The spacing between adjacent rotational levels can be calculated from Equation 10.5:

$$\Delta E = E_\ell - E_{\ell-1} = \frac{\hbar^2}{2I_{CM}} \{\ell(\ell+1) - (\ell-1)\ell\} = \frac{\hbar^2}{I_{CM}}\ell \qquad (10.6)$$

where ℓ is the quantum number of the higher energy state. In going from one rotational state to the next, the molecule loses (or gains) energy, ΔE. The loss (or gain) typically is accompanied by photon emission (or absorption) at the frequency $\omega = \Delta E/\hbar$. Thus, photons should be observed at the frequencies $\omega_0 = \hbar/I_{CM}$, $2\omega_0$, $3\omega_0$, ... These predictions are in excellent agreement with experiment.[4] The wavelengths and frequencies for the absorption spectrum of the CO molecule are given in Table 10.1. The lowest frequencies lie in the *microwave range* of the electromagnetic spectrum, as is typical of the rotational spectra of all molecules. From the data, one can deduce the moment of inertia and the bond length of the molecule, as shown in the following example.

EXAMPLE 10.1   Rotation of the CO Molecule

The $\ell = 0$ to $\ell = 1$ rotational transition of the CO molecule occurs at a frequency of 1.15×10^{11} Hz. (a) Use this information to calculate the moment of inertia of the molecule about its center of mass.

Solution   From Equation 10.6, we see that the energy difference between the $\ell = 0$ and $\ell = 1$ rotational levels is $\hbar^2/I_{CM}$. Equating this to the energy of the absorbed photon, we get

$$\frac{\hbar^2}{I_{CM}} = \hbar\omega$$

The angular frequency ω of the absorbed radiation is

$$\omega = 2\pi f = 2\pi(1.15 \times 10^{11} \text{ Hz}) = 7.23 \times 10^{11} \text{ rad/s}$$

so that I_{CM} becomes

$$I_{CM} = \frac{\hbar}{\omega} = \frac{1.055 \times 10^{-34} \text{ J} \cdot \text{s}}{7.23 \times 10^{11} \text{ rad/s}} = 1.46 \times 10^{-46} \text{ kg} \cdot \text{m}^2$$

(b) Calculate the bond length of the molecule.

Solution   Equation 10.2 can be used to calculate the bond length once the reduced mass of the molecule is

[4] These simple statements must be refined when molecular vibration is taken into account, as discussed in Section 10.3.

found. Since the carbon and oxygen atomic masses are 12 u and 16 u, respectively, the reduced mass of the CO molecule is, from Equation 10.3,

$$\mu = \frac{(12\ u)(16\ u)}{12\ u + 16\ u} = 6.857\ u = 1.14 \times 10^{-26}\ kg$$

where the conversion $1\ u = 1.66 \times 10^{-27}\ kg$ has been used. Then

$$R_0 = \sqrt{\frac{I_{CM}}{\mu}} = \sqrt{\frac{1.46 \times 10^{-46}\ kg \cdot m^2}{1.14 \times 10^{-26}\ kg}}$$

$$= 1.13 \times 10^{-10}\ m = 0.113\ nm$$

This example hints at the immense power of spectroscopic measurements to determine molecular properties!

Molecular Vibration

A molecule is a flexible structure whose atoms are bonded together by what can be considered "effective springs." If disturbed, the molecule can vibrate, taking on vibrational energy. This energy of vibration may be altered if the molecule is exposed to radiation of the proper frequency.

Consider again a diatomic molecule. The potential energy $U(r)$ versus atomic separation r for such a molecule is sketched in Figure 10.7a. The equilibrium separation of the atoms is denoted there by R_0; for small displacements from equilibrium, the atoms vibrate, as if connected by a spring with unstretched length R_0 and force constant K (Fig. 10.7b).[5] Atomic displacements in the direction of the molecular axis give rise to oscillations along the line joining the atoms. For these *longitudinal* vibrations, the system is effectively one-dimensional, with the coordinates of each atom measured along the molecular axis.

(a)

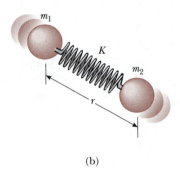

(b)

Figure 10.7   (a) A plot of the potential energy of a diatomic molecule versus atomic separation. The parameter R_0 is the equilibrium separation of the atoms. (b) Model of a diatomic molecule whose atoms are bonded by an effective spring of force constant K. The fundamental vibration is along the molecular axis.

---

[5] In Section 5.6 of Chapter 5 we showed that the effective force constant K is given by the curvature of $U(r)$ evaluated at the equilibrium separation R_0; that is, $K = \partial^2 U/\partial r^2|_{R_0}$.

We denote by ξ_1 and ξ_2 the displacements from equilibrium of m_1 and m_2, respectively. In terms of these displacements, the effective spring is stretched a net amount $\xi_1 - \xi_2$, and the elastic energy of the two-atom pair is

$$U = \tfrac{1}{2}K(\xi_1 - \xi_2)^2$$

Harmonic approximation to molecular vibration

The kinetic energy of the pair is $(p_1{}^2/2m_1) + (p_2{}^2/2m_2)$, but some of this represents translational energy for the molecule as a whole. To isolate the vibrational component, we examine the problem in the center-of-mass (CM) frame, where the total momentum of the molecule is zero. In the CM, the atomic momenta are equal in magnitude but oppositely directed; that is, $p_2 = -p_1$, and the kinetic energy (now of vibration only) is simply

$$KE_{\text{vib}} = p_1{}^2 \left(\frac{1}{2m_1} + \frac{1}{2m_2} \right) = \frac{p_1{}^2}{2\mu}$$

where μ is again the reduced mass defined in Equation 10.3.

The preceding relations describe a one-dimensional oscillator with vibration coordinate $\xi = \xi_1 - \xi_2$. The Schrödinger equation for this case is

$$-\frac{\hbar^2}{2\mu} \frac{d^2\psi}{d\xi^2} + \tfrac{1}{2}K\xi^2\psi(\xi) = E_{\text{vib}}\psi(\xi) \tag{10.7}$$

which is that for the quantum oscillator studied in Chapter 5, with x there replaced by ξ. The allowed energies of vibration are the oscillator levels (cf. Equation 5.29),

$$E_{\text{vib}} = (\nu + \tfrac{1}{2})\hbar\omega \qquad \nu = 0, 1, 2, \ldots \tag{10.8}$$

Allowed energies for vibration

where ν is an integer called the **vibrational quantum number.** The angular frequency ω is the classical frequency of vibration, and is related to the force constant K by

$$K = \mu\omega^2 \tag{10.9}$$

In the lowest vibrational state, $\nu = 0$, we find $E_{\text{vib}} = \tfrac{1}{2}\hbar\omega$, the so-called *zero point energy.* The accompanying vibration—the zero point motion—is present even when the molecule is not excited. From Equation 10.8 we see also that the energy difference between any two successive vibrational levels is the same and is given by $\Delta E_{\text{vib}} = \hbar\omega$. A typical value for ΔE_{vib} can be found from the entries in Table 10.2 and turns out to be about 0.3 eV.

Table 10.2   Fundamental Vibrational Frequencies and Effective Force Constants for Some Diatomic Molecules

Molecule	Frequency (Hz), $\nu = 0$ to $\nu = 1$	Force Constant (N/m)
HF	8.72×10^{13}	970
HCl	8.66×10^{13}	480
HBr	7.68×10^{13}	410
HI	6.69×10^{13}	320
CO	6.42×10^{13}	1860
NO	5.63×10^{13}	1530

From G. M. Barrows, *The Structure of Molecules,* New York, W. A. Benjamin, 1963.

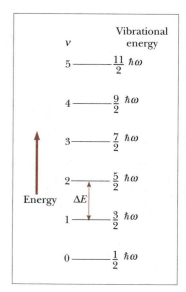

Figure 10.8 Allowed vibrational energies of a diatomic molecule, where ω is the fundamental frequency of vibration given by $\omega = \sqrt{K/\mu}$. Note that the spacings between adjacent vibrational levels are equal.

A level diagram for the vibrational energies of a diatomic molecule is given in Figure 10.8. Normally, most molecules are in the lowest energy state because the thermal energy at ordinary temperatures ($k_B T \approx 0.025$ eV) is insufficient to excite the molecule to the next available vibrational state. Electromagnetic radiation, however, can stimulate transitions to the first excited level. Such a transition would be accompanied by the absorption of a photon to conserve energy. Once excited, the molecule can return to the lower vibrational state by emitting a photon of the same energy. For molecular vibrations, photon absorption and emission occur in the *infrared* region of the spectrum. Absorption frequencies for the $\nu = 0$ to $\nu = 1$ transitions of several diatomic molecules are listed in Table 10.2, together with the effective force constants K calculated from Equation 10.9. Since larger force constants describe stiffer springs, K indicates the strength of the molecular bond. Notice that the CO molecule, which is bonded by several electrons, is much more rigid than such single-bonded molecules as HCl.

EXAMPLE 10.2   Vibration of the CO Molecule

The CO molecule shows a strong absorption line at the frequency 6.42×10^{13} Hz. (a) Calculate the effective force constant for this molecule.

Solution The absorption process is accompanied by a molecular transition from the $\nu = 0$ to the $\nu = 1$ vibrational level. Since the energy difference between these levels is $\Delta E_{\text{vib}} = \hbar\omega$, the absorbed photon must have carried this much energy. It follows that the photon frequency is just $\Delta E_{\text{vib}}/\hbar = \omega$, the frequency of the CO oscillator! From the information given, we calculate

$$\omega = 2\pi f = 2\pi(6.42 \times 10^{13}) = 4.03 \times 10^{14} \text{ rad/s}$$

Using 12 u and 16 u for the carbon and oxygen atomic masses, we find $\mu = 6.857$ u $= 1.14 \times 10^{-26}$ kg, as in Example 10.1. Then

$$K = \mu\omega^2 = (1.14 \times 10^{-26})(4.03 \times 10^{14})^2$$
$$= 1.86 \times 10^3 \text{ N/m}$$

Thus, infrared spectroscopy furnishes useful information on the elastic properties (bond strengths) of molecules.

(b) What is the classical amplitude of vibration for a CO molecule in the $\nu = 0$ vibrational state?

Solution The total vibrational energy for the molecule is $E_{\text{vib}} = \frac{1}{2}\hbar\omega$. At maximum displacement, the CO molecule has transformed all this into elastic energy of the spring,

$$\tfrac{1}{2}\hbar\omega = \tfrac{1}{2}KA^2$$

where A is the vibration amplitude. Using $K = \mu\omega^2$ and $\mu = 1.14 \times 10^{-26}$ kg, we get

$$A = \left(\frac{\hbar}{\mu\omega}\right)^{1/2} = \left\{\frac{1.055 \times 10^{-34}}{(1.14 \times 10^{-26})(4.03 \times 10^{14})}\right\}^{1/2}$$

$$= 4.79 \times 10^{-12} \text{ m} = 0.00479 \text{ nm}$$

Comparing this with the bond length of 0.113 nm, we see that the vibration amplitude is only about 4% of the bond length.

A proper treatment of molecular rotation and vibration should begin with Schrödinger's wave equation in three dimensions. However, because such problems involve many particles, approximation methods are necessary. For diatomic molecules, the particle mass m is replaced by the reduced mass μ of the system, and the potential $U(\mathbf{r})$ describes the interaction between *atoms*. Since the force between atoms depends only on their separation, U is spherically symmetric, and the wavefunctions are just spherical harmonics $Y_{\ell m}(\theta, \phi)$ multiplied by solutions $R(r)$ to the radial wave equation (Section 7.2). In this context, the angular momentum quantum numbers ℓ and m_ℓ specify the rotational state of the molecule. Molecular vibration is described by the radial wave $R(r)$ and reflects the choice of interatomic potential $U(r)$. Near equilibrium, $U(r)$ closely resembles the potential of a spring, leading to vibrational energies characteristic of the quantum oscillator. Example 10.3 shows how these ideas are implemented in a real case.

In actuality, the atoms of a molecule exert complicated forces on one another; these forces are harmonic only when the atoms are close to their equilibrium positions. More vibrational energy implies larger vibration amplitude, and with it a breakdown of the harmonic approximation—the effective spring must give way to a more accurate representation of the true interatomic force. One way to do this consists of replacing the harmonic potential $\frac{1}{2}K\xi^2$ in Equation 10.7 with the *Morse potential* shown in Figure 10.9a. Near equilibrium $(r = R_0)$ the Morse potential is harmonic; further away, $U(r)$ mimics the asymmetry characteristic of interatomic forces, becoming zero when the atoms are widely separated $(r \to \infty)$ but rising sharply as the atoms come close together. In consequence, oscillations take place about an average position $\langle r \rangle$ that *increases* with the energy of vibration. This is the origin of *thermal expansion*, in which the increased vibrational energy results from raising the temperature of the sample.

Unlike the harmonic oscillator, the interval separating successive levels of the Morse oscillator diminishes steadily at higher energies, as illustrated in Figure 10.9b. The Morse oscillator is explored in more detail in Problems 15 and 16.

Anharmonic effects

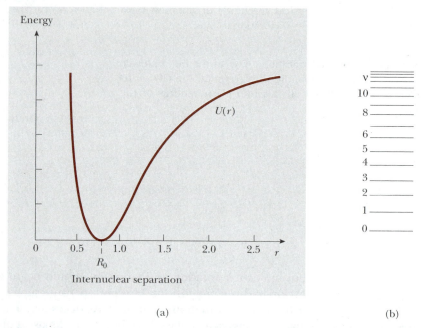

(a)   (b)

Figure 10.9   (a) Potential energy $U(r)$ for the Morse oscillator. In the equilibrium region around R_0, $U(r)$ is harmonic, but rises sharply as the atoms are brought closer together. (b) Allowed energies of vibration for the Morse oscillator. Notice that the separation between adjacent levels *decreases* with increasing energy.

EXAMPLE 10.3   Rotations and Vibrations of H_2: A Numerical Study

MP   In this example we use the MP DESKTOP to explore a few of the lowest lying rotation–vibration states of H_2, using the harmonic approximation to the force between H atoms in this diatomic molecule.

Launch the MP DESKTOP application by running the main program file *MP.EXE*. Next, select *Task* from the menu bar and *EigenStates-3D* from the drop-down task list. The DESKTOP is now ready to solve the (eigenvalue) problem posed in the task field just below the menu bar:

$$-\frac{d^2g}{dr^2} + \left[U(r) + \frac{\ell(\ell+1)}{r^2} \right] g(r) = E \cdot g(r)$$

This is the *radial wave equation* for the pseudo-wavefunction $g(r) = rR(r)$, expressed in dimensionless units of length and energy.

NOTE: Recall that for *EigenStates-3D* the unit of length (δ) and the unit of energy (ϵ) must satisfy $\epsilon \cdot \delta^2 = \hbar^2/2m$ in order to map the equation in the task field to the radial wave equation.

Select the potential energy $U(r)$ by clicking the Potential button located just below the menu bar, and

choosing *EX10-3.U3* (for Example 10.3) from the list of potential files. A second dialog box opens for the function $U(r) = \Gamma(r - \alpha)^2$. In the harmonic approximation, the H atoms forming the molecule are joined by an effective spring, and the potential energy near equilibrium—the *isotropic oscillator potential*—has the parabolic form indicated. The parameter Γ is proportional to the stiffness constant of the effective spring, and α is the equilibrium separation of the atoms. The defaults are chosen to model the H_2 molecule, with distance measured in tenths of angstroms. The corresponding unit of energy is

$$\frac{\hbar^2}{2\mu(0.1\ \text{Å})^2} = \frac{(1.973\ \text{keV}\cdot\text{Å}/c)^2}{(938.28 \times 10^3\ \text{keV}/c^2)(0.1\ \text{Å})^2}$$
$$= 4.15 \times 10^{-4}\ \text{keV}$$

or about 0.415 eV. Notice that we have used $\mu = \frac{1}{2}(938.28)\text{MeV}/c^2$ ($\frac{1}{2}$ proton mass) for the reduced mass of the H_2 molecule. Accept the information as presented by clicking the *OK* button.

The effective potential $U_{\text{eff}}(r) = U(r) + \ell(\ell+1)/r^2$ is plotted in the main window of the screen; since $\ell = 0$ to start, this is the isotropic oscillator potential $U(r)$. (The

Figure 10.10   (Example 10.3) The $\ell = 0$, $\nu = 0$ rotation–vibration state of the H_2 molecule. The vertical line marks the peak of the waveform corresponding to the most probable separation of the H atoms in this state.

asymmetric appearance of the well is an artifact of the logarithmic scale used for r.) The current energy E is recorded in the energy box just below the menu bar. In this example, we will determine allowed energies of vibration using the AutoSearch feature. Click the *Control Panel* entry on the *Options* menu. At the ensuing dialog box click the *Enable AutoSearch* checkbox in the *AutoSearch* field. A checkmark appears in the box showing this feature is activated. Click the *OK* button to finalize the procedure and return to the main screen. In the energy box, input $E = 0$ as a lower limit for the energy of vibration (why?), and press *Enter*. Within seconds AutoSearch locates an eigenvalue and overlays the potential with the wavefunction for this energy. We see the wavefunction is nodeless, and so must describe the vibrational ground state ($\nu = 0$). Clicking the Accuracy button ⌈L⌉ to the right of the energy box improves the precision of the search: the button label **L**(ow), **M**(edium), **H**(igh), or **S**(uper) indicates the current precision level. Increase the precision until E is known confidently to at least three decimal places. You should find $E = 0.656$, or about $(0.656)(0.415 \text{ eV}) = 0.272 \text{ eV}$. The result is actually slightly higher than would be expected for a one-dimensional oscillator with the same spring constant. Can you explain why?

To study the ground state wavefunction, activate the MP Analyzer by clicking ⌈🔍⌉. The wavefunction is read at

the current location of the graphics cursor. Position the cursor on the peak using the mouse or the direction keys [→], [←]. (Keyboard users note that the [Ins] key toggles cursor movement between coarse and fine settings.) The peak position represents the most probable distance separating the H atoms in the molecule. You should find this peak at $r \approx 7.39$, or 0.739 Å (Figure 10.10).

The quantum number ℓ in the effective potential specifies the rotational state of the molecule. Click ⌈🔍⌉ to close the Analyzer, then increase ℓ to 1 by clicking the Orbital button ⌈l= 0⌉. The display is updated to reflect the change in $U_{\text{eff}}(r)$, and AutoSearch has already found the new eigenvalue. This, the lowest vibrational state for $\ell = 1$, has energy $E = (0.695)(0.415 \text{ eV}) = 0.288 \text{ eV}$. The increase, about 0.016 eV, is the contribution to E from molecular rotation. Next, increase ℓ to 2 and determine the energy of this rotation–vibration state. You should find $E = (0.771)(0.415 \text{ eV}) = 0.320 \text{ eV}$; of this, $0.320 \text{ eV} - 0.272 \text{ eV} = 0.048 \text{ eV}$ is energy of rotation. These rotational energies conform to the rule of Equation 10.5, with $\hbar^2/I_{\text{CM}} = 0.016 \text{ eV}$. You should verify that this is a reasonable value for the H_2 molecule (see Eq. 10.2).

In contrast to rotation, the excitation energy for vibration is much larger. Reset ℓ to zero (click ⌈l= 0⌉ while

Figure 10.11   (Example 10.3) The $\ell = 0$, $\nu = 1$ rotation–vibration wavefunction of the H_2 molecule. The single node identifies this as the next-to-lowest vibrational state.

depressing the *Shift* key), and locate the next ($\nu = 1$) vibrational state. Since AutoSearch examines the range $E \pm \Delta E$, you will have to increase E by at least ΔE (the default is $\Delta E = 1$) to avoid returning to the vibrational ground state. You should find $E = (1.969)(0.415 \text{ eV}) = 0.817 \text{ eV}$ for the energy of the $\ell = 0$, $\nu = 1$ state. Thus, the excitation energy for vibration is $0.817 \text{ eV} - 0.272 \text{ eV} = 0.545 \text{ eV}$, more than an order of magnitude larger than that for rotation. The wavefunction for this state is shown in Figure 10.11, following some horizontal rescaling (click ◀▮▶) and scrolling (using the scroll bars to the right and beneath the graphics viewport).

Exercise 1   *Bond Stretching.* Determine the most probable separation of the H atoms for the $\ell = 1$, $\nu = 0$ rotation–vibration state of H_2. Compare with that for the $\ell = 0$, $\nu = 0$ state found in Example 10.3 to confirm that the separation increases in response to the centrifugal forces accompanying molecular rotation. This phenomenon, known as *bond stretching*, is understandably more pronounced in the more massive and flexible structures such as HCl, whose effective spring constant and characteristic vibration frequency are low. An analytical treatment of bond stretching is given as Problem 12.

10.3   MOLECULAR SPECTRA

In general, an excited molecule will rotate and vibrate simultaneously. To a first approximation, these motions are independent, with total energy given by the sum of Equations 10.5 and 10.8:

Rotation–vibration spectrum of a diatomic molecule

$$E_{\text{rot-vib}} = \frac{\hbar^2}{2I_{\text{CM}}} \ell(\ell + 1) + (\nu + \tfrac{1}{2})\hbar\omega \qquad (10.10)$$

Together, the levels prescribed by Equation 10.10 constitute the simplest approximation to the *rotation–vibration spectrum* of a diatomic molecule. For each allowed value of vibrational quantum number ν, there is a complete rotational spectrum, corresponding to $\ell = 0, 1, 2, \ldots$. Since successive rotational levels are separated by energies much smaller than the vibrational energy $\hbar\omega$, the rotation–vibration levels of a typical molecule appear as shown in Figure 10.12. Each level represents a possible internal condition for the molecule and is indexed by the *two* quantum numbers ℓ and ν, specifying the state of rotation and vibration, respectively. Normally, the molecule would take on the configuration with lowest energy, in this case the one designated by $\nu = 0$ and $\ell = 0$. External influences, such as temperature or the presence of electromagnetic radiation, can change the molecular condition, resulting in a transition from one rotation–vibration level to another. The attendant change in molecular energy must be compensated by absorption or emission of energy in some other form. When electromagnetic radiation is involved, the transitions—referred to as *optical transitions*—are subject to other conservation laws as well, since photons carry both momentum and energy.

Any optical transition between molecular levels with energies E_1 and E_2 must be accompanied by photon emission or absorption at the frequency

$$f = \frac{|E_2 - E_1|}{h} \qquad \text{or} \qquad \Delta E = \pm hf \qquad (10.11)$$

Since hf is the photon energy, this is energy conservation for the system of molecule plus photon. Equation 10.11 expresses a kind of *resonance* between the two; unless photons of the correct frequency (energy) are available, no transition is possible. Similar restrictions apply to other quantities that we know must be conserved in the process of transition. In particular, the initial and final states for an optical transition also must differ by exactly one angular momentum unit:

$$|\ell_2 - \ell_1| = 1 \qquad \text{or} \qquad \Delta\ell = \pm 1 \qquad (10.12)$$

The inference to be drawn from Equation 10.12 is that the photon carries angular momentum in the amount of $\hbar$, that is, the photon is a spin 1 particle with spin quantum number $s = 1$. Equation 10.12 then expresses angular momentum conservation for the system molecule plus photon. Equations 10.11 and 10.12 are the **selection rules** for optical transitions.

For the lower vibrational levels, there is also a restriction on the vibrational quantum number ν:

$$|\nu_2 - \nu_1| = 1 \qquad \text{or} \qquad \Delta\nu = \pm 1 \qquad (10.13)$$

Equation 10.13 reflects the harmonic character of the interatomic force rather than any photon property; indeed, for higher vibrational energies the concept of an effective spring joining the atoms is inaccurate, and Equation 10.13 ceases to be valid.

Selection rules greatly restrict the number of photon frequencies or wavelengths observed in molecular spectra, since transitions are prohibited unless all rules are obeyed simultaneously.[6] For instance, a pure rotational transition

Figure 10.12   The rotation–vibration levels for a typical molecule. Note that the vibrational levels are separated by much larger energies so that a complete rotational spectrum can be associated with each vibrational level.

---

[6] The selection rules given are for harmonic oscillations and rotations of a rigid rotor. In practice, violations of these rules are observed for predictable reasons, such as anharmonicity or rotation-vibration coupling.

normally would not be observed, since this requires $\Delta \nu = 0$ in violation of Equation 10.13. In the same way, a pure vibrational transition ($\Delta \ell = 0$) is forbidden, and we conclude that optical transitions usually involve both molecular vibration and rotation.

The spectrum of a particular molecule can be predicted by considering a collection of such molecules, initially undisturbed. At ordinary temperatures there is insufficient thermal energy to excite any but the $\nu = 0$ vibrational mode, although the molecules will be in various states of rotation. Since a pure rotational transition is forbidden, optical *absorption* must result from transitions in which ν increases by one unit but ℓ either increases or decreases, also by one unit (Figure 10.13a). Thus, the molecular absorption spectrum consists of two sequences of lines, represented by $\Delta \ell = \pm 1$, with $\Delta \nu = +1$ for both

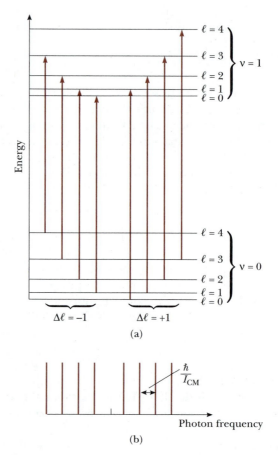

Figure 10.13 (a) Absorptive transitions between the $\nu = 0$ and $\nu = 1$ vibrational states of a diatomic molecule. The transitions obey the selection rule $\Delta \ell = \pm 1$ and fall into two sequences: those for which $\Delta \ell = +1$ and those for which $\Delta \ell = -1$. The transition energies are given by Equation 10.14. (b) Expected lines in the optical absorption spectrum of a molecule. The lines on the right side of center correspond to transitions in which ℓ changes by $+1$, and the lines to the left of center correspond to transitions for which ℓ changes by -1. These same lines appear in the emission spectrum.

cases. The energies of the absorbed photons are readily calculated from Equation 10.10:

$$\Delta E = \hbar\omega + \frac{\hbar^2}{2I_{CM}}2(\ell + 1) \qquad \ell = 0, 1, \ldots \qquad (\Delta\ell = +1)$$

$$\Delta E = \hbar\omega - \frac{\hbar^2}{2I_{CM}}2\ell \qquad \ell = 1, 2, \ldots \qquad (\Delta\ell = -1)$$

$$(10.14)$$

Here ℓ is the rotational quantum number of the initial state. The first of Equations 10.14 generates a series of equally spaced lines at frequencies *above* the characteristic vibration frequency ω, and the second generates a series *below* this frequency. Adjacent lines are separated in (angular) frequency by the fundamental unit $\hbar/I_{CM}$. Notice that ω itself is excluded, since ℓ cannot be zero if the transition is one for which ℓ decreases ($\Delta\ell = -1$). Figure 10.13b shows the expected frequencies in the absorption spectrum for the molecule; these same frequencies appear in the emission spectrum.

The actual absorption spectrum of the HCl molecule is shown in Figure 10.14 for comparison. The similarities are evident and reinforce the validity of our model. One peculiarity is apparent, however: each line in the HCl spectrum is split into a doublet. This doubling occurs because the sample is a mixture of two chlorine isotopes, ^{35}Cl and ^{37}Cl, whose different masses give rise to two distinct values for I_{CM}. Notice, too, that not all spectral lines appear with the same intensity, because even the allowed transitions occur at different *rates* (number of photons absorbed per second). Transition rates are governed chiefly by the populations of the initial and final states, and these depend on the degeneracy of the levels as well as the temperature of the system.

Finally, we note that molecules, like atoms, often simply *scatter* radiation, without having first to absorb and later reemit it. (Indeed, photons of any energy can be scattered, so no resonance is involved.) In *Rayleigh scattering*, the

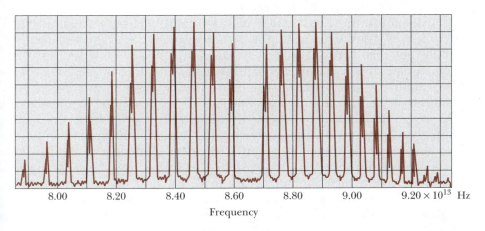

Figure 10.14   The absorption spectrum of the HCl molecule. Each line is split into a doublet because chlorine has two isotopes, ^{35}Cl and ^{37}Cl, which have different nuclear masses. (This is an adaptation of data taken by T. Faulkner and T. Nestrick at Oakland University, Rochester, MI)

photon energy is unchanged by the collision. Rayleigh scattering is stronger at shorter photon wavelengths, and it is this selectivity in the scattering that accounts for the blue color of the daytime sky. But *Raman scattering* also can occur, with the photon losing—or even gaining—energy in the collision. For such inelastic processes, energy is still conserved overall, the photon energy loss or gain being compensated by a suitable change in the internal (typically rotational) state of the molecule. Where rotation is involved, the energy exchanged in Raman scattering is consistent with the selection rule $\Delta \ell = \pm 2$. Raman scattering is relatively weak, and observable only if the incident radiation is sufficiently intense.

10.4   ELECTRON SHARING AND THE COVALENT BOND

In this section we examine the allowed energies and wavefunctions for electrons in a molecule. There are two aspects to the problem: one deals with the complexity of electron motion in the field of several nuclei, the other with the effect of all other electrons on the motion of any one. The two may be divided by first treating a one-electron molecular ion such as H_2^+, and subsequently examining the effects of adding one more electron to give the neutral hydrogen molecule H_2.

The Hydrogen Molecular Ion

The hydrogen molecular ion H_2^+ consists of one electron in the field of two protons (Fig. 10.15). The electron is drawn to each proton by the force of Coulomb attraction, resulting in a total potential energy of

$$U(\mathbf{r}) = -\frac{ke^2}{|\mathbf{r}|} - \frac{ke^2}{|\mathbf{r} - \mathbf{R}|} \tag{10.15}$$

In this expression, one proton is assumed to be at the origin of coordinates $\mathbf{r} = 0$, and the other at $\mathbf{R}$. The equilibrium separation R of the protons in H_2^+ is about 0.1 nm. The stationary state wavefunctions and energies for the electron in H_2^+ are solutions to the time-independent Schrödinger equation with the potential $U(\mathbf{r})$ of Equation 10.15.

As a guide, it is useful to consider the problem in the two limiting cases $R \rightarrow 0$ and $R \rightarrow \infty$, for which the results are already known. As $R \rightarrow 0$, the two protons coalesce to form a nucleus of helium; consequently, the wavefunctions and energies in this *united atom limit* are those of He^+, given by the formulas of Chapter 7 with atomic number $Z = 2$. In particular, the ground and first excited states are both spherically symmetric waves with energies of -54.4 eV and -13.6 eV, respectively. In the *separated atom limit* $R \rightarrow \infty$, the electron sees only the field of a single proton, and the levels are those of atomic hydrogen, -13.6 eV for the ground state and -3.4 eV for the first excited state. Energies for these extreme cases can be marked on the **correlation diagram** of Figure 10.16. The degeneracy of these levels excluding spin is important and is given by the numbers in parentheses. Notice especially that the ground state of the united atom is nondegenerate, but the ground state for the separated atoms is doubly degenerate. The extra degeneracy stems from the presence of the sec-

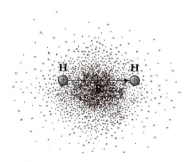

Figure 10.15 The hydrogen molecular ion H_2^+. The lone electron is attracted to both protons by the electrostatic force between opposite charges. The equilibrium separation $|\mathbf{R}|$ of the protons in H_2^+ is about 0.1 nm. (*Adapted from D. Ebbing, General Chemistry 5th ed., Boston, Houghton Mifflin Co., 1996.*)

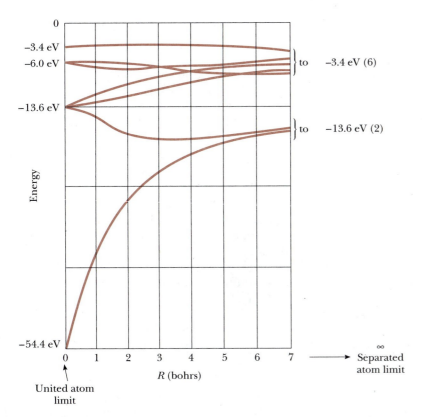

Figure 10.16   Correlation diagram for H_2^+, showing the two lowest electron energy levels as a function of proton separation, in bohrs. [Recall that 1 bohr (a_0) = 0.529 Å]. At $R = 0$ the levels are those of the united atom (ion) He^+, and at $R = \infty$ the levels are those of neutral H. The degeneracy of the various levels (excluding spin) is given by the numbers in parentheses.

ond proton in H_2^+: in the separated atom limit, the electron can occupy a hydrogen-like orbital around *either* proton. The same argument shows that *all of the atomic levels in the separated atom limit have an extra twofold degeneracy*, corresponding to hydrogen-like wavefunctions centered at the site of either proton. The completed correlation diagram for H_2^+ includes the energy levels at any proton separation R.

To find out what happens at intermediate values of R, consider first the case where R is finite but large, so that the influence of the other proton, though weak, cannot be ignored. This appearance of a second attractive force center means that the electron will not remain with its parent proton indefinitely. Given sufficient time, the electron inevitably will *tunnel* through the intervening potential barrier to occupy an atomic orbital around the far proton. Some time later, the electron will tunnel back to the original proton. In principle, the situation is no different from the inversion of the ammonia molecule discussed in Section 6.2: in effect, H_2^+ is a double oscillator, for which stationary states are not wavefunctions concentrated at a single force center but divided equally between them. To be sure, the tunneling time increases with proton separation, and may be very long indeed — something like 1 s for protons 1 nm apart (an eternity by atomic standards). But the time itself is inconsequential, since we are interested here in the stationary states that have, so to speak, "all the time in the world" to form.

From the symmetry of H_2^+, the tunneling probability from one proton to the other must be the same in either direction, and so we expect the electron to spend equal amounts of time in the vicinity of each. It follows that the stationary state wave is an equal mixture of atomic-like wavefunctions centered on each proton as, for example,

Molecular orbitals from atomic wavefunctions

$$\psi_+(\mathbf{r}) = \psi_a(\mathbf{r}) + \psi_a(\mathbf{r} - \mathbf{R}) \qquad (10.16)$$

The subscript a on the right is a collective label for all the quantum numbers needed to specify an atomic state; for the ground state of atomic hydrogen, a is $(1,0,0)$, meaning $n = 1$, $\ell = 0$, and $m_\ell = 0$. Equation 10.16 describes the *symmetric* combination of two atomic wavefunctions centered at the proton sites $\mathbf{r} = 0$ and $\mathbf{r} = \mathbf{R}$. The wavefunction and probability density given by $\psi_+(\mathbf{r})$ are sketched in Figure 10.17a and b; both are symmetric about the molecular center located at $\mathbf{r} = \mathbf{R}/2$, or halfway between the two protons.

However, we could also have written the antisymmetric form

$$\psi_-(\mathbf{r}) = \psi_a(\mathbf{r}) - \psi_a(\mathbf{r} - \mathbf{R}) \qquad (10.17)$$

This again leads to a symmetric probability distribution, but one with a *node* at the molecular center (Fig. 10.17c and d). In fact, Equations 10.16 and 10.17 *both* approximate true molecular wavefunctions when R is large, but they describe states having distinctly different energies. In particular, ψ_- has a somewhat higher energy than ψ_+, since the electron in the antisymmetric state is more confined, being relegated largely to the vicinity of one or the other nucleus. As with the particle in a box, this greater degree of confinement

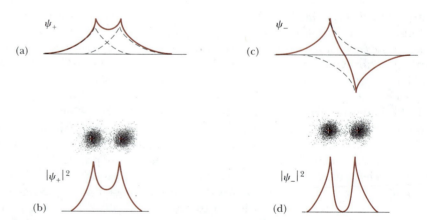

Figure 10.17 The wavefunction (a) and probability density (b) for the approximate molecular wave ψ_+ formed from the symmetric combination of atomic orbitals centered at $\mathbf{r} = 0$ and $\mathbf{r} = \mathbf{R}$. (c) and (d): Wavefunction and probability density for the approximate molecular wave ψ_- formed from the antisymmetric combination of these same orbitals.

comes at the expense of increased kinetic energy. Only in the limit $R \to \infty$ do the two energies merge to form the (doubly degenerate) atomic level of the separate atoms. As $R \to 0$, continuity requires the lower of the two energies to approach -54.4 eV, the ground state energy of the united atom He^+; similarly, the higher energy goes over to the first excited state energy of He^+, -13.6 eV.

We have come upon the quantum origins of the covalent bond. The energetically preferred state ψ_+ is referred to as the **bonding orbital,** since the electron in this state spends much of its time in the space between the two protons, shuttling to and fro and acting as a kind of "glue" that holds the molecule together. The more energetic state ψ_- is the **antibonding orbital** and decreases the molecular stability. Notice that a bonding–antibonding orbital pair is associated with *every* orbital of the separated atoms, not just with the ground state.

EXAMPLE 10.4   Modeling a Diatomic Molecule with the MP DESKTOP

MP In this example we use the MP DESKTOP to study the lowest bonding and antibonding orbitals of the divided square well, which serves as a crude model for electrons in a diatomic molecule.

Launch the DESKTOP application by running the main program file *MP.EXE*. Next, select *Task* from the menu bar and *EigenStates-1D* from the drop-down task list. The DESKTOP is now ready to solve the (eigenvalue) problem posed in the task field just below the menu bar:

$$-\frac{d^2\psi}{dx^2} + U(x)\psi(x) = E \cdot \psi(x)$$

This is Schrödinger's *time-independent* equation expressed in dimensionless units of length and energy.

NOTE: Recall that the length unit (δ) and the energy unit (ϵ) must satisfy $\epsilon \cdot \delta^2 = \hbar^2/2m$ to map the equation in the task field to the Schrödinger equation. For electrons, atomic units (length measured in bohrs and energy in rydbergs) are a convenient choice satisfying this criterion and are the ones adopted in this example.

To specify the potential energy, click the Potential button $\boxed{\mathsf{U(x)}}$ located just below the menu bar. A dialog box opens, presenting a list of potential files. Select *EX10-4.U* (for Example 10.4) from this list, then select *OK*. Another dialog box opens, detailing the potential function $U(x)$ in use. We see that $U(x)$ describes a finite square well of height ϵ on $[-\alpha, \alpha]$, with a central square barrier of width

2β. The barrier divides the original well into two, each representing an "atom" of our (diatomic) molecule. (Since the atomic potentials are identical, this molecule is homonuclear.) With distance measured in bohrs ($a_0 = 0.529$ Å) and energy in rydbergs (1 Ry = 13.60 eV), the defaults model atomic wells 100 eV high and 2 Å (3.780 bohrs) wide. The ground state of one such well was investigated in Example 5.10; as shown there, the lowest energy of our "atomic" electron is about 6.55 eV. In the molecule, the atoms are separated by 2β, or 1 bohr for $\beta = 0.500$ (the default). Accept the information as presented by clicking the *OK* button.

The potential $U(x)$ appears as a solid curve in the main window of the screen; the current energy E is recorded in the energy box just below the menu bar. In the energy box, input a "guess" for the energy of the molecular ground state. A good choice would be the energy of the atomic ground state, 6.55 eV (0.482 Ry), so type in $E = 0.482$, and press *Enter*. The wavefunction for this energy is computed and is superimposed on the potential. Make fine adjustments to E as necessary to generate a smooth wavefunction. Click the Accuracy button $\boxed{\mathsf{L}}$ to the right of the energy box to improve the precision of the search: the button label **L**(ow), **M**(edium), **H**(igh), or **S**(uper) indicates the current precision level. Continue the search until E is known as accurately as possible. You should find $E = 0.47032$ Ry, or about 6.40 eV.

NOTE: The direction keys [↑], [↓] furnish a quick and convenient way to make small changes to the energy. Just press [↑] ([↓]) to increase (decrease) the digit at the cursor by one unit; then press *Enter* to update the display.

Figure 10.18   (Example 10.4) The lowest-lying bonding orbital of a diatomic "molecule," modeled as a divided square well. The energy lies somewhat below that of the "atomic" ground state.

Figure 10.19   (Example 10.4) The first excited state of a molecule is the antibonding orbital derived from the atomic ground state. Notice the comparatively small wave amplitude in the "interatomic" region.

(The digit at the cursor can also be changed by clicking the mouse at the location of the cursor.) Alternatively, the search for eigenvalues can be automated by engaging the AutoSearch feature as described in Example 10.3. AutoSearch has difficulty, however, when the allowed energies are very close together, as is the case for the bonding–antibonding pairs of molecular orbitals studied here. This can be overcome by narrowing the search range in the *AutoSearch* field of the *Desktop Control Panel.*

In the molecule, the electron wavefunction is distributed (equally) between *both* atomic sites. Notice the symmetry of the molecular ground state (Fig. 10.18), and the wave amplitude in the interatomic region. This, the lowest lying bonding orbital of our molecule, derives from the atomic ground state.

Next, adjust E upward in search of the antibonding orbital paired with the bonding orbital just found. You should find $E = 0.49235$ Ry, or about 6.70 eV in this case. Again note the symmetry of the wavefunction and the wave amplitude in the interatomic region (Fig. 10.19). This is the first excited state of our molecule: it has exactly one node (as it should) and is the antibonding orbital derived from the atomic ground state.

Notice that the energies of the bonding and antibonding orbitals straddle the atomic level from which they derive. Furthermore, their energy difference reflects the degree of interaction between the atoms forming the molecule. (In the present case, this "splitting" is about $6.70 - 6.40 = 0.30$ eV.) As the transparency of the barrier separating the atomic wells decreases, the bonding and antibonding levels approach each other. In the limit of zero transparency (no interaction), both molecular levels merge with their parent atomic level, yet the molecular wavefunctions continue to be distributed (equally) between both atomic sites!

Exercise 2   *Exploring Barrier Transparency.* Verify the statements of the preceding paragraph by doubling the barrier width to 2 bohrs, leaving all else unchanged. Find the energies of the lowest bonding and antibonding orbitals, and compare with the energy of the atomic ground state. Describe any changes to the molecular wavefunctions arising from the increased barrier width.

The energy of the bonding orbitals in H_2^+ can be estimated from Schrödinger's equation. For the electron in H_2^+ this is

$$-\frac{\hbar^2}{2m}\nabla^2\psi + U(\mathbf{r})\psi = E\psi$$

with $U(\mathbf{r})$ the potential energy of the electron in the field of the two protons, as in Equation 10.15. Let us multiply every term by ψ^* and integrate over the whole space. Then the right-hand side becomes just E (assuming ψ is normalized),

$$E\int |\psi|^2\, dV = E$$

leaving for the left-hand side an expression that we can use to compute the value of E:

$$E = \int \psi^*\left[-\frac{\hbar^2}{2m}\nabla^2 - \left\{\frac{ke^2}{|\mathbf{r}|} + \frac{ke^2}{|\mathbf{r} - \mathbf{R}|}\right\}\right]\psi dV \qquad (10.18)$$

If ψ is the true molecular wavefunction, this equation furnishes the exact value of particle energy in the state ψ; if ψ is an approximate wavefunction, Equation 10.18 yields an approximation to E.

Substituting for ψ the symmetric combination ψ_+ (suitably normalized—see the Appendix at end of this chapter) gives a first approximation to the energy of the bonding orbitals in H_2^+. For the case where ψ_a refers to the hydrogen atom ground state, the molecular energy E_+ can be shown to be

Bonding orbital of H_2^+

$$E_+ = -1 - \frac{2}{1 + \Delta} \left\{ \frac{1}{R} - \frac{1}{R} (1 + R) e^{-2R} + (1 + R) e^{-R} \right\} \quad (10.19)$$

with Δ the *overlap integral* of atomic wavefunctions found in the Appendix at the end of this chapter:

$$\Delta = \left(1 + R + \frac{R^2}{3} \right) e^{-R}$$

These expressions for E_+ and Δ are given in *atomic units*, where the rydberg (1 Ry = -13.6 eV) is adopted as the unit of energy to go along with the bohr unit of length.

In the limit of large R, $\Delta \to 0$ and E_+ approaches the energy of an isolated hydrogen atom $E_a = -1$ Ry, as it should. As R decreases, E_+ becomes more and more negative, finally reaching -3 Ry at $R = 0$ where the nuclei coalesce. (The correct value of -4 Ry for the ground state of He^+ is not reproduced owing to the approximation inherent in our use of Equation 10.16 for the molecular wavefunction, which fails completely as $R \to 0$.) Since E_+ decreases steadily with R, the molecule appears to be unstable and should collapse to $R = 0$ under the bonding tendency of the orbiting electron. But this overlooks the Coulomb energy of the protons, which must be included to obtain the total molecular energy. Two protons separated by a distance R repel each other with energy ke^2/R, or $2/R$ Ry. The Coulomb repulsion of the protons offsets the bonding attraction of the electron to stabilize the molecule at that separation for which the total energy is minimal. The total molecular energy given by our model is sketched as a function of R in Figure 10.20. The minimum comes at $R_0 = 2.49$ bohrs, or 0.132 nm, and agrees reasonably well with the observed *bond length* for H_2^+, 0.106 nm.

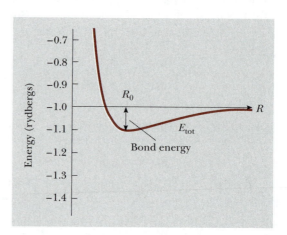

Figure 10.20   The total molecular energy for the bonding orbital of H_2^+, as given by the approximate wavefunction ψ_+ of Equation 10.16. The predicted bond length occurs at the point of stable equilibrium, around $R = 2.5$ bohrs. The predicted bond energy is about 1.77 eV.

At the equilibrium separation of 2.49 bohrs, we find $E_+ = -2.13$ Ry, and a total molecular energy of $E_+ + 2/R_0 = -1.13$ Ry. The negative of this, 1.13 Ry or 15.37 eV, represents the work required to separate the molecule into its constituents and is called the **dissociation energy.** The measured dissociation energy for H_+ is about 16.3 eV. The **bond energy,** or the work required to separate H_2^+ into H and H^+, is the difference between the molecular energy at $R = \infty$ and at equilibrium. Our model predicts a bond energy of 15.37 eV $-$ 13.6 eV $=$ 1.77 eV, which is somewhat less than the actual value of 2.65 eV. The difference can be attributed to our use of Equation 10.16 for the molecular wavefunction, an approximation that is best for larger values of R.

To obtain the energy of the antibonding orbitals for H_2^+, we replace ψ in Equation 10.18 with the antisymmetric wave ψ_- of Equation 10.17 (again, normalized as in the Appendix). With ψ_a the hydrogen ground state, we find for the energy of the lowest antibonding orbital in H_2^+

$$E_- = -1 - \frac{2}{1 - \Delta}\left\{\frac{1}{R} - \frac{1}{R}(1 + R)e^{-2R} - (1 + R)e^{-R}\right\} \quad (10.20)$$

Antibonding orbital of H_2^+

For large separations, E_- coincides with E_+ at -1 Ry. With decreasing R, however, E_- steadily *increases* and always lies *above* E_+. In particular, E_- for any finite value of R is *higher* than the energy of H + H^+ in isolation, and no stable molecular ion will be formed in this state. Inclusion of the nuclear repulsion only enhances the instability. This is the antibonding effect for H_2^+.

Although it must be admitted that E_- seriously overestimates the true molecular energy for small values of R, the conclusion reached vis-a-vis the instability of this state is, nonetheless, correct. The true antibonding orbital energy approaches -1 Ry in the united atom limit and exhibits a broad, shallow minimum near $R = 3$ bohrs (see Fig. 10.16). Equilibrium cannot be sustained, however, when the Coulomb repulsion of the nuclei is included; that is, the curve of total molecular energy shows no minimum for the antibonding state.

The Hydrogen Molecule

The addition of one more electron to H_2^+ gives the hydrogen molecule, H_2. The second electron provides even more "glue" for covalent bonding and should produce a stronger bond than that seen in H_2^+. Indeed, the bond energy for H_2 is 4.5 eV, compared with 2.65 eV for H_2^+. (Naively, we might have expected the bond energy to double with the number of electrons, but the electrons repel each other—an antibonding effect.) The second electron also results in a measured bond length for H_2 equal to 0.074 nm, which is noticeably shorter than the 0.1 nm found for the single-bonded molecular ion H_2^+.

EXAMPLE 10.5 The Hydrogen Molecular Bond

Estimate the bond length and the bond energy of H_2, assuming that each electron moves independently of the other and is described by the approximate wavefunction of Equation 10.16.

Solution If the electrons do not interact, the energy of each must be the bonding energy E_+ of Equation 10.19.

The total energy of the molecule in this *independent particle model* is then

$$E_{\text{tot}} = \frac{2}{R} + 2E_+$$

$$= \frac{2}{R} - 2 - \frac{4}{1 + \Delta}\left\{\frac{1}{R} - \frac{1}{R}(1 + R)e^{-2R} + (1 + R)e^{-R}\right\}$$

By trial and error, we discover that the minimum energy occurs at about $R_0 = 1.44$ bohrs, or 0.076 nm, which becomes the bond length predicted by our model. The molecular energy at this separation is $E_0 = -3.315$ Ry. Comparing this with the total energy of the separated hydrogen atoms, -2 Ry, leaves for the bond energy 1.315 Ry, or 17.88 eV. Due to our neglect of electron repulsion, this model grossly exaggerates the bond energy, although it does get the bond length about right. Even the latter agreement is probably fortuitous, however, since Equation 10.16 is not expected to be accurate even for noninteracting electrons at nuclear separations as small as 1.5 bohrs.

So far we have not recognized that the two electrons in H_2, being identical fermions, are subject to the exclusion principle. If both electrons occupy the bonding orbital of H_2—as we have been assuming—*their spins must be opposite.* If their spins were parallel, one electron would be forced into the antibonding orbital, where it has higher energy. Figure 10.21 shows the electronic energy of the molecule for the two cases. These curves were obtained from laborious calculations that account in an approximate way for the effect of electron–electron repulsion. The results imply that no stable molecule is formed in the parallel spin case.

Figure 10.21   Total molecular energy for the bonding and antibonding orbitals of H_2. For both electrons to be in the bonding orbital, their spins must be opposite. The bond energy for H_2 is 4.5 eV and the bond length is 0.074 nm. Since the energy of the antibonding orbital exceeds that of the isolated H atoms, no stable molecule can be formed in this state.

O P T I O N A L

10.5   BONDING IN COMPLEX MOLECULES

The bonds we found in H_2^+ and H_2 are known as **sigma** type molecular bonds, denoted σ. They arise from the overlap of atomic s states and are characterized by an electron density that is axially symmetric about the line joining the two atoms. Bonding between atoms having more electrons often involves the overlap of p states or other atomic orbitals and leads to other types of molecular bonds with their own distinctive characteristics.

Consider the nitrogen molecule, N_2, the simplest molecule to exhibit bonds other than the σ type. The electron configuration of the nitrogen atom is $1s^2 2s^2 2p^3$. The molecular bond in N_2 is due primarily to the three valence electrons in the $2p$ subshell; the inner s electrons are too tightly bound to their parent atoms to participate in the sharing necessary for bond formation. The $2p$ subshell is made up of three atomic orbitals ($m_\ell = 1, 0, -1$), and each of these gives rise to a bonding and an antibonding orbital for the molecule. The N_2 molecule has six valence electrons to accommodate, three from each N atom. With two electrons in each of the three bonding orbitals, the N_2 molecule is especially stable, with a bond energy of 9.8 eV and a bond length of 0.11 nm.

Of the three bonding–antibonding orbital pairs in N_2, however, only one is a sigma bond. Recall that atomic p states are lobed structures with highly directional characteristics. The electron density in the p_z orbital ($m_\ell = 0$) is concentrated along the z axis; similarly, the p_x and p_y orbitals are marked by electron densities along the x and y axes, respectively (Fig. 7.13). (The p_x and p_y orbitals are not eigenstates of L_z but are formed from the $m_\ell = \pm 1$ states according to the prescription of Equation 7.50 in Chapter 7). When two N atoms are brought together, one of the three axes— say the z axis—will become the molecular axis for N_2 in order to maximize atomic overlaps and produce the strongest bond. This is the sigma bond, because it has axial symmetry. The p_x orbitals, however, also overlap to form a different kind of bond— the **pi bond** (π)—which has a plane of symmetry in the nodal plane of the p orbitals. The same is true of the p_y orbitals. Figure 10.22 illustrates the different kinds of bonds

Sigma and pi bonds for N_2

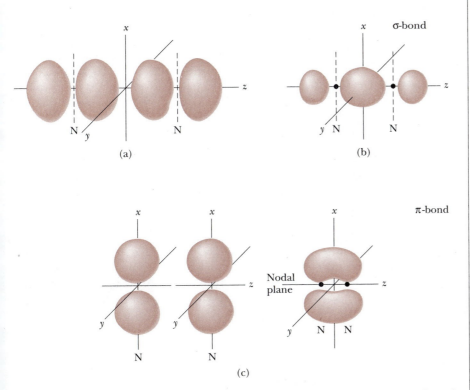

Figure 10.22   (a) and (b): Formation of a sigma bond in N_2 from the overlap of the $2p_z$ orbitals on adjacent N atoms. (c) Formation of a pi bond by overlap of the $2p_x$ orbitals on adjacent N atoms. A similar bond is formed by overlap of the $2p_y$ orbitals.

Bonding in heteronuclear molecules

in the N_2 molecule. The pi bonds are weaker than the sigma bond because they involve less electron overlap.

So far we have been discussing only molecules formed from like atoms. These *homonuclear* molecules exemplify the pure covalent bond. The joining of two different atomic species to form a *heteronuclear* molecule produces *polar covalent* bonds. The hydrogen fluoride molecule, HF, is a good example. The F atom has nine electrons in the configuration $1s^2 2s^2 2p^5$. Of the five $2p$ electrons, four completely fill two of the $2p$ orbitals, leaving one $2p$ electron to be shared with the H atom (the filled orbitals are especially stable and do not significantly affect the molecular bonding in HF). When the two atoms are brought together, the $2p$ orbital of F overlaps the $1s$ orbital of H to form a bonding–antibonding orbital pair for the HF molecule. The result is an ***s-p*** **bond,** with both electrons occupying the more stable bonding orbital. The *s-p* bond in HF is *polar* since the shared electrons spend more time in the vicinity of the F atom due to its high electronegativity. Equivalently, the bond is partly ionic, and the HF molecule shows a permanent *electric dipole moment.* The bond energy for HF is 5.90 eV, and the bond length is 0.092 nm.

The case of HF suggests correctly that purely covalent bonds are found only in homonuclear molecules; heteronuclear molecules always form bonds with some degree of ionicity, as measured by a dipole moment. The water molecule, H_2O, is another example of an *s-p*-bonded structure with a dipole moment, as is the ammonia molecule, NH_3. In NH_3, each of the three $2p$ electrons in the N atom forms an *s-p* bond with an H atom. Since the p_x, p_y, and p_z atomic orbitals are directed along mutually perpendicular axes, we would expect the three *s-p* bonds to be at right angles. In fact, the measured *bond angle* in NH_3 is somewhat larger—about 107°—because of the electrostatic repulsion of the H nuclei.

In closing, we mention briefly the bonds formed by the carbon atom. These bonds result from ***s-p*** **hybridization**, a concept that accounts for the almost endless variety of organic compounds. The C atom has six electrons in the configuration $1s^2 2s^2 2p^2$. We might conclude that only the two $2p$ electrons are prominent in molecular bonding and that carbon is divalent. The existence of hydrocarbons like CH_4 (methane), however, shows that the C atom shares all four of its second-shell electrons, suggesting that the binding energy of the $2s$ electrons in the carbon atom is not much different from that of the $2p$ electrons. But most surprising is the fact that *all four bonds are equivalent!* There are not two *s-p* bonds and two *s-s* bonds as we might have anticipated, but four structurally identical molecular bonds. These bonds arise from the overlap of the $1s$ orbital in H with atomic orbitals in C formed from a *mixture* of carbon $2s$ and $2p$ orbitals. In CH_4 these mixed, or *hybrid, orbitals* are represented by wavefunctions such as

$$\psi = \psi_{2s} + [\psi_{2p}]_x + [\psi_{2p}]_y + [\psi_{2p}]_z \tag{10.21}$$

Other combinations arise by subtracting, rather than adding, the individual wavefunctions, but all mix one $2s$ and three $2p$ atomic orbitals to give four sp^3 *hybrids*. In other carbon compounds, hybridization may involve only one or two of the $2p$ orbitals; these are described as sp and sp^2 hybrids, respectively. It is this complexity that gives rise to the rich variety of organic materials.

SUMMARY

Two or more atoms may combine to form molecules because of a net attractive force between them. The resulting molecular bonds are classified according to the bonding mechanisms and are of the following types:

1. **Ionic bonds.** Certain molecules form ionic bonds because of the Cou-

lomb attraction between oppositely charged ions. Sodium chloride (NaCl) is one example of an ionically bonded molecule.

2. **Covalent bonds.** The covalent bond in a molecule is formed by the sharing of valence electrons of its constituent atoms. For example, the two electrons of the H_2 molecule are equally shared between the nuclei.
3. **Hydrogen bonds.** This type of bonding corresponds to the attraction of two negative ions by an intermediate hydrogen atom (a proton).
4. **Van der Waals bonds.** This is a weak electrostatic bond between atoms that do not form ionic or covalent bonds. It is responsible for the condensation of inert gas atoms and nonpolar molecules into the liquid phase.

The energy of a molecule in a gas consists of contributions from the electronic interactions, the translation of the molecule, rotations, and vibrations. The **allowed values of the rotational energy** of a diatomic molecule are given by

$$E_{rot} = \frac{\hbar^2}{2I_{CM}} \ell(\ell + 1) \qquad \ell = 0, 1, 2, \ldots \qquad (10.5)$$

where I_{CM} is the *moment of inertia* of the molecule about its center of mass and ℓ is an integer called the *rotational quantum number*.

The allowed values of the vibrational energy of a diatomic molecule are given by

$$E_{vib} = (\nu + \tfrac{1}{2})\hbar\omega \qquad \nu = 0, 1, 2, \ldots \qquad (10.8)$$

where ν is the *vibrational quantum number*. The quantity ω is the classical frequency of vibration, and is related to μ, the *reduced mass* of the molecule, and K, the *force constant* of the effective spring bonding the molecule, by the relation $\omega = \sqrt{K/\mu}$.

The internal state of motion of a molecule is some combination of rotation and vibration. Any change in the molecular condition is described as a transition from one rotation–vibration level to another. When accompanied by the emission or absorption of photons, these are called **optical transitions.** Besides conserving energy, optical transitions must conform to the **selection rules**

$$\Delta\ell = \pm 1 \qquad \text{and} \qquad \Delta\nu = \pm 1 \qquad (10.12, 10.13)$$

In the most common case, the absorption spectrum of a diatomic molecule consists of *two* sequences of lines, corresponding to $\Delta\ell = \pm 1$, with $\Delta\nu = +1$ for both sequences. Measurements made on such spectra can be used to determine the length and strength of the molecular bonds.

The electronic states of a molecule are classified as **bonding** or **antibonding.** A bonding state is one for which the electron density is large in the space between the nuclei. An electron in a bonding orbital can be thought of as shuttling rapidly from one nucleus to the other, drawing them together as a kind of "glue." The bonding state is energetically preferred over the anti-bonding state, where the electron spends more of its time outside the bonding region. In a diatomic molecule composed of like atoms (homonuclear molecule), each atomic orbital gives rise to a bonding–anti-bonding orbital pair for the molecule.

Overlap Integrals of Atomic Wavefunctions

The normalization of molecular wavefunctions that are formed as combinations of atomic orbitals involves evaluation of so-called *overlap integrals* having the general form

$$\int \psi_a^*(\mathbf{r}) \psi_b(\mathbf{r} - \mathbf{R}) \; dV \tag{10.22}$$

Such integrals are evaluated here for the case where ψ_a and ψ_b are wavefunctions for the hydrogen ground state:

$$\psi_a(r) = \psi_b(r) = \frac{1}{\sqrt{\pi}} \, a_0^{-3/2} e^{-r/a_0} \tag{10.23}$$

The analysis is most transparent when carried out in *atomic units*, where distances are measured in bohrs (a_0) and energies in rydbergs ($ke^2/2a_0$). In such units r/a_0 becomes simply r, and ke^2/r becomes $2/r$, and so on.

Now the atomic wave of Equation 10.23 is an s state and, therefore, spherically symmetric; that is, ψ_a is a function only of $|\mathbf{r}| = r$. Likewise, $\psi_b(\mathbf{r} - \mathbf{R})$ will depend only on

$$|\mathbf{r} - \mathbf{R}| = \sqrt{r^2 + R^2 - 2\mathbf{r} \cdot \mathbf{R}}$$

If we orient the coordinate axes so that the z axis lies along the direction of $\mathbf{R}$, then $\mathbf{r} \cdot \mathbf{R} = rR \cos \theta$, where θ is the polar angle in spherical coordinates. Since the volume element in spherical coordinates is $dV = r^2 \, dr \, d\Omega$, where the element of solid angle $d\Omega = \sin \theta \, d\theta \, d\phi$, the integration over angle can be carried out using the substitution $u^2 = r^2 + R^2 - 2rR \cos \theta$ (so that $u \, du = rR \sin \theta \, d\theta$). For $\theta = 0$, $u = |r - R| \, (= u_1)$, while $\theta = \pi$ implies $u = |r + R| \, (= u_2)$. Then

$$\int \psi_b(\mathbf{r} - \mathbf{R}) \, d\Omega = \frac{1}{\sqrt{\pi}} a_0^{-3/2} \left(\frac{2\pi}{rR} \right) \int_{u_1}^{u_2} u e^{-u} \, du$$

$$= -\frac{2\sqrt{\pi}}{rR} a_0^{-3/2} (1 + u) e^{-u} \Bigg|_{u_1}^{u_2}$$

$$= \frac{2\sqrt{\pi}}{rR} a_0^{-3/2} \{ (1 + |r - R|) e^{-|r - R|} - (1 + |r + R|) e^{-|r + R|} \}$$

and the full overlap integral becomes

$$\int \psi_a^*(\mathbf{r}) \psi_b(\mathbf{r} - \mathbf{R}) \; dV = \frac{2}{R} \left\{ \int_0^\infty re^{-r}(1 + |r - R|) e^{-|r - R|} \, dr \right.$$

$$\left. - \int_0^\infty re^{-r}(1 + |r + R|) e^{-|r + R|} \, dr \right\}$$

Since r and R are both nonnegative, $|r + R| = r + R$, and the second of the two integrals on the right can be integrated immediately using

$$\int_0^\infty x^n e^{-x}\, dx = n!$$

to get

$$\frac{2}{R} e^{-R} \int_0^\infty r(1 + r + R) e^{-2r}\, dr = \frac{1}{R} e^{-R} \int_0^\infty \left(\frac{x}{2}\right)\left(1 + \frac{x}{2} + R\right) e^{-x}\, dx$$

$$= \left(\frac{1}{R} + \frac{1}{2}\right) e^{-R}$$

The first integral must be evaluated separately in the two intervals $r < R$ and $r > R$. For $r < R$, we have $|r - R| = R - r$ and we find

$$\frac{2}{R} e^{-R} \int_0^R r(1 + R - r)\, dr = [R(1 + R) - 2R^2/3] e^{-R}$$

while for $r > R$ we obtain (with $x = r - R$)

$$\frac{2}{R} \int_R^\infty r(1 + r - R) e^{(-2r + R)}\, dr = \frac{2}{R} e^{-R} \int_0^\infty (x + R)(1 + x) e^{-2x}\, dx$$

$$= \left(\frac{1}{R} + \frac{3}{2}\right) e^{-R}$$

Combining all the above results gives the overlap integral as

$$\int \psi_a^*(\mathbf{r}) \psi_b(\mathbf{r} - \mathbf{R})\, dV = \left(1 + R + \frac{R^2}{3}\right) e^{-R} \qquad (10.24)$$

With only minor alterations, the same method may be used to obtain the expressions given in the text for the energies E_+ and E_- of the lowest bonding and antibonding states, respectively, of H_2^+.

EXAMPLE 10.6 Normalizing a Molecular Wavefunction

Normalize the molecular wavefunction $\psi_\pm$ of Equations 10.16 and 10.17, assuming the atomic wavefunctions ψ_a are already normalized.

Solution The normalization integral for $\psi_\pm$ is

$$\int |\psi_\pm|^2\, dV = \int |\psi_a(\mathbf{r})|^2\, dV + \int |\psi_a(\mathbf{r} - \mathbf{R})|^2\, dV$$

$$\pm \int \psi_a^*(\mathbf{r}) \psi_a(\mathbf{r} - \mathbf{R})\, dV$$

$$\pm \int \psi_a(\mathbf{r}) \psi_a^*(\mathbf{r} - \mathbf{R})\, dV$$

Since the atomic wavefunctions are already normalized, the first two integrals on the right are both unity. If $\mathbf{R}$ is large, the last two integrals can be neglected, since the integrand for each is the product of one atomic function centered at $\mathbf{r} = 0$ with another centered at $\mathbf{r} = \mathbf{R}$. The resulting product is small everywhere provided $\mathbf{R}$ is sufficiently large, and these integrals—the *overlap integrals* described earlier—are very nearly zero. Thus, in the lowest approximation

$$\int |\psi_\pm|^2\, dV \approx 2$$

and approximately normalized wavefunctions can be had by multiplying the waves of Equations 10.16 and 10.17 by the factor $1/\sqrt{2}$.

In a more refined treatment, the overlap integrals have to be evaluated. For the hydrogen atom in the ground state, the overlap integrals are both given by Equation 10.24. Denoting these integrals by Δ, we find

$$\int |\psi_\pm|^2 \, dV \approx 2 \pm 2\Delta$$

and the proper normalization factor for the wavefunction $\psi_\pm$ becomes $1/\sqrt{2(1 \pm \Delta)}$.

SUGGESTIONS FOR FURTHER READING

1. Molecular bonding is discussed extensively at an introductory level in Chapter 24 of J. Brackenridge and R. Rosenberg, *The Principles of Physics and Chemistry*, New York: McGraw-Hill Book Company, Inc., 1970. This work also includes a number of nice illustrations of molecular bonds in various compounds.
2. For an in-depth treatment of the bonding in $H_2{}^+$ and the H$_2$ molecule, see J. C. Slater, *Quantum Theory of Molecules and Solids, Volume 1*, New York: McGraw-Hill Book Company, Inc., 1963.
3. An elaborate discussion of chemical bonding in organic compounds may be found in Chapter 9 of K. Krane, *Modern Physics*, 2nd ed. New York: John Wiley and Sons, Inc., 1996.

QUESTIONS

1. List three ways (modes) a diatomic molecule can store energy internally. Which of the three modes is easiest to excite, that is, requires the least energy? Which requires the most energy?
2. How do the effective force constants for the molecules listed in Table 10.2 compare with those found for typical laboratory springs? Comment on the significance of your findings.
3. Describe hybridization, why it occurs, and what it has to do with molecular bonding. Give an explicit example of where it occurs.
4. Discuss the mechanisms responsible for the different types of bonds that can occur to form stable molecules.
5. Explain the role of the Pauli exclusion principle in determining the stability of molecules, using H_2 and the (hypothetical) species H_3 as examples.
6. Discuss the relationship between tunneling and the covalent bond. From the viewpoint of possible chemical species, what would be the similarities and differences between the world in which we live and one devoid of tunneling?
7. Distinguish between the dissociation energy of a molecule and its bond energy. Which of the two is expected to be greater?
8. In treating the rotational levels of a diatomic molecule, why do we ignore rotation about the internuclear line? Can you think of a case involving a triatomic molecule in which similar considerations might apply?
9. Explain why the noble gases tend to be monatomic rather than diatomic.

PROBLEMS

10.2 Molecular Rotation and Vibration

1. Use the data in Table 10.2 to calculate the reduced mass of the NO molecule; then compute a value for μ using Equation 10.3. Compare the two results.
2. The CO molecule undergoes a rotational transition from the $\ell = 1$ level to the $\ell = 2$ level. Using Table 10.1, calculate the values of the reduced mass and the bond length of the molecule. Compare your results with those of Example 10.1.
3. Use the data in Table 10.2 to calculate the maximum amplitude of vibration for (a) the HI molecule and (b) the HF molecule. Which molecule has the weaker bond?
4. The $\ell = 5$ to $\ell = 6$ rotational absorption line of a diatomic molecule occurs at a wavelength of 1.35 cm (in the vapor phase). (a) Calculate the wavelength and frequency of the $\ell = 0$ to $\ell = 1$ rotational absorption line. (b) Calculate the moment of inertia of the molecule.
5. The HF molecule has a bond length of 0.092 nm. (a) Calculate the reduced mass of the molecule. (b) Sketch the potential energy versus internuclear separation in the vicinity of $r = 0.092$ nm.

6. The HCl molecule is excited to its first rotational energy level, corresponding to $\ell = 1$. If the distance between its nuclei is 0.1275 nm, what is the angular velocity of the molecule about its center of mass?

7. The $\nu = 0$ to $\nu = 1$ vibrational transition of the HI molecule occurs at a frequency of 6.69×10^{13} Hz. The same transition for the NO molecule occurs at a frequency of 5.63×10^{13} Hz. Calculate (a) the effective force constant and (b) the amplitude of vibration for each molecule. (c) Explain why the force constant of the NO molecule is so much larger than that of the HI molecule.

8. Consider the HCl molecule, which consists of a hydrogen atom of mass 1 u bound to a chlorine atom of mass 35 u. The equilibrium separation between the atoms is 0.128 nm, and it requires 0.15 eV of work to increase or decrease this separation by 0.01 nm. (a) Calculate the four lowest rotational energies (in eV) that are possible, assuming the molecule rotates rigidly. (b) Find the molecule's "spring constant" and its classical frequency of vibration. (*Hint:* Recall that $U = \frac{1}{2}Kx^2$.) (c) Find the two lowest vibrational energies and the classical amplitude of oscillation corresponding to each of these energies. (d) Determine the longest wavelength radiation that the molecule can emit in a pure rotational transition and in a pure vibrational transition.

9. The hydrogen molecule comes apart (dissociates) when it is excited internally by 4.5 eV. Assuming that this molecule behaves exactly like a harmonic oscillator with classical frequency $\omega = 8.277 \times 10^{14}$ rad/s, find the vibrational quantum number corresponding to its 4.5-eV dissociation energy.

10. As a model for a diatomic molecule, consider two identical point masses m connected by a rigid massless rod of length R_0. Suppose that this molecule rotates about an axis perpendicular to the rod through its midpoint. Use the Bohr quantization rule for angular momentum to obtain the allowed rotational energies in this approximation. Compare your result with the correct quantum mechanical treatment of this model.

11. For the model of a diatomic molecule described in Problem 10, derive an expression for the minimum energy required to excite the molecule into rotation about the *internuclear line*, that is, about the axis joining the two masses. Assume that the masses are uniform spheres of radius r. Apply your result to the hydrogen molecule, H_2, by taking m equal to the proton mass and r equal to the nuclear size, about 10 fm.

12. The rotational motion of molecules has an effect on the equilibrium separation of the nuclei, a phenomenon known as *bond stretching*. To model this effect, consider a diatomic molecule with reduced mass μ, oscillator frequency ω_0, and internuclear separation R_0 when the angular momentum is zero. The effective potential energy for nonzero values of ℓ is then (see Section 7.5)

$$U_{\text{eff}} = \frac{1}{2}\mu\omega_0^2(r - R_0)^2 + \ell(\ell + 1)\frac{\hbar^2}{2\mu r^2}$$

(a) Minimize the effective potential $U_{\text{eff}}(r)$ to find an equation for the equilibrium separation of the nuclei, R_ℓ, when the angular momentum is ℓ. Solve this equation approximately, assuming $\ell \ll \mu\omega_0 R_0^2/\hbar$. (b) Near the corrected equilibrium point, R_ℓ, the effective potential again is nearly harmonic and can be written approximately as

$$U_{\text{eff}} \approx \frac{1}{2}\mu\omega^2(r - R_\ell)^2 + U_0$$

Find expressions for the corrected oscillator frequency ω and the energy offset U_0 by matching U_{eff} and its first two derivatives at the equilibrium point R_ℓ. Show that the fractional change in frequency is given by

$$\frac{\Delta\omega}{\omega_0} \approx \frac{3\ell(\ell + 1)\hbar^2}{2\mu^2\omega_0^2 R_0^4}$$

13. As an alternative to harmonic interactions, the *Morse potential*

$$U(r) = U_0\{1 - e^{-\beta(r - R_0)}\}^2$$

can be used to describe the vibrations of a diatomic molecule. The parameters R_0, U_0, and β are chosen to fit the data for a particular atom pair. (a) Show that R_0 is the equilibrium separation, and that the potential energy far from equilibrium approaches U_0. (b) Show that near equilibrium ($r \approx R_0$) the Morse potential is harmonic, with force constant $K = m\omega^2 = 2U_0\beta^2$. (c) The lowest vibrational energy for the Morse oscillator can be shown to be

$$E_{\text{vib}} = \frac{1}{2}\hbar\omega - \frac{(\hbar\omega)^2}{16U_0}$$

Obtain from this an expression for the dissociation energy of the molecule. (d) Apply the results of parts (b) and (c) to deduce the Morse parameters U_0 and β for the hydrogen molecule. Use the experimental values 573 N/m and 4.52 eV for the effective spring constant and dissociation energy, respectively. (The measured value of R_0 for H_2 is 0.074 nm.)

14. The allowed energies of vibration for the Morse oscillator can be shown to be

$$E_{\text{vib}} = (\nu + \tfrac{1}{2})\hbar\omega - (\nu + \tfrac{1}{2})^2\frac{(\hbar\omega)^2}{4U_0}$$

$$\text{where} \quad \nu = 0, 1, \ldots$$

Obtain from this an expression for the interval separating successive levels of the Morse oscillator, and show that this interval diminishes steadily at higher energies, as illustrated in Figure 10.9b. From your result, deduce an upper limit for the vibrational quantum number ν. What is the largest vibrational energy permitted for the Morse oscillator?

15. **MP** *The Morse Oscillator Spectrum.* Use the MP DESK-TOP to find the first three pure vibrational states ($\ell = 0$, $\nu = 0, 1, 2$) of the H_2 molecule in the Morse oscillator approximation to the interatomic potential. Adopt the units of Example 10.3 so that you may easily compare results with the isotropic oscillator model of H_2. The number of Morse vibrational states is limited. See if you can find the highest-lying pure vibrational state for this case. What is the vibrational quantum number ν for this state? (The Morse potential may be found in the file *PR10-15.U3;* the default parameters are those for H_2—see Problem 13.)

16. **MP** Consider higher rotational states of the Morse oscillator described in Problem 15. Use the MP DESKTOP to find the energies of the two lowest rotational levels ($\ell = 1$ and $\ell = 2$) associated with the vibrational ground state of the H_2 molecule in this model. (See Example 10.3 and Exercise 1 for guidelines on using the DESKTOP in such applications.) Compare your results with the predictions of Equation 10.10 for this case. Are the rotations and vibrations of the H_2 molecule really independent?

10.4 Electron Sharing and the Covalent Bond

17. A one-dimensional model for the electronic energy of a diatomic homonuclear molecule is described by the potential well and barrier shown in Figure P10.17. In the simplest case, the barrier width is shrunk to zero ($w \to 0$) while the barrier height increases without limit ($U \to \infty$) in such a way that the product Uw approaches

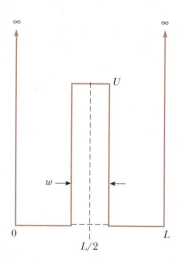

Figure P10.17

a finite value S called the barrier strength. Such a barrier—known as a delta function barrier—can be shown to produce a discontinuous slope in the wavefunction at the barrier site $L/2$ given by

$$\left.\frac{d\psi}{dx}\right|_{L/2+} - \left.\frac{d\psi}{dx}\right|_{L/2-} = \frac{2mS}{\hbar^2}\psi(L/2)$$

Solve the wave equation in the well subject to this condition to obtain expressions for the energies of the ground state and first excited state as functions of the barrier strength S. Examine carefully the limits $S \to 0$ and $S \to \infty$ and comment on your findings.

18. In Section 10.4 it is stated that the approximate electronic energy for the bonding state of H_2^+ as a function of the internuclear separation R is (in atomic units)

$$E_+ = -1 - \frac{2}{1 + \Delta}$$

$$\left\{\frac{1}{R} - \frac{1}{R}(1 + R)e^{-2R} + (1 + R)e^{-R}\right\}$$

To this we add the Coulomb energy of the two protons, $2/R$, to get the total energy of this molecular ion, E_{tot}. (a) Write a simple computer program to evaluate E_{tot} for any given value of R. Use your program to verify that the equilibrium separation of the protons in H_2^+ is $R_0 = 2.49$ bohrs, according to this model. (b) Use this model to predict a value for the effective spring constant that governs the vibrations of the H_2^+ molecular ion, and compare your result with the values reported in Table 10.2. (*Hint:* Use the well-known finite difference approximation to the second derivative,

$$\frac{d^2f}{dx^2} \approx (1/\Delta x)^2 \{f(x + \Delta x) - 2f(x) + f(x - \Delta x)\}$$

where Δx is a small increment.)

19. Repeat the calculations of Problem 18 for the case of the neutral hydrogen molecule, H_2. Take for the molecular energy of H_2

$$E_{tot} = \frac{2}{R} + 2E_+$$

where E_+ is the bonding energy of H_2^+ given in Problem 18. Compare your result for the effective spring constant with the experimental value for H_2, about 573 N/m.

20. **MP** *Modeling a Heteronuclear Molecule.* Consider again the divided square well of Example 10.4 as a model for electrons in a diatomic molecule. Moving the divider off-center destroys the identity of the "atomic" wells to either side and transforms the model to one of a heteronuclear molecule. Do this by editing

$U(x)$, replacing $HVS(\beta - x)$ with $HVS(2*\beta - x)$ and $HVS(x + \beta)$ with $HVS(x)$ [$HVS(x)$ is the Heaviside step function: $HVS(x) = 0$ for $x < 0$ and $HVS(x) = 1$ for $x > 0$]. This replacement moves the divider rightward a distance equal to half its width (β), enlarging the leftside well by β and reducing the rightside well an equal amount. The default values leave atomic wells 100 eV high with widths 1.736 Å and 2.264 Å, and separated by a barrier 0.529 Å wide and 100 eV high. Use the MP DESKTOP to find the bonding–antibonding orbital pair that derives from the atomic ground states for this heteronuclear molecule. Contrast the wavefunctions with those found in Example 10.4 for a homonuclear molecule, specifically noting any similarities or differences.

21. **MP** Use the MP DESKTOP to examine the behavior of an electron initially confined to one side of the divided square well described in Example 10.4. Take for the initial state $\Psi(x, 0)$ a sum of the bonding and antibonding orbitals that derive from the atomic ground state. [From the *Task* menu select *QM Wavegroups-1D*. You will have to use *Fourier* mode along with the *Sum Eigenstates* option and a *Time Step* of 1: see the latter part of Example 6.9 (Chapter 6) for details on configuring the DESKTOP for such applications.] The relevant frequencies $\omega(k)$ (energies) can be found in the file *PR10-21.W;* the amplitudes $a(k)$ are taken from *UNIFORM.A*. How much time, in seconds, elapses before the electron can be said to have crossed the divider? What are the implications of your result?

11

The Solid State

Chapter Outline

Matter in the solid state has been a subject of enormous fascination since the beginnings of civilization. Primitive people were attracted to the solid state by its beauty, as in radiant, symmetric gemstones, and by its utility, as in metal tools. These two attributes, *utility* and *beauty*, are just as important in physics today. Industrial applications have made solid-state physics, or condensed-matter physics, the largest subfield of physics, as evidenced by the number of pages allocated in physics journals and the number of physicists employed in this field. The beautiful symmetry and regularity of crystalline solids have both allowed and stimulated rapid progress in the physics of crystalline solids in the 20th century. Interestingly, although rapid theoretical progress has occurred with the most random (gases) and the most regular (crystalline solids) atomic arrangements, much less has been done with liquids and amorphous (irregular) solids until quite recently. Applications such as solar cells, memory elements, fiber-optic waveguides, and xerography have driven the relatively recent rush of interest in low-cost amorphous materials.

In this chapter we first describe how molecules combine to form crystalline and amorphous solids. We then introduce one of the simplest classical models of conductors, the free-electron gas model, to gain physical insight into the processes of electrical and thermal conduction. Next we introduce the quantum theory of metals to explain the deficiencies of the classical model. We describe the band theory of solids to explain the differences between insulators, conductors, and semiconductors and include a brief discussion of *p-n* junctions and semiconductor devices. Finally we discuss the general principles of lasers and some of the specifics of gas and semiconductor lasers.

11.1   BONDING IN SOLIDS

A crystalline solid consists of a large number of atoms arranged in a regular array, forming a periodic structure. The bonding schemes for molecules discussed in Chapter 10 are also appropriate for describing the bonding mechanisms in solids. For example, the ions in the NaCl crystal are ionically bonded, while the carbon atoms in the diamond structure form covalent bonds. Another type of bonding mechanism is the metallic bond, which is responsible for the cohesion of copper, silver, sodium, and other metals. Finally, the weakest type of bonding, van der Waals bonding, is responsible for the cohesion of organic solids and rare gas crystals. Refer to Section 10.1 for a survey of molecular bonding mechanisms.

Ionic Solids

Many crystals form by ionic bonding, where the dominant interaction is the Coulomb interaction between the ions. Consider the NaCl crystal shown in Figure 11.1, where each Na^+ ion has six nearest-neighbor Cl^- ions and each

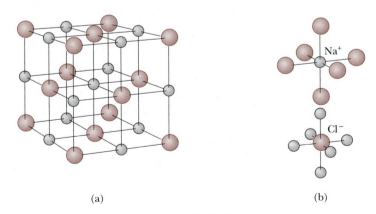

(a)	(b)

Figure 11.1   (a) The crystal structure of NaCl. The unit cell shown here represents four NaCl units and is the smallest repeating unit having the symmetry of the crystal. (b) In the NaCl structure, each positive sodium ion is surrounded by six negative chlorine ions, and each chlorine ion is surrounded by six sodium ions.

Cl^- ion has six nearest-neighbor Na^+ ions. Each Na^+ ion is attracted to the six Cl^- ions. The corresponding attractive potential energy is $-6ke^2/r$, where r is the Na^+-Cl^- separation and k is the Coulomb constant $k = 1/4\pi\epsilon_0$. In addition, there are 12 Na^+ ions at a distance of $\sqrt{2}r$ from the Na^+, which produce a weaker repulsive force on the Na^+ ion. Furthermore, beyond these 12 Na^+ ions one finds more Cl^- ions that produce an attractive force, and so on. The net effect of all these interactions is an attractive electrostatic potential energy,

$$U_{\text{attractive}} = -\alpha k \frac{e^2}{r} \tag{11.1}$$

where α is a constant called the *Madelung constant*. The value of α depends only on the specific type of crystal structure. For example, $\alpha = 1.7476$ for the NaCl structure. When the atoms are brought close together, the subshells resist overlap because of the Pauli exclusion principle, which introduces a repulsive electrostatic potential energy, B/r^m, where m is of the order of 10. The total potential energy per ion pair is therefore of the form

$$U_{\text{total}} = -\alpha k \frac{e^2}{r} + \frac{B}{r^m} \tag{11.2}$$

Figure 11.2 is a plot of the total potential energy per ion pair versus ion separation. The potential energy has its minimum value U_0 at the equilibrium separation where $r = r_0$. It is left to an end-of-chapter problem (Problem 1) to show that B may be expressed in terms of r_0 and m and that the minimum energy U_0 is

Atomic cohesive energy

$$U_0 = -\alpha k \frac{e^2}{r_0}\left(1 - \frac{1}{m}\right) \tag{11.3}$$

The absolute value of U_0, called the **ionic cohesive energy** of the solid, is the energy required to pull the solid apart into a collection of infinitely separated

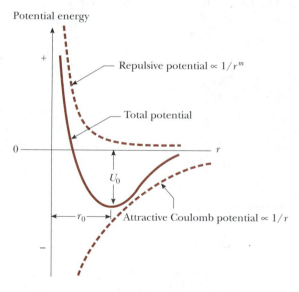

Figure 11.2   Potential energy per ion pair versus ion separation for an ionic solid. U_0 is the ionic cohesive energy, and r_0 is the equilibrium separation between ions.

Table 11.1   Some Properties of Ionic Crystals

Crystal	Equilibrium Separation (Å)	Atomic Cohesive Energy (eV per ion pair)	Melting Point (K)
LiF	2.01	8.32	1143
NaCl	2.82	6.31	1074
CsI	3.95	5.36	621

positive and negative ions. The ionic cohesive energy is 7.84 eV per Na^+ and Cl^- pair. In calculating the atomic cohesive energy, which is the energy needed to pull the solid apart into a collection of infinitely separated Na and Cl neutral atoms, we must keep in mind that one gains 5.14 eV in going from Na^+ to Na, and one must supply 3.61 eV in order to convert Cl^- to Cl. The atomic cohesive energy of NaCl may be computed as

$$+ 7.84 \text{ eV} - 5.14 \text{ eV} + 3.61 \text{ eV} = + 6.31 \text{ eV}$$

and thus the atomic cohesive energy is 6.31 eV per neutral Na and Cl pair. The atomic cohesive energy is an important comparative measure of the strengths of different solids. Table 11.1 lists measured atomic cohesive energies and melting points for a few ionic compounds.

Ionic crystals have the following general properties:

- They form relatively stable and hard crystals.
- They are poor electrical conductors because there are no available free electrons.
- They have fairly high melting and boiling points since appreciable thermal energy must be added to the crystal to overcome the large cohesive energy.
- They are transparent to visible radiation but absorb strongly in the infrared region. This occurs because the electrons form such tightly bound shells in ionic solids that visible radiation does not contain sufficient energy to promote electrons to the next allowed shell and so is not absorbed. The strong infrared absorption (at 20 to 150 μm) occurs because the vibrations of the rather massive ions have a low natural frequency and experience resonant absorption in the low-energy infrared region.
- They are often quite soluble in polar liquids such as water. The water molecule, which has a permanent electric dipole moment, exerts an attractive force on the charged ions, which breaks the ionic bonds and dissolves the solid.

Covalent Solids

As we found in Chapter 10, the covalent bond is very strong and is comparable to the ionic bond. Solid carbon, in the form of diamond, is a crystal whose atoms are covalently bonded. Because atomic carbon has the electron configuration $1s^2 2s^2 2p^2$, it lacks four electrons with respect to a filled shell ($2p^6$). Hence each carbon atom in diamond bonds covalently to four other carbon atoms to form a stable closed-shell structure.

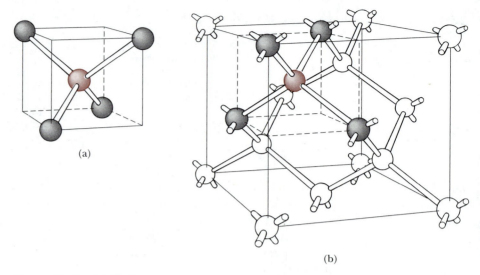

Figure 11.3   (a) Each carbon atom in diamond is covalently bonded to four other carbons and forms a tetrahedral structure. (b) The crystal structure of diamond, showing the tetrahedral bond arrangement. *(After W. Shockley,* Electrons and Holes in Semiconductors, *New York, Van Nostrand, 1950)*

In the diamond structure, each carbon atom is covalently bonded to four other carbon atoms at the corners of a cube, as illustrated in Figure 11.3a. Figure 11.3b shows the crystal structure of diamond. Note that each carbon atom forms covalent bonds with four nearest-neighbor atoms. The basic structure of diamond is called *tetrahedral* (each carbon atom is at the center of a regular tetrahedron), and the angle between the bonds is 109.5°. Other covalent crystals, such as silicon and germanium, have similar structures.

Table 11.2 gives the properties of some covalent solids. Note that the cohesive energies are greater than for ionic solids, which accounts for the hardness of covalent solids. Diamond is particularly hard and has an extremely high melting point (about 4000 K). In general, covalently bonded solids are very hard, have large bond energies and high melting points, and are good insulators, since the electrons are tightly bound in localized bonds. Because electrons are so tightly bound in diamond and other covalent solids, there is insufficient

Table 11.2   Properties of Some Covalent Crystals

Crystal	Cohesive Energy (eV/atom)*	Melting Point (K)
C (diamond)	7.37	~4000
SiC	6.15	2870
Si	4.63	1687
Ge	3.85	1211

* Since covalent cohesive energies are given in electron volts per atom, they should be multiplied by 2 for a proper comparison to Table 11.1.

energy in visible light to raise electrons to excited states. Consequently, many covalent solids do not absorb visible light and so appear transparent.

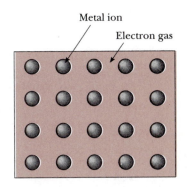

Figure 11.4 The free-electron model of a metal. The negative electron gas acts as a kind of glue to hold the positive metal ions together.

Metallic Solids

Metallic bonds are generally weaker than ionic or covalent bonds. The valence electrons in a metal are relatively free to move throughout the material. There is a large number of such mobile electrons in a metal, typically one or two electrons per atom. The metal structure can be viewed as a lattice of positive ions surrounded by a "gas" of nearly free electrons (Fig. 11.4). The binding mechanism in a metal is the attractive force between the positive ions and the electron gas.

Metals have a cohesive energy in the range of 1 to 4 eV, which is smaller than the cohesive energies of covalent solids but still large enough to produce strong solids (see Table 11.3). Since visible photons also have energies in this range, light interacts strongly with the free electrons in metals. Hence visible light is absorbed and re-emitted quite close to the surface of a metal, which accounts for the shiny nature of metallic surfaces. In addition to the high electrical conductivity of metals produced by the free electrons, the nondirectional nature of the metallic bond allows many different types of metallic atoms to be dissolved in a host metal in varying amounts. The resulting solid solutions or alloys may be designed to have particular properties, such as high strength, low density, ductility, resistance to corrosion, and so on.

Molecular Crystals

A fourth class of binding can occur even when electrons are not available to participate in bond formation, as in the cases of saturated organic molecules (CH_4) and inert gas atoms with closed electron shells. The weak electric forces at work in this type of bonding include **van der Waals forces** and arise from the attractive force between electric dipoles. (See the discussion of van der Waals bonds in Section 10.1.) Some simple molecules containing hydrogen, such as water, have permanent dipole moments (due to the built-in separation of positive and negative centers of charge) and form relatively strong bonds known as **hydrogen bonds.** Ice, for example, is hydrogen bonded with a cohesive energy of about 0.52 eV/molecule or 12 kcal/mol. (Note that 1 eV/molecule = 23 kcal/mol.) Two molecules without permanent dipole moments also experience fluctuation-induced dipole-dipole attractions, but these

Table 11.3 Properties of Some Metallic Crystals

Crystal	Cohesive Energy (eV/atom)	Melting Point (K)
Fe	4.28	2082
Cu	3.49	1631
Ag	2.95	1235
Pb	2.04	874

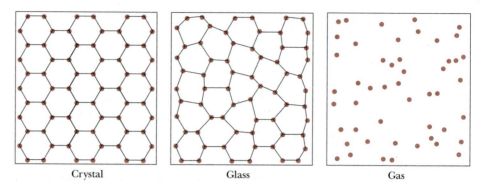

Figure 11.5   Two-dimensional sketches of the atoms in a crystal, a glass, and a gas. Note that the crystal has equal bond lengths and angles, the glass has a distribution of nearly equal bond lengths and angles, and the gas has a completely random spatial distribution of atoms. (*After R. Zallen*, The Physics of Amorphous Solids, *New York, John Wiley and Sons, 1983*)

van der Waals forces are much weaker. Such weak forces hold together many organic solids with comparatively low melting and boiling points as well as inert gas crystals. Two typical cases are solid methane, with a cohesive energy of about 0.10 eV/molecule and a melting point of 91 K, and solid argon, with a cohesive energy of 0.078 eV/molecule and a melting point of 84 K.

Amorphous Solids

The apparent simplicity and regularity of crystalline solids suggest that it will be most profitable to focus attention on perfect crystals. This is indeed the direction we shall take in most of this chapter. However, real crystals are far from perfect, since they contain irregularities (dislocations) and impurities that can have profound effects on their strength, conductivity, and other properties. This gives us some motivation for a brief discussion of crystal defects and the nature of amorphous solids.

It is well known that crystals of a given material form when the liquid state of that material is cooled sufficiently. For example, water forms crystals when cooled to 0°C. In general, more perfect crystals form if a liquid is cooled slowly, allowing the molecules to relax gradually to states having minimum potential energy. Conversely, rapid cooling of a liquid causes many dislocations in the resulting lattice structure. If the cooling is rapid enough, almost any liquid will form an amorphous solid (or glass) without long-range order, as shown in Figure 11.5. Although an amorphous solid has no long-range order, there is extensive short-range order in that bond lengths and bond angles for nearest neighbors are nearly the same throughout the solid. Even metals can be made to solidify in an amorphous form (called a *metallic glass*) if the metal in the liquid state can be cooled by 1000 K in about one millisecond. Figure 11.6 illustrates one clever technique for achieving cooling rates of 10^6 K/s, known as *melt spinning*.

The main experimental techniques for determining the distribution of atoms in crystalline and amorphous solids are x-ray diffraction, electron diffraction, and neutron diffraction. The primary experimental data are measure-

Figure 11.6 Preparing an amorphous metal ribbon by melt spinning.

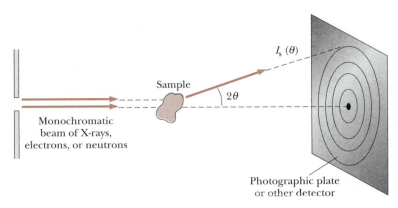

Figure 11.7 A basic experimental setup for determining the distribution of atoms in a powder sample. Since the scattered intensity depends only on θ, rings of constant intensity are obtained on the photographic plate. *(After R. Zallen,* The Physics of Amorphous Solids, *New York, John Wiley and Sons, 1983)*

ments of the scattered intensity $I_s(\theta)$, as sketched in Figure 11.7. Figure 11.8 shows actual photographic records of the electron intensity scattered from amorphous and crystalline iron samples. Note that the regular spacing of atoms over many atomic separations in the crystalline sample leads to many sharp diffraction lines. This is in contrast to the metallic glass sample, in which the distribution of atomic separations and bond angles over one or two atomic separations leads to fewer, more diffuse lines.

The distribution of atoms in space, which is described by the *radial distribution function* (RDF), $\rho(r)$, may be obtained by taking the Fourier transform of $I_s(\theta)$.[1] Note that $\rho(r)\,dr$ is the probability of finding neighboring atoms between r and $r + dr$ from any atom chosen as the origin. Thus we expect $\rho(r)$ to consist of many sharp maxima for crystals, because their structures consist of extremely regular shells of nearest-neighbor atoms, next-nearest-neighbor atoms, and so

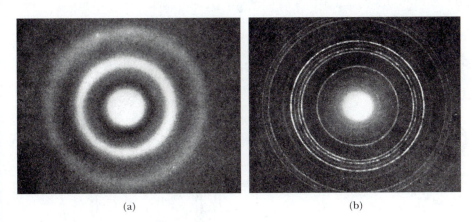

(a)   (b)

Figure 11.8 Electron diffraction patterns of (a) amorphous iron and (b) crystalline iron. *(From T. Ichikawa,* Phys. Status, Solidi A*19:707, 1973)*

[1] See, for example, R. Zallen, *The Physics of Amorphous Solids*, New York, John Wiley and Sons, 1983, Chapter 2.

Figure 11.9 Representative radial distribution functions (RDF) for a substance in crystalline, amorphous, and gas phases. *(After R. Zallen,* The Physics of Amorphous Solids, *New York, John Wiley and Sons, 1983)*

forth. On the other hand, we expect fewer and broader maxima for glasses because of their short-range order and distribution of bond lengths. Figure 11.9 shows schematic illustrations of these differences in the radial distribution function for the crystal, glass, and gas phases. Note that the RDFs for the crystal and glass phases are superimposed on a background equivalent to the gas RDF, $\rho_{gas}(r) = Ar^2$, where A is a constant. The RDFs for both crystal and glass approach the gas RDF for large r values because the distribution of atoms becomes effectively continuous over large distances and smooths out to an average value of $\bar{n}$ atoms per unit volume, as in the gas phase. Since the volume of a spherical shell is $4\pi r^2\, dr$, the number of atoms between r and $r + dr$ in the gas phase is given by $\rho_{gas}(r)\, dr = \bar{n}(4\pi r^2\, dr)$. Thus $\rho_{gas}(r) = Ar^2$, where $A = 4\pi\bar{n}$.

Figure 11.10 shows actual RDFs for crystalline and amorphous germanium obtained from x-ray diffraction measurements. Note that the amorphous RDF

Figure 11.10 RDFs for amorphous and crystalline germanium, calculated from x-ray scattering. *(From R. J. Temkin, W. Paul, and G. A. N. Connell,* Adv. Phys. *22:581, 1973)*

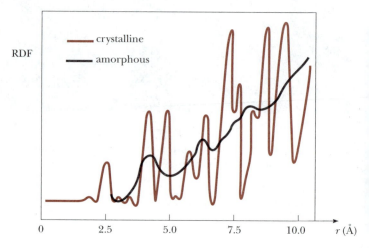

Table 11.4   Examples of Technological Applications of Amorphous Solids

Type	Material	Use
Oxide glass	$(SiO_2)_{0.8}(Na_2O)_{0.2}$	Transparent window glass
Oxide glass	$(SiO_2)_{0.9}(GeO_2)_{0.1}$	Ultratransparent optical fibers
Organic polymer	Polystyrene	Strong, low-density plastics
Chalcogenide glass	Se, As_2Se_3	Photoconductive films used for xerography
Amorphous silicon	$Si_{0.9}H_{0.1}$	Solar cells
Metallic glass	$Fe_{0.8}B_{0.2}$	Ferromagnetic low-loss ribbons used as transformer cores

merges with the smeared background after about four oscillations, indicating only short-range order. However, the crystalline RDF still varies sharply after 14 oscillations. Also, note that the crystalline RDF peaks are not as sharp as might be expected but are broadened by thermal and zero-point motions of the atoms about their lattice positions.

Amorphous solids have many useful and interesting physical properties such as extremely high optical transparency, extremely high strength, and low density. You may wish to read Zallen (cited in the Suggestions for Further Reading at the end of this chapter) for more information about amorphous solids. We conclude our brief introduction to this rapidly expanding field with Table 11.4, which lists some of the technological applications of amorphous solids.

11.2   CLASSICAL FREE-ELECTRON MODEL OF METALS

Shortly after Thomson's discovery of the electron, Drude and Thomson proposed the free-electron theory of metals. According to this theory, the physical properties of a metal may be explained by modeling the metal as a classical gas of conduction electrons moving through a fixed lattice of ion cores. Thomson, Drude, and Lorentz used this picture of a highly mobile electron "fluid" to explain the high electrical and thermal conductivities of metals, shown in Table 11.5, with considerable success. In particular, the model predicts the correct

Table 11.5   Thermal Conductivity, K, and Electrical Conductivity, σ, of Selected Substances at Room Temperature

Substance	K in $W \cdot m^{-1} K^{-1}$	σ in $(\Omega \cdot m)^{-1}$
Copper	390	59×10^6
Aluminum	210	35×10^6
Iron	63	10×10^6
Steel	50	1.4×10^6
Nichrome	14	0.9×10^6
Quartz	13	
NaCl	7.0	$<10^{-4}$

functional form of Ohm's law and the empirical connection between the electrical and thermal conductivities of a metal, known as the Wiedemann-Franz relation.[2] However, the model does not accurately predict the experimental values of electrical and thermal conductivities when classical electronic mean free paths are used in the calculations. In fact, we shall see that the shortcomings of the classical model can be remedied only by taking into account the wave nature of the electron. This involves replacing the Maxwell-Boltzmann rms velocity for electrons with the Fermi velocity (see Section 9.4) as well as replacing the classical mean free path of electrons with the quantum mean free path, which may be hundreds to thousands of times the interatomic distance. Such remarkably long mean free paths occur because the electron is a wave and as such is able to pass freely through a nearly perfect lattice, scattering only when it encounters impurity atoms or other deviations from crystalline regularity.

Ohm's Law

Ohm's law was first established as an experimental result applicable to a wide range of metals and semiconductors. It states simply that the current density in a material is directly proportional to the applied electric field. That is,

$$\mathbf{J} = \sigma\mathbf{E} \tag{11.4}$$

where $\mathbf{J}$ is the current density (A/m^2), σ is the electrical conductivity of the material ($\Omega^{-1}m^{-1}$), and $\mathbf{E}$ is the electric field (V/m). Rather than viewing Equation 11.4 simply as a proportionality between applied field and resulting current, it is more instructive to interpret this equation as a definition of σ. Interpreted in this way, Equation 11.4 tells us that a single constant, σ, which depends on the material and temperature *but not on applied field or voltage*, completely characterizes the electrical conduction in an ohmic solid.

We can show that Ohm's law follows from considering a conductor to consist of a gas of conduction electrons moving through a background of immobile, heavy ions. The electrons in a metal move randomly along straight-line trajectories, which are constantly interrupted by collisions with lattice ions (see Fig. 11.11a). The root-mean-square speed at room temperature is fairly high ($\sim 10^5$ m/s) and may be calculated from the classical equipartition theorem:

$$\tfrac{1}{2}m_e\overline{v^2} = \tfrac{3}{2}k_BT$$

or

$$v_{rms} = \sqrt{\overline{v^2}} = \left(\frac{3k_BT}{m_e}\right)^{1/2} \tag{11.5}$$

Each step of the motion (the path between collisions) shown in Figure 11.11a is a "free path"; the average free path, or *mean free path, L*, is related to the *mean free time, τ*, the average time between collisions, and to v_{rms} by

$$L = v_{rms}\tau \tag{11.6}$$

[2] The Wiedemann-Franz relation states that the ratio of thermal conductivity to electrical conductivity for metals is proportional to the temperature and that the value of the proportionality constant is independent of the metal considered.

Figure 11.11   (a) Random successive displacements of an electron in a metal without an applied electric field. (b) A combination of random displacements and displacements produced by an external electric field. The net effect of the electric field is to add together multiple displacements of length $v_d\tau$ opposite the field direction. For purposes of illustration, this figure greatly exaggerates the size of v_d compared with v_{rms}.

In Drude's original model, L was taken to be several angstroms, consistent with the view that an electron would generally travel one interatomic spacing before bumping into a large ion.

When an electric field is applied to the sample, an electric force is exerted on an electron during each step, resulting in a displacement that is small compared to the mean free path. The cumulative effect of these displacements may be viewed in terms of a small average *drift speed,* v_d, superimposed on the rather high random thermal speed, as shown in Figure 11.11b.

We may derive an expression for v_d by assuming that the total displacement of an electron after b collisions is

$$s = \frac{a}{2}\,(t_1{}^2 + t_2{}^2 + t_3{}^2 + \cdots t_b{}^2) \qquad (11.7)$$

where t_1, t_2, and so on are the successive times between collisions, and a is the acceleration (eE/m_e) produced by the electric field. Note that we have ignored the random initial velocities of the electron in Equation 11.7, since these average to zero. We may write Equation 11.7 in terms of averages as

$$s = \frac{a}{2}\,(b)\,(\overline{t^2}) \qquad (11.8)$$

Since the average value of t^2 is $\overline{t^2} = 2\tau^2$ (see Problem 9), Equation 11.8 becomes

$$s = ab\tau^2$$

or

$$s = \frac{eE}{m_e}\,b\tau^2 \qquad (11.9)$$

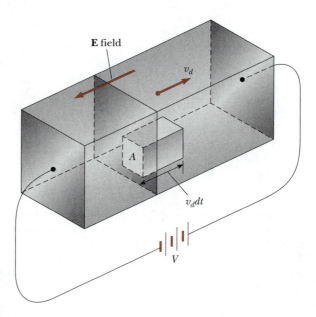

Figure 11.12   The connection between current density and drift velocity. The charge that passes through A in time dt is the charge contained in the small parallelepiped, $neAv_d\, dt$.

Comparing Equation 11.9 to the expression for the displacement in terms of the drift speed, $s = bv_d\tau$, we find

$$v_d = \frac{eE\tau}{m_e} \tag{11.10}$$

To find the current density, J, when n electrons per unit volume all move with speed v_d, we note that in a time dt the electrons move a distance $v_d\, dt$, so that $nAv_d\, dt$ electrons cross an area A perpendicular to the direction of electron flow (Fig. 11.12). Since each electron has a charge e, the charge crossing the area A in the time dt is $neAv_d\, dt$, and the current density is

$$J = \frac{neAv_d\, dt}{A\, dt} = nev_d \tag{11.11}$$

Substituting $v_d = eE\tau/m_e$ into Equation 11.11 yields

$$J = nev_d = \frac{ne^2\tau}{m_e}\, E \tag{11.12}$$

The proportionality of J to E given by Equation 11.12 shows that the classical free-electron model predicts the observed Ohm's law dependence of J on E. For this case, the conductivity, which is *independent of E*, is

$$\sigma = \frac{ne^2\tau}{m_e} \tag{11.13}$$

or, using $\tau = L/v_{\text{rms}}$,

$$\sigma = \frac{ne^2 L}{m_e v_{\text{rms}}} \qquad (11.14)$$

Substituting the Maxwell-Boltzmann rms thermal speed into Equation 11.14 yields

$$\sigma = \frac{ne^2 L}{(3k_B T m_e)^{1/2}} \qquad (11.15)$$

Table 11.6 lists measured values of σ for various metals, for comparison with calculations using Equation 11.15. Since the resistivity, ρ, is the reciprocal of the conductivity, Equation 11.15 may also be written

$$\rho = \frac{(3k_B T m_e)^{1/2}}{ne^2 L} \qquad (11.16)$$

Table 11.6   Electrical Conductivity of Metals at 300 K

Substance	Measured σ in $(\Omega \cdot m)^{-1}$
Copper	59×10^6
Aluminum	35×10^6
Sodium	22×10^6
Iron	10×10^6
Mercury	1.0×10^6

Classical expressions for conductivity and resistivity

EXAMPLE 11.1   The Classical Free-Electron Model of Conductivity in Solids

(a) Show that the rms thermal speed of electrons at 300 K is many orders of magnitude higher than the drift velocity, v_d. To find v_d, assume that a copper wire with a cross-section of 2 mm $\times$ 2 mm can safely carry a current of 10 A. (b) Estimate τ, the average time between collisions for copper at room temperature, assuming that the mean free path is the interatomic distance, 2.6 Å. (c) Calculate the conductivity of copper at room temperature as a test of the classical free-electron gas theory, and compare this to the measured value found in Table 11.6.

Solution   (a) By Equation 11.5,

$$v_{\text{rms}} = \left(\frac{3k_B T}{m_e}\right)^{1/2} = \left[\frac{3(1.38 \times 10^{-23} \text{ J/K})(300 \text{ K})}{9.11 \times 10^{-31} \text{ kg}}\right]^{1/2}$$
$$= 1.17 \times 10^5 \text{ m/s}$$

To calculate v_d we use $J = nev_d$, or $v_d = J/ne$. The number of free (conduction) electrons per cubic centimeter, n, in copper is

$$n = \left(\frac{1 \text{ free electron}}{\text{atom}}\right)\left(\frac{6.02 \times 10^{23} \text{ atoms}}{\text{mol}}\right)$$

$$\times \left(\frac{8.96 \text{ g}}{\text{cm}^3}\right)\left(\frac{1 \text{ mol}}{63.5 \text{ g}}\right)$$
$$= 8.49 \times 10^{22} \text{ electrons/cm}^3$$

Thus

$$v_d = \frac{J}{ne} = \frac{(10 \text{ A})/(4 \times 10^{-6} \text{ m}^2)}{(8.49 \times 10^{28} \text{ m}^{-3})(1.6 \times 10^{-19} \text{ C})}$$
$$= 1.8 \times 10^{-4} \text{ m/s}$$

(b) Using Equation 11.6,

$$\tau = \frac{L}{v_{\text{rms}}} = \frac{2.6 \times 10^{-10} \text{ m}}{1.2 \times 10^5 \text{ m/s}} = 2.2 \times 10^{-15} \text{ s}$$

Thus, in this model the electrons experience several hundred trillion collisions per second!

(c) Equation 11.15 gives for $T = 300$ K

$$\sigma = \frac{ne^2 L}{(3k_B T m_e)^{1/2}} = 5.3 \times 10^6 \ (\Omega \cdot m)^{-1}$$

From Table 11.6 we see that this value of the conductivity is about 10 times smaller than the measured value!

Although the classical electron gas model does predict Ohm's law, we see from Example 11.1 that it results in conductivity values that differ from measured values by an order of magnitude. Worse yet, the *measured resistivity* of most metals is found to be proportional to the absolute temperature (see Fig. 11.13), yet the classical electron gas model predicts a much weaker dependence of ρ on T. From Equation 11.16 we see that the classical model incorrectly predicts

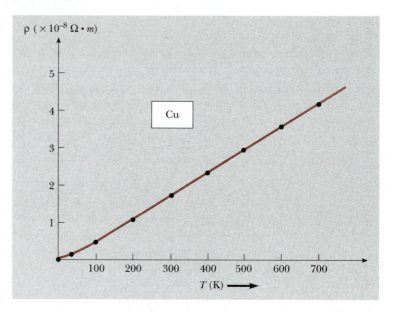

Figure 11.13 The resistivity of pure copper as a function of temperature.

that resistivity should be proportional to the square root of the absolute temperature. In Section 11.3 we will give a different model of electron scattering, which explains the observed linear dependence of ρ on T at high temperatures.

Classical Free-Electron Theory of Heat Conduction

The thermal conductivity of a substance, K, is defined in a similar way to electrical conductivity, σ. Whereas for electrical conductivity we had current density equal to conductivity times voltage gradient, or

$$J = -\sigma \frac{\Delta V}{\Delta x} \tag{11.17}$$

we now have thermal current density (W/m²) equal to thermal conductivity times temperature gradient, or

$$\frac{\Delta Q}{A\,\Delta t} = -K\frac{\Delta T}{\Delta x} \tag{11.18}$$

Here ΔQ is the thermal energy conducted through cross-sectional area A in time Δt between two planes with a temperature gradient of $\Delta T/\Delta x$. Since good electrical conductors are also good conductors of heat, it is natural to assume that the highly mobile electron gas is responsible for transporting charge as well as heat energy through the metal via a random collision process. A remarkable triumph of the classical free-electron model was to show that if free electrons were responsible for both electrical and thermal conduction in metals, then the ratio of K/σ should be a universal constant, *the same for all metals*, and *dependent only on the absolute temperature*. Since such an unusually simple connection between K and σ had already been observed experimentally (the

Wiedemann-Franz law), this prediction confirmed that the motion of the electron gas was basically responsible for both electrical and thermal conductivity.

Since the Wiedemann-Franz law is such a remarkably simple and powerful result, it bears consideration in a bit more detail. The kinetic theory of gases may be applied to the free-electron gas to calculate the flux of thermal energy (W/m^2) carried by electrons moving from a region at temperature $T + \Delta T$ to a region at temperature T.[3] Using this result, one immediately finds that the thermal conductivity depends simply on the heat capacity per unit volume, C_v, v_{rms}, and the mean free path, L, of the electrons, as follows:

$$K = \tfrac{1}{3}C_v v_{rms} L \tag{11.19}$$

If we assume that the electrons behave as a classical ideal gas and obey Maxwell-Boltzmann statistics, the heat capacity per *mole* is found to be $3R/2$ or $3N_A k_B/2$, where N_A is Avogadro's number. To convert this to heat capacity per unit volume for use in Equation 11.19, we must multiply by the ratio of the electron density to Avogadro's number. Thus

$$C_v = (\tfrac{3}{2}N_A k_B)\left(\frac{n}{N_A}\right) = \tfrac{3}{2}k_B n$$

and Equation 11.19 becomes

$$K = \frac{k_B n v_{rms} L}{2} \tag{11.20}$$

If we simply use the interatomic distance for L in Equation 11.20, we will find incorrect values, as in the case of electrical conductivity. However, if we form the ratio of K to σ, we will avoid the problem of having to assign values to n and L, as these quantities cancel. Thus

$$\frac{K}{\sigma} = \frac{k_B n v_{rms} L/2}{n e^2 L/m_e v_{rms}} = \frac{k_B (v_{rms})^2 m_e}{2e^2}$$

Substituting $v^2_{rms} = 3k_B T/m_e$ into this expression finally gives the desired classical result, known as the Wiedemann-Franz law:

$$\frac{K}{\sigma} = \frac{3k_B^2}{2e^2}\,T \tag{11.21}$$

Wiedemann-Franz law

Equation 11.21 shows that the ratio of K/σ is proportional to T and depends only on the universal constants k_B and e, in complete agreement with the empirical law of Wiedemann and Franz announced in 1853. The ratio $K/\sigma T$ is known as the **Lorentz number,** and according to our classical theory, it has the same value, $3k_B^2/2e^2 = 1.1 \times 10^{-8}$ W·Ω/K^2, *for any metal at any temperature.* Table 11.7 shows that, indeed, the ratio of $K/\sigma T$ is nearly constant from metal to metal and with varying temperatures, but that the value of $3k_B^2/2e^2$ does not agree precisely with the measured Lorentz numbers. Overall, however, the agreement of the simple classical theory with experiment is good and is taken to constitute strong evidence that the free-electron gas accounts for both electrical and thermal conductivity in metals.

Table 11.7   Experimental Lorentz Numbers $K/\sigma T$ in Units of 10^{-8} W·Ω/K^2*[*]

Metal	273 K	373 K
Ag	2.31	2.37
Au	2.35	2.40
Cd	2.42	2.43
Cu	2.23	2.33
Ir	2.49	2.49
Mo	2.61	2.79
Pb	2.47	2.56
Pt	2.51	2.60
Sn	2.52	2.49
W	3.04	3.20
Zn	2.31	2.33

[*] From C. Kittel, *Introduction to Solid State Physics*, 2nd ed., New York, John Wiley and Sons, 1965.

[3] See C. Kittel, *Introduction to Solid State Physics*, 6th ed., New York, John Wiley and Sons, 1986, Chapter 5.

11.3   QUANTUM THEORY OF METALS

Although the classical electron gas model broadly describes the electrical and thermal properties of metals, there are notable deficiencies in the predictions of this model—namely in the numerical values of K and σ, in the temperature dependence of σ, and in the prediction of an excess heat capacity for metals (see Section 9.4.) These deficiencies can be rectified by replacing the Maxwell-Boltzmann distribution with the Fermi-Dirac distribution for the conduction electrons in the metal and by calculating the electron mean free path while explicitly taking the wave nature of the electron into account. Since quantum mechanical calculations of electron mean free path are complicated, we shall rely on qualitative arguments and inferences from measured quantities to give physical insight into the surprising transparency of metals to conduction electrons.

Replacement of v_{rms} with v_{F}

As we have seen in Chapter 9, in a metal the electrons obey the Pauli exclusion principle, and essentially all energy levels are filled up to the Fermi energy, E_{F}. Because of restrictions imposed by the Pauli principle on the quantum states of scattered electrons, one might anticipate a "bottleneck" in electron transport through a metal because of a lack of empty final states. That this is not the case is illustrated in Figure 11.14, which shows the velocity distribution in three dimensions of conduction electrons in a metal. Essentially all the electrons have velocities within a radius of $\sqrt{2E_{\text{F}}/m_{\text{e}}}$ in velocity space [$\frac{1}{2}m_{\text{e}}(v_x^2 + v_y^2 + v_z^2) \leq E_{\text{F}}$]. Figure 11.14b shows that the application of an electric field causes all the electron velocities to increase by an increment v_d. Since the shift of the electron originally at A creates an empty final state for the electron originally at B, there is no bottleneck. Furthermore, the net effect of the E-field is to leave an intact

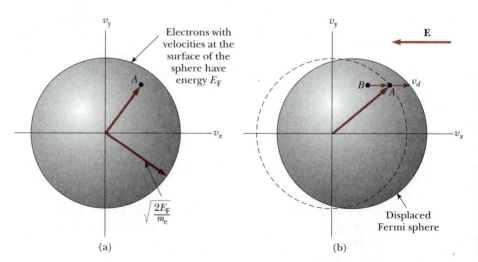

Figure 11.14   (a) Allowed velocity vectors, or positions in velocity space, of conduction electrons in a metal without an applied electric field. (b) The net effect of an applied electric field is a small displacement of the Fermi sphere, the displaced electrons having a speed of approximately v_{F}.

core of states for electrons with $E < E_F$ and to produce a displacement of those electrons near the Fermi surface having $v \approx v_F$. Thus only those electrons with $v \approx v_F$ are free to move and participate in electrical and thermal conduction, and we can presumably use the classical expressions for σ and K,

$$\sigma = \frac{ne^2 L}{m_e v_{\text{rms}}} \quad \text{and} \quad K = \tfrac{1}{3} C_v v_{\text{rms}} L$$

with the Maxwell-Boltzmann rms speed replaced by the Fermi speed; thus,

$$\sigma = \frac{ne^2 L}{m_e v_F} \tag{11.22}$$

and

$$K = \tfrac{1}{3} C_v v_F L \tag{11.23}$$

Wiedemann-Franz Law Revisited

Let us test the validity of the quantum replacement of v_{rms} with v_F by recalculating $K/\sigma T$ and checking for improved agreement with experiment. First we note that using a Maxwell-Boltzmann molar heat capacity of $3R/2$ for electrons in a metal is incorrect, since conduction electrons obey Fermi-Dirac statistics. As already mentioned in Chapter 9, only a fraction of the electrons within $k_B T$ of E_F change energy as the temperature changes, leading to a molar heat capacity that is a small fraction of $3R/2$:

$$C \approx 2\left(\frac{k_B T}{E_F}\right)(\tfrac{3}{2}R)$$

An exact calculation is rather complicated[4] but leads to a similar result:

$$C = \frac{\pi^2}{3}\left(\frac{k_B T}{E_F}\right)(\tfrac{3}{2}R) \tag{11.24}$$

Changing C to a heat capacity per unit volume and substituting into $K = \tfrac{1}{3} C v_F L$ yields

$$K = \tfrac{1}{3}\left(\frac{\pi^2}{3}\right)\left(\frac{k_B T}{E_F}\right)(\tfrac{3}{2}R)\left(\frac{n}{N_A}\right)v_F L$$

Using $R = N_A k$ and $E_F = \tfrac{1}{2} m_e v_F^2$, we find

$$K = \frac{\pi^2}{3}\left(\frac{k_B^2 T}{m_e v_F}\right)nL \tag{11.25}$$

Now, forming the Lorentz number $K/\sigma T$ yields

$$\frac{K}{\sigma T} = \frac{(\pi^2/3)(k_B^2 T/m_e v_F)(nL)}{(ne^2 L/m_e v_F)T}$$

---

[4] See Kittel, op. cit., Chapter 6.

or

Quantum form of the
Wiedemann-Franz law

$$\frac{K}{\sigma T} = \frac{\pi^2 k_B^2}{3e^2} = 2.45 \times 10^{-8} \text{ W} \cdot \Omega/\text{K}^2 \qquad (11.26)$$

A comparison of this result with Table 11.7 shows improved and excellent agreement between theoretical and experimental Lorentz values, thus justifying the replacement of v_{rms} with v_F.

Quantum Mean Free Path of Electrons

Changes from the classical value of electron mean free path arise from the wave properties of the electron. Consider, once again, our familiar copper sample. If we simply use the interatomic distance between copper atoms as the mean free path L for electrons and v_F for v_{rms} in $\sigma = ne^2 L/m_e v_{rms}$, we find a value for σ that is about 200 times smaller than the measured value of the conductivity. This discrepancy implies that we are using the wrong value for L and that the scattering sites for electrons are not adjacent ion cores but more widely separated scattering centers. These scattering centers consist of departures from perfect lattice regularity, such as thermal displacements of ions from equilibrium lattice sites, dislocations, and impurity atoms. Because quantum calculations of L that include these effects are complicated, let us make a rough estimate of L for copper at room temperature by using the measured value of σ. Solving $\sigma = ne^2 L/m_e v_F$ for L yields

$$L = \frac{m_e v_F \sigma}{ne^2} \qquad (11.27)$$

Using a Fermi energy of 7.05 eV for copper (see Table 9.1) gives

$$v_F = \left(\frac{2E_F}{m_e}\right)^{1/2} = \left(\frac{(2)(7.05 \times 1.6 \times 10^{-19} \text{ J})}{9.11 \times 10^{-31} \text{ kg}}\right)^{1/2} = 1.57 \times 10^6 \text{ m/s}$$

Substituting this value into Equation 11.27 gives

$$L = \frac{(9.11 \times 10^{-31} \text{ kg})(1.57 \times 10^6 \text{ m/s})(5.9 \times 10^7 \, \Omega^{-1} \, \text{m}^{-1})}{(8.49 \times 10^{28} \text{ electrons/m}^3)(1.60 \times 10^{-19} \text{ C})^2}$$
$$= 3.9 \times 10^{-8} \text{ m} = 390 \text{ Å}$$

This is about 150 times the distance between atoms. Using $\tau = L/v_F$, we can also calculate the mean free time between electronic collisions in copper at room temperature:

$$\tau = \frac{3.90 \times 10^{-8} \text{ m}}{1.57 \times 10^6 \text{ m/s}} = 2.50 \times 10^{-14} \text{ s}$$

We can account for the unexpectedly long electron mean free path by taking the wave nature of the electron into account. Quantum mechanical calculations show that electron waves with a broad range of energies can pass through a *perfect* lattice of ion cores unscattered, without resistance, and with an infinite mean free path. The actual resistance of a metal is due to the random thermal displacements (thermal vibrations) of ions about lattice points and to other deviations from a perfect lattice, such as impurity atoms and defects that scatter electron waves. The lack of electron scattering by a perfect lattice can be understood by noting that the electron wave generally travels through the metal unattenuated, just as does light through a transparent crystal. Strong reflections of electron waves are set up only for specific electron energies, and when this occurs, the electron wave cannot travel freely through the crystal. As we shall see in the next section, these strong reflections occur when the lattice spacing is equal to an integral number of electronic wavelengths, resulting in a discrete set of forbidden energy bands for electrons.

Causes of resistance in a metal

The observed proportionality of resistivity to absolute temperature at high temperature is the result of the scattering of electrons by lattice ions vibrating thermally about their equilibrium lattice positions. As we saw in Chapter 9, lattice vibrations have a quantized energy $\hbar\omega$, where ω is the angular frequency of vibration of the lattice ion. These quantized lattice vibrations are called **phonons.** The number of phonons with energy $\hbar\omega$ that are available at a given temperature T is denoted by n_p and given by Equation 9.32:

$$n_p \propto \frac{1}{e^{\hbar\omega/k_B T} - 1} \tag{9.32}$$

At high temperatures, $k_B T$ is much greater than $\hbar\omega$, and this equation becomes

$$n_p \propto \frac{k_B T}{\hbar\omega}$$

Thus the number of phonons available to scatter electrons is directly proportional to T. Finally, since the number of electron scatterers is proportional to the temperature, so is the resistivity.

In addition to the temperature-dependent part of the resistivity, there is also a temperature-independent contribution to the resistivity of the metal, which clearly manifests itself for temperatures less than 10 K. This residual resistivity ρ_i, which remains as $T \rightarrow 0$, is produced from electron waves scattering from impurities and structural imperfections in a given sample (see Figure 11.15). The parallel nature of the curves shown in Figure 11.15 implies that the resistivity caused by thermal motion of the lattice, ρ_L, is independent of the impurity concentration and that ρ_i is independent of temperature. This result is formalized in a result known as **Matthiessen's rule,** which states that the resistivity of a metal may usually be written in the form

$$\rho = \rho_i + \rho_L \tag{11.28}$$

where ρ_i depends only on the concentration of crystal imperfections and ρ_L depends only on T.

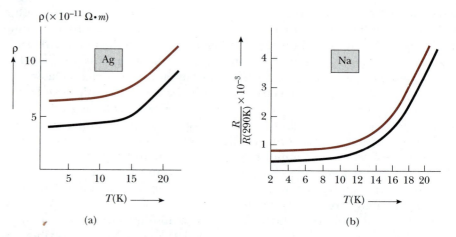

Figure 11.15   (a) Resistivity of silver at low temperature. The colored curve is for a silver sample with a higher concentration of impurity atoms. *(From W. J. de Haas, G. J. van den Berg, J. de Boer, Physica 2:453, 1935)* (b) Resistance of sodium as a function of temperature. The colored curve is for a sodium sample with a higher concentration of imperfections. *(From D. K. C. MacDonald and K. Mendelssohn, Proc. Roy. Soc. (London) A202:103, 1950)*

11.4   BAND THEORY OF SOLIDS

In Sections 11.2 and 11.3 we discussed the classical and quantum models of metals, making the simplifying assumption that the metal consists of a gas of free electrons. Since good insulators have enormous resistivity compared to metals (a factor of 10^{24} greater), their electronic configurations must be quite different. In fact, most outer-shell electrons in an insulator are not free but are involved in ionic or covalent bonds, as discussed earlier. Furthermore, in order to completely understand the electronic properties of solids, one must consider the effect of the lattice ions.

In the general case, there are two approaches to the determination of electronic energies in a solid. One is to follow the behavior of energy levels of isolated atoms as they are brought closer and closer to form a solid. The other is to show that energy bands arise when the Schrödinger equation is solved for electrons subject to a periodic potential representing the lattice ions. We shall follow the isolated-atom approach here first, since it is simpler and immediately leads to an explanation of the differences between conductors, insulators, and semiconductors. Later in this section we shall explicitly consider the effect of a periodic potential on electron waves, to show how energy bands arise.

Isolated-Atom Approach to Band Theory

If two identical atoms are very far apart, they do not interact, and their electronic energy levels can be considered to be those of isolated atoms. Suppose the two atoms are sodium, each having an outermost $3s$ electron with a specific energy. As the two sodium atoms are brought closer together, their wavefunctions overlap, and the two degenerate isolated $3s$ energy levels are split into two different levels, as shown in Figure 11.16a. We can understand this splitting by

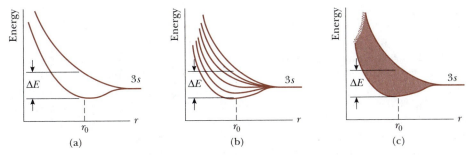

Figure 11.16   (a) The splitting of the $3s$ levels when two sodium atoms are brought together. (b) The splitting when six sodium atoms are brought together. (c) The formation of a $3s$ band when a large number of sodium atoms are assembled to form a solid. Note that r_0 is the actual lattice constant.

considering the appropriate electronic wavefunctions for the case of widely separated atoms and the case of neighboring atoms. (For a more complete and careful treatment of this splitting, see the discussion of covalent bonding in Section 10.4.) Figure 11.17 shows idealized isolated atom wavefunctions ψ_1 and ψ_2 as well as the mixed wavefunctions $\psi_1 + \psi_2$ and $\psi_1 - \psi_2$ for two atoms close together. Note that an electron in the state $\psi_1 + \psi_2$ has a substantial probability of being found midway between the ion cores, while in the state $\psi_1 - \psi_2$ the probability density vanishes at the midpoint. Since the electron spends part of its time midway between the two attractive ion cores in the state $\psi_1 + \psi_2$, the electron is more tightly bound (has lower energy) in the state $\psi_1 + \psi_2$ than in $\psi_1 - \psi_2$. This leads to the two different $3s$ energy levels shown in Figure 11.16a.[5]

When a large number of atoms is brought together to form a solid, a similar

Figure 11.17   Idealized sodium wavefunctions in various stages of combination. (a) Wavefunctions of electrons bound to two ion cores at large separation. (b) One linear combination of wavefunctions that is appropriate for electrons bound to ion cores at small separation is $\psi_1 + \psi_2$. (c) The other linear combination is $\psi_1 - \psi_2$.

---

[5] The width of the energy band ΔE depends on the amount of charge distribution overlap between adjacent atoms and hence on the interatomic separation, r_0.

Figure 11.18 The energy bands of sodium. The solid contains N atoms. Note the energy gaps between allowed bands.

phenomenon occurs. As the atoms are brought close together, the various iso-lated-atom energy levels begin to split. This splitting is shown in Figure 11.16b for six atoms in proximity. In this case there are six energy levels corresponding to six different linear combinations of isolated-atom wavefunctions. The width of an energy band (designated ΔE in Fig. 11.16) depends only on the number of atoms close enough to interact strongly, which is always a small number. If we consider the total number of atoms ($N \approx 10^{23}$ atoms/cm^3), we find a very large number of levels (determined by N) spaced within the width ΔE, so the levels may be regarded as a *continuous band* of energy levels (see Fig. 11.16c). In the case of sodium, it is common to refer to the continuous distribution of allowed energy levels as the $3s$ band, because it originates from the $3s$ levels of individual sodium atoms. In general, a crystalline solid has numerous allowed energy bands that arise from different atomic energy levels. Figure 11.18 shows the allowed energy bands of sodium. Note that energy gaps, or forbidden energy bands, for the electrons occur between the allowed bands. Forbidden energy bands arise from the forbidden regions between different atomic levels and will always be present unless individual atomic levels broaden so much that they overlap.

If the solid contains N atoms, each energy band has N energy levels. The $1s$, $2s$, and $2p$ bands of sodium are each full, as indicated by the gray-shaded areas in Figure 11.18. A level whose orbital angular momentum is ℓ can hold $2(2\ell + 1)$ electrons. The factor of 2 arises from the two possible electron-spin orientations, while the factor $2\ell + 1$ corresponds to the number of possible orientations of the orbital angular momentum. The capacity of each band for a system of N atoms is $2(2\ell + 1)N$ electrons. Hence the $1s$ and $2s$ bands each contain $2N$ electrons ($\ell = 0$), while the $2p$ band contains $6N$ electrons ($\ell = 1$). Because sodium has only one $3s$ electron and there is a total of N atoms in the solid, the $3s$ band contains only N electrons and is only half full. The $3p$ band, which is above the $3s$ band, is completely empty.

Conduction in Metals, Insulators, and Semiconductors

The enormous variation in electrical conductivity of metals, insulators, and semiconductors may be explained qualitatively in terms of energy bands. We shall see that the position and the electronic occupation of the highest band or, at most, of the highest two bands determine the conductivity of a solid.

Metals.   We can understand electrical conduction in metals by considering the half-filled uppermost $3s$ band of sodium. Figure 11.19 shows this typical half-filled metallic band at $T = 0$ K, where the shaded region represents levels that are filled with electrons. Since electrons obey Fermi-Dirac statistics, all levels below the Fermi energy, E_F, are filled with electrons, while all levels above E_F are empty. In the case of sodium, the Fermi energy lies in the middle of the band. At temperatures greater than 0 K, some electrons are thermally excited to levels above E_F (as shown by the Fermi-Dirac contribution to the left of Fig. 11.19), but overall there is little change from the 0-K case. *Note, however, that if an electric field is applied to the metal, electrons with energies near the Fermi energy require only a small amount of additional energy from the field to reach nearby empty energy states.* Thus electrons are free to move with only a small applied field in

Figure 11.19 A half-filled band of a conductor such as the $3s$ band of sodium. At $T = 0$ K, the Fermi energy lies in the middle of the band. The Fermi-Dirac probability that an energy state E is occupied at $T > 0$ K is shown at the left.

a metal because there are many unoccupied states close to occupied energy states.

Insulators. Now consider the two highest-energy bands of a material having the lower band completely filled with electrons and the higher completely empty at 0 K (see Fig. 11.20). It is common to refer to the separation between the outermost filled and empty bands as the *energy gap*, E_g, of the material. The lower band filled with electrons is called the *valence band,* and the upper, empty band is the *conduction band.* The Fermi energy is at the midpoint of the energy gap, as shown in Figure 11.20. Since the energy gap for an insulator is large (~ 10 eV) compared to $k_B T$ at room temperature ($k_B T = 0.025$ eV at 300 K), the Fermi-Dirac distribution predicts that very few electrons will be thermally excited into the upper band at normal temperatures. *Thus, although an insulator has many vacant states in the conduction band that can accept electrons, there are so few electrons actually occupying conduction-band states that the overall contribution to electrical conductivity is very small, resulting in a high resistivity for insulators.*

Semiconductors. Now consider a material that has a much smaller energy gap, on the order of 1 eV. Such materials are called semiconductors. Table 11.8 shows the energy gaps for some representative materials. At $T = 0$ K, all electrons are in the valence band, and there are no electrons in the conduction band. Thus semiconductors are *poor* conductors at low temperatures. At ordinary temperatures, however, the situation is quite different. As shown in Figure 11.21, the populations of the valence and conduction bands are altered. Since the Fermi level, E_F, is located at about the middle of the gap for a semiconductor, and since E_g is small, appreciable numbers of electrons are thermally excited from the valence band to the conduction band. Since there are many empty nearby states in the conduction band, a small applied potential can easily raise the energy of the electrons in the conduction band, resulting in a moderate current. Because thermal excitation across the narrow gap is more probable at higher temperatures, the conductivity of semiconductors depends strongly on temperature and *increases* rapidly with temperature. This contrasts

Table 11.8 Energy-Gap Values for Some Semiconductors*

| Crystal | E_g(eV) | |
	0 K	300 K
Si	1.17	1.14
Ge	0.744	0.67
InP	1.42	1.35
GaP	2.32	2.26
GaAs	1.52	1.43
CdS	2.582	2.42
CdTe	1.607	1.45
ZnO	3.436	3.2
ZnS	3.91	3.6

*These data were taken from C. Kittel, *Introduction to Solid State Physics,* 6th ed., New York, John Wiley and Sons, 1986.

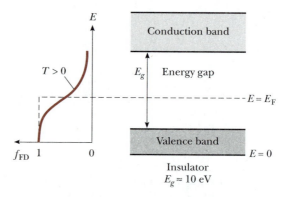

Figure 11.20   An insulator at $T = 0$ K has a filled valence band and an empty conduction band. The Fermi level lies midway between these bands. The Fermi-Dirac probability that an energy state E is occupied at $T > 0$ K is shown to the left.

sharply with the conductivity of a metal, which *decreases slowly* with temperature (see Section 11.3).

It is important to point out that there are both negative and positive charge carriers in a semiconductor. When an electron moves from the valence band into the conduction band, it leaves behind a vacant crystal site, or so-called **hole,** in the otherwise filled valence band. This hole (electron-deficient site) appears as a positive charge, $+e$. The hole acts as a charge carrier in the sense that a valence electron from a nearby site can transfer into the hole, thereby filling it and leaving a hole behind in the electron's original place. Thus the hole migrates through the valence band. In a pure crystal containing only one element or compound, there are equal numbers of conduction electrons and holes. Such combinations of charges are called electron-hole pairs, and a pure semiconductor that contains such pairs is called an *intrinsic semiconductor* (see Fig. 11.22). Note that in the presence of an electric field, the holes move in the direction of the field and the conduction electrons move opposite the field.

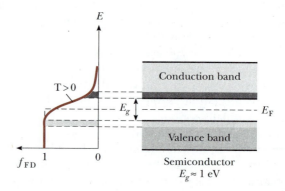

Figure 11.21   The band structure of a semiconductor at ordinary temperatures ($T \approx$ 300 K). Note that the energy gap is much smaller than in an insulator and that many electrons occupy states in the conduction band.

electrons

holes

Energy

Conduction band

Narrow forbidden gap

Valence band

Applied **E** field

Figure 11.22   An intrinsic semiconductor. The electrons move toward the left and the holes move toward the right when the applied electric field is to the right as shown.

EXAMPLE 11.2

Estimate the electric field strength required to produce conduction in diamond, an excellent insulator at room temperature. Assume a mean free path of 5×10^{-8} m and an energy gap of 7 eV in diamond.

Solution   If an electron in diamond is to conduct, it must be supplied with 7 eV of energy from the electric field. Since the electron generally loses most of its excess energy in each collision, the field must supply 7 eV in a *single*

mean free path. Thus, for an electron to gain 7 eV in a distance L, we require an electric field of

$$E = \frac{V}{L} = \frac{7 \text{ V}}{5 \times 10^{-8} \text{ m}} = 1.4 \times 10^8 \text{ V/m}$$

This is an enormous field compared to the field required to produce conduction in metals.

Energy Bands from Electron Wave Reflections

We now take a completely different approach to understanding the origin of energy bands in solids. This approach involves modifying the free-electron wavefunctions to take into account the scattering of electron waves by the periodic crystal lattice.

Recall that a completely free electron moving in the $+x$ direction is represented by a traveling wave with wavenumber $k = 2\pi/\lambda$, described by

$$\Psi_{\text{free}} = A e^{i(kx - \omega t)}$$

According to de Broglie, the wave carries momentum $p = \hbar k$ and energy $E = \hbar \omega$. Furthermore, the energy of the free electron as a function of k is

$$E = \frac{p^2}{2m_e} = \frac{\hbar^2 k^2}{2m_e}$$

and a plot of E versus k yields a parabola (see Fig. 11.23). We note that allowed energy values are distributed continuously from 0 to infinity and that there are no breaks or gaps in energy at particular k values.

Now let us consider what happens to a traveling electron wave when it passes through a one-dimensional crystal lattice with atomic spacing a. Let us start

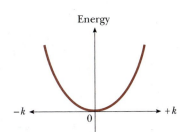

Energy

$-k$ 0 $+k$

Figure 11.23   Energy versus wavenumber k for a free electron, where $k = 2\pi/\lambda$.

with electron waves of very long wavelength, or small k, and low energy, as in Figure 11.24a. In this case, waves reflected from successive atoms and traveling to the left are all slightly out of phase and cancel out on average. Thus the electron wave does not get reflected, and the electron moves through the lattice like a free particle. If, however, we make the electron wavelength shorter and shorter, when we reach the condition $\lambda = 2a$, waves reflected from adjacent atoms will be *in phase*, as shown in Figure 11.24b, and there will be a strong reflected wave. This occurs because the path length difference is $2a$ for a wave reflected directly from atom 1 compared to a wave that travels from 1 to 2 and then is reflected back to 1, as shown in Figure 11.24c. The longest wavelength at which constructive interference of reflected waves occurs is thus

$$2a = \lambda$$

In general, constructive interference also occurs at other, shorter wavelengths,

$$2a = \pm n\lambda \qquad n = 1, 2, 3, \ldots \tag{11.29}$$

where the negative sign arises from reflection from the atom to the left of atom 1. In terms of wavenumber k, Equation 11.29 predicts strong reflected electron waves when

$$k = \pm \frac{n\pi}{a} \qquad n = 1, 2, 3, \ldots \tag{11.30}$$

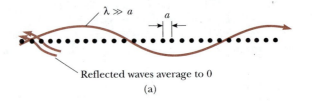

Reflected waves average to 0

(a)

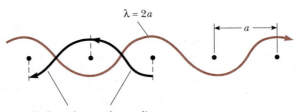

Reflected waves from adjacent
atoms are in phase and reinforce each other

(b)

Constructive interference
Path difference $= 2a = n\lambda$

(c)

Figure 11.24   (a) Reflection of electron waves when $\lambda \gg a$. (b) Reflection of electron waves when $\lambda = 2a$. (c) Constructive interference of waves reflected from atoms 1 and 2.

Thus, for $k = \pm n\pi/a$ the electron wavefunctions are not just waves traveling to the right but are composed of equal parts of waves traveling to the right (incident) and to the left (reflected). Since the waves traveling left and right can be added or subtracted, we actually have two different possible standing wave types, denoted by Ψ_- and Ψ_+:

$$\Psi_- = Be^{i(kx-\omega t)} - Be^{i(-kx-\omega t)} \tag{11.31}$$

$$\Psi_+ = Be^{i(kx-\omega t)} + Be^{i(-kx-\omega t)} \tag{11.32}$$

where $Be^{i(kx-\omega t)}$ is a wave traveling to the right and $Be^{i(-kx-\omega t)}$ is a wave traveling to the left. Using Euler's identity, $e^{i\theta} = \cos\theta + i\sin\theta$, one can easily show that for $k = \pm\pi/a$, Ψ_- and Ψ_+ take the more useful forms

$$\Psi_- = \pm 2Bie^{-i\omega t}\sin\frac{\pi x}{a} \tag{11.33}$$

$$\Psi_+ = 2Be^{-i\omega t}\cos\frac{\pi x}{a} \tag{11.34}$$

The key point is that *Ψ_- has a slightly higher energy than Ψ_{free}, and Ψ_+ has a slightly lower energy than Ψ_{free}, at $k = \pi/a$. This leads to a discontinuity in energy, or a band gap, at $k = \pi/a$.* Figure 11.25 shows this band gap. To show that the Ψ^- state has a higher energy than the Ψ^+ state, we must consider both kinetic and potential energies of the electron in these states. The kinetic energy is the same in both states, since both have the same momentum, $p = \hbar k = \hbar\pi/a$. Thus the average kinetic energy of the states Ψ_+ and Ψ_- is

$$\langle K \rangle = \left\langle \frac{p^2}{2m_e} \right\rangle = \frac{\hbar^2 k^2}{2m_e} = \frac{\hbar^2 \pi^2}{2m_e a^2}$$

The potential energies of the two states are different, however, since Ψ_+ and Ψ_- distribute the electrons differently with respect to the positions of the pos-

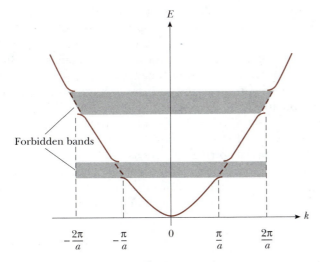

Figure 11.25 Energy versus wavenumber for a one-dimensional lattice with atomic separation a. The E versus k curve for the free-electron case is shown dashed. Note the forbidden bands corresponding to impossible energy states for the electron.

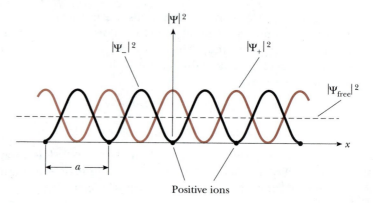

Figure 11.26   Probability densities of standing waves and traveling (free-) electron waves in a one-dimensional lattice. Note that the state Ψ_+ places the electron over the positive ion and consequently has the lowest potential energy.

itive ion cores. Figure 11.26 shows the distributions of electronic charge for the case of a traveling wave and the two standing-wave cases, Ψ_+ and Ψ_-. They are calculated from the probability densities as follows:

$$\Psi_{free}^* \, \Psi_{free} = [Ae^{i(kx-\omega t)}]^*[Ae^{i(kx-\omega t)}] = |A|^2$$

$$\Psi_+^* \Psi_+ = \left[2Be^{-i\omega t} \cos \frac{\pi x}{a} \right]^* \left[2Be^{-i\omega t} \cos \frac{\pi x}{a} \right] = 4|B|^2 \cos^2 \frac{\pi x}{a}$$

$$\Psi_-^* \Psi_- = \left[\pm 2Bie^{-i\omega t} \sin \frac{\pi x}{a} \right]^* \left[\pm 2Bie^{-i\omega t} \sin \frac{\pi x}{a} \right] = 4|B|^2 \sin^2 \frac{\pi x}{a}$$

Note that the traveling-wave solution distributes electronic charge uniformly along the lattice. On the other hand, Ψ_- concentrates electronic charge midway between the ions, while Ψ_+ concentrates electronic charge right over the positive ion cores, minimizing the electron's potential energy.[6] Thus we would expect the average potential energy of the electron in the state Ψ_+ to be somewhat below the free-electron energy and the average potential energy in the state Ψ_- to be a bit above the free-electron energy. This difference in potential energy between Ψ_+ and Ψ_- leads to a difference in total energy, since the kinetic energies of Ψ_+ and Ψ_- are the same. Thus the total energy of the state Ψ_+ at $k = \pi/a$ shown in Figure 11.25 lies slightly below the free-electron energy parabola, and the total energy of the state Ψ_- lies slightly above the free-electron energy. Also, Figure 11.25 shows that the overall effect of the lattice is to introduce forbidden gaps in the free-energy continuum at values of $k = \pm n\pi/a$.

---

[6] Recall that the electron's electrical potential energy is the work done by an external force to move the electron to a particular point. To move the electron to a *positive* ion requires the minimum external work. Thus the electron has the minimum potential energy in this position.

11.5   SEMICONDUCTOR DEVICES

The *p-n* Junction

In order to make devices, one must be able to fabricate semiconductors with well-defined regions of different conductivity. Both the type (positive or negative) and number of carriers in a semiconductor may be tailored to the needs of a particular device by the addition of specific impurities in a process called *doping*. For example, when an atom with five valence electrons, such as arsenic, is added to a semiconductor from Group 4 of the periodic table, four valence electrons participate in the covalent bonds and one electron is left over (see Fig. 11.27a). This extra electron is nearly free and has an energy level that lies within the energy gap, just below the conduction band, as shown in Figure 11.27b. Such a pentavalent atom in effect donates an electron to the semiconductor and is referred to as a *donor atom*. Since the energy spacing between the donor levels and the bottom of the conduction band is very small (typically about 0.05 eV), a small amount of thermal energy causes an electron in these levels to move into the conduction band. (Recall that the average thermal energy of an electron at room temperature is $k_B T \approx 0.025$ eV.) Semiconductors doped with donor atoms are called *n-type semiconductors* because the charge carriers are negatively charged electrons.

If the semiconductor is doped with atoms having three valence electrons, such as indium and aluminum, the three electrons form covalent bonds with neighboring atoms, leaving an electron deficiency, or hole, in the fourth bond (see Fig. 11.28a). The energy levels of such impurities also lie within the energy gap, just above the valence band, as indicated in Figure 11.28b. Electrons from the valence band have enough thermal energy at room temperature to fill these impurity levels, leaving behind a hole in the valence band. Because a trivalent atom in effect accepts an electron from the valence band, such impurities are referred to as *acceptors*. A semiconductor doped with trivalent (acceptor) impurities is known as a *p-type semiconductor* because the charge carriers are posi-

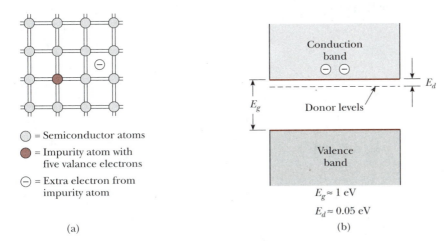

= Semiconductor atoms

= Impurity atom with five valance electrons

= Extra electron from impurity atom

$E_g \approx 1$ eV

$E_d \approx 0.05$ eV

(a)   (b)

Figure 11.27   (a) A two-dimensional representation of a semiconductor containing a donor atom (colored spot). (b) An energy-band diagram of a semiconductor in which the donor levels lie within the forbidden gap, just below the bottom of the conduction band.

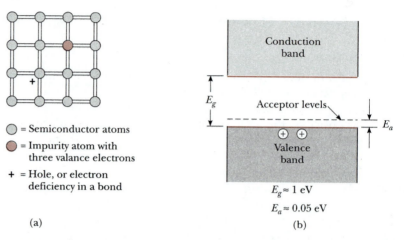

= Semiconductor atoms

= Impurity atom with
  three valance electrons

+ = Hole, or electron
    deficiency in a bond

$E_g \approx 1$ eV

$E_a \approx 0.05$ eV

(a)

(b)

Figure 11.28 (a) A two-dimensional representation of a semiconductor containing an acceptor atom (colored spot). (b) An energy-band diagram of a semiconductor in which the acceptor levels lie within the forbidden gap, just above the top of the valence band.

tively charged holes. When conduction is dominated by acceptor or donor impurities, the material is called an *extrinsic semiconductor*. The typical range of doping densities for n- or p-type semiconductors is 10^{13} to 10^{19} cm^{-3}.

Now consider what happens when a p-type semiconductor is joined to an n-type semiconductor to form a *p-n junction*. We find that the completed junction consists of three distinct semiconductor regions, as shown in Figure 11.29a: a p-type region, a *depletion* region, and an n-type region. The depletion region may be visualized as arising when the two halves of the junction are brought

Figure 11.29 (a) The physical arrangement of a p-n junction. (b) Built-in electric field versus x for the p-n junction. (c) Built-in potential versus x for the p-n junction.

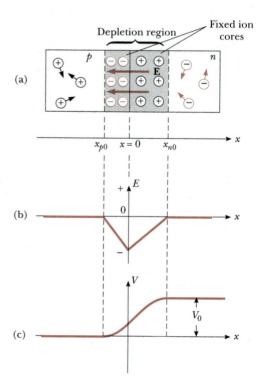

together and mobile donor electrons diffuse to the p side of the junction, leaving behind immobile positive ion cores. (Conversely, holes diffuse to the n side and leave a region of fixed negative ion cores.) The region extending several microns from the junction is called the *depletion region* because it is depleted of *mobile* charge carriers. It also contains a built-in electric field on the order of 10^3 to 10^5 V/cm, which serves to sweep mobile charge out of this region and keep it truly depleted. This electric field creates a potential barrier V_0 that prevents the further diffusion of holes and electrons across the junction and ensures zero current through the junction when no external voltage is applied.

Perhaps the most notable feature of the p-n junction is its ability to pass current in only one direction. Such *diode* action is easiest to understand in terms of the potential diagram in Figure 11.29c. If a positive external voltage is applied to the p side of the junction, the overall barrier is decreased, resulting in a current that increases exponentially with increasing forward voltage or bias. For reverse bias (a positive external voltage applied to the n side of the junction), the potential barrier is increased, resulting in a very small reverse current that quickly reaches a saturation value, I_0, with increasing reverse bias. The current-voltage relation for an ideal diode is

$$I = I_0(e^{qV/k_BT} - 1) \tag{11.35}$$

where q is the electronic charge, k_B is Boltzmann's constant, and T is the temperature in kelvins. Figure 11.30 shows a plot for an I-V characteristic for a real

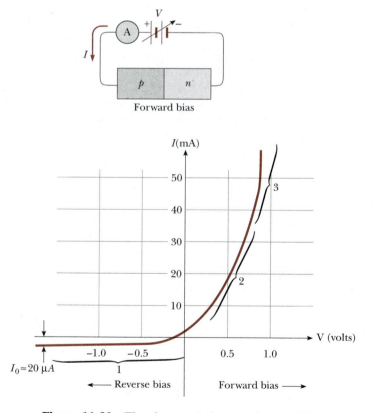

Figure 11.30   The characteristic curve for a real diode.

diode, along with a schematic of a diode under forward bias. Region 1 shows reverse bias operation, region 2 ordinary forward bias, and region 3 the extreme forward bias needed for a *p-n* junction laser. We treat *p-n* junction lasers in Section 11.6.

EXAMPLE 11.3   Forward and Reverse Currents in a Diode

Calculate the ratio of the forward current to the reverse current for a diode described by Equation 11.35 for an applied voltage of $+1.0$ V (forward bias) and -1.0 V (reverse bias).

Solution

$$\frac{I_{forward}}{I_{reverse}} = \frac{I_0(e^{+qV/k_B T} - 1)}{I_0(e^{-qV/k_B T} - 1)}$$

Assuming room-temperature operation (300 K) and taking $k_B = 8.62 \times 10^{-5}$ eV/K gives

$$\frac{I_f}{I_r} = \frac{I_0(e^{1/(8.6 \times 10^{-5})(300)} - 1)}{I_0(e^{-1/(8.6 \times 10^{-5})(300)} - 1)} = \frac{(e^{38.7} - 1)}{(e^{-38.7} - 1)}$$

$$= -6.40 \times 10^{16}$$

Light-Emitting and -Absorbing Diodes—LEDs and Solar Cells

Light emission and absorption in semiconductors is similar to light emission and absorption by gaseous atoms, except that discrete atomic energy levels must be replaced by bands in semiconductors. As shown in Figure 11.31a, an electron excited electrically into the conduction band can easily recombine with a hole (especially if the electron is injected into a *p*-type region), emitting a photon of band-gap energy. Examples of *p-n* junctions that convert electrical input to light output are light-emitting diodes (LEDs) and injection lasers. Conversely, an electron in the valence band may absorb a photon and be promoted to the conduction band, leaving a hole behind (see Fig. 11.31b). Devices in which light-generated electrons and holes are separated by a junction field and collected as current are called solar cells or photovoltaic devices. They are described in detail in an essay by John Meakin at the end of this chapter.

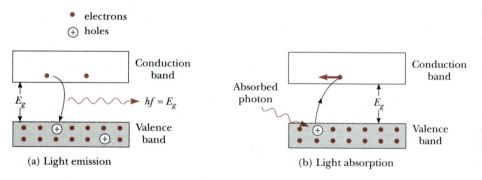

Figure 11.31   (a) Light emission from a semiconductor. (b) Light absorption by a semiconductor.

The Junction Transistor

The development of the transistor by John Bardeen, Walter Brattain, and William Shockley in 1948 totally revolutionized the world of electronics and led to a Nobel prize, shared by the three men, in 1956. By 1960 the transistor had replaced the vacuum tube in many electronic applications, giving rise to a multibillion-dollar industry that produces such popular devices as pocket radios, handheld calculators, computers, television receivers, CD players, and electronic games.

The junction transistor consists of a semiconducting material with a very narrow n region sandwiched between two p regions, called the *pnp transistor*, or a p region sandwiched between two n regions, called the *npn transistor*. Because the operations of the two transistors are essentially the same, we shall describe only the *pnp* transistor.

Figure 11.32 shows the structure of the *pnp* transistor together with its circuit symbol. The outer regions of the transistor are called the *emitter* and the *collector*, and the narrow central region is called the *base*. Note that the configuration contains two junctions. One junction is the interface between the emitter and the base, and the other is that between the base and the collector.

Suppose a voltage is applied to the transistor such that the emitter is at a higher potential than the collector. (This is accomplished with battery V_{ec} in Figure 11.33.) If we think of the transistor as two diodes back to back, we see that the emitter-base junction is forward-biased and the base-collector junction is reverse-biased.

Because the p-type emitter is heavily doped relative to the base, nearly all of the current consists of holes moving across the emitter-base junction. Most of these holes do not recombine in the base because it is very narrow. The holes are finally accelerated across the reverse-biased base-collector junction, producing the current I_c in Figure 11.33.

Although only a small percentage of holes recombine in the base, those that do limit the emitter current to a small value, because positive charge carriers accumulate in the base and prevent holes from flowing into this region. To prevent this limitation of current, some of the positive charge on the base must

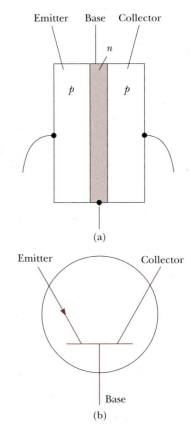

(a)

(b)

Figure 11.32 (a) The *pnp* transistor consists of an *n*-region (base) sandwiched between two *p*-regions (the emitter and the collector). (b) The circuit symbol for the *pnp* transistor.

Figure 11.33 A small base current, I_b, controls a large collector current, I_c. Note that $I_e = I_b + I_c$, $I_b \ll I_c$, and $I_e = \beta I_b$, where $\beta \approx 10$ to 100.

be drawn off; this is accomplished by connecting the base to a second battery, V_{eb} in Figure 11.33. Those positive charges that are not swept across the collector-base junction leave the base through this added pathway. This base current, I_b, is very small, but a small change in it can significantly change the collector current, I_c. If the transistor is properly biased, the collector (output) current is directly proportional to the base (input) current, and the transistor acts as a current amplifier. This condition may be written

$$I_c = \beta I_b \tag{11.36}$$

where β, the current gain, is typically in the range from 10 to 100. Thus the transistor may be used to amplify a small time-varying signal. The small voltage to be amplified is placed in series with the battery V_{eb} as shown in Figure 11.33. The time-dependent input signal produces a small variation in the base current. This variation results in a large change in collector current and hence a large change in voltage across the output resistor.

The Integrated Circuit

The integrated circuit (IC), invented independently by Jack Kilby at Texas Instruments in late 1958 and by Robert Noyce at Fairchild Camera and Instrument in early 1959, has justly been called the most remarkable technology ever to hit mankind. ICs have indeed started a second industrial revolution and are found at the hearts of computers, watches, cameras, automobiles, aircraft, robots, space vehicles, and all sorts of communication and switching networks. In simplest terms, an integrated circuit is a collection of interconnected transistors, diodes, resistors, and capacitors fabricated on a single piece of silicon, affectionately known as a chip. State-of-the-art chips easily contain several million components in a 1-cm^2 area (Fig. 11.34).

Interestingly, integrated circuits were invented partly in an effort to achieve

Figure 11.34 This powerful single-chip 64-bit microprocessor (the R10000 RISC) is designed for a broad range of computer applications, from personal computers to supercomputers. The processor has an area of 298 mm^2 and contains approximately 6 million transistors. *(Courtesy of MIPS Technologies, Inc.)*

Figure 11.35   Jack Kilby's first integrated circuit, tested on September 12, 1958. *(Courtesy of Texas Instruments, Inc.)*

circuit miniaturization and partly to solve the interconnection problem spawned by the transistor. In the era of vacuum tubes, the relatively great power and size of individual components severely limited the number of components that could be interconnected in a given circuit. With the advent of the tiny, low-power, highly reliable transistor, design limits on the number of components disappeared and were replaced by the challenge of wiring together hundreds of thousands of components. The magnitude of this problem can be appreciated when one considers that second-generation computers (consisting of discrete transistors) contained several hundred thousand components requiring more than a million hand-soldered joints to be made and tested.

Figures 11.35 and 11.36 show pictures of the first ICs that contained multiple electronic devices on a single piece of silicon. Noyce's IC also shows interconnections between devices *fabricated as part of the chip manufacturing process*. In addition to solving the interconnection problem, ICs possess the advantages of

Figure 11.36   One of Robert Noyce's first integrated circuits. Note that the leads connecting different components are built in as part of the chip. *(Courtesy of Fairchild Semiconductor Corporation)*

miniaturization and fast response, two attributes that are critical for high-speed computers. The fast response is actually also a product of the miniaturization and close packing of components, because the response time of a circuit depends on the time it takes for electrical signals traveling at about 1 ft/ns to pass from one component to another. This time is clearly reduced by the closer packing of components. Additional information on the history, theory of operation, and fabrication of chips may be found in the Suggestions for Further Reading at the end of this chapter.

11.6   LASERS

Lasers provide an interesting example of the basic principles of the interaction of radiation with atoms. Because of this and because of the large number of engineering applications of lasers—such as CD-ROMs, precision surveying and length measurement, precision cutting and shaping of materials, and communication along optical fibers—it is worthwhile to discuss the principles of operation of lasers in some detail.

Absorption, Spontaneous Emission, and Stimulated Emission

To understand the operation of a laser, we must become familiar with the processes that describe the emission and absorption of radiation by atoms. Although it would be stretching the truth to say that Einstein anticipated the laser, this remarkable genius laid the foundations for the device with his explanation of the processes of atomic absorption, spontaneous emission, and stimulated emission in 1917.[7] Einstein showed that an atom can absorb a photon of energy hf from a radiation field and make a transition from a lower energy state E_1 to a higher state E_2, where $E_2 - E_1 = hf$. Let us set the probability of absorption per unit time per atom equal to $B_{12}u(f, T)$, where B_{12} is Einstein's coefficient of absorption and $u(f, T)$ is the energy density per unit frequency in the radiation field. The probability of absorption of radiation is thus assumed to be proportional to the density of radiation. Once excited, an atom in state 2 is observed to have a definite probability per second of making a *spontaneous transition* (i.e., one not induced by radiation) back to level 1. The spontaneous transition rate may be characterized by the coefficient A_{21} (transitions per unit time) or by the spontaneous emission lifetime, $t_s = 1/A_{21}$ (time per transition).

Spontaneous transition

In addition to spontaneous emission, which is independent of the radiation density, another kind of emission can occur that is *dependent* on the radiation density. If a photon of energy hf interacts with an atom when it is in level 2, the electric field associated with this photon can stimulate or induce atomic emission such that the emitted electromagnetic wave (photon) vibrates in phase with the stimulating wave (photon) and travels in the same direction. Two such photons are said to be *coherent*. We may set the stimulated transition rate per

[7] A. Einstein, "Zur Quanten Theorie der Strahlung (On the Quantum Theory of Radiation)," *Phys. Zeit.* 18:121–128, 1917.

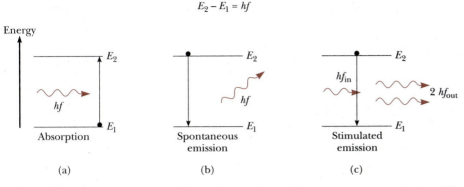

$$E_2 - E_1 = hf$$

Figure 11.37 The processes of (a) absorption and (b) spontaneous emission. The lifetime of the upper state is t_s, and the photon is emitted in a random direction. (c) Stimulated emission. In this process, the emitted photons are in phase with the stimulating photon, and all have the same direction of travel.

atom equal to $B_{21}u(f, T)$, where B_{21} is Einstein's coefficient of stimulated emission. Figure 11.37 summarizes the three processes of absorption, spontaneous emission, and stimulated emission.[8]

Next, following Einstein, consider a mixture of atoms and radiation in thermal equilibrium at temperature T. The populations N_1 and N_2 of the energy levels E_1 and E_2 obey the Boltzmann relation (discussed in Chapter 9).

$$\frac{N_2}{N_1} = e^{-(E_2 - E_1)/k_B T} = e^{-hf/k_B T} \tag{11.37}$$

This equilibrium ratio of populations represents a case of *dynamic equilibrium* in which the number of atomic transitions per second up (from E_1 to E_2) is balanced by the number of spontaneous *and* stimulated transitions per second down (from E_2 to E_1). Since the number of atoms making transitions is proportional to the population of the starting level as well as to the transition probability, we may write

Number of atoms going from $1 \longrightarrow 2$ per unit time $= N_1 u(f, T) B_{12}$

where $u(f, T)$ is the radiation energy density per unit frequency ($J/m^3 \cdot Hz$).

Number of atoms going from $2 \longrightarrow 1$ per unit time by stimulated emission $= N_2 u(f, T) B_{21}$

---

[8] Although spontaneous emission has no classical analog and is a truly quantum effect (arising from quantization of the radiation field), absorption and stimulated emission do have classical analogs. If the atom-photon interaction is modeled as a charged harmonic oscillator driven by an electromagnetic wave, the classical oscillator can either absorb energy from the wave or give up energy, depending on the relative phases of oscillator and wave.

For spontaneous emission there is no dependence on the radiation field, and we merely have

> Number of atoms going from 2 $\longrightarrow$ 1 per unit time spontaneously
> $= N_2 A_{21}$

Since the number of upward transitions per unit time must equal the number of downward transitions per unit time, we have

$$N_1 u(f, T) B_{21} = N_2 [B_{21} u(f, T) + A_{21}] \qquad (11.38)$$

Using Equation 11.37, this becomes

$$u(f, T) = \frac{A_{21}}{B_{12} e^{hf/k_B T} - B_{21}} \qquad (11.39)$$

Equation 11.39 is an expression for the radiation density per unit frequency of a system of atoms and radiation in equilibrium at temperature T expressed in terms of Einstein coefficients A_{21}, B_{12}, and B_{21}. But Planck's blackbody radiation law (Chapter 2) provides us with an expression for a system of radiation and oscillators in thermal equilibrium:

$$u(f, T) = \frac{8\pi h f^3}{c^3} \frac{1}{(e^{hf/k_B T} - 1)} \qquad (2.10)$$

Comparing Equations 11.39 and 2.10, we find the interesting results

Relation of Einstein coefficients of stimulated emission, absorption, and spontaneous emission

$$B_{21} = B_{12} = B \qquad (11.40)$$

and

$$\frac{A_{21}}{B} = \frac{8\pi h f^3}{c^3} \qquad (11.41)$$

Equation 11.40 states that any atom that has a finite probability per unit time of absorption has an *equal* probability of stimulated emission. Equation 11.41 shows that the process of spontaneous emission dominates the process of stimulated emission at higher frequency.

Population Inversion and Laser Action

Since $B_{12} = B_{21}$, there is an equal probability that a photon will cause an upward (absorption) or downward (stimulated emission) atomic transition. When light is incident on a system of atoms in thermal equilibrium, there is usually a net absorption of energy since, according to the Boltzmann distribution, there are more atoms in the ground state than in excited states. However, if one can invert the situation so that there are more atoms in an excited state than in a **Population inversion**   lower state—a condition called *population inversion*—amplification of photons can result. Under the proper conditions, a single input photon can result in a cascade of stimulated photons, all of which are in phase, traveling in the same direction and of the same frequency as the input photon. Since the device that

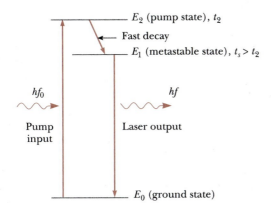

Figure 11.38   A three-level laser system. t_2 is the lifetime of the state E_2, and t_s is the lifetime of the state E_1.

does this is a light amplifier, it is called a **laser,** an acronym for **l**ight **a**mplification by **s**timulated **e**mission of **r**adiation.

Although there are many different types of lasers, most lasers have certain essential features:

- An energy source capable of producing either pulsed or continuous population inversions. In the case of the helium-neon gas laser, the energy source is an electrical discharge that imparts energy by electron-atom collisions. In the case of the ruby crystal laser, the population inversion is produced by intense flashes of broad-band illumination from flash lamps. The process of excitation with intense illumination is called *optical pumping*.
- A lasing medium with at least *three* energy levels: a ground state; an intermediate (metastable) state with a relatively long lifetime, t_s; and a high-energy pump state (Fig. 11.38). To obtain population inversion, t_s must be greater than t_2, the lifetime of the pump state E_2. Note, too, that amplification cannot be obtained with only two levels, because such a system cannot support a population inversion. At most, with extremely intense optical pumping we can increase the population of the upper state in a two-level system until it equals the population of the lower state. Further pumping serves only to excite as many downward transitions as upward transitions, since the probability of absorption is equal to the probability of stimulated emission for a *given* transition. For a population inversion to be produced, energy absorption must occur for a transition *different* from the transition undergoing stimulated emission—thus the need for at least a three-level system.
- A method of containing the initially emitted photons within the laser so that they can stimulate further emission from other excited atoms. In practice this is usually achieved by placing mirrors at the ends of the lasing medium so that photons make multiple passes through the laser. Thus the laser may be thought of as an optical resonator or oscillator with two opposing reflectors at right angles to the laser beam. The oscillation consists of a plane wave bouncing back and forth between the reflectors. The oppositely traveling plane waves, in turn, generate a highly monochromatic

Figure 11.39 Cavity modes in a laser. (a) Longitudinal modes have $L = m\lambda/2$, where m is an integer, λ is the laser wavelength in the laser material, and L is the distance between the two mirrors. (b) Modes with transverse components can exist if both ends and sides of the laser are made reflective. Arrows show the direction of propagation of light rays.

standing wave, which is strongest at resonance when an integral number of half wavelengths just fits between the reflectors (Fig. 11.39). To extract a highly collimated beam from the laser, one of the parallel mirrors is made slightly transmitting so that a small amount of energy leaks out of the cavity.

These three features of lasers lead to several unique characteristics of laser light that make it a much more powerful technological tool than light from ordinary sources. These unique characteristics are high monochromaticity, strong intensity, spatial coherence and low beam divergence. Table 11.9 shows typical values of these quantities for a low-power gas laser. In comparison with a single emission line in a strong conventional light source, the He-Ne laser is about 100 times more monochromatic, generates about 100 times more beam power, and has about 1000 times more exit irradiance (power per unit

Table 11.9   Typical He-Ne Gas Laser Characteristics

Monochromaticity	Beam Power	Beam Diameter at Laser Exit	Beam Divergence
632.81 ± 0.002 nm	10 mW	0.50 mm	1.0 mrad

Irradiance at Laser Exit	Irradiance for a Beam Focused to a 10-μm Diameter
5100 mW/cm^2	1.27×10^7 mW/cm^2

area). The beam divergence, or spread, is measured in milliradians (mrad). A 1.0-mrad laser would have a beam diameter of only 1 cm at a distance of 10 m.

Semiconductor Lasers

One of the most important light sources for fiber-optics communication is the semiconductor laser. This laser is eminently suited to fiber-optics communication because of its small size (maximum dimension several tenths of a millimeter), high efficiency (25 to 30%), simplicity, long lifetime (*p-n* junctions have estimated working lifetimes of about 100 years), ease of modulation (through control of the junction current), and fast response (extending well into the GHz range). Figure 11.40 shows a typical gallium arsenide (GaAs) *p-n* junction laser along with nominal operating values. The first semiconductor laser was made of GaAs and operated at 0.842 μm in the IR.[9,10] Shortly thereafter, Holonyak and Bevacqua reported the operation of a visible GaAs *p-n* junction laser,[11] and the laser was on its way to becoming an important part of semiconductor device technology.

Semiconductor lasers are similar to other lasers in that they require a population inversion between quantum levels in order to produce laser action. They are notably different from gas lasers in that (a) discrete energy levels in

Typical values
L = 0.3 mm   I (at 300 K) = 20 A
a = 0.1 mm   Input power = 136 W
d = 2.0 μm   Light output power = 9 W
λ = 840 nm

Figure 11.40   A gallium arsenide *p-n* junction laser.

[9] R. N. Hall, G. E. Fenner, J. D. Kingsley, T. J. Soltys, and R. O. Carlson, "Coherent Light Emission from GaAs Junctions," *Phys. Rev. Lett.* 9:336, 1962.

[10] M. I. Nathan, W. P. Dumke, G. Burns, F. H. Dill, and G. Lasher, "Stimulated Emission of Radiation from GaAs *p-n* Junctions," *Appl. Phys. Lett.* 1:62, 1962.

[11] N. Holonyak, Jr., and S. F. Bevacqua, *Appl. Phys. Lett.* 1:82, 1962.

gas lasers become *energy bands* in semiconductors, and (b) optical pumping and electrical discharge in conventional lasers are replaced by *injection pumping* in semiconductors. In injection pumping, large forward currents are passed through the laser diode, and electrons and holes are injected into p and n regions, respectively, where they recombine and emit radiation. Let us briefly sketch the main points needed to understand semiconductor lasers.

Energy bands in semiconductors

In semiconductors, the discrete energy levels found in isolated atoms widen into bands because of the perturbations caused by neighboring atoms. We are basically interested in the two outermost energy bands, known as the valence (v) and conduction (c) bands. In a semiconductor such as silicon in thermal equilibrium at room temperature, the valence band is filled with electrons and the conduction band, where electrons can move freely, contains no electrons. Since the two levels are separated by a forbidden energy region or gap, $E_g = 1.1$ eV, very few electrons possess enough thermal energy at room temperature to cross the gap and carry a current in the material. Hence silicon, like other semiconductors, is normally a poor conductor of electricity. However, light, heat, or electrical energy added to a semiconductor can promote electrons into the conduction band and increase the conductivity of a semiconductor. The hole (vacancy) left by the electron in the v-band plays an equally important role in conduction in a semiconductor, since it behaves as a positive charge moving opposite in direction to the conduction electron. In fact, the hole actually completes the current path through the semiconductor. This process of the creation of an electron-hole pair is called *pair generation*. An electron in the c-band can also "fall into" a hole and recombine with it, emitting light of band-gap energy in the process. Both of these important processes are indicated schematically in Figure 11.41. If the thermal-equilibrium situation of empty c-band and full v-band is disturbed by intense optical or electrical excitation, a semiconductor population inversion can be set up, with the v-band empty of electrons (full of holes) to a depth X and the c-band full of electrons to a height Y. As can be seen in Figure 11.42, once a few spontaneously emitted photons with energy E_g induce stimulated emission at E_g, stimulated emission eventually

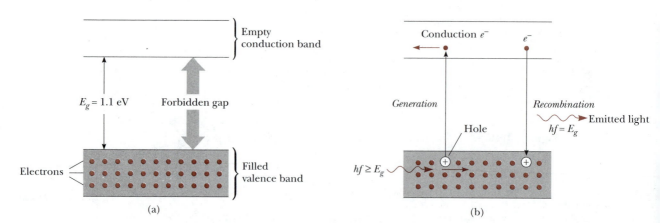

Figure 11.41   (a) Energy bands in pure silicon at 0 K. (b) Generation and recombination of electron-hole pairs in silicon.

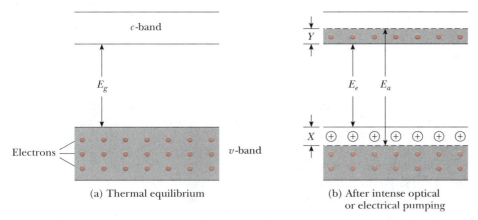

c-band

E_g

Electrons

v-band

(a) Thermal equilibrium

Y

E_e E_a

X

(b) After intense optical
or electrical pumping

Figure 11.42   A population inversion in a semiconductor. Note that E_e, the emission energy, is equal to E_g and that E_a, the absorption energy, is equal to $E_g + X + Y$. The absorption energy is $E_g + X + Y$, and not just $E_g + X$, because electrons obey the Pauli exclusion principle. Thus electrons must absorb enough energy to occupy states above Y.

far exceeds absorption since photons with energy E_g lack sufficient energy to be absorbed. Ultimately the dominance of the rate of stimulated emission over the rate of absorption leads to line narrowing and laser action.

Other Types of Semiconductor Lasers.   For simplicity we have described only the basic GaAs injection laser. To leave the reader with the impression that this is the most important laser would be a serious distortion of the truth, for the present field of semiconductor lasers is enormously rich and varied. While the GaAs laser is a single-homojunction laser (one junction fabricated from a single compound), many current lasers are double-heterojunction (DH) lasers, such as the PbTe-PbSnTe and the GaAs-AlGaAs lasers.[12] When two junctions are formed in the laser, the injected carriers are confined to a narrower region so that a lower threshold current is required for laser action, permitting continuous laser operation at room temperature. In addition, by varying the amount of Sn or Al in the ternary alloy, one can vary the band-gap energy and, consequently, the laser wavelength. Also, GaAs-AlGaAs multilayer junctions are relatively easy to fabricate since AlGaAs layers may be grown epitaxially on GaAs layers. DH lasers can be formed with many other compounds, including quaternary alloys, and tunable semiconductor lasers can be made whose emission wavelengths can be varied by changes in the applied pressure or magnetic field. These techniques have resulted in semiconductor lasers that span the broad range from 0.3 to 40 μm. These and other developments in the field of semiconductor lasers, which is a truly fascinating combination of optics, material science, and device physics, are treated more extensively in some of the Suggestions for Further Reading cited at the end of this chapter.

---

[12] PbSnTe and AlGaAs are called ternary (three-component) alloys. They are crystalline solid solutions with Sn and Al randomly replacing some Pb and Ga atoms, respectively.

SUMMARY

Bonding mechanisms in solids can be classified in a manner similar to the schemes used to describe bonding of atoms to form molecules. *Ionic solids*, such as NaCl, are formed by the net attractive Coulomb force between positive and negative ion neighbors. Ionic crystals are fairly hard, have high melting points, and have strong cohesive energies of several electron volts. *Covalent solids*, such as diamond, form when valence electrons are shared by adjacent atoms to produce bonds of great stability. Covalent solids are generally very hard, have high melting points, and possess the highest cohesive energies, ranging up to 8 eV per atom. Metallic solids are held together by the metallic bond, which arises from the attractive force between the positive ion cores and the electron "gas." This electron gas acts as a kind of "cement" to hold the ion cores together. Generally, metallic solids are softer and have lower cohesive energies than covalent solids. The most weakly bound solids are *molecular crystals*, which are held together by van der Waals or dipole-dipole forces. Molecular crystals, such as solid argon, have low melting points and low cohesive energies on the order of 0.1 eV per molecule.

The classical free-electron model of metals is generally successful in explaining the electrical and thermal properties of metals. According to this theory, a metal is modeled by a gas of conduction electrons moving through a fixed lattice of ion cores. Taking into account the collisions of electrons with the lattice, the classical free-electron model predicts that metals will obey Ohm's law, $\mathbf{J} = \sigma\mathbf{E}$, with the electrical conductivity σ given by

$$\sigma = \frac{ne^2 L}{(3k_B T m_e)^{1/2}} \tag{11.15}$$

where n is the concentration of conduction electrons, L is the electronic mean free path, T is the absolute temperature, and m_e is the mass of the electron. The classical model may also be applied to heat conduction by electrons in metals to show that K/σ, where K is the thermal conductivity, is a universal constant for all metals and depends only on the absolute temperature (Wiedemann-Franz law.) The classical model yields for this ratio

$$\frac{K}{\sigma} = \frac{3k_B^2}{2e^2} T \tag{11.21}$$

Although the classical electron gas model broadly describes the electrical and thermal properties of metals, there are notable deficiencies in its predictions—namely in the numerical values of σ and K and in the temperature dependence of σ. These deficiencies may be rectified by taking into account the wave nature and hence the indistinguishability of electrons, by replacing the Maxwell-Boltzmann distribution with the Fermi-Dirac distribution. When this is done, Equation 11.21 for K/σ takes the new form

$$\frac{K}{\sigma} = \frac{\pi^2 k_B^2}{3e^2} T = 2.45 \times 10^{-8} T \tag{11.26}$$

and shows excellent agreement with experimental values. The correct quantum expression for the conductivity of a metal is

$$\sigma = \frac{ne^2 L}{m_e v_F}$$

where v_F is the Fermi speed and L is the quantum mean free path, *which is usually several hundred times longer than the metal's interatomic distance.* This unusually long mean free path occurs because electron waves can generally pass through a perfect lattice of ions without scattering, since they scatter only from widely spaced crystal imperfections such as thermal displacements of ions and impurity atoms. The observed temperature dependence of σ can also be explained as the result of electron scattering from thermal lattice vibrations, or phonons.

When many atoms are brought together, the individual atomic levels form a set of allowed bands separated by forbidden energy regions, or gaps. One can understand the properties of metals, insulators, and semiconductors in terms of the *band theory of solids.* The highest occupied band of a *metal* is partially filled with electrons. Therefore, many electrons are free to move throughout the metal and contribute to the conduction current. In an *insulator* at $T = 0$ K, the valence band is completely filled with electrons, while the conduction band is empty. The region between the valence band and the conduction band is called the *energy gap* of the material. The energy gap of an insulator is of the order of 10 eV. Because this gap is large compared to $k_B T$ at ordinary temperatures, very few electrons are thermally excited into the conduction band, which explains the small electrical conductivity of an insulator.

A *semiconductor* is a material with a small energy gap, of the order of 1 eV, and a valence band that is filled at 0 K. Because of this small energy gap, a significant number of electrons can be thermally excited from the valence band into the conduction band as temperature increases, thereby producing a semiconductor conductivity that *increases* with temperature. The electrical properties of a semiconductor can be modified by adding donor atoms with five valence electrons (such as arsenic) or by adding acceptor atoms with three valence electrons (such as boron). A semiconductor doped with donor atoms is called an *n-type semiconductor*, while one doped with acceptor atoms is called *p-type*. The energy levels of these donor and acceptor atoms fall within the energy gap of the material, the donor levels falling just below the conduction band and the acceptor levels just above the valence band.

When a *p*-type material is joined to an *n*-type material, a *p-n junction* is formed with the current-voltage characteristics of a diode. If a *p-n junction* is operated under strong forward bias, electrons can be injected into the *p*-type material, where they recombine with holes, emitting photons with band-gap energy. This is the principle of operation of the *light-emitting diode* (LED) and the *injection laser*. If a *p-n* junction absorbs a photon with energy greater than or equal to the band gap, an electron may be promoted to the conduction band (leaving a hole in the valence band), thereby producing a current. Devices in which light-generated electrons and holes are separated by a built-in junction field and collected as current are called *solar cells*. Two other semiconductor devices that have revolutionized life in the latter half of the 20th century are the *junction transistor* and the *integrated circuit* (IC). The transistor is a small, low-power, highly reliable amplifying device that basically consists of two *p-n* junctions placed back to back. The integrated circuit, which is the heart of the small computer, consists of thousands of interconnected transistors, diodes, resistors, and capacitors fabricated on a single chip of silicon. Interestingly, integrated circuits were invented partly in an effort to achieve greater miniaturization and partly to solve the interconnection problem spawned by the transistor.

The processes of atomic absorption, spontaneous emission, and stimulated emission are crucial to an understanding of laser action. By considering a group of atoms and radiation in thermal equilibrium and balancing the rate of upward transitions against the rate of downward transitions, one can show that

$$B_{21} = B_{12} = B \tag{11.40}$$

and

$$\frac{A_{21}}{B} = \frac{8\pi h f^3}{c^3} \tag{11.41}$$

where B_{12} is Einstein's coefficient of absorption, B_{21} is Einstein's coefficient of stimulated emission, and A_{21} is the probability of spontaneous emission per second per atom.

Amplification of the light from a particular atomic transition and generation of an intense, monochromatic, coherent, and highly collimated beam (laser action) may be achieved if certain conditions are met:

- The lasing medium must contain at least three energy levels: a ground state, an intermediate state with a long lifetime, and a high-energy pump state.

- There must be an electrical or optical energy source capable of "pumping" atoms into excited states faster than they leave, so that a population inversion is produced.

- There must be a method of confining the first wave of emitted photons within the laser so that they can stimulate further emission.

Actual lasers include solid-state crystalline lasers such as the ruby laser, gas lasers such as the He-Ne laser, and semiconductor lasers such as the GaAs laser. Although large, pure single crystals of ruby are easy to fabricate, a high pumping power is required to induce lasing in ruby, and consequently this laser must be operated on a pulsed basis. Gas lasers offer an attractive alternative to crystalline solid-state lasers because they may be operated continuously. Continuous operation is possible because of the modest pumping-power requirements of gas lasers, which have small concentrations of optically active atoms. The most important light source for optical signal processing is the semiconductor laser. This laser is essentially a *p-n* junction operated under strong forward bias. Typical advantages of these lasers are small size, high efficiency, simplicity, ease of modulation, and fast response.

SUGGESTIONS FOR FURTHER READING

1. R. Zallen, *The Physics of Amorphous Solids*, New York, John Wiley and Sons, 1983. A comprehensive treatment of amorphous solids at a senior undergraduate level. Great care is exercised in introducing new concepts verbally before launching into mathematical arguments.

2. T. R. Reid, *The Chip: How Two Americans Invented the Microchip and Launched a Revolution*, New York, Simon and Schuster, 1984. An informal, nonmathematical, extremely interesting chip history.

3. B. G. Streetman, *Solid State Electronic Devices*, 2nd ed., Englewood Cliffs, N.J., Prentice-Hall, 1980. A comprehensive treatment of devices at the junior-senior undergraduate level.

4. W. Shockley, *Electrons and Holes in Semiconductors*, Melbourne, Fla., Krieger Publishing Co., 1950. The classic "granddaddy" text on semiconductors. Although somewhat dated, the book is a marvelous compendium of Shockley's insights and analogies in the field of semiconductor physics. It starts out at a very clear introductory level and about halfway through reaches advanced undergraduate and graduate levels.

5. C. Kittel, *Introduction to Solid State Physics*, 7th ed., New York, John Wiley and Sons, 1996. A detailed, comprehensive, up-to-date treatment of solid-state physics at the senior or graduate level. Many useful tables of the properties of solids.

Readings on Lasers

6. C. K. N. Patel, "High-Power Carbon Dioxide Lasers," *Sci. American*, August 1968, 22–23.

7. A. L. Schawlow, "Laser Light," *Sci. American*, September 1968, 120–126.

More advanced treatments of lasers may be found in:

8. J. I. Pankove, *Optical Processes in Semiconductors*, New York, Dover Publications, 1971.

9. S. M. Sze, *Physics of Semiconductor Devices*, New York, John Wiley and Sons, 2nd ed., 1981.

QUESTIONS

1. Explain how the energy bands of metals, semiconductors, and insulators account for the following general optical properties. (a) Metals are opaque to visible light. (b) Semiconductors are opaque to visible light but transparent to infrared (many IR lenses are made of germanium). (c) Many insulators, such as diamond, are transparent to visible light.

2. Table 11.8 shows that the energy gaps for semiconductors decrease with increasing temperature. What do you suppose accounts for this behavior?

3. The resistivity of most metals increases with increasing temperature, whereas the resistivity of an intrinsic semiconductor decreases with increasing temperature. What explains this difference in behavior?

4. Discuss the differences among the band structures of metals, insulators, and semiconductors. How does the band-structure model enable you to better understand the electrical properties of these materials?

5. Discuss the electrical, physical, and optical properties of ionically bonded solids.

6. Discuss the electrical and physical properties of covalently bonded solids.

7. When a photon is absorbed by a semiconductor, an electron-hole pair is said to be created. Give a physical explanation of this concept, using the energy-band model as the basis for your description.

8. In a semiconductor such as silicon, pentavalent atoms such as arsenic are donor atoms, while trivalent atoms such as indium are acceptors. Inspect the periodic table shown on the endpapers, and determine which other elements would be considered either donors or acceptors.

9. Explain how a *p-n* junction diode operates as a rectifier, as a solar cell, and as an LED.

10. What are the essential assumptions made in the classical free-electron theory of metals? How does the energy-band model differ from the classical free-electron theory in describing the properties of metals?

11. Explain the similarities of electrical conduction and heat conduction in metals.

12. Discuss the differences between crystalline solids, amorphous solids, and gases.

13. Discuss the physical sources of electrical resistance in a metal. Does the resistance depend upon the strong reflections set up when the lattice spacing is equal to an integral number of electronic wavelengths?

14. Does it seem reasonable that the average of the square of a quantity is greater than the square of the average? (Recall from Section 11.2 that $\bar{t} = \tau$ and $\overline{t^2} = 2\tau^2$.)

15. How are the equations governing heat flow and charge flow similar for metals?

16. Radiative emission by an atom can be spontaneous or stimulated. Distinguish between these two processes, identifying the characteristic features of each. Describe the role(s) each one plays in the operation of a laser.

17. Why must a lasing medium possess at least three energy levels?

18. Since the light from semiconductor lasers can be modulated easily, these lasers are of great use in light-wave communication systems. Show that laser communication systems possess a great advantage over microwave communication systems in terms of the amount of information that can be carried. (*Hint:* If each speech channel requires a bandwidth of 5 kHz, calculate the number of channels carried by a laser beam of 10^{16} Hz and by a microwave beam of 10^{10} Hz.)

19. Discuss the main criteria that must be met to achieve laser action in a three-level system.

20. What are the three fundamental ways in which light (photons) interacts with matter (atoms)? Briefly explain each.

21. Assuming that K is a positive constant in Equation 11.18, explain why there is a negative sign in this equation.

PROBLEMS

11.1 Bonding in Solids

1. Show that the minimum potential energy of an ion pair in an *ionic* solid is

$$U_0 = -\alpha k \frac{e^2}{r_0}\left(1 - \frac{1}{m}\right)$$

where r_0 is the equilibrium separation between ions, α is the Madelung constant, and the repulsive potential energy between ions is B/r^m. $|U_0|$ is called the *ionic cohesive energy* and is the energy required to separate to infinity an ion pair originally at a separation of r_0 in the crystal. *Hint:*

$$\left.\frac{dU}{dr}\right|_{r=r_0} = 0$$

2. Show that the angle between carbon bonds shown in Figure 11.3a is 109.5°.

3. Calculate the ionic cohesive energy for NaCl, using Equation 11.3. Take $\alpha = 1.7476$, $r_0 = 0.281$ nm, and $m = 8$.

4. LiCl, which has the same crystal structure as NaCl, has $r_0 = 0.257$ nm and has a measured ionic cohesive energy of 199 kcal/mole. (a) Find the integral value of m that best agrees with the measured value 199 kcal/mol. (b) Is the value of U_0 very sensitive to a change in m? Compute the percent change in U_0 when m increases by unity.

5. Consider a one-dimensional chain of alternating positive and negative ions. Show that the potential energy of an ion in this hypothetical crystal is

$$U(r) = -k\alpha \frac{e^2}{r}$$

where $\alpha = 2\ln 2$ (the Madelung constant), and r is the interionic spacing. [*Hint:* Make use of the series expansion for $\ln(1 + x)$.]

6. (a) Show that the force on an ion in an ionic solid can be written

$$F = -k\alpha \frac{e^2}{r^2}\left[1 - \left(\frac{r_0}{r}\right)^{m-1}\right]$$

where r_0 is the equilibrium separation and α is the Madelung constant. (*Hint:* Make use of the value of B found in Problem 1.) (b) Imagine that an ion in the solid is displaced a small distance x from r_0, and show that the ion experiences a restoring force $F = -Kx$, where

$$K = \frac{k\alpha e^2}{r_0^3}(m - 1)$$

(c) Use the result of (b) to estimate the frequency of vibration of an Na^+ ion in NaCl. Take $m = 8$ and use the value $\alpha = 1.7476$. (d) At what wavelength would the Na^+ ion absorb incident radiation? Is the absorption in the UV, visible, or infrared part of the spectrum?

7. (a) Calculate the ionic cohesive energy for KCl, which has the same crystal structure as NaCl. Take $r_0 = 0.314$ nm and $m = 9$. (b) Calculate the atomic cohesive energy of KCl by using the facts that the ionization energy of potassium is 4.34 eV (that is, $K + 4.34$ eV $\rightarrow K^+ + e$) and that the electron affinity of chlorine is 3.61 eV ($Cl^- + 3.61$ eV $\rightarrow Cl + e$).

8. The Madelung constant for the NaCl structure may be found by summing an infinite alternating series of terms giving the electrostatic potential energy between an Na^+ ion and its 6 nearest Cl^- neighbors, its 12 next-nearest Na^+ neighbors, and so on (see Fig. 11.1a). (a) From this expression, show that the first *three* terms of the infinite series for the Madelung constant for the NaCl structure yield $\alpha = 2.13$. (b) Does this infinite series converge rapidly? Calculate the fourth term as a check.

11.2 Classical Free-Electron Model of Metals

9. Consider a group of N electrons, all of which experience a collision with lattice ions at $t = 0$. One can show that the number of electrons that suffer their next collision between t and $t + dt$ follows the exponentially decreasing distribution

$$n(t)\,dt = \frac{Ne^{-t/\tau}}{\tau}\,dt$$

where τ is the mean free time. (a) Show that $\int_0^\infty n(t)\,dt = N$, as expected. (b) Show that this distribution leads to $\bar{t} = \tau$. *Hint:*

$$\bar{t} = \frac{n_1 t_1 + n_2 t_2 + \cdots + n_f t_f}{N} = \frac{1}{N}\int_0^\infty t \cdot n(t)\, dt$$

(c) Show, similarly, that $\overline{t^2} = 2\tau^2$, as stated in the derivation of Equation 11.9.

10. Silver has a density of 10.5×10^3 kg/m³ and a resistivity of 1.60×10^{-8} Ω·m at room temperature. (a) On the basis of the classical free-electron gas model, and assuming that each silver atom contributes one electron to the electron gas, calculate the average time between collisions of the electrons. (b) Calculate the mean free path from τ and the electron's thermal velocity. (c) How does the mean free path compare to the lattice spacing?

11. The contribution of a single electron or hole to the electric conductivity of a semiconductor can be expressed by an important property called the *mobility*. The mobility, μ, is defined as the particle drift speed per unit electric field or, in terms of a formula, $\mu = v_d/E$. Note that mobility describes the ease with which charge carriers drift in an electric field and that the mobility of an electron, μ_n, may be different from the mobility of a hole, μ_p, in the same material. (a) Show that the current density may be written in terms of mobility as $J = ne\mu_n E$, where $\mu_n = e\tau/m_e$. (b) Show that, when both electrons and holes are present, the conductivity may be expressed in terms of μ_p and μ_n as $\sigma = ne\mu_n + pe\mu_p$, where n is the electron concentration and p is the hole concentration. (c) If a germanium sample has $\mu_n = 3900$ cm²/V·s, calculate the drift speed of an electron when a field of 100 V/cm is applied. (d) A pure (intrinsic) sample of germanium has $\mu_n = 3900$ cm²/V·s and $\mu_p = 1900$ cm²/V·s. If the hole concentration is equal to 3.0×10^{13} cm⁻³, calculate the resistivity of the sample.

12. Gallium arsenide (GaAs) is a semiconductor material of great interest for its high power-handling capabilities and fast response time. (a) Calculate the drift speed of electrons in GaAs for a field of 10 V/cm if the electron mobility is $\mu_n = 8500$ cm²/V·s. (See Problem 11 for the definition of μ.) (b) What percent of the electron's thermal speed at 300 K is this drift speed? (c) Assuming an effective electron mass equal to the free-electron mass, calculate the average time between electron collisions. (d) Calculate the electronic mean free path.

11.3   Quantum Theory of Metals

13. Assuming that conduction electrons in silver are described by the Fermi free-electron gas model with $E_F = 5.48$ eV, repeat the calculations of Problem 10.

14. Sodium is a monovalent metal having a density of 0.971 g/cm³, an atomic weight of 23.0 g/mol, and a resistivity of 4.20 μΩ·cm at 300 K. Use this information to calculate (a) the free-electron density (b) the Fermi energy E_F at 0 K (c) the Fermi velocity v_F (d) the average time between electronic collisions (e) the mean free

path of the electrons, assuming that E_F at 0 K is the same as E_F at 300 K and (f) the thermal conductivity. For comparison to (e), the nearest-neighbor distance in sodium is 0.372 nm.

11.4   Band Theory of Solids

15. The energy gap for Si at 300 K is 1.14 eV. (a) Find the lowest-frequency photon that will promote an electron from the valence band to the conduction band of silicon. (b) What is the wavelength of this photon?

16. From the optical absorption spectrum of a certain semiconductor, one finds that the longest wavelength of radiation absorbed is 1.85 μm. Calculate the energy gap for this semiconductor.

17. The simplest way to model the energy-level splitting that is produced when two originally isolated atoms are brought close together is with two finite wells. In this model, the Coulomb potential experienced by the outermost electron in each atom is approximated by a one-dimensional finite square well of depth U and width a. The energy levels for two atoms close together may be found by solving the time-independent Schrödinger equation for a potential consisting of two finite wells separated by a distance b. (a) Start this problem by "warming up" with a solution to the single finite well shown in Figure P11.17a. Justify the solutions listed for regions I, II, and III and apply the standard boundary conditions (ψ and $d\psi/dx$ continuous at $x = 0$ and $x = a$) to obtain a transcendental equation for the bound-state energies. (b) Write a computer program to solve for the bound-state energies when an electron is confined to a well with $U = 100$ eV and $a = 1$ Å. You should find two bound states at approximately 19 and 70 eV. (c) Now consider the finite wells separated by a distance b as shown in Figure P11.17b. Impose the conditions of continuity in ψ and $d\psi/dx$ at $x = 0$ and $x = a$ to obtain

$$\frac{D}{E'} = \frac{2e^{-2Ka}[\cos ka + \frac{1}{2}K/k - k/K)\sin ka]}{(k/K + K/k)\sin ka}$$

Show that the boundary conditions at $x = a + b$ yield

$$\frac{F}{G} =$$

$$\frac{(D/E')e^\beta[\cos\alpha - (K/k)\sin\alpha] + e^{-\beta}[\cos\alpha + (K/k)\sin\alpha]}{(D/E')e^\beta[\sin\alpha + (K/k)\cos\alpha] + e^{-\beta}[\sin\alpha - (K/k)\cos\alpha]}$$

where $\alpha = k(a + b)$ and $\beta = K(a + b)$. The boundary conditions at $x = 2a + b$ yield

$$\frac{F}{G} = \frac{\cos k(2a + b) + (K/k)\sin k(2a + b)}{\sin k(2a + b) - (K/k)\cos k(2a + b)}$$

Thus, the two expressions for F/G may be set equal and the expression for D/E' used to obtain a transcendental

(a)

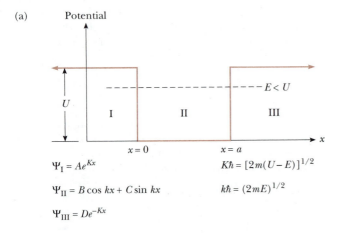

$\Psi_{\mathrm{I}} = Ae^{Kx}$ $K\hbar = [2m(U-E)]^{1/2}$

$\Psi_{\mathrm{II}} = B\cos kx + C\sin kx$ $k\hbar = (2mE)^{1/2}$

$\Psi_{\mathrm{III}} = De^{-Kx}$

(b)

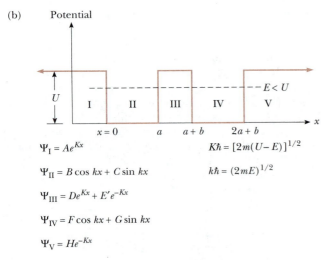

$\Psi_{\mathrm{I}} = Ae^{Kx}$ $K\hbar = [2m(U-E)]^{1/2}$

$\Psi_{\mathrm{II}} = B\cos kx + C\sin kx$ $k\hbar = (2mE)^{1/2}$

$\Psi_{\mathrm{III}} = De^{Kx} + E'e^{-Kx}$

$\Psi_{\mathrm{IV}} = F\cos kx + G\sin kx$

$\Psi_{\mathrm{V}} = He^{-Kx}$

Figure P11.17   (a) Potential and eigenfunctions for a single finite well of depth U, where $E < U$. (b) Potential and eigenfunctions for two finite wells. The width of each well is a, and the wells are separated by a distance b.

equation for the energy E. (d) Write a computer program to solve for the bound-state energies when $U = 100$ eV and $a = b = 1$ Å. You should observe twofold splitting of the bound-state energies found at 19 and 70 eV for the single well. (Your results can be checked using the MP DESKTOP, as described in Example 10.4. Simply load the potential file *EX10-4.U.* and adjust the parameter values to fit the present problem.)

11.5   Semiconductor Devices

18. One can roughly calculate the weak binding energy of a donor electron as well as its orbital radius in a semiconductor on the basis of the Bohr theory of the atom. Recall that, for a single electron bound to a nucleus of

charge Z, the binding energy of the ground state is given by

$$\frac{ke^2Z^2}{2a_0} = 13.6Z^2 \text{ eV} \qquad (3.36)$$

and the ground-state radius by

$$r_1 = a_0/Z \qquad (3.35)$$

For the case of a phosphorus donor atom in silicon, the outermost donor electron is attracted by a nuclear charge of $Z = 1$. However, because the phosphorus nucleus is embedded in the polarizable silicon, the effective nuclear charge seen by the electron is reduced to Z/κ, where κ is the dielectric constant. (a) Calculate the binding energy of a donor electron in $\mathrm{Si}(\kappa = 12)$ and $\mathrm{Ge}(\kappa = 16)$ and compare to the thermal energy available at room temperature. (b) Calculate the radius of the first Bohr orbit of a donor electron in Si and Ge. How does the Bohr orbit radius compare to the nearest-neighbor distance in $\mathrm{Si}(2.34$ Å$)$ and in $\mathrm{Ge}(2.43$ Å$)$?

11.6   Lasers

19. *Cavity modes in lasers.* Under high resolution and at threshold current, the emission spectrum from a GaAs laser is seen to consist of many sharp lines, as shown in Figure P11.19. Although one line generally dominates at higher currents, let us consider the origin and spacing of the multiple lines, or modes, shown in this figure. Recall that standing waves or resonant modes corresponding to the sharp emission lines are formed when an integral number m of half wavelengths fits between the cleaved GaAs surfaces. If L is the distance between

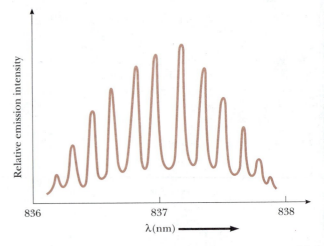

Figure P11.19   A high-resolution emission spectrum of the GaAs laser, operated just below the laser threshold.

cleaved faces, n is the index of refraction of the semi-conductor, and λ is the wavelength in air, we have

$$m = \frac{L}{\lambda/2n}$$

(a) Show that the wavelength separation between adjacent modes (the change in wavelength for the case $\Delta m = +1$) is

$$|\Delta \lambda| = \frac{\lambda^2}{2L\left(n - \lambda\dfrac{dn}{d\lambda}\right)}$$

(*Hint:* Since m is large and we want the small change in wavelength that corresponds to $\Delta m = 1$, we can consider m to be a continuous function of λ and differentiate $m = 2Ln/\lambda$). (b) Using the following typical values for GaAs, calculate the wavelength separation between adjacent modes. Compare your results to Figure P11.19.

$$\left.\begin{array}{l} \lambda = 837 \text{ nm} \\ 2L = 0.60 \text{ mm} \\ n = 3.58 \\ dn/d\lambda = 3.8 \times 10^{-4} \text{ (nm)}^{-1} \end{array}\right\} \text{ at } \lambda = 837 \text{ nm}$$

(c) Estimate the mode separation for the He-Ne gas laser using $\lambda = 632.8$ nm, $2L = 0.6$ m, $n \approx 1$, and $dn/d\lambda = 0$. On the basis of this calculation and the result of part (b), what is the controlling factor in mode separation in both solids and gases?

20. *How monochromatic is a laser?* The natural line width emitted by a collection of Cr^{3+} ions in ruby in thermal equilibrium at 290 K is about 0.4 nm for the 694.4-nm line. This is the *thermal line width*. Show that the emission from a ruby laser is much more monochromatic than 694.4 ± 0.2 nm if laser emission is confined to a single cavity mode. Take the length of the ruby between polished faces to be 10 cm, $n = 1.8$, and $dn/d\lambda \approx 0$. (See Problem 19.)

21. **MP** *Energy Bands from Atomic Levels.* In this problem we use the MP DESKTOP to study how energy bands in a solid arise from the atomic levels of individual atoms. The potential may be found in the file *PR11-21.U;* it consists of multiple square wells (eight in all), each representing the potential of an electron in the vicinity of an atomic nucleus. The "nuclei" are identical and equally spaced, as befits a regular monatomic crystalline "solid." (a) Begin by finding the bound states for a single well 300 eV (≈ 22 Ry) high and 1.00 bohr wide (the de-

faults). Use *AutoSearch* to facilitate the search process. You should find just two bound states, at about 5 Ry and 17 Ry. (b) Now add another well to the right of the first. Return to the *Desktop Function Editor* and click the *Next interval* button to show $U(x)$ on the range $[-6,-4]$. Set $\beta = \alpha + 1$ for a well 1.00 bohr wide and click the *OK* button; in this case, the wells are also separated by 1.00 bohr. Find the bound states of this double well. There are four in all, two each clustered around the single well energies found in part (a). (AutoSearch often misses levels that are close together, so you may have to narrow the search range to locate all members in a cluster. Do this by activating the *Control Panel* from the *Options* menu and editing the value in the *AutoSearch* field: $E \pm 0.1$ should work reasonably well.) Note the symmetry and number of nodes for each bound state wavefunction. (c) Continue adding wells as before, one at a time, until all eight wells are present (the last subinterval is $[+6,+10]$). With eight wells, how many bound states should there be? Describe the grouping of the energy levels, and locate the top and bottom edges of each "band." Determine the size of the band gap. Examine the wavefunction for the state at the bottom of the next-to-lowermost band. Count the nodes and relate this number to the number of wells in the chain. Is this a general result?

22. **MP** *Impurity states in solids.* Refering to the crystalline model of Problem 21, we introduce an "impurity" center by modifying the potential in the interval $[-2,0]$ to read

$$U(x) = (\epsilon - \Gamma)*(\text{HVS}(\alpha - x) + \text{HVS}(x - \beta)) + \Gamma$$

Taking $\Gamma = 30/13.60$ Ry produces a well 30 eV shallower but having the same width as the surrounding wells. Use the MP DESKTOP to find the bound states for this "doped" crystal. (The modified potential may be found in the file *PR11-22.U.*) Locate the band edges and describe any differences you see in the band wavefunctions. Count the states in each band and compare this number with the "pure" crystal results of Problem 21. What happened to the other states? (*Hint:* Search the band gaps and describe the wavefunctions associated with any states found there.) What do you suppose would happen if the impurity well were 30 eV *deeper* than the surrounding wells? Use the MP DESKTOP to confirm your suspicions.

Essay

THE INVENTION OF THE LASER

S. A. Marshall

Roosevelt University
Chicago, Ill.

The laser had its conceptual beginnings in the early 1950s, when many scientists were engaged in radio-frequency and microwave spectroscopy. This field evolved vigorously after World War II, due in part to wartime breakthroughs in radio frequency and microwave technology. These breakthroughs did much to tip the scales of warfare in favor of the Western Allies, principally in their air and naval defense efforts, and contributed measurably to successes in air warfare over Great Britain and antisubmarine warfare in the North Atlantic. The principal centers for radio and microwave research were in Great Britain and the United States.

EARLY MICROWAVE SPECTROSCOPY

The first report of spectroscopy conducted at microwave frequencies was that of Cleeton and Williams at the University of Michigan in 1934.[1] The object of their study was the ammonia gas molecule, NH_3. Through spectroscopic studies, fine structure had been observed in ammonia's infrared rotational bands. This structure was interpreted as being due to the tunneling of the nitrogen atom through the plane of the three hydrogen atoms (see Section 6.2). Theoretical studies of the ammonia molecule's infrared spectrum had suggested that its doublet's wavenumber separation would be of the order of one wavenumber, cm^{-1}, which falls outside the infrared but within the microwave region.[2]

To obtain a source of radiation of this energy, magnetrons were borrowed from the Westinghouse Corporation's research laboratory, but they generated radiation of much lower frequencies than those required for the proposed work on ammonia. Using the Westinghouse magnetrons as models, Cleeton and Williams constructed several magnetrons by scaling down the dimensions. To measure wavelength, they constructed a device similar in operation to an echelon (a diffraction grating consisting of a series of plates of equal thickness, stacked like a staircase). They succeeded in observing the inversion at a wavelength of approximately one centimeter, just where the ammonia infrared data had suggested.

After this early beginning, microwave spectroscopy remained dormant until the end of World War II, largely because microwave technology had not yet developed to the point where spectroscopy in the centimeter wavelength range could be performed. While working on a project for the war effort, some scientists at the MIT Radiation Laboratory observed an unexplained attenuation in test microwave signals directed across the Charles River from Cambridge to Boston. The phenomenon was most prominent at a time when a local garbage scow released waste products into that river. Later, after the advent of microwave spectroscopy, the cause of this microwave signal attenuation was found to be the effluence of ammonia gas from waste discharge!

Microwave spectroscopy is a consequence of transitions between energy levels, which usually generate spectra in the microwave region. Radio frequency spectroscopy, by contrast, is the result of the fact that nuclear electric quadrupole moment and nuclear-magnetic dipole moment energy separations lie in the radio-frequency region of the electromagnetic spectrum. In the case of nuclear electric quadrupole moment transitions, an interaction between nuclear electric quadrupole moments and local electric field gradients causes the partial removal of nuclear spin angular momentum degener-

[1] C. E. Cleeton and N. H. Williams, *Phys. Rev.* 46:235, 1934.
[2] D. M. Dennison and J. D. Hardy, *Phys. Rev.* 39:938, 1932.

acy. Nuclear magnetic resonance, on the other hand, is due to the removal of nuclear-spin angular momentum degeneracy by an external magnetic field.

By the mid-1950s, the physical science and engineering journals were filled with research papers dealing with radio frequency and microwave spectroscopy. During this decade, significant research in either of these fields might be conceived, executed, and submitted for publication within a single month. Toward the end of the 1950s, microwave spectroscopy was headed for a crisis of sorts, because the supply of unstudied and relatively uncomplicated gas molecules with spectra in the centimeter wavelength region was being depleted. One way to open up new areas of investigation was to extend the wavelength range of microwave spectroscopy to the millimeter and submillimeter region, thereby including molecules of greater moment of inertia. However, there seemed to be no way to obtain monochromatic radiation of constant amplitude from vacuum tubes at wavelengths significantly below 3 mm.[3]

The development of coherent sources of radiation in the centimeter, millimeter, and submillimeter wavelength region took place in two steps, both of which involved the ammonia molecule. The first occurred in 1952 with a report by H. Lyons of the Central Radio Propagation Laboratory of the National Bureau of Standards in Washington, D.C.[4] Lyons described the performance of a clock—a quartz-crystal-controlled oscillator slaved to the peak absorption of the (3, 3) rotation inversion line of the ammonia molecule. The notation (3, 3) signifies the $J = 3$ and $K = 3$ angular momentum states of a symmetric top rotor, with J representing the angular-momentum quantum number of the molecule's rotation about its principal axis and K describing rotation about an axis perpendicular to the principal axis.[5] To first order of approximation, the ammonia molecule does not exhibit an electric dipole moment because the nitrogen atom tunnels through the plane of the three hydrogen atoms, thereby causing its electric dipole moment to reverse its direction relative to the molecule's principal axis of symmetry (see Section 6.2). In second order, however, its electric dipole moment is nonvanishing so that, in fact, it exhibits a robust absorption at 23 870.127 MHz in the microwave region. This clock became known as the ammonia clock. When compared with the primary standard quartz-crystal-controlled oscillator clocks at the National Bureau of Standards, the ammonia clock proved to have a stability of the order of a few parts in 10^9. The ammonia clock used the molecule's rotation inversion absorption line as a reference for its own quartz-crystal-controlled oscillator, thereby providing an oscillator of hitherto unprecedented stability.

The second step was a noteworthy advance in 1953, when J. Weber of the University of Maryland provided, for the first time, a theoretical strategy for obtaining coherent amplification of electromagnetic radiation by stimulated emission.[6] The first step, he said, is to create a thermodynamical nonequilibrium distribution of energy levels in a crystalline or gaseous substance. Using the Einstein coefficients for spontaneous and induced transitions,[7] Weber demonstrated the necessary conditions for the coherent amplification of radiation. Oscillations could be sustained if proper control of feedback and gain were provided to compensate for radiation losses. In his paper, Weber estimated the performance of a gas at low pressure. As a choice of an active species, he suggested the gaseous ammonia (3, 3) rotation inversion line, whose absorption is one of the strongest in the microwave region. For a gas pressure of 10^{-2} torr and a frequency of 30 GHz, he estimated that gains of the order of 0.02 dB per meter of ammonia could

[3] W. Gordy, *Annals of the New York Academy of Sciences* 55:774, 1952.

[4] H. Lyons, *Annals of the New York Academy of Sciences* 55:831, 1952.

[5] C. H. Townes and A. L. Schawlow, *Microwave Spectroscopy*, New York, McGraw-Hill Book Co., 1955. See p. 312.

[6] J. Weber, *Trans. IRE Prof. Grp. Electronic Devices, PGED-3*, vol. 1, 1953.

[7] A. Einstein, *Mitt. Phys. Ges. Zurich* 16(18):47, 1916; *Z. Für Phys.* 18:121, 1917.

be achieved, resulting in power levels of 2 mW per 100 cm³ of ammonia. He pointed out that amplification would die out within a period of the order of the molecule's relaxation time unless the system were maintained in a nonequilibrium thermodynamic condition.

THE FIRST MASERS

Within a year or so of Weber's publication, two molecular devices for generating microwave radiation were reported by N. G. Basov and A. M. Prokhorov of the Lebedev Institute in Moscow[8] and by J. P. Gordon, H. J. Zeiger, and C. H. Townes of Columbia University in New York.[9] Both devices used the ammonia molecule as the active species. The acronym *maser* was coined to stand for "**m**icrowave **a**mplification by **s**timulated **e**mission of **r**adiation." In particular, the Columbia University maser made use of molecular beam technology.

As described in Section 6.2, the ammonia molecule can be represented by one of two configurations (see Figure 6.14). In one (the right-handed molecule), the nitrogen lies above the plane of the hydrogen atoms. In the other (the left-handed molecule), nitrogen lies below the plane of hydrogen atoms. Note that no rotation can make these two molecules identical, since inversion through center-of-mass coordinates is not a proper rotation. Both right-handed and left-handed descriptions are valid for the ammonia molecule, so an acceptable description of the ground state of ammonia must be a linear combination of the right-handed and left-handed configurations. There are two such linear combinations. If these two combinations are chosen to be of equal amplitude and orthogonal to each other, then one will be a symmetric representation (the σ state) and the other will be an antisymmetric representation (the α state). These two states behave differently when subjected to a gradient electric field.

In the ammonia maser at Columbia University, molecules were introduced into a high-vacuum chamber and then collimated into a beam. The beam of molecules then traversed an electric field that spatially separated the molecules of the two states. Molecules in the lower-energy state were rejected, while molecules in the excited state were sent to the exit beam. These excited molecules were permitted to drift into a region of stimulating electromagnetic radiation, where amplification took place.

Shortly afterward, the Central Radio Propagation Laboratory of the National Bureau of Standards used the ammonia maser amplifier, a self-sustained oscillator, as a standard clock. Its power output was of the order of one microwatt, and its stability was estimated at a few parts in 10^{10}, an order of magnitude better than that of the crystal-controlled oscillator locked to the ammonia (3, 3) inversion line. The improvement was due to the maser's much narrower line width, since collision broadening of the energy levels is virtually eliminated in the unidirectional, focused molecular beam. Furthermore, Doppler broadening of energy levels is greatly reduced through the use of molecular velocity selection.

Following the development of the ammonia maser, G. Makov, C. Kikuchi, J. Lambe, and R. W. Terhune reported another device.[10] This was the solid-state ruby maser, which used three of the chromium ion's four electronic spin levels of its ground electronic state. Ruby is the mineralogical name for crystalline corundum (Al_2O_3) having trivalent chromium ions as principal impurities. The maser achieved continuous electron-spin population inversion by applying a strategy—suggested by Bloembergen and similar to that known as optical pumping—that achieves preferential population of excited optical

[8] N. G. Basov and A. M. Prokhorov, *Zh. Eksp. Theo. Fiz.* 27:431, 1954.

[9] J. P. Gordon, H. J. Zeiger, and C. H. Townes, *Phys. Rev.* 99:1264, 1955.

[10] G. Makov, C. Kikuchi, J. Lambe, and R. Terhune, *Phys. Rev.* 109:1399, 1958.

states.[11] Since the electronic spin states' energy levels of the trivalent chromium ion in ruby depend on the strength of an applied magnetic field, the ruby maser proved to be tunable. It quickly gained use as a high-gain, low-noise amplifier, a boon to radio astronomy. The ruby crystal used by Kikuchi and his associates as the active element in their maser was donated by the University of Michigan to the Smithsonian Institution in Washington, D.C., and today resides there as a permanent exhibit.

Another important milestone in maser history came in 1960 when H. M. Goldenberg, D. Kleppner, and N. F. Ramsey reported on the development and operation of a hydrogen maser having the astonishing stability of 1 part in 30×10^{12} at a frequency of 1 420 405 751.786 Hz.[12] Since there are about 31 million seconds in a year, this maser will neither gain nor lose more than a few seconds in 100 000 years. The Michelson-Morley experiment was repeated to ever-increasing accuracy with this very precise clock, thereby verifying the conclusion of the original experiment—that the speed of light when measured in an inertial reference frame is invariant and is a universal constant.

FROM MASERS TO LASERS

Almost immediately upon the development of the ammonia maser, A. L. Schawlow and C. H. Townes suggested that the concept be extended to optical wavelengths.[13] The acronym *laser*, by analogy with *maser*, was coined to stand for *l*ight *a*mplification by *s*timulated *e*mission of *r*adiation. The rush to invent was on, and the first laser to be reported was T. H. Maiman's ruby laser at the Hughes Corporation Research Laboratories.[14] The ruby laser consisted of a single crystal rod of synthetic corundum with chromium ions substituted for aluminum ions to the extent of a few parts per thousand. The ends of the rod were polished to be planar and parallel to each other and then metallized to form a radiation cavity resonator. The rod was surrounded by a helical flash lamp (Fig. 1). When a bank of capacitors in the flash lamp discharged, they caused the ruby rod to emit a single pulse of coherent and monochromatic radiation of wavelength 694.3 nm. The pulse duration was of the order of 10^{-8} seconds, and the laser had a power output of some 10^7 W/pulse. Due perhaps to media publicity, *laser* became a household word almost overnight, and the verb *to lase* was coined to stand for the action of a laser.

The laser has evolved into an indispensable tool in both research and engineering. Its numbers, varieties, and applications have experienced astonishing growth. Lasers are used in medicine as precision scalpels. Researchers have focused intense laser beams on targets in attempts to initiate controlled thermonuclear fusion (see Chapter 14). Scientists bounce laser light off of total internal reflectors on the Moon's surface to measure moonquakes. Lasers are used for precision machining and for such mundane tasks as reading bar codes on railroad rolling stock and on grocery-store items.

Nature has produced her own masers. Radio astronomers have observed interesting radiations emanating from extraterrestrial space, believed to be associated with interstellar H_2O molecules, electrically neutral OH and SiO free radicals, and perhaps other species. On the basis of the directions in space from which these radiations emanate, their narrow spectral bandwidths, and their radiation intensities, physicists have deduced that the radiations originate in naturally occurring masers. Although some maser-burst

[11] N. Bloembergen, *Phys. Rev.* 104:324, 1956.

[12] H. M. Goldenberg, D. Kleppner, and N. F. Ramsey, *Phys. Rev. Lett.* 5:361, 1960; *Phys. Rev.* 126: 603, 1962.

[13] A. L. Schawlow and C. H. Townes, *Phys. Rev.* 112:1940, 1958.

[14] T. H. Maiman, *Nature* 187:493, 1960.

Figure 1   (a) A flash lamp surrounds a ruby rod. Chromium ions in the rod absorb light from the flash lamp, which pumps electrons to an excited state. A spontaneous emission from the excited state stimulates other ions to emit. The stimulated emission at 694.3 nm bounces back and forth between the reflecting ends of the rod, building up a coherent laser beam. *(Adapted from D. Ebbing,* General Chemistry, *5th ed., Boston, Houghton Mifflin Co., 1996)* (b) A photograph of the first ruby laser. *(Courtesy of Hughes Aircraft Company)*

radiations are known to issue from the Sun, many cosmic masers are most likely either galactic or extragalactic in origin.[15,16]

Acknowledgments

The authors would like to express indebtedness to Prof. J. Weber of the University of Maryland for his critical reading of the manuscript and to Mary Beth Riedner of Roosevelt University for providing valuable assistance.

Suggestions for Further Reading

N. F. Ramsey, *Science* 248:1612, 1990. A review of the hydrogen atom maser.

[15] M. Miyoshi, *Nature* 371:395, 1994.
[16] M. Elitzur, *Scientific American* 272:78, 1995.

Essay

PHOTOVOLTAIC CONVERSION

John D. Meakin

Department of Mechanical Engineering and Institute of Energy Conversion, University of Delaware

The photovoltaic effect occurs when light is absorbed by a semiconductor. The energy of the photons is transferred to electrons in the valence band of the semiconductor, promoting them into the conduction band and resulting in the formation of electron-hole pairs. Only photons with energies exceeding the band-gap energy of the semiconductor can be effective in this process. If the semiconductor has a small band gap, a large fraction of the incident photons are able to create electron-hole pairs. This yields a large current, but the voltage generated across the solar cell is low. With a large-band-gap semiconductor, the voltage generated is high but the current is low. Analysis has shown that to convert sunlight, in which most of the photons have energies between 1 eV and 3 eV, the optimum band gap for maximum power conversion is approximately 1.5 eV.

In an isolated semiconductor, the excited electron eventually recombines with a hole in the valence band, emitting its excess energy as photons (photoemission) or phonons (heat), and no useful generation of electric energy takes place. To extract the energy of the photoexcited carriers as usable electricity requires the existence of a charge-separating junction such as a *p-n* diode. The passage of the excess charge carriers, known as minority carriers (electrons in the *p*-region, holes in the *n*-region), across the junction prevents electron-hole recombination. Thus a photogenerated electron in *p*-type material has a limited lifetime in the presence of the stable population of holes but an unlimited lifetime after crossing the junction into the *n*-type material (vice versa for holes in *n*-type material).

In the traditional crystalline silicon solar cell, the charge-separating region is formed when the diffusion of specific impurities, or dopants, into a wafer of silicon creates regions of opposite conductivity type. Diffusion of carriers across the transition between the *n*- and *p*-regions—that is, across the junction—occurs until an equilibrium is established in which the electric field created prevents further diffusion of charges. This internal electric field constitutes the charge-separating barrier, which is key to the operation of the solar cell. Electrons created by light in the *p*-type material migrate to the junction and are then swept into the *n*-type region. (An analogous situation exists for holes created on the *n*-type side of the junction.) In an efficient cell, this collection occurs before the electrons recombine with the stable population of holes in the *p*-region. The energy absorbed from the light is thus converted to electrical energy, which can be fed into an external circuit.

Figure 1a illustrates a diode under illumination and shows the production of electron-hole pairs, which are then separated by the built-in electric field. The *I-V* behavior of an ideal photodiode is obtained by adding a light-generated current I_{sc} to the diode equation presented in Chapter 11:

$$I = I_0[\exp(qV/k_\mathrm{B}T) - 1] - I_{sc}$$

where I and V are the external current and voltage, respectively; q is the electronic charge; k_B is Boltzmann's constant; I_{sc} is the short-circuit current; and I_0 is the reverse-saturation current of the diode. The corresponding *I-V* curve (Fig. 1b) is seen to be that of a diode but displaced along the current axis by the "light-generated current," I_{sc}.

If the outside circuit has zero resistance, a maximum current I_{sc} flows as shown on the *I-V* curve. As the load resistance rises, a voltage is generated across the cell, and the external current eventually falls to zero at the open-circuit voltage V_{oc}. In this situation, electrons accumulate on the *n* side of the junction (holes on the *p* side), biasing the junction in the sense opposite that of the built-in field. The maximum voltage the cell can develop corresponds to the forward light-generated current that exactly matches

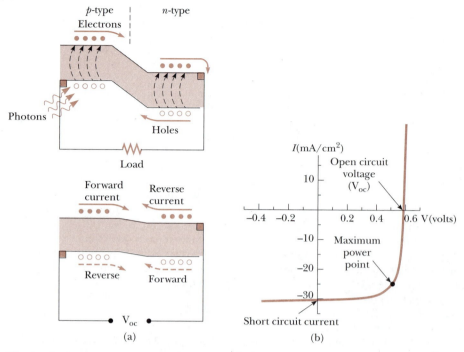

Figure 1   (a) A *p-n* junction under illumination, showing, in the upper diagram, the generation and collection of electrons and holes near the short-circuit current point. As the external load increases, charges accumulate in the cell, reducing the height of the charge-separating barrier until at open circuit there is no net current flow. (b) A typical current-voltage curve for a silicon solar cell under 1 kW/m² insolation.

the reverse current. The maximum power generation occurs where the current-voltage product, $I \times V$, is largest. Under ideal conditions, a single-junction solar cell in direct sunlight should convert about 25% of the incident solar energy to electricity. This figure ignores all the losses that inevitably occur in an actual cell, such as reflection from the front surface and various other electrical and optical losses. The theoretical efficiency limit is increased by operating the cell under concentrated illumination, which can be achieved with mirrors or lenses. Efficiencies of more than 30% are possible from sunlight concentrated a few hundred times, but the concentrating system must track the Sun as it orbits across the sky.

In 1954 the first practical solar cells, which had a structure similar to that shown in Figure 2, were made from single-crystal wafers of semiconductor-grade silicon. These cells converted about 6% of the total incident sunlight to electrical power; that is, they had an efficiency of 6%. Research Si cells have now attained over 23% efficiency under direct sunlight and almost 30% under sunlight concentrated by a factor of a few hundred times. Commercial cells are moving toward 20% efficiency, yielding multicell modules of about 15% efficiency. Why, then, are solar cells not being used more extensively to generate electricity? To answer this question, we must look more closely at the economics of electricity generation from solar cells.

The most significant costs are the initial capital outlays; operation and maintenance costs are small, and there are no fuel costs. A simple analysis reveals the scale of investment that would be economically competitive. A module rated at 1 kW (i.e., able to produce 1 kW of electricity under sunlight of 1 kW/m²—typical of noon under a clear sky) generates, on average, about 5 kWh/day or about 4.10^4 kWh of electricity over its

Figure 2   A schematic drawing of a silicon solar cell. Recent developments include a highly textured front surface to induce light trapping and grids in grooves to reduce the unilluminated area of the cell.

lifetime of, say, 20 years. In America this can be worth as little as $1000 or more than $5000, depending on the location and the type of user. About half of the initial investment will be for costs other than the modules, which must therefore be available for a few hundred dollars per kilowatt-hour. More sophisticated analyses that take inflation and carrying charges into account yield similar order-of-magnitude estimates. Although the first cells produced for space use cost about $1 million/kW, extensive technical and manufacturing progress coupled with much less stringent specifications for terrestrial use have brought the present price down to about $3000 to $5000/kW in large quantities. The price should decline to less than $2000/kW by the year 2000. Is it likely that a further reduction in order of magnitude can be achieved? To answer this question, we must return to the physics of solar cell operation.

The rate at which sunlight is absorbed by a semiconductor of thickness t and absorption coefficient α is

$$F = e^{-\alpha t}$$

where F is the fraction of radiation absorbed by the semiconductor. The absorption coefficient α, measured in cm^{-1}, varies with the wavelength of the light. It is very low for photon energies below the semiconductor band gap and rises as photon energies increase. Crystalline Si has a relatively low absorption coefficient, between 10^2 and 10^4 cm^{-1}, which means that a thickness of about 200 μm is necessary to absorb most of the sunlight. The carriers generated by the photons must then be able to travel about that far to reach the charge-separating junction before recombination occurs, requiring, in turn, a long minority lifetime.

Efficient carrier collection in the traditional crystalline Si cell requires very pure and perfect material, and bringing the price of Si cells down by a further factor of ten cannot be assured, although ongoing developments may yield substantial reductions. These developments include methods to produce large-area, thin sheets directly from a silicon melt to avoid the wafering step used with single crystals. Large-grained polycrystalline material, which is less expensive than single crystals, is also in use. Another promising approach is to texture the front and/or rear surface of the cell so that light undergoes multiple reflections within the cell and can become totally trapped. This approach lengthens the light path for a given cell thickness without increasing the distance that carriers must travel and has yielded ~15% efficient cells that are as thin as 50 μm.

There are many semiconducting compounds in addition to the semiconducting elements of Group IV, Si and Ge. The essential properties that a material must possess to be a candidate for a low-cost solar cell are an appropriate band gap—say, between 1.0 and 1.7 eV—and a very high absorption coefficient. The latter makes it possible to use very thin layers, which in turn relaxes the limitations on purity and perfection, as short collection distances (short carrier lifetimes) now apply. Any compound is a potential semiconductor if the ratio of valence electrons to atoms, e/a, is 4, as is the case for the Group IV elements.

Among the classes of semiconducting compounds are the II-VI (i.e., a compound AB where A is from Group II and B from Group VI), III-V, and I-III-VI$_2$ families. At present, solar cells based on members of each of these groups are being developed with the expectation that a combination of higher efficiency and lower cost will result.

Representative of the III-V compounds are GaAs and InP, which in single-crystal form have yielded solar cells of about 30% efficiency. As a general rule, the III-Vs have excellent carrier properties, and indeed there is a major effort to develop them for high-speed integrated circuits and the like, but their properties are seriously degraded in thin-film form, and single crystals may have to be used to maintain performance. If this should continue to be the case, it may be necessary to use a concentrating system in which an optical assembly focuses the energy from a large aperture onto a much smaller solar cell, thus reducing the total area of solar cell needed for a given output. (See the recommended reading.)

The II-VI compounds are represented by CdTe, which has an ideal band gap of about 1.5 eV and which has yielded solar cells of about 16% efficiency. Techniques for forming large areas of thin film are under development, including electroplating, various types of vapor deposition, and a spray pyrolysis process.

The I-III-VI$_2$ material that is receiving the most attention is CuInSe$_2$. It has a rather low band gap of 1.0 eV, but (as we shall discuss shortly) that may be an advantage for two-junction or tandem cells. Research groups have demonstrated efficiencies of well over 10%, and recently 15% has been reported. Indicative of the sophistication now being practiced is the use of Cu(GaIn)Se$_2$, in which the band gap can be controlled through the Ga:In ratio. All investigators have confirmed a very high degree of stability for these cells, which is particularly important for thin-film cells that in the past have often exhibited unacceptably short usable lifetimes.

The reason for the interest in all of the preceding compounds is their high absorptivity, which means that very thin layers, as little as 1 μm thick, will completely absorb the useful solar spectrum. The carriers generated by the light then need only travel equally short distances to the junction, and relatively impure and imperfect materials, such as polycrystalline thin films, can be used successfully.

In parallel with the development of polycrystalline materials has been extensive investigation and application of amorphous materials based on an alloy of Si and H, conventionally designated a-Si:H. Thin films of this material are deposited by creating a plasma in a gas containing the two elements, most frequently SiH$_4$. The resulting solid contains tetrahedrally bonded Si but in a disordered, noncrystalline array. The hydrogen appears to heal any unsatisfied Si—Si bonds, eliminating energy states that would otherwise introduce electronic states in the band gap. The hydrogenated a-Si can therefore be doped, in contrast to a hydrogen-free amorphous silicon.

Progress in the science and technology of amorphous thin films has been remarkable; in about a decade, a-Si:H solar cells developed from a laboratory curiosity into a familiar component in solar-powered calculators. Panels with areas of 1 m^2 and greater are now entering the market for various uses, such as battery charging and so on. Intense research is focused on the light-induced decay in conversion efficiency, the Staebler-Wronski effect, which is currently the major limitation on widespread use of a-Si:H modules.

Scientifically and technically, the field of photovoltaic conversion remains exciting and challenging. New materials are being developed, and more efficient configurations are being reduced to practice. In place of one cell to harvest the entire solar spectrum, stacked and tandem structures are appearing. In these systems, the top cell removes the high-energy photons, allowing the longer wavelengths (lower-energy photons) to be efficiently harvested in a second or even third cell and thus creating the photovoltaic equivalent of a multistage turbine. The a-Si:H makes an attractive top cell, as its band gap of about 1.6 eV efficiently uses the short-wavelength light while transmitting the longer wavelengths into a bottom cell. Polycrystalline $CuInSe_2$ is available as a material for the bottom cell, as its band gap, 1.0 eV, is close to the ideal value for a two-junction device. Multijunction a-Si:H cells are also being developed with the use of amorphous Si alloys containing C, Ge, or Sn to give various band gaps. A stack of thin cells has been shown to be more resistant to the Staebler-Wronski effect than a single thick cell.

Interest in the nonpolluting use of fossil fuels is encouraging development of an intriguing variant on traditional solar cells known as thermophotovoltaics (TPV). The ideal incident radiation for a given band gap would be at a single wavelength with an energy just above the gap. Such radiation cannot be derived from burning a fuel, but a spectrum optimally centered at the best wavelength can be generated from a heated radiator. The fuel can then be burned under ideal conditions and electricity generated by TPV. One application would be a hybrid fossil fuel electric car that has the advantages of readily portable fuel and electric motive power.

The present world production of solar cells, about 60 MW, is only 0.01% of the conventional electrical generating capacity installed in America alone. However, production has doubled each year for some time, and 10% penetration of the market for new capacity by the year 2000 is quite possible. The electric utilities are showing growing interest in dispersed generation, which relieves the load on the distribution system and can aid demand-side management.

The costs of nuclear energy and traditional fossil fuels, such as oil, gas, and coal, control the economic incentive to develop alternative sources of electricity for use in the developed nations. However, growing concern about the greenhouse effect and acid rain appear likely to motivate more rapid utilization of photovoltaics and other renewable energy resources.

Solar cells are ideally suited for remote locations and Third World areas without grid systems, where the only alternatives are diesel generators and expendable batteries. Under such conditions it is already economical to use solar cells. For example, the American Coast Guard's remote buoys all use solar cells. Space satellites, such as those used to distribute TV signals, continue to be powered entirely by solar cells.

In spite of uncertainty about the rate of growth of electricity generation, particularly in developing nations, there can be no doubt that solar cells will make a very significant contribution and will continue to attract both scientific and commercial interest.

Suggestions for Further Reading

E. Cook, *Man, Energy, Society*, New York, W. H. Freeman, 1976.
P. D. Maycock and E. N. Stirewalt, *Photovoltaics*, Brick House, 1981.
M. A. Green, *Solar Cells*, Englewood Cliffs, N.J., Prentice-Hall, 1983.

12

Superconductivity

Chapter Outline

Most of the material covered in this chapter has to do with the phenomenon of superconductivity. As we shall see, magnetic fields play an important role in the field of superconductivity, so it is important to understand the magnetic properties of materials before discussing the properties of superconductors.

All magnetic effects in matter can be explained on the basis of the current loops associated with atomic magnetic dipole moments. These atomic magnetic moments arise both from the orbital motion of the electrons and from an intrinsic property of the electrons known as *spin*. Our description of magnetism in matter is based in part on the experimental fact that the presence of bulk matter generally modifies the magnetic field produced by currents. For example, when a material is placed inside a current-carrying solenoid, the material sets up its own magnetic field, which adds vectorially to the field that was already present.

The phenomenon of superconductivity has always been very exciting, both for its fundamental scientific interest and because of its many applications.[1] The discovery in the 1980s of high-temperature superconductivity in certain metallic oxides sparked even greater excitement in the scientific and business communities. Many scientists consider this major breakthrough to be as important as the invention of the transistor. For this reason, it is important that all students of science and engineering understand the basic electromagnetic properties of superconductors and become aware of the scope of their current applications.

Superconductors have many unusual electromagnetic properties, and most applications take advantage of such properties. For example, once a current is produced in a superconducting ring maintained at a sufficiently low temperature, that current persists with no measurable decay. The superconducting ring exhibits no electrical resistance to direct currents, no heating, and no losses. In addition to the property of zero resistance, certain superconductors expel applied magnetic fields so that the field is always zero everywhere inside the superconductor.

As we shall see, classical physics cannot explain the behavior and properties of superconductors. In fact, the superconducting state is now known to be a special quantum condensation of electrons. This quantum behavior has been verified through such observations as the quantization of magnetic flux produced by a superconducting ring.

In this chapter we also give a brief historical review of superconductivity, beginning with its discovery in 1911 and ending with recent developments in high-temperature superconductivity. In describing some of the electromagnetic properties displayed by superconductors, we use simple physical arguments whenever possible. We explain the essential features of the theory of superconductivity with the realization that a detailed study is beyond the scope of this text. Finally, we discuss many of the important applications of superconductivity and speculate on potential applications.

12.1   MAGNETISM IN MATTER

The magnetic field produced by a current in a coil of wire gives a hint as to what might cause certain materials to exhibit strong magnetic properties. In general, any current loop has a magnetic field and a corresponding magnetic moment. Similarly, the magnetic moments in a magnetized substance are associated with internal currents on the atomic level. One can view such currents as arising from electrons orbiting around the nucleus and protons orbiting about each other inside the nucleus. However, as we shall see, the intrinsic

[1] M. Brian Maple, a research physicist at the University of California at San Diego, was asked what he found so fascinating about superconductivity. He responded as follows. "For me the fascination of superconductivity is associated with the words perfect, infinite, and zero. A superconductor has the property of being a perfect conductor, or having infinite conductivity, or zero resistance. Whenever you see something that's perfect, infinite, or zero, truly zero, that's obviously a special state of affairs."

magnetic moment associated with the electron is the main source of magnetism in matter.

We begin with a brief discussion of the magnetic moments due to electrons. The mutual forces between these magnetic dipole moments and their interaction with an external magnetic field are of fundamental importance to an understanding of the behavior of magnetic materials. We shall describe three categories of materials—paramagnetic, ferromagnetic, and diamagnetic. **Paramagnetic** and **ferromagnetic** materials are those that have atoms with permanent magnetic dipole moments. **Diamagnetic** materials are those whose atoms have no permanent magnetic dipole moments.

Magnetic Moments of Atoms

As we learned in Section 8.2, the total magnetic moment of an atom has orbital and spin contributions. For atoms or ions containing many electrons, the electrons in closed shells pair up with their spins and orbital angular momenta opposite each other, a situation that results in a net magnetic moment of zero. However, atoms with an odd number of electrons must have at least one "unpaired" electron and a spin magnetic moment of at least one Bohr magneton, μ_B, where

$$\mu_B = \frac{e\hbar}{2m_e} = 9.274 \times 10^{-24} \, \text{J/T} \tag{12.1}$$

The total magnetic moment of an atom is the vector sum of the orbital and spin magnetic moments, and an unpaired outer electron can contribute both an orbital moment and a spin moment. For example, if the unpaired electron is in an s state, $L = 0$ and consequently the orbital moment is zero. However, if the unpaired electron is in a p or d state, $L \neq 0$ and the electron contributes both an orbital moment and a spin moment. The orbital moment is about the same order of magnitude as the Bohr magneton. Table 12.1 gives a few examples of total magnetic moments for different elements. Note that helium and neon have zero moments because their closed shells cause individual moments to cancel.

The nucleus of an atom also has a magnetic moment associated with its constituent protons and neutrons. However, the magnetic moment of a proton or neutron is small compared with the magnetic moment of an electron and can usually be neglected. Because the masses of the proton and neutron are much greater than that of the electron, their magnetic moments are smaller than that of the electron by a factor of approximately 10^3.

Table 12.1 Magnetic Moments of Some Atoms and Ions

Atom (or Ion)	Magnetic Moment (10^{-24} J/T)
H	9.27
He	0
Ne	0
Ce^{3+}	19.8
Yb^{3+}	37.1

Magnetization and Magnetic Field Strength

The magnetization of a substance is described by a quantity called the **magnetization vector, M.** *The magnitude of the magnetization vector is equal to the magnetic moment per unit volume of the substance.* As you might expect, the total magnetic field in a substance depends on both the applied (external) field and the magnetization of the substance.

Consider a region where there exists a magnetic field $\mathbf{B}_0$ produced by a current-carrying conductor. If we now fill that region with a magnetic substance, the total magnetic field $\mathbf{B}$ in that region is $\mathbf{B} = \mathbf{B}_0 + \mathbf{B}_m$, where $\mathbf{B}_m$ is the field produced by the magnetic substance. This contribution can be expressed in terms of the magnetization vector as $\mathbf{B}_m = \mu_0\mathbf{M}$; hence the total magnetic field in the region becomes

$$\mathbf{B} = \mathbf{B}_0 + \mu_0\mathbf{M} \tag{12.2}$$

It is convenient to introduce a field quantity $\mathbf{H}$, called the **magnetic field strength.** This vector quantity is defined by the relationship $\mathbf{H} = (\mathbf{B}/\mu_0) - \mathbf{M}$, or

$$\mathbf{B} = \mu_0(\mathbf{H} + \mathbf{M}) \tag{12.3}$$

In SI units, the dimensions of both $\mathbf{H}$ and $\mathbf{M}$ are amperes per meter.

To better understand these expressions, consider the space enclosed by a solenoid that carries a current I. (We call this space the *core* of the solenoid.) If this space is a vacuum, then $\mathbf{M} = 0$ and $\mathbf{B} = \mathbf{B}_0 = \mu_0\mathbf{H}$. Since $B_0 = \mu_0 nI$ in the core, where n is the number of turns per unit length of the solenoid, then $H = B_0/\mu_0 = \mu_0 nI/\mu_0$, or

$$H = nI \tag{12.4}$$

That is, the magnetic field strength in the core of the solenoid is due to the current in its windings.

If the solenoid core is now filled with some substance and the current I is kept constant, $\mathbf{H}$ inside the substance remains unchanged and has magnitude nI. This is because the magnetic field strength $\mathbf{H}$ is due solely to the current in the solenoid. However, the total field $\mathbf{B}$ changes when the substance is introduced. From Equation 12.3, we see that part of $\mathbf{B}$ arises from the term $\mu_0\mathbf{H}$ associated with the current in the solenoid; the second contribution to $\mathbf{B}$ is the term $\mu_0\mathbf{M}$, due to the magnetization of the substance filling the core.

Classification of Magnetic Substances

In a large class of substances, specifically paramagnetic and diamagnetic substances, the magnetization vector $\mathbf{M}$ is proportional to the magnetic field strength $\mathbf{H}$. For these substances we can write

$$\mathbf{M} = \chi\mathbf{H} \tag{12.5}$$

where χ (Greek letter chi) is a dimensionless factor called the **magnetic susceptibility.** If the sample is paramagnetic, χ is positive, in which case $\mathbf{M}$ is in the same direction as $\mathbf{H}$. If the substance is diamagnetic, χ is negative and $\mathbf{M}$ is opposite $\mathbf{H}$. It is important to note that this linear relationship between $\mathbf{M}$ and $\mathbf{H}$ does not apply to ferromagnetic substances. Table 12.2 gives the susceptibilities of some substances. Substituting Equation 12.5 for $\mathbf{M}$ into Equation 12.3 gives

$$\mathbf{B} = \mu_0(\mathbf{H} + \mathbf{M}) = \mu_0(\mathbf{H} + \chi\mathbf{H}) = \mu_0(1 + \chi)\mathbf{H}$$

or

$$\mathbf{B} = \mu_m\mathbf{H} \tag{12.6}$$

where the constant μ_m is called the **magnetic permeability** of the substance and has the value

$$\mu_m = \mu_0(1 + \chi) \tag{12.7}$$

Table 12.2   Magnetic Susceptibilities of Some Paramagnetic and Diamagnetic Substances at 300 K

Paramagnetic Substance	χ	Diamagnetic Substance	χ
Aluminum	2.3×10^{-5}	Bismuth	-1.66×10^{-5}
Calcium	1.9×10^{-5}	Copper	-9.8×10^{-6}
Chromium	2.7×10^{-4}	Diamond	-2.2×10^{-5}
Lithium	2.1×10^{-5}	Gold	-3.6×10^{-5}
Magnesium	1.2×10^{-5}	Lead	-1.7×10^{-5}
Niobium	2.6×10^{-4}	Mercury	-2.9×10^{-5}
Oxygen (STP)	2.1×10^{-6}	Nitrogen (STP)	-5.0×10^{-9}
Platinum	2.9×10^{-4}	Silver	-2.6×10^{-5}
Tungsten	6.8×10^{-5}	Silicon	-4.2×10^{-6}

Substances may also be classified in terms of how their magnetic permeabilities μ_m compare to μ_0 (the permeability of free space), as follows:

Paramagnetic $\mu_m > \mu_0$

Diamagnetic $\mu_m < \mu_0$

Ferromagnetic $\mu_m \gg \mu_0$

Since χ is very small for paramagnetic and diamagnetic substances (Table 12.2), μ_m is nearly equal to μ_0 in these cases. For ferromagnetic substances, however, μ_m is typically several thousand times larger than μ_0 but is not a constant. Although Equation 12.6 provides a simple relationship between **B** and **H**, it must be interpreted with care in the case of ferromagnetic substances. As mentioned earlier, **M** is not a linear function of **H** for ferromagnetic substances. This is because the value of μ_m *is not a characteristic of the substance* but rather depends on the previous state and treatment of the ferromagnetic material.

EXAMPLE 12.1   An Iron-Filled Toroid

A toroid carrying a current of 5.00 A is wound with 60 turns/m of wire. The core is made of iron, which has a magnetic permeability of $5000\mu_0$ under the given conditions. Find H and B inside the iron.

Solution   Using Equations 12.4 and 12.6, we get

$$H = nI = \left(60.0 \frac{\text{turns}}{\text{m}}\right)(5.00 \text{ A}) = 300 \frac{\text{A} \cdot \text{turns}}{\text{m}}$$

$$B = \mu_m H = 5000\mu_0 H$$

$$= 5000\left(4\pi \times 10^{-7} \frac{\text{Wb}}{\text{A} \cdot \text{m}}\right)\left(300 \frac{\text{A} \cdot \text{turns}}{\text{m}}\right) = 1.88 \text{ T}$$

This value of B is 5000 times larger than the value in the absence of iron!

Exercise   Determine the magnitude and direction of the magnetization inside the iron core.

Answer $M = 1.5 \times 10^6$ A/m; **M** is in the direction of **H**.

Ferromagnetism

Spontaneous magnetization occurs in some substances whose atomic constituents have permanent magnetic dipole moments. The magnetic moments exhibit long-range order, which can take on various forms, as shown in Figure

(a) Ferromagnetic   (b) Antiferromagnetic   (c) Ferrimagnetic

Figure 12.1   Magnetic ordering in three types of solids. (a) In a ferromagnetic substance, all spins are aligned. (b) In an antiferromagnetic substance, spins in two sublattices have the same magnitude but are opposite in direction. (c) In a ferrimagnetic substance, spins in the two sublattices have different magnitudes *and* opposite directions.

12.1. The magnetic moments of a ferromagnet tend to be aligned as in Figure 12.1a, and hence a ferromagnetic substance has a net magnetization. This permanent alignment is due to a strong coupling between neighboring moments, which can be understood only in quantum mechanical terms. In an antiferromagnetic substance (Fig. 12.1b), the magnetic moments all have the same magnitude. However, because the magnetic moments in the sublattices are oppositely directed, the net magnetization of an antiferromagnet is zero. In a ferrimagnetic substance (Fig. 12.1c), the magnetic moments of the atoms in the two sublattices are oppositely directed, but their magnitudes are not the same. Hence a ferrimagnetic substance has a net magnetization.

Iron, cobalt, and nickel are ferromagnetic at sufficiently low temperatures. Rare earths such as gadolinium and terbium are ferromagnetic below room temperature, while other rare earths are ferromagnetic at very low temperatures. At extremely high temperatures, all transition and rare-earth metals become paramagnetic.

All ferromagnetic materials contain microscopic regions called **domains,** within which all magnetic moments are aligned. Each of these domains has a volume of about 10^{-12} to 10^{-8} m^3 and contains 10^{17} to 10^{21} atoms. The boundaries between domains having different orientations are called **domain walls.** In an unmagnetized sample, the domains are randomly oriented such that the net magnetic moment is zero, as shown in Figure 12.2a. When the sample is placed in an external magnetic field, the domains tend to align with the field, which results in a magnetized sample, as in Figure 12.2b. Observations show that domains initially oriented along the external field grow in size at the expense of the less favorably oriented domains. When the external field is removed, the sample may retain a net magnetization in the direction of the original field.[2] At ordinary temperatures, thermal agitation is not sufficiently high to disrupt this preferred orientation of magnetic moments.

(a)

$\mathbf{B}_0$

(b)

Figure 12.2   (a) Random orientation of atomic magnetic dipoles in an unmagnetized substance. (b) When an external magnetic field $\mathbf{B}_0$ is applied, the magnetic dipoles tend to align with the field, giving the sample a net magnetization $\mathbf{M}$.

[2] It is possible to observe the domain walls directly and follow their motion under a microscope. In this technique, a liquid suspension of powdered ferromagnetic substance is applied to the sample. The fine particles tend to accumulate at the domain walls and shift with them.

Figure 12.3 A toroidal winding arrangement used to measure the magnetic properties of a substance. The material under study fills the core of the toroid, and the circuit containing the galvanometer measures the magnetic flux.

A typical experimental arrangement used to measure the magnetic properties of a ferromagnetic material consists of a toroid-shaped sample wound with N turns of wire, as in Figure 12.3. This configuration is sometimes referred to as a **Rowland ring.** A secondary coil connected to a galvanometer is used to measure the magnetic flux. The magnetic field **B** within the core of the toroid is measured by increasing the current in the toroid coil from zero to I. As the current changes, the magnetic flux through the secondary coil changes by BA, where A is the cross-sectional area of the toroid. Because of this changing flux, an emf is induced in the secondary coil that is proportional to the rate of change in magnetic flux. If the galvanometer in the secondary circuit is properly calibrated, one can obtain a value for **B** corresponding to any value of the current in the toroidal coil. The magnetic field **B** is measured first in the empty coil and then with the same coil filled with the magnetic substance. The magnetic properties of the substance are then obtained from a comparison of the two measurements.

Now consider a toroid whose core consists of unmagnetized iron. If the current in the windings is increased from zero to some value I, the field intensity H increases linearly with I according to the expression $H = nI$. Furthermore, the total field B also increases with increasing current, as shown in Figure 12.4. At point O, the domains are randomly oriented, corresponding to $B_m = 0$. As the external field increases, the domains become more aligned until all are nearly aligned at point a. At this point, the iron core is approaching saturation. (The condition of saturation corresponds to the case where all domains are aligned in the same direction.) Next, suppose the current is reduced to zero, thereby eliminating the external field. The B-versus-H curve, called a **magnetization curve,** now follows the path ab shown in Figure 12.4. Note that at point b, the field **B** is not zero, although the external field is equal to zero. This is explained by the fact that the iron core is now magnetized due to the alignment of a large number of domains (that is, $\mathbf{B} = \mathbf{B}_m$). At this point, the iron is said to have a remanent magnetization and could be considered to be a "permanent" magnet. If the external field is reversed in direction and increased in strength by reversal of the current, the domains reorient until the sample is again unmagnetized at point c, where $\mathbf{B} = 0$. A further increase in the reverse current causes the iron to be magnetized in the opposite direction, approaching saturation at point d. A similar sequence of events occurs as the current is reduced to zero and then increased in the original (positive) direction. In this case, the magnetization curve follows the path def. If the current is increased sufficiently, the magnetization curve returns to point a, where the sample again has its maximum magnetization.

The effect just described, called **magnetic hysteresis,** shows that the magnetization of a ferromagnetic substance depends on the history of the substance as well as the strength of the applied field. (The word *hysteresis* literally means "to lag behind.") One often says that a ferromagnetic substance has a "memory" since it remains magnetized after the external field is removed. The closed loop in Figure 12.4 is referred to as a hysteresis loop. Its shape and size depend on the properties of the ferromagnetic substance and on the strength of the maximum applied field. The hysteresis loop for "hard" ferromagnetic materials (those used in permanent magnets) is characteristically wide as in Figure 12.5a, corresponding to a large remanent magnetization. Such materials cannot be easily demagnetized by an external field. This is in contrast with "soft" ferro-

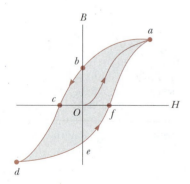

Figure 12.4 A hysteresis curve for a ferromagnetic material.

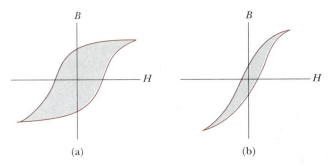

Figure 12.5   Hysteresis curves for (a) a hard ferromagnetic material and (b) a soft ferromagnetic material.

magnetic materials, such as iron, which have very narrow hysteresis loops and small remanent magnetizations (Fig. 12.5b). Such materials are easily magnetized and demagnetized. An ideal soft ferromagnet would exhibit no hysteresis and hence would have no remanent magnetization. One can demagnetize a ferromagnetic substance by carrying the substance through successive hysteresis loops and gradually decreasing the applied field, as in Figure 12.6.

The magnetization curve is useful for another reason. *The area enclosed by the magnetization curve represents the work required to take the material through the hysteresis cycle.* The source of the external field—that is, the emf in the circuit of the toroidal coil—supplies the energy acquired by the sample in the magnetization process. When the magnetization cycle is repeated, dissipative processes within the material due to realignment of the domains result in a transformation of magnetic energy into internal thermal energy, which raises the temperature of the substance. For this reason, devices subjected to alternating fields (such as transformers) use cores made of soft ferromagnetic substances, which have narrow hysteresis loops and correspondingly small energy losses per cycle.

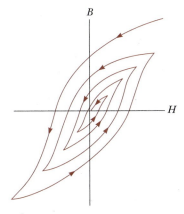

Figure 12.6   Demagnetizing a ferromagnetic material by carrying it through successive hysteresis loops.

Paramagnetism

A paramagnetic substance has a positive but small susceptibility $(0 < \chi \ll 1)$, which is due to the presence of atoms (or ions) with permanent magnetic dipole moments. These dipoles interact only weakly with each other and are randomly oriented in the absence of an external magnetic field. When the substance is placed in an external magnetic field, its atomic dipoles tend to line up with the field. However, this alignment process must compete with thermal motion, which tends to randomize the dipole orientations.

Experimentally, one finds that the magnetization of a paramagnetic substance is proportional to the applied field and inversely proportional to the absolute temperature under a wide range of conditions. That is,

$$M = C \frac{B}{T} \qquad (12.8)$$

Curie's law

This is known as **Curie's law** after its discoverer, Pierre Curie (1859–1906), and the constant C is called **Curie's constant** (Problem 13). This shows that the magnetization increases with increasing applied field and with decreasing temperature. When $B = 0$, the magnetization is zero, corresponding to a random orientation of dipoles. At very high fields or very low temperatures, the

Figure 12.7   A plot of the magnetization versus absolute temperature for a ferromagnetic substance. The magnetic moments are aligned (ordered) below the Curie temperature T_c, where the substance is ferromagnetic. The substance becomes paramagnetic, that is, disordered above T_c.

magnetization approaches its maximum (saturation) value, corresponding to a complete alignment of its dipoles, and Equation 12.8 is no longer valid.

Interestingly, when the temperature of a ferromagnetic substance reaches or exceeds a critical temperature, called the **Curie temperature,** the substance loses its spontaneous magnetization and becomes paramagnetic (Fig. 12.7). Below the Curie temperature, the magnetic moments are aligned and the substance is ferromagnetic. Above the Curie temperature, the thermal energy is large enough to cause a random orientation of dipoles; hence the substance becomes paramagnetic. For example, the Curie temperature for iron is 1043 K. Table 12.3 lists Curie temperatures and saturation magnetization values for several ferromagnetic substances.

Diamagnetism

A diamagnetic substance is one whose atoms have no permanent magnetic dipole moment. When an external magnetic field is applied to a diamagnetic substance such as bismuth or silver, a weak magnetic dipole moment is induced in the direction opposite the applied field (Lenz's law). Although the effect of diamagnetism is present in all matter, it is weak compared to paramagnetism or ferromagnetism.

We can obtain some understanding of diamagnetism by considering two electrons of an atom orbiting the nucleus in opposite directions but with the same speed. The electrons remain in these circular orbits because of the attractive electrostatic force (the centripetal force) of the positively charged nucleus. Because the magnetic moments of the two electrons are equal in magnitude and opposite in direction, they cancel each other, and the dipole moment of the atom is zero. When an external magnetic field is applied, the electrons experience an additional force $q\mathbf{v} \times \mathbf{B}$. This added force modifies the central force and thereby increases the orbital speed of the electron whose magnetic moment is antiparallel to the field and decreases the speed of the electron whose magnetic moment is parallel to the field. As a result, the magnetic moments of the electrons no longer cancel, and the substance acquires a net dipole moment that opposes the applied field. It is important to note that

Table 12.3   Curie Temperatures and Saturation Magnetizations for Several Ferromagnetic Substances

Substance	T_c (K)	M_s (10^6 A/m)
Iron	1043	1.75
Cobalt	1404	1.45
Nickel	631	0.512
Gadolinium	289	2.00
Terbium	230	1.44
Dysprosium	85	2.01
Holmium	20	2.55

Source of data: D. W. Gray, ed., *American Institute of Physics Handbook*, New York, McGraw-Hill, 1963.

this is a classical explanation. Quantum mechanics is needed for a complete explanation of diamagnetism.

**EXAMPLE 12.2   Saturation Magnetization
of Iron**

Estimate the maximum magnetization in a long cylinder of iron, assuming there is one unpaired electron spin per atom.

Solution   The maximum magnetization, called the saturation magnetization, is attained when all the magnetic moments in the sample are aligned. If the sample contains n atoms per unit volume, then the saturation magnetization M_s has the value

$$M_s = n\mu$$

where μ is the magnetic moment per atom. Since the molecular weight of iron is 56 g/mol and its density is

7.9 g/cm^3, the value of n is 8.5×10^{28} atoms/m^3. Assuming each atom contributes one Bohr magneton (due to one unpaired spin) to the magnetic moment, we get

$$M_s = \left(8.5 \times 10^{28} \, \frac{\text{atoms}}{\text{m}^3} \right) \left(9.27 \times 10^{-24} \, \frac{\text{A} \cdot \text{m}^2}{\text{atom}} \right)$$
$$= 7.9 \times 10^5 \, \text{A/m}$$

This is about one-half the experimentally determined saturation magnetization for annealed iron, which indicates that there are actually two unpaired electron spins per atom.

12.2   A BRIEF HISTORY OF SUPERCONDUCTIVITY

The era of low-temperature physics began in 1908 when the Dutch physicist Heike Kamerlingh Onnes first liquefied helium, which boils at 4.2 K at standard pressure. Three years later, in 1911, Kamerlingh Onnes and one of his assistants discovered the phenomenon of superconductivity while studying the resistivity of metals at low temperatures.[3] They first studied platinum and found that its resistivity, when extrapolated to 0 K, depended on purity. They then decided to study mercury because very pure samples could easily be prepared by distillation. Much to their surprise, the resistance of the mercury sample dropped sharply at 4.15 K to an unmeasurably small value. It was quite natural that Kamerlingh Onnes would choose the name **superconductivity** for this new phenomenon of perfect conductivity. Figure 12.8 shows the experimental results for mercury and platinum. Note that platinum does not exhibit superconducting behavior, as indicated by its finite resistivity as T approaches 0 K. In 1913 Kamerlingh Onnes was awarded the Nobel prize in physics for the study of matter at low temperatures and the liquefaction of helium.

We now know that the resistivity of a superconductor is truly zero. Soon after the discovery by Kamerlingh Onnes, many other elemental metals were found to exhibit zero resistance when their temperatures were lowered below a certain characteristic temperature of the material, called the **critical temperature,** T_c.

The magnetic properties of superconductors are as dramatic and as difficult to understand as their complete lack of resistance. In 1933 W. Hans Meissner and Robert Ochsenfeld studied the magnetic behavior of superconductors and found that when certain ones are cooled below their critical temperatures in the presence of a magnetic field, *the magnetic flux is expelled from the interior of the*

---

[3] H. Kamerlingh Onnes, *Leiden Comm.*, 120b, 122b, 124c, 1911.

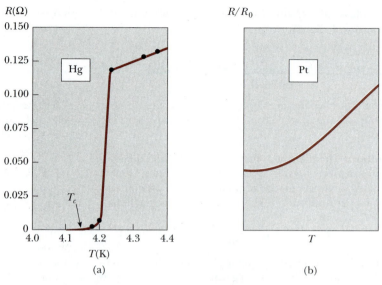

Figure 12.8   Plots of resistance versus temperature for (a) mercury (the original data published by Kamerlingh Onnes) and (b) platinum. Note that the resistance of mercury follows the path of a normal metal above the critical temperature, T_c, and then suddenly drops to zero at the critical temperature, which is 4.15 K for mercury. In contrast, the data for platinum show a finite resistance R_0 even at very low temperatures.

superconductor.[4] Furthermore, these materials lost their superconducting behavior above a certain temperature-dependent **critical magnetic field,** $B_c(T)$. In 1935 Fritz and Heinz London developed a phenomenological theory of superconductivity,[5] but the actual nature and origin of the superconducting state were first explained by John Bardeen, Leon N. Cooper, and J. Robert Schrieffer in 1957.[6] A central feature of this theory, commonly referred to as the BCS theory, is the formation of bound two-electron states called **Cooper pairs.** In 1962 Brian D. Josephson predicted a tunneling current between two superconductors separated by a thin (<2 mm) insulating barrier, where the current is carried by these paired electrons.[7] Shortly thereafter, Josephson's predictions were verified, and today there exists a whole field of device physics based on the Josephson effect. Early in 1986 J. Georg Bednorz and Karl Alex Müller reported evidence for superconductivity in an oxide of lanthanum, barium, and copper at a temperature of about 30 K.[8] This was a major breakthrough in superconductivity because the highest known value of T_c at that time was about 23 K in a compound of niobium and germanium. This remarkable discovery, which marked the beginning of a new era of high-temperature superconductivity, received worldwide attention in both the scientific community and the business world. Recently, researchers have reported critical tempera-

[4] W. H. Meissner and R. Ochsenfeld, *Naturwisschaften* 21:787, 1933.

[5] F. London and H. London, *Proc. Roy. Soc. (London)* A149:71, 1935.

[6] J. Bardeen, L. N. Cooper, and J. R. Schrieffer, *Phys. Rev.* 108:1175, 1957.

[7] B. D. Josephson, *Phys. Letters* 1:251, 1962.

[8] J. G. Bednorz and K. A. Müller, *Z. Phys.* B64:189, 1986.

tures as high as 150 K in more complex metallic oxides, but the mechanisms responsible for superconductivity in these materials remain unclear.

Until the discovery of high-temperature superconductors, the use of superconductors required coolant baths of liquefied helium (rare and expensive) or liquid hydrogen (very explosive). On the other hand, superconductors with $T_c > 77$ K require only liquid nitrogen, which boils at 77 K and is comparatively inexpensive, abundant, and relatively inert. If superconductors with T_c's above room temperature are ever found, technology will be drastically altered.

12.3   SOME PROPERTIES OF TYPE I SUPERCONDUCTORS

Critical Temperature and Critical Magnetic Field

Table 12.4 lists the critical temperatures of some superconducting elements, classified as **type I superconductors.** Note the absence of copper, silver, and gold, which are excellent electrical conductors at ordinary temperatures but do not exhibit superconductivity.

In the presence of an applied magnetic field **B**, the value of T_c decreases with increasing magnetic field, as indicated in Figure 12.9 for several type I superconductors. When the magnetic field exceeds the critical field, **B**$_c$, the superconducting state is destroyed and the material behaves as a normal conductor with finite resistance.

The magnitude of the critical magnetic field varies with temperature according to the approximate expression

$$B_c(T) = B_c(0) \left[1 - \left(\frac{T}{T_c} \right)^2 \right]$$ (12.9)

Table 12.4   Critical Temperatures and Critical Magnetic Fields (at $T = 0$ K) of Some Elemental Superconductors

Superconductor	T_c (K)	$B_c(0)$ (T)
Al	1.196	0.0105
Ga	1.083	0.0058
Hg	4.153	0.0411
In	3.408	0.0281
Nb	9.26	0.1991
Pb	7.193	0.0803
Sn	3.722	0.0305
Ta	4.47	0.0829
Ti	0.39	0.010
V	5.30	0.1023
W	0.015	0.000115
Zn	0.85	0.0054

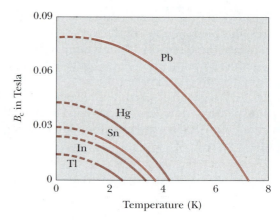

Figure 12.9 Critical magnetic field versus critical temperature for several type I superconductors. Extrapolations of these fields to 0 K give the critical fields listed in Table 12.4. For a given metal, the material is superconducting at fields and temperatures below its critical temperature and behaves as a normal conductor above that curve.

As you can see from this equation and Figure 12.9, the value of B_c is a maximum at 0 K. The value of $B_c(0)$ is found by determining B_c at some finite temperature and extrapolating back to 0 K, a temperature that cannot be achieved in the laboratory. The value of the critical field limits the maximum current that can be sustained in a type I superconductor.

Note that $B_c(0)$ is the maximum magnetic field that is required to destroy superconductivity in a given material. If the applied field exceeds $B_c(0)$, the metal never becomes superconducting at any temperature. Values for the critical field for type I superconductors are quite low, as Table 12.4 shows. For this reason, type I superconductors are not used to construct high-field magnets, called superconducting magnets, because the magnetic fields generated by modest currents destroy, or "quench," the superconducting state.

Magnetic Properties of Type I Superconductors

One can use simple arguments based on the laws of electricity and magnetism to show that the magnetic field inside a superconductor cannot change with time. According to Ohm's law, the electric field inside a conductor is proportional to the resistance of the conductor. Thus, since $R = 0$ for a superconductor, *the electric field in its interior must be zero.* Now recall that Faraday's law of induction can be expressed as

$$\oint \mathbf{E} \cdot d\mathbf{s} = -\frac{d\Phi_B}{dt} \tag{12.10}$$

That is, the line integral of the electric field around any closed loop is equal to the negative rate of change in the magnetic flux Φ_B through the loop. Since **E** is zero everywhere inside the superconductor, the integral over any closed path inside the superconductor is zero. Hence $d\Phi_B/dt = 0$, which tells us that *the magnetic flux in the superconductor cannot change.* From this we can conclude that B ($= \Phi_B/A$) *must remain constant inside the superconductor.*

Prior to 1933 it was assumed that superconductivity was a manifestation of perfect conductivity. If a perfect conductor is cooled below its critical temperature in the presence of an applied magnetic field, the field should be trapped in the interior of the conductor even after the field is removed. The final state of a perfect conductor in an applied magnetic field should depend upon which occurs first, the application of the field or the cooling below the critical temperature. If the field is applied after cooling below T_c, the field should be expelled from the superconductor. On the other hand, if the field is applied before cooling, the field should not be expelled from the superconductor after cooling below T_c.

When experiments were conducted in the 1930s to examine the magnetic behavior of superconductors, the results were quite different. In 1933 Meissner and Ochsenfeld[4] discovered that, when a metal became superconducting in the presence of a weak magnetic field, the field was expelled so that **B** equaled 0 everywhere in the interior of the superconductor. Thus the same final state, **B** = 0, was achieved whether the field was applied before or after the material was cooled below its critical temperature. Figure 12.10 illustrates this effect for a material in the shape of a long cylinder. Note that the field penetrates the cylinder when its temperature is greater than T_c (Fig. 12.10a). As the temperature is lowered below T_c, however, the field lines are spontaneously expelled from the interior of the superconductor (Fig. 12.10b). Thus a type I superconductor is more than a perfect conductor (resistivity $\rho = 0$); it is also a perfect diamagnet (**B** = 0). The phenomenon of the expulsion of magnetic fields from the interior of a superconductor is known as the Meissner effect. The property that **B** = 0 in the interior of a type I superconductor is as fundamental as the property of zero resistance and shows the important role that magnetism plays in superconductivity. If the applied field is sufficiently large ($B > B_c$), the superconducting state is destroyed and the field penetrates the sample.

Because a superconductor is a perfect diamagnet, it repels a permanent magnet. In fact, one can perform a dazzling demonstration of the Meissner effect by floating a small permanent magnet above a superconductor and

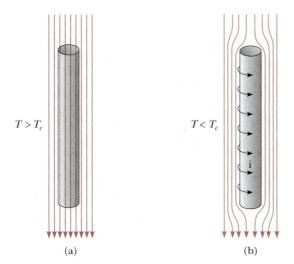

$T > T_c$ (a)

$T < T_c$ i (b)

Figure 12.10   A type I superconductor in the form of a long cylinder in the presence of an external magnetic field. (a) At temperatures above T_c, the field lines penetrate the sample because it is in its normal state. (b) When the rod is cooled to $T < T_c$ and becomes superconducting, magnetic flux is excluded from its interior by the induction of surface currents.

Figure 12.11 A small permanent magnet levitated above a pellet of the $YBa_2Cu_3O_{7-\delta}$ superconductor cooled to the temperature of liquid nitrogen, 77 K. *(Courtesy of IBM Research)*

achieving magnetic levitation. Figure 12.11 is a dramatic photograph of magnetic levitation. The details of this demonstration are provided in Questions 16 through 19.

You should recall from your study of electricity that a good conductor expels static electric fields by moving charges to its surface. In effect, the surface charges produce an electric field that exactly cancels the externally applied field inside the conductor. In a similar manner, a superconductor expels magnetic fields by forming surface currents. To illustrate this point, consider again the superconductor in Figure 12.10. Let us assume that the sample is initially at a temperature $T > T_c$, as in Figure 12.10a, so that the field penetrates the cylinder. As the cylinder is cooled to a temperature $T < T_c$, the field is expelled as in Figure 12.10b. In this case, surface currents are induced on the superconductor, producing a magnetic field that exactly cancels the externally applied field inside the superconductor. As you might expect, the surface currents disappear when the external magnetic field is removed.

EXAMPLE 12.3   Critical Current in a Pb Wire

A lead wire has a radius of 3.00 mm and is at a temperature of 4.20 K. Find (a) the critical magnetic field in lead at this temperature and (b) the maximum current the wire can carry at this temperature and still remain superconducting.

Solution   (a) We can use Equation 12.9 to find the critical field at any temperature if $B_c(0)$ and T_c are known. From Table 12.4 we see that the critical magnetic field of lead at 0 K is 0.0803 T and its critical temperature is 7.193 K. Hence Equation 12.9 gives

$$B_c(4.2 \text{ K}) = (0.0803 \text{ T}) \left[1 - \left(\frac{4.20}{7.193} \right)^2 \right] = 0.0529 \text{ T}$$

(b) According to Ampère's law, if a wire carries a steady current I, the magnetic field generated at an exterior point a distance r from the wire is

$$B = \frac{\mu_0 I}{2\pi r}$$

When the current in the wire equals a certain critical cur-

rent I_c, the magnetic field at the wire surface equals the critical magnetic field B_c. (Note that $B = 0$ inside, because all the current is on the wire surface.) Using the preceding expression and taking r equal to the radius of the wire, we find

$$I = \frac{2\pi r B}{\mu_0} = 2\pi \frac{(3.00 \times 10^{-3} \text{ m}) (0.0529 \text{ T})}{4\pi \times 10^{-7} \text{ N/A}^2} = 794 \text{ A}$$

Penetration Depth

As we have seen, magnetic fields are expelled from the interior of a type I superconductor by the formation of surface currents. In reality, these currents are not formed in an infinitesimally thin layer on the surface. Instead, they penetrate the surface to a small extent. Within this thin layer, which is about 100 nm thick, the magnetic field **B** decreases exponentially from its external value to zero, according to the expression

$$B(x) = B_0 e^{-x/\lambda} \tag{12.11}$$

where it is assumed that the external magnetic field is parallel to the surface of the sample. In this equation, B_0 is the value of the magnetic field at the surface, x is the distance from the surface to some interior point, and λ is a parameter called the **penetration depth.** The variation of magnetic field with distance inside a type I superconductor is plotted in Figure 12.12. The superconductor occupies the region on the positive side of the x axis. As you can see, the magnetic field becomes very small at depths a few times λ below the surface. Values for λ are typically in the range 10 to 100 nm.

Penetration depth varies with temperature according to the empirical expression

$$\lambda(T) = \lambda_0 \left[1 - \left(\frac{T}{T_c} \right)^2 \right]^{-1/2} \tag{12.12}$$

where λ_0 is the penetration depth at 0 K. From this expression we see that λ becomes infinite as T approaches T_c. Furthermore, as T approaches T_c, while the sample is in the superconducting state, an applied magnetic field penetrates more and more deeply into the sample. Ultimately, the field penetrates the entire sample (λ becomes infinite), and the sample becomes normal.

Magnetization

When a bulk sample is placed in an external magnetic field $\mathbf{B}_0$, the sample acquires a magnetization **M** (see Section 12.1). The magnetic field **B** inside the sample is related to $\mathbf{B}_0$ and **M** through the relationship $\mathbf{B} = \mathbf{B}_0 + \mu_0 \mathbf{M}$. When the sample is in the superconducting state, $\mathbf{B} = 0$; therefore it follows that the magnetization is

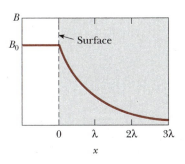

Figure 12.12 The magnetic field **B** inside a superconductor versus distance x from the surface of the superconductor. The field outside the superconductor (for $x < 0$) is B_0, and the superconductor is to the right of the dashed line.

$$\mathbf{M} = -\frac{\mathbf{B}_0}{\mu_0} = \chi\mathbf{H} \tag{12.13}$$

where χ $(=-1)$ is the magnetic susceptibility. That is, whenever a material is in a superconducting state, its magnetization opposes the external magnetic field and the magnetic susceptibility has its maximum negative value. Again we see that a type I superconductor is a perfect diamagnetic substance.

Figure 12.13a is a plot of magnetic field inside a type I superconductor versus external field (parallel to a long cylinder) at $T < T_c$. (Long cylinders are used to minimize end effects.) The magnetization versus external field at some constant temperature is plotted in Figure 12.13b. Note that when $B_0 > B_c$, the magnetization is approximately zero.

With the discovery of the Meissner effect, Fritz and Heinz London were able to develop phenomenological equations for type I superconductors based on equilibrium thermodynamics. They could explain the critical magnetic field in terms of the energy increase of the superconducting state, an increase resulting from the exclusion of flux from the interior of the superconductor. According to equilibrium thermodynamics, a system prefers to be in the state having the lowest free energy. Hence, the superconducting state must have a lower free energy than the normal state. If E_s represents the energy of the superconducting state per unit volume and E_n the energy of the normal state per unit volume, then $E_s < E_n$ below T_c and the material becomes superconducting. The exclusion of a field B causes the total energy of the superconducting state to increase by an amount equal to $B^2/2\mu_0$ per unit volume. The critical field value is defined by the equation

$$E_s + \frac{B_c^2}{2\mu_0} = E_n \tag{12.14}$$

Because the London theory also gives the temperature dependence of E_s, an exact expression for $B_c(T)$ could be obtained. Note that the field exclusion energy $B_c^2/2\mu_0$ is just the area under the curve in Figure 12.13b.

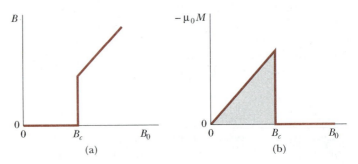

Figure 12.13   The magnetic field–dependent properties of a type I superconductor. (a) A plot of internal field versus applied field, where $\mathbf{B} = 0$ for $B_0 < B_c$. (b) A plot of magnetization versus applied field. Note that $\mathbf{M} \approx 0$ for $B_0 > B_c$.

12.4   TYPE II SUPERCONDUCTORS

By the 1950s researchers knew there was another class of superconductors, which they called type II superconductors. These materials are characterized by two critical magnetic fields, designated B_{c1} and B_{c2} in Figure 12.14. When the external magnetic field is less than the lower critical field B_{c1}, the material is entirely superconducting and there is no flux penetration, just as with type I superconductors. When the external field exceeds the upper critical field B_{c2}, the flux penetrates completely and the superconducting state is destroyed, just as for type I materials. For fields lying between B_{c1} and B_{c2}, however, the material is in a mixed state, referred to as the **vortex state.** (This name is given because of swirls of currents that are associated with this state.) While in the vortex state, the material can have zero resistance and has partial flux penetration. Vortex regions are essentially filaments of normal material that run through the sample when the external field exceeds the lower critical field, as illustrated in Figure 12.15. As the strength of the external field increases, the number of filaments increases until the field reaches the upper critical value, and the sample becomes normal.

One can view the vortex state as a cylindrical swirl of supercurrents surrounding a cylindrical normal-metal core that allows some flux to penetrate the interior of the type II superconductor. Associated with each vortex filament is a magnetic field that is greatest at the core center and falls off exponentially outside the core with the characteristic penetration depth λ. The supercurrents are the "source" of **B** for each vortex. In type II superconductors, the radius of the normal-metal core is smaller than the penetration depth.

Table 12.5 gives critical temperatures and B_{c2} values for several type II superconductors. The values of B_{c2} are very large in comparison with those of B_c for type I superconductors. For this reason, type II superconductors are well suited for the construction of high-field superconducting magnets. For example, using the alloy NbTi, superconducting solenoids may be wound to produce magnetic fields in the range 5 to 10 T. Furthermore, they require no power to maintain the field. Iron-core electromagnets rarely exceed 2 T and consume

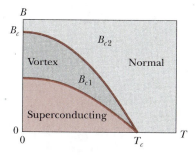

Figure 12.14 Critical magnetic fields as a function of temperature for a type II superconductor. Below B_{c1}, the material behaves as a type I superconductor. Above B_{c2}, the material behaves as a normal conductor. Between these two fields, the superconductor is in the vortex (mixed) state.

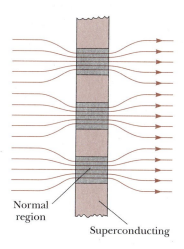

Figure 12.15 A schematic diagram of a type II superconductor in the vortex state. The sample contains filaments of normal (unshaded) regions through which magnetic field lines can pass. The field lines are excluded from the superconducting (shaded) regions.

Table 12.5	**Critical Temperatures and Upper Critical Magnetic Fields (at $T = 0$ K) of Some Type II Superconductors**	
Superconductor	**T_c (K)**	**$B_{c2}(0)$ (T)**
Nb_3Al	18.7	32.4
Nb_3Sn	18.0	24.5
Nb_3Ge	23	38
NbN	15.7	15.3
NbTi	9.3	15
$Nb_3(AlGe)$	21	44
V_3Si	16.9	23.5
V_3Ga	14.8	20.8
PbMoS	14.4	60

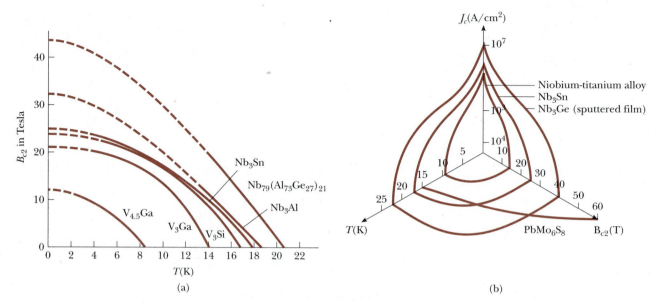

Figure 12.16   (a) Upper critical field, B_{c2}, as a function of temperature for several type II superconductors. *(From S. Foner, et al., Physics Letters 31A:349, 1970)* (b) A three-dimensional plot showing the variation of critical current density, J_c, with temperature, and the variation of the upper critical field with temperature for several type II superconductors.

power to maintain the field. Notice also that type II superconductors are compounds formed from elements of the transition and actinide series. Plots of B_{c2} variations with temperature appear in Figure 12.16a. The three-dimensional plot in Figure 12.16b shows the variation of critical temperature with both B_{c2} and critical current density, J_c.

Figure 12.17a shows internal magnetic field versus external field for a type II superconductor, while Figure 12.17b shows the corresponding magnetization versus external field.

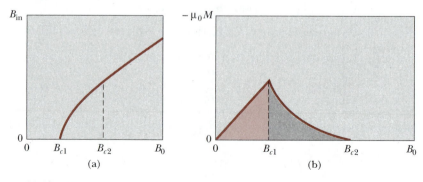

Figure 12.17   The magnetic behavior of a type II superconductor. (a) A plot of internal field versus external field. (b) A plot of magnetization versus external field.

When a type II superconductor is in the vortex state, sufficiently large currents can cause the vortices to move perpendicular to the current. This vortex motion corresponds to a change in flux with time and produces resistance in the material. By adding impurities or other special inclusions, one can effectively pin the vortices and prevent their motion, to produce zero resistance in the vortex state. The critical current for type II superconductors is the current that, when multiplied by the flux in the vortices, gives a Lorentz force that overcomes the pinning force.

EXAMPLE 12.4 A Type II Superconducting Solenoid

A solenoid is to be constructed from wire made of the alloy Nb_3Al, which has an upper critical field of 32.0 T at $T = 0$ K and a critical temperature of 18.0 K. The wire has a radius of 1.00 mm, the solenoid is to be wound on a hollow cylinder of diameter 8.00 cm and length 90.0 cm, and there are to be 150 turns of wire per centimeter of length. (a) How much current is required to produce a magnetic field of 5.00 T at the center of the solenoid?

Solution The magnetic field at the center of a tightly wound solenoid is $B = \mu_0 n I$, where n is the number of turns per unit length along the solenoid, and I is the current in the solenoid wire. Taking $n = 150$ turns/cm $= 1.50 \times 10^4$ turns/m, and $B = 5.00$ T, we

find

$$I = \frac{B}{\mu_0 n} = \frac{5.00 \text{ T}}{(4\pi \times 10^{-7} \text{ N/A}^2)(1.50 \times 10^4 \text{ m}^{-1})} = 265 \text{ A}$$

(b) What maximum current can the solenoid carry if its temperature is to be maintained at 15.0 K and it is to remain superconducting? (Note that B near the solenoid windings is approximately equal to B on its axis.)

Solution Using Equation 12.9, with $B_c(0) = 32.0$ T, we find $B_c = 9.78$ T at a temperature of 15.0 K. For this value of B, we find $I_{max} = 518$ A.

12.5 OTHER PROPERTIES OF SUPERCONDUCTORS

Persistent Currents

Because the dc resistance of a superconductor is zero below the critical temperature, once a current is set up in the material, it persists *without any applied voltage* (which follows from Ohm's law and the fact that $R = 0$). These **persistent currents,** sometimes called supercurrents, have been observed to last for several years with no measurable losses. In one experiment conducted by S. S. Collins in Great Britain, a current was maintained in a superconducting ring for 2.5 years, stopping only because a trucking strike delayed delivery of the liquid helium that was necessary to maintain the ring below its critical temperature.[9]

To better understand the origin of persistent currents, consider a loop of wire made of a superconducting material. Suppose the loop is placed, in its normal state ($T > T_c$), in an external magnetic field, and then the temperature is lowered below T_c so that the wire becomes superconducting, as in Figure 12.18a. As with a cylinder, the flux is excluded from the interior of the wire because of the induced surface currents. However, note that *flux lines still pass through the hole in the loop*. When the external field is turned off, as in Figure 12.18b, *the flux through this hole is trapped because the magnetic flux through the loop*

---

[9] This charming story was provided by Steve Van Wyk.

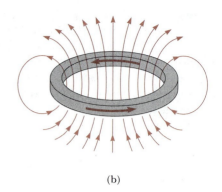

(a)   (b)

Figure 12.18   (a) When a superconducting loop at $T > T_c$ is placed in an external magnetic field and the temperature is then lowered to $T < T_c$, flux passes through the hole in the loop even though it does not penetrate the interior of the material forming the loop. (b) After the external field is removed, the flux through the hole remains trapped, and an induced current appears in the material forming the loop.

cannot change.[10] The superconducting wire prevents the flux from going to zero through the advent of a large spontaneous current induced by the collapsing external magnetic field. If the dc resistance of the superconducting wire is truly zero, this current should persist forever. Experimental results using a technique known as nuclear magnetic resonance indicate that such currents will persist for more than 10^5 years! The resistivity of a superconductor based on such measurements has been shown to be less than 10^{-26} $\Omega \cdot$ m. This reaffirms the fact that R is zero for a superconductor. (See Problem 43 for a simple but convincing demonstration of zero resistance.)

Now consider what happens if the loop is cooled to a temperature $T < T_c$ before the external field is turned on. When the field is turned on while the loop is maintained at this temperature, *flux must be excluded from the entire loop, including the hole,* because the loop is in the superconducting state. Again, a current is induced in the loop to maintain zero flux through the loop and through the interior of the wire. In this case, the current disappears when the external field is turned off.

Coherence Length

Another important parameter associated with superconductivity is the **coherence length,** ξ. The coherence length is the smallest dimension over which superconductivity can be established or destroyed. Table 12.6 lists typical values of the penetration depth, λ, and ξ at 0 K for selected superconductors.

A superconductor is type I if $\xi < \lambda$; most of the pure metals that are superconductors fall into this category. An increase in the ratio λ / ξ favors type II superconductivity. A detailed analysis shows that coherence length and penetration depth both depend on the mean free path of the electrons in the normal state. The mean free path of a metal can be reduced by the addition of impurities to the metal, which causes the penetration depth to increase while co-

---

[10] Alternatively, one can apply Equation 12.12, taking the line integral of the **E** field over the loop. Since $\rho = 0$ everywhere along a path on the superconductor, **E** = 0, the integral is zero, and $d\Phi_B / dt = 0$.

Table 12.6   Penetration Depths and Coherence Lengths of Selected Superconductors at $T = 0$ K[a]

Superconductor	λ (nm)	ξ (nm)
Al	16	160
Cd	110	760
Pb	37	83
Nb	39	38
Sn	34	23

[a] These are calculated values from C. Kittel, *Introduction to Solid State Physics*, New York, John Wiley, 1986.

herence length decreases. Thus one can cause a metal to change from type I to type II by introducing an alloying element. For example, pure lead is a type I superconductor but changes to type II (with almost no change in T_c) when alloyed with 2% indium (by weight).

Flux Quantization

The phenomenon of flux exclusion by a superconductor applies only to a simply connected object—that is, one with no holes or their topological equivalent. However, when a superconducting ring is placed in a magnetic field and the field is removed, flux lines are trapped and are maintained by a persistent circulating current, as shown in Figure 12.18b. Realizing that superconductivity is fundamentally a quantum phenomenon, Fritz London suggested that the trapped magnetic flux should be quantized in units of h/e.[11] (The electronic charge e in the denominator arises because London assumed that the persistent current is carried by single electrons.) Subsequent delicate measurements on very small superconducting hollow cylinders showed that the flux quantum is one-half the value postulated by London.[12] That is, the magnetic flux Φ is quantized not in units of h/e but in units of $h/2e$:

$$\Phi = \frac{nh}{2e} = n\Phi_0 \qquad (12.15)$$

where n is an integer and

$$\Phi_0 = \frac{h}{2e} = 2.0679 \times 10^{-15} \text{ T} \cdot \text{m}^2 \qquad (12.16)$$

Magnetic flux quantum

is the **magnetic flux quantum.**

[11] F. London, *Superfluids*, vol. I, New York, John Wiley, 1954.

[12] The effect was discovered by B. S. Deaver, Jr., and W. M. Fairbank, *Phys. Rev. Letters* 7:43, 1961, and independently by R. Doll and M. Nabauer, *Phys. Rev. Letters* 7:51, 1961.

12.6 ELECTRONIC SPECIFIC HEAT

The thermal properties of superconductors have been extensively studied and compared with those of the same materials in the normal state, and one very important measurement is specific heat. When a small amount of thermal energy is added to a normal metal, some of the energy is used to excite lattice vibrations, and the remainder is used to increase the kinetic energy of the conduction electrons. The electronic specific heat C is defined as the ratio of the thermal energy absorbed by the electrons to the increase in temperature of the system.

Figure 12.19 shows how the electronic specific heat varies with temperature for both the normal state and the superconducting state of gallium, a type I superconductor. At low temperatures, the electronic specific heat of the material in the normal state, C_n, varies with temperature as AT, as explained in Chapter 11. The electronic specific heat of the material in the superconducting state, C_s, is substantially altered below the critical temperature. As the temperature is lowered starting from $T > T_c$, the specific heat first jumps to a very high value at T_c and then falls below the value for the normal state at very low temperatures. Analyses of such data show that at temperatures well below T_c, the electronic part of the specific heat is dominated by a term that varies as $\exp(-\Delta/kT)$. A result of this form suggests the existence of an energy gap in the energy levels available to the electrons. We shall see that the energy gap is a measure of the thermal energy necessary to move electrons from a set of ground states (superconducting) to a set of excited states (normal), and that the energy gap is actually 2Δ.

Figure 12.19 Electronic specific heat versus temperature for superconducting gallium (in zero applied magnetic field) and normal gallium (in a 0.020-T magnetic field). For the superconducting state, note the discontinuity that occurs at T_c and the exponential dependence on $1/T$ at low temperatures. *(Taken from N. Phillips, Phys. Rev. 134: 385, 1964)*

12.7   BCS THEORY

According to classical physics, part of the resistivity of a metal is due to collisions between free electrons and thermally displaced ions of the metal lattice, and part is due to scattering of electrons from impurities or defects in the metal. Soon after the discovery of superconductivity, scientists recognized that this classical model could never explain the superconducting state, because the electrons in a material always suffer some collisions, and therefore resistivity can never be zero. Nor could superconductivity be understood through a simple microscopic quantum mechanical model, where one views an individual electron as an independent wave traveling through the material. Although many phenomenological theories based on the known properties of superconductors were proposed, none could explain why electrons enter the superconducting state and why electrons in this state are not scattered by impurities and lattice vibrations.

Several important developments in the 1950s led to better understanding of superconductivity. In particular, many research groups reported that the critical temperatures of isotopes of a metal decreased with increasing atomic mass. This observation, called the **isotope effect,** was early evidence that lattice motion played an important role in superconductivity. For example, in the case of mercury, $T_c = 4.161$ K for the isotope ^{199}Hg, 4.153 K for ^{200}Hg, and 4.126 K for ^{204}Hg. The characteristic frequencies of the lattice vibrations are expected to change with the mass M of the vibrating atoms. In fact, the lattice vibrational frequencies are expected to be proportional to $M^{-1/2}$ [analogous to the angular frequency ω of a mass-spring system, where $\omega = (k/M)^{1/2}$]. On this basis, it became apparent that any theory of superconductivity for metals must include electron-lattice interactions, which is somewhat surprising because electron-lattice interactions increase the resistance of normal metals.

The full microscopic theory of superconductivity presented in 1957 by Bardeen, Cooper, and Schrieffer has had good success in explaining the features of superconductors. The details of this theory, now known as the BCS theory, are beyond the scope of this text, but we can describe some of its main features and predictions.

The central feature of the BCS theory is that two electrons in the superconductor are able to form a bound pair called a **Cooper pair** if they somehow experience an attractive interaction. This notion at first seems counterintuitive since electrons normally repel one another because of their like charges. However, a net attraction can be achieved if the electrons interact with each other via the motion of the crystal lattice as the lattice structure is momentarily deformed by a passing electron.[13] To illustrate this point, Figure 12.20 shows two electrons moving through the lattice. The passage of electron 1 causes nearby ions to move inward toward the electron, resulting in a slight increase in the

---

[13] For a lively description of this process, see D. Teplitz, ed., *Electromagnetism: Path to Research*, New York, Plenum Press, 1982. In particular, see Chapter 1, "Electromagnetic Properties of Superconductors," by Brian B. Schwartz and Sonia Frota-Pessoa. Note that the electron that causes the lattice to deform remains in that region for a very short time, $\approx 10^{-16}$ s, compared to the much longer time it takes the lattice to deform, $\approx 10^{-13}$ s. Thus the sluggish ions continue to move inward for a time interval about 1000 times longer than the response time of the electron, so the region is effectively positively charged between 10^{-16} s and 10^{-13} s.

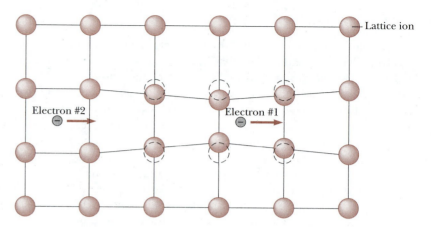

Figure 12.20   The basis for the attractive interaction between two electrons via the lattice deformation. Electron 1 attracts the positive ions, which move inward from their equilibrium positions (dashed circles). This distorted region of the lattice has a net positive charge, and hence electron 2 is attracted to it.

concentration of positive charge in this region. Electron 2 (the second electron of the Cooper pair), approaching before the ions have had a chance to return to their equilibrium positions, is attracted to the distorted (positively charged) region. The net effect is a weak delayed attractive force between the two electrons, resulting from the motion of the positive ions. As one researcher has beautifully put it, "the following electron surfs on the virtual lattice wake of the leading electron."[14] In more technical terms, one can say that the attractive force between two Cooper electrons is an *electron-lattice-electron interaction*, where the crystal lattice serves as the mediator of the attractive force. Some scientists refer to this as a *phonon-mediated mechanism*, because quantized lattice vibrations are called *phonons*.

A Cooper pair in a superconductor consists of two electrons having opposite momenta and spin, as described schematically in Figure 12.21. In the superconducting state, the linear momenta can be equal and opposite, corresponding to no net current, or slightly different and opposite, corresponding to a

Figure 12.21   A schematic diagram of a Cooper pair. The electron moving to the right has a momentum **p** and its spin is up, while the electron moving to the left has a momentum −**p** and its spin is down. Hence the total momentum of the system is zero and the total spin is zero.

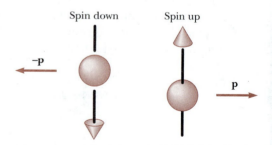

Spin down   Spin up

−p   p

[14] Many authors choose to refer to this cooperative state of affairs as a **collective state.** As an analogy, one author wrote that the electrons in the paired state "move like mountain-climbers tied together by a rope: should one of them leave the ranks due to the irregularities of the terrain (caused by the thermal vibrations of the lattice atoms) his neighbors would pull him back."

net superconducting current. Because Cooper pairs have zero spin, they can all be in the same state. This is in contrast with electrons, which are fermions (spin $\frac{1}{2}$) that must obey the Pauli exclusion principle. In the BCS theory, a ground state is constructed in which *all electrons form bound pairs.* In effect, all Cooper pairs are "locked" into the *same quantum state.* One can view this state of affairs as a condensation of all electrons into the same state. Also note that, because the Cooper pairs have zero spin (and hence zero angular momentum), their wavefunctions are spherically symmetric (like the *s*-states of the hydrogen atom.) In a "semiclassical" sense, the electrons are always undergoing head-on collisions and as such are always moving in each other's wakes. Because the two electrons are in a bound state, their trajectories always change directions in order to keep their separation within the coherence length.

The BCS theory has been very successful in explaining the characteristic superconducting properties of zero resistance and flux expulsion. From a qualitative point of view, one can say that in order to reduce the momentum of any single Cooper pair by scattering, it is necessary to simultaneously reduce the momenta of all the other pairs — in other words, it is an all-or-nothing situation. One cannot change the velocity of one Cooper pair without changing those of all of them.[15] Lattice imperfections and lattice vibrations, which effectively scatter electrons in normal metals, have no effect on Cooper pairs! In the absence of scattering, the resistivity is zero and the current persists forever. It is rather strange, and perhaps amazing, that the mechanism of lattice vibrations that is responsible (in part) for the resistivity of normal metals also provides the interaction that gives rise to their superconductivity. Thus, copper, silver, and gold, which exhibit small lattice scattering at room temperature, are not superconductors, whereas lead, tin, mercury, and other modest conductors have strong lattice scattering at room temperature and become superconductors at low temperatures.

As we mentioned earlier, the superconducting state is one in which the Cooper pairs act collectively rather than independently. The condensation of all pairs into the same quantum state makes the system behave as a giant quantum mechanical system or macromolecule that is quantized on the macroscopic level. *The condensed state of the Cooper pairs is represented by a single coherent wavefunction Ψ that extends over the entire volume of the superconductor.*

The stability of the superconducting state is critically dependent on strong correlation between Cooper pairs. In fact, the theory explains superconducting behavior in terms of the energy levels of a kind of "macromolecule" and the existence of an energy gap E_g between the ground and excited states of the system, as in Figure 12.22a. Note that in Figure 12.22b there is no energy gap for a normal conductor. In a normal conductor, the Fermi energy E_F represents the largest kinetic energy the free electrons can have at 0 K.

The energy gap in a superconductor is very small, of the order of $k_B T_c$ ($\approx 10^{-3}$ eV) at 0 K, as compared with the energy gap in semiconductors (≈ 1 eV) or the Fermi energy of a metal (≈ 5 eV). The energy gap represents the energy needed to break apart a Cooper pair. The BCS theory predicts that

---

[15] A Cooper pair is somewhat analogous to a helium atom, $^{4}_{2}\text{He}$, in that both are bosons with zero spin. It is well known that the superfluidity of liquid helium may be viewed as a condensation of bosons in the ground state. Likewise, superconductivity may be viewed as a superfluid state of Cooper pairs, all in the same quantum state.

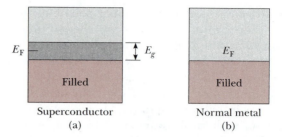

Figure 12.22   (a) A simplified energy-band structure for a superconductor. Note the energy gap between the lower filled states and the upper empty states. (b) The energy-band structure for a normal conductor has no energy gap. At $T = 0$ K, all states below the Fermi energy E_F are filled, and all states above it are empty.

at $T = 0$ K,

$$E_g = 3.53 \, k_B T_c \tag{12.17}$$

Thus superconductors that have large energy gaps have relatively high critical temperatures. The exponential dependence of the electronic heat capacity discussed in the preceding section, $\exp(-\Delta/k_B T)$, contains an experimental factor, $\Delta = E_g/2$, that may be used to determine the value of E_g. Furthermore, the energy-gap values predicted by Equation 12.17 are in good agreement with the experimental values in Table 12.7. (The tunneling experiment used to obtain these values is described later.) As we noted earlier, the electronic heat capacity in zero magnetic field undergoes a discontinuity at the critical temperature. Furthermore, at finite temperatures, thermally excited individual electrons interact with the Cooper pairs and reduce the energy gap continuously from a peak value at 0 K to zero at the critical temperature, as shown in Figure 12.23 for several superconductors.

Because the two electrons of a Cooper pair have opposite spin angular momenta, an external magnetic field raises the energy of one electron and lowers

Table 12.7   Energy Gaps for Some Superconductors at $T = 0$ K

Superconductor	E_g (meV)
Al	0.34
Ga	0.33
Hg	1.65
In	1.05
Pb	2.73
Sn	1.15
Ta	1.4
Zn	0.24
La	1.9
Nb	3.05

Figure 12.23   The points on this graph represent reduced values of the observed energy gap $E_g(T)/E_g(0)$ as a function of the reduced temperature T/T_c for tin, tantalum, and niobium. The solid curve gives the values predicted by the BCS theory. (*Data are from electron tunneling measurements by P. Townsend and J. Sutton*, Phys. Ref. *28*:591, 1962)

the energy of the other (Problem 29). If the magnetic field is made strong enough, it becomes energetically favorable for the pair to break up into a state where both spins point in the same direction and the superconducting state is destroyed. The value of the external field that causes the breakup corresponds to the critical field.

EXAMPLE 12.5   The Energy Gap for Lead

Use Equation 12.17 to calculate the energy gap for lead, and compare the answer with the experimental value in Table 12.7.

Solution   Because $T_c = 7.193$ K for lead, Equation 12.17 gives

$$E_g = 3.53k_B T_c = (3.53)(1.38 \times 10^{-23} \text{ J/K})(7.193 \text{ K})$$
$$= 3.50 \times 10^{-22} \text{ J} = 0.00219 \text{ eV} = 2.19 \times 10^{-3} \text{ eV}$$

The experimental value is 2.73×10^{-3} eV, corresponding to a difference of about 20%.

12.8   ENERGY GAP MEASUREMENTS

Single-Particle Tunneling

The energy gaps in superconductors can be measured very precisely in single-particle tunneling experiments (those involving normal electrons), first reported by Giaever in 1960.[16] As described in Chapter 6, tunneling is a phenom-

---

[16] I. Giaever, *Phys. Rev Letters* 5:147, 464, 1960.

enon in quantum mechanics that enables a particle to penetrate and go through a barrier even though classically it has insufficient energy to go over the barrier. If two metals are separated by an insulator, the insulator normally acts as a barrier to the motion of electrons between the two metals. However, if the insulator is made sufficiently thin (less than about 2 nm), there is a small probability that electrons will tunnel from one metal to the other.

First consider two normal metals separated by a thin insulating barrier, as in Figure 12.24a. If a potential difference V is applied between the two metals, electrons can pass from one metal to the other, and a current is set up. For small applied voltages, the current-voltage relationship is linear (the junction obeys Ohm's law). However, if one of the metals is replaced by a superconductor maintained at a temperature below T_c, as in Figure 12.24b, something quite unusual occurs. As V increases, no current is observed until V reaches a threshold value that satisfies the relationship $V_t = E_g/2e = \Delta/e$, where Δ is half the energy gap. (The voltage source provides the energy required to break a Cooper pair and free an electron to tunnel. The factor of one half comes from the fact that we are dealing with single-particle tunneling, and the energy required is one-half the binding energy of a pair, 2Δ.) That is, if $eV \geqslant 0.5E_g$, then tunneling can occur between the normal metal and the superconductor.

Thus, single-particle tunneling provides a direct experimental measurement of the energy gap. The value of Δ obtained from such experiments is in good

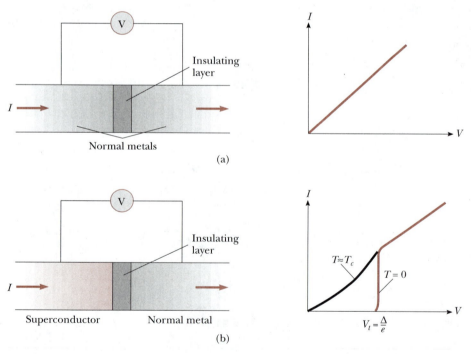

Figure 12.24   (a) The current-voltage relationship for electron tunneling through a thin insulator between two normal metals. The relationship is linear for small currents and voltages. (b) The current-voltage relationship for electron tunneling through a thin insulator between a superconductor and a normal metal. The relationship is nonlinear and strongly temperature-dependent. (*Adapted from N. W. Ashcroft and N. D. Mermin,* Solid State Physics, *Philadelphia, Saunders College Publishing, 1975*)

agreement with the results of electronic heat capacity measurements. The *I-V* curve in Figure 12.24b shows the nonlinear relationship for this junction. Note that as the temperature increases toward T_c, some tunneling current occurs at voltages smaller than the energy-gap threshold voltage. This is due to a combination of thermally excited electrons and a decrease in the energy gap.

Absorption of Electromagnetic Radiation

Another experiment used to measure the energy gaps of superconductors is the absorption of electromagnetic radiation. In superconductors, photons can be absorbed by the material when their energy is greater than the gap energy. Electrons in the valence band of the semiconductor absorb incident photons, exciting the electrons across the gap into the conduction band. In a similar manner, superconductors absorb photons if the photon energy exceeds the gap energy, 2Δ. If the photon energy is less than 2Δ, no absorption occurs. When photons are absorbed by the superconductor, Cooper pairs are broken apart. Photon absorption in superconductors typically occurs in the range between microwave and infrared frequencies, as the following example shows.

EXAMPLE 12.6   Absorption of Radiation by Lead

Find the minimum frequency of a photon that can be absorbed by lead at $T = 0$ K to break apart a Cooper pair.

Solution   From Table 12.7 we see that the energy gap for lead is 2.73×10^{-3} eV. Setting this equal to the photon energy hf, and noting that 1 eV $= 1.60 \times 10^{-19}$ J, we find

$$hf = 2\Delta = 2.73 \times 10^{-3} \text{ eV} = 4.37 \times 10^{-22} \text{ J}$$

$$f = \frac{4.37 \times 10^{-22} \text{ J}}{6.626 \times 10^{-34} \text{ J} \cdot \text{s}} = 6.60 \times 10^{11} \text{ Hz}$$

Exercise   What is the maximum wavelength of radiation that can be absorbed by lead at 0 K?

Answer $\lambda = c/f = 0.455$ mm, which is in the far infrared region.

12.9 JOSEPHSON TUNNELING

In the preceding section we described single-particle tunneling from a normal metal through a thin insulating barrier into a superconductor. Now we consider tunneling between two superconductors separated by a thin insulator. In 1962 Brian Josephson proposed that, in addition to single particles, Cooper pairs can tunnel through such a junction. Josephson predicted that pair tunneling can occur without any resistance, producing a direct current when the applied voltage is zero and an alternating current when a dc voltage is applied across the junction.

At first, physicists were very skeptical about Josephson's proposal because it was believed that single-particle tunneling would mask pair tunneling. However, when the phase coherence of the pairs is taken into account, one finds that, under the appropriate conditions, the probability of tunneling of pairs across the junction is comparable to that of single-particle tunneling. In fact, when the insulating barrier separating the two superconductors is made sufficiently thin (say, ≈ 1 nm) Josephson tunneling is as easy to observe as single-particle tunneling.

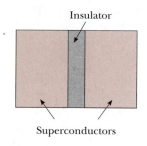

Insulator

Superconductors

Figure 12.25 A Josephson junction consists of two superconductors separated by a very thin insulator. Cooper pairs can tunnel through this insulating barrier in the absence of an applied voltage, setting up a direct current.

In the remainder of this section, we describe three remarkable effects associated with pair tunneling.

The dc Josephson Effect

Consider two superconductors separated by a thin oxide layer, typically 1 to 2 nm thick, as in Figure 12.25. Such a configuration is known as a Josephson junction. In a given superconductor, the pairs could be represented by a wavefunction $\Psi = \Psi_0 e^{i\phi}$, where ϕ is the phase and is the same for every pair. If the superconductor to the left of the insulating layer has a phase ϕ_1 and that to the right has a phase ϕ_2, Josephson showed that at zero voltage there appears across the junction a supercurrent satisfying the relationship

$$I_s = I_{max} \sin(\phi_2 - \phi_1) = I_{max} \sin \delta \qquad (12.18)$$

where I_{max} is the maximum current across the junction under zero-voltage conditions. The value of I_{max} depends on the surface area of each superconductor-oxide interface and decreases exponentially with the thickness of the oxide layer.

Rowell and Anderson made the first confirmation of the dc Josephson effect in 1963. Since then, all of Josephson's other theoretical predictions have been verified.

The ac Josephson Effect

When a dc voltage V is applied across the Josephson junction, a most remarkable effect occurs: *the dc voltage generates an alternating current I*, given by

$$I = I_{max} \sin(\delta + 2\pi ft) \qquad (12.19)$$

where δ is a constant equal to the phase at $t = 0$ and f is the frequency of the Josephson current:

$$f = \frac{2eV}{h} \qquad (12.20)$$

A dc voltage of 1 μV results in a current frequency of 483.6 MHz. Precise measurements of the frequency and voltage have enabled physicists to obtain the ratio e/h to unprecedented precision.

The ac Josephson effect can be demonstrated in various ways. One method is to apply a dc voltage and detect the electromagnetic radiation generated by the junction. Another method is to irradiate the junction with external radiation of frequency f'. With this method, a graph of direct current versus voltage has steps when the voltages correspond to Josephson frequencies f that are integral multiples of the external frequency f'—that is, when $V = hf/2e = nhf'/2e$ (Fig. 12.26). Because the two sides of the junction are in different quantum states, the junction behaves as an atom undergoing transitions between these states as it absorbs or emits radiation. In effect, when a Cooper pair crosses the junction, a photon of frequency $f = 2eV/h$ is either emitted or absorbed by the system.

Quantum Interference

Quantum interference is the behavior of the direct-tunneling current in the presence of an external magnetic field. When a Josephson junction is subjected to a magnetic field, the maximum critical current in the junction depends on

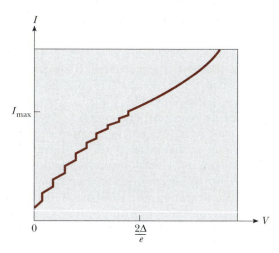

Figure 12.26   A plot of dc current as a function of bias voltage for a Josephson junction placed in an electromagnetic field. At a frequency of 10 GHz, as many as 500 steps have been observed. (In the presence of an applied electromagnetic field, there is no clear onset of single-particle tunneling.)

the magnetic flux through the junction. The tunneling current under these conditions is predicted to be periodic in the number of flux quanta through the junction.

Figure 12.27a shows the maximum tunneling current as a function of magnetic field in an Sn-SnO-Sn junction at 1.90 K. Note that the current depends

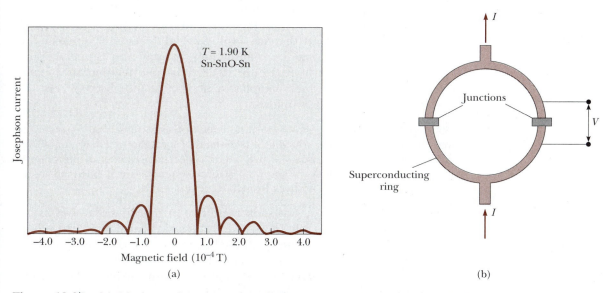

(a)   (b)

Figure 12.27   (a) Maximum Josephson tunneling current as a function of applied magnetic field in an Sn-SnO-Sn junction. The zero current points are directly related to the number of flux quanta through the junction. (*From R. D. Parks, ed.,* Superconductivity, *vol. 1, New York, Dekker, 1969*) (b) A schematic diagram of a SQUID constructed from two Josephson junctions in parallel.

periodically on the magnetic flux. For typical junctions, the field periodicity is about 10^{-4} T. This is not surprising in view of the quantum nature of the magnetic flux.

If a superconducting circuit is constructed with two Josephson junctions in parallel, such as in the ring in Figure 12.27b, one can observe interference effects similar to the interference of light waves in Young's double-slit experiment. In this case, the total current depends periodically on the flux inside the ring. Because the ring can have an area much greater than that of a single junction, the magnetic field sensitivity is greatly increased.

A device that contains one or more Josephson junctions in a loop is called a SQUID, an acronym for *su*perconducting *qu*antum *i*nterference *d*evice. SQUIDs are very useful for detecting very weak magnetic fields, of the order of 10^{-14} T, which is a very small fraction of the Earth's field ($B_{Earth} \approx 0.5 \times 10^{-4}$ T). Commercially available SQUIDs are capable of detecting a change in flux of about $10^{-5}\ \Phi_0 \approx 10^{-20}$ T·m² in a bandwidth of 1 Hz (where Φ_0 is the magnetic flux quantum given by Eq. 12.16). For example, they are being used to scan "brain waves," a medical term for fields generated by current-carrying neurons. The essay by Clark Hamilton following this chapter contains a more detailed discussion of SQUIDs and other superconducting devices.

12.10   HIGH-TEMPERATURE SUPERCONDUCTIVITY

For many years, scientists have searched for materials that are superconductors at higher temperatures, and until 1986 the alloy Nb₃Ge had the highest known critical temperature, 23.2 K. Theorists made various predictions that the maximum critical temperature for a superconductor in which the electron-lattice interaction was important would be in the neighborhood of 30 K. Early in 1986 J. Georg Bednorz and Karl Alex Müller of IBM Research Laboratory in Zurich made a remarkable discovery that has revolutionized the field of superconductivity. They found that an oxide of lanthanum, barium, and copper became superconducting at about 30 K. Figure 12.28 shows the temperature dependence of the resistivity of their samples. A portion of the abstract from this unpretentious paper reads as follows:

> Upon cooling, the samples show a linear decrease in resistivity, then an approximately logarithmic increase, interpreted as a beginning of localization. Finally, an abrupt decrease by up to three orders of magnitude occurs, reminiscent of the onset of percolative superconductivity. The highest onset temperature is observed in the 30 K range. It is markedly reduced by high current densities.[8]

The superconducting phase was soon identified at other laboratories as the compound $La_{2-x}Ba_xCuO_4$, where $x \approx 0.2$. By replacing barium with strontium, researchers soon raised the value of T_c to about 36 K. Inspired by these developments, scientists worldwide worked feverishly to discover materials with even higher T_c values, and research in the superconducting behavior of metallic oxides accelerated at a tremendous pace. The year 1986 marked the beginning of a new era of high-temperature superconductivity. Bednorz and Müller were awarded the Nobel prize in 1987 (the fastest-ever recognition by the Nobel committee) for their very important discovery.

Early in 1987 research groups at the University of Alabama at Huntsville and

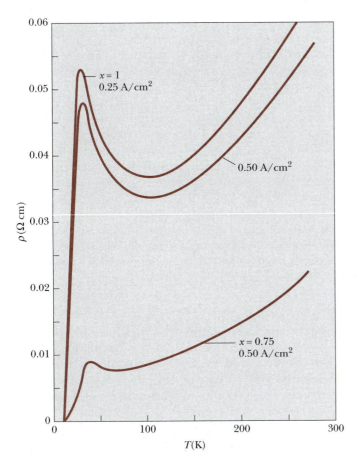

Figure 12.28   Temperature dependence of the resistivity of Ba-La-Cu-O for samples containing different concentrations of barium and lanthanum. The influence of current density is shown in the upper two curves. (*Taken from J. G. Bednorz and K. A. Müller, Z. Phys. B, 64:189, 1986*)

the University of Houston announced the discovery of superconductivity near 92 K in a mixed-phase sample containing yttrium, barium, copper, and oxygen.[17] The discovery was confirmed in other laboratories around the world, and the superconducting phase was soon identified to be the compound $YBa_2Cu_3O_{7-\delta}$. Figure 12.29 shows a plot of resistance versus temperature for this compound. This was an important milestone in high-temperature superconductivity, because the transition temperature of this compound is above the boiling point of liquid nitrogen (77 K), a coolant that is readily available, inexpensive, and much easier to handle than liquid helium.

At an American Physical Society meeting on March 18, 1987, a special panel discussion on high-temperature superconductivity introduced the world to the newly discovered novel superconductors. This all-night session, which attracted a standing-room-only crowd of about 3000, produced great excitement in the

[17] M. K. Wu, J. R. Ashburn, C. J. Torng, P. H. Hor, R. L. Meng, L. Gao, Z. J. Huang, Y. Q. Wang, and C. W. Chu, *Phys. Rev. Letters* 58:908, 1987.

Figure 12.29 Temperature dependence of the resistance of a sample of $YBa_2Cu_3O_{7-\delta}$ showing the transition temperature near 92 K. (*Taken from M. W. Wu et al.*, Phys. Rev. Letters *58:908, 1987*)

scientific community and has been referred to as "the Woodstock of physics." Realizing the possibility of routinely operating superconducting devices at liquid-nitrogen temperature and perhaps eventually at room temperature, thousands of scientists from a variety of disciplines entered the arena of superconductivity research. The exceptional interest in these novel materials arises from at least four factors:

- The metallic oxides are relatively easy to fabricate and hence can be investigated at smaller laboratories and universities.
- They have very high T_c values and very high upper critical magnetic fields, estimated to be greater than 100 T in several materials (Table 12.8).

Table 12.8   Some Properties of High-T_c Superconductors in the Form of Bulk Ceramics

Superconductor	T_c (K)	$B_{c2}(0)$ (T)[a]
La-Ba-Cu-O	30	
$La_{1.85}Sr_{0.15}CuO_5$	36.2	>36
La_2CuO_4	40	
$YBa_2Cu_3O_{7-\delta}$	92	≈160
$ErBa_2Cu_3O_{9-\delta}$	94	>28
$DyBa_2Cu_3O_7$	92.5	
Bi-Sr-Ca-Cu-O	120	
Tl-Ba-Ca-Cu-O	125	
$HgBa_2Ca_2Cu_3O_8$[b]	134	

[a] These are projected extrapolations based on data up to about 30 T.

[b] The compound $HgBa_2Ca_2Cu_3O_8$ has a critical temperature of 164 K under a hydrostatic pressure of 45 GPa.

- Their properties and the mechanisms responsible for their superconducting behavior represent a great challenge to theoreticians.
- They may be of considerable technological importance.

Since 1986, several complex metallic oxides in the form of ceramics have been investigated, and critical temperatures above 100 K (triple-digit superconductivity) have been observed. Early in 1988 researchers reported superconductivity at 120 K in a Bi-Sr-Ca-Cu-O compound and at 125 K in a Tl-Ba-Ca-Cu-O compound. As of early 1995, the record high critical temperature was 134 K, in the compound $HgBa_2Ca_2Cu_3O_8$. The increase in T_c since 1986 is dramatized in Figure 12.30. As you can see from this graph, the new high-T_c materials are all copper oxides of one form or another.

The various superconducting compounds that have been extensively studied to date can be classified in terms of what are called perovskite crystal structures. The first class (Fig. 12.31a) consists of the cubic perovskites ($a = b = c$), such as $BaPb_{1-x}Bi_xO_3$, one of the original "high-T_c materials" ($T_c \approx 10$ K). The second class (Fig. 12.31b) is single-layer perovskites, which have a tetragonal distortion ($a = b \neq c$); La_2CuO_4 is an example ($T_c \approx 38$ K). Note that the lattice parameters a and b are measured in the copper-oxygen planes, while c is perpendicular to these planes. The third class is multilayer perovskites (Fig.

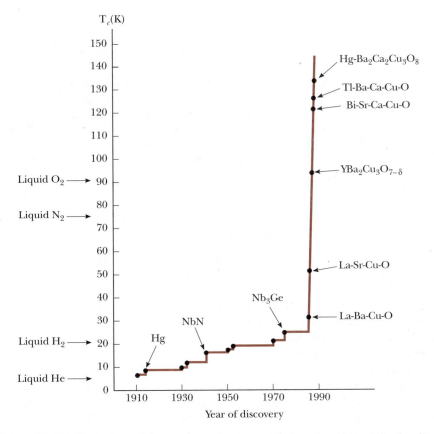

Figure 12.30   Evolution of the superconducting critical temperature since the discovery of the phenomenon.

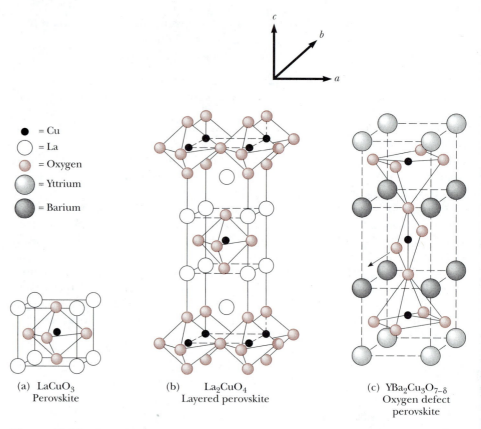

(a) LaCuO$_3$
Perovskite

(b) La$_2$CuO$_4$
Layered perovskite

(c) YBa$_2$Cu$_3$O$_{7-\delta}$
Oxygen defect
perovskite

Figure 12.31 Crystal structures for some of the new superconducting materials, the general class structure of which is perovskite. (a) The fundamental perovskite unit. (b) The single-layered perovskite having a tetragonal distortion ($a = b \neq c$). The first high-temperature superconductor ($T_c = 30$ K) falls in this class. (c) The double-layered perovskite with its orthorhombic structure ($a \neq b \neq c$). This structure is also related to the fundamental perovskite cube but has missing oxygen atoms.

12.31c), such as YBa$_2$Cu$_3$O$_7$ ($T_c \approx 92$ K), with an orthorhombic structure ($a \neq b \neq c$). Compounds in this class are sometimes called 1-2-3 materials because of their relative metallic composition.

The structures of the more complex copper oxides are not illustrated in Figure 12.31, but a most interesting observation can be made about them. *There appears to be a direct correlation between the number of copper-oxygen layers and critical temperature,* with critical temperature clearly increasing as copper-oxygen layers are added. There is some reason to expect that adding even more copper-oxygen layers to these complex oxides will raise the critical temperature even higher. On the basis of such results, some experts in the field are predicting T_c values above 200 K.

The maximum supercurrents in these structures are high in the copper-oxygen planes and much lower in the direction perpendicular to those planes. In effect, conduction can be viewed as being two-dimensional. Recently the Los Alamos Superconductivity Technology Center manufactured flexible superconducting tape having a critical-current density of more than 10^{10} A/m^2 at 77 K with a critical temperature of about 90 K. The high critical-current density was achieved when a metallurgical technique known as texturing was used to achieve grain alignment. The tape consists of three layers: a substrate of flexible nickel, a second layer of textured cubic zirconia, and a third layer of superconducting $YBa_2Cu_3O_7$.

Mechanisms of High-T_c Superconductivity

Although the framework of the conventional BCS theory has been quite successful in explaining the behavior and properties of the "old-generation" superconductors, theoreticians are still trying to understand the nature of superconductivity in the "new-generation" high-T_c cuprate oxides. The various models and mechanisms that have been proposed are far too technical to describe here.

Recently, researchers at IBM measured the symmetry of the wavefunction that describes Cooper pairs in $YBa_2Cu_3O_7$ to be a d-wave ($L = 2$).[18] They made their measurements on a sample consisting of a ring of $YBa_2Cu_3O_7$ deposited on a tricrystal substrate of $SrTiO_3$. By forming an odd number of Josephson junctions at the grain boundaries of the substrate, they observed a spontaneous magnetic flux equal to an odd integer times the magnetic flux quantum (that is, $\Phi_0/2$, $3\Phi_0/2$, and so on). This experiment provides strong evidence that antiferromagnetism in the cuprate oxides plays an important role in the establishment of the superconducting state in these high-T_c compounds and the formation of corresponding Cooper pairs.

12.11   APPLICATIONS OF SUPERCONDUCTIVITY   O P T I O N A L

High-temperature superconductors may lead to many important technological advances, such as highly efficient, lightweight superconducting motors. However, many significant materials-science problems must be overcome before such applications become reality. Perhaps the most difficult technical challenge is to mold the brittle ceramic materials into useful shapes, such as wires and ribbons for large-scale applications and thin films for small devices (e.g., SQUIDs). Another major problem is the relatively low current density that bulk ceramic compounds can carry. Assuming that such problems will be overcome, it is interesting to speculate on some of the future applications of these newly discovered materials.

An obvious application using the property of zero resistance to direct currents is low-loss electrical power transmission. A significant fraction of electrical power is lost as heat when current is passed through normal conductors. If power transmission lines could be made superconducting, these dc losses could be eliminated and substantial savings in energy costs would result.

The new superconductors could have a major impact in the field of electronics. Because of its switching properties, the Josephson junction can be used as a computer element. In addition, if one could use superconducting films to interconnect computer chips, chip size could be reduced and consequently speeds would be enhanced. Thus information could be transmitted more rapidly and more chips could be contained on a circuit board with far less heat generation.

---

[18] C. C. Tsuei and J. R. Kirtley, *Phys. Rev. Lett.* 73:593, 1994.

This prototype train, constructed in Japan, has superconducting magnets built into its base. A powerful magnetic field both levitates the train a few inches above the track and propels it smoothly at speeds of 300 mph or more. (*Joseph Brignolo/The Image Bank*)

The phenomenon of magnetic levitation can be exploited in the field of transportation. In fact, a prototype train that runs on superconducting magnets has been constructed in Japan. The moving train levitates above a normal conducting metal track through eddy-current repulsion. One can envision a future society of vehicles of all sorts gliding above a freeway, making use of superconducting magnets. Some scientists are speculating that the first major market for levitating devices will be the toy industry.

Another very important application of superconductivity is the superconducting magnet, a crucial component in particle accelerators. Currently, all high-energy particle accelerators use liquid-helium-based superconducting technology. Again, significant savings in cooling and operating costs would result if a liquid-nitrogen-based technology were developed.

An important application of superconducting magnets is a diagnostic tool called magnetic resonance imaging (MRI). This technique has played a prominent role in diagnostic medicine because it uses relatively safe radio-frequency radiation, rather than x-rays, to produce images of body sections. Because the technique relies on intense magnetic fields generated by superconducting magnets, the initial and operating costs of MRI systems are high. A liquid-nitrogen-cooled magnet could reduce such costs significantly.

In the field of power generation, companies and government laboratories have worked for years to develop superconducting motors and generators. In fact, a small-scale superconducting motor using the newly discovered ceramic superconductors has already been constructed at Argonne National Laboratory in Illinois.

We have already mentioned some small-scale applications of superconductivity—namely SQUIDs and magnetometers that make use of the Josephson effect and quantum interference. Josephson junctions can also be used as voltage standards and as infrared detectors. SQUIDs have been fabricated from films of $YBa_2Cu_3O_{7-\delta}$, and Josephson junctions using this compound have been operated at liquid-nitrogen temperature and above.

SUMMARY

Paramagnetic and **ferromagnetic** materials are those that have atoms with permanent magnetic dipole moments that tend to align in the direction of an external magnetic field impressed on these materials. **Diamagnetic** materials are those whose atoms have no permanent magnetic dipole moments. When a diamagnetic material is placed in an external magnetic field, weak dipole moments opposing the field are induced in the material.

Superconductors are materials that have zero dc resistance below a certain temperature T_c, called the **critical temperature.** A second characteristic of a type I superconductor is that it behaves as a perfect diamagnet. At temperatures below T_c, any applied magnetic flux is expelled from the interior of a type I superconductor. This phenomenon of flux expulsion is known as the **Meissner effect.** The superconductivity of a type I superconductor is destroyed when an applied magnetic field exceeds the **critical magnetic field,** B_c, which is less than 0.2 T for the elemental superconductors.

A type II superconductor is characterized by two critical fields. When an applied field is less than the lower critical field, B_{c1}, the material is entirely superconducting and there is no flux penetration. When the applied field exceeds the upper critical field, B_{c2}, the superconducting state is destroyed and the flux penetrates the material completely. However, when the applied field lies between B_{c1} and B_{c2}, the material is in a vortex state that is a combination of superconducting regions threaded by regions of normal resistance.

Once set up in a superconducting ring, **persistent currents** (also called supercurrents) circulate for several years with no measurable losses and with zero applied voltage. This is a direct consequence of the fact that the dc resistance is truly zero in the superconducting state.

A central feature of the BCS theory of superconductivity for metals is the formation of a bound state called a **Cooper pair,** consisting of two electrons having opposite momenta and opposite spins. The two electrons can form a bound state through a weak attractive interaction in which the crystal lattice of the metal serves as a mediator. In effect, one electron is weakly attracted to the other after the lattice is momentarily deformed by the first electron. In the ground state of the superconducting system, all electrons form Cooper pairs and all Cooper pairs are in the same quantum state. Thus the superconducting state is represented by a single coherent wavefunction that extends over the entire volume of the sample.

The BCS model predicts an energy gap $E_g = 3.53kT_c$, in contrast to a normal conductor, which has no energy gap. This value represents the energy needed to break up a Cooper pair and is of the order of 1 meV for the elemental superconductors.

High-temperature superconductors have critical temperatures as high as 150 K. They are all copper oxides, and the critical temperatures appear to be linked to the number of copper-oxygen layers in the structures. The new-generation materials, known to be type II superconductors, have highly anisotropic resistivities and high upper critical fields. However, in the form of bulk ceramic samples, the materials have limited critical currents and are quite brittle. Although the BCS model appears to be consistent with most empirical observations on high-temperature superconductors, the basic mechanisms giving rise to the superconducting state remain unknown.

QUESTIONS

1. Why is **M** = 0 in a vacuum? What is the relationship between **B** and **H** in a vacuum?
2. Explain why some atoms have permanent magnetic dipole moments and others do not.
3. Explain the significance of the Curie temperature for a ferromagnetic substance.
4. Explain why it is desirable to use hard ferromagnetic materials to make permanent magnets.
5. Describe how you would measure the two major characteristics of a superconductor.
6. Why is it not possible to explain zero resistance using a classical model of charge transport through a solid?
7. Discuss the differences between type I and type II superconductors. Discuss their similarities.
8. The specific heat of a superconductor in the absence of a magnetic field undergoes an anomaly at the critical temperature and decays exponentially toward zero below this temperature. What information does this behavior provide?

9. What is the isotope effect, and why does it play an important role in testing the validity of the BCS theory?
10. What are Cooper pairs? Discuss their essential properties, such as their momentum, spin, binding energy, and so on.
11. How would you explain the fact that lattice imperfections and lattice vibrations can scatter electrons in normal metals but have no effect on Cooper pairs?
12. Define single-particle tunneling and the conditions under which it can be observed. What information can one obtain from a tunneling experiment?
13. Define Josephson tunneling and the conditions under which it can be observed. What is the difference between Josephson tunneling and single-particle tunneling?
14. What are the features of high-temperature superconductors that limit their possible applications?
15. Discuss four features of high-T_c superconductors that make them superior to the old-generation superconductors. In what way are the new superconductors inferior to the old ones?

The following questions deal with the Meissner effect. A small permanent magnet is placed on top of a high-temperature superconductor (usually $YBa_2Cu_3O_{7-\delta}$), starting at room temperature. As the superconductor is cooled with liquid nitrogen (77 K), the permanent magnet levitates—a most dazzling phenomenon. Assume, for simplicity, that the superconductor is type I.

16. Because of the Meissner effect, a magnetic field is expelled from the superconductor below its critical temperature. How does this explain the levitation of the permanent magnet? What must happen to the superconductor in order to account for this behavior?

17. As soon as the permanent magnet has levitated, it gains potential energy. What accounts for this increase in mechanical energy? (This is a tricky one.)

18. If the experiment is repeated by first cooling the superconductor below its critical temperature and then placing the permanent magnet on top of it, will the magnet still levitate? If so, will there be any difference in its elevation compared to the previous case?

19. If a calibrated thermocouple is attached to the superconductor to measure its temperature, describe how you could use this demonstration to obtain a value for T_c. (*Hint:* Start the observation below T_c with a levitated permanent magnet.)

PROBLEMS

12.1 Magnetism in Matter

1. What is the relative permeability of a material that has a magnetic susceptibility of 1.00×10^{-4}?

2. An iron-core toroid is wrapped with 250 turns of wire per meter of its length. The current in the winding is 8.00 A. Taking the magnetic permeability of iron to be $\mu_m = 5000\mu_0$, calculate (a) the magnetic field strength **H** and (b) the magnetic flux density **B**. (*Note:* **B** is frequently called the magnetic field as well, but its units [tesla] clearly distinguish it from **H**, which has units of amperes per meter.)

3. A toroid with a mean radius of 20 cm and 630 turns of wire is filled with powdered steel whose magnetic susceptibility χ is 100. If the current in the windings is 3.00 A, find B (assumed uniform) inside the toroid.

4. A toroid has an average radius of 9.0 cm and carries a current of 0.50 A. How many turns are required to produce a magnetic field strength of $700 \, A \cdot turns/m$ within the toroid?

5. A magnetic field of magnitude 1.3 T is to be set up in an iron-core toroid. The toroid has a mean radius of 10 cm and magnetic permeability of $5000\mu_0$. What current is required if there are 470 turns of wire in the winding?

6. A toroid has an average radius of 10 cm and a cross-sectional area of $1.0 \, cm^2$. There are 400 turns of wire on the soft iron core, which has a permeability of $800\mu_0$. Calculate the current necessary to produce a magnetic flux of 5.0×10^{-4} Wb through a cross-section of the core.

7. A coil of 500 turns is wound on an iron ring ($\mu_m = 750\mu_0$) of 20-cm mean radius and 8.0-cm^2 cross-sectional area. Calculate the magnetic flux Φ_B in this Rowland ring when the current in the coil is 0.50 A.

8. A uniform ring of radius R and total charge Q rotates with constant angular speed ω about an axis perpendicular to the plane of the ring and passing through its center. What is the magnetic moment of the rotating ring?

9. In the text, we found that an alternative description for the magnetic field **B** in terms of magnetic field strength **H** and magnetization **M** is $B = \mu_0 H + \mu_0 M$. Relate the magnetic susceptibility χ to $|H|$ and $|M|$ for paramagnetic or diamagnetic materials.

10. A magnetized cylinder of iron has a magnetic field $B = 0.040$ T in its interior. The magnet is 3.0 cm in diameter and 20 cm long. If the same magnetic field is to be produced by a 5.0-A current carried by an air-core solenoid with the same dimensions as the cylindrical magnet, how many turns of wire must be on the solenoid?

11. In Bohr's 1913 model of the hydrogen atom, the electron is in a circular orbit of radius 5.3×10^{-11} m, and its speed is 2.2×10^6 m/s. (a) What is the magnitude of the magnetic moment due to the electron's motion? (b) If the electron orbits counterclockwise in a horizontal circle, what is the direction of this magnetic moment vector?

12. At saturation, the alignment of spins in iron can contribute as much as 2.0 T to the total magnetic field B. If each electron contributes a magnetic moment of $9.27 \times 10^{-24} \, A \cdot m^2$ (one Bohr magneton), how many electrons per atom contribute to the saturated field of iron? (*Hint:* There are 8.5×10^{28} iron atoms in a cubic meter.)

13. *Curie's law of paramagnetism.* Curie's law, Equation 12.8, may be derived on the basis of a few assumptions. Recall that paramagnetism is produced by the net alignment of electronic magnetic moments of size $\cong \mu_B$ along the direction of an applied magnetic field, **B**. The paramagnetic magnetization $M = \mu_B(n_1 - n_2)$, where n_1 is the number of moments per unit volume parallel to the applied field and n_2 is the number of moments antiparallel. (a) Show that the ratio

$$\frac{n_1}{n_2} = e^{2\mu_B B/k_B T} \cong 1 + \frac{2\mu_B B}{k_B T}$$

by using the Maxwell-Boltzmann distribution and assuming that dipoles oriented parallel to **B** have energy $-\mu_B B$ and those antiparallel to **B** have energy $+\mu_B B$. (b) Show that the exponent $2\mu_B B/k_B T \ll 1$ for $B = 1.0$ T and $T = 300$ K. Thus

$$\frac{n_1 - n_2}{n_2} \cong \frac{2\mu_B B}{k_B T}$$

(c) Show that $n_1 - n_2 \cong n\mu_B B/k_B T$, where n is the total number of magnetic moments, and show that the paramagnetic magnetization is given by Equation 12.8, $M = C(B/T)$, with $C = n\mu_B^2/k_B$.

12.3   Some Properties of Type I Superconductors
12.4   Type II Superconductors

14. A wire made of Nb_3Al has a radius of 2.0 mm and is maintained at 4.2 K. Using the data in Table 12.5, find (a) the upper critical field for this wire at this temperature, (b) the maximum current that can pass through the wire before its superconductivity is destroyed, and (c) the magnetic field 6.0 mm from the wire surface when the current has its maximum value.

15. A superconducting solenoid is to be designed to generate a 10-T magnetic field. (a) If the winding has 2000 turns/m, what is the required current? (b) What force per meter does the magnetic field exert on the inner windings?

16. Determine the current generated in a superconducting ring of niobium metal 2.0 cm in diameter if a 0.020-T magnetic field directed perpendicular to the ring is suddenly decreased to zero. The inductance of the ring is $L = 3.1 \times 10^{-8}$ H.

17. Determine the magnetic field energy, in joules, that needs to be added to destroy superconductivity in 1.0 cm^3 of lead near 0 K. Use the fact that $B_c(0)$ for lead is 0.080 T.

18. The penetration depth for lead at 0.0 K is 39 nm. Find the penetration depth in lead at (a) 1.0 K, (b) 4.2 K, and (c) 7.0 K.

19. Find the critical magnetic field in mercury at (a) 1.0 K and (b) 4.0 K.

12.5   Other Properties of Superconductors

20. *Persistent currents.* In an experiment carried out by S. C. Collins between 1955 and 1958, a current was maintained in a superconducting lead ring for 2.5 years with no observed loss. If the inductance of the ring was 3.14×10^{-8} H, and the sensitivity of the experiment was 1 part in 10^9, determine the maximum resistance of the ring. (*Hint:* Treat this as a decaying current in an RL circuit, and recall that $e^{-x} \cong 1 - x$ for small x.)

21. *Speed of electron flow.* Current is carried throughout the body of niobium-tin, a type II superconductor. If a niobium-tin wire of cross-section 2.0 mm^2 can carry a maximum supercurrent of 1.0×10^5 A, estimate the average

speed of the superconducting electrons. (Assume that the density of conducting electrons is $n_s = 5.0 \times 10^{27}/m^3$.)

22. *Diamagnetism.* When a superconducting material is placed in a magnetic field, the surface currents established make the magnetic field inside the material truly zero. (That is, the material is perfectly diamagnetic.) Suppose a circular disk, 2.0 cm in diameter, is placed in a 0.020-T magnetic field with the plane of the disk perpendicular to the field lines. Find the equivalent surface current if it all lies at the circumference of the disk.

23. *Surface currents.* A rod of a superconducting material 2.5 cm long is placed in a 0.54-T magnetic field with its cylindrical axis along the magnetic field lines. (a) Sketch the directions of the applied field and the induced surface current. (b) Estimate the magnitude of the surface current.

12.7   BCS Theory
12.8   Energy Gap Measurements

24. Calculate the energy gap for each superconductor in Table 12.4 as predicted by the BCS theory. Compare your answers with the experimental values in Table 12.7.

25. Calculate the energy gap for each superconductor in Table 12.5 as predicted by the BCS theory. Compare your values with those found for type I superconductors.

26. *High-temperature superconductor.* Estimate the energy gap E_g for the high-temperature superconductor $YBa_2Cu_3O_{7-\delta}$, which has a critical temperature of 92 K, assuming BCS theory holds.

27. *Isotope effect.* Because of the isotope effect, $T_c \propto M^{-\alpha}$. Use these data for mercury to determine the value of the constant α. Is your result close to what you might expect on the basis of a simple model?

Isotope	T_c (K)
^{199}Hg	4.161
^{200}Hg	4.153
^{204}Hg	4.126

28. *Cooper pairs.* A Cooper pair in a type I superconductor has an average separation of about 1.0×10^{-4} cm. If these two electrons can interact within a volume of this diameter, how many other Cooper pairs have their centers within the volume occupied by one pair? Use the appropriate data for lead, which has $n_s = 2.0 \times 10^{22}$ electrons/cm^3.

29. *Dipole in a magnetic field.* The potential energy of a magnetic dipole of moment $\boldsymbol{\mu}$ in the presence of a magnetic field **B** is $U = -\boldsymbol{\mu} \cdot \mathbf{B}$. When an electron is placed in a magnetic field, its magnetic moment can be aligned either with or against the field. The energy separation between these two states is $\Delta E = 2\mu B$, where the magnetic moment of the electron is $\mu = 5.79 \times 10^{-5}$ eV/T.

(a) If a Cooper pair is subjected to a 38-T magnetic field (the critical field for Nb_3Ge), calculate the energy separation between the spin-up and spin-down electrons. (b) Calculate the energy gap for Nb_3Ge as predicted by the BCS theory at 0 K, using the fact that $T_c = 23$ K. (c) How do your answers to (a) and (b) compare? What does this result suggest, based on what you have learned about critical fields?

12.9   Josephson Tunneling

30. Estimate the area of a ring that would fit one of your fingers, and calculate the magnetic flux through the ring due to the Earth's magnetic field (5.8×10^{-5} T). If this flux were quantized, how many fluxons would the ring enclose?

31. A Josephson junction is fabricated using indium for the superconducting layers. If the junction is biased with a dc voltage of 0.50 mV, find the frequency of the alternating current generated. (For comparison, note that the energy gap of indium at 0 K is 1.05 meV.)

32. If a magnetic flux of $1.0 \times 10^{-4}\Phi_0$ ($\frac{1}{10\,000}$ of the flux quantum) can be measured with a SQUID (Fig. P12.32), what is the smallest magnetic field change ΔB that can be detected with this device for a ring having a radius of 2.0 mm?

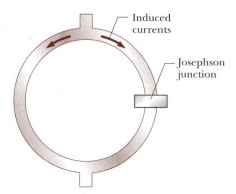

Figure P12.32   A superconducting *quantum interference device* (SQUID).

33. A superconducting circular loop of very fine wire has a diameter of 2.0 mm and a self-inductance of 5.0 nH. The flux through the loop, Φ, is the sum of the applied flux, Φ_{app}, and the flux due to the supercurrent, $\Phi_{sc} = LI$, where L is the self-inductance of the loop. Because the flux through the loop is quantized, we have

$$n\Phi_0 = \Phi_{app} + \Phi_{sc} = \Phi_{app} + LI$$

where Φ_0 is the flux quantum. (a) If the applied flux is zero, what is the smallest current that can meet this quantization condition? (b) If the applied field is perpendicular to the plane of the loop and has a magnitude of 3.0×10^{-9} T, find the smallest current that circulates around the loop.

Additional Problems

34. A solenoid of diameter 8.0 cm and length 1 m is wound with 2000 turns of superconducting wire. If the solenoid carries a current of 150 A, find (a) the magnetic field at the center, (b) the magnetic flux through the center cross-section, and (c) the number of flux quanta through the center.

35. *Energy storage.* A novel method has been proposed to store electrical energy. A huge underground superconducting coil, 1.0 km in diameter, carries a maximum current of 50 kA through each winding of a 150-turn Nb_3Sn solenoid. (a) If the inductance of this huge coil is 50 H, what is the total energy stored? (b) What is the compressive force per meter acting between two adjacent windings 0.25 m apart?

36. *Superconducting power transmission.* Superconductors have been proposed for power transmission lines. A single coaxial cable (Fig. P12.36) could carry 1.0×10^3 MW (the output of a large power plant) at 200 kV, dc, over a distance of 1000 km without loss. The superconductor would be a 2.0-cm-radius inner Nb_3Sn cable carrying the current I in one direction, while the outer surrounding superconductor of 5.0-cm radius would carry the return current I. In such a system, what is the magnetic field (a) at the surface of the inner conductor and (b) at the inner surface of the outer conductor? (c) How much energy would be stored in the space between the conductors in a 1000-km superconducting line? (d) What is the force per meter length exerted on the outer conductor?

Figure P12.36

37. *Penetration depth.* The penetration depth of a magnetic field into a superconductor is found, from London's equations, to be

$$\lambda = \sqrt{\frac{m_e}{\mu_0 n_s e^2}}$$

where m_e is the mass of the electron, e is the charge of the electron, and n_s is the number of superconducting electrons per unit volume. (a) Estimate n_s in lead at 0 K if the magnetic penetration depth near 0 K is $\lambda_0 = 4.0 \times 10^{-8}$ m. (b) Starting with Equation 12.12, determine n_s in lead as a function of temperature for $T \leq T_c$.

38. *"Floating" a wire.* Is it possible to "float" a superconducting lead wire of radius 1.0 mm in the magnetic field of the Earth? Assume that the horizontal component of the Earth's magnetic field is 4.0×10^{-5} T.

39. *Magnetic field inside a wire.* A type II superconducting wire of radius R carries current uniformly distributed through its cross-section. If the total current carried by the wire is I, show that the magnetic energy per unit length inside the wire is $\mu_0 I^2/16\pi$.

40. *Magnetic levitation.* If a very small but very strong permanent magnet is lowered toward a flat-bottomed type I superconducting dish, at some point the magnet will levitate above the superconductor because the superconductor is a perfect diamagnet and expels all magnetic flux. Therefore the superconductor acts like an identical magnet lying an equal distance below the surface (Fig. P12.40). At what height does the magnet levitate if its mass is 4.0 g and its magnetic moment $\mu = 0.25$ A·m^2? (*Hint:* The potential energy between two dipole magnets a distance r apart is $\mu_0\mu^2/4\pi r^3$).

Figure P12.40 A magnetic levitation experiment.

41. *Designing a superconducting solenoid.* A superconducting solenoid was made with Nb$_3$Zr wire wound on a 10-cm-diameter tube. The solenoid winding consisted of a double layer of 0.50-mm-diameter wire with 984 turns (corresponding to a coil length of 25 cm). (a) Calculate the inductance, L, of the solenoid, assuming its length is great relative to its diameter. (b) The magnitude of the persistent current in this solenoid in the superconducting state has been reported to decrease by 1 part in 10^9 per hour. If the resistance of the solenoid is R, then the current in the circuit should decay according to $I = I_0 e^{-Rt/L}$. This resistance, although small, is due to magnetic flux migration in the superconductor. Determine an upper limit of the coil's resistance in the supercon-

ducting state. (These data were reported by J. File and R. G. Mills, *Phys. Rev. Letters* 10:93, 1963.)

42. *Entropy difference.* The entropy difference per unit volume between the normal and superconducting states is

$$\frac{\Delta S}{V} = -\frac{\partial}{\partial T}\left(\frac{B^2}{2\mu_0}\right)$$

where $B^2/2\mu_0$ is the magnetic energy per unit volume required to destroy superconductivity. Determine the entropy difference between the normal and superconducting states in 1.0 mol of lead at 4.0 K if the critical magnetic field $B_c(0) = 0.080$ T and $T_c = 7.2$ K.

43. *A convincing demonstration of zero resistance.* A direct and relatively simple demonstration of zero dc resistance can be carried out using the four-point probe method. The experimental set-up, shown in Figure P12.43, consists of a disc of YBa$_2$Cu$_3$O$_{7-\delta}$ (a high-T_c superconductor) to which four wires are attached by indium solder (or some other suitable contact material). A constant current, maintained through the sample by a dc voltage between points a and b, is measured with a dc ammeter. (The current is varied with the variable resistance R.) The potential difference V_{cd} between c and d is measured with a digital voltmeter. When the disc is immersed in liquid nitrogen, it cools quickly to 77 K, which is below its critical temperature (92 K); the current remains approximately constant, but V_{cd} *drops abruptly to zero.* (a) Explain this observation on the basis of what you know about superconductors and your understanding of Ohm's law. (b) The data in Table 12.9 represent actual values of V_{cd} for different values of I taken on a sample at room temperature. Make an I-V plot of the data, and determine whether the sample behaves in a linear

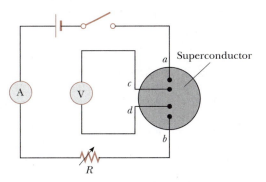

Figure P12.43 A circuit diagram used in the four-point probe measurement of the dc resistance of a sample. A dc digital ammeter is used to measure the current, and the potential difference between c and d is measured with a dc digital voltmeter. Note that there is no voltage source in the inner loop circuit where V_{cd} is measured.

manner. From the data, obtain a value of the dc resistance of the sample at room temperature. (c) At room temperature, it is found that $V_{cd} = 2.234$ mV for $I = 100.3$ mA, but after the sample is cooled to 77 K, $V_{cd} = 0$ and $I = 98.1$ mA. What do you think might explain the slight decrease in current?

Table 12.9   Current Versus Potential Difference V_{cd} in a Bulk Ceramic Sample of $YBa_2Cu_3O_{7-\delta}$ at Room Temperature[a]

I (mA)	V_{cd} (mV)
57.8	1.356
61.5	1.441
68.3	1.602
76.8	1.802
87.5	2.053
102.2	2.398
123.7	2.904
155	3.61

[a] The current was supplied by a 6-V battery in series with a variable resistor R. The values of R ranged from 10 Ω to 100 Ω. The data are from the author's laboratory.

Essay   SUPERCONDUCTING DEVICES*

Superconductivity was discovered in 1911 by Dutch physicist Heike Kamerlingh Onnes. In experiments carried out to measure the resistance of frozen mercury, he made the startling discovery that resistance vanished completely at a temperature a few degrees above absolute zero. Since that time, superconductivity has been observed in a large number of elemental metals and alloys. The first proposed application of this new phenomenon was the construction of powerful magnets, since superconductors can carry current with no heat generation. Unfortunately it was soon discovered that modest magnetic fields destroy the superconductivity in materials such as mercury, tin, and lead. In fact, the conditions under which superconductivity exists in any material can be described by a three-dimensional diagram such as that shown in Figure 1a. The axes are temperature, magnetic field, and current density. The curved surface in this figure represents the transition between normal and superconducting behavior, and its intersections with the three axes are the critical current density (J_c), critical field (H_c), and critical temperature (T_c). The primary objective of research in superconducting materials since 1911 has been to expand the volume of superconducting behavior in this diagram. This effort has generated a wide variety of what are now called conventional superconductors, with critical temperatures up to 23 K and critical fields up to 40 K (the Earth's magnetic field is 0.00005 T). The primary drawback of these conventional superconductors is the expense and inconvenience of their refrigeration systems. Nevertheless, the fabrication of superconducting magnets for magnetic resonance imaging, accelerator magnets, fusion research, and other applications is a $200 million/year business.

Clark A. Hamilton

National Institute of Standards and Technology Electromagnetic Technology Division

In 1987 Karl Alex Müller and J. Georg Bednorz received the Nobel prize in physics for their discovery in 1986 of a new class of higher-temperature superconducting ceramics consisting of oxygen, copper, barium, and lanthanum. Expanding on their work, researchers at the University of Alabama at Huntsville and the University of Houston developed the compound $YBa_2Cu_3O_x$, which has a transition temperature above 90 K. For the first time, superconductivity became possible at liquid-nitrogen temperatures.

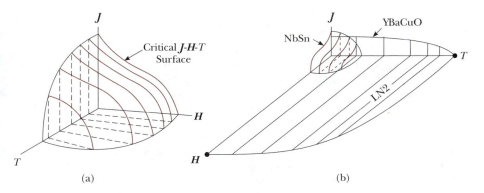

(a)   (b)

Figure 1 (a) A three-dimensional diagram illustrating the surface separating normal and superconducting behavior as a function of temperature (T), magnetic field (H), and current density (J). (b) A comparison of the J-H-T diagrams for Nb_3Sn and $YBa_2Cu_3O_x$. *(Courtesy of Alan Clark, National Institute of Standards and Technology.)*

---

* Contribution of the National Institute of Standards and Technology, not subject to copyright.

An explosion of research on the new materials has resulted in the discovery of many materials with transition temperatures near 100 K and hints of superconductivity at much higher temperatures. Figure 1b compares the transition parameters of the old and new materials and illustrates the potential of the new superconductors. With their high transition temperatures and critical fields projected to hundreds of tesla, the parameter space where superconductivity exists has been greatly expanded. Unfortunately, the critical-current density of the new materials over most of the *H-T* plane is very low. It is expected that this problem will be solved when the new materials can be made in sufficiently pure form.

SUPERCONDUCTIVITY

In 1957 John Bardeen, L. N. Cooper, and J. R. Schrieffer published a theory which shows that superconductivity results from an interaction between electrons and the lattice in which they flow.[1,2,3] Coulomb attraction between electrons and lattice ions produces a lattice distortion that can attract other electrons. At sufficiently low temperatures, the attractive force overcomes thermal agitation, and the electrons begin to condense into pairs. As the temperature is reduced further, more and more electrons condense into the paired state. In so doing, each electron pair gives up a condensation energy 2Δ. The electron pairs are all described by a common quantum mechanical wavefunction, and thus a long-range order exists among all of the electron pairs in the superconductor. As a result of this long-range order, the pairs can flow through the lattice without suffering the collisions that lead to resistance in normal conductors.

We can model a superconductor as a material in which current is carried by two different electron populations. The first population consists of normal electrons, which behave as the electrons in any normal metal do. The second population consists of electron pairs, which behave coherently and are described by a common wavefunction $\Psi(x, t)$. The amplitude of $\Psi(x, t)$ is usually constant throughout the superconductor, but its phase may vary, depending on the current. The condensation energy of the electron pairs produces an energy gap between normal and superconducting electrons, which is $2\Delta/e \approx 2$ meV for conventional superconductors. This model is known as the two-fluid model.

JOSEPHSON JUNCTIONS

In addition to the work on large-scale applications of superconductivity (primarily magnets), considerable research has been done on small-scale superconducting devices. Most of these devices take advantage of the unusual properties of a Josephson junction, which is simply two superconductors separated by an insulating barrier, as shown in Figure 2. When the barrier is thin enough, electron pairs can tunnel from one superconductor to the other. Brian Josephson shared the Nobel prize in physics in 1973 for his prediction of the behavior of such a device. The remainder of this essay will discuss the operation of Josephson devices and circuits.

The equations for the current associated with electron pairs across a Josephson junction can be derived by solving two coupled wave equations with a potential energy difference eV that depends on the voltage V across the junction.[1] The result is the Josephson equations:

$$I_J = I_0 \sin \phi \tag{1}$$

$$\frac{d\phi}{dt} = \frac{2eV}{\hbar} \tag{2}$$

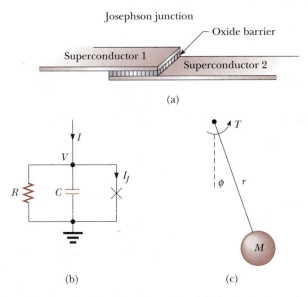

Figure 2   The geometry of a Josephson junction (a), its equivalent circuit (b), and a mechanical analog for the junction behavior (c).

where I_J is the electron-pair current through the junction, ϕ is the phase difference between the superconducting wavefunctions on the two sides of the barrier, and I_0 is a constant related to the barrier thickness. The full behavior of the junction can be modeled by the equivalent circuit of Figure 2b. The resistance R describes the flow of normal electrons; X represents a pure Josephson element having a flow of electron pairs, I_J, given by Equations 1 and 2; and C is the normal geometrical capacitance. Thus the total junction current is

$$I = \frac{V}{R} + C\frac{dV}{dt} + I_0 \sin\phi \qquad (3)$$

Substituting Equation 2, this becomes

$$I = \frac{C\hbar}{2e}\frac{d^2\phi}{dt^2} + \frac{\hbar}{2eR}\frac{d\phi}{dt} + I_0 \sin\phi \qquad (4)$$

Under different drive and parameter conditions, Equation 4 leads to an almost infinite variety of behavior and is the subject of more than 100 papers.

THE PENDULUM ANALOGY

In understanding Equation 4, it is useful to make the analogy between the current in a Josephson junction and the torque applied to a pendulum that is free to rotate 360° in a viscous medium, as shown in Figure 2c. The applied torque, τ, on such a pendulum can be written

$$\tau = Mr^2\frac{d^2\phi}{dt^2} + k\frac{d\phi}{dt} + Mgr\sin\phi \qquad (5)$$

where $k\dfrac{d\phi}{dt}$ is the damping torque, $Mr^2\dfrac{d^2\phi}{dt^2}$ is the inertial term, and $Mgr\sin\phi$ is the gravitational restoring torque. Since Equations 4 and 5 have identical form, we can say

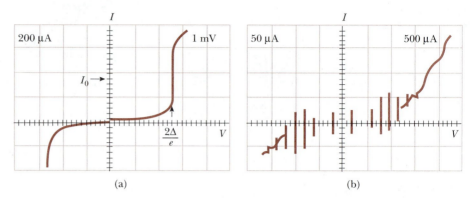

Figure 3   (a) The *I-V* curve of a Josephson junction. (b) The *I-V* curve when the junction is exposed to 96 GHz radiation. The quantized voltage levels occur at the values $V_n = nhf/2e$.

that the torque on the pendulum is the analog of current in a Josephson junction, and the rate of rotation of the pendulum is the analog of voltage across the junction.

We are now in a position to understand the Josephson junction *I-V* curve shown in Figure 3a. For currents less than I_0, the junction phase assumes a value such that all of the current is carried by the Josephson element, as described by Equation 1. Except for a small transient, the voltage remains zero. In the analog, the pendulum assumes a fixed angle that balances the torque and the gravitational restoring force. When the torque exceeds the critical value *Mgr*, the pendulum rotation accelerates until the viscous damping term can absorb the applied torque. Similarly, when the current applied to a Josephson junction exceeds the critical value I_0, the junction switches to a voltage state in which the junction current oscillates and has an average value determined by the normal state resistance. As the current decreases, the voltage follows the normal state curve until the current reaches a value I_{min}. Below I_{min} the voltage falls abruptly to zero as the junction switches back into the superconducting state. In the analog, with decreasing torque, the pendulum rotation slows down to the point where the applied torque is insufficient to push it over the top, and it then settles into a nonrotating state.

The nonlinearity in the normal state curve for the junction can be explained as follows. For small voltages, the resistance is high because there are comparatively few normal-state electrons to tunnel across the barrier. At voltages greater than the superconducting energy gap voltage, $2\Delta/e$, there is sufficient energy from the bias source to break superconducting pairs, and the substantial increase in the number of normal carriers causes an abrupt increase in current.

JOSEPHSON VOLTAGE STANDARD

One of the first applications of the Josephson effect comes from the fact that the Josephson current oscillates at a frequency proportional to the applied voltage, where the frequency is $f = 2eV/h$. Since this relation is independent of all other parameters and has no known corrections, it forms the basis of voltage standards throughout the world. In practice, these standards operate by applying a microwave current through the junction and allowing the Josephson frequency to phase-lock to some harmonic of the applied frequency. With the correct choice of design parameters, the only stable operating points of the junction will be at one of the voltages $V = nhf/2e$, where *n* is an integer. These voltages are very accurately known, because frequency can be measured with great accuracy and $2e/h = 483.594$ GHz/mV is a defined value for the purpose of voltage standards. Figure 3b is the *I-V* curve of a junction that is driven at 96 GHz. About ten quantized voltage levels span the range from -1 to 1 mV. Each of these levels represents a phase lock between a harmonic of the applied frequency at 96 GHz and the quantum phase difference, ϕ, across the junction. Practical voltage standards use up to 20 000 junctions in series to generate reference voltages up to 10 V.

FLUX QUANTIZATION

Consider a closed superconducting loop surrounding magnetic flux Φ. We know from Faraday's law that any change in Φ must induce a voltage around the loop. Since the voltage across a superconductor is zero, the flux cannot change and is therefore trapped in the loop. There is a further restriction in that the phase of the quantum wavefunction around the loop must be continuous. It can be shown that the variation of the quantum phase along a superconducting wire is proportional to the product of the current and the inductance L of the wire,

$$\Delta \phi = \frac{2\pi LI}{h/2e} \tag{6}$$

In order for ϕ to be continuous around the loop, $\Delta \phi$ must be a multiple of 2π. It is thus a simple matter to show that the product LI must be quantized in units of $h/2e$. Since LI is just the magnetic flux through the loop, the flux is quantized; the quantity $\Phi_0 = h/2e = 2.07 \times 10^{-15}$ T·m^2 is the flux quantum.

SQUIDs

One of the most interesting of all superconducting devices is formed by inserting two Josephson junctions in a superconducting loop, as shown in Figure 4a.[2,3] From a circuit viewpoint, this looks like two junctions in parallel, and it therefore has an *I-V* curve like that of a single junction (Fig. 4b). The current through the loop is just the sum of the currents in the two junctions.

$$I = I_{01} \sin \phi_1 + I_{02} \sin \phi_2 \tag{7}$$

The equation for continuity of phase around the loop is

$$\Delta \phi = 2\pi n = \phi_1 - \phi_2 + 2\pi \frac{L(I_1 - I_2)}{\Phi_0} + 2\pi \frac{\Phi}{\Phi_0} \tag{8}$$

When Equations 7 and 8 are solved to maximize I, the total critical current I_{max} is found to be a periodic function of the magnetic flux Φ applied to the loop as shown in Figure 4c. This pattern can be interpreted as the consequence of interference between the superconducting wavefunctions on the two sides of the junction. These devices are therefore called superconducting *quantum interference devices*, or SQUIDs.

The most important application of SQUIDs is the measurement of small magnetic fields. This is done by using a large pick-up loop to inductively couple flux into the SQUID loop. A sensitive electronic circuit senses the changes in critical current I_{max} caused by changing flux in the SQUID loop. This arrangement can resolve flux changes of as little as 10^{-6} Φ_0. For a typical SQUID with an inductance $L = 100$ pH, this corresponds to an energy resolution of $\Delta E = \Delta \Phi^2/2L = 2 \times 10^{-32}$ J, which approaches the theoretical limit set by Planck's constant.

A rapidly expanding application of SQUIDs is the measurement of magnetic fields caused by the small currents in the heart and brain. Systems with as many as seven SQUIDs are being used to map these biomagnetic fields. This new nonintrusive tool may one day play an important role in locating the source of epilepsy and other brain dis-

Figure 4   (a) The circuit diagram for a SQUID. (b) Its *I-V* curve. (c) The dependence of critical current on applied magnetic flux.

orders. In such applications it is important to discriminate against distant sources of magnetic noise. This is done by using two pick-up coils wound in opposite directions, so that the SQUID is sensitive only to the *gradient* of the magnetic field.

DIGITAL APPLICATIONS

The possibility of digital applications for Josephson junctions is suggested immediately by the bistable nature of the *I-V* curve in Figure 3a. At a fixed-bias current, the junction can be in either the zero-voltage state or the finite-voltage state, representing a binary 0 or 1. In the mid-1970s a major effort was mounted to develop large-scale computing systems that used Josephson devices.[4] The motivation for this is twofold. First, Josephson junctions can switch states in only a few picoseconds; second, the power dissipation of Josephson devices is nearly three orders of magnitude less than that of semiconductor devices. The low power dissipation results from the low logic levels (0 and 2 mV) of Josephson devices. Dissipation becomes increasingly important in ultra-high-speed computers because the need to limit the propagation delay forces the computer into a small volume. As the computer shrinks, heat removal becomes increasingly difficult.

A variety of Josephson logic circuits have been proposed. We will consider just one, based on the SQUID circuit of Figure 4a. Suppose that the SQUID carries a current at the level of the lower dashed line in Figure 4. Current in the two input lines (I_a, I_b) moves the bias point horizontally along this line in Figure 4c. For example, a current in either line moves the operating point to (c), while a current in both lines moves the operating point across the threshold curve to (d). When the threshold curve is crossed, the SQUID switches into the voltage state. The output of the device thus is the "AND" function of the two inputs. Other logic combinations can be achieved by adjusting the

Figure 5   A photograph of a Josephson integrated circuit consisting of two coupled SQUIDs. The minimum feature size is 4 microns.

coupling of the input lines. Memory can be made by using coupled SQUIDs, where one SQUID senses the presence or absence of flux in a second SQUID. Figure 5 is a photograph of two coupled thin-film SQUIDs that form a flip-flop circuit. These devices are made by photolithographic processes similar to those used in the semiconductor industry.

Josephson logic and memory chips with thousands of junctions have been built and successfully operated. However, there are still significant problems to be overcome in developing a complete Josephson computer. The greatest obstacle is the dependence of Josephson logic on threshold switching. This results in circuits that operate over only a small range of component parameters. Such circuits are said to have small *margins*. Until a way is found to improve the margins or until fabrication processes with very tight control are developed, the Josephson computer will remain on the drawing board.

Suggestions for Further Reading

1. R. P. Feynman, R. B. Leighton, and M. Sands, "The Feynman Lectures on Physics," vol. 3, Addison Wesley, Reading, Mass., 1966, 21.1–21.18.
2. D. N. Langenberg, D. J. Scalapino, and B. N. Taylor, "The Josephson Effects," *Sci. American*, May 1966.
3. D. G. McDonald, "Superconducting Electronics," *Physics Today*, February 1981.
4. J. Matisoo, "The Superconducting Computer," *Sci. American*, May 1980.

13

Nuclear Structure

Chapter Outline

In 1896, the year that marks the birth of nuclear physics, the French physicist Henri Becquerel (1852–1908) discovered radioactivity in uranium compounds. Following his discovery, scientists did a great deal of research in attempts to understand the nature of the radiation emitted by radioactive nuclei. Pioneering work by Ernest Rutherford showed that the emitted radiation was of three types, which he called alpha, beta, and gamma rays. Rutherford classified the rays according to the natures of the electric charges they possessed and their abilities to penetrate matter and ionize air. Later experiments showed that alpha rays are helium nuclei, beta rays are electrons, and gamma rays are high-energy photons.

In 1911 Rutherford and his students Geiger and Marsden performed a number of important scattering experiments involving alpha particles. Those experiments established that the nucleus of an atom can be regarded as essentially a point mass and point charge and that most of the atomic mass is contained in the nucleus. Subsequent studies revealed the presence of a previously unknown type of force, the nuclear force, which is predominant at distances of less than approximately 10^{-14} m and is zero at great distances. That is, the nuclear force is a short-range force.

Other milestones in the development of nuclear physics include:

- The observation of nuclear reactions in 1930 by Cockroft and Walton, using artificially accelerated particles
- The discovery of the neutron in 1932 by Chadwick and the conclusion that neutrons make up about half of the nucleus
- The discovery of artificial radioactivity in 1933 by Joliot and Irene Curie
- The discovery of nuclear fission in 1938 by Meitner, Hahn, and Strassmann
- The development of the first controlled fission reactor in 1942 by Fermi and his collaborators

In this chapter we discuss the properties and structure of the atomic nucleus. We first describe the basic properties of nuclei, then discuss nuclear forces and binding energy, nuclear models, and the phenomenon of radioactivity. Finally, we examine the processes by which nuclei decay.

13.1   SOME PROPERTIES OF NUCLEI

All nuclei are composed of two types of particles: protons and neutrons. The only exception is the ordinary hydrogen nucleus, which is a single proton. In describing the atomic nucleus, we must use the following quantities:

- The **atomic number**, Z (sometimes called the charge number), which equals the number of protons in the nucleus
- The **neutron number**, N, which equals the number of neutrons in the nucleus
- The **mass number**, A, which equals the number of nucleons (neutrons plus protons) in the nucleus

In representing nuclei, it is convenient to have a system of symbols to show how many protons and neutrons are present. The symbol used is $^{A}_{Z}X$, where X represents the chemical symbol for the element. For example, $^{56}_{26}Fe$ (iron) has a mass number of 56 and an atomic number of 26; it therefore contains 26 protons and 30 neutrons. When no confusion is likely to arise, we omit the subscript Z because the chemical symbol can always be used to determine Z.

The nuclei of all atoms of a particular element contain the same number of protons but often contain different numbers of neutrons. Nuclei that are related in this way are called **isotopes.**

The isotopes of an element have the same Z value but different N and A values.

Isotopes

The natural abundances of isotopes can differ substantially. For example, $^{11}_{6}C$, $^{12}_{6}C$, $^{13}_{6}C$, and $^{14}_{6}C$ are four isotopes of carbon. The natural abundance of the $^{12}_{6}C$ isotope is about 98.9%, whereas that of the $^{13}_{6}C$ isotope is only about 1.1%. Some isotopes do not occur naturally but can be produced in the laboratory through nuclear reactions. Even the simplest element, hydrogen, has isotopes: $^{1}_{1}H$, the ordinary hydrogen nucleus; $^{2}_{1}H$, deuterium; and $^{3}_{1}H$, tritium.

Ernest   Rutherford   (1871–1937), a physicist from New Zealand, was awarded a Nobel prize in 1908 for discovering that atoms can be broken apart by alpha rays and for studying radioactivity. "On consideration, I realized that this scattering backward must be the result of a single collision, and when I made calculations I saw that it was impossible to get anything of that order of magnitude unless you took a system in which the greater part of the mass of the atom was concentrated in a minute nucleus. It was then that I had the idea of an atom with a minute massive center carrying a charge." (*Photo courtesy of AIP Niels Bohr Library*)

Charge and Mass

The proton carries a single positive charge, equal in magnitude to the electron charge (where $e = 1.602\ 177\ 3 \times 10^{-19}$ C). The neutron is electrically neutral, as its name implies. Because the neutron has no charge, it is difficult to detect.

Atomic mass (the mass of an atom containing a nucleus and Z electrons) can be measured with great precision with the help of the mass spectrometer and the analysis of nuclear reactions. The proton is approximately 1836 times as massive as the electron, and the masses of the proton and the neutron are almost equal. It is convenient to define, for atomic masses, the atomic mass unit, u, in such a way that the mass of the isotope ^{12}C is exactly 12 u. That is, the mass of an atom is measured relative to the mass of an atom of the neutral carbon-12 isotope (the nucleus plus six electrons). Thus the mass of ^{12}C is exactly 12 u, where 1 u = $1.660\ 540 \times 10^{-27}$ kg. The proton and neutron each have a mass of approximately 1 u, and the electron has a mass that is only a small fraction of an atomic mass unit:

$$\text{Mass of proton} = 1.007\ 276\ 5 \text{ u}$$

$$\text{Mass of neutron} = 1.008\ 664\ 9 \text{ u}$$

$$\text{Mass of electron} = 0.000\ 548\ 579\ 90 \text{ u}$$

Because the rest energy of a particle is given by $E = mc^2$, it is often convenient to express the atomic mass unit in terms of its rest-energy equivalent. For one atomic mass unit, we have

$$E = mc^2 = (1.660\ 540 \times 10^{-27} \text{ kg}) \frac{(2.997\ 924\ 6 \times 10^8 \text{ m/s})^2}{(1.602\ 177\ 3 \times 10^{-19} \text{ J/eV})}$$

$$= 931.494\ 3 \text{ MeV}$$

As noted in Chapter 1, physicists often express mass in terms of the unit MeV/c^2, so here the mass of 1 u is

$$1 \text{ u} \equiv 931.494\ 3 \ \frac{\text{MeV}}{c^2}$$

Table 13.1 gives the rest masses of the proton, neutron, and electron. The masses and some other properties of selected isotopes are provided in Appendix B.

Table 13.1 Rest Masses of the Proton, Neutron, and Electron in Various Units

Particle	Mass		
	kg	u	MeV/c^2
Proton	$1.672\ 623 \times 10^{-27}$	1.007 276	938.272 3
Neutron	$1.674\ 929 \times 10^{-27}$	1.008 665	939.565 6
Electron	$9.109\ 390 \times 10^{-31}$	$5.48\ 579\ 9 \times 10^{-4}$	0.510 999 1

EXAMPLE 13.1   The Atomic Mass Unit

Use Avogadro's number to show that the atomic mass unit is 1 u = 1.66 × 10⁻²⁷ kg.

Solution   We know that exactly 12 g of ¹²C contains Avogadro's number of atoms. Avogadro's number, N_A, has the value 6.02 × 10²³ atoms/mol. Thus the mass of one carbon atom is

$$\text{Mass of one }^{12}\text{C atom} = \frac{0.012 \text{ kg}}{6.02 \times 10^{23} \text{ atoms}}$$

$$= 1.99 \times 10^{-26} \text{ kg}$$

Since one atom of ¹²C is defined to have a mass of 12 u, we find that

$$1 \text{ u} = \frac{1.99 \times 10^{-26} \text{ kg}}{12} = 1.66 \times 10^{-27} \text{ kg}$$

Size and Structure of Nuclei

The size and structure of nuclei were first investigated in Rutherford's scattering experiments, discussed in Section 3.2. In those experiments, Rutherford directed positively charged nuclei of helium atoms (alpha particles) at a thin piece of metal foil. As the alpha particles moved through the foil, they often passed near a nucleus of the metal. Because of the positive charge on both the incident particles and the nuclei, particles were deflected from their straight-line paths by the Coulomb repulsive force. In fact, some particles were even deflected backward, through an angle of 180° from the incident direction. Those particles were apparently moving directly toward a nucleus, in a head-on collision course.

Rutherford employed an energy calculation and found an expression for the distance, d, at which a particle approaching a nucleus is turned around by Coulomb repulsion. In such a head-on collision, the kinetic energy of the incoming alpha particle must be converted completely to electrical potential energy when the particle stops at the point of closest approach and turns around (Fig. 13.1). If we equate the initial kinetic energy of the alpha particle to the electrical potential energy of the system (alpha particle plus target nucleus), we have

$$\frac{1}{2} mv^2 = k \frac{q_1 q_2}{r} = k \frac{(2e)(Ze)}{d}$$

Solving for d, the distance of closest approach, we get

$$d = \frac{4kZe^2}{mv^2}$$

Figure 13.1   An alpha particle on a head-on collision course with a nucleus of charge Ze. Because of the Coulomb repulsion between the like charges, the alpha particle approaches to a distance d from the nucleus, called the distance of closest approach.

From this expression, Rutherford found that 7-MeV alpha particles approached nuclei to within 3.2 × 10⁻¹⁴ m when the foil was made of gold. Thus the radius of the gold nucleus must be less than this value. For silver atoms, the distance of closest approach was found to be 2 × 10⁻¹⁴ m. From these results, Rutherford concluded that the positive charge in an atom is concentrated in a small sphere with a radius of approximately 10⁻¹⁴ m, which he called the nucleus. Because such small dimensions are common in nuclear physics, a convenient

Figure 13.2 James Chadwick (1891–1974). Chadwick, a student of Rutherford, was one of the greatest experimental nuclear physicists. He was awarded the 1935 Nobel prize for the discovery of the neutron. *(Photo by Bortzells Esselte, courtesy AIP Emilo Segrè Visual Archives)*

length unit is the femtometer (fm), sometimes called the **fermi,** defined as

$$1 \text{ fm} \equiv 10^{-15} \text{ m}$$

In the early 1920s it was known from the work of Rutherford, Bohr, and Moseley that the nucleus contained Z protons with a charge $+Ze$ but had a mass approximately equal to that of A protons, where $A \cong 2Z$. To account for the nuclear mass, Rutherford proposed that the nucleus contained $(A - Z)$ neutral proton-electron combination particles in addition to the Z protons. He called the hypothetical combination particle a neutron. This hypothesis prompted an experimental search to confirm the existence of the neutron in a free state outside the nucleus—a search that culminated in 1932. In that year, the important experimental work of the British physicist James Chadwick (1891–1974) clearly showed the existence of a particle with a charge of zero and a mass of approximately 1 u. Chadwick (Fig. 13.2) was awarded the 1935 Nobel prize for this basic discovery.

Although Rutherford's idea that the neutron was a neutral proton-electron combination led to the important search for the neutron, both it and the idea that electrons exist inside the nucleus and can leak out in beta decay have since been abandoned. The current view is that the neutron is a fundamental particle, not a proton-electron combination, and that the beta particles (electrons) emitted in radioactive nuclear decay are created at the moment of decay. As mentioned in Chapter 4 (see Example 4.11), if electrons were confined in the nucleus, the uncertainty principle would require them to possess unrealistically large kinetic energies compared to the energies actually observed in beta decay.

Another argument against electrons in the nucleus concerns the magnetic moments of nuclei, which are observed to be approximately 2000 times smaller than the magnetic moment of the electron. As we saw in Chapter 8, the magnetic moment of the electron is of the order of a Bohr magneton, $\mu_B = e\hbar/2m_e$, where m_e is the electronic mass. Thus, if the same kind of theory accounts for the magnetic moment of a nucleus and there are no electrons in the nucleus, we expect the nuclear magnetic moment to be of the order of a nuclear magneton, $\mu_n = e\hbar/2m_p$, where m_p is the mass of a proton. Since m_p is about 2000 times m_e, this model accounts for the observed size difference between nuclear and electronic magnetic moments. If there were electrons in the nucleus, their much larger moments would dominate and nuclei would have magnetic moments of the order of $e\hbar/2m_e$, in disagreement with actual observations.

Table 13.2 lists important properties of the electron, proton, and neutron. This table presents yet another reason for discarding the idea that a neutron is a bound electron-proton pair: the spin of the neutron is $\frac{1}{2}$ and cannot be composed from the electron and proton spins, both of which are $\frac{1}{2}$.

Since the time of Rutherford's scattering experiments, a multitude of other experiments have shown that most nuclei are approximately spherical and have an average radius of

Nuclear radius

$$r = r_0 A^{1/3} \tag{13.1}$$

where A is the mass number and r_0 is a constant equal to 1.2×10^{-15} m. Because the volume of a sphere is proportional to the cube of the radius, it follows from Equation 13.1 that the volume of a nucleus (assumed to be spherical) is

Table 13.2   Masses, Spins, and Magnetic Moments of the Proton, Neutron, and Electron

Particle	Mass (MeV/c^2)	Spin	Magnetic Moment
Proton	938.28	$\frac{1}{2}$	$2.7928\mu_n$
Neutron	939.57	$\frac{1}{2}$	$-1.9135\mu_n$
Electron	0.510 99	$\frac{1}{2}$	$-1.0012\mu_B$

directly proportional to A, the total number of nucleons. This suggests that *all nuclei have nearly the same density.* When the nucleons combine to form a nucleus, they combine as though they were tightly packed spheres (Fig. 13.3). This fact has led to an analogy between the nucleus and a drop of liquid, in which the density of the drop is independent of its size. We shall discuss the liquid-drop model in Section 13.3.

Figure 13.3   A nucleus can be modeled as a cluster of tightly packed spheres, each of which is a nucleon.

EXAMPLE 13.2   The Volume and Density of a Nucleus

Find (a) an approximate expression for the mass of a nucleus of mass number A, (b) an expression for the volume of this nucleus in terms of the mass number, and (c) a numerical value for its density.

Solution   (a) The mass of the proton is approximately equal to that of the neutron. Thus, if the mass of one of these particles is m, the mass of the nucleus is approximately Am.

(b) Assuming that the nucleus is spherical and using Equation 13.1, we find that the volume is

$$V = \tfrac{4}{3}\pi r^3 = \tfrac{4}{3}\pi r_0^3 A$$

(c) The nuclear density can be found as follows:

$$\rho_n = \frac{\text{mass}}{\text{volume}} = \frac{Am}{\tfrac{4}{3}\pi r_0^3 A} = \frac{3m}{4\pi r_0^3}$$

Taking $r_0 = 1.2 \times 10^{-15}$ m and $m = 1.67 \times 10^{-27}$ kg, we find that

$$\rho_n = \frac{3(1.67 \times 10^{-27} \text{ kg})}{4\pi(1.2 \times 10^{-15} \text{ m})^3} = 2.3 \times 10^{17} \frac{\text{kg}}{\text{m}^3}$$

The nuclear density is approximately 2.3×10^{14} times as great as the density of water ($\rho_{\text{water}} = 1.0 \times 10^3$ kg/m³)!

Nuclear Stability

The nucleus consists of a closely packed collection of protons and neutrons. Like charges (the protons) in proximity exert very large repulsive electrostatic forces on each other, which should cause the nucleus to fly apart. However, nuclei are stable because of the presence of the **nuclear force.** This is an attractive force, with a very short range (about 2 fm), that acts between all nuclear particles. The protons attract each other via the nuclear force, and at the same time they repel each other through the Coulomb force. The attractive nuclear force also acts between pairs of neutrons and between neutrons and protons.

There are approximately 260 stable nuclei; hundreds of other nuclei have been observed but these are unstable. Figure 13.4 is a plot of N versus Z for some stable nuclei. Note that light nuclei are most stable if they contain equal numbers of protons and neutrons — that is, if $N = Z$. Also note that heavy nuclei are more stable if the number of neutrons exceeds the number of protons. We can understand this by recognizing that, as the number of protons increases,

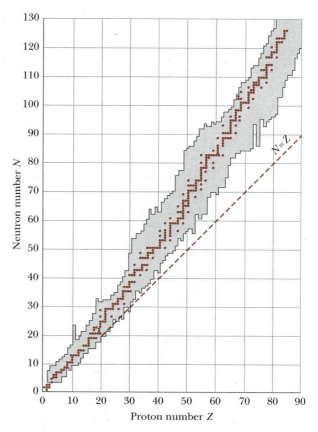

Figure 13.4 A plot of neutron number N versus atomic number Z for the stable nuclei (solid points). The dashed line, corresponding to the condition $N = Z$, is called the line of stability. The shaded area shows radioactive nuclei.

the strength of the Coulomb force increases, which tends to break the nucleus apart. As a result, more neutrons are needed to keep the nucleus stable, since neutrons experience only the attractive nuclear forces. Eventually, the repulsive forces between protons cannot be compensated for by the addition of more neutrons; this occurs when $Z = 83$. Elements that contain more than 83 protons do not have stable nuclei.

Interestingly, most stable nuclei have even values of A. In fact, certain values of Z and N correspond to unusually high stability in nuclei. These values of N and Z, called **magic numbers,** are

$$Z \text{ or } N = 2, 8, 20, 28, 50, 82, 126 \tag{13.2}$$

For example, the helium nucleus (two protons and two neutrons), with $Z = 2$ and $N = 2$, is very stable. The unusual stability of nuclei with progressively larger magic numbers suggests a shell structure for the nucleus that is similar to atomic shell structure. In Section 13.3 we briefly treat the shell model of the nucleus, which explains magic numbers.

Nuclear Spin and Magnetic Moment

In Chapter 8 we discussed the fact that an electron has an intrinsic angular momentum associated with its spin. A nucleus, like an electron, has an intrinsic angular momentum that arises from relativistic properties. The magnitude of the nuclear angular momentum is $\sqrt{I(I+1)}\hbar$, where I is a quantum number called the nuclear spin and may be an integer or a half-integer. Nuclear spin is the total angular momentum of all the nucleons, including both orbital and spin angular momentum of each nucleon. The maximum component of the angular momentum projected along the z axis is $I\hbar$. Figure 13.5 illustrates the possible orientations of the nuclear spin and its projections along the z axis for the case where $I = \frac{3}{2}$.

The nuclear angular momentum has a nuclear magnetic moment associated with it, similar to that of the electron. The magnetic moment of a nucleus is measured in terms of the **nuclear magneton** μ_n, a unit of moment defined as

$$\mu_n \equiv \frac{e\hbar}{2m_\mathrm{p}} = 5.05 \times 10^{-27} \text{ J/T} \qquad (13.3)$$

where m_p is the proton mass. This definition is analogous to that of the Bohr magneton, μ_B, which corresponds to the spin magnetic moment of a free electron (Section 8.2). Note that μ_n is smaller than μ_B by a factor of approximately 2000 due to the large difference in masses of the proton and electron.

The magnetic moment of a free proton is approximately $2.7928\mu_n$. Unfortunately, there is no general theory of nuclear magnetism that explains this value. Another surprising point is that a neutron also has a magnetic moment, with a value of $-1.9135\mu_n$. The existence of a magnetic moment is evidence that the uncharged neutron has an internal charge distribution. The minus sign indicates that this moment is opposite the spin angular momentum of the neutron.

Figure 13.5 The possible orientations of the nuclear spin and its projections along the z axis for the case $I = \frac{3}{2}$.

Nuclear Magnetic Resonance and MRI

Nuclear magnetic moments (as well as electronic magnetic moments) precess when placed in an external magnetic field. The frequency at which they precess, called the **Larmor precessional frequency** ω_L, is directly proportional to the magnetic field. This frequency is described schematically in Figure 13.6a, where the magnetic field is along the z axis. For example, the Larmor frequency of a proton in a 1-T magnetic field is 42.577 MHz. The potential energy of a magnetic dipole moment in an external magnetic field is given by $-\boldsymbol{\mu} \cdot \mathbf{B}$. When the projection of $\boldsymbol{\mu}$ is along the field, the potential energy of the dipole moment is $-\mu B$; that is, it has its minimum value. When the projection of $\boldsymbol{\mu}$ is against the field, the potential energy is μB and it has its maximum value. Figure 13.6b shows these two energy states for a nucleus with a spin of $\frac{1}{2}$.

It is possible to observe transitions between these two spin states through a technique called **nuclear magnetic resonance (NMR)**. A dc magnetic field is introduced to align the magnetic moments (Fig. 13.6a), along with a second, weak, oscillating magnetic field oriented perpendicular to $\mathbf{B}$. When the frequency of the oscillating field is adjusted to match the Larmor precessional frequency, a torque acting on the precessing moments causes them to "flip"

Larmor frequency

Nuclear magnetic resonance

Figure 13.6 (a) When a nucleus is placed in an external magnetic field, **B**, the magnetic moment precesses about the magnetic field with a frequency proportional to the field. (b) A proton, whose spin is $\frac{1}{2}$, can occupy one of two energy states when placed in an external magnetic field. The lower energy state E_1 corresponds to the case where the spin is aligned with the field, and the higher energy state E_2 corresponds to the case where the spin is opposite the field. The reverse is true for an electron in a magnetic field.

between the two spin states. These transitions result in a net absorption of energy by the spin system, an absorption that can be detected electronically. Figure 13.7 is a diagram of the apparatus used in NMR. The absorbed energy is supplied by the generator producing the oscillating magnetic field. Nuclear magnetic resonance and a related technique called electron spin resonance are extremely important methods of studying nuclear and atomic systems and how those systems interact with their surroundings. Figure 13.8 shows a typical NMR spectrum.

Figure 13.7 An experimental arrangement for nuclear magnetic resonance. The radiofrequency magnetic field of the coil, provided by the variable-frequency oscillator, must be perpendicular to the dc magnetic field. When the nuclei in the sample meet the resonance condition, the spins absorb energy from the field of the coil, and this absorption changes the Q of the circuit in which the coil is included. Most modern NMR spectrometers use superconducting magnets at fixed field strengths and operate at frequencies of approximately 200 MHz.

Figure 13.8   An NMR spectrum of ^{31}P in a bridged metallic complex containing platinum. The lines that flank the central strong peak are due to the interaction between ^{31}P and other distant ^{31}P nuclei. The outermost set of lines is due to the interaction between ^{31}P and neighboring platinum nuclei. The spectrum was recorded at a fixed magnetic field of about 4 T, and the mean frequency was 200 MHz.

A widely used diagnostic technique called **magnetic resonance imaging (MRI)** is based on nuclear magnetic resonance. In MRI, the patient is placed inside a large solenoid that supplies a spatially varying magnetic field. Because of the gradient in the magnetic field, protons in different parts of the body precess at different frequencies, and so the resonance signal can provide information on the positions of the protons. A computer is used to analyze the position information to provide data for the construction of a final image. Figure 13.9 shows some MRI scans taken on a human head.

The main advantage of MRI over other imaging techniques in medical diagnostics is that it causes minimal damage to cellular structures. Photons associated with the radio-frequency signals used in MRI have energies of only about 10^{-7} eV. Because molecular bond strengths are much greater (approximately 1 eV), the radio-frequency radiation causes little cellular damage. In comparison, x-rays and gamma rays have energies ranging from 10^4 to 10^6 eV and can cause considerable cellular damage.

Figure 13.9   A computer-digitized MRI scan of a brain with a glioma tumor. (*1996 Scott Camazine*)

13.2   BINDING ENERGY AND NUCLEAR FORCES

The total mass of a nucleus is always less than the sum of the masses of its nucleons. Because mass is a measure of energy, *the total energy of the bound system (the nucleus) is less than the combined energy of the separated nucleons.* This difference in energy is called the **binding energy** of the nucleus and can be thought of as the energy that must be added to a nucleus to break it apart into its components. Therefore, in order to separate a nucleus into its constituent protons and neutrons, energy must be put into the system.

Conservation of energy and the Einstein mass-energy equivalence relationship show that the binding energy of any nucleus of mass M_A is

$$E_b \, (\text{MeV}) = [ZM(\text{H}) + Nm_n - M(_Z^A\text{X})] \times 931.494 \, \frac{\text{MeV}}{\text{u}}   \qquad (13.4)$$

Binding energy of a nucleus

where $M(H)$ is the atomic mass of the neutral hydrogen atom, $M(_Z^A X)$ represents the atomic mass of the element $(_Z^A X)$, m_n is the mass of the neutron, and the masses are all in atomic mass units. Note that the mass of the Z electrons included in the first term in Equation 13.4 cancels with the mass of the Z electrons included in the term $M(_Z^A X)$, within a small difference associated with the atomic binding energy of the electrons. Since atomic binding energies are typically several electron volts and nuclear binding energies are several MeV, this difference is negligible.

EXAMPLE 13.3 The Binding Energy of the Deuteron

Calculate the binding energy of the deuteron, which consists of a proton and a neutron, given that the atomic mass of the deuteron is 2.014 102 u.

Solution We know that the atomic mass of hydrogen and the nuclear mass of the neutron are

$$M(H) = 1.007\ 825\ u$$

$$m_n = 1.008\ 665\ u$$

We can use atomic masses for these calculations, since the electron masses cancel. Therefore,

$$M(H) + m_n = 2.016\ 490\ u$$

To calculate the mass difference, we subtract the deuteron mass from this value:

$$\Delta m = [M(H) + m_n] - M(D)$$

$$= 2.016\ 490\ u - 2.014\ 102\ u$$

$$= 0.002\ 388\ u$$

Since 1 u corresponds to an equivalent energy of 931.494 MeV, the mass difference corresponds to the binding energy

$$E_b = (0.002\ 388\ u)(931.5\ MeV/u) = 2.224\ MeV$$

This result tells us that, in order to separate a deuteron into its constituent parts (a proton and a neutron), it is necessary to add 2.224 MeV of energy to the deuteron. One way of supplying the deuteron with this energy is by bombarding it with energetic particles.

If the binding energy of a nucleus were zero, the nucleus would separate into its constituent protons and neutrons without the addition of any energy; that is, it would spontaneously break apart.

A plot of binding energy per nucleon, E_b/A, as a function of mass number for various stable nuclei is shown in Figure 13.10. Except for the lighter nuclei, the average binding energy per nucleon is about 8 MeV. For the deuteron, the average binding energy per nucleon is $E_b/A = 2.224/2\ MeV = 1.112\ MeV$. Note that the curve in Figure 13.10 peaks in the vicinity of $A = 60$. That is, nuclei with mass numbers greater or less than 60 are not as strongly bound as those near the middle of the periodic table. The higher values of binding energy near $A = 60$ imply that energy is released when a heavy nucleus with $A \approx 200$ splits or fissions into several lighter nuclei that lie near $A = 60$. Energy is released in fission because the final state, consisting of two lighter fragments, is more tightly bound, or lower in energy, than the original nucleus. Similarly, energy can be released when two light nuclei with $A \leq 20$ combine or fuse to form one heavier nucleus. These two important processes, *fission* and *fusion*, are considered in detail in Chapter 14.

Another important feature of Figure 13.10 is that the binding energy per nucleon is approximately constant for $A > 20$. In this case, the nuclear forces between a particular nucleon and all the other nucleons in the nucleus are said

Saturation of nuclear forces

to be saturated; that is, a particular nucleon forms attractive bonds with only a limited number of other nucleons. Because of the short-range character of the

Figure 13.10   A plot of binding energy per nucleon versus mass number for the stable nuclei shown in Figure 13.4.

nuclear force, these other nucleons can be viewed as being the nearest neighbors in the close-packed structure shown in Figure 13.3. If every nucleon could interact with every other nucleon, each nucleon would form $(A - 1)$ bonds, and the binding energy per nucleon would be proportional to $(A - 1)$ rather than constant as observed.

The general features of the nuclear binding force have been revealed in a wide variety of experiments. We summarize them as follows.

- The attractive nuclear force is a different kind of force from the common forces of electromagnetism and gravitation, and since it dominates the repulsive Coulomb force between protons in the nucleus, it is stronger than the electromagnetic force.
- The nuclear force is a short-range force that rapidly falls to zero when the separation between nucleons exceeds several fermis. Evidence for the limited range of nuclear forces comes from scattering experiments and from the saturation of nuclear forces already mentioned. The neutron-proton (n-p) potential energy plot of Figure 13.11a, obtained by scattering neutrons from a target containing hydrogen, shows the short range of the nuclear force. The depth of the n-p potential energy is 40 to 50 MeV and contains a strong repulsive component that prevents the nucleons from approaching much closer than 0.4 fm. Another interesting feature of the nuclear force is that its size depends on the relative spin orientations of the nucleons, as shown by scattering experiments performed with spin-polarized beams and targets.
- From n-n, n-p, and p-p scattering experiments and other indirect evidence, it is found that the nuclear force is independent of the electric charge of the interacting nucleons. As might be expected from this "charge-blind" character of the nuclear force, the nuclear force does not affect electrons, and so it enables energetic electrons to serve as point-like probes of the

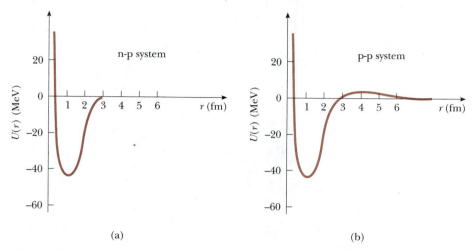

Figure 13.11   (a) Potential energy versus separation for the neutron-proton system. (b) Potential energy versus separation for the proton-proton system. The difference in the two curves is due mainly to the Coulomb repulsion in the case of the proton-proton interaction.

Charge independence of the nuclear force

charge density of nuclei. The charge independence of the nuclear force also means that the main difference between the n-p and p-p interactions is that the p-p potential energy consists of a *superposition* of nuclear and Coulomb interactions, as shown in Figure 13.11b. At distances less than 2 fm, p-p and n-p potential energies are nearly identical, but at distances greater than this, the p-p potential has a positive energy barrier, with a maximum of about 1 MeV at 4 fm.

Although the preceding "laundry list" of observed properties of the nuclear force may be interesting, it leaves unanswered the deeper question of what mechanism or model accounts for the powerful nuclear force. One model that successfully explains the nuclear force is the *exchange force model*. According to this model, two nucleons—a proton and a neutron, for example—experience an attractive force when one spontaneously emits a particle and the other absorbs that particle. From a classical point of view, the emitting and absorbing nucleons are both strongly attracted to the exchanged particle as if connected to it by elastic bands or springs. Since both nucleons are strongly attracted to the same particle, they appear to be strongly attracted to each other.

Particle exchange model

You may wonder how the exchange of a particle of mass m between the proton and neutron can occur without violating conservation of energy and conservation of momentum. A violation of energy conservation must occur if a proton at rest with energy $m_p c^2$ emits a particle of mass m with energy mc^2 and yet remains a proton with energy $m_p c^2$. Also, conservation of momentum would cause a paired proton and neutron to recoil away from each other as the proton emitted a particle and the neutron absorbed it, instead of producing motion of the proton toward the neutron as required for an attractive nuclear force.

The answer to both dilemmas seems to be a trick: the exchanged particle exists for such a short time and is confined to such a small region of space that it is a *virtual particle*. A virtual particle cannot be directly detected or measured during the momentum- and energy-violating processes taking place inside the nucleus. Perhaps the best way to understand this is through the uncertainty principles $\Delta E \, \Delta t \geq \hbar/2$ and $\Delta p_x \, \Delta x \geq \hbar/2$. These relationships imply that energy conservation can be violated by an amount ΔE for a short time interval Δt and that momentum conservation can be violated by an amount Δp_x over a small nuclear region of size Δx. Although it may appear that a virtual particle can never be detected, if an incident particle strikes the nucleus in just the right way, it can supply the missing momentum and energy to make a virtual particle real, thereby permitting the exchanged particle to be measured in the laboratory.

A simple application of the energy-time uncertainty principle enables us to estimate the mass of the exchanged particle that should be sought in high-energy collisions. Since the amount by which energy conservation is violated in the exchange process is $\Delta E = mc^2$, where m is the mass of the exchanged particle, we have

$$\Delta t \cong \frac{\hbar}{\Delta E} = \frac{\hbar}{mc^2} \tag{13.5}$$

Realizing that Δt is the time it takes the exchanged particle to travel a distance d between nucleons, and that the maximum value of d is $d = c\,\Delta t$ for an exchanged particle traveling at the speed of light, we find that

$$d_{max} = c\,\Delta t = c\left(\frac{\hbar}{mc^2}\right) = \frac{\hbar}{mc} \tag{13.6}$$

This expression shows that *the range of the force, d_{max}, is inversely proportional to the mass of the exchanged particle*, a general feature of exchange-force models. Finally, because the range of the nuclear force is approximately 2 fm, we can estimate the rest energy of the exchanged particle that is responsible for the nuclear force:

$$mc^2 \cong \frac{\hbar c}{d_{max}} = \frac{(1.05 \times 10^{-34}\,\text{J} \cdot \text{s})(3.00 \times 10^8\,\text{m/s})}{2.0 \times 10^{-15}\,\text{m}}$$

$$= 1.6 \times 10^{-11}\,\text{J} = 100\,\text{MeV}$$

Twelve years after the Japanese physicist Hideki-Yukawa (1907–1981) proposed the exchange theory of the nuclear force just described, a strongly interacting particle with a mass of about 140 MeV/c^2, called the pi meson, was found in cosmic ray interactions, closely confirming his predictions. We discuss the pi meson and other exchanged particles more fully in Chapter 15.

13.3   NUCLEAR MODELS

Although the details of nuclear forces are still not completely understood, several nuclear models with adjustable parameters have been proposed, and they help us understand various features of nuclear experimental data and the mechanisms responsible for binding energy. The models we shall discuss are (1) the liquid-drop model, which accounts for the nuclear binding energy; (2) the independent-particle model, which accounts for more detailed features of nuclei, including large differences in stability between nuclei with similar Z and A values; and (3) the collective model.

Liquid-Drop Model

The **liquid-drop model**, proposed by C. F. von Weizsächer in 1935, treats the nucleons as though they were molecules in a drop of liquid. The nucleons interact strongly with each other and undergo frequent collisions as they jiggle around within the nucleus. This jiggling motion is analogous to the thermally agitated motions of molecules in a drop of liquid.

Three major effects influence the binding energy of the nucleus in the liquid-drop model:

- **The volume effect.** Earlier we showed that the binding energy per nucleon is approximately constant, indicating that the nuclear force saturates (Fig. 13.10). Therefore the total binding energy of the nucleus is proportional to A and to the nuclear volume. The contribution to the binding energy of this volume effect is C_1A, where C_1 is an adjustable constant.
- **The surface effect.** Because nucleons on the surface of the drop have fewer neighbors than those in the interior, surface nucleons reduce the binding energy by an amount proportional to the number of surface nucleons. Because the number of surface nucleons is proportional to the surface area of the nucleus, $4\pi r^2$, and $r^2 \propto A^{2/3}$ (Eq. 13.1), the surface term can be expressed as $-C_2A^{2/3}$, where C_2 is a constant.
- **The Coulomb repulsion effect.** Each proton repels every other proton in the nucleus. The corresponding potential energy per pair of interacting particles is ke^2/r, where k is the Coulomb constant. The total Coulomb energy represents the work required to assemble Z protons from infinity to a sphere of volume V. This energy is proportional to the number of proton pairs $Z(Z-1)/2$ and inversely proportional to the nuclear radius. Consequently, the reduction in energy that results from the Coulomb effect is $-C_3Z(Z-1)/A^{1/3}$, where C_3 is a constant.

Another effect that decreases the binding energy is significant for heavy nuclei with large excesses of neutrons. Since it is observed that for a given A value, nuclei with $Z = N$ have the largest binding energy, the binding energy must be corrected by a symmetry term that favors $Z = N$ and decreases the binding energy symmetrically for $N > Z$ or $N < Z$. The symmetry term is generally written as $-C_4(N-Z)^2/A$, where C_4 is an adjustable constant.

Adding these contributions, we get as the total binding energy

Semiempirical binding-energy formula

$$E_b = C_1A - C_2A^{2/3} - C_3\frac{Z(Z-1)}{A^{1/3}} - C_4\frac{(N-Z)^2}{A} \tag{13.7}$$

This equation is often referred to as the Weizsächer **semiempirical binding energy formula,** because it has some theoretical justification but contains four constants that are adjusted to fit this expression to experimental data. For nuclei with $A \geq 15$, the constants have the values

$$C_1 = 15.7 \text{ MeV} \qquad C_2 = 17.8 \text{ MeV}$$
$$C_3 = 0.71 \text{ MeV} \qquad C_4 = 23.6 \text{ MeV}$$

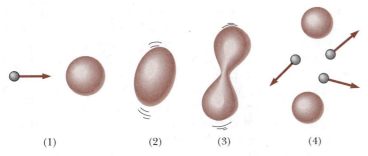

Figure 13.12 Steps leading to fission according to the liquid-drop model of the nucleus.

Equation 13.7, together with these constants, fits the known nuclear mass values very well. However, the liquid drop model does not account for some finer details of nuclear structure, such as certain stability rules and angular momentum. On the other hand, it does provide a qualitative description of nuclear fission, shown schematically in Figure 13.12. If the drop vibrates with a large amplitude (which may be initiated by collision with another particle), it distorts, and under the right conditions, it breaks apart. We shall discuss the process of fission further in Chapter 14.

Independent-Particle Model

The **independent-particle model,** often called the *shell model,* is based upon the assumption that *each nucleon moves in a well-defined orbital state within the nucleus in an averaged field produced by the other nucleons.* This model is similar to the shell model of the atom except for the character of the force term. In the shell model, the nucleons exist in quantized energy states, and there are few collisions between nucleons. Obviously, the assumptions of this model differ greatly from those made in the liquid-drop model.

Each quantized orbital state for a proton or neutron is analogous to one of the orbital states of an electron in an atomic subshell, with the exception that the ordering of nuclear states in energy is more closely modeled by a spherical-well potential than by the Coulomb potential that is appropriate for atomic electrons. (See Problem 32 in Chapter 7 for a solution of the spherical well.) The quantized states occupied by the nucleons can be described by a set of quantum numbers. Since both the proton and the neutron have spin $\frac{1}{2}$, we can apply the Pauli exclusion principle to describe the allowed states (as we did for electrons in Chapter 8). That is, each orbital state can contain only two protons (or two neutrons) having *opposite* spin (Fig. 13.13). The protons have a set of allowed states, which differ slightly from those of the neutrons because they move in different potential wells. The proton levels are higher in energy than the neutron levels because the protons experience a superposition of Coulomb potential and nuclear potential, while the neutrons experience only a nuclear potential. These two different wells are sketched in Figure 13.13.

Using only the fact that nuclear energy levels exist and that they fill according to the Pauli exclusion principle, one can explain the tendency for N to equal Z in stable, light nuclei. To achieve maximum stability for a given A, all the

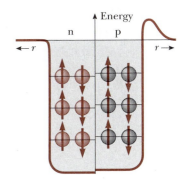

Figure 13.13 A square-well potential containing 12 nucleons. The gray circles represent protons, and the colored circles represent neutrons. The energy levels for the protons are slightly higher than those for the neutrons because of the Coulomb potential in the case of the protons. The difference in the levels increases as Z increases. Note that only two nucleons with opposite spin can occupy a given level, as required by the Pauli exclusion principle.

nucleons must be packed sequentially into the lowest energy levels, leaving no gaps in filled levels. Since any energy state is filled when it contains two protons (or two neutrons), another proton (or neutron) added to the nucleus produces an increase in energy and a decrease in stability of the nucleus. Thus, comparing $^{12}_{5}B$ and $^{12}_{6}C$, the seventh neutron in boron must occupy a higher energy level than the sixth neutron in carbon (two nucleons per energy state), and consequently $^{12}_{5}B$ has higher energy and is less stable than $^{12}_{6}C$.

To understand the observed characteristics of nuclear ground states, such as angular momentum and the high stability of magic-number nuclei, it is necessary to include *nuclear spin-orbit effects*. Unlike the spin-orbit interaction for an electron in an atom, which is magnetic in origin, the spin-orbit effect for nucleons in a nucleus is due to the nuclear force. It is much stronger than in the atomic case and has opposite sign, so that in nuclei, spin-orbit split states with higher angular momentum lie lower in energy. When the nuclear spin-orbit effect is added to a spherical finite potential, the magic numbers are predicted, because the spin-orbit potential produces especially large jumps between certain energy levels. In effect, the spin-orbit interaction substantially raises the energy levels containing 9, 21, 29, 51, 83, and 127 nucleons, thereby making the levels that contain 8, 20, 28, 50, 82, and 126 nucleons unusually stable.

Finally, it is possible to understand how individual nucleons can be considered to be moving in well-defined orbitals within the cramped confines of the nucleus, which is literally swarming with other nucleons. Under these circumstances, it would seem that a given nucleon would undergo many collisions and not move in a well-defined orbital. However, for the low-energy nuclear ground state, the exclusion principle inhibits energy-changing collisions by preventing colliding nucleons from occupying already filled low-lying energy states. In effect, the exclusion principle prevents nucleon collisions within the densely packed nucleus at low energy and justifies the shell-model approach.

Collective Model

A third model of nuclear structure, known as the **collective model,** combines some features of the liquid-drop model and the independent-particle model. The nucleus is considered to have some "extra" nucleons moving in quantized orbits in addition to the filled core of nucleons. The extra nucleons are subject to the field produced by the core, as in the independent-particle scheme. Deformations can be set up in the core as a result of a strong interaction between the core and the extra nucleons, thereby initiating vibrational and rotational motions, as in the liquid-drop model. The collective model has been very successful in explaining many nuclear phenomena.

13.4   RADIOACTIVITY

In 1896 Henri Becquerel (1852–1908) accidentally discovered that uranyl potassium sulfate crystals emitted an invisible radiation that could darken a photographic plate when the plate was covered to exclude light. After a series of experiments, he concluded that the radiation emitted by the crystals was of a new type, one that required no external stimulation and was so penetrating that it could darken protected photographic plates and ionize gases. This process of spontaneous emission of radiation by uranium was soon to be called **radioactivity.** Subsequent experiments by other scientists showed that other substances were more powerfully radioactive. Marie (1867–1934) and Pierre Curie (1859–1906) conducted the most significant investigations of this type. After several years of careful and laborious chemical separation processes on tons of pitchblende, a radioactive ore, the Curies reported the discovery of two previously unknown elements, both radioactive, which they named polonium and radium. Subsequent experiments, including Rutherford's famous work on alpha-particle scattering, suggested that radioactivity was the result of the decay, or disintegration, of unstable nuclei.

Three types of radiation can be emitted by a radioactive substance: alpha (α), in which the emitted particles are ^{4}He nuclei; beta (β), in which the emitted particles are either electrons or positrons; and gamma (γ), in which the emitted "rays" are high-energy photons. A **positron** is a particle like the electron in all respects except that the positron has a charge of $+e$ (in other words, the positron is the antimatter twin of the electron). The symbol e$^-$ is used to designate an electron, and e$^+$ designates a positron.

B I O G R A P H Y

——

MARIE SKLODOWSKA CURIE
(1867–1934)

Marie Sklodowska Curie was born in Poland shortly after the unsuccessful Polish revolt against Russia in 1863. After high school, she worked diligently to help meet the educational expenses of her older brother and sister in Paris. At the same time, she managed to save enough money for her own move to Paris, and entered the Sorbonne in 1891. Although she lived very frugally during this period (she once fainted from hunger in the classroom), she graduated at the top of her class.

In 1895 Marie Sklodowska married the French physicist Pierre Curie (1859–1906), who was already known for the discovery of piezoelectricity. (A piezoelectric crystal exhibits a potential difference under pressure.) Using piezoelectric materials to measure the activity of radioactive substances, Marie Curie demonstrated the radioactive nature of the elements uranium and thorium. In 1898 she and her husband discovered a new radioactive element contained in uranium ore, which they called polonium after her native land. By the end of 1898, the Curies succeeded in isolating trace amounts of an even more radioactive element, which they named radium. In an effort to produce weighable quantities of radium, they took on the painstaking job of isolating radium from pitchblende, an ore rich in uranium. After four years of purifying and repurifying tons of ore, and using their own life savings to finance their work, the Curies succeeded in preparing about 0.1 g of radium. In 1903, along with Henri Becquerel, they received the Nobel prize in physics for their studies of radioactive substances.

After her husband's death in an accident in 1906, Marie Curie assumed his professorship at the Sorbonne. Unfortunately, she experienced prejudice in the scientific community because she was a woman. For example, after being nominated to the French Academy of Sciences, she was refused membership after losing by one vote.

In 1911 Marie Curie was awarded a second Nobel prize, this one in chemistry, for the discovery of radium and polonium. She spent the last few decades of her life supervising the Paris Institute of Radium.

(Courtesy of AIP Niels Bohr Library)

It is possible to distinguish these three forms of radiation using the scheme described in Figure 13.14. The radiation from a radioactive sample is directed into a region in which there is a magnetic field. The beam splits into three components, two bending in opposite directions and the third experiencing no change in direction. From this simple observation, we can conclude that the radiation of the undeflected beam carries no charge (the gamma ray), the component deflected upward corresponds to positively charged particles (alpha particles), and the component deflected downward corresponds to negatively charged particles (e^-). If the beam includes a positron (e^+), it is deflected upward.

The three types of radiation have quite different penetrating powers. Alpha particles barely penetrate a sheet of paper, beta particles can penetrate a few millimeters of aluminum, and gamma rays can penetrate several centimeters of lead.

The rate at which a particular decay process occurs in a radioactive sample is proportional to the number of radioactive nuclei present (that is, those nuclei that have not yet decayed). If N is the number of radioactive nuclei present at some instant, the rate of change of N is

$$\frac{dN}{dt} = -\lambda N \tag{13.8}$$

where λ, called the **decay constant,** is the probability per unit time that a nucleus will decay. The minus sign indicates that dN/dt is negative; that is, N is decreasing in time.

If we write Equation 13.8 in the form

$$\frac{dN}{N} = -\lambda \, dt$$

Figure 13.14   The radiation from a radioactive source can be separated into three components through the use of a magnetic field to deflect the charged particles. The photographic plate at the right records the events. The gamma ray is not deflected by the magnetic field.

we can integrate the expression to give

$$\int_{N_0}^{N} \frac{dN}{N} = -\lambda \int_{0}^{t} dt$$

$$\ln\left(\frac{N}{N_0}\right) = -\lambda t$$

or

$$N = N_0 e^{-\lambda t} \tag{13.9}$$

Exponential decay

where the constant N_0 represents the number of radioactive nuclei at $t = 0$. Equation 13.9 shows that the number of radioactive nuclei in a sample decreases exponentially with time.

The **decay rate** R, or the number of decays per unit time, can be obtained by differentiating Equation 13.9 with respect to time:

$$R = \left|\frac{dN}{dt}\right| = N_0 \lambda e^{-\lambda t} = R_0 e^{-\lambda t} \tag{13.10}$$

Decay rate

where $R_0 = N_0 \lambda$ is the decay rate at $t = 0$ and $R = \lambda N$. The decay rate of a sample is often referred to as its **activity.** Note that both N and R decrease exponentially with time. The plot of N versus t shown in Figure 13.15 illustrates the exponential decay law.

Another parameter that is useful in characterizing the decay of a particular nucleus is the **half-life,** $T_{1/2}$:

> The **half-life** of a radioactive substance is the time it takes half of a given number of radioactive nuclei to decay.

Setting $N = N_0/2$ and $t = T_{1/2}$ in Equation 13.9 gives

$$\frac{N_0}{2} = N_0 e^{-\lambda T_{1/2}}$$

Writing this in the form $e^{\lambda T_{1/2}} = 2$ and taking the natural logarithm of both sides, we get

$$T_{1/2} = \frac{\ln 2}{\lambda} = \frac{0.693}{\lambda} \tag{13.11}$$

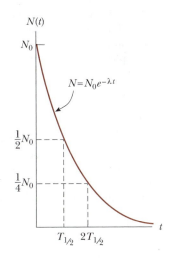

Figure 13.15 A plot of the exponential decay law for radioactive nuclei. The vertical axis represents the number of radioactive nuclei present at any time t, and the horizontal axis is time. The time $T_{1/2}$ is the half-life of the sample.

This is a convenient expression for relating half-life to the decay constant. Note that after an elapsed time of one half-life, $N_0/2$ radioactive nuclei remain (by definition); after two half-lives, half of these have decayed and $N_0/4$ radioactive nuclei are left; after three half-lives, $N_0/8$ are left; and so on. In general, after n half-lives, the number of radioactive nuclei remaining is $N_0/2^n$. Thus we see that nuclear decay is independent of the past history of a sample.

A frequently used unit of activity is the **curie** (Ci), defined as

$$1 \text{ Ci} \equiv 3.7 \times 10^{10} \text{ decays/s}$$

The curie

This value was originally selected because it is the approximate activity of 1 g of radium. The SI unit of activity is the **becquerel** (Bq):

$$1 \text{ Bq} \equiv 1 \text{ decay/s}$$

Therefore, $1 \text{ Ci} = 3.7 \times 10^{10}$ Bq. The curie is a rather large unit, and the more frequently used activity units are the millicurie mCi (10^{-3} Ci) and the microcurie, μCi (10^{-6} Ci)

EXAMPLE 13.4   How Many Nuclei Are Left?

The isotope carbon-14, $^{14}_6$C, is radioactive and has a half-life of 5730 years. If you start with a sample of 1000 carbon-14 nuclei, how many will still be around in 22 920 years?

Solution In 5730 years half the sample will have decayed, leaving 500 carbon-14 nuclei. In another 5730 years (for a total elapsed time of 11 460 years), the number will be reduced to 250 nuclei. After another 5730 years (total time 17 190 years), 125 will remain. Finally, after four half-lives (22 920 years), only about 62 will remain.

These numbers represent ideal circumstances. Calculation of radioactive decay is an averaging process conducted with a very large number of atoms, and the actual outcome depends on statistics. Our original sample in this example contained only 1000 nuclei, certainly not a very large number. Thus, if we actually counted the number remaining from this small sample after one half-life, it probably would not be 500.

EXAMPLE 13.5   The Activity of Radium

The half-life of the radioactive nucleus $^{226}_{88}$Ra is about 1.6×10^3 years. (a) What is the decay constant of $^{226}_{88}$Ra?

Solution We can calculate the decay constant λ by using Equation 13.11 and the fact that

$$T_{1/2} = 1.6 \times 10^3 \text{ years}$$
$$= (1.6 \times 10^3 \text{ years})(3.16 \times 10^7 \text{ s/year})$$
$$= 5.0 \times 10^{10} \text{ s}$$

Therefore,

$$\lambda = \frac{0.693}{T_{1/2}} = \frac{0.693}{5.0 \times 10^{10} \text{ s}} = 1.4 \times 10^{-11} \text{ s}^{-1}$$

Note that this result gives the probability that any single $^{226}_{88}$Ra nucleus will decay in 1 s.

(b) If a sample contains 3.0×10^{16} such nuclei at $t = 0$, determine its activity at this time.

Solution We can calculate the activity of the sample at $t = 0$ using $R_0 = \lambda N_0$, where R_0 is the decay rate at $t = 0$ and N_0 is the number of radioactive nuclei present at

$t = 0$. Since $N_0 = 3.0 \times 10^{16}$, we have

$$R_0 = \lambda N_0 = (1.4 \times 10^{-11} \text{ s}^{-1})(3.0 \times 10^{16})$$
$$= 4.2 \times 10^5 \text{ decays/s}$$

Since $1 \text{ Ci} = 3.7 \times 10^{10}$ decays/s, the activity, or decay rate, at $t = 0$ is

$$R_0 = 11.1 \ \mu\text{Ci}$$

(c) What is the decay rate after the sample is 2.0×10^3 years old?

Solution We can use Equation 13.10 and the fact that $2.0 \times 10^3 \text{ years} = (2.0 \times 10^3 \text{ years})(3.15 \times 10^7 \text{ s/year}) = 6.3 \times 10^{10}$ s:

$$R = R_0 e^{-\lambda t}$$
$$= (4.2 \times 10^5 \text{ decays/s})e^{-(1.4 \times 10^{-11} \text{ s}^{-1})(6.3 \times 10^{10} \text{ s})}$$
$$= 1.7 \times 10^5 \frac{\text{decays}}{\text{s}}$$

EXAMPLE 13.6   The Activity of Carbon

A radioactive sample contains 3.50 μg of pure $^{11}_6$C, which has a half-life of 20.4 min. (a) Determine the number of nuclei in the sample at $t = 0$.

Solution The atomic mass of $^{11}_6$C is approximately 11.0, and so 11.0 g contains Avogadro's number (6.02×10^{23}) of nuclei. Therefore 3.50 μg contains N nuclei, where

$$\frac{N}{6.02 \times 10^{23} \text{ nuclei/mol}} = \frac{3.50 \times 10^{-6} \text{ g}}{11.0 \text{ g/mol}}$$
$$N = 1.92 \times 10^{17} \text{ nuclei}$$

(b) What is the activity of the sample initially and after 8.00 h?

Solution Since $T_{1/2} = 20.4$ min $= 1224$ s, the decay constant is

$$\lambda = \frac{0.693}{T_{1/2}} = \frac{0.693}{1224 \text{ s}} = 5.66 \times 10^{-4} \text{ s}^{-1}$$

Therefore the initial activity of the sample is

$$R_0 = \lambda N_0 = (5.66 \times 10^{-4} \text{ s}^{-1})(1.92 \times 10^{17})$$
$$= 1.08 \times 10^{14} \frac{\text{decays}}{\text{s}}$$

We can use Equation 13.10 to find the activity at any time t. For $t = 8.00$ h $= 2.88 \times 10^4$ s, we see that $\lambda t = 16.3$, and so

$$R = R_0 e^{-\lambda t} = (1.09 \times 10^{14} \text{ decays/s}) e^{-16.3}$$

$$= 8.96 \times 10^6 \frac{\text{decays}}{\text{s}}$$

Table 13.3 lists values for activity versus time in this situation.

Table 13.3   Activity Versus Time for the Sample Described in Example 13.6

t (h)	R (decays/s)
0	1.08×10^{14}
1	1.41×10^{13}
2	1.84×10^{12}
3	2.39×10^{11}
4	3.12×10^{10}
5	4.06×10^9
6	5.28×10^8
7	6.88×10^7
8	8.96×10^6

Exercise 1   Calculate the number of radioactive nuclei that remain after 8.00 h.

Answer 1.58×10^{10} nuclei

EXAMPLE 13.7   A Radioactive Isotope of Iodine

A sample of the isotope ^{131}I, which has a half-life of 8.04 days, has an activity of 5 mCi at the time of shipment. Upon receipt of the ^{131}I in a medical laboratory, its activity is 4.2 mCi. How much time has elapsed between the two measurements?

Solution   We can make use of Equation 13.10 in the form

$$\frac{R}{R_0} = e^{-\lambda t}$$

Taking the natural logarithm of each side, we get

$$\ln \left(\frac{R}{R_0} \right) = -\lambda t$$

$$(1) \qquad t = -\frac{1}{\lambda} \ln \left(\frac{R}{R_0} \right)$$

To find λ, we use Equation 13.11:

$$(2) \qquad \lambda = \frac{0.693}{T_{1/2}} = \frac{0.693}{8.04 \text{ days}}$$

Substituting (2) into (1) gives

$$t = -\left(\frac{8.04 \text{ days}}{0.693} \right) \ln \left(\frac{4.2 \text{ mCi}}{5.0 \text{ mCi}} \right) = 2.02 \text{ days}$$

13.5   DECAY PROCESSES

As we stated in the preceding section, a radioactive nucleus spontaneously decays by means of one of three processes: alpha decay, beta decay, or gamma decay. Let us discuss these processes in more detail.

Alpha Decay

If a nucleus emits an alpha particle ($^4_2$He), it loses two protons and two neutrons. Therefore the atomic number Z decreases by 2, the mass number A decreases by 4, and the neutron number decreases by 2. The decay can be written as

$$^A_Z X \longrightarrow ^{A-4}_{Z-2} Y + ^4_2 He \qquad (13.12) \qquad \textbf{Alpha decay}$$

where X is called the **parent nucleus** and Y the **daughter nucleus.** For example, ^{238}U and ^{226}Ra are both alpha emitters and decay according to the schemes

$$^{238}_{92}\text{U} \longrightarrow ^{234}_{90}\text{Th} + ^{4}_{2}\text{He} \qquad (13.13)$$

$$^{226}_{88}\text{Ra} \longrightarrow ^{222}_{86}\text{Rn} + ^{4}_{2}\text{He} \qquad (13.14)$$

The half-life for ^{238}U decay is 4.47×10^9 years, and that for ^{226}Ra decay is 1.60×10^3 years. In both cases, note that the mass number of the daughter nucleus is 4 less than that of the parent nucleus. Likewise, the atomic number is reduced by 2. The differences are accounted for in the emitted alpha particle (the ^4He nucleus). Observe that alpha decay processes release energy because the decay products, especially the alpha particle, are more tightly bound than the parent nucleus (see Fig. 13.10.)

Figure 13.16 depicts the decay of ^{226}Ra. When one element changes into another, as in the process of alpha decay, the process is called spontaneous decay. As a general rule, (1) the sum of the mass numbers A must be the same on both sides of the equation, and (2) the net charge must be the same on both sides of the equation. In addition, the total relativistic energy and momentum must be conserved. If we call M_X the atomic mass of the parent, M_Y the mass of the daughter, and M_α the mass of the alpha particle, we can define the **disintegration energy** Q:

Disintegration energy Q

$$Q = (M_X - M_Y - M_\alpha)c^2 \qquad (13.15)$$

Note that atomic mass rather than nuclear mass can be used here, because the electronic masses cancel in an evaluation of the mass differences. Q is in joules when the masses are in kilograms and c is the usual 3.00×10^8 m/s. However, when masses are expressed in the more convenient unit u, the value of Q can be calculated in mega-electron volts with the expression

$$Q = (M_X - M_Y - M_\alpha) \times 931.494 \text{ MeV/u} \qquad (13.16)$$

Figure 13.16 Alpha decay of radium. The radium nucleus is initially at rest. After the decay, the radon nucleus has kinetic energy K_{Rn} and momentum $\mathbf{p}_{\text{Rn}}$, and the alpha particle has kinetic energy K_α and momentum $\mathbf{p}_\alpha$.

The disintegration energy Q appears in the form of kinetic energy in the daughter nucleus and the alpha particle. The quantity given by Equation 13.16 is sometimes referred to as the Q value of the nuclear reaction. In the case of the ^{226}Ra decay described in Figure 13.16, if the parent nucleus is at rest when it decays, the residual kinetic energy of the products is 4.87 MeV. Most of the kinetic energy is associated with the alpha particle because this particle is much less massive than the recoiling daughter nucleus, ^{222}Rn. That is, because momentum must be conserved, the lighter alpha particle recoils with a much higher speed than the daughter nucleus. Generally, light particles carry off most of the energy in nuclear decays.

It is fairly easy to calculate the fraction of the disintegration energy carried off by the alpha particle by applying conservation of energy and momentum:

$$Q = K_Y + K_\alpha \qquad (13.17)$$

$$p_Y = p_\alpha \qquad (13.18)$$

where the subscript Y stands for the daughter nucleus. Since the total kinetic energy released in alpha decay (several MeV) is small compared to the rest energies of the alpha particle (3726 MeV) and the daughter nucleus (206.793 BeV for ^{222}Rn), we can use the classical expressions for momentum and kinetic energy in Equations 13.17 and 13.18 to show that

$$K_\alpha = \frac{M_Y}{M_Y + M_\alpha} Q \qquad (13.19)$$

where M_Y and M_α are the atomic masses of the daughter nucleus and the alpha particle (see Problem 43 at the end of the chapter).

Interestingly, if one assumed that ^{238}U (or other alpha emitters) decayed by emitting a proton or neutron, the mass of the decay products would exceed that of the parent nucleus, corresponding to negative Q values. These negative Q values indicate that such decays do not occur spontaneously.

EXAMPLE 13.8 The Energy Liberated When Radium Decays

The ^{226}Ra nucleus undergoes alpha decay according to Equation 13.12. Calculate the Q value for this process. Take the atomic masses to be 226.025 406 u for ^{226}Ra, 222.017 574 u for ^{222}Rn, and 4.002 603 u for $^{4}_{2}$He, as found in Appendix B.

Solution Using Equation 13.16, we see that

$$Q = (M_X - M_Y - M_\alpha) \times 931.494 \frac{\text{MeV}}{\text{u}}$$

$$= (226.025\,406\,\text{u} - 222.017\,574\,\text{u}$$

$$- 4.002\,603\,\text{u}) \times 931.494 \frac{\text{MeV}}{\text{u}}$$

$$= (0.005\,229\,\text{u}) \times \left(931.494 \frac{\text{MeV}}{\text{u}}\right) = 4.87 \text{ MeV}$$

It is left to Problem 43 to show that the kinetic energy of the alpha particle is about 4.8 MeV, whereas the recoiling daughter nucleus has only about 0.1 MeV of kinetic energy.

We now turn to the mechanism of alpha decay. Imagine that an alpha particle is somehow formed within the nucleus. Figure 13.17 is a plot of potential energy versus distance r from the nucleus for a typical alpha particle–nucleus system, where R is the range of the nuclear force. The curve represents the combined effects of (1) the Coulomb repulsive energy, which gives the positive peak for $r > R$, and (2) the nuclear attractive force, which causes the curve to be negative for $r < R$. As we saw in Example 13.8, a typical disintegration energy is about 5 MeV, which is the approximate kinetic energy of the alpha particle, represented by the lower dashed line in Figure 13.17. According to classical physics, the alpha particle is trapped in the potential well. How, then, does it ever escape from the nucleus?

The answer to this question was first provided by George Gamow in 1928 and, independently, by R. W. Gurney and E. U. Condon in 1929, using quantum mechanics. Briefly, the view of quantum mechanics is that there is always some probability that the particle can penetrate (tunnel through) the barrier (Section 6.2). Recall that the probability of locating the particle depends on its wavefunction ψ and that the probability of tunneling is measured by $|\psi|^2$. Figure 13.18 is a sketch of the wavefunction for a particle of energy E meeting a square barrier of finite height, a shape that approximates the nuclear barrier. Note that the wavefunction exists both inside and outside the barrier. Although the amplitude of the wavefunction is greatly reduced on the far side of the barrier, its finite value in this region indicates a small but finite probability that the particle can penetrate the barrier. As the energy E of the particle increases, its

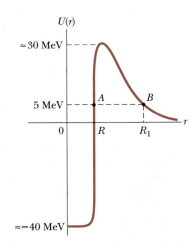

Figure 13.17 Potential energy versus separation for the alpha particle–nucleus system. Classically, the energy of the alpha particle is not great enough to overcome the barrier, and so the particle should not be able to escape the nucleus.

Figure 13.18 The nuclear potential energy is modeled as a square barrier. The energy of the alpha particle is E, which is less than the height of the barrier. According to quantum mechanics, the alpha particle has some chance of tunneling through the barrier, as indicated by the finite size of the wavefunction for $r > R_1$.

probability of escaping also increases. Furthermore, the probability increases as the width of the barrier decreases.

EXAMPLE 13.9 Probability for Alpha Decay

Apply the tunneling methods of Chapter 6 to compute the probability of escape from a $^{226}_{88}$Ra nucleus of an alpha particle with disintegration energy 5 MeV.

Solution The escape probability is none other than the transmission coefficient $T(E)$ for the Coulomb barrier shown in Figure 13.17. Equation 6.10 of Chapter 6 gives $T(E)$ approximately as

$$T(E) \simeq e^{-(2/\hbar)\sqrt{2m}\int\sqrt{U(r)-E}\,dr}$$

The integral is taken over the *classically forbidden region* where $E < U$. For alpha decay, this region is bounded below by the nuclear radius R and above by $R_1 = 2Zke^2/E$ [from $E = U(R_1) = 2Zke^2/R_1$] (see Fig. 13.17). In this expression, Z is the atomic number of the *daughter* nucleus.

The tunneling integral to be evaluated is

$$\int \sqrt{U(r) - E}\,dr = \sqrt{E}\int_R^{R_1} \sqrt{\frac{R_1}{r} - 1}\,dr$$

$$= R_1\sqrt{E}\int_{R/R_1}^1 \sqrt{\frac{1}{z} - 1}\,dz$$

where $z = r/R_1$. An exact value for this integral can be had with some effort, but a useful approximation is found readily by noting that R/R_1 is a small number ($R \sim 10$ fm; $R_1 \sim 50$ fm for $E \sim 5$ MeV). Thus, as a first estimate we set the lower limit to zero and change variables with $z = \cos^2\theta$, obtaining

$$\int_0^1 \sqrt{\frac{1}{z} - 1}\,dz = 2\int_0^{\pi/2} \sin^2\theta\,d\theta$$

$$= \int_0^{\pi/2} [1 - \cos 2\theta]\,d\theta = \frac{\pi}{2}$$

To improve upon this, we break the original integral into two and approximate the second, using $1/z \gg 1$ for z small, to get

$$\int_0^1 \sqrt{\frac{1}{z} - 1}\,dz - \int_0^{R/R_1} \sqrt{\frac{1}{z} - 1}\,dz \simeq \frac{\pi}{2}$$

$$- \int_0^{R/R_1} \frac{dz}{\sqrt{z}} = \frac{\pi}{2} - 2\sqrt{\frac{R}{R_1}}$$

Combining this result with $R_1 = 2Zke^2/E$ gives, for the decay probability,

$$T(E) = \exp\{-4\pi Z\sqrt{(E_0/E)} + 8\sqrt{Z(R/r_0)}\}$$

The parameter $r_0 = \hbar^2/M_\alpha ke^2$ is a kind of Bohr radius for the alpha particle, with the value 7.25 fm, and E_0 is an energy unit analogous to the Rydberg in atomic physics:

$$E_0 = \frac{ke^2}{2r_0} = \frac{14.40 \text{ eV·Å}}{(2)(7.25 \times 10^{-5} \text{ Å})} = 0.0993 \text{ MeV}$$

For the alpha decay of radium, the daughter nucleus is radon with atomic number $Z = 86$ and mass number $A = 222$. Equation 13.1 predicts the radius R of the radon nucleus to be

$$R = (1.2 \times 10^{-5} \text{ fm})(222)^{1/3} = 7.27 \text{ fm}$$

Then the decay probability for alpha disintegration at $E = 5$ MeV is

$$\exp\{-4\pi(86)\sqrt{(0.0993/5)} + 8\sqrt{86(7.27/7.25)}\}$$

$$= \exp\{-78.008\} = 1.32 \times 10^{-34}$$

This probability is quite small, but the actual number of disintegrations per second is much larger because of the many collisions the alpha particle makes with the nuclear barrier. This collision frequency f is the reciprocal of the transit time for the alpha particle crossing the nucleus; that is, $f = v/2R$, where v is the speed of the alpha particle inside the nucleus. In most cases f is about 10^{21} collisions per second (see Problem 16 in Chapter 6), leading to a predicted decay rate $\simeq 10^{-13}$ disintegrations per second for this case, in reasonable agreement with the observed value of $\lambda = 1.4 \times 10^{-11}$ s^{-1}.

Beta Decay

When a radioactive nucleus undergoes beta decay, the daughter nucleus has the same number of nucleons as the parent nucleus, but the atomic number is changed by 1:

$$_Z^A X \longrightarrow \,_{Z+1}^{A}Y + e^- \qquad (13.20)$$

$$_Z^A X \longrightarrow \,_{Z-1}^{A}Y + e^+ \qquad (13.21)$$

Beta decay

Again, note that the nucleon number and total charge are both conserved in these decays. However, as we shall see later, *these processes are not described completely by such expressions.* We shall explain this shortly.

Two typical beta decay processes are

$$_6^{14}C \longrightarrow \,_7^{14}N + e^-$$

$$_7^{12}N \longrightarrow \,_6^{12}C + e^+$$

Notice that in beta decay a neutron changes into a proton (or vice versa). It is also important to note that the electron or positron in these decays is not present beforehand in the nucleus but is created at the moment of decay out of the rest energy of the decaying nucleus.

Now consider the energy of the system before and after the decay. As with alpha decay, energy must be conserved. Experimentally, it is found that the beta particles are emitted with a continuous range of kinetic energies up to some maximum value, $K_{\max}$ (Fig. 13.19). The kinetic energy of the particles must be balanced by the decrease in mass of the residual system—that is, the Q value. However, because all decaying nuclei have the same initial mass, *the Q value must be the same for each decay.* In view of this, why do the emitted particles have different kinetic energies? The law of conservation of energy seems to be violated! Further analysis shows that, according to the decay processes given by Equations 13.20 and 13.21, the principles of conservation of both angular momentum (spin) and linear momentum are also violated!

After a great deal of experimental and theoretical study, Pauli in 1930 proposed that a third particle must be present to carry away the "missing" energy and momentum. Fermi later named this particle the **neutrino** ("little neutral one") because it had to be electrically neutral and have little or no rest mass. Although it eluded detection for many years, the neutrino (symbolized by ν) was finally detected experimentally in 1956. It has the following properties:

- Zero electric charge
- A rest mass that is much smaller than that of the electron and is possibly zero. Recent experiments suggest that the mass of the neutrino is less than 7 eV/c^2.
- A spin of $\frac{1}{2}$, which satisfies the law of conservation of angular momentum when applied to beta decay
- Very weak interaction with matter, which makes it very difficult to detect

Properties of the neutrino

For the general form of the beta decays considered earlier, we can now write

$$_Z^A X \longrightarrow \,_{Z+1}^{A}Y + e^- + \bar{\nu} \qquad (13.22)$$

$$_Z^A X \longrightarrow \,_{Z-1}^{A}Y + e^+ + \nu \qquad (13.23)$$

where the symbol $\bar{\nu}$ represents the **antineutrino,** the antiparticle to the neutrino. As in the case of alpha decay, the decays just listed are analyzed through

Figure 13.19 A typical beta decay curve. The maximum kinetic energy observed for the beta particles corresponds to the Q value for the reaction.

Enrico Fermi, an Italian-American physicist, received his doctorate from the University of Pisa in 1922, then did postdoctoral work in Germany under Max Born. He returned to Italy in 1924 and in 1926 became a professor of physics at the University of Rome. He received the Nobel prize for physics in 1938 for his work with the production of transuranic radioactive elements (those more massive than uranium) by neutron bombardment.

Fermi first became interested in physics at the age of 14, after reading an old physics book in Latin. (He was an excellent scholar and could recite Dante's *Divine Comedy* and much of Aristotle's writings from memory.) His great ability to solve problems in theoretical physics and his skill for simplifying very complex situations made him somewhat of an oracle. Fermi was also a gifted experimentalist and teacher. During one of his early lecture trips to the United States, a car that he had purchased became disabled, and he pulled into

B I O G R A P H Y

———

ENRICO FERMI
(1901–1954)

a nearby gas station. After Fermi repaired the car with ease, the station owner offered him a job on the spot.

Fermi and his family immigrated to the United States, and he became a naturalized citizen in 1944. He

taught first at Columbia University, then at the University of Chicago. As part of the Manhattan Project during World War II, Fermi was commissioned to design and build a structure, called an atomic pile, in which a self-sustained chain reaction might occur. The structure, built in the squash court of the University of Chicago, contained uranium in combination with graphite blocks to slow the neutrons to thermal speeds. Cadmium rods inserted in the pile were used to absorb neutrons and control the reaction rate. History was made at 3:45 P.M. on December 2, 1942, as the cadmium rods were slowly withdrawn and a self-sustained chain reaction was observed. Fermi's earth-shaking achievement—the world's first nuclear reactor—marked the beginning of the atomic age.

Fermi died of cancer in 1954 at the age of 53. One year later, the 100th element was discovered and named *fermium* in his honor.

(National Accelerator Laboratory)

conservation of energy and momentum, but we must use relativistic expressions because the kinetic energies of the electron and neutrino are not small compared to their rest energies.

Electron capture   A process that competes with e^+ decay is called **electron capture.** This occurs when a parent nucleus captures one of its own orbital atomic electrons and emits a neutrino. The final product after decay is a nucleus whose charge is $Z - 1$:

$$^A_Z X + e^- \longrightarrow \, ^{\,A}_{Z-1} X + \nu \qquad (13.24)$$

In most cases it is an inner K-shell electron that is captured, and this is referred to as **K capture.** One example of this process is the capture of an electron by $^7_4 Be$ to become $^7_3 Li$:

$$^7_4 Be + e^- \longrightarrow \, ^7_3 Li + \nu$$

Finally, it is instructive to mention the Q values for the beta-decay processes. The Q values for e^- decay and electron capture are $Q = (M_X - M_Y)c^2$, while the Q values for e^+ decay are given by $Q = (M_X - M_Y - 2m_e)c^2$. These relationships are useful for determining whether or not possible beta-decay processes are energetically allowed.

Carbon Dating

The beta decay of ^{14}C is commonly used to date organic samples. Cosmic rays in the upper atmosphere cause nuclear reactions that create ^{14}C. In fact, the ratio of ^{14}C to ^{12}C in the carbon dioxide molecules of our atmosphere has a constant value of approximately 1.3×10^{-12}. All living organisms have this same ratio of ^{14}C to ^{12}C, because they continuously exchange carbon dioxide with their surroundings. When an organism dies, however, it no longer absorbs ^{14}C from the atmosphere, and so the $^{14}C/^{12}C$ ratio decreases as a result of the beta decay of ^{14}C, which has a half-life of 5730 years. It is therefore possible to measure the age of a material by measuring its activity per unit mass caused by the decay of ^{14}C. Using this technique, scientists have been able to identify samples of wood, charcoal, bone, and shell as having lived from 1000 to 25 000 years ago. This knowledge has helped us reconstruct the history of living organisms—including humans—during that time span.

A particularly interesting example is the dating of the Dead Sea Scrolls, a group of manuscripts discovered by a shepherd in 1947. Translation showed them to be religious documents, including most of the books of the Old Testament. Because of their historical and religious significance, scholars wanted to know their age. Carbon dating performed on the material in which the scrolls were wrapped established their age at approximately 1950 years.

EXAMPLE 13.10   The Age of Ice Man

In 1991 a German tourist discovered the well-preserved remains of a human, later dubbed "Ice Man," trapped in a glacier in the Italian Alps. Radioactive dating of a sample of Ice Man established an age of 5300 years. Why did scientists date the sample with the isotope ^{14}C rather than ^{11}C, a beta emitter with a half-life of 20.4 min?

Reasoning   Carbon-14 has a long half-life, 5730 years, so the fraction of ^{14}C nuclei remaining after one half-life is high enough to enable accurate measurements of changes in the sample's activity. The ^{11}C isotope, which has a very short half-life, is not useful because its activity decreases to a vanishingly small value over the age of the sample, making it impossible to detect.

If a sample to be dated is not very old—say, about 50 years—then the scientist should select the isotope of some other element whose half-life is comparable to the age of the sample. For example, if the sample contained hydrogen, one could measure the activity of ^{3}H (tritium), a beta emitter with a half-life of 12.3 years. As a general rule, the expected age of the sample should be great enough to allow measurement of a change in activity but not so great that the activity is undetectable.

EXAMPLE 13.11   Radioactive Dating

An archaeologist finds a 25.0-g piece of charcoal in the ruins of an ancient city. The sample shows a ^{14}C activity of 250 decays/min. How long has the tree from which this charcoal came been dead?

Solution   First calculate the decay constant for ^{14}C, which has a half-life of 5730 years.

$$\lambda = \frac{0.693}{T_{1/2}} = \frac{0.693}{(5730 \text{ years})(3.16 \times 10^7 \text{ s/year})}$$
$$= 3.83 \times 10^{-12} \text{ s}^{-1}$$

The number of ^{14}C nuclei can be calculated in two steps. (1) The number of ^{12}C nuclei in 25.0 g of carbon is

$$N(^{12}C) = \frac{6.02 \times 10^{23} \text{ nuclei/mol}}{12.0 \text{ g/mol}}(25.0 \text{ g})$$
$$= 1.26 \times 10^{24} \text{ nuclei}$$

Knowing that the ratio of ^{14}C to ^{12}C in the live sample was 1.3×10^{-12}, we see that the number of ^{14}C nuclei in 25.0 g *before* decay is

$$N_0(^{14}C) = (1.3 \times 10^{-12})(1.26 \times 10^{24})$$
$$= 1.6 \times 10^{12} \text{ nuclei}$$

Hence the initial activity of the sample is

$$R_0 = N_0\lambda = (1.6 \times 10^{12} \text{ nuclei})(3.83 \times 10^{-12} \text{ s}^{-1})$$
$$= 6.13 \frac{\text{decays}}{\text{s}} = 370 \frac{\text{decays}}{\text{min}}$$

(2) We can now calculate the age of the charcoal, using

Equation 13.10, which relates the activity R at any time t to the initial activity R_0:

$$R = R_0 e^{-\lambda t} \quad \text{or} \quad e^{-\lambda t} = \frac{R}{R_0}$$

Because it is given that $R = 250$ decays/min and because we found that $R_0 = 370$ decays/min, we can calculate t by taking the natural logarithm of both sides of the last equation:

$$-\lambda t = \ln\left(\frac{R}{R_0}\right) = \ln\left(\frac{250}{370}\right) = -0.39$$

$$t = \frac{0.39}{\lambda} = \frac{0.39}{3.84 \times 10^{-12}\ \text{s}^{-1}}$$

$$= 1.0 \times 10^{11}\ \text{s} = 3.2 \times 10^3\ \text{years}$$

Gamma Decay

Very often, a nucleus that undergoes radioactive decay is left in an excited energy state. The nucleus can then undergo a second decay to a lower energy state, perhaps to the ground state, by emitting a high-energy photon:

Gamma decay

$$_{Z}^{A}\text{X}^* \longrightarrow _{Z}^{A}\text{X} + \gamma \tag{13.25}$$

where X* indicates a nucleus in an excited state. The typical half-life of an excited nuclear state is 10^{-10} s. Photons emitted in such a de-excitation process are called gamma rays. Such photons have very high energy (in the range of 1 MeV to 1 GeV) relative to the energy of visible light (about 1 eV). Recall that the energy of photons emitted (or absorbed) by an atom equals the difference in energy between the two electronic states involved in the transition. Similarly, a gamma-ray photon has an energy hf that equals the energy difference ΔE between two nuclear energy levels. When a nucleus decays by emitting a gamma ray, the nucleus doesn't change its atomic mass A or atomic number Z.

A nucleus may reach an excited state as the result of a violent collision with another particle. It is also very common for a nucleus to be in an excited state after undergoing an alpha or beta decay. The following sequence of events represents a typical situation in which gamma decay occurs.

$$_{5}^{12}\text{B} \longrightarrow _{6}^{12}\text{C}^* + e^- + \bar{\nu} \tag{13.26}$$

$$_{6}^{12}\text{C}^* \longrightarrow _{6}^{12}\text{C} + \gamma \tag{13.27}$$

Figure 13.20 shows the decay scheme for ^{12}B, which undergoes beta decay with a half-life of 20.4 ms to either of two levels of ^{12}C. It can either (1) decay directly to the ground state of ^{12}C by emitting a 13.4-MeV electron or (2) undergo e^- decay to an excited state of ^{12}C*, followed by gamma decay to the ground state. The latter process results in the emission of a 9.0-MeV electron and a 4.4-MeV photon.

Table 13.4 summarizes the pathways by which a radioactive nucleus can undergo decay.

Figure 13.20   The ^{12}B nucleus undergoes e^- decay to two levels of ^{12}C. The decay to the excited level, ^{12}C*, is followed by gamma decay to the ground state.

Table 13.4   Decay Processes

Alpha decay	$_Z^A X \rightarrow \,_{Z-2}^{A-4} X + \,_2^4 He$
Beta decay (e^-)	$_Z^A X \rightarrow \,_{Z+1}^{A} X + e^- + \bar{\nu}$
Beta decay (e^+)	$_Z^A X \rightarrow \,_{Z-1}^{A} X + e^+ + \nu$
Electron capture	$_Z^A X + e^- \rightarrow \,_{Z-1}^{A} X + \nu$
Gamma decay	$_Z^A X^* \rightarrow \,_Z^A X + \gamma$

13.6   NATURAL RADIOACTIVITY

Radioactive nuclei are generally classified into two groups: (1) unstable nuclei found in nature, which give rise to what is called **natural radioactivity,** and (2) nuclei produced in the laboratory through nuclear reactions, which exhibit **artificial radioactivity.**

There are three series of naturally occurring radioactive nuclei (Table 13.5). Each series starts with a specific long-lived radioactive isotope whose half-life exceeds that of any of its descendants. The three natural series begin with the isotopes ^{238}U, ^{235}U, and ^{232}Th, and the corresponding stable end products are three isotopes of lead: ^{206}Pb, ^{207}Pb, and ^{208}Pb. The fourth series in Table 13.5 begins with ^{237}Np and has as its stable end product ^{209}Bi. The element ^{237}Np is a transuranic element (one having an atomic number greater than that of uranium) not found in nature. This element has a half-life of "only" 2.14×10^6 years.

Figure 13.21 shows the successive decays for the ^{232}Th series. Note that ^{232}Th first undergoes alpha decay to ^{228}Ra. Next, ^{228}Ra undergoes two successive β decays to ^{228}Th. The series continues and finally branches when it reaches ^{212}Bi. At this point, there are two decay possibilities. The end of the decay series is the stable isotope ^{208}Pb. Each step in the decay sequence shown in Figure 13.21 may be characterized by a decrease of either 4 (for alpha decays) or 0 (for beta or gamma decays) in the mass number A, together with appropriate changes in Z.

The two uranium series are somewhat more complex than the ^{232}Th series. Also, there are several naturally occurring radioactive isotopes, such as ^{14}C and ^{40}K, that are not part of the aforementioned decay series.

Radioactive series in nature constantly replenish our environment with radioactive elements that would otherwise have disappeared long ago. For ex-

Figure 13.21 Successive decays for the ^{232}Th series.

Table 13.5   The Four Radioactive Series

Series	Starting Isotope	Half-Life (years)	Stable End Product
Uranium	$_{92}^{238}U$	4.47×10^9	$_{82}^{206}Pb$
Actinium (natural)	$_{92}^{235}U$	7.04×10^8	$_{82}^{207}Pb$
Thorium	$_{90}^{232}Th$	1.41×10^{10}	$_{82}^{208}Pb$
Neptunium	$_{93}^{237}Np$	2.14×10^6	$_{83}^{209}Bi$

ample, because the Solar System is approximately 5×10^9 years old, the supply of ^{226}Ra (whose half-life is only 1600 years) would have been depleted by radioactive decay long ago if it were not for the decay series that starts with ^{238}U.

SUMMARY

A nuclear species can be represented by ^A_ZX, where A is the **mass number,** the total number of nucleons, and Z is the **atomic number,** the total number of protons. The total number of neutrons in a nucleus is the **neutron number** N, where $A = N + Z$. Elements with the same Z but different A and N values are called **isotopes.**

Assuming that a nucleus is spherical, its radius is

$$r = r_0 A^{1/3} \tag{13.1}$$

where $r_0 = 1.2$ fm (1 fm $= 10^{-15}$ m).

Nuclei are stable because of the **nuclear force** between nucleons. This short-range force dominates the Coulomb repulsive force at distances of less than about 2 fm and is nearly independent of charge. Light nuclei are most stable when the number of protons equals the number of neutrons. Heavy nuclei are most stable when the number of neutrons exceeds the number of protons. In addition, the most stable nuclei have Z and N values that are both even. Nuclei with unusually high stability have Z or N values of 2, 8, 20, 28, 50, 82, and 126, called **magic numbers.**

Nuclei have a total angular momentum of magnitude $\sqrt{I(I + 1)}\ \hbar$, where I is called the **nuclear-spin quantum number.** The magnetic moment of a nucleus is measured in terms of the **nuclear magneton** μ_n, where

$$\mu_n \equiv \frac{e\hbar}{2m_\mathrm{p}} = 5.05 \times 10^{-27}\ \mathrm{J/T} \tag{13.3}$$

When a nuclear moment is placed in an external magnetic field, it precesses about the field with a frequency that is proportional to the field.

The difference between the combined mass of the separate nucleons and that of the compound nucleus containing those nucleons, when multiplied by c^2, gives the **binding energy** E_b of the nucleus. We can calculate the binding energy of any nucleus of mass $M(^A_Z$X$)$ with the expression

$$E_b\ (\mathrm{MeV}) = [ZM(\mathrm{H}) + Nm_\mathrm{n} - M(^A_Z\mathrm{X})] \times 931.494\ \frac{\mathrm{MeV}}{\mathrm{u}} \tag{13.4}$$

where all masses are atomic masses and m_n is the mass of the neutron.

The **liquid-drop model** of nuclear structure treats the nucleons as molecules in a drop of liquid. The three main contributions influencing binding energy are the volume effect, the surface effect, and the Coulomb repulsion. Summing such contributions results in the **semiempirical binding energy formula.**

The **independent-particle model,** or shell model, of nuclear structure assumes that each nucleon moves in a well-defined quantized orbit within the nucleus. The stability of certain nuclei—particularly those with magic numbers—can be explained with this model. Nuclei can move in well-defined orbits inside the densely packed nucleus because two nucleons will collide only if the energy of each state after the collision corresponds to one of the unoccupied nuclear states.

The **collective model** of the nucleus combines some features of the liquid-drop model and some features of the independent-particle model. It has been very successful in describing a variety of nuclear phenomena.

A radioactive substance decays by alpha decay, beta decay, or gamma decay. An alpha particle is a ^{4}He nucleus; a beta particle is either an electron (e^-) or a positron (e^+), accompanied by the emission of an antineutrino ($\bar{\nu}$) or a neutrino (ν), respectively; a gamma particle is a high-energy photon.

If a radioactive material contains N_0 radioactive nuclei at $t = 0$, the number N of nuclei remaining after a time t has elapsed is

$$N = N_0 e^{-\lambda t} \tag{13.9}$$

where λ is the **decay constant,** the probability per unit time that a nucleus will decay. The **decay rate,** or **activity,** of a radioactive substance is

$$R = \left| \frac{dN}{dt} \right| = R_0 e^{-\lambda t} \tag{13.10}$$

where $R_0 = N_0 \lambda$ is the activity, or number of decays per unit time at $t = 0$. The **half-life** $T_{1/2}$ is defined as the time required for half of a given number of radioactive nuclei to decay, where

$$T_{1/2} = \frac{0.693}{\lambda} \tag{13.11}$$

Alpha decay can occur because, according to quantum mechanics, alpha particles can tunnel through the Coulomb barriers of nuclei. A nucleus undergoing beta decay emits either an electron (e^-) and an antineutrino ($\bar{\nu}$) or a positron (e^+) and a neutrino (ν). In electron capture, the nucleus of an atom absorbs one of its own orbital electrons and emits a neutrino. In gamma decay, a nucleus in an excited state decays to a lower energy state and emits a gamma ray.

QUESTIONS

1. Why are heavy nuclei unstable?
2. A proton precesses with a frequency ω_p in the presence of a magnetic field. If the intensity of the magnetic field is doubled, what happens to the precessional frequency?
3. Explain why nuclei that are well off the line of stability in Figure 13.4 tend to be unstable.
4. Consider two heavy nuclei, X and Y, with similar mass numbers. If X has the higher binding energy, which nucleus tends to be more unstable?
5. Discuss the differences between the liquid-drop model and the independent-particle model of the nucleus.
6. How many values of I_z are possible for $I = \frac{5}{2}$? for $I = 3$?
7. In nuclear magnetic resonance, how does increasing the

dc magnetic field change the frequency of the ac field that excites a particular transition?
8. Would the liquid-drop model or the independent-particle model be more appropriate for predicting the behavior of a nucleus in a fission reaction? Which would be more successful at predicting the magnetic moment of a given nucleus? Which could better explain the gamma-ray spectrum of an excited nucleus?
9. If a nucleus has a half-life of one year, does that mean it will be completely decayed after two years? Explain.
10. What fraction of a radioactive sample has decayed after two half-lives have elapsed?
11. Two samples of the same radioactive nuclide are pre-

pared, each having the same size. Sample A has twice the initial activity of sample B. How does the half-life of A compare with the half-life of B? After each sample has passed through five half-lives, what is the ratio of their activities?

12. Explain why the half-lives of radioactive nuclei are essentially independent of temperature.

13. The radioactive nucleus $^{226}_{88}$Ra has a half-life of approximately 1.6×10^3 years. Given that the Solar System is about 5 billion years old, why do we still find this nucleus in nature?

14. A free neutron undergoes beta decay with a half-life of about 10 min. Can a free proton undergo a similar decay?

15. Explain how you can carbon-date the age of a sample.

16. What is the difference between a neutrino and a photon?

17. Use Equation 13.24 to explain why the neutrino must have a spin of $\frac{1}{2}$.

18. A nucleus such as ^{226}Ra that is initially at rest undergoes alpha decay. Which has more kinetic energy after the decay, the alpha particle or the daughter nucleus?

19. Can a nucleus emit alpha particles with different energies? Explain.

20. Explain why many heavy nuclei undergo alpha decay but do not spontaneously emit neutrons or protons.

21. If an alpha particle and an electron have the same kinetic energy, which undergoes greater deflection when passed through a magnetic field?

22. If photographic film is kept in a wooden box, alpha particles from a radioactive source outside the box cannot expose the film, but beta particles can. Explain.

23. Pick any beta-decay process, and show that the neutrino must have zero charge.

24. Suppose it could be shown that the intensity of cosmic rays at the Earth's surface was much greater 10 000 years ago. How would this difference affect what we accept as valid carbon-dated ages of ancient samples of once-living matter?

25. Why is carbon dating unable to provide accurate estimates of very old material?

26. Element X has several isotopes. What do these isotopes have in common? How do they differ?

27. Explain the main differences among alpha, beta, and gamma rays.

28. How many protons are in the nucleus $^{222}_{86}$Rn? How many neutrons? How many orbiting electrons are in the neutral atom?

PROBLEMS

Table 13.6 will be useful for many of these problems. A more complete list of atomic masses appears in Appendix B.

Table 13.6   Some Atomic Masses

Element	Atomic Mass (u)	Element	Atomic Mass (u)
$^{4}_{2}$He	4.002 603	$^{27}_{13}$Al	26.981 539
$^{7}_{3}$Li	7.016 003	$^{30}_{15}$P	29.978 310
$^{9}_{4}$Be	9.012 182	$^{40}_{20}$Ca	39.962 591
$^{10}_{5}$B	10.012 937	$^{42}_{20}$Ca	41.958 63
$^{12}_{6}$C	12.000 000	$^{43}_{20}$Ca	42.958 766
$^{13}_{6}$C	13.003 355	$^{56}_{26}$Fe	55.934 939
$^{14}_{7}$N	14.003 074	$^{64}_{30}$Zn	63.929 145
$^{15}_{7}$N	15.000 109	$^{63}_{29}$Cu	63.929 599
$^{15}_{8}$O	15.003 065	$^{93}_{41}$Nb	92.906 377
$^{17}_{8}$O	16.999 131	$^{197}_{79}$Au	196.966 543
$^{18}_{8}$O	17.999 160	$^{202}_{80}$Hg	201.970 617
$^{18}_{9}$F	18.000 937	$^{216}_{84}$Po	216.001 888
$^{20}_{10}$Ne	19.992 436	$^{220}_{86}$Rn	220.011 368
$^{23}_{11}$Na	22.989 768	$^{234}_{90}$Th	234.043 593
$^{23}_{12}$Mg	22.994 124	$^{238}_{92}$U	238.050 785

13.1   Some Properties of Nuclei

1. Find the radii of (a) a nucleus of $^{4}_{2}$He and (b) a nucleus of $^{238}_{92}$U. (c) What is the ratio of these radii?

2. The compressed core of a star formed in the wake of a supernova explosion consists of only neutrons and is called a *neutron star*. Calculate the mass of 10 cm^3 of a neutron star.

3. Consider the hydrogen atom to be a sphere of radius equal to the Bohr radius a_0, given by Equation 3.29 in Chapter 3, and calculate the approximate value of the ratio of the nuclear mass density to the atomic mass density.

4. The Larmor precessional frequency is

$$f = \frac{\Delta E}{h} = \frac{2\mu B}{h}$$

Calculate the radio-wave frequency at which resonance absorption occurs for (a) free neutrons in a magnetic field of 1 T, (b) free protons in a magnetic field of 1 T, and (c) free protons in the Earth's magnetic field at a location where the magnetic field strength is 50 μT.

5. (a) Use energy methods to calculate the distance of closest approach for a head-on collision between an alpha particle with an initial energy of 0.5 MeV and a gold nucleus (^{197}Au) at rest. (Assume that the gold nucleus

remains at rest during the collision.) (b) What minimum initial speed must the alpha particle have in order to approach to a distance of 300 fm?

6. A neutron star is a nucleus composed entirely of neutrons and has a radius of about 10 km. Such a star forms when a larger star cools down and collapses under the influence of its own gravitational field. The compression continues until the protons and electrons merge to form neutrons. (a) Assuming that Equation 13.1 is valid, estimate the mass number A and mass M of a neutron star of radius $R = 10$ km. (b) Find the acceleration due to gravity at the surface of such a star. (c) Find the rotational kinetic energy of the star if it rotates about its axis 30 times/s. (Assume that the star is a sphere of uniform density.)

7. Consider a hydrogen atom with the electron in the $2p$ state. The magnetic field at the nucleus produced by the orbiting electron has a value of 12.5 T. The proton can have its magnetic moment aligned in either of two directions perpendicular to the plane of the electron's orbit. Because of the interaction of the proton's magnetic moment with the electron's magnetic field, there will be a difference in energy between the states with the two different orientations of the proton's magnetic moment. Find that energy difference in electron volts.

8. Using a reasonable scale, sketch an energy-level diagram for (a) a proton and (b) a deuteron, both in a magnetic field B. (c) What are the absolute values of the changes in energy that accompany the possible transitions between the levels shown in your diagrams?

9. Copper, as it occurs naturally, consists of two stable isotopes, ^{63}Cu and ^{65}Cu. What is the relative abundance of the two forms? In your calculations, take the atomic weight of copper to be 63.55, and take the masses of the two isotopes to be 62.95 u and 64.95 u.

10. (a) Find the radius of the $^{12}_{6}$C nucleus. (b) Find the force of repulsion between a proton at the surface of a $^{12}_{6}$C nucleus and the remaining five protons. (c) How much work (in MeV) must be done to overcome this electrostatic repulsion and put the last proton into the nucleus? (d) Repeat (a), (b), and (c) for $^{238}_{92}$U.

13.2 Binding Energy and Nuclear Forces

11. In Example 13.3, the binding energy of the deuteron was calculated to be 2.224 MeV. This corresponds to a value of 1.112 MeV/nucleon. What is the binding energy per nucleon for the heaviest isotope of hydrogen, ^{3}H (called *tritium*)?

12. Using the atomic mass of $^{56}_{26}$Fe given in Table 13.6, find its binding energy. Then compute the binding energy per nucleon and compare your result with Figure 13.10.

13. The $^{139}_{57}$La isotope of lanthanum is stable. A radioactive isobar (see Problem 14) of this lanthanum isotope, $^{139}_{59}$Pr, appears below the line of stable nuclei in Figure

13.4 and decays by β^+ emission. Another radioactive isobar of ^{139}La, $^{139}_{55}$Cs, decays by β^- emission and appears above the line of stable nuclei in Figure 13.4. (a) Which of these three isobars has the highest neutron-to-proton ratio? (b) Which has the greatest binding energy per nucleon? (c) Which of the two radioactive nuclei (^{139}Pr or ^{139}Cs) do you expect to be heavier?

14. Two nuclei with the same mass number are known as *isobars*. If the two nuclei also have interchanged atomic and neutron numbers, such as $^{23}_{11}$Na and $^{23}_{12}$Mg, the nuclei are called *mirror isobars*. Binding-energy measurements on these nuclei can be used to obtain evidence of the charge independence of nuclear forces (that is, proton-proton, proton-neutron, and neutron-neutron forces are approximately equal). (a) Calculate the difference in binding energy for the two mirror nuclei $^{15}_{8}$O and $^{15}_{7}$N. (b) Calculate the difference in binding energy per nucleon for the mirror isobars $^{23}_{11}$Na and $^{23}_{12}$Mg. How do you account for the difference?

15. Calculate the binding energy per nucleon for the nuclei (a) $^{20}_{10}$Ne, (b) $^{40}_{20}$Ca, (c) $^{93}_{41}$Nb, and (d) $^{197}_{79}$Au.

16. Calculate the minimum energy required to remove a neutron from the $^{43}_{20}$Ca nucleus. (*Hint:* Use a formula analogous to Eq. 13.4.)

17. Using the graph in Figure 13.10, estimate how much energy is released when a nucleus of mass number 200 is split into two nuclei, each of mass number 100.

13.3 Nuclear Models

18. (a) In the liquid-drop model of nuclear structure, why does the surface-effect term $- C_2 A^{2/3}$ have a minus sign? (b) The binding energy of the nucleus increases as the volume-to-surface ratio increases. Calculate this ratio for both spherical and cubical shapes, and explain which is more plausible for nuclei.

19. Treat the nucleus as a sphere of uniform volume charge density ρ. (a) Derive an expression for the total energy required to assemble a sphere of charge corresponding to a nucleus of atomic number Z and radius R. (b) Using the result of (a), find an expression for the electrostatic potential energy U in terms of the mass number A for the case when $N = Z$ (the nucleus is on the line of stability). (c) Evaluate the result of (b) for the nucleus $^{30}_{15}$P.

20. (a) Use Equation 13.7 and the given values for the constants C_1, C_2, C_3, and C_4 to calculate the binding energy per nucleon for the isobars $^{64}_{29}$Cu and $^{64}_{30}$Zn. (b) Compare these values to the binding energy per nucleon calculated with Equation 13.4.

13.4 Radioactivity

21. Measurements on a radioactive sample show that its activity decreases by a factor of 5 during a 2-h interval. (a) Determine the decay constant of the radioactive nu-

cleus. (b) Calculate the value of the half-life for this isotope.

22. The half-life of ^{131}I is 8.04 days. (a) Calculate the decay constant for this isotope. (b) Find the number of ^{131}I nuclei necessary to produce a sample with an activity of 0.5 μCi.

23. A freshly prepared sample of a certain radioactive isotope has an activity of 10 mCi. After an elapsed time of 4 h, its activity is 8 mCi. (a) Find the decay constant and half-life of the isotope. (b) How many atoms of the isotope were contained in the freshly prepared sample? (c) What is the sample's activity 30 h after it is prepared?

24. Tritium has a half-life of 12.33 years. What percentage of the ^{3}H nuclei in a tritium sample will decay during a period of 5 years?

25. A sample of radioactive material is said to be *carrier-free* when no stable isotopes of the radioactive element are present. Calculate the mass of strontium in a carrier-free 5-mCi sample of ^{90}Sr whose half-life is 28.8 years.

26. How many radioactive atoms are present in a sample that has an activity of 0.2 μCi and a half-life of 8.1 days?

27. A laboratory stock solution is prepared with an initial activity due to ^{24}Na of 2.5 mCi/mL, and 10 mL of the stock solution is diluted (at $t_0 = 0$) to a working solution with a total volume of 250 mL. After 48 h, a 5-mL sample of the working solution is monitored with a counter. What is the measured activity? (*Note:* 1 mL = 1 milliliter, and the half-life of ^{24}Na is 15.0 h.)

28. Start with Equation 13.10, and find the following useful forms for the decay constant and half-life:

$$\lambda = \frac{1}{t} \ln\left(\frac{R_0}{R}\right) \qquad T_{1/2} = \frac{(\ln 2)\, t}{\ln(R_0/R)}$$

29. The radioactive isotope ^{198}Au has a half-life of 64.8 h. A sample containing this isotope has an initial activity ($t = 0$) of 40 μCi. Calculate the number of nuclei that will decay in the time interval from $t_1 = 10$ h to $t_2 = 12$ h.

30. During the manufacture of a steel engine component, radioactive iron (^{59}Fe) is included in the total mass of 0.2 kg. The component is placed in a test engine when the activity due to this isotope is 20 μCi. After a 1000-h test period, oil is removed from the engine and found to contain enough ^{59}Fe to produce 800 disintegrations/min per liter of oil. The total volume of oil in the engine is 6.5 L. Calculate the total mass worn from the engine component per hour of operation. (The half-life for ^{59}Fe is 45.1 days.)

31. The activity of a sample of radioactive material was measured over 12 h, and the following *net* count rates were obtained at the times indicated:

Time (h)	Counting Rate (counts/min)
1	3100
2	2450
4	1480
6	910
8	545
10	330
12	200

(a) Plot the activity curve on semilog paper. (b) Determine the disintegration constant and the half-life of the radioactive nuclei in the sample. (c) What counting rate would you expect for the sample at $t = 0$? (d) Assuming the efficiency of the counting instrument to be 10%, calculate the number of radioactive atoms in the sample at $t = 0$.

32. A by-product of some fission reactors is the isotope $^{239}_{94}$Pu, which is an alpha emitter with a half-life of 24 000 years:

$$^{239}_{94}\text{Pu} \longrightarrow \,^{235}_{92}\text{U} + \alpha$$

Consider a sample of 1 kg of pure $^{239}_{94}$Pu at $t = 0$. Calculate (a) the number of $^{239}_{94}$Pu nuclei present at $t = 0$, (b) the initial activity in the sample, and (c) the time required for the activity to decrease to 1 decay/s.

33. A piece of charcoal has a mass of 25 g and is known to be about 25 000 years old. (a) Determine the number of decays per minute expected from this sample. (b) If the radioactive background in the counter without a sample is 20 counts/min and we assume 100% efficiency in counting, explain why 25 000 years is close to the limit of dating with this technique.

34. A fission reactor accident evaporates 5×10^6 Ci of ^{90}Sr ($T_{1/2} = 27.7$ years) into the air. The ^{90}Sr falls out over an area of 10^4 km^2. How long will it take the activity of the ^{90}Sr to reach the agriculturally "safe" level of 2 μCi/m^2?

35. What specific activity (see Problem 36), in disintegrations/min·g, would be expected for carbon samples from 2000-year-old bones? (Note that the ratio of ^{14}C to ^{12}C in living organisms is 1.3×10^{-12} and the half-life of ^{14}C is 5730 years.)

36. In addition to the radioactive nuclei included in the natural decay series, there are several other radioactive nuclei that occur naturally. One is ^{147}Sm, which is 15% naturally abundant and has a half-life of approximately 1.3×10^{10} years. Calculate the number of decays per second per gram (due to this isotope) in a sample of natural samarium. The atomic weight of samarium is 150.4. (Activity per unit mass is called *specific activity*.)

37. A radioactive nucleus with decay constant λ decays to a stable daughter nucleus. (a) Show that the number of daughter nuclei, N_2, increases with time according to the expression

$$N_2 = N_{01}(1 - e^{-\lambda t})$$

where N_{01} is the initial number of parent nuclei. (b) Starting with 10^6 parent nuclei at $t = 0$, with a half-life of 10 h, plot the number of parent nuclei and the number of daughter nuclei as functions of time over the interval 0 to 30 h.

38. Many radioisotopes have important industrial, medical, and research applications. One of these is ^{60}Co, which has a half-life of 5.2 years and decays by the emission of a beta particle (energy 0.31 MeV) and two gamma photons (energies 1.17 MeV and 1.33 MeV). A scientist wishes to prepare a ^{60}Co sealed source that will have an activity of at least 10 Ci after 30 months of use. (a) What minimum initial mass of ^{60}Co is required? (b) At what rate will the source emit energy after 30 months?

39. The concept of radioactive half-life was described in Section 13.4, and Equation 13.11 gives the relationship between $T_{1/2}$ and λ. Another parameter that is often useful in the description of radioactive processes is the *mean life*, τ. Although the half-life of a radioactive isotope is accurately known, it is not possible to predict the time when any individual atom will decay. The mean life is a measure of the average length of existence of all the atoms in a particular sample. Show that $\tau = 1/\lambda$. (*Hint:* Remember that τ is essentially an average value, and use the fact that the number of atoms that decay between t and $t + dt$ is equal to dN. Furthermore, note that these dN atoms have a finite time of existence, t.)

40. Potassium as it occurs in nature includes a radioactive isotope, ^{40}K, which has a half-life of 1.27×10^9 years and a relative abundance of 0.0012%. These nuclei decay by two different pathways—89% by β^- emission and 11% by β^+ emission. Calculate the total activity in Bq associated with 1 kg of KCl *due to β^- emission*.

13.5 The Decay Processes

41. Find the energy released in the alpha decay of $^{238}_{92}$U:

$$^{238}_{92}\text{U} \longrightarrow ^{234}_{90}\text{Th} + ^{4}_{2}\text{He}$$

You will find the mass values in Table 13.6 useful.

42. When, after a reaction or disturbance of any kind, a nucleus is left in an excited state, it can return to its normal (ground) state by emission of a gamma-ray photon (or several photons). Equation 13.25 describes this process. The emitting nucleus must recoil in order to conserve both energy and momentum. (a) Show that the recoil energy of the nucleus is

$$E_r = \frac{(\Delta E)^2}{2Mc^2}$$

where ΔE is the difference in energy between the excited and ground states of a nucleus of mass M. (b) Calculate the recoil energy of the ^{57}Fe nucleus when it decays by gamma emission from the 14.4-keV excited state. For this calculation, take the mass to be 57 u. (*Hint:* When writing the equation for conservation of energy, use $(Mv)^2/2M$ for the kinetic energy of the recoiling nucleus. Also, assume that $hf \ll Mc^2$.)

43. Equation 13.12 represents the decay of an unstable nucleus at rest by alpha emission. The disintegration energy Q given by Equation 13.15 must be shared by the alpha particle and the daughter nucleus in order to conserve both energy and momentum in the decay process. (a) Show that Q and K_α, the kinetic energy of the alpha particle, are related by the expression

$$Q = K_\alpha \left(1 + \frac{M_\alpha}{M_Y}\right)$$

where M_Y is the mass of the daughter nucleus. (b) Use the result of (a) to find the energy of the alpha particle emitted in the decay of ^{226}Ra. (c) What is the kinetic energy of the daughter?

44. Find the kinetic energy of an alpha particle emitted during the alpha decay of $^{220}_{86}$Rn. Assume that the daughter nucleus, $^{216}_{84}$Po, has zero recoil velocity.

45. Determine which of the following suggested decays can occur spontaneously:
    (a) $^{40}_{20}\text{Ca} \rightarrow e^+ + ^{40}_{19}\text{K} + \nu$
    (b) $^{98}_{44}\text{Ru} \rightarrow ^{4}_{2}\text{He} + ^{94}_{42}\text{Mo}$
    (c) $^{144}_{60}\text{Nd} \rightarrow ^{4}_{2}\text{He} + ^{140}_{58}\text{Ce}$

46. (a) Why is the following inverse beta decay forbidden for a free proton?

$$p \longrightarrow n + e^+ + \nu$$

(b) Why is the same reaction possible if the proton is bound in a nucleus? For example, the following reaction occurs:

$$^{13}_{7}\text{N} \longrightarrow ^{13}_{6}\text{C} + e^+ + \nu$$

(c) How much energy is released in the reaction given in (b)? [Take the masses to be $m(e^+) = 0.000549$ u, $M(^{13}\text{C}) = 13.003355$ u, and $M(^{13}\text{N}) = 13.005739$ u, and see Problem 49.]

47. Use the Heisenberg uncertainty principle to make a reasonable argument against the hypothesis that free electrons can be present in a nucleus. Use relativistic expressions for the momentum and energy, and include appropriate assumptions and approximations.

48. The simplest example of beta decay is the beta decay of

the free neutron. Free neutrons have a half-life of 10.2 min and decay according to the reaction

$$n \longrightarrow p + e^- + \bar{\nu}$$

(a) Find the maximum kinetic energy the electron may have when a neutron decays at rest. (*Hint:* In this case, the energy of the antineutrino is approximately zero, the momentum of the proton equals the momentum of the electron, and the kinetic energy of the proton is negligible compared to that of the electron.) (b) Find the proton's kinetic energy in this case. (c) Find the maximum kinetic energy and momentum that the antineutrino may have and the proton's kinetic energy in this case. (*Hint:* Assume that the proton's kinetic energy is negligible compared to that of the $\bar{\nu}$.)

49. *Q values for beta decay:* One must be careful in calculating *Q* values for beta decay. Atomic masses cannot always be used without correction, because electron masses do not always cancel as in alpha decay. Show that the correct expressions for beta decay are

Electron emission $Q = [M(_Z^A\text{X}) - M(_{Z+1}^A\text{Y})]c^2$

Positron emission $Q = [M(_Z^A\text{X}) - M(_{Z-1}^A\text{Y}) - 2m_e]c^2$

Electron capture $Q = [M(_Z^A\text{X}) - M(_{Z-1}^A\text{Y})]c^2$

where $M(_Z^A\text{X})$ and $M(_{Z\pm1}^A\text{Y})$ are the atomic masses of the parent and daughter, respectively, and m_e is the mass of the electron.

13.6   Natural Radioactivity

50. Starting with $_{92}^{235}\text{U}$, the sequence of decays shown in Figure P13.50 is observed, ending with the stable isotope $_{82}^{207}\text{Pb}$. Enter the correct isotope symbol in each open square.

Figure P13.50

Additional Problems

51. A ^{3}H nucleus beta-decays into ^{3}He by creating an electron and an antineutrino according to the reaction

$$_1^3\text{H} \longrightarrow _2^3\text{He} + e^- + \bar{\nu}$$

Use Appendix B to determine the total energy released in this reaction.

52. The nucleus $_8^{15}$O decays by electron capture. Write (a) the basic nuclear process and (b) the decay process. (c) Determine the energy of the neutrino. Disregard the daughter's recoil.

53. In a piece of rock from the Moon, the ^{87}Rb content is assessed to be 1.82×10^{10} atoms per gram of material, and the ^{87}Sr content is found to be 1.07×10^9 atoms per gram. (a) Determine the age of the rock. (b) Could the material in the rock actually be much older? What assumption is implicit in the use of the radioactive dating method? (The relevant decay is ^{87}Rb $\rightarrow$ ^{87}Sr $+ e^-$. The half-life of the decay is 4.8×10^{10} years.)

54. (a) Can ^{57}Co decay by β^+ emission? Explain. (b) Can ^{14}C decay by β^- emission? Explain. (c) If either answer is yes, what is the range of kinetic energies available for the β particle?

55. When a material of interest is irradiated by neutrons, radioactive atoms are produced continually, and some decay according to their given half-lives. (a) If radioactive atoms are produced at a constant rate R and their decay is governed by the conventional radioactive decay law, show that the number of radioactive atoms accumulated after an irradiation time t is

$$N = \frac{R}{\lambda}(1 - e^{-\lambda t})$$

(b) What is the maximum number of radioactive atoms that can be produced?

56. The ground state of $_{43}^{93}$Tc (molar mass 92.9102) decays by electron capture and β^+ emission to energy levels of the daughter (molar mass in ground state 92.9068) at 2.44 MeV, 2.03 MeV, 1.48 MeV, and 1.35 MeV. (a) For which of these levels are electron capture and β^+ decay allowed? (b) Identify the daughter and sketch the decay scheme, assuming that all excited states de-excite by direct gamma decay to the ground state.

14

Nuclear Physics Applications

Chapter Outline

This chapter is concerned primarily with nuclear reactions in which particles and nuclei collide and change into other nuclei and with the two means by which energy can be derived from nuclear reactions: *fission*, in which a large nucleus splits, or fissions, into two smaller nuclei, and *fusion*, in which two small nuclei fuse to form a larger one. In either case, a release of energy occurs that can then be used either destructively through bombs or constructively through production of electric power. We also examine the interaction of radiation with matter and several devices for detecting radiation.

14.1 NUCLEAR REACTIONS

It is possible to change the structures of nuclei by bombarding them with energetic particles. Such collisions that change the identities of the target nuclei are called **nuclear reactions.** Rutherford was the first to observe them in 1919, using naturally occurring radioactive sources for the bombarding particles. Since then, thousands of nuclear reactions have been observed, especially after the development of charged-particle accelerators in the 1930s. With today's technology in particle accelerators, it is possible to achieve particle energies of approximately 1000 GeV. These high-energy particles are used to create new particles whose properties are helping to solve the mystery of the nucleus.

Consider a reaction in which a target nucleus X is bombarded by a particle a, resulting in a nucleus Y and a particle b:

Nuclear reaction

$$a + X \longrightarrow Y + b \tag{14.1}$$

Sometimes this reaction is written in the more compact form

$$X(a, b)Y$$

As an example, consider the reaction $^7Li(p, \alpha)^4He$, or

$$^1_1H + ^7_3Li \longrightarrow ^4_2He + ^4_2He$$

Cockroft and Walton were first to observe this reaction in 1932, using protons accelerated to 600 keV in an accelerator they had designed and built. A nuclear reaction such as this, and in fact any reaction, can occur only if it satisfies certain *conservation laws*. The conservation laws for nuclear reactions are:

- *Conservation of mass number, A.* The total number of nucleons must be the same after the reaction as before. For the reaction under discussion, $A_{before} = 1 + 7 = A_{after} = 4 + 4$.
- *Conservation of charge, q.* Here the charged nuclear particles are protons, and $q_{before} = 1 + 3 = q_{after} = 2 + 2$.
- *Conservation of energy, linear momentum, and angular momentum.* These quantities are conserved because a nuclear reaction involves only internal forces between a target nucleus and a bombarding nucleus, and there are no external forces to upset these conservation principles.

Let us apply the conservation of energy to a reaction of the form of Equation 14.1 to compute *the total kinetic energy released (or absorbed) in the reaction*, which is called the *reaction energy, Q*. Assume that the target nucleus X is originally at rest, the bombarding particle a has kinetic energy K_a, and the reaction products b and Y have kinetic energies K_b and K_Y. Conserving energy,

$$M_X c^2 + K_a + M_a c^2 = M_Y c^2 + K_Y + M_b c^2 + K_b$$

As the total kinetic energy released in the reaction, Q, is equal to the difference between the kinetic energy of the final particles and that of the initial particle, we find

Q of a reaction

$$Q = (K_Y + K_b) - K_a = (M_X + M_a - M_Y - M_b)c^2 \tag{14.2}$$

A reaction for which Q is positive converts nuclear mass to kinetic energy of the products Y and b and is called an *exothermic reaction*. A reaction for which Q is negative requires some minimum input kinetic energy from the bombarding particle in order to occur. Such a reaction is called *endothermic*.

For an endothermic reaction to proceed, the incident particle must have a minimum kinetic energy called the *threshold energy, K_{th}*. Since K_{th} must not only supply $|Q|$, the excess mass-energy of the products, but also supply some kinetic energy to the products to conserve momentum, K_{th} is greater than $|Q|$.

For low-energy reactions, where the kinetic energies of all the interacting particles are small compared to their rest energies, we can apply the nonrelativistic expressions $K = \frac{1}{2}mv^2$ and $p = mv$ to find the threshold energy. It is left

Table 14.1   Q Values for Nuclear Reactions Involving Light Nuclei

Reaction[a]	Measured Q Value (MeV)
^{2}H(n, γ)^{3}H	6.257 ± 0.004
^{2}H(d, p)^{3}H	4.032 ± 0.004
^{6}Li(p, α)^{3}H	4.016 ± 0.005
^{6}Li(d, p)^{7}Li	5.020 ± 0.006
^{7}Li(p, n)^{7}Be	-1.645 ± 0.001
^{7}Li(p, α)^{4}He	17.337 ± 0.007
^{9}Be(n, γ)^{10}Be	6.810 ± 0.006
^{9}Be(γ, n)^{8}Be	-1.666 ± 0.002
^{9}Be(d, p)^{10}Be	4.585 ± 0.005
^{9}Be(p, α)^{6}Li	2.132 ± 0.006
^{10}B(n, α)^{7}Li	2.793 ± 0.003
^{10}B(p, α)^{7}Be	1.148 ± 0.003
^{12}C(n, γ)^{13}C	4.948 ± 0.004
^{13}C(p, n)^{13}N	-3.003 ± 0.002
^{14}N(n, p)^{14}C	0.627 ± 0.001
^{14}N(n, γ)^{15}N	10.833 ± 0.007
^{18}O(p, n)^{18}F	-2.453 ± 0.002
^{19}F(p, α)^{16}O	-8.124 ± 0.007

From C. W. Li, W. Whaling, W. A. Fowler, and C. C. Lauritsen, *Physical Review* 83:512, 1951.

[a] The symbols n, p, d, α, and γ denote the neutron, proton, deuteron, alpha particle, and photon, respectively.

as a problem (Problem 9) to show that when momentum and energy are conserved in a negative Q reaction, the threshold energy is

$$K_{th} = -Q\left(1 + \frac{M_a}{M_X}\right) \qquad (14.3)$$

Finally, we have included a selected list of measured Q values for reactions involving light nuclei in Table 14.1.

EXAMPLE 14.1

(a) Calculate the Q value for the reaction observed by Cockcroft and Walton.

Solution   (a) The reaction is

$$^1_1H + {}^7_3Li \longrightarrow {}^4_2He + {}^4_2He$$

Using $Q = (M_{Li} + M_H - 2M_{He})c^2$ and substituting mass values from Appendix B, we find

$Q = (7.016\ 003\ u + 1.007\ 825\ u$
$\qquad - 2(4.002\ 603\ u))(931.50\ \text{MeV/u}) = 17.3\ \text{MeV}$

(b) Find the kinetic energy of the products if 600-keV protons are incident.

Solution   (b) Since $Q = K_{products} - K_{incident\ particle}$,

$K_{products} = Q + K_{incident\ particle} = 17.3\ \text{MeV} + 0.6\ \text{MeV}$
$\qquad\qquad = 17.9\ \text{MeV}$

This means that the two alpha particles share 17.9 MeV of kinetic energy.

Figure 14.1 (a) A beam of particles incident on a thin foil target of thickness *x*. The view shown is from the edge of the target. (b) A front view of the target, where the circles represent the target nuclei.

14.2   REACTION CROSS-SECTION

Since we deal in this chapter with the interactions between nuclei and matter, it is useful to introduce a quantity called the *cross-section*, which is a measure of the probability that a particular nuclear reaction will occur.

When a beam of particles is incident on a target in the form of a thin foil, not every particle interacts with a target nucleus. The probability that an inter-action will occur depends on the ratio of the "effective" area of the target nucleus to the area of the foil. The situation is analogous to throwing darts at a large wall upon which many inflated balloons are hanging. If the darts are thrown at random in the direction of the wall and the balloons are spread out so as not to touch each other, there is some chance that you will hit one balloon on any given throw. Furthermore, if you throw darts at a rate R_0, the rate R at which balloons burst will be less than R_0. In fact, the probability of hitting a balloon will equal R/R_0. The ratio R/R_0 will depend on the number of balloons N on the wall, the area σ of each balloon, and the area A of the wall. Since the total cross-sectional area of the balloons is $N\sigma$, the probability R/R_0 equals the ratio $N\sigma/A$.

With this analogy, we can now understand the concept of cross-section as it pertains to nuclear events. Suppose a beam of particles is incident upon a thin-foil target, as in Figure 14.1a. Each target nucleus X has an effective area σ called the **cross-section.** You can think of σ as an effective area of the nucleus at right angles to the direction of motion of the bombarding particles, as in Figure 14.1b. It is assumed that the reaction X(a, b)Y will occur only if the incident particle strikes the area σ. Therefore, the probability that a collision will occur is proportional to σ. That is, the probability increases as σ increases. In the general case, the size of σ for a specific reaction may also depend on the energy of the incident particle. Let us consider the concept of cross-section in more detail. In what follows, we will take the foil thickness to be x and its area to be A. Furthermore, we will use the following notation.

R_0 = Rate at which incident particles strike the foil (particles/s)

R = Rate at which reaction events occur (reactions/s)

n = Number of target nuclei per unit volume (particles/m^3), each with cross-section σ

Since the total number of target nuclei in the foil is nAx, the total area exposed to the incident beam must be σnAx. It is assumed that the foil is sufficiently thin that some nuclei are not "hidden" by others as in Figure 14.1b. The ratio of the rate of interactions to the rate of incident particles, R/R_0, must equal the ratio of the area σnAx to the total area A of the foil, in analogy with the dart-balloon collision events. That is,

Reaction rate is proportional to cross-section and target density

$$\frac{R}{R_0} = \frac{\sigma nAx}{A} = \sigma nx \tag{14.4}$$

This result shows that the probability that a nuclear reaction will occur is proportional to the cross-section σ, the density of target nuclei n, and the thickness of the target, x. A value for σ for a specific reaction can therefore be obtained by measuring R, R_0, n, and x and using Equation 14.4.

We can use the same reasoning to arrive at an expression for the number of particles that penetrate (that is, do not undergo reactions within) a foil. Suppose that N_0 particles are incident on a foil of thickness dx, and dN is the number of particles that interact with the target nuclei (Fig. 14.2). The ratio of the number of interacting particles to the number of incident particles, dN/N, equals the ratio of the total target cross-section, $nA\sigma\, dx$, to the total foil area A. That is,

$$-\frac{dN}{N} = \frac{nA\sigma\, dx}{A} = n\sigma\, dx$$

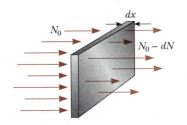

Figure 14.2 If N_0 is the number of particles incident on a target of thickness dx in some time interval, the number emerging from the target is $N_0 - dN$.

The minus sign indicates that particles are being removed from the beam, so dN must be negative. Integrating this expression and taking $N = N_0$ at $x = 0$,

$$\int_{N_0}^{N} \frac{dN}{N} = -n\sigma \int_{0}^{x} dx$$

$$\ln\left(\frac{N}{N_0}\right) = -n\sigma x$$

$$N = N_0 e^{-n\sigma x} \tag{14.5}$$

Number of particles transmitted through a target of thickness x

That is, if N_0 is the number of incident particles, the number that emerge from the slab is N. Thus the number of beam particles that penetrate a target decreases exponentially with target thickness.

Nuclear cross-sections, which have dimensions of area, are typically of the order of the square of the nuclear radius, which is about 10^{-28} m^2. For this reason, it is common to use the unit of 10^{-28} m^2 for measuring nuclear cross-sections. This small unit is known, strangely enough, as the *barn* (b)[1] and is defined as

$$1 \text{ barn} = 10^{-28} \text{ m}^2 \tag{14.6}$$

The barn

In reality, the concept of cross-section in nuclear and atomic physics has little to do with the actual geometric area of target nuclei. The model we have used is simply a convenient one for describing the probability of occurrence of any nuclear reaction. In fact, cross-sections vary with both the specific reaction considered and with the incident particle's kinetic energy over much more than several times the target nucleus's geometrical area. For example, the cross-section for an antineutrino to interact with a proton via the nuclear weak interaction is only about 10^{-19} b in the reaction

$$\bar{\nu} + \text{p} \longrightarrow \text{e}^+ + \text{n} \qquad (\sigma = 10^{-19} \text{ b})$$

The cross-sections for inelastic scattering of neutrons from iodine and xenon via the nuclear strong interaction, however, are about 4 barns in the following reactions.

$$\text{n} + {}^{127}\text{I} \longrightarrow {}^{127}\text{I*} + \text{n} \qquad (\sigma = 4 \text{ b})$$

$$\text{n} + {}^{129}\text{Xe} \longrightarrow {}^{129}\text{Xe*} + \text{n} \qquad (\sigma = 4 \text{ b})$$

---

[1] "Barn" was introduced by the American physicists M. G. Holloway and C. P. Baker in 1942 in a humorous twist. It served the purpose of a code word in concealing war work on reaction probabilities and was appropriate because a cross-section of 10^{-28} m^2 really is "as big as the broad side of a barn" for nuclear processes.

Figure 14.3 The neutron-capture cross-section for cadmium. *(From H. H. Goldsmith, H. W. Ibser, and B. T. Feld, Rev. Mod. Phys. 19:261, 1947.)*

(An inelastic scattering reaction is one in which the incident particle loses energy to the target nucleus, emerging from the reaction with less kinetic energy but leaving the target in an excited state, here denoted by the asterisk.)

Finally, an example of a nuclear reaction with a large scattering cross-section (10^4 b), and with a strongly energy-dependent cross-section as well, is the neutron capture reaction

$$n + {}^{113}Cd \longrightarrow {}^{114}Cd + \gamma \qquad (\sigma = 10^4 \text{ b})$$

In this reaction, which has a high probability only for low-energy neutrons, the cadmium target incorporates an extra neutron into its nucleus, is raised to an excited state, and emits its excess energy by gamma decay. Figure 14.3 shows the strong dependence of cross-section on incident neutron energy in this case.

14.3 INTERACTIONS INVOLVING NEUTRONS

In order to understand the process of nuclear fission and the physics of the nuclear reactor, you should first understand the manner in which neutrons interact with nuclei. As we mentioned earlier, because of their charge neutrality, neutrons are not subject to Coulomb forces. Since neutrons interact very weakly with electrons, matter appears quite "open" to neutrons. In general, one finds that typical cross-sections for neutron-induced reactions increase as the neutron energy decreases. Free neutrons undergo β decay with a mean lifetime of about 10 min. On the other hand, neutrons in matter are usually absorbed by nuclei before they decay.

When a fast neutron (one with an energy greater than about 1 MeV) moves through matter, it undergoes many scattering events with nuclei. In each such event, the neutron gives up some of its kinetic energy to a nucleus. The neutron continues to undergo collisions until its energy is of the order of the thermal energy $k_B T$, where k_B is Boltzmann's constant and T is the absolute temperature. A neutron with this energy is called a **thermal neutron.** At this low energy, there is a high probability that the neutron will be captured by a nucleus, as can be seen for the case of Cd, shown in Figure 14.3. This **neutron-capture** process can be written

$$_0^1 n + {_Z^A}X \longrightarrow {_Z^{A+1}}X + \gamma \qquad (14.7)$$

Neutron-capture process

Although we do not indicate it here, the nucleus X is in an excited state X* for a very short time before it undergoes gamma decay. Also, the product nucleus $_Z^{A+1}X$ is usually radioactive and may decay by α or β emission.

The neutron-capture cross-section associated with the preceding process depends on the nature of the target nucleus and the energy of the incident neutron. For some materials and for fast neutrons, the cross-section is so small that elastic collisions dominate. Materials for which this occurs are called **moderators,** since they slow down (or moderate) the originally energetic neutrons very effectively. Boron, graphite, and water are a few examples of moderator materials.

During an elastic collision between two particles, the maximum kinetic energy is transferred from one particle to the other when they have the same mass. Consequently, a neutron loses all of its kinetic energy when it collides head-on with a proton, in analogy with the collision between a moving billiard ball and a stationary one. If the collision is oblique, the neutron loses only part of its kinetic energy. For this reason, materials which are abundant in hydrogen, such as paraffin and water, are good moderators for neutrons.

At some point, many of the neutrons in the moderator become thermal neutrons, which are neutrons in thermal equilibrium with the moderator material. Their average kinetic energy at room temperature is

$$K_{av} = \frac{3}{2} k_B T \approx 0.04 \text{ eV}$$

which corresponds to a neutron root-mean-square speed of about 2800 m/s. Thermal neutrons have a distribution of velocities just as the molecules in a container of gas do. A high-energy neutron, whose energy is several MeV, will thermalize (that is reach K_{av}) in less than 1 ms when incident on a moderator such as graphite (carbon) or water.

EXAMPLE 14.2 Neutron Capture by Aluminum

An aluminum foil of thickness 0.30 mm is bombarded by energetic neutrons. The aluminum nuclei undergo neutron capture according to the process ^{27}Al(n, γ) ^{28}Al, with a measured capture cross-section of 2.0×10^{-3} b = 2.0×10^{-31} m^2. Assuming the flux of incident neutrons to be 5.0×10^{12} neutrons/cm$^2 \cdot$s, calculate the number of neutrons captured per second by 1.0 cm^2 of the foil.

Solution To solve this problem, we must first evaluate the density of nuclei n (which equals the density of atoms).

Since the density of aluminum is 2.7 g/cm^3 and $A = 27$ for aluminum, we have

$$n = \frac{6.02 \times 10^{23} \text{ nuclei/mol}}{27 \text{ g/mol}} \times 2.7 \frac{\text{g}}{\text{cm}^3}$$

$$= 6.02 \times 10^{22} \frac{\text{nuclei}}{\text{cm}^3} = 6.02 \times 10^{28} \frac{\text{nuclei}}{\text{m}^3}$$

Substituting this value into Equation 14.4, together with the given data for x and σ, gives

$$\frac{R}{R_0} = \sigma n x$$

$$= (2.0 \times 10^{-31} \text{ m}^2)(6.02 \times 10^{28} \text{ nuclei/m}^3)$$

$$(0.30 \times 10^{-3} \text{ m}) = 3.6 \times 10^{-6}$$

Since the rate of incident particles per unit area is $R_0 = 5.0 \times 10^{12}$ neutrons/cm$^2 \cdot$s,

$$R = \left(5.0 \times 10^{12} \frac{\text{neutrons}}{\text{cm}^2 \cdot \text{s}}\right)(3.6 \times 10^{-6})$$

$$= 1.8 \times 10^7 \frac{\text{neutrons}}{\text{cm}^2 \cdot \text{s}}$$

Therefore the number of neutrons captured by 1.0 cm^2/s is only 1.8×10^7 neutrons, whereas the incident number is 5.0×10^{12} neutrons. That is, only about four neutrons out of 1 million are captured!

14.4 NUCLEAR FISSION

Nuclear fission occurs when a very heavy nucleus such as ^{235}U splits, or fissions, into two particles of comparable mass. In such a reaction, *the total mass of the product particles is less than the original mass.* Fission is initiated by the capture of a thermal neutron by a heavy nucleus and involves the release of about 200 MeV per fission. This energy release occurs because the smaller fission-product nuclei are more tightly bound by about 1 MeV per nucleon than the original heavy nucleus.

The process of nuclear fission was first observed in 1939 by Otto Hahn and Fritz Strassmann, following some basic studies by Fermi concerning the interaction of thermal neutrons with uranium. After bombarding uranium ($Z = 92$) with neutrons, Hahn and Strassmann performed a chemical analysis and discovered among the products two medium-mass elements, barium and lanthanum. Shortly thereafter, Lise Meitner and Otto Frisch explained what had happened. The uranium nucleus could split into nearly equal fragments after absorbing a neutron. Measurements showed that about 200 MeV of energy was released in each fission event.

The fission of ^{235}U by thermal neutrons can be represented by

Fission of ^{235}U

$$^{1}_{0}n + {}^{235}_{92}U \longrightarrow {}^{236}_{92}U^* \longrightarrow X + Y + \text{neutrons} \tag{14.8}$$

where ^{236}U* is an intermediate excited state that lasts for only about 10^{-12} s before splitting into X and Y. The resulting nuclei X and Y are called **fission fragments.** There are many combinations of X and Y that satisfy the requirements of conservation of mass-energy, charge, and nucleon number. Figure 14.4 shows the actual mass distribution of fragments in the fission of ^{235}U. The process results in the production of several neutrons, typically two or three. On the average, about 2.5 neutrons are released per event. A typical reaction of this type is

$$^{1}_{0}n + {}^{235}_{92}U \longrightarrow {}^{141}_{56}Ba + {}^{92}_{36}Kr + 3{}^{1}_{0}n \tag{14.9}$$

Of the 200 MeV or so released in this reaction, most goes into the kinetic energy of the heavy fragments barium and krypton.

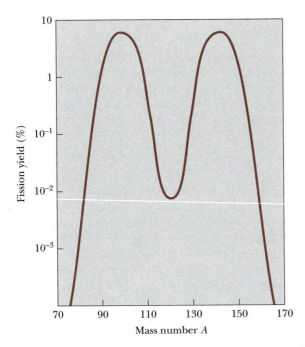

Figure 14.4   The distribution of fission products versus mass number for the fission of ^{235}U bombarded with slow neutrons. Note that the ordinate has a logarithmic scale.

The breakup of the uranium nucleus can be compared to what happens to a drop of water when excess energy is added to it. All the atoms in the drop have energy, but not enough to break up the drop. However, if enough energy is added to set the drop into vibration, it undergoes elongation and compression until the amplitude of vibration becomes large enough to cause the drop to break. In the uranium nucleus, a similar process occurs. Figure 14.5 shows

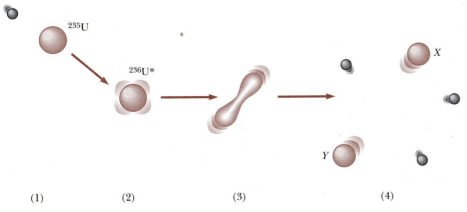

(1)   (2)   (3)   (4)

Figure 14.5   The stages in a nuclear fission event as described by the liquid-drop model of the nucleus.

various stages of the nucleus as the result of neutron capture. The sequence of events for ^{235}U can be described as follows:

1. The ^{235}U nucleus captures a thermal (slow-moving) neutron.
2. This capture results in the formation of ^{236}U*, and the excess energy of this nucleus causes it to undergo violent oscillations.
3. The ^{236}U* nucleus becomes highly distorted, and the force of repulsion between protons in the two halves of the dumbbell shape tends to increase the distortion.
4. The nucleus splits into two fragments, emitting several neutrons in the process.

As can be seen in Figure 14.4, the most probable fission events correspond to fission fragments with mass numbers $A \approx 140$ and $A \approx 95$. These fragments, which share the protons and neutrons of the mother nucleus, both fall on the neutron-rich side of the stability line in Figure 13.4 (Chapter 13). Since fragments that have a large excess of neutrons are unstable, the neutron-rich fragments almost instantaneously release two or three neutrons. The remaining fragments are still rich in neutrons and proceed to decay to more stable nuclei through a succession of beta decays. In the process of such decays, gamma rays are also emitted by nuclei in excited states.

Let us estimate the disintegration energy Q released in a typical fission process. From Figure 13.8 we see that the binding energy per nucleon for heavy nuclei ($A \approx 240$) is about 7.6 MeV, whereas in the intermediate mass range, the binding energy per nucleon is about 8.5 MeV. Taking the mass number of the mother nucleus to be $A = 240$, we see that the energy released per nucleon is estimated to be

$$Q = (240 \text{ nucleons}) \left(8.5 \frac{\text{MeV}}{\text{nucleon}} - 7.6 \frac{\text{MeV}}{\text{nucleon}} \right) = 200 \text{ MeV}$$

About 85% of this energy appears in the form of kinetic energy in the heavy fragments. This energy is very large compared to the energy released in chemical processes. For example, the energy released in the combustion of one molecule of octane used in gasoline engines is about one-millionth the energy released in a single fission event!

EXAMPLE 14.3   The Fission of Uranium

In addition to the barium-lanthanum reaction observed by Meitner and Frisch and the barium-krypton reaction of Equation 14.9, two other ways in which ^{235}U can fission when bombarded with a neutron are (1) by forming ^{140}Xe and ^{94}Sr and (2) by forming ^{132}Sn and ^{101}Mo. In each case, neutrons are also released. Find the number of neutrons released in each event.

Solution   By balancing mass numbers and atomic numbers, we find that these reactions can be written

$$^{1}_{0}\text{n} + ^{235}_{92}\text{U} \longrightarrow ^{140}_{54}\text{Xe} + ^{94}_{38}\text{Sr} + 2(^{1}_{0}\text{n})$$

$$^{1}_{0}\text{n} + ^{235}_{92}\text{U} \longrightarrow ^{132}_{50}\text{Sn} + ^{101}_{42}\text{Mo} + 3(^{1}_{0}\text{n})$$

Thus two neutrons are released in the first event and three in the second.

EXAMPLE 14.4   The Energy Released in the Fission of ^{235}U

Calculate the total energy released if 1.00 kg of ^{235}U undergoes fission, taking the disintegration energy per event to be $Q = 208$ MeV (a more accurate value than the estimate given before).

Solution   We need to know the number of nuclei in 1 kg of uranium. Since $A = 235$, the number of nuclei is

$$N = \frac{6.02 \times 10^{23} \text{ nuclei/mol}}{235 \text{ g/mol}} (1.00 \times 10^3 \text{ g})$$

$$= 2.56 \times 10^{24} \text{ nuclei}$$

Hence the disintegration energy is

$$E = NQ = (2.56 \times 10^{24} \text{ nuclei}) \left(208 \frac{\text{MeV}}{\text{nucleus}} \right)$$

$$= 5.32 \times 10^{26} \text{ MeV}$$

Since 1 MeV is equivalent to 4.14×10^{-20} kWh, we find that $E = 2.37 \times 10^7$ kWh. This shows that 1 kg of ^{235}U supplies a large amount of energy.

EXAMPLE 14.5   A Rough Mechanism for the Fission Process

Consider the fission reaction $^{235}_{92}\text{U} + \text{n} \rightarrow {}^{141}_{56}\text{Ba} + {}^{92}_{36}\text{Kr} + 2{}^{1}_{0}\text{n}$. Show that the model of the fission process in which an excited $^{235}_{92}\text{U}$ nucleus elongates enough to overcome attractive nuclear forces and separates into two charged fragments can be used to estimate the energy released in this fission process. In this model, we assume that the incident neutron provides a few MeV of excitation energy to separate the Ba and Kr nuclei within the uranium nucleus so that the two fragments $^{141}_{56}\text{Ba}$ and $^{92}_{36}\text{Kr}$ are driven apart by Coulomb repulsion. We also neglect the small amount of kinetic energy (several MeV) carried off by the neutrons produced in the reaction.

Solution   First we calculate the separation, r, of the Ba and Kr nuclei at which the nuclear force between them falls to zero. This is $r \cong r_{\text{Ba}} + r_{\text{Kr}}$, where r_{Ba} and r_{Kr} are the nuclear radii of Ba and Kr given by Equation 13.1. Thus

$$r_{\text{Ba}} = (1.2 \times 10^{-15} \text{ m})(141)^{1/3} = 6.2 \times 10^{-15} \text{ m}$$

$$r_{\text{Kr}} = (1.2 \times 10^{-15} \text{ m})(92)^{1/3} = 5.4 \times 10^{-15} \text{ m}$$

$$r \cong 12 \times 10^{-15} \text{ m}$$

Next calculate the Coulomb potential energy for two charges of $Z_1 = 56$ and $Z_2 = 36$ separated by a distance of 12 fm. The potential energy of the two nuclei on the brink of separation is

$$U = \frac{k(Z_1 e)(Z_2 e)}{r} = \frac{(1.440 \text{ eV} \cdot \text{nm})(56)(36)}{12 \times 10^{-15} \text{ m}} = 240 \text{ MeV}$$

As the two fragments separate, this potential energy is converted to an amount of kinetic energy consistent with the total measured energy release of about 200 MeV. This shows that the simple fission mechanism suggested here is plausible.

14.5   NUCLEAR REACTORS

In the last section we saw that, when ^{235}U fissions, an average of 2.5 neutrons are emitted per event. These neutrons can in turn trigger other nuclei to fission, with the possibility of a chain reaction (Fig. 14.6). Calculations show that if the

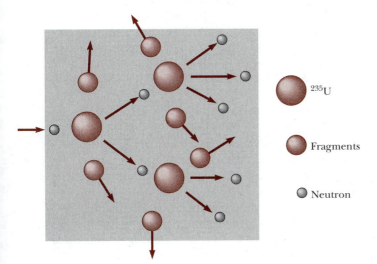

Figure 14.6   A nuclear chain reaction initiated by the capture of a neutron. (Many pairs of different isotopes are produced.)

^{235}U

Fragments

Neutron

Figure 14.7   A sketch of the world's first reactor, which was composed of layers of graphite interspersed with uranium. (Because of wartime secrecy, there are no photographs of the completed reactor.) The first self-sustained chain reaction was achieved on December 2, 1942. Word of the success was telephoned immediately to Washington, D.C., in the form of this coded message: "The Italian navigator has landed in the New World and found the natives very friendly." The historic event took place in an improvised laboratory in the racquet court under the west stands of the University of Chicago's Stagg Field, and the Italian navigator guiding the work was Fermi.

chain reaction is not controlled (that is, if it does not proceed slowly), it can result in a violent explosion, with the release of an enormous amount of energy. For example, if the energy in 1 kg of ^{235}U were released, it would be equivalent to detonating about 20 000 tons of TNT! This, of course, is the principle behind the first nuclear bomb, an uncontrolled fission reaction.

Chain reaction   A nuclear reactor is a system designed to maintain what is called a **self-sustained chain reaction.** This important process was first achieved in 1942 by Fermi at the University of Chicago, with natural uranium as the fuel (Fig. 14.7). Most reactors in operation today also use uranium as fuel. Natural uranium contains only about 0.7% of the ^{235}U isotope, with the remaining 99.3% being ^{238}U. This fact is important for the operation of a reactor, because ^{238}U almost never fissions. Instead, it tends to absorb neutrons, producing neptunium and plutonium. For this reason, reactor fuels must be artificially enriched to contain at least a few percent ^{235}U.

To achieve a self-sustained chain reaction, one (on average) of the neutrons emitted in ^{235}U fission must be captured by another ^{235}U nucleus and cause it to undergo fission. A useful parameter for describing the level of reactor op-
Reproduction constant   eration is the **reproduction constant** K, defined as *the average number of neutrons from each fission event that actually cause another fission event.* As we have seen, K can have a maximum value of 2.5 in the fission of uranium. In practice, however, K is less than this because of several factors, to be discussed.

A self-sustained chain reaction is achieved when $K = 1$. Under this condition the reactor is said to be **critical.** When $K < 1$, the reactor is subcritical and the reaction dies out. When K is substantially greater than unity, the reactor is said to be supercritical, and a runaway reaction occurs. In a nuclear reactor run by a utility company to furnish power, it is necessary to maintain a value of K slightly greater than unity.

Figure 14.8 shows the basic ingredients of a nuclear reactor core. The fuel elements consist of enriched uranium. The function of the remaining parts of the reactor and some aspects of its design will now be described.

Neutron Leakage

In any reactor, a fraction of the neutrons produced in fission leak out of the core before inducing other fission events. If the fraction leaking out is too great, the reactor will not operate. The percentage lost is large if the reactor is very small because leakage is a function of the ratio of surface area to volume. Therefore, a critical feature of the design of a reactor is to choose the correct surface-area-to-volume ratio so that a sustained reaction can be achieved.

Regulating Neutron Energies

Recall that the neutrons released in fission events are very energetic, having kinetic energies of about 2 MeV. It is necessary to slow these neutrons to thermal energies to allow them to be captured and produce fission of other ^{235}U nuclei, because the probability of neutron-induced fission increases with decreasing energy, as shown in Figure 14.9. The energetic neutrons are slowed down by a moderator substance surrounding the fuel, as shown in Figure 14.8.

Neutron Capture

In the process of being slowed down, neutrons may be captured by nuclei that do not fission. The most common event of this type is neutron capture by ^{238}U, which constitutes over 90% of the uranium in the fuel elements. The probability of neutron capture by ^{238}U is very high when the neutrons have high kinetic energies and very low when they have low kinetic energies. Thus the slowing down of the neutrons by the moderator serves the secondary purpose of making them available for reaction with ^{235}U and decreasing their chances of being captured by ^{238}U.

Control of Power Level

It is possible for a reactor to reach the critical stage ($K = 1$) after all the neutron losses just described are minimized. However, some method of control is needed to maintain a K value near unity. If K were to rise above this value, the heat produced in the runaway reaction would melt the reactor. To control the power level, control rods are inserted into the reactor core (see Fig. 14.8). These rods are made of materials, such as cadmium, that absorb neutrons very efficiently (see Fig. 14.3). By adjusting the number and position of these control rods in the reactor core, the K value can be varied and any power level within the design range of the reactor can be achieved.

Although there are several types of reactor systems that convert the kinetic energy of fission fragments to electrical energy, the most common type in use

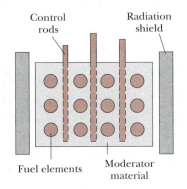

Figure 14.8 A cross-section of a reactor core showing the control rods, fuel elements, and moderating material surrounded by a radiation shield.

Figure 14.9 The cross-section for neutron-induced fission of ^{235}U. The average cross-section for room-temperature neutrons is about 500 b.

Figure 14.10   Main components of a pressurized-water reactor.

in the United States is the pressurized-water reactor (Fig. 14.10). Its main parts are common to all reactor designs. Fission events in the reactor core supply heat to the water contained in the primary (closed) loop, which is maintained at high pressure to keep it from boiling. This water also serves as the moderator. The hot water is pumped through a heat exchanger, and the heat is transferred to the water contained in the secondary loop. The hot water in the secondary loop is converted to steam, which drives a turbine-generator system to create electric power. Note that the water in the secondary loop is isolated from the water in the primary loop to prevent contamination of the secondary water and steam by radioactive nuclei from the reactor core.

Safety and Waste Disposal

One of the dangers inherent in a nuclear reactor is the possibility of interrupted water flow in the secondary cooling system. If this should occur, it is conceivable that the temperature of the reactor could increase to the point where the fuel elements would melt and in turn melt the bottom of the reactor and the ground below. Additionally, the large amounts of heat generated could lead to a high-pressure steam explosion. To decrease the chances of such an event, every reactor is built with a backup cooling system that takes over if the regular cooling system fails. It is important to note that a well-designed reactor cannot explode or produce meltdown (the so-called "China syndrome").

Another problem in fission reactors is the disposal of radioactive material when the core is replaced. This waste material contains long-lived, highly radioactive isotopes that must be stored over long periods of time in such a way that there is no chance of environmental contamination. At present, sealing radioactive wastes in deep salt mines seems to be the method of choice. However, because there is no official government storage facility, waste continues to build up in the backyards of hospitals and power plants.

Another consequence of nuclear power plants is the disposal of excess heat in the cooling system. Water from a nearby river is often used to cool the reactor. This raises the river temperatures downstream from the reactor and affects the organisms living in and adjacent to the river. Another technique for cooling the reactor water is the use of evaporation towers.

Even including the Chernobyl incident in Ukraine in 1987, the adverse impact of nuclear reactors on the environment is probably less than other means of power generation. Thus both increased use of nuclear power and increased emphasis on energy conservation methods appear to be logical components of a sensible energy policy.

14.6   NUCLEAR FUSION

In Chapter 13 we found that the binding energy for light nuclei (those having mass numbers less than 20) is much smaller than the binding energy for heavier nuclei. This suggests a process that is the reverse of fission. When *two light nuclei combine to form a heavier nucleus*, the process is called **nuclear fusion**. Because the mass of the final nucleus is less than the combined rest masses of the original nuclei, a loss of mass occurs, accompanied by a release of energy. The following are examples of such energy-liberating fusion reactions occuring in the Sun:

$$
{}_1^1\text{H} + {}_1^1\text{H} \longrightarrow {}_1^2\text{H} + {}_1^0\text{e}^+ + \nu
$$
$$
{}_1^1\text{H} + {}_1^2\text{H} \longrightarrow {}_2^3\text{He} + \gamma
$$

(14.10)

This second reaction is followed by one of the following reactions:

$$
{}_1^1\text{H} + {}_2^3\text{He} \longrightarrow {}_2^4\text{He} + {}_1^0\text{e}^+ + \nu
$$
$$
{}_2^3\text{He} + {}_2^3\text{He} \longrightarrow {}_2^4\text{He} + {}_1^1\text{H} + {}_1^1\text{H}
$$

These are the basic reactions in what is called the **proton-proton cycle,** believed to be one of the basic cycles by which energy is generated in the Sun and other stars that have an abundance of hydrogen. Most of the energy production takes place in the Sun's interior, where the temperature is approximately 1.5×10^7 K. As we will see later, such high temperatures are required to drive these reactions that they are called **thermonuclear fusion reactions.** The hydrogen (fusion) bomb, which was first exploded in 1952, is an example of an uncontrolled thermonuclear fusion reaction.

Thermonuclear reactions

All of the reactions in the proton-proton cycle are exothermic—that is, they involve a release of energy. An overall view of the proton-proton cycle is that four protons combine to form an alpha particle and two positrons, with the release of 25 MeV of energy.

Fusion Reactions

The enormous amount of energy released in fusion reactions suggests the possibility of harnessing this energy for useful purposes here on Earth. A great deal of effort is currently directed toward developing a sustained and control-

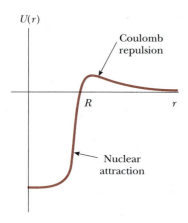

Figure 14.11   Potential energy as a function of separation between two deuterons. The Coulomb repulsive force is dominant at long range, whereas the nuclear attractive force is dominant at short range, where R is of the order of several fermi.

lable thermonuclear reactor—a fusion power reactor. Controlled fusion is often called the ultimate energy source because of the availability of its fuel source: water. For example, if deuterium were used as the fuel, 0.12 g of it could be extracted from 1 gal of water at a cost of about 4 cents. Such rates would make the fuel costs of even an inefficient reactor almost insignificant. An additional advantage of fusion reactors is that comparatively few radioactive by-products are formed. For the proton-proton cycle described earlier in this section, the end product of the fusion of hydrogen nuclei is safe, nonradioactive helium. Unfortunately, a thermonuclear reactor that can deliver a net power output spread out over a reasonable time interval is not yet a reality even though research has been in progress since the 1950s, and many difficulties must be resolved before a successful device is constructed.

We have seen that the Sun's energy is based, in part, upon a set of reactions in which hydrogen is converted to helium. Unfortunately, the proton-proton interaction is not suitable for use in a fusion reactor, because the event requires very high pressures and densities. The process works in the Sun only because of the extremely high density of protons in the Sun's interior.

The fusion reactions that appear most promising for a fusion power reactor involve deuterium ($^2_1$H) and tritium ($^3_1$H):

$$^2_1\text{H} + {}^2_1\text{H} \longrightarrow {}^3_2\text{He} + {}^1_0\text{n} \qquad Q = 3.27 \text{ MeV}$$

$$^2_1\text{H} + {}^2_1\text{H} \longrightarrow {}^3_1\text{H} + {}^1_1\text{H} \qquad Q = 4.03 \text{ MeV} \qquad (14.11)$$

$$^2_1\text{H} + {}^3_1\text{H} \longrightarrow {}^4_2\text{He} + {}^1_0\text{n} \qquad Q = 17.59 \text{ MeV}$$

As noted earlier, deuterium is available in almost unlimited quantities from our lakes and oceans and is very inexpensive to extract. Tritium, however, is radioactive ($T_{1/2} = 12.3$ years) and undergoes beta decay to ^{3}He. As a result, tritium does not occur naturally to any great extent and must be artificially produced.

One of the major problems in obtaining energy from nuclear fusion is the fact that the Coulomb repulsion force between two charged nuclei must be overcome before they can fuse. The potential energy as a function of particle separation for two deuterons (each with charge $+e$) is shown in Figure 14.11. The potential energy is positive in the region $r > R$, where the Coulomb repulsive force dominates ($R \approx 1$ fm), and negative in the region $r < R$, where the strong nuclear force dominates. The fundamental problem, then, is to give the two nuclei enough kinetic energy to overcome this repulsive potential barrier. This can be accomplished by heating the fuel to extremely high temperatures (about 10^8 K, far greater than the interior temperature of the Sun). At these high temperatures, the atoms are ionized and the system consists of a collection of electrons and nuclei, commonly referred to as a plasma. The following example shows how to estimate the temperature required to produce fusion.

High temperatures are required to overcome the large Coulomb barrier

EXAMPLE 14.6   The Fusion of Two Deuterons

The separation between two deuterons must be about 1.0×10^{-14} m for the attractive nuclear force to overcome the repulsive Coulomb force. (a) Calculate the height of the potential barrier due to the repulsive force.

Solution   The potential energy associated with two charges separated by a distance r is

$$U = k\frac{q_1 q_2}{r}$$

where k is the Coulomb constant. For the case of two deuterons, $q_1 = q_2 = +e$, so

$$U = k\frac{e^2}{r} = \left(8.99 \times 10^9 \frac{\text{N} \cdot \text{m}^2}{\text{C}^2}\right)\frac{(1.60 \times 10^{-19} \text{ C})^2}{1.0 \times 10^{-14} \text{ m}}$$

$$= 2.3 \times 10^{-14} \text{ J} = 0.14 \text{ MeV}$$

(b) Estimate the effective temperature required for a deuteron to overcome the potential barrier, assuming an energy of $\frac{3}{2}k_BT$ per deuteron (where k_B is Boltzmann's constant).

Solution   Since the total Coulomb energy of the pair of deuterons is 0.14 MeV, the Coulomb energy per deuteron is 0.07 MeV = 1.1×10^{-14} J. Setting this equal to the average thermal energy per deuteron gives

$$\tfrac{3}{2}k_BT = 1.1 \times 10^{-14} \text{ J}$$

where k_B is equal to 1.38×10^{-23} J/K. Solving for T gives

$$T = \frac{2 \times (1.1 \times 10^{-14} \text{ J})}{3 \times (1.38 \times 10^{-23} \text{ J/K})} = 5.3 \times 10^8 \text{ K}$$

Example 14.6 suggests that deuterons must be heated to about 5×10^8 K to achieve fusion. This estimate of the required temperature is too high, however, because the particles in the plasma have a Maxwellian speed distribution, and therefore some fusion reactions are caused by particles in the high-energy "tail" of this distribution. Furthermore, even the particles without enough energy to overcome the barrier have some probability of tunneling through the barrier. When these effects are taken into account, a temperature of "only" 4×10^8 K appears adequate to fuse two deuterons.

The temperature at which the power generation rate exceeds the loss rate (due to mechanisms such as radiation losses) is called the **critical ignition temperature.** This temperature for the deuterium-deuterium (D-D) reaction is 4×10^8 K. According to $E \cong k_BT$, this temperature is equivalent to approximately 35 keV. It turns out that the critical ignition temperature for the deuterium-tritium (D-T) reaction is about 4.5×10^7 K, or only 4 keV. Figure 14.12 is a plot of the power generated by fusion versus temperature for the two reactions. The straight line represents the power lost, via the radiation mechanism known as **bremsstrahlung,** versus temperature. This is the principal mechanism of energy loss, in which radiation (primarily x-rays) is emitted as the result of electron-ion collisions within the plasma.[2] The intersections of the P_{lost} line with the P_f curves give the critical ignition temperatures.

Critical ignition temperature

[2] Cyclotron radiation is another loss mechanism; it is especially important in the case of the D-D reaction.

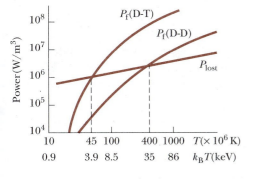

Figure 14.12 Power generated (or lost) versus temperature for the deuterium-deuterium and deuterium-tritium fusion reactions. When the generation rate P_f exceeds the loss rate P_{lost}, ignition takes place.

In addition to the high temperature requirements, there are two other critical parameters that determine whether or not a thermonuclear reactor will be successful: the **ion density,** n, and **confinement time,** τ. *The confinement time is the period for which the interacting ions are maintained at a temperature equal to or greater than the ignition temperature.* The British physicist J. D. Lawson has shown that the ion density and confinement time must both be large enough to ensure that more fusion energy is released than is required to heat the plasma. In particular, **Lawson's criterion** states that a net energy output is possible under the following conditions.

$$n\tau \geq 10^{14} \text{ s/cm}^3 \quad \text{(D-T)}$$
$$n\tau \geq 10^{16} \text{ s/cm}^3 \quad \text{(D-D)} \tag{14.12}$$

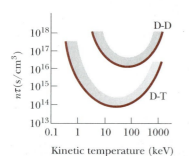

Figure 14.13 The Lawson number $n\tau$ versus kinetic temperature for the D-T and D-D fusion reactions.

Figure 14.13 is a graph of $n\tau$ versus the so-called **kinetic temperature** $k_B T$ for the D-T and D-D reactions.

Lawson arrived at his criterion by comparing the energy required to heat the plasma with the energy generated by the fusion process. The energy E_h required to heat the plasma is proportional to the ion density n; that is, $E_h = D_1 n$. The energy generated by the fusion process, E_f, is proportional to $n^2\tau$, or $E_f = D_2 n^2 \tau$. This can be understood by realizing that the fusion energy released is proportional to both the rate at which interacting ions collide, n^2, and the confinement time, τ. Net energy is produced when the energy generated by fusion, E_f, exceeds E_h. When the constants D_1 and D_2 are calculated for different reactions, the condition that $E_f \geq E_h$ leads to Lawson's criterion.[3]

In summary, the three basic requirements of a successful thermonuclear power reactor are

- The plasma temperature must be very high—about 4.5×10^7 K for the D-T reaction and about 4×10^8 K for the D-D reaction.
- The ion density must be high. A high density of interacting nuclei is necessary to increase the collision rate between particles.
- The confinement time of the plasma must be long. To meet Lawson's criterion, the product $n\tau$ must be large. For a given value of n, the probability of fusion between two particles increases as τ increases.

Current efforts are aimed at meeting Lawson's criterion at temperatures exceeding the critical ignition temperature. Although the minimum plasma densities have been achieved, the problem of confinement time is more difficult. How can a plasma be confined at 10^8 K for 1 s? Two basic techniques for confining plasmas are under investigation: magnetic field confinement and inertial confinement.

Magnetic Field Confinement

Many fusion-related plasma experiments use **magnetic field confinement** to contain the charged plasma. Figure 14.14a shows a device called a **tokamak,** first developed in Russia. A combination of two magnetic fields is used to con-

---

[3] Note that Lawson's criterion neglects the energy needed to set up the strong magnetic field that is used to confine the hot plasma. This energy is expected to be about 20 times greater than the energy required to heat the plasma. Consequently, it is necessary to have a magnetic energy recovery system or to make use of superconducting magnets.

Figure 14.14   (a) A schematic diagram of a tokamak used in magnetic confinement. The total magnetic field **B** is the superposition of the toroidal field $\mathbf{B}_t$ and the poloidal field $\mathbf{B}_p$. The plasma is trapped within the spiraling field lines as shown. (*Adapted from McGraw-Hill Encyclopedia of Science and Technology, New York, McGraw-Hill Book Co., 1987.*) (b) A cutaway view of the TFTR. (*Courtesy of the Plasma Physics Laboratory at Princeton University*)

fine and stabilize the plasma: (1) a strong toroidal field, produced by the current in the windings, and (2) a weaker "poloidal" field, produced by the toroidal current. The toroidal current heats the plasma in addition to confining it. The resultant helical field lines spiral around the plasma and keep it from touching the walls of the vacuum chamber. If the plasma comes into contact with the walls, its temperature is reduced and heavy impurities sputtered from the walls "poison" it and lead to large power losses. One of the major breakthroughs in the 1980s was in the area of auxiliary heating to reach ignition temperatures. Experiments have shown that injecting a beam of energetic neutral particles into the plasma is a very efficient method of heating the plasma to ignition temperatures (5 to 10 keV). Radio-frequency heating will probably be needed for reactor-size plasmas. Figure 14.14b shows a cutaway view of the Princeton Tokamak Fusion Test Reactor.

To date, the best fusion results have come from the Joint European Torus (JET), a tokamak near Oxford, England, and the Tokamak Fusion Test Reactor (TFTR) in Princeton, New Jersey. In 1991 JET, the world's largest tokamak, exceeded the Lawson break-even point and reached the condition of $n\tau = 3.5 \times 10^{13}$ s/cm^3 at an ion temperature of 25 keV. When small amounts of tritium were introduced into the deuterium plasma, about 2 MW of power was produced, although 20 MW of input power was required to heat the plasma. In 1994 both the JET and TFTR groups ran the first full-scale experiments on plasmas containing 50:50 mixtures of tritium and deuterium. The world record for power generation is now held by the TFTR, which, operating at 3×10^8 K, generated a burst of 9 MW on May 27, 1994, for about 1 s, with an input heating power of about 30 MW. In these tests it was found that no unusual losses of alpha particles were produced in the D-T fusion and that the alphas were absorbed by the plasma, producing additional plasma heating, as had been hoped.

Inertial Confinement

The second technique for confining a plasma, called **inertial confinement**, makes use of a D-T target that has a very high particle density of 5×10^{25} particles/cm^3, or a mass density of about 200 g/cm^3. In this scheme, the confinement time is very short (typically 10^{-11} to 10^{-9} s), and so, because of their own inertia, the particles do not have a chance to move appreciably from their initial positions. Thus Lawson's criterion can be satisfied by combining a high particle density with a short confinement time.

Laser fusion is the most common form of inertial confinement. A small D-T pellet about 1 mm in diameter is struck simultaneously by several focused, symmetrically incident high-intensity laser beams, resulting in a large pulse of input energy that causes the surface of the fuel pellet to evaporate (Fig. 14.15). The escaping particles produce a reaction force on the core of the pellet, resulting in a strong, inwardly moving, compressive shock wave. This shock wave increases the pressure and density of the core and produces a corresponding increase in temperature. When the temperature of the core reaches ignition temperature, fusion reactions cause the pellet to explode. The process can be viewed as a miniature hydrogen bomb explosion.

Two of the leading laser fusion laboratories in the United States are the Omega facility at the University of Rochester and the Nova facility at Lawrence Livermore National Laboratory in California. Both facilities use neodymium glass lasers; the Omega facility focuses 60 laser beams on a target chamber about 3 m in diameter. Figure 14.16a shows the target at Omega, and Figure 14.16b shows the tiny, spherical D-T pellets used. Nova, operating at higher input power levels than Omega, is capable of injecting a power of 200 kJ into a

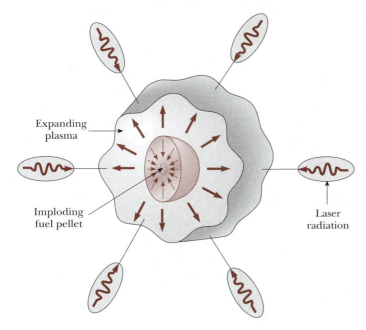

Figure 14.15 In the inertial confinement scheme, a D-T fuel pellet fuses when struck by several high-intensity laser beams simultaneously and symmetrically.

(a)   (b)

Figure 14.16 (a) A view of the Omega target bay from high above the target chamber shows critical components associated with the UV transport system. (b) An example of a plastic microballoon used as an inertial confinement fusion target. The microballoon has a diameter of 900 μm and is filled with an equimolar mix of deuterium and tritium. *(This work was supported by the U.S. Department of Energy (DOE), the University of Rochester, and the New York State Energy Research and Development Authority. The support of DOE does not constitute an endorsement by DOE of the views expressed in this text. Staff Photographer: Eugene Kowaluk)*

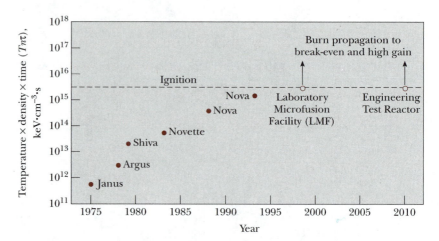

Figure 14.17   A plot of the product of temperature, density, and confinement time for laser fusion versus year. The filled circles indicate currently achieved results.

0.5-mm D-T pellet in 1 ns. With these high input powers, Nova has achieved $n\tau \approx 5 \times 10^{14}$ s/cm³ and ion temperatures of 5.0 keV, values close to those required for D-T ignition. Figure 14.17 illustrates progress in laser fusion experiments with time. This steady progress has led the U.S. Department of Energy and other groups to plan a national facility that will involve a laser fusion device with an input energy in the 5- to 10-MJ range.

Fusion Reactor Design

In the D-T fusion reaction

$$\ce{^2_1H} + \ce{^3_1H} \longrightarrow \ce{^4_2He} + \ce{^1_0n} \qquad Q = 17.6 \text{ MeV}$$

the alpha particle carries 20% of the energy and the neutron carries 80%, or about 14 MeV. While the charged alphas are primarily absorbed in the plasma and produce desired additional plasma heating, the 14-MeV neutrons pass through the plasma and must be absorbed in a surrounding blanket material to extract their large kinetic energy and generate electric power. A frequently proposed scheme is to use molten lithium metal as the neutron-absorbing material and to circulate the lithium in a closed heat-exchanging loop to produce steam and drive turbines, as in a conventional power plant. Figure 14.18 is a diagram of such a fusion reactor system. It is estimated that a blanket of lithium about 1 m thick would capture nearly 100% of the neutrons from the fusion of a small D-T pellet, not only absorbing the neutron kinetic energy but limiting the dangerous neutron flux to nearby workers.

The capture of neutrons by lithium is described by the reaction

$$\ce{^1_0n} + \ce{^6_3Li} \longrightarrow \ce{^3_1H} + \ce{^4_2He}$$

where the kinetic energies of the charged tritium and alpha particle products are converted to heat in the lithium. An extra advantage of using lithium as the energy transfer medium is the production of tritium, $\ce{^3_1H}$, which may be separated from the lithium and returned as fuel to the reactor. The process is indicated in the generic fusion reactor shown in Figure 14.18.

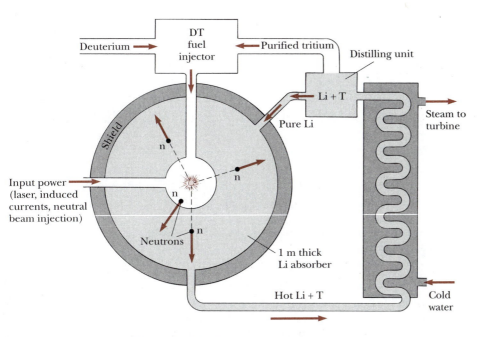

Figure 14.18   A generic fusion reactor.

Advantages and Problems of Fusion

If fusion power can be harnessed, it will offer several advantages over fission-generated power: (1) the low cost and abundance of the fuel (deuterium), (2) the impossibility of runaway accidents, and (3) a lesser radiation hazard than with fission. Some of the anticipated problem areas and disadvantages include (1) its as-yet-unestablished feasibility, (2) the very high proposed plant costs, (3) the scarcity of lithium, (4) the limited supply of helium needed to cool the superconducting magnets used to produce strong confining fields (this problem may be significantly reduced by the development of high-temperature superconductors), (5) structural damage and induced radioactivity caused by neutron bombardment, and (6) the anticipated high degree of thermal pollution. If these basic problems and the engineering design factors can be resolved, nuclear fusion may become a feasible source of energy by the middle of the 21st century.

14.7   RECENT FUSION ENERGY DEVELOPMENTS[4]

The search for alternative energy sources has led to a highly successful international research effort to develop fusion as a practical energy source. This process is illustrated in Figure 14.19, where deuterium and tritium ions, in a high-temperature plasma and confined by a strong magnetic field, undergo fusion reactions and produce both 3.5-MeV alpha particles that are trapped in

[4] This section was prepared by Dr. Ronald C. Davidson, Director of the Princeton Plasma Physics Laboratory and Professor of Astrophysical and Physical Sciences at Princeton University.

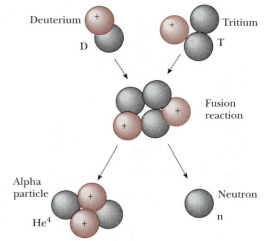

Figure 14.19 A schematic diagram of the D-T fusion reaction. Eighty percent of the energy released is in the 14-MeV neutrons.

the plasma and energetic 14-MeV neutrons that carry away the heat from the reaction.

As shown in Figure 14.20, the power produced in laboratory fusion experiments has increased by a factor of more than 60 million since construction of the largest U.S. tokamak, the Tokamak Fusion Test Reactor (TFTR) at the

PLT	Princeton Large Torus
PDX	Princeton Divertor Experiment
JET	Joint European Torus
JT-60	Japan
ITER	International Thermonuclear Experimental Reactor
DIII & DIII-D	General Atomics Tokamak Experiments
ATC & TFTR	Princeton Plasma Physics Laboratory
ALCATOR A, C	Massachusetts Institute of Technology

Figure 14.20 Progress in the fusion power achieved by tokamaks since construction authorization of the Tokamak Fusion Test Reactor in 1976.

Figure 14.21   A plot of the total fusion power versus time achieved in the historic experiments on the Tokamak Fusion Test Reactor in December 1993.

Princeton Plasma Physics Laboratory. Scientific breakthroughs and world re-cords in plasma performance have seesawed among the large tokamaks in Eu-rope, Japan, and the United States. The United States moved into the lead in December 1993, when 6.2 MW of fusion power was generated during the his-toric experiments on the TFTR.

Figure 14.14b shows the TFTR device, including the high-power neutral beam injectors around the periphery of the tokamak that are used to heat and fuel the plasma to ion temperatures approaching 4×10^8 K and ion densities of 0.6×10^{14} cm^{-3}. Once the high-energy (100-keV) neutral tritium and deu-terium atoms enter the plasma chamber, they are ionized and collisionally heat the background plasma. The high-temperature plasma in TFTR is confined and insulated from contact with the chamber wall by a strong toroidal mag-netic field of magnitude 5 T, which is sustained for about 2 s by external power supplies.

Properties of the plasma in TFTR are measured by an array of noninvasive diagnostic instruments, ranging from detectors that sense microwave and light emission from the plasma to detectors that measure energetic ions escaping from the plasma and the 14-MeV neutrons produced in the deuterium-tritium fusion process. Typical TFTR data on fusion-power production are illustrated in Figure 14.21, where the fusion power (in the neutrons and alpha particles) is plotted versus time for three different experiments with varying concentra-tions of tritium and deuterium in the fuel mixture. In each case, the neutral beam injectors are "on" between $t = 0$ and $t = 0.75$ s. Note from the figure that a 50% tritium–50% deuterium mixture is optimal for maximizing the fusion power production at 6.2 MW. Figure 14.22 is a plot of the peak ion temperature T_i in the center of the plasma versus time, for the same time interval as shown in Figure 14.21 for fusion power data. Note that T_i increases to about 35 keV (corresponding to 3.9×10^8 K) in these experiments.

Also in the high-power experiments with deuterium-tritium plasmas, there are preliminary indications that the plasma electrons are heated collisionally by energetic 3.5-MeV alpha particles created in the fusion process. The elec-trons experience a temperature increase of about 0.75 keV ($\cong 0.8 \times 10^7$ K) above the baseline electron temperature of 9 keV ($\cong 10^8$ K) obtained in other-

Figure 14.22   A plot of the peak ion temperature T_i versus time achieved in the highest-fusion-power experiment (6.2 MW) in Figure 14.21.

wise similar deuterium plasmas. Heating of the background plasma by the alpha particles will be very important in later power reactors and in experimental facilities such as the International Thermonuclear Experimental Reactor (ITER), which is presently being designed. The ITER device will be a tokamak much larger in physical size than TFTR. It is designed to produce 1000 MW of fusion power—about a factor of 100 larger than TFTR's ultimate design capability of 10 MW. In ITER, the energy content of the alpha particles will be so intense that they will heat and sustain the plasma, allowing the auxiliary heating sources to be turned off once the reaction is initiated. Such a state of "sustained burn" of a plasma is referred to as ignition.

14.8   INTERACTION OF PARTICLES WITH MATTER

In this section we shall consider processes involving the interaction between energetic particles and matter. We shall deal mainly with the interaction of charged particles and photons with matter, since these are of primary importance in such factors as shielding characteristics, the design of particle detectors, and biological effects.

Heavy Charged Particles

Range of a particle

A heavy charged particle, such as an alpha particle or a proton, moving through a solid, liquid, or gas travels a well-defined distance before coming to rest. This distance is called the **range** of the particle. As the particle passes through the medium, it loses energy, primarily through the excitation and ionization of atoms in the medium. Some energy is also lost through elastic collisions with nuclei. The highly energetic particle loses its energy in many small increments, and to a good approximation one can treat the problem as a continuous loss of energy. At the end of its range, the particle is left with just the average thermal energy of its surroundings. The range depends on the charge, mass, and energy of the particle, the density of the medium through which it travels, and the ionization potential and atomic number of the atoms in the medium.

The ranges of alpha particles and protons in air as a function of their energy are plotted in Figure 14.23. Note that for a given energy, the proton has a range about ten times that of the alpha particle. This is because the proton has less charge and so does not interact with the medium as strongly as the alpha particle. Furthermore, since the alpha particle is more massive, it travels at a lower average speed. As a result, the slower-moving alpha particle loses its energy more readily, since it has more time to interact with atoms in the medium.

Figure 14.24 shows the rate of energy loss per unit length (energy loss rate), $-dE/dx$, versus the energy of the charged particle. This **energy loss rate** (also called the *stopping power*) is approximately proportional to the inverse of the energy at low energies ($v \ll c$) and reaches a maximum at some point. At very high energies (as $v \to c$), the energy loss per unit length is approximately energy-independent. At very low energies, the decrease in energy loss is a result of the fact that the particle is moving too slowly to produce ionization and loses its energy mainly by elastic collisions.

The energy loss rate is approximately proportional to the density of the medium through which the charged particles travel. This is explained by recognizing that the primary energy loss mechanism (ionization) involves the excitation of electrons in the medium, and the density of electrons increases with increasing density of the medium. For example, the range in aluminum of protons having an energy of 1 to 10 MeV is about 1/1600 of their range in air.

Fission fragments with energies of about 80 MeV have a range of only about 0.02 m in air, which is about 1000 times smaller than the range of protons with the same energy. This difference is due to the greater charge of the fission fragments and to the fact that the energy loss rate is proportional to the square of the charge.

Electrons

When an electron of energy less than about 1 MeV passes through matter, it loses its energy by the same processes that occur for heavy charged particles. However, the range of the electron is not as well defined. Since an electron is much smaller than a proton, there are large statistical variations in

Figure 14.23 The range of alpha particles and protons in air under standard conditions. (*From E. Segre, ed.,* Experimental Nuclear Physics, *New York, Wiley, 1953, Vol. 1.*)

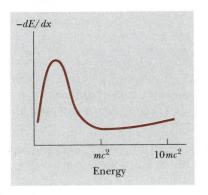

Figure 14.24 The energy loss $-dE/dx$ versus the energy of a charged particle of mass m moving through a medium.

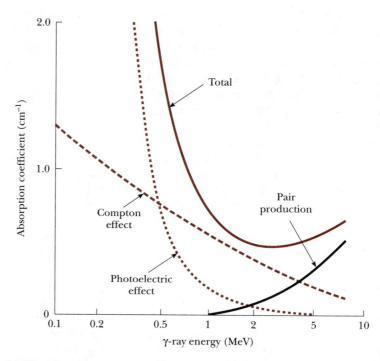

Figure 14.25 The absorption of gamma rays in lead. Shown are the three important absorption processes, which add up to the total absorption constant μ, measured in cm^{-1}. (*From W. Heitler*, The Quantum Theory of Radiation, *3rd ed., Oxford, Clarendon Press, 1954.*)

the electron's path length as a result of a phenomenon called **straggling.** It takes only a few large-deflection collisions to stop an electron in matter. Furthermore, electrons scatter much more readily than heavy charged particles with the same energy. As we mentioned earlier, a decelerated charged particle emits electromagnetic radiation called *bremsstrahlung.* This energy loss process is more important for electrons than for heavy charged particles, since low-mass electrons undergo greater accelerations when passing through matter. For example, the energy loss rate due to bremsstrahlung for 10-MeV electrons passing through lead is about equal to the loss rate due to ionization.

Photons

Since photons are uncharged, they are not as effective as charged particles in producing ionization or excitation in matter. Nevertheless, photons can be removed from a beam by either scattering or absorption in the medium. Figure 14.25 illustrates the three important processes that contribute to the total absorption of gamma rays in lead:

Gamma ray absorption

• At low photon energies (less than about 0.5 MeV), the predominant process that removes photons from a beam is the photoelectric effect, in which a photon is absorbed and transfers all of its energy to an electron.

- At intermediate photon energies, the predominant process is Compton scattering, in which the photon transfers some of its energy to an electron. Both the photoelectric effect and Compton scattering were discussed in Chapter 2.
- At high energies, the predominant process is **pair production,** in which an electron-positron pair is created as a photon passes near a nucleus in the medium. Since the rest energy of an electron-positron pair is $2mc^2 = 1.02$ MeV (twice the rest energy of an electron), the gamma-ray photon must have at least this much energy to produce a pair. Pair production will be discussed further in Chapter 15.

If a beam of photons is incident on a medium, its intensity decreases exponentially with increasing depth of penetration into the medium. This reduction in intensity is referred to as **attenuation** of the beam. As in the case of neutrons (Eq. 14.5), the intensity in a medium varies according to the relation

$$I(x) = I_0 e^{-\mu x} \tag{14.13}$$

Absorption of photons in a medium

where I_0 is the incident photon intensity (measured in photons/$m^2 \cdot s$), x is the distance the beam travels in the medium, $I(x)$ is the intensity of the beam after it travels a distance x, and μ is a parameter called the **linear absorption coefficient** of the medium. This coefficient depends on the energy of the photon as well as on the properties of the medium. The variation of μ with energy is shown in Figure 14.25 for gamma rays in lead and in Tables 14.2 and 14.3 for x-rays and gamma rays in various media.

Linear absorption coefficient

Table 14.2   Linear Absorption Coefficients for X-Rays in Various Media

λ (pm)	Linear Absorption Coefficient (cm^{-1})				
	Air	Water	Aluminum	Copper	Lead
10		0.16	0.43	3.2	43
20		0.18	0.76	13	55
30		0.29	1.3	38	158
40		0.44	3.0	87	350
50	8.6×10^{-4}	0.66	5.4	170	610
60	1.3×10^{-3}	1.0	9.2	286	1000
70	1.95×10^{-3}	1.5	14	430	1600
80	2.73×10^{-3}	2.1	20	625	
90	3.64×10^{-3}	2.8	30	875	
100	4.94×10^{-3}	3.8	41	1200	
150	1.56×10^{-2}	12	124		
200	3.64×10^{-2}	28	275		
250	6.63×10^{-2}	51	524		

Table 14.3   Linear Absorption Coefficients for Gamma Rays in Various Media

Photon Energy (MeV)	Linear Absorption Coefficient (cm⁻¹)[a]			
	Water	Aluminum	Iron	Lead
0.1	0.167	0.432	2.69	59.8
0.15	0.149	0.359	1.43	20.8
0.2	0.139	0.324	1.08	10.1
0.3	0.118	0.278	0.833	3.79
0.4	0.106	0.249	0.722	2.35
0.5	0.0967	0.227	0.651	1.64
0.6	0.0894	0.210	0.598	1.29
0.8	0.0786	0.184	0.525	0.946
1.0	0.0706	0.166	0.468	0.772
1.5	0.0576	0.135	0.380	0.581
2.0	0.0493	0.166	0.332	0.510
3.0	0.0396	0.0953	0.282	0.463
4.0	0.0339	0.0837	0.259	0.470
5.0	0.0302	0.0767	0.247	0.486
6.0	0.0277	0.0718	0.240	0.514
8.0	0.0242	0.0656	0.234	0.532
10.0	0.0221	0.0626	0.236	0.568

[a] Calculated using data in I. Kaplan, *Nuclear Physics*, Reading, Mass., Addison-Wesley, 1962, Table 15.6.

EXAMPLE 14.7   Half-Value Thickness

The half-value thickness of an absorber is defined as the thickness that will reduce the intensity of a beam of particles by a factor of 2. Calculate the half-value thickness for lead, assuming an x-ray beam of wavelength 20 pm (1 pm = 10^{-12} m = 10^{-3} nm).

Solution   The intensity varies with distance traveled in the medium according to Equation 14.13:

$$I(x) = I_0 e^{-\mu x}$$

In this case, we are looking for a value of x such that $I(x) = I_0/2$. That is, we require that

$$\frac{I_0}{2} = I_0 e^{-\mu x}$$

or

$$e^{-\mu x} = \tfrac{1}{2}$$

Taking the natural logarithm of both sides of this equation gives

$$\mu x = \ln 2$$

or

$$x = \frac{\ln 2}{\mu}$$

Since $\mu = 55$ cm⁻¹ for x-rays in lead at a wavelength of 20 pm (Table 14.2),

$$x = \frac{\ln 2}{55 \text{ cm}^{-1}} = \frac{0.693}{55 \text{ cm}^{-1}}$$

$$= 1.26 \times 10^{-2} \text{ cm} = 0.126 \text{ mm}$$

Hence we conclude that lead is a very good absorber for x-rays.

14.9 RADIATION DAMAGE IN MATTER

Radiation passing through matter can cause severe damage. The degree and type of damage depend upon several factors, including the type and energy of the radiation and the properties of the matter. For example, the metals used in nuclear reactor structures can be severely weakened by high fluxes of energetic neutrons, because these high fluxes often lead to metal fatigue. The damage in such situations is in the form of atomic displacements, which often result in major alterations in the material properties. Materials can also be damaged by ionizing radiations, such as gamma rays and x-rays. For example, defects called color centers can be produced in inorganic crystals by irradiating the crystals with x-rays. One extensively studied color center has been identified as an electron trapped in a Cl^- ion vacancy.

Radiation damage in biological organisms is due primarily to ionization effects in cells. The normal operation of a cell may be disrupted when highly reactive ions or radicals form as the result of ionizing radiation. For example, hydrogen and hydroxyl radicals produced from water molecules can induce chemical reactions that may break bonds in proteins and other vital molecules. Furthermore, the ionizing radiation may directly affect vital molecules by removing electrons from their structures. Large doses of radiation are especially dangerous, because damage to a great number of molecules in a cell may kill the cell. Although the death of a single cell is usually not a problem, the death of many cells may irreversibly damage the organism. Also, cells that do survive the radiation may become defective. These defective cells can produce more defective cells when they divide, leading to cancer.

In biological systems, it is common to separate radiation damage into two categories: somatic and genetic. *Somatic damage* is the damage associated with any body cell except the reproductive cells. At high dose rates, such damage can lead to cancer or seriously alter the characteristics of specific organisms. *Genetic damage* affects only reproductive cells. Damage to the genes in reproductive cells can lead to defective offspring. Clearly, we must be concerned about the effects of diagnostic treatments such as x-rays and other forms of radiation exposure.

Several units are used to quantify the amount, or dose, of any radiation that interacts with a substance.

> The **roentgen** (R) is that amount of ionizing radiation that produces an electric charge of $\frac{1}{3} \times 10^{-9}$ C in 1 cm^3 of air under standard conditions.

The roentgen

Equivalently, the roentgen is that amount of radiation that deposits an energy of 8.76×10^{-3} J into 1 kg of air.

For most applications, the roentgen has been replaced by the rad (which is an acronym for radiation absorbed dose):

> One **rad** is that amount of radiation that deposits 10^{-2} J of energy into 1 kg of absorbing material.

The rad

Although the rad is a perfectly good physical unit, it is not the best unit for measuring the degree of biological damage produced by radiation, because

Table 14.4   RBE[a] for Several Types of Radiation

Radiation	RBE Factor
X-rays and gamma rays	1.0
Beta particles	1.0–1.7
Alpha particles	10–20
Slow neutrons	4–5
Fast neutrons and protons	10
Heavy ions	20

[a] RBE = relative biological effectiveness.

damage depends not only on the dose but on the type of the radiation. For example, a given dose of alpha particles causes about ten times more biological damage than an equal dose of x-rays. The **RBE** (relative biological effectiveness) factor for a given type of radiation is the *number of rads of x-radiation or gamma radiation that produces the same biological damage as 1 rad of the radiation being used.* Table 14.4 gives the RBE factors for several types of radiation. The values are only approximate, because they vary with particle energy and with the form of the damage. The RBE factor should be considered only a first-approximation guide to the actual effects of radiation.

Finally, the **rem** (radiation equivalent in man) is the product of the dose in rad and the RBE factor:

$$\text{Dose in rem} \equiv \text{dose in rad} \times \text{RBE} \tag{14.14}$$

According to this definition, 1 rem of any type of radiation produces the same amount of biological damage as 1 rem of any other type. From Table 14.4 we see that a dose of 1 rad of fast neutrons represents an effective dose of 10 rem. On the other hand, 1 rad of gamma radiation is equivalent to a dose of 1 rem.

Low-level radiation from a natural source, such as cosmic rays or radioactive rocks and soil, delivers to each of us a dose of about 0.13 rem/year and is called *background radiation.* It is important to note that background radiation varies with geography. The upper limit of radiation dose (apart from background radiation) recommended by the U.S. government is about 0.5 rem/year. Many occupations involve much higher radiation exposures, and so an upper limit of 5 rem/year has been set for combined whole-body exposure. Higher upper limits are permissible for certain parts of the body, such as the hands and forearms. A dose of 400 to 500 rem results in a mortality rate of about 50% (which means that half the people exposed to this radiation level die). The most dangerous forms of exposure are ingestion and inhalation of radioactive isotopes, especially those elements the body retains and concentrates, such as ^{90}Sr. In some cases, a dose of 1000 rem can result from ingesting 1 mCi of radioactive material.

14.10   RADIATION DETECTORS

Various devices have been developed for detecting radiation. They are used for a variety of purposes, including medical diagnoses, radioactive dating measurements, measurement of background radiation, and measurement of the mass, energy, and momentum of particles created in high-energy nuclear reactions.

Ion chamber

An **ion chamber** is a general class of detector that makes use of the electron-ion pairs generated by the passage of radiation through a gas to produce an electrical signal. It consists basically of two charged plates maintained at different potentials by a voltage supply (Fig. 14.26). The plates attract electrons or ions, depending on plate polarity, and cause a current pulse i to flow that is proportional to the number of electron-ion pairs produced or to the particle energy if the particle comes to rest in the chamber. When an ion chamber is used in this way, both to detect the presence of an energetic charged particle and to measure its energy, it is called a **proportional counter.**

Figure 14.26 A schematic of an ion chamber. A charged particle stopping in the chamber creates electron-ion pairs, which are collected by the plates. This sets up a current *i* in the external circuit that is proportional to the particle's kinetic energy.

The **Geiger counter** (Fig. 14.27) is perhaps the most common form of ion chamber used to detect radiation. It can be considered the prototype of all counters that use the ionization of a medium as the basic detection process. It consists of a cylindrical metal tube filled with gas at low pressure and a long wire along the axis of the tube. The wire is maintained at a high positive potential (about 10^3 V) with respect to the tube. When a high-energy particle or photon enters the tube through a thin window at one end, some of the atoms of the gas are ionized. The electrons removed from these atoms are attracted

Geiger counter

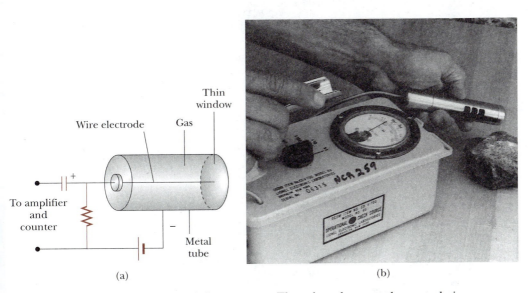

Figure 14.27 (a) A diagram of a Geiger counter. The voltage between the central wire and the metal tube is usually about 1000 V. (b) The use of a Geiger counter to measure the activity in a radioactive mineral. (*Henry Leap and Jim Lehman*)

toward the positive wire, and in the process they ionize other atoms in their path. This results in an avalanche of electrons that produces a current pulse at the output of the tube. After the pulse is amplified, it can either be used to trigger an electronic counter or be delivered to a loudspeaker that clicks each time a particle is detected. While a Geiger counter easily detects the presence of an energetic charged particle, the energy lost by the particle in the counter is *not* proportional to the current pulse produced in the avalanche process. Thus, although it is rugged, simple, and portable, a Geiger counter cannot be used to measure the energy of a particle.

Semiconductor diode detector

A **semiconductor diode detector** is essentially a reverse-bias *p-n* junction. Recall from Chapter 11 that a *p-n* junction diode passes current readily when forward-biased and prohibits a current under reverse-bias conditions. As an energetic particle passes through the junction, electrons are excited into the conduction band, and holes form in the valence band. The internal electric field sweeps the electrons toward the positive (*p*) side of the junction and the holes toward the negative (*n*) side. This movement of electrons and holes creates a pulse of current that is measured with an electronic counter. In a typical device, the duration of the pulse is about 10^{-8} s.

Scintillation counter

A **scintillation counter** (Fig. 14.28) usually uses a solid or liquid material whose atoms are easily excited by radiation. These excited atoms emit visible light when they return to their ground state. Common materials used as scintillators are transparent crystals of sodium iodide and certain plastics. If such a material is attached to one end of a device called a **photomultiplier** (PM) tube, the photons emitted by the scintillator can be converted to an electric signal. The PM tube consists of numerous electrodes, called dynodes, whose potentials increase in succession along the length of the tube as shown in Figure

Figure 14.28   A diagram of a scintillation counter connected to a photomultiplier tube.

14.28. The top of the tube contains a photocathode, which emits electrons by the photoelectric effect. As one of these emitted electrons strikes the first dynode, the electron has sufficient kinetic energy to eject several other electrons. When these electrons are accelerated to the second dynode, many more electrons are ejected and a multiplication process occurs. The end result is 1 million or more electrons striking the last dynode. Hence one particle striking the scintillator produces a sizable electric pulse at the output of the PM tube, and in turn this pulse is sent to an electronic counter. Both the scintillator and diode detector are much more sensitive than a Geiger counter, mainly because of the higher density of the detecting medium. Both can also be used to measure particle energy if the particle stops in the detector.

Track detectors are devices that can be used to view the tracks of charged particles directly. High-energy particles produced in particle accelerators may have energies ranging from 10^9 to 10^{12} eV. Thus they cannot be stopped entirely, and their energy cannot be measured in the small detectors already mentioned. Instead, the energy and momentum of these energetic particles are found from the curvature of the path of the particle in a known magnetic field, hence the necessity of track detectors.

Track detectors

A **photographic emulsion** is the simplest track detector. A charged particle ionizes the atoms in an emulsion layer. The path of the particle corresponds to a family of points at which chemical changes have occurred in the emulsion. When the emulsion is developed, the particle's track becomes visible.

Photographic emulsion

A **cloud chamber** contains a vapor that has been supercooled to just below its usual condensation point. An energetic particle passing through the chamber ionizes the vapor along its path. These ions serve as centers for condensation of the supercooled vapor. The track can be seen with the naked eye and can be photographed. A magnetic field can be applied to determine the charges of the particles as well as their momentum and energy.

Cloud chamber

A device called a **bubble chamber,** invented in 1952 by Donald A. Glaser, makes use of a liquid (usually liquid hydrogen) maintained near its boiling point. Ions produced by incoming charged particles leave bubble tracks, which can be photographed (Fig. 14.29). Because the density of the detecting medium of a bubble chamber is much higher than the density of the gas in a cloud chamber, the bubble chamber has a much higher sensitivity.

Bubble chamber

A **spark chamber** is a counting device that consists of an array of conducting parallel plates and is capable of recording a three-dimensional track record. Even-numbered plates are grounded, and odd-numbered plates are maintained at a high potential (about 10 kV). The spaces between the plates contain a noble gas at atmospheric pressure. When a charged particle passes through the chamber, ionization occurs in the gas, resulting in a large surge of current and a visible series of sparks along the particle path. These paths may be photographed or electronically detected and sent to a computer for path reconstruction and the determination of particle mass, momentum, and energy.

Wire chambers are similar to spark chambers in their ability to record 3-D tracks and provide fast electronic readout to a computer for track reconstruction and display. A wire chamber consists of closely spaced parallel wires that collect the electrons created by a passing ionizing particle. A second grid, with its wires perpendicular to the first, allows the x, y position of the particle in the plane of the two sets of wires to be determined. Finally, several such x, y locating grids arranged parallel to each other along the z axis can be used to determine the particle's trajectory in three dimensions.

Figure 14.29 A bubble-chamber photograph showing tracks of subatomic particles. (*Courtesy of Lawrence Radiation Laboratory, University of California, Berkeley*)

Neutron detectors are a bit more difficult to construct than charged particle detectors, because neutrons do not interact electrically with atoms in their passage through matter. Fast neutrons, however, can be detected by filling an ion chamber with hydrogen gas and detecting the ionization produced by high-speed recoiling protons produced in neutron-proton collisions. Slow neutrons with energies less than 1 MeV do not transfer sufficient energy to protons to be detected in this way but can be detected with an ion chamber filled with BF_3 gas. In this case, the boron nuclei disintegrate in a slow neutron-capture process, emitting highly ionizing alphas that are easily detected in the ion chamber.

SUMMARY

A **nuclear reaction** can occur when a target nucleus X at rest is bombarded by a particle a, resulting in a nucleus Y and a particle b:

$$a + X \longrightarrow Y + b \quad \text{or} \quad X(a, b)Y \quad (14.1)$$

The energy released in such a reaction, called the **reaction energy** Q, is

$$Q = (M_X + M_a - M_Y - M_b)c^2 \quad (14.2)$$

If N_0 beam particles are incident on a target of thickness x, the number N of beam particles that emerge from the target is

$$N = N_0 e^{-n\sigma x} \quad (14.5)$$

where n is the density of nuclei in the target and σ is the **cross-section.** Nuclear cross-sections have dimensions of area and are usually measured in *barns* (b), where $1 \text{ b} = 10^{-28} \text{ m}^2$.

The probability that neutrons will be captured as they move through matter generally increases with decreasing neutron energy. A *thermal neutron* (one whose energy is approximately $k_B T$) has a high probability of being captured by a nucleus according to the following **neutron-capture process:**

$$_0^1 n + {}_Z^A X \longrightarrow {}_Z^{A+1} X + \gamma \quad (14.7)$$

Energetic neutrons are slowed down readily in materials called **moderators**. These materials have small cross-sections for neutron capture, and so neutrons lose their energy mainly through elastic collisions.

Nuclear fission occurs when a very heavy nucleus, such as ^{235}U, splits into two smaller fragments. Thermal neutrons can create fission in ^{235}U:

$$_0^1 n + {}_{92}^{235} U \longrightarrow {}_{92}^{236} U^* \longrightarrow X + Y + \text{neutrons} \quad (14.8)$$

where X and Y are the fission fragments and $^{236}U^*$ is a nucleus in an excited state. On the average, 2.5 neutrons are released per fission event. The fragments and neutrons have a great deal of kinetic energy after the fission event. The fragments then undergo a series of beta and gamma decays to various stable isotopes. The energy released per fission event is about 200 MeV.

The **reproduction constant** K is the average number of neutrons released from each fission event that cause another event. In a power reactor, it is necessary to maintain K slightly greater than 1. The value of K is affected by such factors as reactor geometry, the mean neutron energy, and the probability of neutron capture. Proper design of the reactor geometry is necessary to minimize neutron leakage from the reactor core. Neutron energies are regulated with a moderator material to slow down energetic neutrons and therefore increase the probability of neutron capture by other ^{235}U nuclei. The power level of the reactor is adjusted with control rods made of a material that absorbs neutrons very efficiently. The value of K can be adjusted by inserting the rods into the reactor core at varied depths.

Nuclear fusion is a process in which two light nuclei fuse to form a heavier nucleus. A great deal of energy is released in such a process. The major obstacle to obtaining useful energy from fusion is the large Coulomb repulsive force between the charged nuclei at close separations. Sufficient energy must be supplied to the particles to overcome this Coulomb barrier. The temperature required to produce fusion is of the order of 10^8 K, and at this temperature all matter occurs as plasma.

In a fusion reactor, the plasma temperature must reach at least the **critical ignition temperature,** the temperature at which the power generated by the fusion reactions exceeds the power lost in the system. The most promising fusion reaction is the D-T reaction, which has a critical ignition temperature of approximately 4.5×10^7 K. Two important parameters involved in fusion reactor design are **ion density** n and **confinement time** τ. The confinement time is the time period for which the interacting particles must be maintained at a temperature equal to or greater than the critical ignition temperature. **Lawson's criterion** states that for the D-T reaction, $n\tau \geqslant 10^{14}$ s/cm^3.

When energetic particles interact with a medium, they lose their energy by several processes. Heavy particles, such as alpha particles, lose most of their energy by excitation and ionization of atoms in the medium. The particles have a finite range in the medium, which depends on the energy, mass, and charge of the particle as well as the properties of the medium. Energetic electrons moving through a medium also lose their energy by excitation and ionization. However, they scatter more readily than heavy particles and therefore undergo fewer collisions before coming to rest.

Photons, or gamma rays, which have no charge, are not as effective as charged particles in producing ionization or excitation. However, they can be absorbed in a medium by several processes: (1) the photoelectric effect, predominant at low photon energies; (2) Compton scattering, predominant at intermediate photon energies; and (3) pair production, predominant at high photon energies.

If a beam of photons with intensity I_0 is incident on a medium, the intensity of the beam decreases exponentially with increasing depth of penetration into the medium, according to the relation

$$I(x) = I_0 e^{-\mu x} \tag{14.13}$$

where x is the distance traveled into the medium, $I(x)$ is the intensity of the beam after it travels a distance x in the medium, and μ is the **linear absorption coefficient** of the medium.

Devices used to detect radiation are the Geiger counter, the semiconductor diode detector, the scintillation counter, the photographic emulsion, the bubble chamber, the spark chamber, and the wire chamber. Both the spark chamber and the wire chamber can easily make three-dimensional track measurements of charged particles, and both may be interfaced to computers for rapid data collection and trajectory reconstruction.

SUGGESTIONS FOR FURTHER READING

1. A popular history and overview of fusion are given in R. Herman, *Fusion: The Search for Endless Energy*, Cambridge, England, Cambridge University Press, 1990.
2. The history of the German and U.S. A-bomb projects (including several interesting photographs) is given in several articles in *Physics Today* 48:8, August 1995, Part 1.
3. The ITER is discussed in R. W. Conn, et al., "The International Thermonuclear Experimental Reactor," *Scientific American*, 226:103, April 1992.
4. For an extensive bibliography and selected reprints on medical applications of nuclear physics, see *Medical Physics*, edited by Russel K. Hobbie, College Park, Md., American Association of Physics Teachers, 1986.

QUESTIONS

1. Explain the functions of a moderator and a control rod in a fission reactor.
2. Why is water a better shield against neutrons than lead or steel?
3. Discuss the advantages and disadvantages of fission reactors from the standpoints of safety, pollution, and resources. Make a comparison with power generated by the burning of fossil fuels.
4. Why would a fusion reactor produce less radioactive waste than a fission reactor?
5. Lawson's criterion states that the product of ion density and confinement time must exceed a certain number before a break-even fusion reaction can occur. Why should these two parameters determine the outcome?
6. Why is the temperature required for the D-T fusion less than that for the D-D fusion? Estimate the relative importance of Coulomb repulsion and nuclear attraction in each case.
7. What factors make a fusion reaction difficult to achieve?
8. Discuss the similarities and differences between fusion and fission.
9. Discuss the advantages and disadvantages of fusion power from the standpoints of safety, pollution, and resources.
10. Discuss three major problems associated with the development of a controlled-fusion reactor.
11. Describe two techniques that are being pursued in an effort to obtain power from nuclear fusion.
12. If two radioactive samples have the same activity measured in curies, will they necessarily create the same damage to a medium? Explain.
13. One method of treating cancer of the thyroid is to insert a small radioactive source directly into the tumor. The radiation emitted by the source can destroy cancerous cells. Very often, the radioactive isotope $^{131}_{53}\text{I}$ is injected into the bloodstream in this treatment. Why do you suppose iodine is used?
14. Why should a radiologist be extremely cautious about x-ray doses when treating pregnant women?
15. The design of a PM tube might suggest that any number of dynodes may be used to amplify a weak signal. What factors do you suppose would limit the amplification in this device?

PROBLEMS

14.1 Nuclear Reactions

1. (a) Calculate the Q value corresponding to the reaction $^{18}\text{O}(\text{p}, \text{n})^{18}\text{F}$, and compare your result with the measured value in Table 14.1. (b) Calculate the threshold energy of the incident proton.

2. Supernova explosions are incredibly powerful nuclear reactions that tear apart the cores of relatively massive stars (greater than four times the Sun's mass). These blasts are produced by carbon fusion, which requires a temperature of about 6×10^8 K to overcome the strong

Coulomb repulsion between carbon nuclei. (a) Estimate the repulsive energy barrier to fusion, using the required ignition temperature for carbon fusion. (In other words, what is the kinetic energy for a carbon nucleus at 6×10^8 K?) (b) Calculate the energy (in MeV) released in each of these "carbon-burning" reactions:

$$^{12}C + {}^{12}C \longrightarrow {}^{20}Ne + {}^4He$$

$$^{12}C + {}^{12}C \longrightarrow {}^{24}Mg + \gamma$$

(c) Calculate the energy (in kWh) given off when 2 kg of carbon completely fuses according to the first reaction.

3. The following reaction, first observed in 1930, led to the discovery of the neutron by Chadwick:

$$^9_4Be(\alpha, n)^{12}_6C$$

Calculate the Q value of this reaction.

4. The following is the first known reaction (achieved in 1934) in which the product nucleus is radioactive:

$$^{27}_{13}Al(\alpha, n)^{30}_{15}P$$

Calculate the Q value of this reaction.

5. There are a few nuclear reactions in which the emerging particle and the product nucleus are identical. One example of this is the reaction

$$^7_3Li(p, \alpha)^4_2He$$

Calculate the Q value of this reaction.

6. Show that the following inverse reactions have the same *absolute value* of Q:

$$^{10}_5B(\alpha, p)^{13}_6C \quad \text{and} \quad ^{13}_6C(p, \alpha)^{10}_5B$$

7. (a) The first nuclear transmutation was achieved in 1919 by Rutherford, who bombarded nitrogen atoms with alpha particles emitted by the isotope ^{214}Bi. The reaction is

$$^4_2He + {}^{14}_7N \longrightarrow {}^{17}_8O + {}^1_1H$$

What is the Q value of the reaction? What is the threshold energy? (b) Cockroft and Walton performed the first nuclear reaction utilizing particle accelerators. In that case, accelerated protons were used to bombard lithium nuclei, producing the reaction

$$^1_1H + {}^7_3Li \longrightarrow {}^4_2He + {}^4_2He$$

Since the masses of the particles involved in the reaction were well known, these results were used to obtain an early proof of the Einstein mass-energy relation. Calculate the Q value of the reaction.

8. (a) One method of producing neutrons for experimental use is based on the bombardment of light nuclei by alpha particles. In one arrangement, alpha particles emitted by plutonium are incident on beryllium nuclei, and this results in the production of neutrons:

$$^4_2He + {}^9_4Be \longrightarrow {}^{12}_6C + {}^1_0n$$

What is the Q value for this reaction? (b) Neutrons are also often produced by small-particle accelerators. In one design, deuterons (^{2}H) that have been accelerated in a Van de Graaf generator are used to bombard other deuterium nuclei, resulting in the reaction

$$^2_1H + {}^2_1H \longrightarrow {}^3_2He + {}^1_0n$$

Is this reaction exothermic or endothermic? Calculate its Q value.

9. When the nuclear reaction represented by Equation 14.2 is endothermic, the disintegration energy Q is negative. In order for the reaction to proceed, the incoming particle must have a minimum kinetic energy, called the *threshold energy*, K_{th}. (a) Show that

$$K_{th} = -Q\left(1 + \frac{M_a}{M_X}\right)$$

by using the fact that $K_{th} = -Q$ in the CM frame and by transforming this result for the CM frame back to the laboratory frame. Note that in the CM frame, a and X have equal and opposite momenta, $p = M_a v = M_X V$. In the lab frame, a has momentum $p_{lab} = M_a(v + V)$ and X is at rest. (b) Calculate the threshold energy of the incident alpha particle in the reaction

$$^4_2He + {}^{14}_7N \longrightarrow {}^{17}_8O + {}^1_1H$$

14.2 Reaction Cross-Section

10. Consider a slab consisting of two layers of material with thicknesses x_1 and x_2 and target densities n_1 and n_2. If N_0 is the number of particles incident on the first layer of the slab in some time interval, determine the number N that emerge from the second layer in that interval. Assume that the cross-section σ is the same for each material. What would you guess is the relationship for three or more layers?

11. The density of the liquid hydrogen target in a bubble chamber is 70 kg/m^3. If 20% of a beam of slow neutrons incident on the bubble chamber has reacted with the hydrogen by the time the beam has traveled 2 m through the hydrogen, what is the cross-section, in barns, for the reaction of these slow neutrons with hydrogen atoms?

12. The density of lead is 11.35 g/cm^3, and its atomic weight is 207.2. Assume that 1.000 cm of lead reduces a beam of 1-MeV gamma rays to 28.65% of its initial intensity. (a) How much lead is required to reduce the beam to 10^{-4} of its initial intensity? (b) What is the effective cross-section of a lead atom for a 1-MeV photon?

13. Neutrons are captured by a cadmium foil. Using the data in Figure 14.3, find (a) the ratio of 10-eV neutrons cap-

tured to 1-eV neutrons captured, (b) the ratio of 1-eV neutrons captured to 0.1-eV neutrons captured, and (c) the ratio of 0.1-eV neutrons captured to 0.01-eV neutrons captured. (d) In what range of energies can cadmium be used as an energy selector?

14. The atomic weight of cadmium is 112.41, and its density is 8.65 g/cm³. Using Figure 14.3, estimate the attenuation distance of a thermal neutron beam in cadmium. (The attenuation distance is the distance traveled after which the intensity of the beam is reduced to $1/e$ of its initial value, where e is the base of the natural logarithms.)

15. A beam of 100-MeV protons from a Van de Graaf generator is incident on a gold foil 5.1×10^{-5} m thick. The beam current is 0.1 μA, and the beam has a cross-sectional area of 1 cm². If the scattering cross-section is 500 b, calculate (a) the ratio N/N_0, (b) the number per second that pass through the foil, and (c) the number lost from the beam each second by scattering. (Gold has an atomic weight of 197 and a density of 19.3 g/cm³.)

16. A tiny sphere is made of a material that absorbs all the photons incident on it. Many such spheres are embedded randomly in a transparent medium. (a) If the radius of one of these spheres is b, what is its cross-section σ in terms of b for the absorption of photons? (b) If the radius of each sphere is $b = 2 \times 10^{-3}$ m, what is the cross-section? (c) If 3×10^4 of these spheres are uniformly embedded in a cylinder of a transparent medium of height 2 m and cross-sectional area 0.5 m², and a light of beam intensity 0.75 W/m² is incident normally on one end of the cylinder, what is the intensity of the beam of light that emerges from the other end?

14.3   Interactions Involving Neutrons

17. The density of cadmium is 8.65 g/cm³, and its cross-section for thermal neutron absorption is 2450 b. Find the ratio of neutron absorption to neutron decay in cadmium for thermal neutrons at a temperature of 27°C. (*Hint:* The number of neutrons absorbed per second is

$$\frac{dN}{dt} = \frac{dN}{dx}\frac{dx}{dt} = \frac{dN}{dx}v$$

where v is the neutron speed and dN/dx is the number of neutrons lost per meter of travel. The half-life of the neutron is 636 s.)

18. In natural silver, the abundance of ^{107}Ag is 51.35%, and that of ^{109}Ag is 48.65%. The neutron-absorption cross-section of ^{107}Ag is 31 b, and that of ^{109}Ag is 87 b. The activation products ^{108}Ag and ^{110}Ag formed by neutron absorption decay by beta emission with half-lives of 144 s and 24.5 s, respectively. A silver sample is removed from a fission reactor, and after some delay it is found that the ratio of ^{108}Ag to ^{110}Ag is 20:1. How long was

the delay? (*Hint:* Assume that the sample has been in the reactor for a time sufficiently long that the decay rate equals the production rate for both ^{108}Ag and ^{110}Ag.)

19. A 1-MeV neutron is emitted in a fission reactor. If it loses half of its kinetic energy in each collision with a moderator atom, how many collisions must it undergo in order to achieve thermal energy (0.039 eV)?

20. A particle cannot generally be localized to distances much smaller than its de Broglie wavelength. This means that a slow neutron appears to be larger to a target particle than does a fast neutron, in the sense that the slow neutron will probably be found over a large volume of space. For a thermal neutron at room temperature (300 K), find (a) the linear momentum and (b) the de Broglie wavelength. Compare this effective neutron size with both nuclear and atomic dimensions.

14.4   Nuclear Fission

21. Find the energy released in the fission reaction

$$^1_0n + ^{235}_{92}U \longrightarrow ^{141}_{56}Ba + ^{92}_{36}Kr + 3(^1_0n)$$

The required masses are

$$M(^1_0n) = 1.008665 \text{ u}$$
$$M(^{235}_{92}U) = 235.043915 \text{ u}$$
$$M(^{141}_{56}Ba) = 140.9139 \text{ u}$$
$$M(^{92}_{36}Kr) = 91.8973 \text{ u}$$

14.5   Nuclear Reactors

22. (a) How many grams of ^{235}U must undergo fission to operate a 1000-MW power plant for one day? (b) If the density of ^{235}U is 18.7 g/cm³, how large a sphere of ^{235}U could you make from this much uranium?

23. In order to minimize neutron leakage from a reactor, the surface-area-to-volume ratio should be a minimum for a given shape. For a given volume V, calculate this ratio for (a) a sphere, (b) a cube, and (c) a parallelepiped of dimensions $a \times a \times 2a$. (d) Which of these shapes would have the minimum leakage? (e) Which would have the maximum leakage?

24. It has been estimated that there are 10^9 tons of natural uranium (mainly ^{238}U) at concentrations exceeding 100 parts per million, of which 0.7% is ^{235}U. If all the world's power needs (7×10^{12} J/s) were to be supplied by ^{235}U fission, how long would this supply last? (This estimate of uranium supply was taken from K. S. Deffeyes and I. D. MacGregor, *Scientific American*, January 1980, p. 66.)

25. An electrical power plant operates on the basis of thermal energy generated in a pressurized-water reactor. The electrical power output of the plant is 1 GW, and its efficiency is 30%. (a) Find the total power generated by the reactor. (b) How much power is discharged to the

environment as waste heat? (c) Calculate the rate of fission events in the reactor core. (d) Calculate the mass of ^{235}U used up in one year. (e) Using the results from (a), determine the rate at which fuel is converted to energy (in kg/s) in the reactor core, and compare your answer with the result from (d).

26. (a) Estimate the volume of space required to store the radioactive wastes that would be produced in one year if all the annual U.S. electricity production (which is about 2.2×10^{12} kWh/year) came from uranium enriched to 3% ^{235}U. (Assume that the conversion efficiency is 30% and that the waste is in the form of a liquid with a density of 1 g/cm^3.) (b) If the waste could be formed into a cube, what would be the length of the cube's sides?

14.6 Nuclear Fusion

27. Consider the deuterium-tritium fusion reaction with the tritium nucleus at rest:

$$^{2}_{1}H + ^{3}_{1}H \longrightarrow ^{4}_{2}He + ^{1}_{0}n$$

(a) From Equation 13.1, estimate the required distance of closest approach. (b) What is the Coulomb potential energy (in electron volts) at this distance? (c) If the deuteron has just enough energy to reach the distance of closest approach, what is the final velocity of the combined deuterium and tritium nuclei in terms of the initial deuteron velocity, v_0? (d) Use energy methods to estimate the minimum initial deuteron energy required to achieve fusion. (e) Why does the fusion reaction occur at much lower deuteron energies than that calculated in (d)?

28. The half-life of tritium is 12 years. If the TFTR fusion reactor contains 50 m^3 of tritium at a density equal to 1.5×10^{14} particles/cm^3, how many curies of tritium are in the plasma? Compare this with a fission reactor supply of 4×10^{10} Ci.

29. (a) Calculate the amounts of energy carried off by the $^{4}_{2}He$ nucleus and the neutron in the D-T fusion reaction. (Assume that the momentum is initially zero.) (b) Does this explain why only 20% of the energy produced can be used for critical ignition?

30. The Sun radiates energy at the rate of 4×10^{23} kW. Assuming that the reaction

$$4(^{1}_{1}H) \longrightarrow ^{4}_{2}He + 2e^{+} + 2\nu + \gamma$$

accounts for all the energy released, calculate (a) the number of protons fused per second and (b) the mass transformed into energy per second.

31. Suppose the target in a laser fusion reactor is a sphere of solid hydrogen with a diameter of 10^{-4} m and a density of 0.2 g/cm^3. Also assume that half of the nuclei are ^{2}H and half are ^{3}H. (a) If 1% of a 200-kJ laser pulse goes into heating this sphere, what temperature will the sphere reach? (b) If all of the hydrogen "burns" according to the D-T reaction, how many joules of energy will be released?

32. In a tokamak fusion reactor, suppose a 500-eV deuteron moves at an angle of 30° to the toroidal magnetic field. Assume that $B_t = 1$ T and $B_p = 0$ (see Fig. 14.14). (a) Calculate the components of velocity parallel and perpendicular to $\mathbf{B}_t$. (b) What is the radius of the spiral motion for the deuteron? (c) How far does the deuteron travel *along* the magnetic field before it completes one revolution around the magnetic field?

33. The carbon cycle, first proposed by Bethe in 1939, is another cycle by which energy is released in stars and hydrogen is converted to helium. The carbon cycle requires higher temperatures than the proton-proton cycle. The series of reactions is

$$^{12}C + ^{1}H \longrightarrow ^{13}N + \gamma$$
$$^{13}N \longrightarrow ^{13}C + e^{+} + \nu$$
$$^{13}C + ^{1}H \longrightarrow ^{14}N + \gamma$$
$$^{14}N + ^{1}H \longrightarrow ^{15}O + \gamma$$
$$^{15}O \longrightarrow ^{15}N + e^{+} + \nu$$
$$^{15}N + ^{1}H \longrightarrow ^{12}C + ^{4}He$$

(a) If the proton-proton cycle requires a temperature of 1.5×10^{7} K, estimate the temperature required for the first step in the carbon cycle. (b) Calculate the Q value for each step in the carbon cycle and the overall energy released. (c) Do you think the energy carried off by the neutrinos is deposited in the star? Explain.

34. (a) Calculate the energy (in kWh) released if 1 kg of ^{239}Pu undergoes complete fission and the energy released per fission event is 200 MeV. (b) Calculate the energy (in MeV) released in the D-T fusion:

$$^{2}_{1}H + ^{3}_{1}H \longrightarrow ^{4}_{2}He + ^{1}_{0}n$$

(c) Calculate the energy (in kWh) released if 1 kg of deuterium undergoes fusion. (d) Calculate the energy (in kWh) released by the combustion of 1 kg of coal if each $C + O_2 \rightarrow CO_2$ reaction yields 4.2 eV. (e) List the advantages and disadvantages of each of these methods of energy generation.

35. The ^{6}Li isotope is only about 7.5% naturally abundant. The remaining 92.5% of lithium is ^{7}Li. It is estimated that about 2×10^{13} g of lithium is available. If ^{6}Li is used as a tritium source in fusion reactors, with an energy release of 22 MeV per ^{6}Li nucleus, estimate the total energy available in D-T fusion. How does this number compare with the world's fossil fuel supply, which is estimated to be about 2.5×10^{23} J?

36. Find the Q value for each of the reactions in the proton-proton cycle (Eq. 14.10), and show that the overall Q value for the cycle is 25.7 MeV.

37. (a) Estimate the net power output of a fusion reactor that burns ten 50:50 3-mg D-T pellets every second. Assume that 30% of the fuel ignites and that a 5×10^{14} W laser pulse lasting 10 ns is needed to initiate burning. (b) What is the equivalent in liters of oil for a day's operation, if 2 L of oil gives off 100 MJ when burned?

38. In a laser fusion reaction, a pellet containing a 50:50 mixture of D and T experiences a density increase of 1000 times when struck by a laser pulse. (a) Find the particle density in the compressed state if the normal density is 0.20 g/cm^3. (b) How long must the D and T nuclei be confined at this density if the reaction is to achieve a break-even condition?

39. Two nuclei with atomic numbers Z_1 and Z_2 approach each other with a total energy E. (a) If the minimum distance of approach for fusion to occur is $r = 10^{-14}$ m, find E in terms of Z_1 and Z_2. (b) Calculate the minimum energy for fusion for the D-D and D-T reactions (the first and third reactions in Eq. 14.11).

40. To understand why containment of a plasma is necessary, consider the rate at which a plasma would be lost if it were not contained. (a) Estimate the rms speed of deuterons in a plasma at 10^8 K. (b) Estimate the time interval for which such a plasma would remain in a 10-cm cube if no steps were taken to contain it.

41. Of all the hydrogen nuclei in the ocean, 0.0156% are deuterium. The oceans have a volume of 317 million cubic miles. (a) If all the deuterium in the oceans were fused to $_2^4He$, how many joules of energy would be released? (b) Current world energy consumption is about 7×10^{12} W. If consumption were 100 times greater, how many years would the energy calculated in (a) last?

42. It has been pointed out that fusion reactors are safe from explosion because there is never enough energy in the plasma to do much damage. (a) Using a particle density of 10^{15} cm^{-3} and a kinetic temperature of 10 keV, calculate the amount of energy stored in the plasma of the TFTR reactor. (b) How many kilograms of water could be boiled by this much energy? (The plasma volume of the TFTR reactor is about 50 m^3.)

43. In order to confine a stable plasma, the magnetic energy density in the magnetic field must exceed the pressure $2nkT$ of the plasma by a factor of at least 10. In the following, assume a confinement time $\tau = 1$ s. (a) Using Lawson's criterion, determine the required ion density. (b) From the ignition temperature criterion for the D-T reaction, determine the required plasma pressure. (c) Determine the magnitude of the magnetic field that is required to contain the plasma.

14.8   Interaction of Particles with Matter

44. The densities and atomic weights of the materials in Table 14.3 are as follows:

Substance	Density (g/cm^3)	Atomic Weight
H_2O	1	18
Al	2.7	27
Fe	7.8	55.8
Pb	11.35	207.2

Compute the number of electrons per cubic centimeter for each material, and plot the gamma-ray linear absorption coefficient versus electron density. Draw three graphs corresponding to gamma-ray energies of 0.1, 1.0, and 10 MeV. What do you conclude from your graphs?

45. Three equally thick layers of aluminum, copper, and lead are used to reduce the intensity of an x-ray beam to one third of its original intensity. The wavelength of the beam is 50 pm. (a) Find the thickness of *each layer* of material. (b) By what fraction would the initial beam intensity be reduced by the lead alone?

46. X-rays of wavelength 25 pm and gamma rays of energy 0.1 MeV have approximately the same absorption coefficient in lead. How do their energies compare?

47. What is the half-value thickness (Example 14.7) of water to x-rays of wavelength 20 pm? Since the human body is more than 90% water, what does your answer indicate about the use of x-rays as a diagnostic technique?

48. In a large-scale nuclear attack, typical radiation intensity from radioactive fallout might be 2000 rad in most places. In the following calculations, assume that one third of the radiation is 10-MeV gamma radiation and that the linear absorption coefficient is the same for aluminum and concrete. (a) What thickness (in meters) of concrete would be needed to reduce the radiation intensity to 1 rad? (b) If a particular shelter were located at a "hot spot" receiving 100 000 rad, what thickness of concrete would be needed to reduce the radiation intensity to 1 rad?

14.9   Radiation Damage in Matter

49. Assume that an x-ray technician takes an average of eight x-rays per day and receives a dose of 5 rem/year as a result. (a) Estimate the dose in rem per x-ray taken. (b) How does this result compare with low-level background radiation?

50. In terms of biological damage, how many rad of heavy ions are equivalent to 100 rad of x-rays?

51. Two workers using an industrial x-ray machine accidentally insert their hands in the x-ray beam for the same

length of time. The first worker inserts one hand in the beam, and the second worker inserts both hands. Which worker receives the larger dose in rad?

52. Calculate the radiation dose, in rad, supplied to 1 kg of water such that the energy deposited equals (a) the rest energy of the water and (b) its thermal energy. (Assume that each molecule has a thermal energy of $k_B T$.)

53. A person whose mass is 75 kg is exposed to a dose of 25 rad. How many joules of energy are deposited in the person's body?

14.10 Radiation Detectors

54. In a Geiger tube, the voltage between the electrodes is typically 1 kV, and the current pulse generated by the detection of a β particle fully charges a 5-pF capacitor. (a) What is the energy amplification of this device for a 0.5-MeV beta ray? (b) How many electrons are avalanched by the initial electron?

55. In a PM tube, assume that there are seven dynodes with potentials of 100 V, 200 V, 300 V, . . . , 700 V. The average energy required to free an electron from the dynode surface is 10 eV. For each incident electron, how many electrons are freed (a) at the first dynode and (b) at the last dynode?

15

Particle Physics and Cosmology

Chapter Outline

The word *atom* is from the Greek *atomos*, which means "indivisible." At one time, atoms were thought to be the indivisible constituents of matter; that is, they were regarded as elementary particles. All the way from the teachings of the ancient Greek philosopher Democritus to the relatively recent works of John Dalton and Dmitri Mendeleev, the idea that all matter consisted of elementary atoms differing in mass and chemical properties has been quite successful in explaining many properties of matter. When discoveries in the early part of the 20th century revealed that the atom is composed of other constituents, another simplification occurred with Bohr's atomic model and the invention of quantum mechanics. The variety of physical and chemical properties of approximately 100 elements was explained in terms of rules governing just three constituents: electrons, protons, and neutrons. With the exception of the free neutron, these particles are very stable.

Beginning in about 1945, many new particles were discovered in experiments involving high-energy collisions between known particles. Such particles are characteristically very unstable and have very short half-lives, ranging between 10^{-6} and 10^{-23} s. So far, more than 400 of these unstable, temporary particles have been catalogued.

Since the early 1960s, many powerful particle accelerators have been constructed throughout the world, making it possible to observe collisions of highly energetic particles under controlled laboratory conditions so as to reveal the subatomic world in finer detail. Until the 1960s, physicists were puzzled by the large number and variety of subatomic particles being discovered. They wondered whether the particles were like the varied animals in a zoo, with no systematic relationship connecting them, or whether a pattern was emerging that would provide a better understanding of the elaborate structure in the subnuclear world. Since about 1970, physicists have tremendously advanced our knowledge of the structure of matter by recognizing that all particles except electrons, photons, and a few related particles are made of smaller particles called **quarks.** Thus protons and neutrons, for example, are not truly elementary but are systems of tightly bound quarks. The quark model has reduced the array of particles to a manageable number and has successfully predicted new quark combinations that have subsequently been observed in many experiments. A kind of wave mechanics for quarks (quantum chromodynamics) has also been developed. This theory, although mathematically difficult, has deepened our understanding of elementary particles and has helped to tame the particle "zoo."

This chapter examines the properties and classifications of the known elementary particles, the interactions that govern their behavior, and the methods of producing elementary particles and measuring their properties. We also discuss the current theory of elementary particles, in which everything is believed to consist of only two families of "matter" particles—quarks and leptons—and one family of force-transmitting, or "field," particles. Finally, we discuss how such models help scientists understand the creation and evolution of the Universe.

15.1 THE FUNDAMENTAL FORCES IN NATURE

To understand the properties of elementary particles, we must be able to describe the forces between them. Particles in nature are subject to four fundamental forces; in order of decreasing strength, these are the strong force, the electromagnetic force, the weak force, and the gravitational force.

The **strong force** is responsible for binding quarks tightly together to form protons, neutrons, and other hadrons. It is extremely short-range and is negligible for separations greater than approximately 10^{-15} m. The nuclear force that binds neutrons and protons in nuclei is currently believed to be a residual effect of the more basic strong force between quarks, much as the molecular force binding electrically neutral atoms together in molecules is a residual electrical interaction.

The **electromagnetic force,** which binds electrons and protons within atoms and molecules to form ordinary matter, is approximately two orders of magnitude weaker than the strong force. It is a long-range force that decreases in strength as the inverse square of the separation between interacting particles.

The **weak force** is a short-range force that accounts for the beta decay of nuclei and the decay of heavier quarks and leptons. Its strength is only about 10^{-6} times that of the strong force. (As we shall discuss later, scientists now believe that the weak and electromagnetic forces are two manifestations of a single force called the electroweak force.)

Finally, the **gravitational force** is a long-range force that has a strength only about 10^{-43} times that of the strong force. Although this familiar interaction holds the planets, stars, and galaxies together, its effect on elementary particles is negligible.

Classically, the entity that is responsible for transmitting a force from one particle to another is the field. For example, a positive electric charge produces an electric field in space, which in turn exerts an attractive force on a nearby negative charge. Furthermore, the field can carry energy and momentum from one particle to the other. According to current quantum field theories, the energy and momentum of all fields are quantized, and the quantum that carries a "chunk" of momentum and energy from one type of particle to another is called a field particle. Thus, in particle physics, interactions between particles are described in terms of the exchange of field particles, or quanta. In the case of the familiar electromagnetic interaction, for instance, the field particles are photons. In the language of modern physics, one can say that the electromagnetic force is *mediated*, or carried, by photons and that photons are the quanta of the electromagnetic field. Likewise, the strong force is mediated by field particles called *gluons*, the weak force is mediated by particles called the $W^{\pm}$ and Z^0 *bosons*, and the gravitational force is mediated by quanta of the gravitational field, called *gravitons*. These interactions, their ranges, their relative strengths, and the corresponding field particles, are summarized in Table 15.1. Note that the third column of Table 15.1 presents another way of classifying interactions—by means of the observed lifetime of a decaying particle. In fact, with only a few exceptions, the lifetimes of decaying particles are excellent indicators of what interaction has caused the decay, with shorter lifetimes being associated with stronger forces. As shown in Table 15.1, particles decaying via

Table 15.1   Particle Interactions

Interaction (Force)	Relative Strength[a]	Typical Lifetimes for Decays via a Given Interaction	Range of Force	Mediating Field Particle
Strong	1	$\leqslant 10^{-20}$ s	Short (≈ 1 fm)	Gluon
Electromagnetic	$\approx 10^{-2}$	$\approx 10^{-16}$ s	Long ($\propto 1/r^2$)	Photon
Weak	$\approx 10^{-6}$	$\geqslant 10^{-10}$ s	Short ($\approx 10^{-3}$ fm)	$W^{\pm}$, Z^0 bosons
Gravitational	$\approx 10^{-43}$	?	Long ($\propto 1/r^2$)	Graviton

[a] For two u quarks at 3×10^{-17} m.

the strong force are usually the shortest-lived; next come those decaying via electromagnetic forces; and finally, particles decaying by the weak interaction have the longest observed lifetimes.

15.2   POSITRONS AND OTHER ANTIPARTICLES

In the 1920s the theoretical physicist Paul Adrien Maurice Dirac (1902–1984) developed a version of quantum mechanics that incorporated special relativity. Dirac's theory automatically explained the origin of the electron's spin and its magnetic moment. However, it also presented a major difficulty. Dirac's relativistic wave equation required solutions corresponding to negative energy states even for free particles.[1] But if negative energy states existed, one would expect an electron in a state of positive energy to make a rapid transition to one of these states, emitting a photon in the process. Dirac avoided this difficulty by postulating that all negative energy states are filled. Electrons that occupy the negative energy states are called the "Dirac sea." Electrons in the Dirac sea are not directly observable, because the Pauli exclusion principle does not allow them to react to external forces. However, if one of the negative energy states is vacant, leaving a hole in the sea of filled states, the hole can react to external forces and would be observable. (This is analogous to the behavior of a hole in the valence band of a semiconductor.)

The second, equivalent interpretation of this theory is that for every particle there is also an antiparticle. The antiparticle would have the same mass as the particle, but their charges would be opposite. For example, the electron's antiparticle (called a positron) would have a mass of 0.511 MeV/c^2 and a positive charge of $+1.6 \times 10^{-16}$ C. Usually we shall denote an antiparticle with a bar over the symbol for the particle. Thus $\bar{e}$ denotes the positron (although sometimes the notation e^+ is preferred), $\bar{p}$ denotes the antiproton, and $\bar{\nu}$ denotes the antineutrino.

In 1932 Carl Anderson discovered the positron, and in 1936 he was awarded the Nobel prize for that achievement. Anderson discovered the positron while examining tracks created by electron-like particles of positive charge in a cloud chamber. (The early experiments used cosmic rays—mostly energetic protons passing through interstellar space—to initiate high-energy reactions in the upper atmosphere which resulted in the production of positrons at ground level.) To discriminate between positive and negative charges, Anderson placed the cloud chamber in a magnetic field, causing moving charges to follow curved paths. He noted that some of the electron-like tracks deflected in a direction corresponding to a positively charged particle.

Since Anderson's initial discovery, the positron has been observed in other experiments. Perhaps the most common process for producing positrons is **pair production,** in which a gamma-ray photon with sufficiently high energy collides with a nucleus and an electron-positron pair is created. Because the total rest energy of the electron-positron pair is $2m_e c^2 = 1.02$ MeV (where m_e is the mass of the electron), the photon must have at least this much energy to create an electron-positron pair. Thus electromagnetic energy in the form of

Pair production

---

[1] P. A. M. Dirac, *The Principles of Quantum Mechanics*, 3rd ed., New York, Oxford University Press, 1947, Chapter 11.

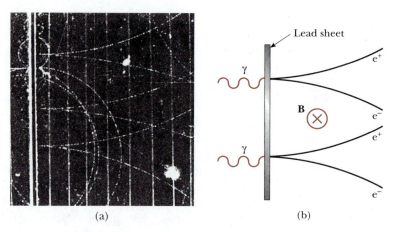

Figure 15.1   (a) Bubble-chamber tracks of electron-positron pairs produced by 300-MeV gamma rays striking a lead sheet. *(Courtesy Lawrence Berkeley Laboratory, University of California)* (b) A sketch of the pertinent pair-production events. Note that the positrons deflect upward while the electrons deflect downward in an applied magnetic field directed into the page.

a gamma ray is converted to mass in accordance with Einstein's famous relationship $E = mc^2$. Figure 15.1 shows tracks of electron-positron pairs created by 300-MeV gamma rays striking a lead sheet.

The reverse process can also occur. Under the proper conditions, an electron and a positron can annihilate each other to produce two gamma-ray photons with a combined energy of at least 1.02 MeV:

Electron-positron annihilation

$$e^- + e^+ \longrightarrow 2\gamma$$

Prior to 1955, on the basis of the Dirac theory, it was expected that every particle had a corresponding antiparticle, but antiparticles such as the antiproton and antineutron had not been detected experimentally. Since the relativistic Dirac theory had had some failures (it predicted the wrong-size magnetic moment for the proton) as well as many successes, it was important to determine whether the antiproton really existed. In 1955 a team led by Emilio Segrè and Owen Chamberlain used the Bevatron particle accelerator at the University of California, Berkeley, to produce both antiprotons and antineutrons, and thus established with certainty the existence of antiparticles. For this work Segrè and Chamberlain received the Nobel prize in 1959.

It is now established that every particle has a corresponding antiparticle of equal mass and spin and of equal and opposite charge, magnetic moment, and strangeness. (The property of strangeness is explained in Section 15.6.) The only exceptions to these rules for particles and antiparticles are the neutral photon, pion, and eta, each of which is its own antiparticle.

An intriguing aspect of antiparticle theory is that if we replace every proton, neutron, and electron in an atom with its antiparticle, we should create a stable antiatom; combinations of antiatoms should form antimolecules and eventually antiworlds. As far as we know, everything would behave in the same way in an antiworld as in our world, and in principle, it is possible that some distant antimatter galaxies exist, separated from normal-matter galaxies by millions of lightyears. Unfortunately, since the photon is its own antiparticle, there is no difference in the light emitted from matter and antimatter galaxies, and so astronomical observations cannot tell the two apart. Although no evidence of antimatter galaxies exists at present, it is awe-inspiring to contemplate the kind of cosmic spectacle that would result if matter and antimatter worlds were to collide: a gigantic eruption of jets of annihilation radiation, transforming the entire galactic mass into ephemeral neutrinos, antineutrinos, and gamma rays, all fleeing the collision point at the speed of light.

EXAMPLE 15.1   Pair Production

When an electron and a positron meet at low speeds in free space, why are *two* 0.511-MeV gamma rays produced rather than *one* gamma ray with an energy of 1.022 MeV?

Reasoning   Gamma rays are photons, and photons carry momentum. If only one photon were produced, momen-

tum would not be conserved, because the total momentum of the electron-positron system is approximately zero, whereas a single photon of energy 1.022 MeV would have a very large momentum. On the other hand, the two gamma-ray photons that are produced travel off in opposite directions, so their total momentum is zero.

15.3   MESONS AND THE BEGINNING OF PARTICLE PHYSICS

In the mid-1930s, physicists had a fairly simple view of the structure of matter. The building blocks were the proton, the electron, and the neutron. Three other particles were known or had been postulated at the time: the photon, the neutrino, and the positron. These six particles were considered the fundamental constituents of matter. With this marvelously simple picture of the world, however, no one could provide an answer to an important question: Since the many protons in proximity in any nucleus should strongly repel each other because of their like charges, what is the nature of the force that holds the nucleus together? Scientists recognized that this mysterious force must be much stronger than anything encountered in nature up to that time.

In 1935 the Japanese physicist Hideki Yukawa (1907–1981) proposed the first theory to successfully explain the nature of the nuclear force, an effort that later earned him the Nobel prize. To understand Yukawa's theory, it is useful to first recall that in the modern view of electromagnetic interactions, *charged particles interact by exchanging photons.* Yukawa used this idea to explain the nuclear force by proposing a new particle whose exchange between nucleons in the nucleus produces the nuclear force. Furthermore, he established that the range of the force is inversely proportional to the mass of this particle and predicted that the mass would be about 200 times the mass of the electron. Since the new particle would have a mass between that of the electron and that of the proton, it was called a **meson** (from the Greek *meso*, "middle").

In an effort to substantiate Yukawa's predictions, physicists began an experimental search for the meson by studying cosmic rays entering the Earth's atmosphere. In 1937 Carl Anderson and his collaborators discovered a particle of mass 106 MeV/c^2, about 207 times the mass of the electron. However, subsequent experiments showed that the particle interacted very weakly with matter and hence could not be the carrier of the strong force. The puzzling situ-

Hideki Yukawa (1907–1981), a Japanese physicist, was awarded the Nobel prize in 1949 for predicting the existence of mesons. This photograph of Yukawa at work was taken in 1950 in his office at Columbia University. *(UPI/Bettmann)*

Figure 15.2 Feynman diagram showing how a photon mediates the electromagnetic force between two interacting electrons. The *t*-arrow shows the direction of increasing time.

ation inspired several theoreticians to propose that there were two mesons with slightly different masses. This idea was confirmed by the discovery in 1947 of the pi (π) meson, or simply *pion*, by Cecil Frank Powell (1903–1969) and Giuseppe P. S. Occhialini (b. 1907). The particle discovered by Anderson in 1937, the one thought to be a meson, is not really a meson. Instead, it takes part in the weak and electromagnetic interactions only, and it is now called the *muon*.

The pion, Yukawa's carrier of the nuclear force, comes in three varieties corresponding to three charge states: π^+, π^-, and π^0. The π^+ and π^- particles have masses of 139.6 MeV/c^2, while the π^0 has a mass of 135.0 MeV/c^2. Pions and muons are very unstable particles. For example, the π^-, which has a mean lifetime of 2.6×10^{-8} s, first decays to a muon and a muon antineutrino. The muon, which has a mean lifetime of 2.2 μs, then decays to an electron, a neutrino, and an electron antineutrino:

$$\pi^- \longrightarrow \mu^- + \bar{\nu}_\mu \tag{15.1}$$

$$\mu^- \longrightarrow e^- + \nu_\mu + \bar{\nu}_e$$

The interaction between two particles can be represented in a simple diagram called a *Feynman diagram*, developed by the American physicist Richard P. Feynman (1918–1988). Figure 15.2 is such a diagram for the electromagnetic interaction between two electrons. In this simple case, a photon is the field particle that mediates the electromagnetic force between the electrons. The photon transfers energy and momentum from one electron to the other in this interaction. The photon is called a *virtual photon* because it is emitted and reabsorbed without being detected. Virtual photons can violate the law of conservation of energy by ΔE for a very short time, Δt, provided that $\Delta E \, \Delta t \approx \hbar/2$. Note that such quantum violations of energy conservation take place only in the short term and that system energy is conserved in the long run when the photon is reabsorbed by the other electron.

Now consider the pion exchange between a proton and a neutron that transmits the nuclear force according to Yukawa (Fig. 15.3). We can reason that the energy ΔE needed to create a pion of mass m_π is given by Einstein's equation $\Delta E = m_\pi c^2$. Again, the very existence of the pion would violate conservation of energy if it lasted for a time greater than $\Delta t \approx \hbar/2\Delta E$ (from the minimum form of the uncertainty principle), where ΔE is the energy of the pion and Δt is the time it takes the pion to travel from one nucleon to the other. Therefore,

$$\Delta t \approx \frac{\hbar}{2 \, \Delta E} = \frac{\hbar}{2 m_\pi c^2} \tag{15.2}$$

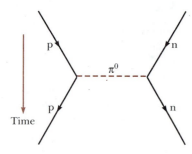

Figure 15.3 Feynman diagram representing a proton interacting with a neutron via the strong force. In this case, the pion mediates the strong force. The *t*-arrow shows the direction of increasing time.

Because the pion cannot travel faster than the speed of light, the maximum distance d it can travel in a time Δt is $c \, \Delta t$. Using Equation 15.2 and $d = c \, \Delta t$, we find this maximum distance to be

$$d \approx \frac{\hbar}{2 m_\pi c} \tag{15.3}$$

We know that the range of the nuclear force is approximately 1.0×10^{-15} m.

Richard Phillips Feynman was a brilliant theoretical physicist who shared the 1965 Nobel prize in physics with Julian S. Schwinger and Shinichiro Tomonaga for their fundamental work in the principles of quantum electrodynamics. Feynman's many important contributions to physics include the invention of simple diagrams to represent particle interactions graphically, the theory of the weak interaction of subatomic particles, a reformulation of quantum mechanics, the theory of superfluid helium, and his contribution to physics education through the magnificent three-volume text *The Feynman Lectures on Physics.*

Feynman did his undergraduate work at MIT and received his PhD in 1942 from Princeton University, where he studied under John Archibald Wheeler. During World War II he worked on the Manhattan Project at Princeton and at Los Alamos, New Mexico. He then joined the faculty at Cornell University in 1945 and in 1950 was appointed professor of physics at California Institute of Technology, where he remained for the rest of his career.

It is well known that Feynman had a passion for finding new and better ways to formulate each problem or, as he would say, "turning it around." In the early part of his career, he was fascinated with electrodynamics and developed an intuitive view of quantum electrodynamics. Convinced that the electron could not interact with its own field, he said, "That was the beginning, and the idea seemed so obvious to me that I fell deeply in love with it. . . ." Often called the outstanding intuitionist of our age, he said in his Nobel acceptance speech, "Often, even in a physicist's sense, I did not have a demonstration of how to get all of these rules and equations, from conventional elec-

B I O G R A P H Y

RICHARD P. FEYNMAN
(1918–1988)

trodynamics. . . . I never really sat down, like Euclid did for the geometers of Greece, and made sure that you could get it all from a single set of axioms."

In 1986 Feynman was a member of the presidential commission to investigate the explosion of the space shuttle *Challenger.* In this capacity, he performed a simple experiment for the commission members that showed that one of the shuttle's O-ring seals was the likely cause of the disaster. After placing a seal in a pitcher of ice water and squeezing it with a clamp, he demonstrated that the seal failed to spring back into shape once the clamp was removed.[1]

Feynman worked in physics with a style commensurate with his personality—that is, with energy, vitality, and humor. The following quotations from some of his colleagues hint at the great impact he made on the scientific community.[2]

Murray Gell-Mann: "A brilliant, vital, and amusing neighbor, Feynman was a stimulating (if sometimes exasperating) partner in discussions of profound issues—we would ex-

change ideas and silly jokes in between bouts of mathematical calculation—we struck sparks off each other, and it was exhilarating."

David Pines: "Reading Feynman is a joy and a delight, for in his papers, as in his talks, Feynman communicated very directly, as though the reader were watching him derive the results at the blackboard."

David L. Goodstein: "He loved puzzles and games. In fact, he saw all the world as a sort of game, whose progress of 'behavior' follows certain rules, some known, some unknown . . . . Find places or circumstances where the rules don't work, and invent new rules that do."

Valentine L. Telegdi: "Feynman was not a theorist's theorist, but a physicist's physicist and a teacher's teacher."

Laurie M. Brown, one of his graduate students at Cornell, noted that Feynman, a playful showman, was "undervalued at first because of his rough manners [but] in the end triumphs through native cleverness, psychological insight, common sense and the famous Feynman humor . . . . Whatever else Dick Feynman may have joked about, his love for physics approached reverence."

[1]Feynman's own account of this inquiry can be found in *Physics Today* 4:26, February 1988.

[2]For more on Feynman's life and contributions, see the articles in a special memorial issue of *Physics Today* 42, February 1989. For a personal account of Feynman, see his popular autobiographical books, *Surely You're Joking Mr. Feynman,* New York, Bantam Books, 1985, and *What Do You Care What Other People Think,* New York, W. W. Norton & Co., 1987.

(Courtesy of Tamiko Thiel Thinking Corporation)

Using this value for d in Equation 15.3, we calculate the rest energy of the pion to be

$$m_\pi c^2 \approx \frac{\hbar c}{2d} = \frac{197 \text{ MeV} \cdot \text{fm}}{2(1 \times 10^{-15} \text{ m})}$$
$$\approx 100 \text{ MeV}$$

This corresponds to a mass of 100 MeV/c^2 (approximately 250 times the mass of the electron), a value in reasonable agreement with the observed pion mass.

The concept we have just described is quite important. In effect, it says that a proton can change into a proton plus a pion as long as it returns to its original state in a very short time. Physicists often say that a nucleon undergoes "fluctuations" as it emits and absorbs pions. As we have seen, these fluctuations are a consequence of a combination of quantum mechanics (through the uncertainty principle) and special relativity (through Einstein's energy-mass relationship, $E = mc^2$). Also, as seen in Chapter 13, these virtual fluctuations can become real particles in collision processes if the incident particle can furnish the "missing" energy and momentum.

This section has dealt with early ideas about particles that carry the nuclear force, namely the pions, and the mediators of the electromagnetic force, photons. Current ideas indicate that the nuclear force is more accurately described as an average or residual effect of the strong color force between quarks, as will be explained in Section 15.10. The graviton, which is the mediator of the gravitational force, has yet to be observed. The $W^\pm$ and Z^0 particles that mediate the weak force were discovered in 1983 by the Italian physicist Carlo Rubbia (b. 1934) and his associates, using a proton-antiproton collider. Rubbia and Simon van der Meer, both at CERN near Geneva, Switzerland, shared the 1984 Nobel prize for the discovery of the $W^\pm$ and Z^0 particles and the development of the proton-antiproton collider. In this accelerator, protons and antiprotons that have a momentum of 270 GeV/c undergo head-on collisions with each other. In some of the collisions $W^\pm$ and Z^0 particles are produced, which in turn are identified by their decay products.

15.4 CLASSIFICATION OF PARTICLES

All particles other than photons can be classified into two broad categories, hadrons and leptons, according to their interactions.

Hadrons

Particles that interact through the strong force are called **hadrons.** The two classes of hadrons, *mesons* and *baryons,* are distinguished by their masses and spins.

Mesons all have zero or integer spin (0 or 1), with masses between that of the electron and that of the proton. All mesons are known to decay finally into electrons, positrons, neutrinos, and photons. The pion is the lightest of known mesons; it has a mass of approximately 140 MeV/c^2 and a spin of 0. Another is the K meson, with a mass of approximately 500 MeV/c^2 and a spin of 0.

Baryons, the second class of hadrons, have masses equal to or greater than the proton mass (*baryon* means "heavy" in Greek), and their spins are always half odd-integer values ($\frac{1}{2}$ or $\frac{3}{2}$). Protons and neutrons are baryons, as are many other particles. With the exception of the proton, all baryons decay in such a way that the end products include a proton. For example, the baryon called the Ξ^- hyperon first decays to the Λ^0 baryon and a π^- in about 10^{-10} s. The Λ^0 then decays to a proton and a π^- in approximately 3×10^{-10} s.

Today it is believed that hadrons are composed of more elemental units called quarks. Table 15.2 lists some of the important properties of hadrons. The symbols B, L_e, L_μ, L_τ, and S stand for baryon, electron, muon, and tau numbers and strangeness, respectively, and are explained in Sections 15.5 and 15.6.

Table 15.2   Some Particles and Their Properties

Category	Particle Name	Symbol	Anti-particle	Mass (MeV/c^2)	B	L_e	L_μ	L_τ	S	Lifetime (s)	Principal Decay Modes[a]
Leptons	Electron	e^-	e^+	0.511	0	+1	0	0	0	Stable	
	Electron-Neutrino (e)	ν_e	$\bar{\nu}_e$	<7 eV/c^2	0	+1	0	0	0	Stable	
	Muon	μ^-	μ^+	105.7	0	0	+1	0	0	2.20×10^{-6}	$e^-\bar{\nu}_e\nu_\mu$
	Muon-Neutrino (μ)	ν_μ	$\bar{\nu}_\mu$	<0.3	0	0	+1	0	0	Stable	
	Tau	τ^-	τ^+	1784	0	0	0	+1	0	$<4 \times 10^{-13}$	$\mu^-\bar{\nu}_\mu\nu_\tau$ $e^-\bar{\nu}_e\nu_\tau$
	Tau-Neutrino (τ)	ν_τ	$\bar{\nu}_\tau$	<30	0	0	0	+1	0	Stable	
Hadrons											
Mesons	Pion	π^+	π^-	139.6	0	0	0	0	0	2.60×10^{-8}	$\mu^+\nu_\mu$
		π^0	Self	135.0	0	0	0	0	0	0.83×10^{-16}	2γ
	Kaon	K^+	K^-	493.7	0	0	0	0	+1	1.24×10^{-8}	$\mu^+\nu_\mu$ $\pi^+\pi^0$
		K^0_S	$\bar{K}^0_S$	497.7	0	0	0	0	+1	0.89×10^{-10}	$\pi^+\pi^-$ $2\pi^0$
		K^0_L	$\bar{K}^0_L$	497.7	0	0	0	0	+1	5.2×10^{-8}	$\pi^\pm e^\mp \bar{\nu}_e$ $3\pi^0$ $\pi^\pm \mu^\mp \bar{\nu}_\mu$
	Eta	η	Self	548.8	0	0	0	0	0	$<10^{-18}$	$2\gamma, 3\pi$
		η'	Self	958	0	0	0	0	0	2.2×10^{-21}	$\eta\pi^+\pi^-$
Baryons	Proton	p	$\bar{\text{p}}$	938.3	+1	0	0	0	0	Stable	
	Neutron	n	$\bar{\text{n}}$	939.6	+1	0	0	0	0	920	$pe^-\bar{\nu}_e$
	Lambda	Λ^0	$\bar{\Lambda}^0$	1115.6	+1	0	0	0	−1	2.6×10^{-10}	$p\pi^-, n\pi^0$
	Sigma	Σ^+	$\bar{\Sigma}^-$	1189.4	+1	0	0	0	−1	0.80×10^{-10}	$p\pi^0, n\pi^+$
		Σ^0	$\bar{\Sigma}^0$	1192.5	+1	0	0	0	−1	6×10^{-20}	$\Lambda^0\gamma$
		Σ^-	$\bar{\Sigma}^+$	1197.3	+1	0	0	0	−1	1.5×10^{-10}	$n\pi^-$
	Xi	Ξ^0	$\bar{\Xi}^0$	1315	+1	0	0	0	−2	2.9×10^{-10}	$\Lambda^0\pi^0$
		Ξ^-	$\bar{\Xi}^+$	1321	+1	0	0	0	−2	1.64×10^{-10}	$\Lambda^0\pi^-$
	Omega	Ω^-	Ω^+	1672	+1	0	0	0	−3	0.82×10^{-10}	$\Xi^0\pi^0$ $\Lambda^0 K^-$

[a] Notations in this column such as $p\pi^-$, $n\pi^0$ mean two possible decay modes. In this case, the two possible decays are $\Lambda^0 \rightarrow p + \pi^-$ and $\Lambda^0 \rightarrow n + \pi^0$.

Leptons

Leptons (from the Greek *leptos,* meaning "small" or "light") are a group of particles that participate in the electromagnetic (if charged) and weak interactions. All leptons have spins of $\frac{1}{2}$. Included in this group are electrons, muons, and neutrinos, which are less massive than the lightest hadron. Although hadrons have size and structure, *leptons appear to be truly elementary point-like particles with no structure.*

Quite unlike hadrons, the number of known leptons is small. Currently, scientists believe there are only six leptons: the electron, the muon, the tau, and—associated with each of these particles—three different neutrinos, the electron neutrino (ν_e), the muon neutrino (ν_μ), and the tau neutrino (ν_τ). We now classify the six known leptons into three pairs called families:

$$\begin{pmatrix} e^- \\ \nu_e \end{pmatrix} \qquad \begin{pmatrix} \mu^- \\ \nu_\mu \end{pmatrix} \qquad \begin{pmatrix} \tau^- \\ \nu_\tau \end{pmatrix}$$

Current evidence suggests that neutrinos travel with the speed of light and therefore have zero mass. As we shall see later, a firm knowledge of the neutrino's mass (it may be small, not exactly zero as in Table 15.2) could have great significance in cosmological models and predictions of the future of the Universe. Finally, note that each of the six leptons has a corresponding antiparticle.

15.5   CONSERVATION LAWS

Conservation laws are important to an understanding of why certain decays and reactions occur and others do not. In general, the laws of conservation of energy, linear momentum, angular momentum, and electric charge provide us with a set of rules that all processes must follow. For example, conservation of electric charge requires that the total charge before a reaction equals the total charge after the reaction.

Certain new conservation laws are important in the study of elementary particles. Two of these laws, concerning baryon number and lepton number, are described in this section, and others will be discussed later in this chapter. Although the two described here have no theoretical foundation, they are supported by abundant empirical evidence.

Baryon Number

Conservation of baryon number

To apply conservation of baryon number, we must assign a baryon number $B = +1$ for all baryons, $B = -1$ for all antibaryons, and $B = 0$ for all other particles. The **law of conservation of baryon number** states that *whenever a nuclear reaction or decay occurs, the sum of the baryon numbers before the process must equal the sum of the baryon numbers after the process.* An equivalent statement is that the net number of baryons remains constant in any process.

If baryon number is absolutely conserved, the proton must be absolutely stable. If it were not for the law of conservation of baryon number, the proton could decay to a positron and a neutral pion. However, such a decay has never been observed. At present, we can say only that the proton has a half-life of at least 10^{31} years (the estimated age of the Universe is only about 10^{10} years). In one recent version of a grand unified theory, physicists predicted that the proton is unstable. According to this theory, baryon number is not absolutely conserved.

EXAMPLE 15.2   Checking Baryon Numbers

Determine whether or not each of the following reactions can occur on the basis of the law of conservation of baryon number.

$$(1) \qquad p + n \longrightarrow p + p + n + \bar{p}$$

$$(2) \qquad p + n \longrightarrow p + p + \bar{p}$$

Solution   For reaction 1, recall that $B = +1$ for baryons and $B = -1$ for antibaryons. Hence the left side of reaction 1 gives a total baryon number of $1 + 1 = 2$. The right side of reaction 1 gives a total baryon number of $1 + 1 + 1 + (-1) = 2$. Thus the reaction can occur, provided the incoming proton has sufficient energy.

The left side of reaction 2 gives a total baryon number of $1 + 1 = 2$. However, the right side gives $1 + 1 + (-1) = 1$. Because the baryon number is not conserved, the reaction cannot occur.

Lepton Number

There are three conservation laws involving lepton numbers, one for each variety of lepton. The **law of conservation of electron-lepton number** states that *the sum of the electron-lepton numbers before a reaction or decay must equal the sum of the electron-lepton numbers after the reaction or decay.*

Conservation of lepton number

The electron and the electron neutrino are assigned a positive lepton number, $L_e = +1$; the antileptons e^+ and $\bar{\nu}_e$ are assigned a negative lepton number, $L_e = -1$; all others (the muon and tau families) have $L_e = 0$. For example, consider the decay of the neutron

$$n \longrightarrow p + e^- + \bar{\nu}_e$$

Before the decay the electron-lepton number is $L_e = 0$; after the decay it is $0 + 1 + (-1) = 0$. Thus the electron-lepton number is conserved. It is important to recognize that the baryon number must also be conserved. We can easily check this by noting that before the decay $B = +1$, and after the decay $B = +1 + 0 + 0 = +1$.

Similarly, when a decay involves muons, the muon-lepton number, L_μ, is conserved. The μ^- and the ν_μ are assigned positive numbers, $L_\mu = +1$; the antimuons μ^+ and $\bar{\nu}_\mu$ are assigned negative numbers, $L_\mu = -1$ and all others have $L_\mu = 0$. Finally, the tau-lepton number, L_τ, is conserved, and similar assignments can be made for the tau lepton and its neutrino.

EXAMPLE 15.3   Checking Lepton Numbers

Determine which of the following decay schemes can occur on the basis of conservation of electron-lepton number.

$$(1) \qquad \mu^- \longrightarrow e^- + \bar{\nu}_e + \nu_\mu$$

$$(2) \qquad \pi^+ \longrightarrow \mu^+ + \nu_\mu + \nu_e$$

Solution   Because decay 1 involves both a muon and an electron, L_μ and L_e must both be conserved. Before the decay, $L_\mu = +1$ and $L_e = 0$. After the decay, $L_\mu = 0 + 0 + 1 = +1$, and $L_e = +1 - 1 + 0 = 0$. Thus both numbers are conserved, and on this basis the decay mode is possible.

Before decay 2 occurs, $L_\mu = 0$ and $L_e = 0$. After the decay, $L_\mu = -1 + 1 + 0 = 0$, but $L_e = +1$. Thus the decay is not possible, because the electron-lepton number is not conserved.

Exercise 1   Determine whether the decay $\mu^- \rightarrow e^- + \bar{\nu}_e$ can occur.

Answer   No. The muon-lepton number is $+1$ before the decay and 0 after the decay.

15.6   STRANGE PARTICLES AND STRANGENESS

Many particles discovered in the 1950s were produced by the nuclear interaction of pions with protons and neutrons in the atmosphere. Three of these particles—namely, the kaon (K), lambda (Λ), and sigma (Σ) particles—exhibited unusual properties in production and decay and hence were called *strange particles*.

One unusual property of these particles is their production in pairs. For example, when a pion collides with a proton, two neutral strange particles are produced with high probability (Fig. 15.4):

$$\pi^- + p \longrightarrow K^0 + \Lambda^0$$

On the other hand, the reaction $\pi^- + p \rightarrow K^0 + n$ never occurred, even though no known conservation laws would have been violated and the energy of the pion was sufficient to initiate the reaction.

The second peculiar feature of strange particles is that, although they are

Figure 15.4   This bubble-chamber photograph shows many events, and the inset is a drawing of identified tracks. The strange particles Λ^0 and K^0 form at the bottom as the π^- interacts with a proton according to $\pi^- + p \rightarrow \Lambda^0 + K^0$. (Note that the neutral particles leave no tracks, as indicated by the dashed lines.) The Λ^0 and K^0 then decay according to $\Lambda^0 \rightarrow \pi^- + p$ and $K^0 \rightarrow \pi^+ + \mu^- + \bar{\nu}_\mu$. *(Courtesy Lawrence Berkeley Laboratory, University of California, Photographic Services)*

produced by the strong interaction at a high rate, they do not decay into particles that interact via the strong force at a very high rate. Instead, they decay very slowly, which is characteristic of the weak interaction, as shown in Table 15.1. Their half-lives are in the range 10^{-10} s to 10^{-8} s; most other particles that interact via the strong force have lifetimes on the order of 10^{-20} s and less.

To explain these unusual properties of strange particles, a law called conservation of strangeness was introduced with a new quantum number S, called **strangeness.** The strangeness numbers for some particles are given in Table 15.2. The production of strange particles in pairs is explained by assigning $S = +1$ to one of the particles and $S = -1$ to the other. All nonstrange particles are assigned strangeness $S = 0$. The **law of conservation of strangeness** states that *whenever a nuclear reaction or decay occurs, the sum of the strangeness numbers before the process must equal the sum of the strangeness numbers after the process.*

Conservation of strangeness number

The slow decay of strange particles can be explained by assuming that the strong and electromagnetic interactions obey the law of conservation of strangeness, but the weak interaction does not. Because the decay reaction involves the loss of one strange particle, it violates strangeness conservation and hence proceeds slowly via the weak interaction.

EXAMPLE 15.4   Is Strangeness Conserved?

(a) Determine whether the following reaction occurs on the basis of conservation of strangeness.

$$\pi^0 + n \longrightarrow K^+ + \Sigma^-$$

Solution   The initial state has strangeness $S = 0 + 0 = 0$. Because the strangeness of the K^+ is $S = +1$ and the strangeness of the Σ^- is $S = -1$, the strangeness of the final state is $+1 - 1 = 0$. Thus strangeness is conserved and the reaction is allowed.

(b) Show that the following reaction does not conserve strangeness.

$$\pi^- + p \longrightarrow \pi^- + \Sigma^+$$

Solution   The initial state has strangeness $S = 0 + 0 = 0$, and the final state has strangeness $S = 0 + (-1) = -1$. Thus strangeness is not conserved.

Exercise 2   Show that the reaction $p + \pi^- \rightarrow K^0 + \Lambda^0$ obeys the law of conservation of strangeness.

15.7   HOW ARE ELEMENTARY PARTICLES PRODUCED AND PARTICLE PROPERTIES MEASURED?

Examination of the bewildering array of entries in Table 15.2 leaves one yearning for firm ground. It is natural to wonder about an entry for a particle that exists for 10^{-20} s and has a mass of 1192.5 MeV/c^2. How is it possible to detect a particle that exists for only 10^{-20} s? Furthermore, how can the mass be measured? If a standard attribute of a particle is some type of permanence or stability, in what sense is a fleeting entity that exists for 10^{-20} s a particle? In this section we attempt to answer such questions and explain how elementary particles are produced and how their properties are measured.

Elementary particles, most of which are unstable and only rarely come about naturally in cosmic ray showers, are created abundantly in man-made collisions of high-energy particles with a suitable target. Since very high-energy beams of incident particles are desirable, stable charged particles such as electrons or protons generally make up the incident beam, because it takes considerable time to accelerate particles to high energies with electromagnetic fields. Simi-

One of the colliding beam areas at CERN. (*Philippe Plailly/Eurelios/Science Photo Library/ Photo Researchers, Inc.*)

larly, targets must be simple and stable, and the simplest target, hydrogen, serves nicely as both a target and a detector. In a liquid hydrogen bubble chamber, which is basically a large container filled with hydrogen near its boiling point, a charged particle traversing the chamber ionizes the atoms along its path, and the ionization causes a visible track of tiny bubbles. The liquid hy-

"Particles, particles, particles."

drogen also serves as an efficient source of target protons, with a proton density sufficient to ensure many incident particle-target collisions within a reasonable time.

Figure 15.4 shows a typical event in which a bubble chamber has served as both target source and detector. In this figure, many parallel tracks of negative pions are visible entering the photograph from the bottom. One of the pions has hit a stationary proton in the hydrogen and produced two strange particles, the Λ^0 and K^0, according to the reaction

$$\pi^- + p \longrightarrow \Lambda^0 + K^0$$

Neither neutral strange particle leaves a track, but their subsequent decays into charged particles can be seen clearly, as indicated in Figure 15.4. A magnetic field directed into the plane of the photograph causes the track of each charged particle to curve, and from the measured curvature one can determine the particle's charge and linear momentum. If the mass and momentum of the incident particle are known, we can then usually calculate the product particle mass, kinetic energy, and speed from conservation of momentum and energy (see Section 1.10 and Example 1.18). Finally, combining a product particle's speed with a measurable decay track length, we can calculate the product particle's lifetime. Figure 15.4 shows that sometimes one can use this lifetime technique even for a neutral particle, which leaves no track. As long as the start and finish of the missing track are known, as well as the particle speed, one can infer the missing track length and find the lifetime of the neutral particle.

Resonance Particles

With clever experimental technique and much effort, decay track lengths as short as 1 micron (10^{-6} m) can be measured. This means that lifetimes as short as 10^{-16} s can be measured with this technique in the case of high-energy particles traveling at about the speed of light. We arrive at this result by assuming that a decaying particle travels 1 micron in the laboratory at a speed of $0.99c$, yielding a laboratory lifetime of $\tau_{lab} = 10^{-6}$ m/$0.99c \approx 0.33 \times 10^{-14}$ s. Now, relativity helps us. Since the proper lifetime, as measured in the decaying particle's rest frame, is shorter than τ_{lab} by a factor of $\sqrt{1 - v^2/c^2}$, we can actually measure lifetimes of duration:

$$\tau_{proper} = \tau_{lab} \sqrt{1 - \frac{v^2}{c^2}} = (0.33 \times 10^{-14} \text{ s}) \sqrt{1 - \frac{(0.99c)^2}{c^2}} = 4 \times 10^{-16} \text{ s}$$

Unfortunately, even with Einstein's help, we are several orders of magnitude away from 10^{-20} to 10^{-23} s lifetimes with the best efforts of the track-length method. How, then, can we detect the presence of particles that exist for as short a time as 10^{-23} s? As we shall see shortly, the masses, lifetimes, and very existence of these very short-lived particles, known as *resonance particles*, can be inferred from peaks (resonances) in the cross-section-versus-energy plots describing their decay products.

Let's consider this in more detail by looking at the case of the resonance particle called the delta plus (Δ^+), which has a mass of 1232 MeV/c^2 and a lifetime of about 10^{-23} s. The Δ^+ is produced in the reaction

$$e^- + p \longrightarrow e^- + \Delta^+ \tag{15.4}$$

which is followed in 10^{-23} s by the decay of the Δ^+ according to

$$\Delta^+ \longrightarrow \pi^+ + n \tag{15.5}$$

Because the Δ^+ lifetime is so short, it leaves no measurable track, and it might seem impossible to distinguish the reactions given in Equations 15.4 and 15.5 from the net direct reaction in which no Δ^+ is produced:

$$e^- + p \longrightarrow e^- + \pi^+ + n \tag{15.6}$$

In fact, we can tell whether a Δ^+ was formed by measuring the momentum and energy of the suspected decay products (pion and neutron) and using the conservation of momentum and energy to see whether these values combine to give a Δ^+ mass of 1232 MeV/c^2.

To understand this in detail, consider the decay of the Δ^+ shown in Figure 15.5. The energy and momentum of the Δ^+ must satisfy the equation

$$E_\Delta{}^2 = (p_\Delta c)^2 + (m_\Delta c^2)^2$$

or

$$m_\Delta c^2 = \sqrt{E_\Delta{}^2 - (p_\Delta c)^2} \tag{15.7}$$

where m_Δ is the Δ^+ mass. Although we cannot directly measure E_Δ and $\mathbf{p}_\Delta$, since the delta particle leaves no track, we can measure E_π, $\mathbf{p}_\pi$, E_n, and $\mathbf{p}_n$ and, using conservation of momentum and energy, find an expression for $m_\Delta c^2$ in terms of these measured quantities. Thus we have $E_\Delta = E_\pi + E_n$ and $\mathbf{p}_\Delta = \mathbf{p}_\pi + \mathbf{p}_n$, and substituting into Equation 15.7,

$$m_\Delta c^2 = \sqrt{(E_\pi + E_n)^2 - (\mathbf{p}_\pi + \mathbf{p}_n)^2 c^2} \tag{15.8}$$

Equation 15.8 holds for all events in which a Δ^+ particle actually formed and decayed. That is, for many different measured values of E_π, $\mathbf{p}_\pi$, E_n, and $\mathbf{p}_n$ corresponding to many repeated experiments, we will always find the same value of the quantity $m_\Delta c^2 = 1232$ MeV within experimental uncertainty *if the decay of a Δ^+ is involved*. On the other hand, if no Δ^+ is involved and the direct reaction $e^- + p \to e^- + \pi^+ + n$ occurs, $\sqrt{(E_\pi + E_n)^2 - (\mathbf{p}_\pi + \mathbf{p}_n)^2 c^2}$ will not equal 1232 MeV but will sweep over a broad range of values, some larger and some smaller than $m_\Delta c^2$, as the experiment is repeated. The typical method

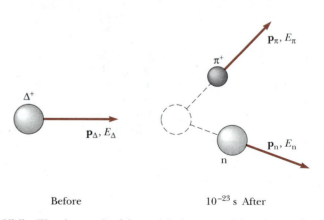

Before 10^{-23} s After

Figure 15.5   The decay of a Δ^+ particle into a positive pion and a neutron.

Figure 15.6   Experimental evidence for the existence of the Δ^+ particle. The sharp peak at 1232 MeV was produced by the events in which a Δ^+ formed and promptly decayed into a π^+ and neutron.

for showing the existence of a resonance particle involves calculating the quantity $Z = \sqrt{(E_\pi + E_n)^2 - (\mathbf{p}_\pi + \mathbf{p}_n)^2 c^2}$ for a large number of events in which a π^+ and a neutron are produced and then drawing a histogram of those events. By plotting the number of events with a given value of Z against this value of Z, we obtain a slowly varying curve with sharp peaks superimposed, showing the existence of resonance particles. Figure 15.6 shows such an experimental histogram for the Δ^+ particle. The broad background (dashed curve) is produced by direct events in which no Δ^+ was created; the sharp peak at $Z = 1232$ MeV, containing many events, was produced by all the events in which a Δ^+ formed and decayed into a pion and neutron with just the right energies and momenta to make up a delta particle. Peaks corresponding to two other resonance particles of larger mass can also be seen in Figure 15.6.

Histograms like Figure 15.6 can tell us not only the mass of a short-lived particle but also the lifetime of the particle from the full width at half maximum of the resonance peak, denoted by Γ in Figure 15.6. Since Γ is twice the uncertainty in energy of the delta, $\Gamma = 2\,\Delta E$, we can use the energy-time uncertainty relation, $\Delta E\,\Delta t \approx \hbar/2$, to infer the lifetime of the delta, Δt:

$$\Delta t \approx \frac{\hbar}{2\,\Delta E} = \frac{\hbar}{\Gamma} \qquad (15.9)$$

The measured width of $\Gamma = 115$ MeV leads to a value for the lifetime of the delta particle:

$$\Delta t \approx \frac{\hbar}{\Gamma} = \frac{6.6 \times 10^{-16}\ \text{eV}\cdot\text{s}}{115 \times 10^{6}\ \text{eV}} = 5.7 \times 10^{-24}\ \text{s}$$

In this incredibly short lifetime, a delta particle moving at the highest possible speed of c travels only 10^{-15} m, or about one nuclear diameter.

Energy Considerations in Particle Production

It is ironic that the highly sophisticated branch of physics known as elementary particle physics relies for its very existence on the most primitive and brutish of experimental methods: smashing an incident particle moving at relativistic speed into a stationary target particle and observing what pieces come flying out! Yet this process is at the heart of experimental particle physics, and in this section we will be concerned with the important issue of finding the threshold energy that is required for the production of new particles in a collision. Since the energy needed to manufacture new particles comes from the kinetic energy of the incident particle, and incident particle energies are quite large (the Fermi National Laboratory produces 1000-GeV protons), we must use relativistic equations in the calculation of these threshold energies.

Consider the specific particle production process

$$m_1 + m_2 \longrightarrow m_3 + m_4 + m_5 \tag{15.10}$$

Here m_1 is the mass of the incident particle, m_2 is the mass of the target particle which is at rest in the laboratory, and m_3, m_4, and m_5 are product particle masses. Figure 15.7a shows such a particle reaction in the laboratory reference frame. Observe that the energies shown in Figure 15.7—E_1, E_2, and so on—are total energies (kinetic energy + rest energy) and that the momenta are labeled $\mathbf{p}_1$, $\mathbf{p}_2$, and so on. At first glance it might seem that the total initial energy of m_1 and m_2, $E_1 + E_2$, could be converted completely to rest energy of particles 3, 4, and 5. However, as we have already seen in the calculation of nuclear reaction thresholds in Chapter 14, some of the initial energy must go into energy of motion of the product particles in order to conserve momentum in the collision. This means that not all of the initial energy can go into creating new masses. The issue is to find out how much of the initial energy can go toward creating masses m_3, m_4, and m_5. That is, we wish to calculate the minimum or threshold kinetic energy K_{th} that m_1 must have in order to create particles with masses of m_3, m_4, and m_5 and also conserve momentum.

An important clue to finding K_{th} most easily comes from understanding that if we could somehow arrange for a zero initial momentum, all of the initial energy could be converted into new particles. What we *can* do is solve for K_{th} in the frame in which the initial momentum is zero—the center-of-mass frame—and then transform that result back into the frame in which the experiment is actually carried out, the laboratory frame. Here's how it goes.

Figure 15.7b shows the same reaction as 15.7a but viewed from the center-of-mass (CM) frame, in which by definition the total momentum is always zero. Thus, in Figure 15.7b, m_1 and m_2 have equal and opposite momenta, and the vector sum of the momenta of m_3, m_4, and m_5 is zero. Actually, Figure 15.7b shows the case of an incident particle with more than threshold energy. A moment's reflection reveals that when m_1 has the minimum, or threshold, energy in the CM frame, all of the initial energy, $E_1' + E_2'$, should be converted to the masses of particles 3, 4, and 5. This occurs when the product particles are created at rest in the CM frame, which is now possible since the total mo-

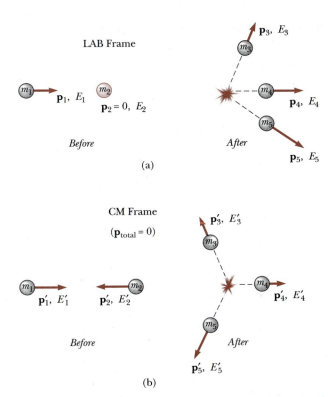

Figure 15.7   (a) The reaction $m_1 + m_2 \rightarrow m_3 + m_4 + m_5$ viewed in the laboratory frame, where m_2 is initially at rest and the energies are total relativistic energies (kinetic plus rest energy). (b) The same reaction viewed from the center-of-mass (CM) frame, in which the total momentum is always zero.

mentum is zero in the CM frame. Figure 15.8 shows the reaction $m_1 + m_2 \rightarrow m_3 + m_4 + m_5$ in the CM frame when m_1 has the threshold kinetic energy and m_3, m_4, and m_5 are created at rest.

To calculate a numerical expression for K_{th}, we make use of the invariant quantity $E^2 - p^2 c^2$, introduced in Chapter 1. Recall that $E^2 - p^2 c^2$ is called an invariant because it has the same numerical value for a system of particles in any inertial reference frame (see Problem 1.42). Applying the invariance of

Figure 15.8   The reaction $m_1 + m_2 \rightarrow m_3 + m_4 + m_5$ in the CM frame when m_1 has the threshold kinetic energy required to produce m_3, m_4, and m_5. For the threshold condition, m_3, m_4, and m_5 are created at rest.

$E^2 - p^2c^2$ to the CM and laboratory frames, we have, for the period before the collision,

$$E_{CM}^2 - p_{CM}^2 c^2 = E_{lab}^2 - p_{lab}^2 c^2 \qquad (15.11)$$

Here E_{CM} and E_{lab} stand for the sum of the energies of particles m_1 and m_2 in the CM and laboratory frames, respectively. Likewise, p_{CM} and p_{lab} stand for the sum of the momenta of particles m_1 and m_2 in the CM and laboratory frames. Since $p_{CM} = 0$, $E_{lab} = E_1 + m_2 c^2$, and $p_{lab} = p_1$, Equation 15.11 becomes

$$E_{CM \text{ before}}^2 = (E_1 + m_2 c^2)^2 - p_1^2 c^2 \qquad (15.12)$$

We can eliminate p_1 from Equation 15.12 by using $E_1^2 = p_1^2 c^2 + m_1^2 c^4$ to obtain

$$E_{CM \text{ before}}^2 = 2E_1 m_2 c^2 + (m_2^2 + m_1^2) c^4 \qquad (15.13)$$

Observe that Equation 15.13 is in a useful form, since we can readily solve for E_1 or K_{th} ($E_1 = K_{th} + m_1 c^2$) in terms of all the masses if we can find an expression for $E_{CM \text{ before}}$ in terms of the product masses m_3, m_4, and m_5. We can find such an expression by conserving relativistic energy in the CM frame:

$$E_{CM \text{ before}} = E_{CM \text{ after}} = E_3' + E_4' + E_5' = (m_3 + m_4 + m_5) c^2 \quad (15.14)$$

Substituting this expression into Equation 15.13 and using $E_1 = K_{th} + m_1 c^2$, after a bit of algebra we obtain our final result:

$$K_{th} = \frac{(m_3 + m_4 + m_5)^2 c^2 - (m_1 + m_2)^2 c^2}{2m_2} \qquad (15.15)$$

Equation 15.15 gives the threshold kinetic energy of an incident particle m_1 colliding with a stationary target m_2 required to produce three product particles of mass m_3, m_4, and m_5. For more than three product particles, inspection of our derivation shows that Equation 15.15 can be generalized as

$$K_{th} = \frac{(m_3 + m_4 + m_5 + m_6 + \cdots)^2 c^2 - (m_1 + m_2)^2 c^2}{2m_2} \qquad (15.16)$$

EXAMPLE 15.5   How To Make a Virtual Particle Real

Consider the scattering of two protons. Assume that the protons interact by exchanging virtual field particles or field quanta. Thus, if the protons are attracted gravitationally, they exchange a graviton; if they are repelled electrically, they exchange a photon. In the present case, we wish to consider the nuclear force between two protons that we assume is carried by the field particle called the pi meson or pion. When an incoming proton scatters from a stationary proton at low energy, a virtual pion with a mass of about 140 MeV/c^2 can blink into existence, transport energy and momentum from one proton to the other, and then blink out of existence in a time so short that violation of energy conservation is not observable. However, if an incident proton interacts with a stationary target at high enough energy, the incident proton may supply enough energy to make a virtual pion real according to the reaction

$$p + p \longrightarrow p + p + \pi^0 \qquad (15.17)$$

where π^0 represents a neutral pion. In this example, we wish to find the threshold energy for π^0 production according to the reaction given by Equation 15.17. The neutral pion has a rest energy of 135 MeV.

Solution Use Equation 15.15 with $m_1 = m_2 = m_3 = m_4 = m_p = 938.3$ MeV/c^2 and $m_5 = m_\pi = 135$ MeV/c^2.

Equation 15.15 becomes

$$K_{th} = \frac{(2m_p + m_\pi)^2 c^2 - (2m_p)^2 c^2}{2m_p}$$

$$= \frac{(4m_p m_\pi + m_\pi^2) c^2}{2m_p} = 2m_\pi c^2 + \frac{m_\pi^2 c^2}{2m_p}$$

$$= 2(135 \text{ MeV}/c^2) c^2 + \frac{(135 \text{ MeV}/c^2)^2 c^2}{2(938.3 \text{ MeV}/c^2)}$$

$$= 280 \text{ MeV}$$

Thus, if stationary protons are bombarded with protons of at least 280 MeV kinetic energy, particle accelerators can produce neutral pion beams that can be used for additional pion studies.

Exercise 3   Since equally strong nuclear forces exist between proton and proton, neutron and neutron, and proton and neutron, we might expect the pion to come in three different charge states, $+e$, $-e$, and 0, and this is indeed the case. Three different pions exist—π^+, π^-, and π^0—with masses $m_{\pi^-} = 139.6$ MeV/c^2, $m_{\pi^+} = 139.6$ MeV/c^2, and $m_{\pi^0} = 135$ MeV/c^2. Two reactions involving the production of the π^+ and π^- are

$$p + p \longrightarrow p + n + \pi^+ \tag{15.18}$$

$$p + p \longrightarrow p + p + \pi^+ + \pi^- \tag{15.19}$$

Calculate the threshold energy for these reactions.

Answers $K_{th} = 292$ MeV for Equation 15.18, $K_{th} = 600$ MeV for Equation 15.19

15.8   THE EIGHTFOLD WAY

As we have seen, we associate conserved quantities such as spin, baryon number, lepton number, and strangeness with particles. Many classification schemes have been proposed that group particles into families. Consider the first eight baryons listed in Table 15.2, all with spins of $\frac{1}{2}$. If we plot their strangeness versus charge using a sloping coordinate system, as in Figure 15.9a, a fascinating pattern is observed. Six of the baryons form a hexagon with the remaining two at its center.

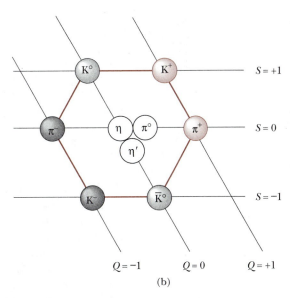

Figure 15.9   (a) The hexagonal eightfold-way pattern for the eight spin-$\frac{1}{2}$ baryons. This strangeness-versus-charge plot uses a sloping axis for the charge number Q but a horizontal axis for the strangeness (S) values. (b) The eightfold-way pattern for the nine spin-0 mesons.

Now consider the family of mesons listed in Table 15.2, which have spins of zero. If we count both particles and antiparticles, there are nine such mesons. Figure 15.9b is a plot of strangeness versus charge for this family. Again a fascinating hexagonal pattern emerges. In this case, the particles on the perimeter of the hexagon lie opposite their antiparticles, and the remaining three (which form their own antiparticles) are at its center. In 1961 Murray Gell-Mann and Yuval Ne'eman independently proposed these and related symmetric patterns, called the **eightfold way.**

The groups of baryons and mesons can be displayed in many other symmetric patterns within the framework of the eightfold way. For example, the family of spin-$\frac{3}{2}$ baryons contains ten particles arranged in a pattern like the tenpins in a bowling alley. When the pattern was proposed in 1961, one of the particles was missing—it had yet to be discovered. Gell-Mann predicted that the missing particle, which he called the omega minus (Ω^-), should have a spin of $\frac{3}{2}$, a charge of -1, a strangeness of -3, and a rest energy of about 1680 MeV. Shortly thereafter, in 1964, scientists at the Brookhaven National Laboratory found the missing particle through careful analyses of bubble-chamber photographs and confirmed all its predicted properties.

The patterns of the eightfold way in particle physics have much in common with the periodic table. Whenever a vacancy (a missing particle or element) occurs in the organized patterns, experimentalists have a guide for their investigations.

15.9   QUARKS

As we have noted, leptons appear to be truly elementary particles because they have no measurable size or internal structure, are limited in number, and do not seem to break down into smaller constituents. Hadrons, on the other hand, are complex particles having size and structure. The existence of the eightfold-way patterns suggests that baryons and mesons—in other words, hadrons—have a more elemental substructure. Furthermore, we know that hadrons decay into other hadrons and are many in number. Table 15.2 lists only the hadrons that are stable against hadronic decay; hundreds of others have been discovered. These facts strongly suggest that hadrons cannot be truly elementary. In this section we show that the complexity of hadrons can be explained by a simpler substructure.

The Original Quark Model

In 1963 Gell-Mann and George Zweig independently proposed that hadrons have a more elemental substructure. According to their model, all hadrons are composite systems of two or three fundamental constituents called **quarks** (pronounced to rhyme with *forks*). (Zweig called them "aces." Gell-Mann borrowed the word *quark* from the passage "Three quarks for Muster Mark" in James Joyce's *Finnegan's Wake.*) In the original quark model, there were three types of quarks designated by the symbols u, d, and s. These were given the arbitrary names *up, down,* and *sideways* or, now more commonly, *strange.*

An unusual property of quarks is that they had to be assigned fractional electronic charges. The u, d, and s quarks have charges of $+2e/3$, $-e/3$, and

Table 15.3   Properties of Quarks and Antiquarks

Quarks

Name	Symbol	Spin	Charge	Baryon Number	Strangeness	Charm	Bottomness	Topness
Up	u	$\frac{1}{2}$	$+\frac{2}{3}e$	$\frac{1}{3}$	0	0	0	0
Down	d	$\frac{1}{2}$	$-\frac{1}{3}e$	$\frac{1}{3}$	0	0	0	0
Strange	s	$\frac{1}{2}$	$-\frac{1}{3}e$	$\frac{1}{3}$	-1	0	0	0
Charmed	c	$\frac{1}{2}$	$+\frac{2}{3}e$	$\frac{1}{3}$	0	$+1$	0	0
Bottom	b	$\frac{1}{2}$	$-\frac{1}{3}e$	$\frac{1}{3}$	0	0	$+1$	0
Top	t	$\frac{1}{2}$	$+\frac{2}{3}e$	$\frac{1}{3}$	0	0	0	$+1$

Antiquarks

Name	Symbol	Spin	Charge	Baryon Number	Strangeness	Charm	Bottomness	Topness
Anti-up	$\bar{u}$	$\frac{1}{2}$	$-\frac{2}{3}e$	$-\frac{1}{3}$	0	0	0	0
Anti-down	$\bar{d}$	$\frac{1}{2}$	$+\frac{1}{3}e$	$-\frac{1}{3}$	0	0	0	0
Anti-strange	$\bar{s}$	$\frac{1}{2}$	$+\frac{1}{3}e$	$-\frac{1}{3}$	$+1$	0	0	0
Anti-charmed	$\bar{c}$	$\frac{1}{2}$	$-\frac{2}{3}e$	$-\frac{1}{3}$	0	-1	0	0
Anti-bottom	$\bar{b}$	$\frac{1}{2}$	$+\frac{1}{3}e$	$-\frac{1}{3}$	0	0	-1	0
Anti-top	$\bar{t}$	$\frac{1}{2}$	$-\frac{2}{3}e$	$-\frac{1}{3}$	0	0	0	-1

$-e/3$, respectively, where e is the charge of a positron. Each quark has a baryon number of $\frac{1}{3}$ and a spin of $\frac{1}{2}$. (The spin $\frac{1}{2}$ means that all quarks are *fermions*, defined in Section 9.2 as any particle having half integral spin.) The u and d quarks have strangeness 0, and the s quark has strangeness -1. Table 15.3 gives other properties of quarks and antiquarks. Associated with each quark is an antiquark of opposite charge, baryon number, and strangeness.

The compositions of all hadrons that were known when Gell-Mann and Zweig presented their models could be completely specified by three simple rules:

- A meson consists of one quark and one antiquark, which gives it a baryon number of 0, as required. Mesons are thus quark-antiquark combinations (quarkonium) bound together by a swarm of gluons, which are the field particles that transmit the strong force between quarks.
- A baryon consists of three quarks and is a sort of quark "molecule" held together by gluons.
- An antibaryon consists of three antiquarks.

Table 15.4 lists the quark compositions of several mesons and baryons. Note that just two of the quarks, u and d, are contained in all hadrons encountered in ordinary matter (protons and neutrons). The third quark, s, is needed only to construct strange particles with a strangeness number of either $+1$ or -1. Figure 15.10 is a pictorial representation of the quark compositions of several particles.

Table 15.4   Quark Compositions of Several Hadrons

Particle	Quark Composition
Mesons	
π^+	$u\bar{d}$
π^-	$\bar{u}d$
K^+	$u\bar{s}$
K^-	$\bar{u}s$
K^0	$d\bar{s}$
Baryons	
p	uud
n	udd
Λ^0	uds
Σ^+	uus
Σ^0	uds
Σ^-	dds
Ξ^0	uss
Ξ^-	dss
Ω^-	sss

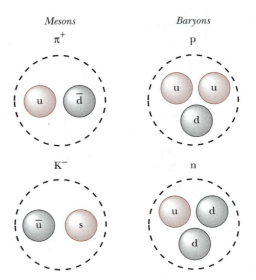

Figure 15.10   Quark compositions of two mesons and two baryons. Note that each meson on the left contains two quarks, while each baryon on the right contains three quarks.

Charm and Other Recent Developments

Although the original quark model was highly successful in classifying particles into families, some discrepancies occurred between predictions of the model and certain experimental decay rates. Consequently, several physicists proposed a fourth quark in 1967. They argued that if there are four leptons (as was thought at the time), then there should also be four quarks because of an underlying symmetry in nature. The fourth quark, denoted by c, was given a property called **charm.** A *charmed* quark would have charge $+2e/3$, but its charm would distinguish it from the other three quarks. The new quark would

The American physicists Richard Feynman *(left)* and Murray Gell-Mann *(right)* were awarded the Nobel prize in 1965 and 1969, respectively, for their theoretical studies dealing with subatomic particles. *(Courtesy of Michael R. Dressler)*

have a charm of $C = +1$, its antiquark would have a charm of $C = -1$, and all other quarks would have $C = 0$, as indicated in Table 15.3. Charm, like strangeness, would be conserved in strong and electromagnetic interactions but not in weak interactions.

In 1974 a new heavy meson called the J/Ψ particle (or simply Ψ) was discovered independently by two groups, one led by Burton Richter at the Stanford Linear Accelerator (SLAC) and the other led by Samuel Ting at the Brookhaven National Laboratory. Richter and Ting were awarded the Nobel prize in 1976 for this work. The J/Ψ particle did not fit into the three-quark model but had the properties of a combination of a charmed quark and its anti-quark ($c\bar{c}$). It was much more massive than the other known mesons ($\sim 3100 \text{ MeV}/c^2$), and its lifetime was much longer than those that decay via the strong force. Soon, related charmed mesons were discovered that corresponded to such quark combinations as $\bar{c}d$ and $c\bar{d}$, all of which have large masses and long lifetimes. In 1975 researchers at Stanford University reported strong evidence for the tau (τ) lepton, with a mass of $1784 \text{ MeV}/c^2$. Such discoveries led to more elaborate quark models and to the proposal of two new quarks, *top* (t) and *bottom* (b). (Some physicists prefer the designators *truth* and *beauty*.) To distinguish these quarks from the old ones, quantum numbers called *topness* and *bottomness* were assigned to these new particles; they are included in Table 15.3. In 1977 researchers at the Fermi National Laboratory, under the direction of Leon Lederman, reported the discovery of a very massive new meson, Υ, whose composition is considered to be $b\bar{b}$. In March 1995 researchers at the Fermi National Laboratory announced the discovery of the top quark (supposedly the last of the quarks), with a mass of $173 \text{ GeV}/c^2$. The researchers identified $t\bar{t}$ pairs produced in the collision of 0.9-TeV protons with 0.9-TeV antiprotons from their decay into W bosons and bottom quarks and the subsequent decay of those particles into leptons, neutrinos, and hadrons. An interesting essay by Melissa Franklin and David Kestenbaum at the end of this chapter explains the operation of the accelerator and the particle detectors used to find the top quark.

You are probably wondering whether such discoveries will ever end. How many "building blocks" of matter really exist? At the present, physicists believe that the fundamental matter particles in nature are 12 fermions, including six quarks and six leptons (together with their antiparticles). Table 15.5 lists some of the properties of these particles as well as the properties of the fundamental field particles.

Despite extensive experimental efforts, no isolated quark has ever been observed. Physicists now believe that quarks are permanently confined inside hadrons because of an exceptionally strong force that prevents them from escaping. This force, called the strong or "color" force, increases with separation distance, eventually approaching a constant value—implying a potential energy that grows linearly with quark separation and the requirement of infinite energy to produce two truly free quarks. One author has described the great strength of the force between quarks as follows:[2]

---

[2] Harald Fritzsch, *Quarks, The Stuff of Matter*, London, Allen Lane, 1983.

Table 15.5   Properties of the Fundamental Point Particles

Particle		Mass in Terms of Proton Mass	Interactions Experienced[a]	Interaction Mediated	Electric Charge in Units of $|e|$	Color Charge
Matter Particles						
Leptons	e^-, μ, τ	$\frac{1}{1836}, \frac{1}{9}, 1.9$	EM, W	None	-1	No
	ν_e, ν_μ, ν_τ	$<10^{-8}, <\frac{1}{3500}, <\frac{1}{30}$	W	None	0	No
Quarks	u, c, t	$\frac{1}{235}, 1.6, 165$	EM, W, S	None	$+\frac{2}{3}$	Yes
	d, s, b	$\frac{1}{135}, \frac{1}{6}, 5.2$	EM, W, S	None	$-\frac{1}{3}$	Yes

All are fermions with spin $\frac{1}{2}$.

Particle		Mass in Terms of Proton Mass	Interactions Experienced[a]	Interaction Mediated	Electric Charge in Units of $|e|$	Color Charge
Field Particles						
Photon	γ	0	EM	EM	0	No
Intermediate bosons	$W^\pm$	85	W, EM	W	±1	No
	Z^0	97	W	W	0	No
Gluons	g	0	S	S	0	Yes

All are bosons with spin 1.

[a] EM = electromagnetic; W = weak; S = strong.

> Quarks are slaves of their own color charge, . . . bound like prisoners of a chain gang. . . . Any locksmith can break the chain between two prisoners, but no locksmith is expert enough to break the gluon chains between quarks. Quarks remain slaves forever.

15.10   COLORED QUARKS, OR QUANTUM CHROMODYNAMICS

Shortly after the concept of quarks was proposed, scientists recognized that certain particles had quark compositions that violated the Pauli exclusion principle. Because all quarks are fermions with spins of $\frac{1}{2}$, they are expected to follow the exclusion principle. One example of a particle that violates the exclusion principle is the Ω^- (sss) baryon, which contains three s quarks with parallel spins, giving it a total spin of $\frac{3}{2}$. Other examples of baryons that have identical quarks with parallel spins are the Δ^{++} (uuu) and the Δ^- (ddd). To resolve this problem, Moo-Young Han and Yoichiro Nambu suggested in 1965 that quarks possess a new property called **color.** This property is similar in many respects to electric charge except that it occurs in three varieties called red, green, and blue. (The antiquarks have the colors antired, antigreen, and antiblue.) To satisfy the exclusion principle, all three quarks in a baryon must have different colors. Just as a combination of actual colors of light can produce the neutral color white, a combination of three quarks with different colors is also white, or colorless. A meson consists of a quark of one color and an antiquark of the corresponding anticolor. The result is that both baryons and mesons are always colorless (or white).

Although the concept of color in the quark model was originally conceived to satisfy the exclusion principle, it also provided an improved theory for explaining certain experimental results. For example, the modified theory correctly predicts the lifetime of the π^0 meson. The theory of how quarks interact with each other is called **quantum chromodynamics** (QCD), to parallel quantum electrodynamics, the theory of interaction between electric charges. In QCD the quark is said to carry a *color charge*, in analogy with electric charge, and the color charge is responsible for the strong force between quarks. The strong force between quarks is often called the *color force.*

As stated earlier, the strong interaction between quarks is carried by massless particles called **gluons** (g), which are analogous to photons for the electromagnetic force. According to QCD, there are eight gluons, all with color charge. When a quark emits or absorbs a gluon, its color changes. For example, a blue quark that emits a gluon may become a red quark, and a red quark that absorbs this gluon becomes a blue quark. The color force between quarks is analogous to the electric force between charges; like colors repel and opposite colors attract. Therefore, two red quarks repel each other, but a red quark is attracted to an antired quark. The attraction between quarks of opposite color to form a meson (q$\bar{\text{q}}$) is indicated in Figure 15.11a. Differently colored quarks also attract each other, but with less intensity than opposite colors of quark and antiquark. For example, the red, blue, and green quarks in a cluster all attract each other to form baryons, as indicated in Figure 15.11b. Thus every baryon contains three quarks of three different colors.

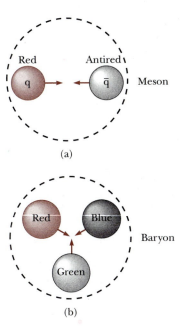

Figure 15.11   (a) A red quark is attracted to an antired quark. This forms a meson whose quark structure is (q$\bar{\text{q}}$). (b) Three differently colored quarks attract each other to form a baryon.

Experimental Evidence for Quarks

As already mentioned, an isolated quark has never been observed and probably never will be, according to QCD. This leaves us with the question of how quarks can be detected and what experimental evidence we have for the existence of quarks.

Experiments at the Stanford Linear Accelerator (SLAC) in the late 1960s established two important results concerning quarks. (1) The recoils of high-energy electrons scattered from protons could be completely modeled by assuming *three* point-like quarks within the proton. (2) Although gluons were not directly detected, it was found that the incident electrons (which carry no color charge) interacted with only about half the mass making up the proton. This is exactly what is expected if approximately half the material in a proton consists of gluons that are electrically neutral and do not interact electromagnetically with the electron (see Table 15.5).

It is interesting to consider what happens to a quark-antiquark pair (q$\bar{\text{q}}$) formed in a high-energy collision as the pair attempts to separate. Because the color force is so strong, all quarks must form colorless baryons (three quarks) or mesons (two quarks). The process goes like this. As two quarks separate, the increase in energy of the color field is so large that additional quark-antiquark pairs are created in streams (or jets) following the original (q$\bar{\text{q}}$) pair. When all the available energy is used up, the quarks cluster into color-neutral or colorless combinations, which continue to separate indefinitely as two jets of hadrons, as in Figure 15.12a. Figure 15.12b shows actual experimental data corresponding to three jets of hadrons emerging from a collision in which a quark and antiquark were produced and a gluon was radiated. Because gluons carry color

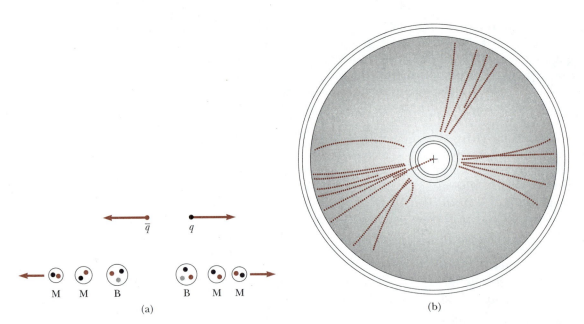

Figure 15.12 (a) As the quark-antiquark pair separates, the potential energy in the color field is transformed into additional q$\bar{q}$ pairs, which quickly condense into colorless mesons (M) and baryons (B). These separate as jets of hadrons. (b) Three jets of hadrons produced by a quark, an antiquark, and a gluon. This figure is from the JADE detector at the German laboratory DESY in Hamburg. (*Adapted from Gordon Kane,* The Particle Garden, *Figure 6.2, p. 100*)

charge, they, like quarks, must also appear as a jet of hadrons. Thus *experimental signatures of quarks and gluons are narrow "jets" of hadrons.*

Explanation of Nuclear Force in Terms of Quarks

Although the color force between two color-neutral hadrons is negligible at large separations, the strong color force between their constituent quarks does not exactly cancel at small separations. This residual strong force is, in fact, the force that binds protons and neutrons to form nuclei. In Section 15.3 we explained the nuclear interaction of a proton with a neutron using Yukawa's theory of pion exchange. According to QCD, a more basic explanation of nuclear force can be given in terms of quarks and gluons, as shown by contrasting Feynman diagrams of the same process in Figure 15.13. Each quark within the neutron and proton is continually emitting and absorbing virtual gluons and creating and annihilating virtual (q$\bar{q}$) pairs. When the neutron and proton approach within 1 fm of each other, these virtual gluons and quarks can be exchanged between the two nucleons, and such exchanges produce the nuclear force. Figure 15.13b depicts one likely possibility or contribution to the process shown in Figure 15.13a. A down quark emits a virtual gluon (represented by a wavy line, g), which creates a u$\bar{u}$ pair. Both the recoiling d quark and the $\bar{u}$ are transmitted to the proton, where the $\bar{u}$ annihilates a proton u quark (with the creation of a gluon) and the d is captured.

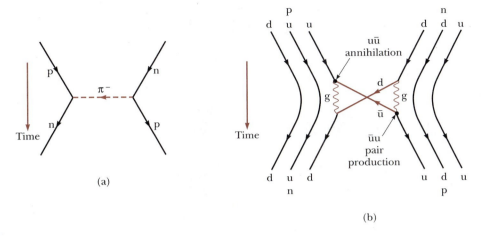

Figure 15.13   (a) A nuclear interaction between a proton and a neutron explained in terms of Yukawa's pion exchange model. (b) The same interaction as in (a), explained in terms of quarks and gluons. Note that the exchanged $\bar{u}d$ quark pair makes up a π^- meson.

15.11   ELECTROWEAK THEORY AND THE STANDARD MODEL

Recall that the weak interaction is an extremely short-range force with an interaction distance of approximately 10^{-18} m. Such a short-range interaction implies that the quantized particles that carry the weak field (the spin-1 W^+, W^-, and Z^0 bosons) are quite massive, as is indeed the case (see Table 15.5). These bosons are especially amazing when we think of them as structureless, point-like particles as massive as krypton atoms. As mentioned earlier, the weak interaction is responsible for neutron decay and the beta decay of other heavier baryons. More important, the weak interaction is responsible for the decay of the c, s, b, and t quarks into lighter, more stable u and d quarks as well as the decay of the massive μ and τ leptons into (lighter) electrons. Thus, *the weak interaction is very important because it governs the stability of the basic matter particles.*

A mysterious feature of the weak interaction is its lack of symmetry, especially in comparison with the high degree of symmetry shown by the strong, electromagnetic, and gravitational interactions. For example, the weak interaction, unlike the strong interaction, is not symmetric under mirror reflection or charge exchange. (Mirror reflection means that all the quantities in a given particle reaction are exchanged as in a mirror reflection—left for right, an inward motion toward the mirror for an outward motion. Charge exchange means that all the electric charges in a particle reaction are converted to their opposites—all positives to negatives and vice versa.) When we say that the weak interaction is not symmetric, we mean that the reaction with all quantities changed occurs less frequently than the direct reaction. For example, the decay of the K^0, which is governed by the weak interaction, is not symmetric under charge exchange, since

$$K^0 \longrightarrow \pi^- + e^+ + \nu_e$$

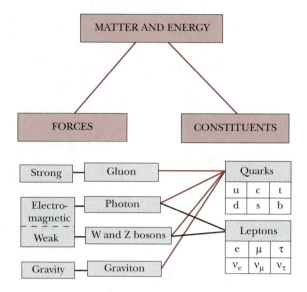

Figure 15.14   The Standard Model of particle physics.

occurs much more frequently than

$$K^0 \longrightarrow \pi^+ + e^- + \bar{\nu}_e$$

In 1979 Sheldon Glashow, Abdus Salam, and Steven Weinberg won a Nobel prize for developing a theory that unifies the electromagnetic and weak interactions. This **electroweak theory** postulates that the weak and electromagnetic interactions have the same strength at very high particle energies. Thus the two interactions are viewed as two different manifestations of a single unifying electroweak interaction. The photon and the three massive bosons ($W^\pm$ and Z^0) play key roles in the electroweak theory. Perhaps the most spectacular of the theory's many concrete predictions are the predictions of the masses of the W and Z particles at about 80 GeV/c^2 and 91 GeV/c^2, respectively. The 1984 Nobel prize was awarded to Carlo Rubbia and Simon van der Meer for their work leading to the discovery of these particles with just those masses at the CERN Laboratory.

The combination of the electroweak theory and QCD for the strong interaction is called the *Standard Model* in high-energy physics. Although the details of the Standard Model are complex, its essential ingredients can be summarized with the help of Figure 15.14. The strong force, mediated by gluons, holds quarks together to form composite particles such as protons, neutrons, and mesons. Leptons participate only in the electromagnetic and weak interactions. The electromagnetic force is mediated by photons, while the weak force is mediated by W and Z bosons. Note that all fundamental forces are mediated by spin-1 bosons whose properties are given, to a large extent, by symmetries involved in the theories.

The Standard Model does not answer all questions, however. A major question is why the photon has no mass while the W and Z bosons do have mass. Because of this difference, the electromagnetic and weak forces are quite distinct at low energies but become similar at very high energies, where the rest energies of the W and Z bosons are insignificant fractions of their total energies. This behavior in the transition from high to low energies, called *symmetry breaking*, leaves open the question of the origin of particle masses. To resolve this problem, a hypothetical particle called the *Higgs boson* has been proposed; it provides a mechanism for breaking the electroweak symmetry. The Standard Model, including the Higgs mechanism, provides a logically consistent explanation of the massive nature of the W and Z bosons. Unfortunately, the Higgs boson has not yet been found, but physicists think its mass should be less than 1 TeV (10^{12} eV).

To determine whether the Higgs boson exists, two quarks of at least 1 TeV of energy must collide, but calculations show that this requires injecting 40 TeV of energy within the volume of a proton. Scientists are convinced that, because of the limited energy available in conventional accelerators using fixed targets, it is necessary to build colliding-beam accelerators called **colliders.** The concept of colliders is straightforward. Particles with equal masses and kinetic energies, traveling in opposite directions in an accelerator ring, collide head-on to produce the required reaction and the formation of new particles. Because the total momentum of the interacting particles is zero, all of their kinetic energy is available for the reaction. The Large Electron-Positron Collider (LEP) at CERN and the Stanford Linear Collider in California collide both electrons and protons. The Super Proton Synchrotron at CERN accelerates protons and antiprotons to energies of 270 GeV, while the world's highest-energy proton accelerator, the Tevatron, at the Fermi National Laboratory in Illinois produces protons at almost 1000 GeV (1 TeV). The Superconducting Super Collider (SSC) was an accelerator under construction in Texas that was designed to produce 20-TeV protons in a ring 52 mi in circumference. After much debate in Congress and an investment of almost $2 billion, the U.S. Department of Energy canceled the SSC project in October 1993. CERN recently approved the development of the Large Hadron Collider (LHC), a proton-proton collider that will provide a center-of-mass energy of 14 TeV and allow an exploration of Higgs-boson physics. The accelerator will be constructed in the same 27-km-circumference tunnel as CERN's Large Electron-Positron Collider, and many countries are expected to participate in the project.

Following the success of the electroweak theory, scientists attempted to combine it with QCD in a *grand unification theory*, or GUT. In this model, the electroweak force is merged with the strong color force to form a grand unified force. One version of the theory considers leptons and quarks as members of the same family that can change into each other by exchanging an appropriate particle. Many GUT theories predict that protons are unstable and will decay with a lifetime of about 10^{31} years. Attempts to detect such proton decays have so far been unsuccessful.

15.12   THE COSMIC CONNECTION

In this section we shall describe one of the most fascinating theories in all of science—the Big Bang theory of the creation of the Universe—and the experimental evidence that supports it. This theory of cosmology states that the Universe had a beginning and, furthermore, that the beginning was so cataclysmic that it is impossible to look back beyond it. According to this theory,

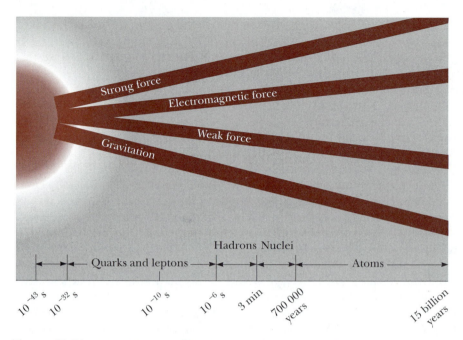

Figure 15.15   A brief history of the Universe from the Big Bang to the present. The four forces became distinguishable during the first microsecond. After that, all the quarks combined to form particles that interact via the strong force. However, the leptons remained separate and have existed as individually observable particles to this day.

the Universe erupted from a point-like singularity about 10 to 20 billion years ago. Such extremes of energy occurred in the first few minutes after the Big Bang that it is believed that all four interactions of physics were unified and that all matter melted down into an undifferentiated "quark soup."

Figure 15.15 shows the evolution of the four fundamental forces from the Big Bang to the present. During the first 10^{-43} s (the ultra-hot epoch during which $T \approx 10^{32}$ K), it is presumed that the strong, electroweak, and gravitational forces were joined to form a completely unified force. In the first 10^{-32} s following the Big Bang (the hot epoch, $T \approx 10^{29}$ K), gravity broke free of this unification while the strong and electroweak forces remained joined, as described by a grand unification theory. This was a period when particle energies were so great ($>10^{16}$ GeV) that very massive particles as well as quarks, leptons, and their antiparticles existed. Then the Universe rapidly expanded and cooled during the warm epoch, when the temperatures ranged from 10^{29} to 10^{15} K; the strong and electroweak forces parted company, and the grand unification scheme was broken. As the Universe continued to cool, the electroweak force split into the weak force and the electromagnetic force about 10^{-10} s after the Big Bang.

During the first half-hour of creation, the temperature was probably about 10^{10} K and the Universe could be viewed as a cosmic thermonuclear bomb fusing protons and neutrons into deuterium and then helium. Until about 700 000 years after the Big Bang, the Universe was dominated by radiation. Ions absorbed and re-emitted photons, thereby ensuring thermal equilibrium of

George Gamow (1904–1968) and two of his students, Ralph Alpher and Robert Herman, were the first to take the first half hour of the Universe seriously. In a mostly overlooked paper published in 1948, they made truly remarkable cosmological predictions. They correctly calculated the abundances of hydrogen and helium after the first half hour (75% H and 25% He) and predicted that radiation from the Big Bang should still be present, with an apparent temperature of about 5 K. In Gamow's own words, the Universe's supply of hydrogen and helium were created very quickly, "in less time than it takes to cook a dish of duck and roast potatoes." The comment is characteristic of this interesting physicist, who is known as much for his explanation of alpha decay and his theories of cosmology as for his delightful popular books, his cartoons, and his wonderful sense of humor. A classic Gamow story holds that, having coauthored a paper with Alpher, he made Hans Bethe an honorary author so that the credits would read "Alpher, Bethe, and Gamow" (to resemble the Greek letters α, β, and γ).

radiation and matter. Energetic radiation also prevented matter from forming clumps or even single hydrogen atoms. When the Universe was about 700 000 years old, it had expanded and cooled to about 3000 K, and protons could bind to electrons to form neutral hydrogen atoms. Because neutral atoms do not appreciably scatter photons, the Universe suddenly became transparent to photons. Radiation no longer dominated the Universe, and clumps of neutral matter steadily grew—first atoms, then molecules, gas clouds, stars, and finally galaxies. The state of affairs for radiation- and matter-dominated periods is shown in Figure 15.16.

Observation of Radiation from the Primordial Fireball

In 1965 Arno A. Penzias and Robert W. Wilson of Bell Laboratories were testing a sensitive microwave receiver and made an amazing discovery. A pesky signal that produced a faint background hiss was interfering with their satellite communications experiments. Despite their valiant efforts, the signal remained. Ultimately it became clear that they were perceiving microwave background radiation (at a wavelength of 7.35 cm) that represented the leftover glow from the Big Bang.

The microwave horn that served as their receiving antenna is shown in Figure 15.17. The intensity of the detected signal remained unchanged as the antenna was pointed in different directions. The fact that the radiation had equal strengths in all directions suggested that the entire Universe was the source of this radiation. Eviction of a flock of pigeons from the 20-foot horn and cooling

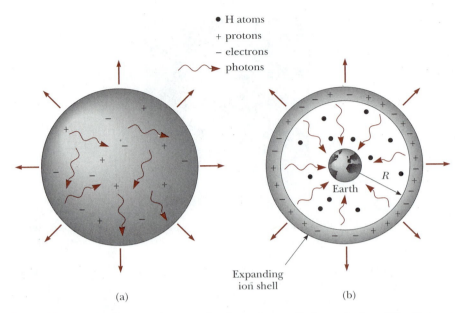

Figure 15.16   Radiation- and matter-dominated stages in the evolution of the Universe. (a) The radiation-dominated stage ($T > 3000$ K, age $< 700\,000$ years) is an expanding Universe filled with protons, electrons, neutrons, photons, and neutrinos. Ions continually scatter photons, ensuring thermal equilibrium of radiation and matter. (b) The matter-dominated state ($T < 3000$ K, age $> 700\,000$ years). Hydrogen atoms form, making the Universe transparent to photons. Photons emitted from an expanding ion shell at $R \approx 14$ billion lightyears are currently seen on Earth enormously Doppler-shifted to the red.

Figure 15.17   Robert W. Wilson *(left)* and Arno A. Penzias with the Bell Telephone Laboratories horn-reflector antenna. *(AT&T Bell Laboratories)*

the microwave detector both failed to remove the "spurious" signal. Through a casual conversation, Penzias and Wilson discovered that a group at Princeton had predicted the residual radiation from the Big Bang and were planning an experiment, seeking to confirm the theory. The excitement in the scientific community was high when Penzias and Wilson announced that they had already observed an excess microwave background that was compatible with a 3-K blackbody source.

Because Penzias and Wilson took their measurements at a single wavelength, they did not completely confirm the radiation as 3-K blackbody radiation. One can imagine the excitement in the late 1960s and 1970s as experimentalists fought obscuring atmospheric absorption and, point by point, added intensity data at different wavelengths. The results shown in Figure 15.18 and careful measurements taken at wavelengths less than 0.1 cm by the Cosmic Background Explorer Satellite (COBE) confirm that the radiation is that of a black body at 2.7 K. This figure is, perhaps, the most clear-cut evidence for the Big Bang theory. It also presents the earliest view of the Universe, a much earlier view than that available with the largest optical and radio telescopes. The 1978 Nobel prize was awarded to Penzias and Wilson for their very important discovery.

Figure 15.18 Radiation spectrum of the Big Bang. The solid line is the spectrum calculated for a black body at 2.7 K.

Spectrum Emitted by a Receding Black Body

Since the observed 3-K blackbody spectrum is the single strongest piece of evidence for the Big Bang theory of the Universe, it is important to examine this topic in more detail. In particular, we wish to show what happens to a blackbody spectrum in an expanding Universe.

Consider blackbody radiation from the Big Bang at the time when the Universe first became transparent to photons, 700 000 years after the Big Bang (ABB). Recall from Chapter 2 that the wavelength distribution for blackbody radiation is

$$u(\lambda, T) \, d\lambda = \frac{8\pi hc}{\lambda^5} \left(\frac{1}{e^{hc/\lambda k_B T} - 1} \right) d\lambda \tag{15.20}$$

where $u(\lambda, T) \, d\lambda$ is the energy per unit volume of the radiation with wavelength between λ and $\lambda + d\lambda$ emitted by a black body at temperature T. Suppose that, between then (700 000 years ABB) and now (approximately 15 000 000 000 years ABB), the Universe has expanded and a photon of original wavelength λ is now Doppler-shifted to λ', where λ' is longer than λ. Our question is: What happens to the form of the original wavelength distribution $u(\lambda, T) \, d\lambda$?

Let $f(\lambda', T') \, d\lambda'$ be the present energy per unit volume of the residual Big Bang radiation as measured from the Earth. Since the ion shell that emitted the radiation is moving away from the Earth at extremely relativistic speed (Fig. 15.16), we must use the relativistic Doppler shift for light from a receding source to relate λ' to λ. As shown in Chapter 1, Example 1.5, this connection is

$$\lambda' = \frac{\sqrt{1 + v/c}}{\sqrt{1 - v/c}} \, \lambda \tag{15.21}$$

where v is the speed of recession of the ion shell. For our purposes, Equation 15.21 may be written more simply as

$$\lambda' = F\lambda \tag{15.22}$$

which shows that the present wavelength λ' is greater than the original wavelength λ by the factor $F = \sqrt{1 + v/c}/\sqrt{1 - v/c}$. (Note that F is much greater than unity, since $v \approx c$.) To find the relation between the present blackbody spectrum $f(\lambda', T')\,d\lambda'$ and the original $u(\lambda, T)\,d\lambda$, we substitute $\lambda = \lambda'/F$ into the expression for $u(\lambda, T)\,d\lambda$:

$$u(\lambda, T)\,d\lambda = \frac{8\pi hc}{\lambda^5}\left(\frac{1}{e^{hc/\lambda k_{\mathrm{B}} T} - 1}\right)d\lambda = \frac{8\pi hc}{(\lambda'/F)^5}\left(\frac{1}{e^{hc/(\lambda'/F)k_{\mathrm{B}} T} - 1}\right)\frac{d\lambda'}{F}$$

or

$$\frac{u(\lambda, T)\,d\lambda}{F^4} = \frac{8\pi hc}{(\lambda')^5}\left(\frac{1}{e^{hc/\lambda' k_{\mathrm{B}} T'} - 1}\right)d\lambda' \tag{15.23}$$

where $T' = T/F$. We can identify the right side of Equation 15.23 as the *present* blackbody spectrum $f(\lambda', T')\,d\lambda'$ because it has the standard functional form of a blackbody spectrum with λ replaced by λ' and T replaced by T'. Thus we can write Equation 15.23 as

$$\frac{u(\lambda, T)\,d\lambda}{F^4} = f(\lambda', T')\,d\lambda' \tag{15.24}$$

where $f(\lambda', T')\,d\lambda'$ has the explicit form

$$f(\lambda', T')\,d\lambda' = \frac{8\pi hc}{(\lambda')^5}\left(\frac{1}{e^{hc/\lambda' k_{\mathrm{B}} T'} - 1}\right)d\lambda' \tag{15.25}$$

with $T' = T/F$. Equation 15.25 shows that the radiation from a receding black body still has the same spectral distribution, but its temperature T' drops by a factor of F and its energy per unit volume decreases by a factor of F^4. In a frame of reference moving with the ions that last scattered the radiation, the characteristic temperature of the radiation is about 3000 K. Equation 15.25 shows that in our frame of reference the radiation appears to have a characteristic temperature of $(3000 \text{ K})/F$, or 3 K, since we know that $F \approx 1000$.

Other Evidence for the Expanding Universe

Most of the key discoveries supporting the theory of an expanding Universe, and indirectly the Big Bang theory of cosmology, were made in the 20th century. In 1913 Vesto Melvin Slipher, an American astronomer, reported that most nebulae were receding from the Earth at speeds up to several million miles per hour. Slipher was one of the first to use the methods of Doppler shifts in spectral lines to measure velocities. The key to this method is to locate in nebular spectra the characteristic lines of some element whose wavelengths have been measured on Earth. The speed of the receding object is then directly

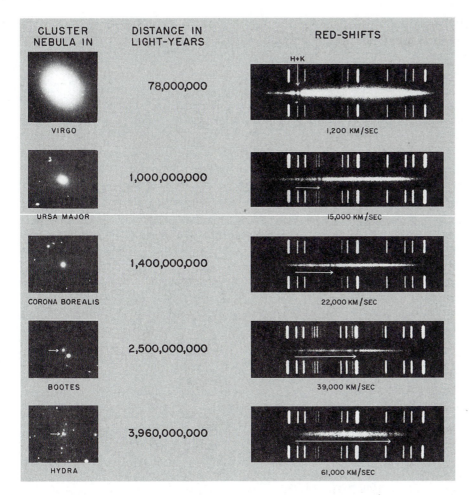

CLUSTER NEBULA IN	DISTANCE IN LIGHT-YEARS	RED-SHIFTS
VIRGO	78,000,000	1,200 KM/SEC
URSA MAJOR	1,000,000,000	15,000 KM/SEC
CORONA BOREALIS	1,400,000,000	22,000 KM/SEC
BOOTES	2,500,000,000	39,000 KM/SEC
HYDRA	3,960,000,000	61,000 KM/SEC

Figure 15.19   Red shift of the K and H lines of calcium for five galaxies. Red shifts are expressed as speeds $c(\Delta\lambda/\lambda)$. Calibration spectra taken in the laboratory appear above and below each galaxy's spectrum.

proportional to the wavelength shift to the red, as shown in Figure 15.19. Slipher's measurements of nebular speeds were quite accurate, but the distances to the nebula were virtually unknown, although there was some indication that more distant objects had higher recessional speeds.

Shortly after Slipher's discovery, a flowering of ideas occurred in theoretical cosmology. In 1916 Einstein published his general theory of relativity, in which he formulated the differential equations connecting the distribution of matter in the Universe with the curvature of space-time. Einstein erroneously solved these equations and found that in order to have a Universe static in time,[3] he had to add a "cosmic repulsion" force between galaxies. This was a rather unusual force, since it was independent of mass and grew stronger with increasing galaxy separation. In the same year, the Dutch mathematician and astronomer

---

[3] The average distance between clumps of matter on a large scale is constant with time.

Figure 15.20 Milton L. Humason. He and Edwin Hubble did the most to demonstrate the expanding Universe experimentally. The two astronomers present an interesting contrast. Humason was originally a mule-team driver and janitor at Mount Wilson. He taught himself astronomy and became a consummate experimentalist.

Willem de Sitter found another static solution to Einstein's equations. This solution predicted an increasing red shift for lines emitted by more distant objects. De Sitter's solution stirred much interest because it seemed capable of being confirmed experimentally.

A major breakthrough in the cosmological problem came in 1922 when Alexander Friedmann,[4] a Russian mathematician, found an error in Einstein's work and showed that two *nonstatic* models were possible: an expanding Universe and a contracting Universe. At first Einstein found it difficult to exchange the eternal, steady, static Universe for an expanding, dynamic Universe that presumably had a beginning in time. He wrote to de Sitter, "This circumstance of an expanding Universe irritates me" and "to admit such possibilities seems senseless."[5] However, by 1933 Einstein was finally convinced of the impeccable logic of Friedmann's arguments. Any remaining doubt about the nonstatic nature of the Universe was erased by the powerful experimental results of Edwin P. Hubble and Milton Humason on the red shifts of distant nebulae (see Figs. 15.20 and 15.21).

Hubble first established that nebulae were actually composed of billions of stars and then went on to measure their distances. By 1923 he had already determined that the Andromeda nebula was more than 1 million lightyears distant, well beyond the 100 000-lightyear extent of our own Milky Way galaxy. Hubble measured the distances to nearby galaxies, or "island Universes," by observing the brightness of stars with fluctuating intensity, called *cepheid variables*. Since the intrinsic intensity I_{int} was known from the period of light fluctuations, Hubble used the observed intensity I_{obs} and the inverse-square law to find the distance, R:

$$I_{obs} = \frac{I_{int}}{R^2} \tag{15.26}$$

In the late 1920s, Hubble made the bold assertion that the whole Universe is expanding. From 1928 to 1936, he and Humason toiled at Mount Wilson to prove this assertion until they reached the limits of the 100-inch telescope. The results of their work and its continuation on a 200-inch telescope in the 1940s showed that the speeds of galaxies increase in direct proportion to their distance R from us (Fig. 15.22). This linear relationship, known as **Hubble's law,**

---

[4] Friedmann was an amateur meteorologist and balloonist who taught the young Gamow relativity theory.

[5] There has been much scientific irritation over expanding Universe theories. Presumably the irritation derives from the fact that the Big Bang theory places the cause of creation permanently beyond the reach of physical investigation. Fred Hoyle has never accepted the Big Bang theory and prefers instead a steady-state model of the Universe in which matter is created out of nothing at a rate necessary to compensate for the increasing distances between visible galaxies. Others have expressed their irritation as follows:

   Sir Arthur Eddington: "I have no axe to grind in this discussion but the notion of a beginning is repugnant to me. . . . I simply do not believe that the present order of things started off with a bang. . . . The expanding Universe is preposterous and incredible; it leaves me cold."

   Walter Nernst: "To deny the infinite duration of time would be to betray the very foundation of science."

   Phillip Morrison: "I find it hard to accept the big bang theory. I would like to reject it."

   Allan Sandage: "It is such a strange conclusion. . . . It cannot really be true."

Figure 15.21   Edwin P. Hubble. Hubble was born to wealth, graduated from the University of Chicago, and was a Rhodes Scholar, lawyer, athlete, soldier, and astronomer with an unerring eye for important problems in astronomy.

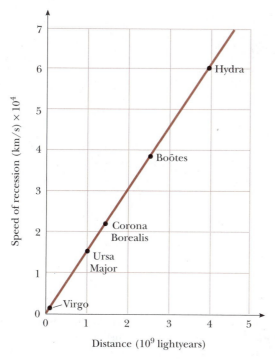

Figure 15.22   Hubble's law: The speed of recession is directly proportional to distance. The five galaxies shown in Figure 15.19 are plotted here.

may be written

$$v = HR \qquad (15.27)$$

where H, called the **Hubble parameter,** has the approximate value

$$H = 17 \times 10^{-3} \text{ m/(s·lightyear)}$$

with an uncertainty of about 50%.

Will the Universe Expand Forever?

In the 1950s and 1960s Allan R. Sandage used the 200-inch telescope at Mount Palomar to measure the speeds of galaxies at distances up to 6 billion lightyears. His measurements showed that these very distant galaxies were moving about 10 000 km/s faster than the Hubble law predicted. According to this result, the Universe must have been expanding more rapidly 1 billion years ago, and consequently the expansion is slowing (Fig. 15.23).[6] Today astronomers and physicists are trying to determine the rate of slowing. If the average mass density of the Universe is less than some critical density ($\rho_c \approx 4$ hydrogen atoms/m^3), the galaxies will slow in their outward rush but still escape to infinity. If the average density exceeds the critical value, the expansion will eventually stop and contraction will begin, possibly leading to a superdense state and another expansion or an oscillating Universe.

[6] The data at large distances have large observational uncertainties and may be systematically in error from selection effects such as abnormal brightness in the most distant visible clusters.

Figure 15.23   Red shift, or speed of recession, versus apparent magnitude, or distance, of 18 faint clusters. Curve *A* is the trend suggested by the six faintest clusters of galaxies. Line *C* corresponds to a Universe with a constant rate of expansion. If the data fall between *B* and *C*, the expansion slows but never stops. If the data fall to the left of *B*, expansion stops and contraction occurs.

EXAMPLE 15.6   Recession of a Quasar

A quasar is a star-like object that is very distant from the Earth. Its speed can be measured from Doppler-shift measurements of the light it emits. A certain quasar recedes from the Earth at a speed of $0.55c$. How far away is it?

Solution   We can find the distance with Hubble's law:

$$R = \frac{v}{H} = \frac{(0.55)(3.00 \times 10^8 \text{ m/s})}{17 \times 10^{-3} \text{ m/(s·lightyear)}}$$

$$= 9.7 \times 10^9 \text{ lightyears}$$

Exercise 4   Assuming that the quasar has moved with the speed $0.55c$ ever since the Big Bang, estimate the age of the Universe.

Answer $t = R/v = 1/H \approx 18$ billion years (in reasonable agreement with other calculations)

EXAMPLE 15.7   Critical Density of the Universe

Estimate the critical mass density of the Universe, ρ_c, using energy considerations.

Solution   Figure 15.24 shows a large section of the Universe with radius R, containing galaxies with a total mass M. A galaxy of mass m and speed v at R will just escape to infinity with zero speed if the sum of its kinetic energy and gravitational potential energy is zero. Thus

$$E_{\text{total}} = 0 = K + U = \tfrac{1}{2}mv^2 - \frac{GmM}{R}$$

$$\tfrac{1}{2}mv^2 = \frac{Gm\frac{4}{3}\pi R^3\rho_c}{R}$$

$$(1) \qquad v^2 = \frac{8\pi G}{3} R^2\rho_c$$

Because the galaxy of mass m obeys the Hubble law, $v = HR$, Equation 1 becomes

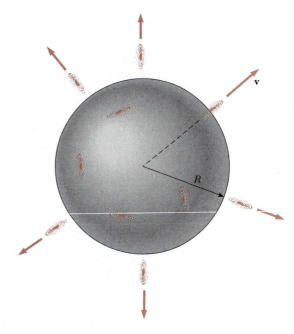

Figure 15.24   (Example 15.7) A galaxy escaping from a large cluster contained within radius R. Only the mass within R slows the mass m.

$$H^2 = \frac{8\pi G}{3} \rho_c$$

$$(2) \qquad \rho_c = \frac{3H^2}{8\pi G}$$

Using $H = 17 \times 10^{-3}$ m/(s·lightyear), where 1 lightyear = 9.46×10^{15} m, and $G = 6.67 \times 10^{-11}$ N·m^2/kg^2 yields $\rho_c = 5.8 \times 10^{-27}$ kg/m^3. As the mass of a hydrogen atom is 1.67×10^{-27} kg, ρ_c corresponds to about 4 hydrogen atoms per cubic meter.

Missing Mass in the Universe (?)

The visible matter in galaxies averages out to 5×10^{-30} kg/m^3. The radiation in the Universe has a mass equivalent of approximately 2% of the visible matter. Nonluminous matter (such as interstellar gas or black holes) can be estimated from the speeds of galaxies orbiting each other in the cluster. The higher the galaxy speeds, the more mass in a galaxy cluster. Results from measurements on the Coma cluster of galaxies indicate, surprisingly, that the amount of invisible matter is 20 to 30 times the amount present in stars and luminous gas clouds. Yet even this large invisible component, if applied to the Universe as a whole, leaves the observed mass density a factor of 10 less than ρ_c. This so-called *missing mass* (or *dark matter*) has been the subject of intense theoretical and experimental work. Exotic massive particles supposedly left over from the early moments of the Universe, such as the photino (mass $\approx 100 \text{ GeV}/c^2$), have been suggested as candidates for the missing mass. More mundane proposals have been that the missing mass is present in galaxies as neutrinos. In fact, neutrinos are so abundant that a tiny neutrino rest mass on the order of 20 eV could furnish the missing mass and "close" the Universe.

Although we are a bit more sure about the beginning of the Universe, we are uncertain about its end. Will the Universe expand forever? Will it collapse and repeat its expansion in an endless series of oscillations? Results and answers to these questions remain inconclusive, and the exciting controversy continues.

15.13   PROBLEMS AND PERSPECTIVES

While particle physicists have been exploring the realm of the very small, cosmologists have been exploring cosmic history back to the first instants of the Big Bang. Observation of events that occur when two particles collide in an accelerator is essential to a reconstruction of the early moments in cosmic history. Perhaps the key to understanding the early Universe is to first understand the world of elementary particles. Cosmologists and physicists now find that they have many common goals and are joining hands to attempt to understand the physical world at its most fundamental level.

Our understanding of physics at short distances is far from complete. Particle physics is faced with many questions. Why is there so little antimatter in the Universe? Do neutrinos have a small rest mass, and if so, how do they contribute to the "dark matter" of the Universe? Is it possible to unify the strong and electroweak theories in a logical and consistent manner? Why do quarks and leptons form three similar but distinct families? Are muons and taus the same as electrons (apart from their different masses), or do they have other subtle differences that have not been detected? Why are some particles charged and others neutral? Why do quarks carry a fractional charge? What determines the masses of the fundamental leptons and quarks? Can isolated quarks exist? The questions go on and on. Given the rapid advances and new discoveries in the field of particle physics, some of these questions may be resolved by the time you read this book.

An important and obvious question that remains is whether leptons and quarks have a substructure. If they have a substructure, we can envision an infinite number of deeper structural levels. However, if leptons and quarks are indeed the ultimate constituents of matter, as physicists today tend to believe, we should be able to construct a final theory of the structure of matter, as Einstein dreamed. In the view of many physicists, the end of the road is in sight, but how long it will take to reach that goal is anyone's guess.

SUMMARY

There are four fundamental forces in nature: strong, electromagnetic, weak, and gravitational. The strong, or color force is the force between quarks that binds them into neutrons, protons, and other hadrons. The weak force is responsible for beta decay. The electromagnetic and weak forces are now considered to be manifestations of a single force called the electroweak force. Every fundamental interaction is said to be transmitted or mediated by the exchange of field particles. The electromagnetic interaction is mediated by the photon; the weak interaction is mediated by the $W^\pm$ and Z^0 bosons; the gravitational interaction is mediated by gravitons; the strong interaction is mediated by gluons.

An antiparticle and a particle have the same mass but opposite charge, and other properties may also have opposite values, such as lepton number and baryon number.

Matter particles are classified as hadrons or leptons. **Hadrons** interact through the strong force. They have size and structure and are not elementary particles. There are two types of hadrons, *baryons* and *mesons*. Mesons have baryon number zero and have either zero or integral spin. Baryons, which generally are the most massive particles, have nonzero baryon numbers and spins of $\frac{1}{2}$ or $\frac{3}{2}$. The neutron and proton are examples of baryons.

Leptons have no structure or size and are considered truly elementary. They interact only through the weak and electromagnetic forces. There are six leptons: the electron, e^-; the muon, μ^-; the tau, τ^-; and their neutrinos, ν_e, ν_μ, and ν_τ.

In all reactions and decays, quantities such as energy, linear momentum, angular momentum, electric charge, baryon number, and lepton number are strictly conserved. Conservation of relativistic energy and momentum yields the following expression of the threshold energy for particle production:

$$K_{th} = \frac{(m_3 + m_4 + m_5 + \cdots)^2 c^2 - (m_1 + m_2)^2 c^2}{2m_2}$$

where K_{th} is the kinetic energy of the incident particle of mass m_1, m_2 is the mass of the stationary-target particle, and the remaining m's are product particle masses.

Certain particles have properties called **strangeness** and **charm.** These unusual properties are conserved only in the reactions and decays that occur via the strong force.

Theories in elementary particle physics have postulated that all hadrons are composed of smaller units known as **quarks.** Quarks have fractional electric charge and baryon numbers of $\frac{1}{3}$, and they come in six "flavors": up (u), down (d), strange (s), charmed (c), top (t), and bottom (b). Each baryon contains three quarks, and each meson contains one quark and one antiquark.

According to the theory of **quantum chromodynamics,** quarks have a property called **color,** and the strong force between quarks is referred to as the **color force.**

The background microwave radiation observed by Penzias and Wilson strongly suggests that the Universe started with a Big Bang about 15 billion years ago. The background radiation is equivalent to that of a black body at approximately 3 K.

Astronomical measurements show that the Universe is expanding. According to **Hubble's law,** distant galaxies are receding from the Earth at a speed $v = HR$ where R is the distance from Earth to the galaxy and H is Hubble's constant, which is approximately 17×10^{-3} m/(s·lightyear).

SUGGESTIONS FOR FURTHER READING

Particle Physics

1. J. Bahcall, "The Solar Neutrino Problem," *Sci. American*, May 1990.
2. H. Breuker et al., "Tracking and Imaging Elementary Particles," *Sci. American*, August 1991.
3. F. Close, *The Cosmic Onion: Quarks and the Nature of the Universe*, The American Institute of Physics, 1986. A timely monograph on particle physics, including lively discussions of the Big Bang theory.
4. H. Fritzsch, *Quarks, The Stuff of Matter*, London, Allen and Lane, 1983. An excellent introductory overview of elementary particle physics.
5. S. Glashow, *Interactions: A Journey Through the Mind of a Particle Physicist*, New York, Warner Books, 1988.
6. H. Harari, "The Structure of Quarks and Leptons," *Sci. American*, April 1983.
7. L. M. Lederman, "The Value of Fundamental Science," *Sci. American*, November 1984.
8. N. B. Mistry, R. A. Poling, and E. H. Thorndike, "Particles with Naked Beauty," *Sci. American*, July 1983.
9. C. Quigg, "Elementary Particles and Forces," *Sci. American*, April 1985.
10. M. Riordan, "The Discovery of Quarks," *Science* 29, May, 1992.
11. M. J. G. Veltman, "The Higgs Boson," *Sci. American*, November 1986.
12. E. Witten, "Reflections on the Fate of Spacetime," *Physics Today*, April 1996. This article discusses the string theory of elementary particles.

Cosmology and the Big Bang

13. G. Gamow, "The Evolutionary Universe," *Sci. American*, September 1956.
14. G. Gamow, "Gravity," *Sci. American*, March 1961.
15. T. Goldman et al., "Gravity and Antimatter," *Sci. American*, March 1988.
16. D. Layzer, *Constructing the Universe*, Scientific American Library, New York, W. H. Freeman and Co., 1984, Chapters 7 and 8.
17. R. A. Muller, "The Cosmic Background Radiation and the New Aether Drift," *Sci. American*, May 1978.
18. A. R. Sandage, "The Red-Shift," *Sci. American*, September 1956.
19. C. Sagan and F. Drake, "The Search for Extraterrestrial Intelligence," *Sci. American*, May 1975.
20. F. Wilczek, "The Cosmic Asymmetry Between Matter and Anti-Matter," *Sci. American*, December 1980.

QUESTIONS

1. Name the four fundamental interactions and the field particle that mediates each.
2. Describe the quark model of hadrons, including the properties of quarks.
3. What are the differences between hadrons and leptons?
4. Describe the properties of baryons and mesons and the important differences between them.
5. Particles known as resonances have very short lifetimes, of the order of 10^{-23} s. From this information, would you guess they are hadrons or leptons? Explain.
6. Kaons all decay into final states that contain no protons or neutrons. What is the baryon number of kaons?
7. The Ξ^0 particle decays by the weak interaction according to the decay mode $\Xi^0 \rightarrow \Lambda^0 + \pi^0$. Would you expect this decay to be fast or slow? Explain.
8. Identify the particle decays listed in Table 15.2 that occur by the weak interaction. Justify your answers.
9. Identify the particle decays listed in Table 15.2 that occur by the electromagnetic interaction. Justify your answers.
10. Two protons in a nucleus interact via the strong interaction. Are they also subject to the weak interaction?
11. Discuss the following conservation laws: energy, linear momentum, angular momentum, electric charge, baryon number, lepton number, and strangeness. Are all of these laws based on fundamental properties of nature? Explain.
12. An antibaryon interacts with a meson. Can a baryon be produced in such an interaction? Explain.
13. Describe the essential features of the Standard Model of particle physics.
14. How many quarks are in (a) a baryon? (b) an antibaryon? (c) a meson? (d) an antimeson? How do you account for the fact that baryons have half-integral spins while mesons have spins of 0 or 1? (*Hint:* Quarks have spin $\frac{1}{2}$.)
15. In the theory of quantum chromodynamics, quarks come in three colors. How would you justify the statement that "all baryons and mesons are colorless"?
16. Which baryon did Murray Gell-Mann predict in 1961? What is the quark composition of this particle?
17. What is the quark composition of the Ξ^- particle? (See Table 15.4.)
18. The $W^{\pm}$ and Z bosons were first produced at CERN in 1983 (by the collision of a beam of protons and a beam of antiprotons at high energy). Why was this an important discovery?
19. How did Edwin Hubble (in 1928) determine that the Universe is expanding?

PROBLEMS

15.1 Fundamental Forces in Nature

1. When a high-energy proton or pion traveling near the speed of light collides with a nucleus, it travels an average distance of 3×10^{-15} m before interacting. From this information, estimate the time required for the strong interaction to occur.

2. The neutral ρ^0 meson decays by the strong interaction into two pions according to $\rho^0 \rightarrow \pi^+ + \pi^-$, with a half-life of about 10^{-23} s. The neutral K^0 meson also decays into two pions according to $K^0 \rightarrow \pi^+ + \pi^-$ but with a much longer half-life, about 10^{-10} s. How do you explain these observations?

15.2 Positrons and Other Antiparticles

3. Two photons are produced when a proton and an antiproton annihilate each other. What are the minimum frequency and the corresponding wavelength of each photon?

4. A photon produces a proton-antiproton pair at rest according to the reaction $\gamma \rightarrow p + \bar{p}$. If a nearby nucleus of mass 100 u (initially at rest) carries off the photon's momentum, what is the frequency of the photon? What is its wavelength?

15.3 Mesons and the Beginning of Particle Physics

5. One of the mediators of the weak interaction is the Z^0 boson, which has a mass of 91 GeV/c^2. Use this information to find an approximate value for the range of the weak interaction.

15.5 Conservation Laws

6. High-energy muons occasionally collide with electrons and produce two neutrinos according to the reaction $\mu^+ + e^- \rightarrow 2\nu$. What kind of neutrinos are these?

7. (a) Show that baryon number and charge are conserved in the following reactions of a pion with a proton.

$$\pi^- + p \longrightarrow K^- + \Sigma^+ \tag{1}$$

$$\pi^- + p \longrightarrow \pi^- + \Sigma^+ \tag{2}$$

(b) The first reaction is observed, but the second never occurs. Explain these observations.

8. The following reactions or decays involve one or more neutrinos. Supply the missing neutrinos (ν_e, ν_μ, or ν_τ).
   (a) $\pi^- \rightarrow \mu^- + ?$
   (b) $K^+ \rightarrow \mu^+ + ?$
   (c) $? + p \rightarrow n + e^+$
   (d) $? + n \rightarrow p + e^-$
   (e) $? + n \rightarrow p + \mu^-$
   (f) $\mu^- \rightarrow e^- + ? + ?$

9. Determine which of the following reactions can occur. For those that cannot occur, determine the conservation law (or laws) that each violates.
   (a) $p \rightarrow \pi^+ + \pi^0$

(b) $p + p \rightarrow p + p + \pi^0$
(c) $p + p \rightarrow p + \pi^+$
(d) $\pi^+ \rightarrow \mu^+ + \nu_\mu$
(e) $n \rightarrow p + e^- + \bar{\nu}_e$
(f) $\pi^+ \rightarrow \mu^+ + n$

15.6 Strange Particles and Strangeness

10. Determine whether strangeness is conserved in each of the following decays and reactions.
    (a) $\Lambda^0 \rightarrow p + \pi^-$
    (b) $\pi^- + p \rightarrow \Lambda^0 + K^0$
    (c) $\bar{p} + p \rightarrow \bar{\Lambda}^0 + \Lambda^0$
    (d) $\pi^- + p \rightarrow \pi^- + \Sigma^+$
    (e) $\Xi^- \rightarrow \Lambda^0 + \pi^-$
    (f) $\Xi^0 \rightarrow p + \pi^-$

11. The following decays are forbidden. Determine a conservation law that each violates.
    (a) $\mu^- \rightarrow e^- + \gamma$
    (b) $n \rightarrow p + e^- + \nu_e$
    (c) $\Lambda^0 \rightarrow p + \pi^0$
    (d) $p \rightarrow e^+ + \pi^0$
    (e) $\Xi^0 \rightarrow n + \pi^0$

12. The following reactions are forbidden. Determine a conservation law that each violates.
    (a) $p + \bar{p} \rightarrow \mu^+ + e^-$
    (b) $\pi^- + p \rightarrow p + \pi^+$
    (c) $p + p \rightarrow p + \pi^+$
    (d) $p + p \rightarrow p + p + n$
    (e) $\gamma + p \rightarrow n + \pi^0$

15.7 How Are Elementary Particles Produced and Particle Properties Measured?

13. In 1959 Emilio Segrè and Owen Chamberlain were awarded the Nobel prize for demonstrating the existence of the antiproton. In a series of experiments started in 1955, using the Bevatron accelerator at Berkeley, they actually produced both antiprotons and antineutrons in the reactions

$$p + p \longrightarrow p + p + p + \bar{p}$$

$$p + p \longrightarrow p + p + n + \bar{n}$$

(a) Calculate the threshold kinetic energy of the incident proton (fixed-target proton) required for the production of an antiproton, $\bar{p}$. (b) For the same initial conditions, calculate the threshold kinetic energy needed for the production of an antineutron, $\bar{n}$.

14. Calculate the threshold for production of strange particles in the following reactions. Assume that the first particle is moving and the second is at rest.

$$\text{(a)} \quad p + p \longrightarrow n + \Sigma^+ + K^0 + \pi^+$$

$$\text{(b)} \quad \pi^- + p \longrightarrow \Sigma^0 + K^0$$

15. *Efficiency of fixed-target accelerators vs. colliding-beam accelerators.* The efficiency is the percentage of the initial kinetic energy that is converted to the mass of new product particles in a given reaction. Calculate the efficiencies of the reactions in Example 15.5, Exercise 3, and Problem 13. If one could arrange these experiments so that the two colliding protons approached each other with equal speed at threshold, the total momentum would be zero and none of the initial kinetic energy would go into kinetic energy of the products. All of the initial kinetic energy would then go into creating new mass, and the efficiency would be 100%. That is why most current experiments involving the production of heavy particles are carried out with colliding-beam accelerators. Although there are formidable experimental difficulties in storing, focusing, and causing two oppositely circulating low-density beams of particles to collide, the great gain in efficiency makes colliders worthwhile.

16. Consider the reaction $p + p \rightarrow p + p + X$. (a) For a fixed-target accelerator capable of producing incident protons with a kinetic energy of 1000 GeV, find the heaviest particle X that can be produced. (b) If colliding protons, each with a kinetic energy of 500 GeV, are available in a collider, what is the mass of the heaviest particle X that can be produced in the same reaction?

15.9   Quarks

17. A Σ^0 particle traveling through matter strikes a proton, and a Σ^+ and a gamma ray, as well as a third particle, emerge. Use the quark models of the Σ^+ and the gamma ray to determine the identity of the third particle.

18. The quark compositions of the K^0 and Λ^0 particles are $d\bar{s}$ and uds, respectively. Show that the charge, baryon number, and strangeness of these particles equal the sums of these numbers for the quark constituents.

19. Neglecting binding energies, estimate the masses of the u and d quarks from the masses of the proton and neutron.

20. Figure P15.20 shows a neutron-proton interaction in which a virtual π^+ is exchanged. Draw a Feynman diagram showing this interaction in terms of quarks and gluons.

21. A neutron undergoing beta decay is believed to emit a virtual W^- that decays into an e^- and an antielectron neutrino, $\bar{\nu}_e$. Draw a Feynman diagram showing this decay in terms of quarks, the W^-, the e^-, and the $\bar{\nu}_e$.

15.12   The Cosmic Connection

22. Using Hubble's law (Eq. 15.27), estimate the wavelength of the 590-nm sodium line emitted from galaxies (a) 2×10^6 lightyears away from Earth, (b) 2×10^8 lightyears away, and (c) 2×10^9 lightyears away. (*Hint:* Use the relativistic Doppler formula for wavelength λ' of light emitted from a receding source):

$$\lambda' = \lambda \sqrt{\frac{1 + v/c}{1 - v/c}}$$

23. A distant quasar is moving away from Earth at such high speed that the blue 434-nm hydrogen line is observed at 650 nm, in the red portion of the spectrum. (a) How fast is the quasar receding? (See the hint in the preceding problem.) (b) Using Hubble's law, determine the distance from Earth to this quasar.

Additional Problems

24. An unstable particle, initially at rest, decays into a proton (rest energy 938.3 MeV) and a negative pion (rest energy 139.5 MeV). A uniform magnetic field of 0.250 T exists perpendicular to the velocities of the created particles. The radius of curvature of each track is found to be 1.33 m. What is the rest mass of the original unstable particle?

25. Calculate the kinetic energies of the proton and pion resulting from the decay of a Λ^0 at rest:

$$\Lambda^0 \longrightarrow p^+ + \pi^-$$

26. A Σ^0 particle at rest decays according to

$$\Sigma^0 \longrightarrow \Lambda^0 + \gamma$$

Find the gamma-ray energy.

27. If a K^0 meson at rest decays in 0.90×10^{-10} s, how far will a K^0 meson travel if it is moving at $0.96c$ through a bubble chamber?

28. A pi meson at rest decays according to $\pi^- \rightarrow \mu^- + \bar{\nu}_\mu$. What is the energy carried off by the neutrino? (Assume that the neutrino moves off with the speed of light.) $m_\pi c^2 = 139.5$ MeV, $m_\mu c^2 = 105.7$ MeV, $m_\nu = 0$.

29. Two protons approach each other with 70.4 MeV of kinetic energy and engage in a reaction in which a proton and positive pion emerge at rest. What third particle, obviously uncharged and therefore difficult to detect, must have been created?

Figure P15.20

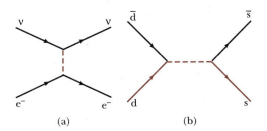

Figure P15.30

30. What processes are described by the Feynman diagrams in Figure P15.30? What is the exchanged particle in each process?

31. Identify the mediators for the two interactions described in the Feynman diagrams in Figure P15.31.

32. The Λ^0 is an unstable particle that decays into a proton and a negatively charged pion. Determine the kinetic energies of the proton and pion if the Λ^0 is at rest when it decays. The rest mass of the Λ^0 is 1115.7 MeV/c^2, the mass of the π^- is 139.5 MeV/c^2, and the mass of the proton is 938.3 MeV/c^2.

33. A free neutron beta-decays by creating a proton, an electron, and an antineutrino according to the reaction $n \rightarrow p + e^- + \bar{\nu}$. Assume that a free neutron beta-decays by creating a proton and electron according to the reaction

$$n \longrightarrow p + e^-$$

and assume that the neutron is initially at rest in the laboratory. (a) Determine the energy released in this reaction. (b) Determine the speeds of the proton and electron after the reaction. (Energy and momentum are conserved in the reaction.) (c) Are any of these particles moving at relativistic speeds? Explain.

34. The quark composition of the proton is uud, while that of the neutron is udd. Show that the charge, baryon number, and strangeness of these particles equal the sums of those same numbers for their quark constituents.

Figure P15.31

Essay

HOW TO FIND A TOP QUARK

Melissa Franklin

Harvard University

David Kestenbaum

Harvard University

I n 1932 Carl Andersen observed a new particle in his cosmic ray experiment at Pike's Peak. This was the muon, the first in a second generation of leptons. "Who ordered that?," exclaimed the Nobel prize winner I. I. Rabi. Nature had spoken, and the theory had to listen. The Standard Model was expanded to include a second lepton generation.

In 1977 when the discovery of the bottom quark ushered in the addition of a third family of quarks, the Standard Model demanded a "top" quark to balance it. This time the new addition had been "ordered," but would nature deliver? In 1994, when the top quark was finally sighted, it marked not only a great triumph for the theory, but also a remarkable achievement for modern detectors and accelerators. This is the story of the accelerator and detector at Fermi National Accelerator Laboratory ("Fermilab") that brought you the top quark.

THE ACCELERATOR

In quantum mechanics a particle is described as a wave with wavelength inversely proportional to the particle momentum. Using waves with smaller and smaller wavelengths enables scientists to probe smaller and smaller structures. Therefore, being able to accelerate particles up to large momenta enables us to probe small structures such as the proton. In order to create new high-mass particles in the laboratory, we need high-energy particle beams, since the available energy to create new states is approximately $\sqrt{2mE}$ in a fixed target experiment, or $2E$ for a collider experiment, where E is the energy of the particle beam and m is the mass of the target particle. For these two reasons it is necessary to understand techniques used to accelerate charged particles.

A particle accelerator has two main elements, the beam and the accelerating structure. All acceleration techniques use only electromagnetic fields. Particles are usually accelerated in evacuated pipes to reduce scattering. The accelerating fields must be carefully controlled, and methods for aligning the beam are necessary. Finally, since the magnets get very hot, they need to be cooled.

There are different types of accelerating techniques. One uses a DC electric field, and another uses either standing or traveling radio frequency (RF) waves to accelerate particles. The DC technique is limited by arcing. It is used for initial acceleration of particles, followed by the RF wave technique, which leads to bunched beams.

The source of charged particles for the beam is either a hot metal filament for electrons or a bottle of hydrogen for protons. The particles must be guided down the beam pipe. This involves steering or bending the beam, focusing it, and breaking it up to form bunches of particles rather than a continuous beam.

The proton accelerator at Fermilab consists of many parts (Fig. 1). First, hydrogen is ionized to form H⁻ ions which are then introduced into an accelerating structure ("PreAcc" in Fig. 1) which produces a DC electric field of 750 kV. After the ions are accelerated in this field, they enter a linear accelerator (LINAC) which uses 201 MHz–805 MHz RF waves supplied to accelerating cavities. The RF signal produces an alternating electric field in a cavity pointed first along the beam and then opposite to it. The ions are accelerated in the forward direction by shielding them from the RF signal in drift tubes while the **E** field is of the wrong sign. The ions are accelerated down the 145-m long LINAC to an energy of 400 MeV. The particles introduced into the first accelerating cavity do not all arrive simultaneously and have a spread of energies. A particle arriving with exactly the right energy and at the right time will be accelerated to the desired energy. Because of the alternating nature of the electric field, particles

An aerial view of the Fermi National Accelerator Laboratory, Batavia, Illinois. The largest circle is the main accelerator. Three experimental lines extend at a tangent from the accelerator. The 16-story twin-towered Wilson Hall is seen at the base of the experimental lines. *(Fermilab Photo Dept.)*

arriving slightly after or before the "ideal" particle will be accelerated in such a way that they tend toward the ideal particle trajectory. This feature is called *phase stability* and is a necessary feature for our accelerators. Both electrons in the H^- ion are then stripped off by a foil, and the protons are introduced into a circular accelerator called the Synchrotron Booster.

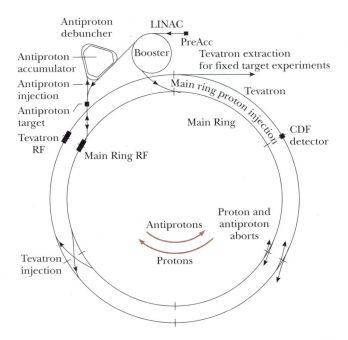

Figure 1   Fermilab accelerator complex.

The Booster's structure also uses RF electromagnetic fields to accelerate the protons by 550 keV per turn. The protons, kept inside the ring by magnetic fields, are now traveling at a speed of $0.37c$. Each time around the ring, the protons pass through accelerating cavities and pick up energy. Dipole magnets (shown schematically in Fig. 2) are used to bend the particles in a circle. The magnetic fields in these magnets are increased from 0.05 to 0.7 T as the particles change energy from 400 MeV to 8 GeV in such a way that the protons stay in the same orbit in the circular beam pipe. When their energy reaches 8 GeV, the protons are extracted using fast-rising magnetic fields and introduced into the 2000-m diameter main proton synchrotron. The 6000-m tunnel houses two accelerating structures, one atop the other. The first, called the Main Ring, built in 1969 with conventional dipole magnets whose field ranges from 0.04 to 0.67 T, can accelerate protons up to 400 GeV.

The superconducting magnet accelerator, the Tevatron, built in 1979, makes use of superconducting magnets that produce strong, uniform magnetic fields. The field in the superconducting dipole magnets ranges from 0.66 to 4.4 T, thus enabling the Tevatron to accelerate particles to 900 GeV in the same radius as the Main Ring. In general, we can describe the relationship between accelerator radius, energy achieved, and magnetic field as follows:

$$\text{Radius} = 44 \frac{E(\text{GeV})}{B(\text{kG})} \, m$$

This assumes that 75% of the ring is filled with dipole magnets. In the case of the Tevatron, there are 774 dipoles, each 6 m long, comprising 74% of the Tevatron cir-

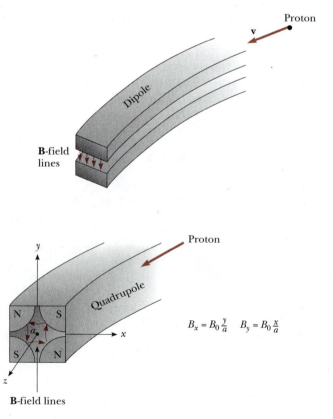

$$B_x = B_0 \frac{y}{a} \quad B_y = B_0 \frac{x}{a}$$

Figure 2   A dipole magnet in the synchrotron exerts a force on the proton toward the center of the ring. A quadrupole magnet exerts a force on protons traveling off axis.

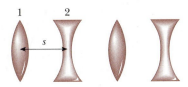

Figure 3 Strong Focusing using quadrupole magnets separated by a distance s and rotated 90° in series. Quadrupole magnets in series with equal and opposite focal lengths f have a total focal length given by $1/f_{12} = s/f^2$.

cumference. This shows that the limitation for accelerators is either building magnets with a high uniform magnetic field or having enough money to make an accelerator with a huge radius.

The synchrotron would be unable to keep the protons inside the beampipe during acceleration without a method of focusing the beam. The concept of Strong Focusing developed in 1952 is crucial. Quadrupole magnets shown in Figure 2 have zero field on axis, but have a field gradient off axis. As a particle wanders away from the axis in the x direction, it experiences a force that pushes it back toward the axis. A particle that wanders off in the y direction experiences a force away from the axis. Strong focusing relies on a series of focusing and defocusing magnets that have an overall effect of focusing the beam, provided the magnets are spaced sufficiently closely together, as shown in Figure 3. The field gradients in the quadrupoles in the Tevatron reach as high as 67 T/m. The synchrotron consists of strings of magnets, something like F-B-B-B-B-D-B-B-B-B-F etc., where F stands for focusing, B stands for bending, and D stands for defocusing magnets.

Particles in the Tevatron are accelerated by RF cavities driven at 53 MHz. The protons are accelerated through a beampipe maintained at a vacuum of 10^{-9} torr to a final energy of 900 ± 0.72 GeV.

The tunnel of the main accelerator at Fermilab. The upper ring of magnets is the 400-GeV accelerator. The lower ring are superconducting magnets for the Tevatron. *(Fermilab Photo Dept.)*

The top quark was discovered at a proton-antiproton collider. The antiprotons are produced by extracting the proton beam at 120 GeV and shining it on a copper target. Approximately 2.5×10^{12} protons are extracted from the Main Ring every 2.4 s, and 3×10^7 antiprotons are produced with a distribution of energy peaked at 8.9 GeV. A smaller 250-m radius ring first debunches the antiprotons into a continuous beam and then "cools" the antiprotons until they form an intense, monoenergetic and focused beam. The method used, called stochastic cooling, was invented by Simon Van der Meer in 1972. The antiprotons are cooled and stored in the accumulator, a ring concentric with the debuncher held at a vacuum of 2×10^{-10} torr for a few hours until enough antiprotons are accumulated for reentry into the Main Ring.

The antiprotons are introduced into the Main Ring in bunches approximately 50 cm long and are accelerated to 900 GeV in the same accelerating structure used for accelerating the protons. Bunches of roughly 3×10^{11} protons and 1×10^{11} antiprotons circulating in opposite directions in the Tevatron collide at two interaction regions containing the particle detectors every 3.5 μs.

THE DETECTOR

Because the mass of the top quark is almost 200 times the mass of a proton, it is very rarely produced, even in collisions where the center of mass energy is 1.8 TeV. Only a few top quark pairs are produced from a typical 50 billion collisions per day. The Standard Model predicts that the top quark (with a lifetime of 10^{-24} s) should decay in a manner similar to the familiar β-decay which converts a neutron to a proton by emitting a W boson which then decays to an electron and neutrino. Each top quark decays to a W and a b-quark. We focus on the case where one W decays either to a $\mu\nu_{\mu}$ or $e\nu_{e}$ and the other W decays to a quark and an antiquark, for example, an up quark and an antidown quark (Fig. 4). Quarks are confined to hadronic bound states; hence the outgoing quarks in this interaction manifest themselves as collimated "jets" of hadrons.

To find the top quark pairs produced, the detector must then be able to identify electrons, muons, neutrinos, light quark jets, and b-quark jets and to measure their momentum and energy.

All detectors function by registering the effects of energy lost by particles as they pass through matter. Fortunately electrons, hadrons, jets, neutrinos, and muons all behave quite differently, which allows us to identify each type.

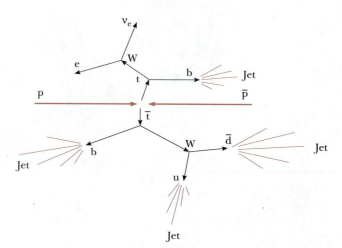

Figure 4   Schematic of top quark pair production and subsequent top quark decays.

Figure 5   Schematic of a quadrant of the CDF detector.

The experiment that first observed the production of top quarks, called the Collider Detector at Fermilab (CDF), was designed for a variety of physics measurements. The detector surrounds the point where the protons and antiprotons collide (called the interaction point), and measures the momentum, energy, and identity of the particles produced in the collision. Figure 5 is a schematic view of one quadrant of the CDF detector. To measure the momentum of high-energy particles, we require a large region with a **B** field to measure their curvature. To measure their energy, we use a calorimeter that contains enough material to bring the particles to rest. Hence, detectors must be large and heavy; CDF, which weighs over 6000 tons, is 12 m high and 32 m long. From each collision, it records over 50 000 pieces of information. CDF, like most collider detectors, surrounds the collision point with tracking chambers, followed by calorimeters and muon chambers.

Charged Particle Tracking

The momentum of charged particles emerging from a collision can be measured in a drift chamber. In its simplest form, a drift chamber is a gas-filled rectangular tube with a high voltage wire running along its axis, similar to a Geiger counter. The wire (called a "sense" wire) is held at a high potential relative to the tube to create an electric field. By adding wires of different voltages, the field can be made roughly constant over the width of the tube. When a charged particle passes through the tube, it ionizes some of the atoms in the gas. The electrons from ionization drift in the field toward the sense wire (Fig. 6). With a suitably chosen electric field strength and gas, the drift velocity can be made roughly constant. Near the wire, the electric field strength increases proportional to $1/r$. In this region the electrons gain enough energy to ionize more atoms in the gas. This process continues, producing an "avalanche" of electrons that appears as a small electronic pulse.

By measuring the drift time, we can determine the position of the original particle. Drift velocities are typically 5 cm/μs, and although the drift time can be measured to less than 1 ns, the position resolution (typically 250 μm) is dominated by variations in the arrival time of the avalanche as a result of collisions with the gas molecules. A single planar drift chamber measures position in one dimension, but if a drift chamber with

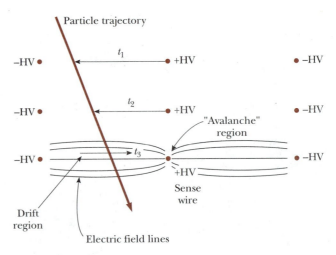

Figure 6   Schematic of a planar drift chamber showing **E** field lines, avalanche region, electron drift region, sense wires (+HV), incident particle trajectory and the times measured between traversal of the particle in each cell and avalanche at each sense wire.

two planes of wires is used (one, for example, with its sense wire along the x direction, the other along the y direction) the particle's position can be reconstructed in two dimensions. If a series of drift chambers are placed in a magnetic field, the particle's momentum can be determined by measuring the curvature of its track.

CDF uses a 1-m-radius, 3-m-long cylindrical drift chamber surrounding the collision point, containing over 4000 sense wires. The wires are grouped into nine layers. Five layers run parallel to the beam line and alternate with the other four, which are tilted at ± 3 degrees to the beam line to provide 3D position information. Another 30 000 wires along the length of the cylinder create drift fields (of about 1 kV/cm) around each sense wire. The entire chamber is immersed in a 1.4-T magnetic field, which is directed parallel to the beam line and is generated by a superconducting solenoid. Momentum is measured more accurately for low momentum tracks with large curvatures than for high momentum tracks which bend only a few degrees in the field. A 20-GeV/c particle traveling in the radial direction can be measured to about 4%, but a 1-GeV/c particle's momentum can be determined to 0.2%.

For extremely short-lived particles, we need a detector with higher resolution than the drift chamber can provide. Although b-hadrons have a lifetime of only 1.5 ps, they can often be identified by the short distance they travel from the collision before decaying. B-hadrons from top decays that have an average momentum of 40 GeV/c transverse to the beam and travel an average of 3.4 mm before decaying. B-hadrons can be identified by tracing back the tracks from the decay to a vertex which is displaced from the collision point (Fig. 7). We use a high precision silicon microstrip detector (an 8-cm-radius, 60-cm-long cylindrical device) which is placed inside the drift chamber and immediately surrounds the beampipe.

Strips of silicon separated by about 20 μm provide the necessary resolution for finding displaced vertices. The semiconducting strips are reverse-biased with voltages applied to the n and p sides. This produces an electric field in the depletion region, while preventing much current flow. A charged particle passing through the detector can create electron-hole pairs that move in the electric field and appear as small pulses on the n and p electrodes. The CDF microstrip detector uses four layers of silicon strips placed at small radii inside the main drift chamber. The combination of the two tracking devices provides enough resolution to pick out tracks and identify displaced vertices from

tt̄ Event
SVX Display

Jet 2

Jet 3

Jet 1 (b-jet)

Primary
vertex

l_1 Secondary
vertex

l_2

| l_1 = 4.5 mm |
| l_2 = 2.2 mm |

Jet 4 (b-jet)

e^+

ν

$m_{\text{top}} = 170 \pm 10 \text{ GeV}/c^2$ 24 September, 1992
run #40758, event #44414

Figure 7   Schematic of top quark candidate event.

b-hadron decays. For a b-hadron traveling in the radial direction, the position of the displaced vertex can be resolved to about 130 μm.

Calorimetry: Measuring the Energy of Particles

We now have a snapshot of the event which tells us the direction and momentum of many of the particles. Outside the tracking chambers and the solenoid coil are calorimeters designed to measure particle energies. The energy of the hadron jets is crucial to the determination of the top quark mass. Most particles lose energy in the calorimeters by producing a shower of lower-energy particles. Figure 8 illustrates the various effects of these particles in the detector. Calorimeters contain alternate layers of an "absorber" like lead or iron, with "active" layers of detectors for sampling the particles in the shower as it develops. Since we would like to contain the entire shower, high-energy particles require deep calorimeters, and since we want to know roughly where the energy was deposited, the calorimeters are segmented into hundreds of read-out towers. Unlike tracking chambers which are sensitive only to charged particles, calorimeters can also provide information about neutral particles like photons and neutrons. As we will see, they also aid in distinguishing electrons from hadrons.

Electrons, because of their low mass, are accelerated by the protons in the lead nuclei, and lose much of their energy through *bremsstrahlung*, or "braking radiation." Because energy loss due to bremsstrahlung radiation for a particle falls as M^{-2}, this process is neglible for the heavy particles. Hadrons, instead, lose most of their energy through inelastic collisions with the lead nuclei, producing secondary hadrons.

In both these cases, the loss of energy results in the production of other, lower-energy particles. Photons from bremsstrahlung will subsequently convert to electron-positron pairs which continue in the same manner. Hadrons also shower, producing lower-energy

Figure 8   Schematic of muons (μ), electrons (e) and jets interacting in a detector.

hadrons. Eventually, all the particles in the "shower" will come to rest. By measuring the energy deposited in the active layers of the calorimeter, a good measurement of the energy of the particle is made. The more times we sample the shower the better our estimate of the energy of the incident particle.

In practice, although we use the same method to measure both electron and hadron energies, we often use two separate devices, an "electromagnetic" calorimeter (in our case containing about 7.5 cm of lead) followed by a "hadronic" calorimeter with much more material (about 80 cm of steel). Each uses about 30 active layers to sample the number of particles in the shower. The sampling layers are slabs of plastic scintillator. The scintillation light they produce is converted to electronic pulses by photomultiplier tubes. By summing these pulses, we obtain a measure of the total energy deposited in the region.

In the CDF, electrons traverse 18 radiation lengths of lead in the electromagnetic calorimeter. A pion traverses about 5 nuclear absorption lengths of iron in the hadronic calorimeter. Electron and hadron showers are distinguishable by the fact that electrons shower early. In lead a radiation length is 0.56 cm while an absorption length is about 10 cm. As a result, an electron will lose most of its energy in a very short amount of material compared to a pion or a proton.

Identifying Muons and Neutrinos

We have discussed some of the ways in which jets, b-hadrons and electrons can be identified and are left now with the problem of finding muons and neutrinos. A muon is just a heavy version of the electron. Recall that high-energy electrons lose energy mostly through bremsstrahlung, a process which is negligible for the heavier muons. As a result, muons lose energy almost entirely through the relatively slow process of ionization. A 20-GeV muon from a top quark decay will travel through the drift chamber, leaving a track, but then pass through the calorimeters depositing only a few GeV of energy. To identify muons, we place small drift chambers outside the calorimeters and look for the track of a particle. Because most other particles stop in the calorimeter, a track in the muon drift chambers establishes the particle as a muon candidate. We already know the muon's momentum from the track it left in the main drift chamber.

Neutrinos are like muons without charge and little or no mass. Because they have no charge, they cannot lose energy through ionization. Neutrinos interact only through the weak interaction. Only one out of a billion neutrinos from a top quark decay will interact in the detector. Neutrinos can be detected only by measuring the missing momentum in the collision.

Top quark pairs are produced predominantly by quark–antiquark interactions. Although proton–antiproton collisions take place at the CDF in their center-of-mass frame, the quarks inside them carry only a fraction of their momentum. As a result, the top quark system is typically boosted along the beamline. Fortunately, the quarks and antiquarks collide with very little momentum transverse to the beam direction. Conservation of momentum requires that the final transverse momentum should then also be near zero. If a large imbalance is detected, we can infer the presence of a neutrino and

Figure 9   Top quark candidates' invariant mass distribution. (*Phys. Rev. Lett.* 74:2626, 1995)

estimate its momentum transverse to the beam. Since undetected particles also result in such an imbalance, this technique requires a "hermetic" detector with few cracks, intercepting most of the particles from the collision.

RECONSTRUCTING THE TOP

For over two years, this process continued, beginning with a bottle of hydrogen, and ending with a few thousand magnetic tapes full of data. After a long period of analyzing the data, selecting electrons, muons, and b-hadrons, a small pool of 50 events were identified containing about 28 top quark pairs. Using the measured momentum and energies, the mass of the top was reconstructed to be 176 ± 13 GeV/c^2 (Fig. 9). The top quark is the heaviest known fundamental particle; why it is so heavy is one of the fundamental questions remaining to be answered.

Suggestions for Further Reading

D. A. Edwards and M. J. Syphers, *An Introduction to the Physics of High Energy Accelerators*, New York, John Wiley & Sons, 1993.

T. Ferbel, *Experimental Techniques in High Energy Physics*, Frontiers in Physics Lecture Note Series, Reading, Mass., Addison-Wesley, 1987.

M. Reiser, *Theory and Design of Charged Particle Beams*, New York, John Wiley & Sons, 1994.

Best Known Values for Physical Constants

There is an uncertainty in the last digit, typically of 1 to 7 parts.

speed of light, c	299 792 458 m/s (exact)
gravitational constant, G	$6.672\ 6 \times 10^{-11}$ N$\cdot$m$^2\cdot$kg^{-2}
Planck's constant, h	$6.626\ 076 \times 10^{-34}$ J$\cdot$s
Boltzmann constant, k_B	$1.380\ 66 \times 10^{-23}$ J$\cdot$K^{-1}
Stefan-Boltzmann constant, σ	$5.670\ 5 \times 10^{-8}$ W$\cdot$m$^{-2}\cdot$K^{-4}
Rydberg constant, R	$1.097\ 373\ 153 \times 10^{7}$ m^{-1}
fine-structure constant, α	1/137.035 989
Bohr radius, a_0	$5.291\ 772\ 5 \times 10^{-11}$ m
Avogadro's number, N_A	$6.022\ 137 \times 10^{23}$ mol^{-1}
mass of neutron, m_n	$1.674\ 929 \times 10^{-27}$ kg
	939.565 63 MeV/c^2
mass of proton, m_p	$1.672\ 623\ 1 \times 10^{-27}$ kg
	938.272 3 MeV/c^2
mass of electron, m_e	$9.109\ 389\ 7 \times 10^{-31}$ kg
	0.510 999 1 MeV/c^2
elementary charge, e	$1.602\ 177\ 3 \times 10^{-19}$ C
permeability of vacuum, μ_0	$4\pi \times 10^{-7}$ N/A^2 (exact)
	$12.566\ 370\ 6 \times 10^{-7}$ N/A^2
permittivity of vacuum, ϵ_0	$1/\mu_0 c^2$ C^2/N$\cdot$m^2 (exact)
	$8.854\ 187\ 81 \times 10^{-12}$ C^2/N$\cdot$m^2
Coulomb constant, k	$1/(4\pi\epsilon_0)$ (exact)
	$8.987\ 551\ 78 \times 10^{9}$ N$\cdot$m^2/C^2
Bohr magneton, μ_B	$9.274\ 015\ 4 \times 10^{-24}$ J/T
	$5.788\ 382\ 6 \times 10^{-5}$ eV/T
Nuclear magneton, μ_n	$5.050\ 786\ 5 \times 10^{-27}$ J/T
	$3.152\ 451\ 7 \times 10^{-8}$ eV/T
Atomic mass unit, u	931.494 3 MeV/c^2
	$1.66\ 054\ 02 \times 10^{-27}$ kg

Values from E. Richard Cohen and Barry N. Taylor, "Fundamental Constants and Metric Practice," Part 2, *Physics Today*, August 1995.

APPENDIX B

Table of Selected Atomic Masses*

$1 \text{ u} = 931.494\ 3 \text{ MeV}/c^2 = 1.66\ 054\ 02 \times 10^{-27} \text{ kg}$

Atomic Number Z	Element	Symbol	Mass Number A	Atomic Mass** (u)	Percent Abundance, or Decay Mode (if radioactive)†	Half-Life (if radioactive)
0	(Neutron)	n	1	1.008665	β^-	10.6 min
1	Hydrogen	H	1	1.007825	99.985	
	Deuterium	D	2	2.014102	0.015	
	Tritium	T	3	3.016049	β^-	12.33 y
2	Helium	He	3	3.016029	0.00014	
			4	4.002603	≈ 100	
3	Lithium	Li	6	6.015122	7.5	
			7	7.016004	92.5	
4	Beryllium	Be	7	7.016929	EC, γ	53.3 days
			8	8.005305	2α	6.7×10^{-17} s
			9	9.012182	100	
5	Boron	B	10	10.012937	19.8	
			11	11.009306	80.2	
6	Carbon	C	11	11.011433	β^+, EC	20.4 min
			12	12.000000	98.89	
			13	13.003355	1.11	
			14	14.003242	β^-	5730 y
7	Nitrogen	N	13	13.005739	β^+	9.96 min
			14	14.003074	99.63	
			15	15.000109	0.37	
8	Oxygen	O	15	15.003066	β^+, EC	122 s
			16	15.994915	99.76	
			18	17.999160	0.204	
9	Fluorine	F	19	18.998403	100	
10	Neon	Ne	20	19.992440	90.51	
			22	21.991386	9.22	
11	Sodium	Na	22	21.994437	β^+, EC, γ	2.602 y
			23	22.989770	100	
			24	23.990964	β^-, γ	15.0 h
12	Magnesium	Mg	24	23.985042	78.99	
13	Aluminum	Al	27	26.981538	100	
14	Silicon	Si	28	27.976926	92.23	
			31	30.975363	β^-, γ	2.62 h
15	Phosphorus	P	31	30.973761	100	
			32	31.973907	β^-	14.28 days
16	Sulfur	S	32	31.972071	95.0	
			35	34.969032	β^-	87.4 days

Atomic Number Z	Element	Symbol	Mass Number A	Atomic Mass** (u)	Percent Abundance, or Decay Mode (if radioactive)†	Half-Life (if radioactive)
17	Chlorine	Cl	35	34.968853	75.77	
			37	36.965903		
18	Argon	Ar	40	39.962383	99.60	
19	Potassium	K	39	38.963707	93.26	
			40	39.963999	β^-, EC, γ, β^+	1.28×10^9 y
20	Calcium	Ca	40	39.962591	96.94	
21	Scandium	Sc	45	44.955910	100	
22	Titanium	Ti	48	47.947947	73.7	
23	Vanadium	V	51	50.943964	99.75	
24	Chromium	Cr	52	51.940512	83.79	
25	Manganese	Mn	55	54.938049	100	
26	Iron	Fe	56	55.934942	91.8	
27	Cobalt	Co	59	58.933200	100	
			60	59.933822	β^-, γ	5.271 y
28	Nickel	Ni	58	57.935348	68.3	
			60	59.930790	26.1	
			64	63.927969	0.91	
29	Copper	Cu	63	62.929601	69.2	
			64	63.929766	β^-, β^+	12.7 h
			65	64.927794	30.8	
30	Zinc	Zn	64	63.929146	48.6	
			66	65.926036	27.9	
31	Gallium	Ga	69	68.925581	60.1	
32	Germanium	Ge	72	71.922076	27.4	
			74	73.921178	36.5	
33	Arsenic	As	75	74.921597	100	
34	Selenium	Se	80	79.916522	49.8	
35	Bromine	Br	79	78.918338	50.69	
36	Krypton	Kr	84	83.911508	57.0	
			89	88.917563	β^-	3.2 min
37	Rubidium	Rb	85	84.911792	72.17	
38	Strontium	Sr	86	85.909265	9.8	
			88	87.905617	82.6	
			90	89.907738	β^-	28.8 y
39	Yttrium	Y	89	88.905849	100	
40	Zirconium	Zr	90	89.904702	51.5	
41	Niobium	Nb	93	92.906376	100	
42	Molybdenum	Mo	98	97.905407	24.1	
43	Technetium	Te	98	97.907215	β^-, γ	4.2×10^6 y
44	Ruthenium	Ru	102	101.904349	31.6	
45	Rhodium	Rh	103	102.905504	100	
46	Palladium	Pd	106	105.903484	27.3	
47	Silver	Ag	107	106.905093	51.83	
			109	108.904756	48.17	
48	Cadmium	Cd	114	113.903359	28.7	
49	Indium	In	115	114.903879	95.7; β^-	5.1×10^{14} y
50	Tin	Sn	120	119.902199	32.4	
51	Antimony	Sb	121	120.903822	57.3	
52	Tellurium	Te	130	129.906223	34.5; β^-	2×10^{21} y

Atomic Number Z	Element	Symbol	Mass Number A	Atomic Mass** (u)	Percent Abundance, or Decay Mode (if radioactive)†	Half-Life (if radioactive)
53	Iodine	I	127	126.904468	100	
			131	130.906118	β^-, γ	8.04 days
54	Xenon	Xe	132	131.904155	26.9	
			136	135.907220	8.9	
55	Cesium	Cs	133	132.905447	100	
56	Barium	Ba	137	136.905822	11.2	
			138	137.905242	71.7	
			144	143.922845	β^-	11.9 s
57	Lanthanum	La	139	138.906349	99.911	
58	Cerium	Ce	140	139.905435	88.5	
59	Praseodymium	Pr	141	140.907648	100	
60	Neodymium	Nd	142	141.907719	27.2	
61	Promethium	Pm	145	144.912745	EC, α, γ	17.7 y
62	Samarium	Sm	152	151.919729	26.6	
63	Europium	Eu	153	152.921227	52.1	
64	Gadolinium	Gd	158	157.924101	24.8	
65	Terbium	Tb	159	158.925343	100	
66	Dysprosium	Dy	164	163.929171	28.1	
67	Holmium	Ho	165	164.930319	100	
68	Erbium	Er	166	165.930290	33.4	
69	Thulium	Tm	169	168.934211	100	
70	Ytterbium	Yb	174	173.938858	31.6	
71	Lutecium	Lu	175	174.940768	97.39	
72	Hafnium	Hf	180	179.946549	35.2	
73	Tantalum	Ta	181	180.947996	99.988	
74	Tungsten (wolfram)	W	184	183.950932	30.7	
75	Rhenium	Re	187	186.955750	62.60, β^-	4×10^{10} y
76	Osmium	Os	191	190.960928	β^-, γ	15.4 days
			192	191.961479	41.0	
77	Iridium	Ir	191	190.960591	37.3	
			193	192.962923	62.7	
78	Platinum	Pt	195	194.964774	33.8	
79	Gold	Au	197	196.966551	100	
80	Mercury	Hg	202	201.970625	29.8	
81	Thallium	Tl	205	204.974412	70.5	
82	Lead	Pb	203	202.973375	β^-	51.9 h
			204	203.973028	β^-, 1.48	1.4×10^{17} y
			206	205.974449	24.1	
			207	206.975880	22.1	
			208	207.976636	52.3	
			210	209.984163	α, β^-, γ	22.3 y
			211	210.988735	β^-, γ	36.1 min
			212	211.991871	β^-, γ	10.64 h
			214	213.999798	β^-, γ	26.8 min
83	Bismuth	Bi	209	208.980384	100	
			211	210.987258	α, β^-, γ	2.15 min
84	Polonium	Po	210	209.982857	α, γ	138.38 days
			214	213.995176	α, γ	164 μs
85	Astatine	At	210	209.987131	β^+	8.1 h

Atomic Number Z	Element	Symbol	Mass Number A	Atomic Mass** (u)	Percent Abundance, or Decay Mode (if radioactive)†	Half-Life (if radioactive)
86	Radon	Rn	222	222.017570	α, γ	3.8235 days
87	Francium	Fr	223	223.019731	α, β^-, γ	21.8 min
88	Radium	Ra	226	226.025402	α, γ	1.60×10^3 y
			228	228.031064	β^-	5.76 y
89	Actinium	Ac	227	227.027747	α, β^-, γ	21.773 y
90	Thorium	Th	228	228.028715	α, γ	1.9131 y
			232	232.038050	100, α, γ	1.41×10^{10} y
91	Protactinium	Pa	231	231.035878	α, γ	3.28×10^4 y
92	Uranium	U	232	232.037129	α, γ	72 y
			233	233.039627	α, γ	1.592×10^5 y
			235	235.043922	0.72; α, γ	7.038×10^8 y
			236	236.045561	α, γ	2.342×10^7 y
			238	238.050784	99.275; α, γ	4.468×10^9 y
			239	239.054289	β^-, γ	23.5 min
93	Neptunium	Np	239	239.052933	β^-, γ	2.35 days
94	Plutonium	Pu	239	239.052156	α, γ	2.41×10^4 y
95	Americium	Am	243	243.061372	α, γ	7.37×10^3 y
96	Curium	Cm	247	247.070346	α	15.6×10^6 y
97	Berkelium	Bk	247	247.070298	α, γ	1.4×10^3 y
98	Californium	Cf	251	251.079579	α	898 y
99	Einsteinium	Es	254	254.088019	α, β^-, γ	276 days
100	Fermium	Fm	253	253.085173	EC, α, γ	3.0 days
101	Mendelevium	Md	258	258.098427	α	55 days
102	Nobelium	No	259	259.101050	EC, α	58 min
103	Lawrencium	Lr	260	260.105570	α	3 min
104	Rutherfordium‡	Rf	261	261.10891	α	1.1 min
105	Hahnium	Ha	262	262.114370	α	0.7 min
106	Seaborgium	Sg	263	263.118218	α	0.9 s
107	Nielsbohrium	Ns	262	262.123120	α	115 ms
108	Hassium	Hs	264	264.128630	α	0.08 ms
109	Meitnerium	Mt	266	266.137830	α	3.4 ms
110	(discovered Nov. 1994)					
111	(discovered Dec. 1994)					

* Data are taken from *Chart of the Nuclides,* 12th ed., General Electric, 1977, from C. M. Lederer and V. S. Shirley, eds., *Table of Isotopes,* 7th ed., John Wiley & Sons, Inc., New York, 1979, and from G. Audi and H. Wapstra, "The 1993 Atomic Mass Evaluation," *Nuclear Physics* A565, 1, 1993.

** The masses given in column 5 are those for the neutral atom, including the *Z* electrons.

† The abbreviation EC stands for "electron capture."

‡ The names for elements 104–109 are controversial. The names given here are those recommended by the discoverers.

Nobel Prizes

All Nobel Prizes in physics are listed (and marked with a P), as well as relevant Nobel Prizes in Chemistry (C). The key dates for some of the scientific work are supplied; they often antedate the prize considerably.

1901   (P) *Wilhelm Roentgen* for discovering x-rays (1895).

1902   (P) *Hendrik A. Lorentz* for predicting the Zeeman effect and *Pieter Zeeman* for discovering the Zeeman effect, the splitting of spectral lines in magnetic fields.

1903   (P) *Antoine-Henri Becquerel* for discovering radioactivity (1896) and *Pierre* and *Marie Curie* for studying radioactivity.

1904   (P) *Lord Rayleigh* for studying the density of gases and discovering argon.

     (C) *William Ramsay* for discovering the inert gas elements helium, neon, xenon, and krypton, and placing them in the periodic table.

1905   (P) *Philipp Lenard* for studying cathode rays, electrons (1898–1899).

1906   (P) *J. J. Thomson* for studying electrical discharge through gases and discovering the electron (1897).

1907   (P) *Albert A. Michelson* for inventing optical instruments and measuring the speed of light (1880s).

1908   (P) *Gabriel Lippmann* for making the first color photographic plate, using interference methods (1891).

     (C) *Ernest Rutherford* for discovering that atoms can be broken apart by alpha rays and for studying radioactivity.

1909   (P) *Guglielmo Marconi* and *Carl Ferdinand Braun* for developing wireless telegraphy.

1910   (P) *Johannes D. van der Waals* for studying the equation of state for gases and liquids (1881).

1911   (P) *Wilhelm Wien* for discovering Wien's law giving the peak of a blackbody spectrum (1893).

     (C) *Marie Curie* for discovering radium and polonium (1898) and isolating radium.

1912   (P) *Nils Dalén* for inventing automatic gas regulators for lighthouses.

1913   (P) *Heike Kamerlingh Onnes* for the discovery of superconductivity and liquefying helium (1908).

1914   (P) *Max T. F. von Laue* for studying x-rays from their diffraction by crystals, showing that x-rays are electromagnetic waves (1912).

     (C) *Theodore W. Richards* for determining the atomic weights of sixty elements, indicating the existence of isotopes.

1915   (P) *William Henry Bragg* and *William Lawrence Bragg,* his son, for studying the diffraction of x-rays in crystals.

1917   (P) *Charles Barkla* for studying atoms by x-ray scattering (1906).

1918   (P) *Max Planck* for discovering energy quanta (1900).

1919   (P) *Johannes Stark,* for discovering the Stark effect, the splitting of spectral lines in electric fields (1913).

1920   (P) *Charles-Édouard Guillaume* for discovering invar, a nickel-steel alloy with low coefficient of expansion.

(C) *Walther Nernst* for studying heat changes in chemical reactions and formulating the third law of thermodynamics (1918).

1921   (P) *Albert Einstein* for explaining the photoelectric effect and for his services to theoretical physics (1905).

(C) *Frederick Soddy* for studying the chemistry of radioactive substances and discovering isotopes (1912).

1922   (P) *Niels Bohr* for his model of the atom and its radiation (1913).

(C) *Francis W. Aston* for using the mass spectrograph to study atomic weights, thus discovering 212 of the 287 naturally occurring isotopes.

1923   (P) *Robert A. Millikan* for measuring the charge on an electron (1911) and for studying the photoelectric effect experimentally (1914).

1924   (P) *Karl M. G. Siegbahn* for his work in x-ray spectroscopy.

1925   (P) *James Franck* and *Gustav Hertz* for discovering the Franck-Hertz effect in electron-atom collisions.

1926   (P) *Jean-Baptiste Perrin* for studying Brownian motion to validate the discontinuous structure of matter and measure the size of atoms.

1927   (P) *Arthur Holly Compton* for discovering the Compton effect on x-rays, their change in wavelength when they collide with matter (1922), and *Charles T. R. Wilson* for inventing the cloud chamber, used to study charged particles (1906).

1928   (P) *Owen W. Richardson* for studying the thermionic effect and electrons emitted by hot metals (1911).

1929   (P) *Louis Victor de Broglie* for discovering the wave nature of electrons (1923).

1930   (P) *Chandrasekhara Venkata Raman* for studying Raman scattering, the scattering of light by atoms and molecules with a change in wavelength (1928).

1932   (P) *Werner Heisenberg* for creating quantum mechanics (1925).

1933   (P) *Erwin Schrödinger* and *Paul A. M. Dirac* for developing wave mechanics (1925) and relativistic quantum mechanics (1927).

(C) *Harold Urey* for discovering heavy hydrogen, deuterium (1931).

1935   (P) *James Chadwick* for discovering the neutron (1932).

(C) *Irène* and *Frédéric Joliot-Curie* for synthesizing new radioactive elements.

1936   (P) *Carl D. Anderson* for discovering the positron in particular and antimatter in general (1932) and *Victor F. Hess* for discovering cosmic rays.

(C) *Peter J. W. Debye* for studying dipole moments and diffraction of x-rays and electrons in gases.

1937   (P) *Clinton Davisson* and *George Thomson* for discovering the diffraction of electrons by crystals, confirming de Broglie's hypothesis (1927).

1938   (P) *Enrico Fermi* for producing the transuranic radioactive elements by neutron irradiation (1934–1937).

1939   (P) *Ernest O. Lawrence* for inventing the cyclotron.

1943   (P) *Otto Stern* for developing molecular-beam studies (1923), and using them to discover the magnetic moment of the proton (1933).

1944   (P) *Isidor I. Rabi* for discovering nuclear magnetic resonance in atomic and molecular beams.

(C) *Otto Hahn* for discovering nuclear fission (1938).

1945   (P) *Wolfgang Pauli* for discovering the exclusion principle (1924).

1946   (P) *Percy W. Bridgman* for studying physics at high pressures.

1947   (P) *Edward V. Appleton* for studying the ionosphere.

1948   (P) *Patrick M. S. Blackett* for studying nuclear physics with cloud-chamber photographs of cosmic-ray interactions.

1949   (P) *Hideki Yukawa* for predicting the existence of mesons (1935).

1950   (P) *Cecil F. Powell* for developing the method of studying cosmic rays with photographic emulsions and discovering new mesons.

1951   (P) *John D. Cockcroft* and *Ernest T. S. Walton* for transmuting nuclei in an accelerator (1932).

(C) *Edwin M. McMillan* for producing neptunium (1940) and *Glenn T. Seaborg* for producing plutonium (1941) and further transuranic elements.

1952   (P) *Felix Bloch* and *Edward Mills Purcell* for discovering nuclear magnetic resonance in liquids and gases (1946).

1953   (P) *Frits Zernike* for inventing the phase-contrast microscope, which uses interference to provide high contrast.

1954   (P) *Max Born* for interpreting the wave function as a probability (1926) and other quantum-mechanical discoveries and *Walther Bothe* for developing the coincidence method to study subatomic particles (1930–1931), producing, in particular, the particle interpreted by Chadwick as the neutron.

1955   (P) *Willis E. Lamb, Jr.* for discovering the Lamb shift in the hydrogen spectrum (1947) and *Polykarp Kusch* for determining the magnetic moment of the electron (1947).

1956   (P) *John Bardeen, Walter H. Brattain,* and *William Shockley* for inventing the transistor (1956).

1957   (P) *T.-D. Lee* and *C.-N. Yang* for predicting that parity is not conserved in beta decay (1956).

1958   (P) *Pavel A. Čerenkov* for discovering Čerenkov radiation (1935) and *Ilya M. Frank* and *Igor Tamm* for interpreting it (1937).

1959   (P) *Emilio G. Segrè* and *Owen Chamberlain* for discovering the antiproton (1955).

1960   (P) *Donald A. Glaser* for inventing the bubble chamber to study elementary particles (1952).

(C) *Willard Libby* for developing radiocarbon dating (1947).

1961   (P) *Robert Hofstadter* for discovering internal structure in protons and neutrons and *Rudolf L. Mössbauer* for discovering the Mössbauer effect of recoilless gamma-ray emission (1957).

1962   (P) *Lev Davidovich Landau* for studying liquid helium and other condensed matter theoretically.

1963   (P) *Eugene P. Wigner* for applying symmetry principles to elementary-particle theory and *Maria Goeppert Mayer* and *J. Hans D. Jensen* for studying the shell model of nuclei (1947).

1964   (P) *Charles H. Townes, Nikolai G. Basov,* and *Alexandr M. Prokhorov* for developing masers (1951–1952) and lasers.

1965   (P) *Sin-itiro Tomonaga, Julian S. Schwinger,* and *Richard P. Feynman* for developing quantum electrodynamics (1948).

1966   (P) *Alfred Kastler* for his optical methods of studying atomic energy levels.

1967   (P) *Hans Albrecht Bethe* for discovering the routes of energy production in stars (1939).

1968   (P) *Luis W. Alvarez* for discovering resonance states of elementary particles.

1969   (P) *Murray Gell-Mann* for classifying elementary particles (1963).

1970   (P) *Hannes Alfvén* for developing magnetohydrodynamic theory and *Louis Eugène Félis Néel* for discovering antiferromagnetism and ferrimagnetism (1930s).

1971   (P) *Dennis Gabor* for developing holography (1947).

        (C) *Gerhard Herzberg* for studying the structure of molecules spectroscopically.

1972   (P) *John Bardeen, Leon N. Cooper,* and *John Robert Schrieffer* for explaining superconductivity (1957).

1973   (P) *Leo Esaki* for discovering tunneling in semiconductors, *Ivar Giaever* for discovering tunneling in superconductors, and *Brian D. Josephson* for predicting the Josephson effect, which involves tunneling of paired electrons (1958–1962).

1974   (P) *Anthony Hewish* for discovering pulsars and *Martin Ryle* for developing radio interferometry.

1975   (P) *Aage N. Bohr, Ben R. Mottelson,* and *James Rainwater* for discovering why some nuclei take asymmetric shapes.

1976   (P) *Burton Richter* and *Samuel C. C. Ting* for discovering the J/psi particle, the first charmed particle (1974).

1977   (P) *John H. Van Vleck, Nevill F. Mott,* and *Philip W. Anderson* for studying solids quantum-mechanically.

        (C) *Ilya Prigogine* for extending thermodynamics to show how life could arise in the face of the second law.

1978   (P) *Arno A. Penzias* and *Robert W. Wilson* for discovering the cosmic background radiation (1965) and *Pyotr Kapitsa* for his studies of liquid helium.

1979   (P) *Sheldon L. Glashow, Abdus Salam,* and *Steven Weinberg* for developing the theory that unified the weak and electromagnetic forces (1958–1971).

1980   (P) *Val Fitch* and *James W. Cronin* for discovering CP *(charge-parity) violation* (1964), which possibly explains the cosmological dominance of matter over antimatter.

1981   (P) *Nicolaas Bloembergen* and *Arthur L. Schawlow* for developing laser spectroscopy and *Kai M. Siegbahn* for developing high-resolution electron spectroscopy (1958).

1982   (P) *Kenneth G. Wilson* for developing a method of constructing theories of phase transitions to analyze critical phenomena.

1983   (P) *William A. Fowler* for theoretical studies of astrophysical nucleosynthesis and *Subramanyan Chandrasekhar* for studying physical processes of importance to stellar structure and evolution, including the prediction of white dwarf stars (1930).

1984   (P) *Carlo Rubbia* for discovering the W and Z particles, verifying the electroweak unification, and *Simon van der Meer,* for developing the method of stochastic cooling of the CERN beam that allowed the discovery (1982–1983).

1985   (P) *Klaus von Klitzing* for the quantized Hall effect, relating to conductivity in the presence of a magnetic field (1980).

1986   (P) *Ernst Ruska* for inventing the electron microscope (1931), and *Gerd Binnig* and *Heinrich Rohrer* for inventing the scanning-tunneling electron microscope (1981).

1987   (P) *J. Georg Bednorz* and *Karl Alex Müller* for the discovery of high temperature superconductivity (1986).

1988   (P) *Leon M. Lederman, Melvin Schwartz,* and *Jack Steinberger* for a collaborative experiment that led to the development of a new tool for studying the weak nuclear force, which affects the radioactive decay of atoms.

1989   (P) *Norman Ramsay* for various techniques in atomic physics; and *Hans Dehmelt* and *Wolfgang Paul* for the development of techniques for trapping single charge particles.

1990   (P) *Jerome Friedman, Henry Kendall,* and *Richard Taylor* for experiments important to the development of the quark model.

1991   (P) *Pierre-Gilles de Gennes* for discovering that methods developed for studying order phenomena in simple systems can be generalized to more complex forms of matter, in particular to liquid crystals and polymers.

1992   (P) *George Charpak* for developing detectors that trace the paths of evanescent subatomic particles produced in particle accelerators.

1993   (P) *Russell Hulse* and *Joseph Taylor* for discovering evidence of gravitational waves.

1994   (P) *Bertram N. Brockhouse* and *Clifford G. Schull* for pioneering work in neutron scattering.

1995   (P) *Martin L. Perl* for discovery of the tau particle, and *Frederick Reines* for first detection of a neutrino.

MP DESKTOP Software for Modern Physics

INTRODUCTION

MP DESKTOP is a sophisticated interactive learning tool with considerable flexibility and scope. Among its many capabilities the DESKTOP can:

- Add any number of harmonic waves with arbitrary amplitude and phase to produce a wave group, and display the time evolution of the resulting waveform.
- Display the time evolution of quantum wave packets in one dimension, subject to arbitrary forces (specified by a potential energy function).
- Find bound state energies and display the corresponding wavefunctions for a particle moving in a one-dimensional potential well of any shape.
- Determine scattering (transmission and reflection) coefficients for a particle incident on a one-dimensional potential barrier of any shape.
- Find stationary state energies and display the corresponding wavefunctions (including probability "clouds") for any spherically symmetric potential in three dimensions.

A number of examples and problems in the text have been written specifically for use with the DESKTOP and illustrate its potential for clarifying some of the more abstract ideas inherent in wave motion and quantum physics. The examples, beginning with Example 4.6, are your guide as you learn to use the program. By working through all the examples, you will come to appreciate the full capabilities of MP DESKTOP. The DESKTOP may then be used to investigate other quantum phenomena of your own choosing. All examples and problems requiring use of the MP DESKTOP are clearly identified with the MP logo **MP** .

To use the software, you will need an IBM PC (or close compatible) with a VGA display and color monitor; a mouse is *strongly* recommended. (A 486 processor with a clock speed of at least 50 MHz is desirable for displaying the time evolution of waveforms.)

INSTALLATION

Start by copying the program files on the floppy disk to your hard drive (about 500 kbytes of disk space are required). The INSTALL utility on the program disk automates this chore. To install:

(1) Make the drive containing the installation disk (drive A) the current drive.
(2) Type 'A:INSTALL' at the DOS prompt, and press ⟨Enter⟩.

The destination drive and directory default to C:\MPDESK. You may rename this directory at the installation prompt.

The application file MP.EXE **must** be executed from the C:\MPDESK subdirectory where it resides. After copying the files, INSTALL will launch the DESKTOP automatically if you desire. Otherwise, or in subsequent sessions, you must do the following:

(1) Make drive C (your computer's hard drive) the current drive.
(2) Type 'cd\MPDESK' and press ⟨*Enter*⟩. (If you installed to a different subdirectory, substitute the name of your subdirectory for 'MPDESK.')
(3) Type 'MP' and press ⟨*Enter*⟩.

THE DESKTOP SCREEN

The main DESKTOP screen is divided into six logical elements organized according to function.

The *Desktop Menu Bar* allows selections to be made from four pull-down menus:

The *Task Menu* is used to select a specific DESKTOP task.

The *Options Menu* governs exactly how the DESKTOP performs various tasks, where it will find its data files, the display colors it uses, etc.

The *View Menu* determines what graphics are visible and the display options in effect.

The *Help Menu* accesses the on-line help database.

The *Desktop Task Field*, just below the menu bar, provides information specific to the current task, and includes command buttons to access the needed function input (e.g., potential energy). Here you will also find the *Desktop Time\Energy Box* for entering values of time (t) or energy (E) as appropriate to task.

The *Graphics Viewport*, the central and dominant element of the DESKTOP screen, is where all graphics images are displayed. The *Horizontal and Vertical Scroll Bars* immediately below and to the right of the Viewport are used for scrolling the image. The Viewport may be split into two separate windows (with the *Split-Screen Control:* see below), but only one—the active window—can be scrolled or otherwise manipulated at any given time. The active window is identified by the scroll bar beneath it. The other (inactive) window may be activated by moving the mouse into the window and clicking the left mouse button, or by clicking the *Swap Window Bar* (blank bar just below the inactive window).

The *Graphics Toolbar* consists of six controls grouped to the right of the Vertical Scroll Bar and above the MP logo. These controls are used to further manipulate the graphics image (scale, zoom in or out, split the Viewport, control plot appearance) and analyze graphics output.

The *Auxiliary Toolbar* is located just below the MP logo and consists of anywhere from zero to two additional controls specific to the selected task.

The *Desktop Status Bar*, extending along the bottom of the screen, keeps the user informed of current program settings. The Status Bar is updated continually during program operation.

GETTING HELP

You can get help while using the MP DESKTOP by choosing a command from the *Desktop Help Menu* or by pressing the ⟨*F1*⟩ key. Some dialog boxes also have a *Help* button. In addition, MP DESKTOP supports *QuickHelp*, which displays summary information for each object on the DESKTOP.

The *Desktop Help Menu:* Selecting *Contents* brings up the main help window listing all help topics organized according to task, with cross-references (hyperlinks) to related topics. (The *Help* command appearing in selected dialog boxes is a context-specific gateway to this on-line reference.) The *Search for Help On. . .* menu option allows you to identify a help topic or topics by entering a keyword. To print any of the on-line help topics, select the *Print* command on the button bar at the top of the help window.

The ⟨*F1*⟩ (Help) key: Another context-specific gateway to the on-line reference material. The help topic displayed will depend on the activity at hand. In many—but not all—situations, pressing the ⟨*F1*⟩ key produces the same action as choosing *Contents* from the Desktop Help Menu.

QuickHelp: Click the *right* mouse button on any DESKTOP object and a pop-up window appears describing the object's function, along with any special usage notes.

A WORD ABOUT UNITS

In order to achieve maximum flexibility, the MP DESKTOP works in units whose numerical values are adapted for convenience to fit the problem at hand. However, the units employed must be interrelated in specific ways according to task.

Classical Waves—Units

- Coordinate [x]: arbitrary. The spatial domain extends from $x = -1$ to $x = +1$, and cannot be changed.
- Wavenumber [k]: the reciprocal of the unit for coordinate.
- Angular frequency [ω]: the angular frequency unit is fixed by the application and reflected in the dispersion relation $\omega(k)$.
- Time [t]: the reciprocal of the frequency unit.

Example: Consider transverse waves on a string. Such waves propagate with speed $v = \sqrt{T/\mu}$ where T is the tension and μ the mass per unit length of string. Thus, the dispersion relation for this case is

$$\omega(k) = v|k|$$

Adopting the meter (m) as the unit of length, we find the DESKTOP display covers the range -1 m $< x < +1$ m, and wavenumbers k are read in units of $1/$m. If $\omega(k)$ is measured in Hz, a wave speed of 5 m/s is reproduced by taking $v = 5$. Alternatively, we could decide to measure $\omega(k)$ in kHz and take $v = 0.005$ to describe the same wave speed. In the first case, time is measured in seconds, and in the second, time is measured in milliseconds.

QM Waves—Units

- Coordinate [*x* or *r*]: fixed by the application and reflected in the potential energy $U(x)$ or $U(r)$. An exception occurs for free particles, where the coordinate unit is arbitrary. The coordinate measure sets all other units, as described below.
- Energy [*E*]: related to the coordinate unit so that Schrödinger's time-independent equation matches that in the Desktop Task Field. In particular, the energy unit (ϵ) and coordinate unit (δ) must satisfy

$$\epsilon\delta^2 = \hbar^2/2m$$

where $\hbar$ is Planck's constant divided by 2π, and m is the particle mass. In describing electrons, atomic units ($\delta = 1$ bohr [$= 0.529$ Å] and $\epsilon = 1$ Ry [$= 13.60$ eV]) are a convenient pair satisfying this criterion.
- Time [*t*]: related to the energy unit (ϵ) as $\hbar/\epsilon$, to make Schrödinger's time-dependent equation match that in the Desktop Task Field.
- Angular frequency [ω]: the reciprocal of the time unit.
- Momentum: related to the coordinate unit (δ) as $\hbar/\delta$.

Example: A Finite Square Well is represented by the potential energy

$$U(x) = U \text{ for } 0 < x < L$$
$$= 0 \text{ elsewhere}$$

For an electron in this well we might adopt atomic units, measuring U in rydbergs and L in bohrs. An electron in a well 100 eV high and 1 Å wide would be described by $U = 100/13.60$ Ry and $L = 1.0/0.529$ bohrs. Then a given energy, say $E = 1.0$, is the equivalent of 1.0 Ry or 13.60 eV, and coordinate values *x* are read in bohrs.

Many other unit choices are possible, provided only that they are properly interrelated. Indeed, one of the well parameters can be absorbed in the units by taking L itself as the coordinate unit. Then the well occupies a region of unit length on the DESKTOP, and the unit of energy is $\epsilon = \hbar^2/2mL^2$. The potential energy $U(x)$ for this case would be written as

$$U(x) = U \text{ for } 0 < x < 1$$
$$= 0 \text{ elsewhere}$$

where now U must be entered in ϵ units. Again, for an electron in a well 100 eV high and 1 Å wide, we calculate $\epsilon = 3.81$ eV. Thus, a well 100 eV high would be described by $U = 100/3.81$, and an energy of $E = 1.0$ now corresponds to 3.81 eV. (To describe protons in the same well, the mass increases by a factor of 1836, U increases by the same factor, and the energy unit is reduced to $3.81/1836 = 2.07$ meV.)

USING THE PROGRAM

Space limitations preclude us from reproducing the entire on-line reference here, but we now describe several of the most important and frequently used program operations.

To Input a Function, such as potential energy, use the *Desktop Function Editor.* DESKTOP functions are saved on disk in the form of data files. These files are created and modified with the Desktop Function Editor. The Function Editor dialog box appears when a data file is first opened or, for files already present in memory, when a function command button is selected. The Function Editor dialog box is divided into several fields:

Interval Field
This field specifies the domain over which the function will be evaluated, beginning with the *Start* entry and continuing to the *Stop* entry.

- Complicated functions often require multiple intervals for their definition. In each interval a new function expression with new parameters may be introduced. Successive intervals are created with the *Next* command. The Start limit for the new interval is taken as the Stop limit of the previous one, and is NOT editable. Expressions and parameter values are duplicated from the previous interval by default, but may be modified as desired.
- Use the *Next* and *Back* commands to navigate among intervals. Once created, an interval can be deleted only by setting the Stop limit for that interval equal to the Start limit. There is no restriction on the number of intervals you can create.
  Note: Interval limits for initial wave amplitudes $[a(k)$ or $\Psi(x)]$ are taken from the corresponding environment domain $[\omega(k)$ or $U(x)]$, and are NOT editable. Accordingly, these functions are limited to a single interval of definition.

Expressions (Function) Field
The function definition over the current interval is entered in the equation edit box, and may contain numerical constants, functions, and parameters. The syntax is that used in many programming languages. Thus

$$(2*x - 1)/(x + 1)$$

is a valid expression prescribing division $(/)$, multiplication $(*)$, addition $(+)$ and subtraction $(-)$. The Function Editor also recognizes exponentiation $(\hat{})$, integer division $(\backslash)$, and the remainder after integer division $(@)$.

- Function names MUST be upper case. The following functions are supported:

ABS	absolute value
EXP	exponential
LOG	natural logarithm
COS	trigonometric cosine
SIN	trigonometric sine
HVS	Heaviside step [HVS$(x) = 0$ for $x < 0$, 1 for $x > 0$]
RE	complex number, real part
IM	complex number, imaginary part

- Parameters are represented by Greek letters. To enter a Greek symbol, hold down the ⟨*Alt*⟩ key and type the appropriate code using the numeric keypad.

Valid symbols and their codes are:

Symbol	Code	Symbol	Code
α	224	τ	231
β	225	Φ	232
Γ	226	θ	233
π	227	Ω	234
Σ	228	δ	235
σ	229	ϕ	237
μ	230	ϵ	238

'i' and 'π' (code 227) are reserved for the imaginary unit and pi, respectively.

Parameters Field
The Parameters field lists those symbols appearing in the function expression which require further definition. Supported operations, functions, etc., are identical to those used in other expressions, but NO variables are allowed. Parameters may be defined in terms of other parameters which appear earlier in the list.

Notes Field
The Notes field is used to record units and/or other information pertinent to the file record.

To **Manipulate Program Graphics,** use the *Graphics Toolbar.* The Graphics Toolbar, located to the right of the Graphics Viewport, consists of six controls for manipulating and analyzing program graphics:

 The *Analyzer Control* is used for quantitative analysis of the graphics image (see below).

 The *Vertical Scale Control* is used to stretch (or shrink) the image vertically. Each click of the mouse stretches/shrinks the image by about 10%. To shrink the image, hold down the ⟨*Shift*⟩ key when clicking on the control. The change is reflected in the size of the scroll box on the Vertical Scroll Bar. To recover the default scaling, click the control while holding down the ⟨*Ctrl*⟩ key.
Note: *Vertical Scale* affects only the topmost plot in the active window; to scale an underlying plot, eliminate the topmost plot using the View Menu, then restore it after scaling is completed.

 The *Horizontal Scale Control* is used to stretch (or shrink) the image horizontally. Images are scaled horizontally by changing the number of function elements displayed in the active window. With each invocation of the control, this number changes by about 33%. The change is reflected in the size of the scroll box on the Horizontal Scroll Bar. To shrink the image, hold down the ⟨*Shift*⟩ key when clicking on the control. To recover the default scaling, click the control while holding down the ⟨*Ctrl*⟩ key. Horizontal Scale affects all plots in the active window.

The *Zoom Control* is used to zoom in or out on the graphics image. Click the Zoom Control (or press ⟨*Alt-Z*⟩) to zoom in on the image. To zoom out, hold down the ⟨*Shift*⟩ key when clicking on the control (or press ⟨*Shift-Alt-Z*⟩). Zooming works by scaling the image vertically and horizontally in the proper proportion so as to preserve the aspect ratio. The change is reflected in the size of the scroll box on both the *Horizontal* and *Vertical Scroll Bars*. To recover the default scaling, click the control while holding down the ⟨*Ctrl*⟩ key. The *Zoom Control* affects all plots in the active window.

 The *Split-Screen Control* is used to split the Graphics Viewport. Click the Split-Screen Control to split the Viewport into two windows; click the control a second time to recover a single window. When restoring the Viewport to its unsplit condition, the visible window becomes the one that was active when the command was invoked. Splitting the Graphics Viewport is useful in the following circumstances:

- To show two views (different scales, ranges) of the same graphics.
- To display spectral frequencies and amplitudes used in the construction of wave groups (*Classical Waves, QM WaveGroups*— Fourier mode only).
- To show probability 'clouds' derived from 3d wavefunctions *(Eigenstates- 3D)*.

 The *Color-for-Phase Control* is used to display modulus and phase information for a complex-valued function. Phase is represented by colors according to the "colorwheel" on the face of the control. Phase angle increases steadily around the wheel in a counterclockwise sense. The dark line on the right marks the zero of phase; phase angles range from $-180°$ to $+180°$. Color-for-phase plotting is the default display style for *QM WaveGroups*. Clicking the control a second time disables color-for-phase plotting, in which case only the real part of the function is displayed.

To **Analyze Program Graphics,** use the *MP Analyzer*. Quantitative analysis of DESKTOP graphics is performed with the Analyzer Control, the top button on the Graphics Toolbar. When active, the Analyzer overlays the graphics image with a graphics cursor (vertical line) and a readout window displaying ordinate and abscissa values at the current cursor. Ordinate values can be read anywhere by moving the cursor with the left and right arrow keys or clicking the left mouse button at the desired location.

- Keyboard control of the graphics cursor is facilitated by the ⟨*Ins*⟩ key. ⟨*Ins*⟩ toggles between coarse and fine movement (tuning). The greatest degree of cursor control is afforded by the keyboard (NOT the mouse) set for fine tuning.
- Reference marks may be added to the display by positioning the cursor at the desired location and pressing the period key ⟨.⟩; when the cursor is next moved, a line is left behind. The technique is useful for marking features of special interest. To erase a mark, overlay it with the graphics cursor and press ⟨.⟩ again.

Deactivate the Analyzer by selecting the Analyzer Control again.

To **Change Default Settings,** access the *Desktop Control Panel.* MP DESKTOP performs certain key operations according to the settings in the Desktop Control Panel. The Control Panel is accessed from the Control Panel entry on the Options Menu after a task has been selected. Controls are grouped together in fields according to purpose. All data entry from the Control Panel follows a fixed decimal format. In this format, only digits 0–9 are echoed and a digit always fills each place in the field. Also, the normal delete and backspace operations are suspended (to preserve the field format).

Coordinate Field
Potentials and wavefunctions are evaluated over a discrete set of points referred to as the coordinate grid. Adjacent grid points are separated by a distance equal to the *Mesh Size.* Function variations on a scale finer than the mesh size will be lost, but a coarse mesh results in fewer (and faster) computations. For *Classical Waves,* Mesh Size defaults to 0.001; for all other tasks the default is 0.005. The Mesh Size together with the function domain fixes the number of grid points. The total number of grid points cannot exceed 4095. With logarithmic scaling, the grid points become further separated as the coordinate value increases. Mesh Size here refers to the increment in the logarithm of the coordinate (which is uniform). Logarithmic scaling is useful when the potential is long range and all coordinate values are positive. This is the default setting for Schrödinger's radial wave equation of three-dimensional quantum mechanics.

Fourier Field
In Fourier mode the initial waveform is built from a superposition of individual waves. For *Classical Waves,* these are harmonic oscillations indexed by a continuous wavenumber k. For *QM Waves,* the building blocks are the stationary states; these are similarly indexed, but k likely is not a wavenumber here (in fact, for bound states, k is an integer quantum number). *Fourier Mesh Size* is the interval in k for the individual waves included in the mixture: the default value is 1. The Fourier Mesh Size is NOT editable unless the *Enable Mode* switch is set. For *QM Waves,* this requires first that a potential energy function $U(x)$ be selected. (For *Classical Waves,* Fourier Mode is enabled by default and cannot be disabled.)

Evolution Field
QM Waves are evolved by integrating the time-dependent Schrödinger equation [TDSE] or, in Fourier Mode, by direct summation of the stationary states. For the latter, *Time Step* simply represents the elapsed time between screen updates. When solving the Schrödinger equation *Time Step* also is the step size used for integration and, to give reliable results, must be smaller than any natural time scale for the system under study.

- For *Classical Waves,* the direct summation method is enabled by default and cannot be disabled.
- The Evolution field of the Control Panel is inactive if an Eigenvalue Task has been selected.

The Evolution setting (SUM, INTegral) is recorded on the Desktop Status Bar.

AutoSearch Field

Eigenvalue searches are often facilitated by the AutoSearch feature, engaged with the *Enable Mode* switch. AutoSearch specifies an automated search for eigenvalues in the range indicated, and centered on the energy value E entered in the Desktop Energy Box.

- AutoSearch may miss energies in the search range which are too close together. In some cases this can be overcome by narrowing the search range.
- The AutoSearch field of the *Control Panel* is inactive if an Evolution Task has been selected.

The Autosearch setting (Auto-ON) is recorded on the Desktop Status Bar.

Answers to Odd Problems

CHAPTER 1

3. In the rest frame, $p_i = p_f = 4.0 \times 10^4$ kg·m/s
In the moving frame, $p_i' = p_f' = 7.5 \times 10^4$ kg·m/s

5. (a) Time upwind/downwind $= (2L/c)(1/(1 - v^2/c^2))$
Time crosswind $= (2L/c)(1/\sqrt{1 - v^2/c^2})$
(b) Time upwind/downwind $= 0.417$ h
Time crosswind $= 0.408$ h
$\Delta t = 0.0088$ h $\approx \frac{1}{2}$ min

7. $0.866c$

9. $0.661c$

11. (a) 8.3×10^{-8} s   (b) 24 m

13. (a) 70 beats/min   (b) 31 beats/min

17. $0.166c = 1.11 \times 10^8$ mi/h

19. (a) 1790 Hz   (b) 2090 Hz   (c) 3 parts per 100 million

21. $0.98c$

25. $0.98c$

27. (a) $0.85c$ (b) No change, because masses cancel.

29. (b) As expected, as $v \rightarrow c$, $a \rightarrow 0$, because the maximum speed is c.   (c) $v = qEct/\sqrt{m^2c^2 + q^2E^2t^2}$
$x = (c/qE)(\sqrt{m^2c^2 + q^2E^2t^2} - mc)$

33. (a) 939 MeV   (b) 3.01×10^3 MeV
(c) 2.07×10^3 MeV

35. $0.75c$

37. (a) $0.0236c$ (b) $6.18 \times 10^{-4}c$

39. (a) 4.4×10^9 kg/s   (b) 1.4×10^{13} years

41. (a) In S, $p = mu/\sqrt{1 - u^2/c^2}$, $E = mc^2/\sqrt{1 - u^2/c^2}$

In S', $p' = \dfrac{(mu - mv)/(1 - uv/c^2)}{\sqrt{1 - (u - v)^2/(1 - uv/c^2)^2 c^2}}$

$E' = \dfrac{mc^2}{\sqrt{1 - (u - v)^2/c^2(1 - uv/c^2)^2}}$

43. (a) 3.26 MeV   (b) 2.75 MeV

45. $\Delta m = m_n - m_p - m_{e^-} = 8.404 \times 10^{-4}$ u
$E = (8.404 \times 10^{-4}$ u$)(931.5$ MeV/u$) = 0.783$ MeV

47. (a) $p = 206$ MeV/c $u = 0.827c$ (b) 498 MeV/c^2

49. (a) 39.2 μs   (b) Accurate to one digit. More precisely, he aged 1.78 μs less on each orbit.

51. $0.789c$

53. (a) $\dfrac{2v}{1 + v^2/c^2}$ (b) $\dfrac{3v + v^3/c^2}{1 + 3v^2/c^2}$

55. (a) $\dfrac{2d}{v + c}$ (b) $\dfrac{2d\sqrt{1 - v^2/c^2}}{v + c}$

57. 15.0 h in the spacecraft frame, 10.0 h less than in the Earth frame.

59. (a) 83.3 m and 62.5 m   (b) 27.0 m and 20.3 m
(c) 6.00 ks   (d) 3.60 ks   (e) 4.80 ks
(f) Both rockets and their crew are destroyed.

CHAPTER 2

1. (c) $\Delta\omega = 8.8 \times 10^{10}$ rad/s; $\Delta\omega/\omega = 2.3 \times 10^{-5}$
(d) If the electron's plane of rotation is parallel to **B**, the magnetic flux Φ_B is always zero so that **E** and F are zero and there is no Δv for the electron.

3. (a) $E_{total} = 2.0$ J; $f = 0.56$ Hz   (b) 5.4×10^{33}
(c) 3.7×10^{-34} J

5. (a) $x = 5(1 - e^{-x})$, where $x = hc/\lambda_{max} k_B T$
(b) Solution to (a) is $x = 4.965$ or
$\lambda_{max} T = 2.90 \times 10^{-3}$ m·K

7. (a) 2.07 eV   (b) 4.14×10^{-5} eV
(c) 1.24×10^{-7} eV

9. 1.60×10^{30} photons/s

11. 3.0×10^{19} photons/s

13. (a) 1.30 eV   (b) 554 nm

15. (a) 1.91 eV   (b) 1.20 V

17. (a) 0.45 eV   (b) 2.03 eV   (c) 612 nm

19. 2.00 eV

21. 1.35 eV

23. (a) 0.00243 nm   (b) 74.4 eV

25. (a) 0.0403, 0.0412, 0.0424, 0.0436, 0.0445, 0.0448, 0.0445 nm   (b) 231, 905, 1760, 2570, 3140, 3330, 3140 eV   (c) $\theta = 180°$

29. (a) $\theta = 41.5°$ (b) 0.679 MeV

31. (a) 9.11×10^4 eV   (b) 8.90 keV   (c) 55.4°

33. (a) 1.95×10^{-3} eV   (b) 0.0861 nm

35. (a) 17.4 keV   (b) 0.0760 nm   (c) 16.3 keV   (d) 1.1 keV

37. (a) 3.77×10^{-5} eV   (b) 3.10 eV   (c) No, because the maximum energy transferred ($\theta = 180°$) is insufficient.

39. (a) $d_1 = d_0/\sqrt{2}$, $d_2 = d_0/\sqrt{5}$
(b) $\Theta_1 = 51.4°$, $\Theta_2 = 57.8°$, $\Theta_3 = 64.3°$

43. (a) 0.571 eV   (b) $V_s = 1.54$ V

45. 22.9°

CHAPTER 3

1. 1.60×10^{-19} C

3. (a) 9.58×10^7 C/kg   (b) proton
(c) 2.19×10^6 m/s $\sim 0.01c$

5. (b) speed = 2.39×10^8 m/s = $0.795c$
Using $p = mv$ gives $r = 0.0769$ m and $y = 0.00408$ m
Using $p = \gamma mv$ gives $r = 0.1267$ m and $y = 0.00243$ m

7. mass = 8.418×10^{-11} g
$mg/E = 25.96 \times 10^{-19}$ C
$\bar{q} = 1.661 \times 10^{-19}$ C

9. 6.00×10^{-15} m

11. 656.112 nm, 486.009 nm, 433.937 nm

13. (a) 3   (b) no

15. (a) $E_n = (-13.6 \text{ eV}) Z^2/n^2 = (-54.4 \text{ eV})/n^2$
(b) 54.4 eV

17. $r_{He^+} = 0.0265$ nm
$r_{Li^{2+}} = 0.0177$ nm
$r_{Be^{3+}} = 0.0132$ nm

19. (a) 1.89 eV   (b) 658 nm   (c) 4.56×10^{14} Hz

21. (a) $\lambda_{max} = 1874.606$ nm $\lambda_{min} = 820.140$ nm
(b) $E_{min} = 0.6627$ eV $E_{max} = 1.515$ eV

23. (a) 0.0529 nm   (b) 1.99×10^{-24} kg·m/s
(c) 1.05×10^{-34} kg·m²/s $= \hbar$ (d) 13.6 eV
(e) -27.2 eV   (f) -13.6 eV

25. (a) 1.60×10^{14} Hz   (b) For $n = 3$,
$f = 2.44 \times 10^{14}$ Hz   For $n = 4$, $f = 1.03 \times 10^{14}$ Hz

29. (b) 3.23×10^{-8} %

33. (a) 3.1×10^{-15} m   (b) -18.9 MeV

35. $r_{positronium} = 2r_{hydrogen}$
$E_{positronium} = \frac{1}{2}E_{hydrogen}$

39. (a) 4.9000134 eV   (b) 1.31×10^6 m/s = 4.38×10^{-3} c (c) 3.60 m/s   (d) 3.68×10^{-11} eV

41. $\lambda = 634$ nm, red

43. (a) B   (b) A   (c) B and C

CHAPTER 4

1. 3.97×10^{-13} m

3. 1.79×10^{-36} m

5. For electrons: (a) 0.0150 eV   (b) 150 eV
(c) 1240 MeV
For alphas: (a) 2.06×10^{-6} eV
(b) 0.0206 eV   (c) 201 MeV

7. 9.05 fm = 9.05×10^{-15} m

9. 1.06×10^{-34} m

11. (a) 124 MeV   (b) The kinetic energy is too large to expect that the electron could be confined to a region the size of the nucleus.

13. 50th plane

17. $v_p = \sqrt{k^2 c^2 + (m_e c^2/\hbar)^2}/k$
$v_g = kc^2/\sqrt{k^2 c^2 + (m_e c^2/\hbar)^2}$; $v_p v_g = c^2$
If $v_p > c$, $v_g < c$.

19. Wave crest at $x = 0$ moves to $x = 0.9$ during time interval $t = 0.00$ to $t = 0.09$, giving a wave speed of
$0.9/0.09 = 10$, in agreement with $v_p = \dfrac{\omega}{k}$.
With $k \rightarrow -k$, the wave moves in the opposite direction with the same speed.

21. $\omega(k) = \sqrt{gk}$. The packet distorts and disperses as the nodes of the waveform move through the central peak, thus indicating that $v_p < v_g$.

23. 4.56×10^{-14} m

25. $\Delta x = \sqrt{\hbar/2m} \cdot \sqrt[4]{H/2g} = 1.84 \times 10^{-16}$ m

27. (a) $\Delta v \geqslant 0.25$ m/s   (b) 2.25 m

29. The intrinsic energy width, ΔE, is given by $\Delta E \geqslant 3.29 \times 10^{-6}$ eV. It can't be measured with a gamma detector that has a resolution of ± 5 eV.

31. 546 eV

33. 2.25 to 1

35. 9.5×10^{27} m

37. $\Delta m/m = 5.62 \times 10^{-8}$

CHAPTER 5

1. (a) Not acceptable—diverges as $x \rightarrow \infty$
(b) Acceptable   (c) Acceptable   (d) Not acceptable—not single-valued   (e) Not acceptable—wavefunction is discontinuous (as is its slope).

3. (a) 0.126 nm   (b) 5.26×10^{-24} kg·m/s
(c) 95 eV

5. $U(x) = \dfrac{\hbar^2}{2mL^2}\left\{\dfrac{4x^2}{L^2} - 6\right\}$ is a parabola opening upward with its vertex at $\left(0, -\dfrac{3\hbar^2}{mL^2}\right)$

9. $\lambda = 2.02 \times 10^{-4}$ nm
$E = 6.14$ MeV
This is the gamma-ray region of the electromagnetic spectrum.

11. $\psi_1(x) = \sqrt{2/L}\,\cos(\pi x/L)$ $P_1(x) = (2/L)\cos^2(\pi x/L)$
$\psi_2(x) = \sqrt{2/L}\,\sin(2\pi x/L)$ $P_2(x) = (2/L)\sin^2(2\pi x/L)$
$\psi_3(x) = \sqrt{2/L}\,\cos(3\pi x/L)$ $P_3(x) = (2/L)\cos^2(3\pi x/L)$

13. (a) 5.13×10^{-3} eV   (b) 9.40 eV   (c) Much smaller electron mass

15. (a) $U = -\dfrac{7}{3}\dfrac{ke^2}{d}$ (b) $\dfrac{h^2}{36\,md^2}$

(c) $d = \dfrac{h^2}{42\,mke^2} = 0.050$ nm

(d) Lithium spacing = $a = 0.28$ nm

17. (b) $P_1 = 0.200$ (c) $P_2 = 0.351$
(d) $E_1 = 0.377$ eV; $E_2 = 1.51$ eV

19. $n = 4.27 \times 10^{28}$; excitation energy $= 4.69 \times 10^{-32}$ J

21.

23. $\psi(x) = A\sin kx$ for $0 \le x \le L$ and $\psi(x) = Ce^{-\alpha x}$ for
$x > L$, where $k^2 = \dfrac{2mE}{\hbar^2}$ and $\alpha^2 = \dfrac{2m(U-E)}{\hbar^2}$.

Allowed energies satisfy $\dfrac{kL}{\sin(kL)} = \left[\dfrac{2mUL^2}{\hbar^2}\right]^{1/2}$,

which has solutions only if $\dfrac{2mUL^2}{\hbar^2} > 1$.

25. (a) $\alpha = \dfrac{m\omega}{2\hbar}$; $E = \dfrac{3\alpha\hbar^2}{m} = \dfrac{3\hbar\omega}{2}$

(b) $C = \left(\dfrac{32\alpha^3}{\pi}\right)^{1/4}$

27. $P_c(x) = (1/\pi)(A^2 - x^2)^{-1/2}$

29. (b) $\langle x \rangle = L/2$, in agreement with the quantum result. $\langle x^2 \rangle = L^2/3$, larger than the quantum result by an amount $L^2/2(n\pi)^2$ which goes to zero for n large (correspondence principle).

33. $\langle x \rangle = 0$ $\langle x^2 \rangle = \dfrac{\hbar}{2m\omega}$

$\Delta x = \left(\dfrac{\hbar}{2m\omega}\right)^{1/2}$

35. $\Delta x = \left(\dfrac{\hbar}{2m\omega}\right)^{1/2}$ (from Prob. 33)

$\Delta p_x = \left(\dfrac{m\hbar\omega}{2}\right)^{1/2}$ (from Prob. 34)

$\Delta x\,\Delta p_x = \tfrac{1}{2}\hbar$, the minimum permitted by the uncertainty principle.

37. $E = 1.019$ (ground state) and $E = 2.338$ (first excited state). Ground state wavefunction is nodeless and symmetric; first excited state is antisymmetric, with one node at the center of the triangular well.

39. $\Psi(x, t) = \dfrac{1}{\sqrt{2}}\left[\psi_1(x)e^{-iE_1 t/\hbar} + \psi_2(x)e^{-iE_2 t/\hbar}\right]$

where E_1, E_2 are the ground and first excited state energies, respectively.

CHAPTER 6

1. (b) $k^2 = 2mE/\hbar^2$ to the left and $k^2 = 2m(U-E)/\hbar^2$ to the right of the step. $E/U = \tfrac{1}{2}$ (c) 1.44 fm

3. $R = 0.146$, $T = 0.854$ for both, provided their energies are the same.

5. (b) (1) 0.90   (2) 0.36   (3) 0.41   (4) ≈ 0

9. $T = 0.333461$
$R = 0.666539$
$T(E) = 0.333596$ (exact value)
Thickness = 0.0234 Å for protons.

11. (a) The matter wave reflected from the front of the well ($x = 0$) suffers a 180° change in phase; that reflected from the rear is not phase-shifted.
(b) $\Psi(x, t) = \psi(x)\,e^{-i\omega t}$, where
$\psi(x) = Ae^{ik'x} + Be^{-ik'x}$ (region 1)
$ = Ce^{-ikx} + De^{ikx}$ (region 2)
$ = Fe^{ik'x} + Ge^{-ik'x}$ (region 3)
with $k = \sqrt{2mE/\hbar^2}$
$k' = \sqrt{2m(E-U)/\hbar^2}$
and $G = 0$ for particle incident from the left.

13. $R = -\dfrac{E_0}{E}\left(1 - \dfrac{E_0}{E}\right)^{-1}$; $T = \left(1 - \dfrac{E_0}{E}\right)^{-1}$;

with $E_0 = -\dfrac{mS^2}{2\hbar^2}$

15. $E = 0.576$ MeV
$v = 0.0175c$
Since $E < 4.05$ MeV, the ejected alpha particle could not have originated in the ground state.

17. 3.85×10^{30} Hz

CHAPTER 7

1. $E/E_0 = 6, 9, 9, 12, 14, 14$ where $E_0 = \dfrac{\pi^2 \hbar^2}{8mL^2}$

3. (a) $E = \dfrac{11\pi^2 \hbar^2}{2mL^2}$ (b) n_1, n_2, n_3 can have values

(1, 1, 3); (1, 3, 1); (3, 1, 1)
(c) $\psi_{113} = A \sin(\pi x/L) \sin(\pi y/L) \sin(3\pi z/L)$
   $\psi_{131} = A \sin(\pi x/L) \sin(3\pi y/L) \sin(\pi z/L)$
   $\psi_{311} = A \sin(3\pi x/L) \sin(\pi y/L) \sin(\pi z/L)$

5. (a) $E_{111} \simeq 1.54$ MeV   (b) The states ψ_{221}, ψ_{122}, and ψ_{212} have the same energy $E = 3E_{111} \simeq 4.63$ MeV. The states ψ_{211}, ψ_{121}, ψ_{112} have the same energy $= 2E_{111} \approx 3.08$ MeV.   (c) Both states are threefold degenerate.

7. $A = \sqrt{8/V}$, where $V = L_1 \cdot L_2 \cdot L_3$ is the volume of the box.

9. $\ell = 4$

11. (a) $\ell \simeq 4.58 \times 10^{65}$
(b) $\Delta L/L \simeq 1/\ell = 2.18 \times 10^{-66}$

13. (a) For $n = 3$, ℓ can have the values of 0, 1, or 2. For $\ell = 0$, $m_\ell = 0$; for $\ell = 1$, $m_\ell = 1, 0, -1$; for $\ell = 2$, $m_\ell = 2, 1, 0, -1, -2$.   (b) $E = -6.04$ eV

17. $L_z = \pm 2\hbar, \pm \hbar$, or zero

19. (a) $n = 6$ (b) -0.378 eV   (c) $\ell = 4$; $L = \sqrt{20}\hbar$
(d) m_ℓ can be $-4, -3, -2, -1, 0, +1, +2, +3$, or $+4$ and $L_z = m_\ell \hbar$, giving $\theta = 153.4°, 132.1°, 116.6°, 102.9°, 90°, 77.1°, 63.4°, 47.9°, 26.6°$, respectively.

23. $\langle r \rangle = 5a_0 = 2.645$ Å

25. $\langle U \rangle = -27.2$ eV and $\langle K \rangle = 13.6$ eV

27. $P = 0.176$ for the $2s$ state and $P = 0.371$ for the $2p$ state.

29. $\Delta r = 0.866a_0$, more than half the average itself.

31. $\psi(0) = 0$; $E_n = -\dfrac{mA^2}{2\hbar^2}\left[\dfrac{1}{n^2}\right]$ $(n = 1, 2, \ldots)$

33. $E = 3.700$ Ry ($= 50.32$ eV) and $\ell = 1$ for first excited states.

CHAPTER 8

1. $f = 9.79 \times 10^9$ Hz

3. (a) 2   (b) 8   (c) 18   (d) 32   (e) 50

5. $\dfrac{dB_z}{dz} = 0.387$ T/m

7. $v/c = 278.6$

9. $|S| = \dfrac{\sqrt{15}}{2}\hbar$ $\theta = 140.8°, 105.0°, 75.0°, 39.2°$

Ω^- does obey the Pauli exclusion principle.

11. $j = \frac{5}{2}, \frac{3}{2}$, and

$m_j = \pm\frac{5}{2}, \pm\frac{3}{2}, \pm\frac{1}{2}$ $(j = \frac{5}{2})$

$m_j = \pm\frac{3}{2}, \pm\frac{1}{2}$ $(j = \frac{3}{2})$

13. (a) $n = 4$, $\ell = 3$, $j = \frac{5}{2}$ (b) $|\mathbf{J}| = \dfrac{\sqrt{35}}{2}\hbar$

(c) $J_z = m_j\hbar$, where $m_j = \pm\frac{5}{2}, \pm\frac{3}{2}, \pm\frac{1}{2}$

15. $\Delta E_{calc} = 3.18 \times 10^{-5}$ eV; $\Delta E_{meas} = 5 \times 10^{-5}$ eV

17. (a) 394.8 eV
$(n_1, n_2, n_3) = (1, 1, 1)$
$(n_1, n_2, n_3) = (1, 1, 2), (1, 2, 1),$ or $(2, 1, 1)$
(b) All eight particles go into the $(n_1, n_2, n_3) = (1, 1, 1)$ state, and so the energy is $8 \times E_{111} = 225.6$ eV.

19. $6s$ wavefunction has five nodes and $E = -0.635$ Ry. ($= -8.64$ eV). Ionization energy is 8.64 eV vs 9.22 eV from Table 8.2. Most probable distance is 1.37 bohrs compared with 1.00 bohr for the (outermost) electron in hydrogen, suggesting that the gold atom is not much larger even though it has 79 electrons.

21. (a) $1s^2 2s^2 2p^4$
(b) $(1, 0, 0, \pm\frac{1}{2})$
$(2, 0, 0, \pm\frac{1}{2})$
$(2, 1, 0, \pm\frac{1}{2})$
$(2, 1, \pm 1, \pm\frac{1}{2})$

23. [Kr] $4d^{10}$ has the lesser energy, but [Kr] $4d^9 5s^1$ has two more unpaired spins. Thus, Hund's rule is violated in this case (since filled subshells are especially stable). The element is Pd.

25. (b) $(\text{Slope})_{theor} = 0.496 \times 10^8$ Hz$^{1/2}$ compared with $(\text{Slope})_{expt} \approx 0.50 \times 10^8$ Hz$^{1/2}$ (c) Intercepts also agree very well, indicating the charge seen by an L shell electron is very nearly $Z - 1$.

CHAPTER 9

3. $s_{\bar{v}} = 1.58$ cm $s_{rms} = 1.45$ cm $s_{mp} = 1.78$ cm

7. (b) $2e^{-2p\mathcal{E}/k_B T}$ (c) $T = 2.83$ K

(d) $\bar{E} = \dfrac{2p\mathcal{E}}{1 + \frac{1}{2}e^{2p\mathcal{E}/k_B T}}$

(f) C is a maximum at $T = 0.055$ K

9. $\bar{v} = 1.51 \times 10^3$ m/s so the classical Doppler shift formula can be used.

$$\dfrac{\Delta f}{f_0} = 1.01 \times 10^{-5}$$

11. (a) Since $(N/V)\hbar^3/8 \, (k_B m T)^{3/2} = 2.74 \times 10^{-11} \ll 1$, the MB distribution may be used.
(b) The same quantity $= 0.014$, which is not $\ll 1$, so the BE distribution should be used.

13. (b) T_E (lead) ≈ 100 K, T_E (aluminum) ≈ 400 K, and T_E (silicon) ≈ 750 K

15. (a) 4.23 eV   (b) $T = 32\ 700$ K

21. 50% probability, independent of T.

23. 5.28 eV
25. (a) 1.1/1   (b) $1.55 \times 10^{-25}/1$

CHAPTER 10

1. $\mu = 1.22 \times 10^{-26}$ kg, $\mu_{calc} = 1.24 \times 10^{-26}$ kg
3. (a) 0.0120 nm   (b) 0.00923 nm; HI has the weaker bond.
5. 0.95 u
7. (a) $K(HI) = 292$ N/m; $K(NO) = 1550$ N/m
    (b) $A(HI) = 0.0123$ nm; $A(NO) = 0.00492$ nm
9. $\nu = 7$ is the highest vibrational level that can be excited without the molecule coming apart.
11. $\Delta E_{rot} = \dfrac{5\hbar^2}{4mr^2} = 0.519$ MeV for H_2
13. (c) $E_{diss} = U_0 - E_{vib}$ (d) $U_0 = 4.79$ eV;
    $\alpha = 19.3$ nm^{-1}
15. $E_0 = 0.268$ eV $E_1 = 0.781$ eV $E_2 = 1.26$ eV
    $E_{16} = 4.77$ eV is the highest-lying vibrational state
    ($\nu = 16$).
17. Ground state: $E = (2x^2\hbar^2/mL^2)$, where x is the smallest root of the equation $\tan x = -(2\hbar^2/mSL)x$. First excited state: $E = (2\pi^2\hbar^2/mL^2)$. As $S \to \infty$, $x \to \pi$ and the two energies coincide; as $S \to 0$, $x \to \pi/2$, and the energies reduce to the ground and first excited states of an infinite well with no barrier.
19. (a) $R_0 = 1.44$ bohrs
    (b) $K = 1.03$ Ry/bohr$^2 = 801$ N/m
21. Crossing time $\approx 7 \times 10^{-15}$ s, showing that this is not a stationary state of the divided well.

CHAPTER 11

3. 7.84 eV
7. (a) 7.12 eV   (b) 6.39 eV
11. (c) 3.9×10^5 cm/s   (d) 0.36 $\Omega\cdot$m
13. (a) 3.80×10^{-14} s   (b) 52.7 nm
    (c) $L/d \approx 200$
15. (a) 2.75×10^{14} Hz   (b) 1090 nm (IR)
19. (b) 0.36 nm   (c) 6.7×10^{-4} nm. The controlling factor is cavity length.
21. (a) $E = 4.741$ Ry and $E = 17.07$ Ry (single well)
    (b) 4.703 Ry
        4.779 Ry } bound state energies
        16.72 Ry } for double well
        17.52 Ry

Ground state is nodeless; 1st, 2nd, and 3rd excited states have 1, 2, and 3 nodes, respectively.
(c) 16 bound states altogether, with the energies grouped into two "bands," each with 8 levels. The lower band extends from 4.671 Ry to 4.813 Ry; upper band from 16.44 Ry to 17.99 Ry. The band gap is 11.63 Ry. The wavefunction for state at bottom of second band has 8 nodes, the same as the number of potential wells.

CHAPTER 12

1. 1.0001
3. 191 mT
5. 277 mA
7. 150 μWb
9. M/H
11. (a) 9.3×10^{-24} A$\cdot$m^2 (b) Because the electron is negatively charged, its (conventional) current is clockwise, and **M** points downward.
15. (a) 3980 A   (b) 39 800 N/m, outward
17. 2.54 mJ/cm^3
19. (a) 38.6 mT   (b) 2.9 mT
21. 63 m/s
23. (a) See Figure 12.10b.   (b) 10.7 kA

25.

Superconductor	T_c (K)	E_g (theo.)
Nb$_3$Al	18.7	5.70×10^{-3} eV
Nb$_3$Sn	18.0	5.48×10^{-3} eV
Nb$_3$Ge	23.0	7.00×10^{-3} eV
NbN	15.7	4.78×10^{-3} eV
NbTi	9.30	2.80×10^{-3} eV
Nb$_3$(AlGe)	21.0	6.40×10^{-3} eV
V$_3$Si	16.9	5.10×10^{-3} eV
V$_3$Ga	14.8	4.50×10^{-3} eV
PbMoS	14.4	4.40×10^{-3} eV

In general, type II energy gaps are greater than type I, as are the critical temperatures.

27. $\alpha \approx 0.34$
29. (a) 4.4 meV   (b) 7.0 meV   (c) Because the answers are comparable, the energy required to break up a Cooper pair is approximately the same in the two substances.
31. 241 GHz
33. (a) 0.414 μA   (b) 0.186 μA
35. (a) 6.24×10^{10} J   (b) 2000 N/m
37. (a) 1.8×10^{28} electrons/m^3
    (b) $(1.8 \times 10^{28}$ m$^{-3}) \left[1 - \left(\dfrac{T}{T_c} \right)^4 \right]$
41. (a) 0.0382 H   (b) 1.06×10^{-14} Ω
43. (a) Since $V = IR$, it follows that $V = 0$ if $R = 0$.   (b) The graph of V versus I is approximately linear, with a slope $= 1/R = \Delta I/\Delta V = 4.31$ Ω^{-1}. Thus, $R = 0.0232$ Ω.

CHAPTER 13

1. (a) 1.9 fm   (b) 7.44 fm   (c) 3.92
3. 8.57×10^{13}
5. (a) 4.55×10^{-13} m   (b) 6.03×10^6 m/s
7. 2.2×10^{-6} eV
9. 30% ^{63}Cu
11. 2.657 MeV/nucleon
13. (a) ^{139}Cs with $N/Z = 1.53$
    (b) ^{139}La, with 8.379 MeV/nucleon
    (c) ^{139}Cs, with a mass of 138.913 u
15. (a) 8.03 MeV/nucleon   (b) 8.55 MeV/nucleon
    (c) 8.66 MeV/nucleon   (d) 7.92 MeV/nucleon
17. 160 MeV
19. (a) $U = 3k\dfrac{(Ze)^2}{5R}$ (b) $2.88 \times 10^{-14}\,(A^{5/3})$ J

    (c) 52.1 MeV
21. (a) 0.805 h^{-1} (b) 0.861 h
23. (a) $\lambda = 1.55 \times 10^{-5}$ s^{-1}; $T_{1/2} = 12.4$ h
    (b) 2.39×10^{13} atoms   (c) 1.87 mCi
25. 36.3×10^{-6} g
27. 0.055 mCi
29. 9.46×10^9
31. (b) $\lambda = 0.25$ h^{-1}; $T_{1/2} = 2.77$ h
    (c) 4×10^3 counts/min   (d) 9.59×10^6
33. (a) 18.3 counts/min
35. 11.8 decays/min·g
41. 2.26 MeV
43. (b) 4.79 MeV   (c) 0.08 MeV
45. (a) $Q = -2.37$ MeV, so reaction *cannot* occur.
    (b) $Q = -1.68$ MeV, so reaction *cannot* occur.
    (c) $Q = 1.86$ MeV, so reaction *can* occur.
47. Using the uncertainty principle, electrons in a nucleus are found to have an energy of about 100 MeV. Since the most energetic electrons emitted in beta decay have energies less than 10 MeV, electrons are not present in the nucleus.
51. 18.6 keV
53. (a) 3.96×10^9 years   (b) It could be older.
55. (b) R/λ

CHAPTER 14

1. (a) -2.4386 MeV   (b) 2.5751 MeV
3. 5.70 MeV
5. 17.35 MeV
7. (a) 1.53 MeV   (b) 17.35 MeV
9. (b) 1.53 MeV
11. 0.0266 b
13. (a) 0.0373   (b) 0.0663   (c) $\simeq 1$
    (d) 0.1 to 10 eV
15. (a) 0.86   (b) 6.1×10^{11} protons/s
    (c) 8.7×10^{10} protons/s
17. 2.25×10^{12}

19. 25
21. 200.6 MeV
23. (a) 4.84 $V^{-1/3}$ (b) 6 $V^{-1/3}$ (c) 6.30 $V^{-1/3}$
    (d) Sphere   (e) Parallelipiped
25. (a) 3333 MW   (b) 2333 MW
    (c) 1.04×10^{20} events/s   (d) 1.34×10^3 kg
    (e) 3.7×10^{-8} kg/s
27. (a) 2.70×10^{-15} m   (b) 720 keV   (c) $v_F = v_0 m_D/(m_D + m_T)$ (d) 1.2 MeV   (e) Possibly by tunneling
29. (a) $K_n = 14.1$ MeV $K_\alpha = 3.45$ MeV   (b) Yes; since the neutron is uncharged, it is not confined by the **B** field, and only K_α contributes directly to achieving critical ignition.
31. (a) 1.9×10^9 K   (b) 355 kJ
33. (a) 52×10^6 K   (b) $1.943 + 1.709 + 7.551 + 7.297 + 2.242 + 4.966 = 25.75$ MeV   (c) Most energy is lost, since neutrinos have such low cross-sections.
35. 5.3×10^{23} J
37. (a) 3000 MW   (b) 5.2×10^6 L
39. (a) $2.3 \times 10^{-19}\, Z_1 Z_2$ J   (b) 0.14 MeV
41. (a) 2.63×10^{33} J   (b) 119 billion years
43. (a) 10^{14}/cm^3 (b) 2.2×10^5 J/m^3 (c) 2.35 T
45. (a) 1.4×10^{-3} cm   (b) $I/I_0 = 0.426$
47. 3.85 cm
49. (a) 0.0025 rem   (b) 38 times background of 0.13 rem/yr
51. The two workers receive the same dose in rad.
53. 18.8 J
55. (a) 10   (b) 10^8

CHAPTER 15

1. 10^{-23} s
3. $f_{min} = 2.26 \times 10^{23}$ Hz, $\lambda_{max} = 1.32 \times 10^{-15}$ m
5. 2.1×10^{-18} m
9. (a) Does not occur (violates baryon number)
    (b) Occurs
    (c) Does not occur (violates baryon number)
    (d) Occurs   (e) Occurs
    (f) Does not occur (violates baryon number and muon-lepton number)
11. (a) Electron and muon-lepton number
    (b) Electron-lepton number
    (c) Strangeness and charge   (d) Baryon number
    (e) Strangeness
13. (a) 5.63 GeV   (b) 5.64 GeV
15. For Example 15.5, $E = 48\%$; for Exercise 3, $E = 48\%$, $E = 46\%$; for Problem 13, $E = 33\%$.
17. A neutron, udd
19. $m_u = 312.3$ MeV/c^2, $m_d = 313.6$ MeV/c^2

21.

23. (a) $0.384c$, 38.4% of the speed of light
(b) 6.7×10^9 lightyears

25. $K_p = 5.4$ MeV, $K_\pi = 32.3$ MeV
27. 9.3 cm
29. A neutron
31. (a) A Z^0 boson   (b) A gluon
33. (a) 0.782 MeV   (b) Proton speed $= 0.001266c =$
380 km/s, electron speed $= 0.9185c = 276 \times 10^6$ m/s
(c) The electron is relativistic, the proton is not.

Index

Page numbers in *italics* indicate figures; page numbers followed by "n" indicate footnotes; page numbers followed by "t" indicate tables.

Periodic Table of the Elements

Transition elements

Legend:
Symbol — Ca
Atomic mass † — 40.08
Atomic number — 20
Electron configuration — $4s^2$

Group I

Z	Symbol	Mass	Config
1	H	1.0080	$1s^1$
3	Li	6.94	$2s^1$
11	Na	22.99	$3s^1$
19	K	39.102	$4s^1$
37	Rb	85.47	$5s^1$
55	Cs	132.91	$6s^1$
87	Fr	(223)	$7s^1$

Group II

Z	Symbol	Mass	Config
4	Be	9.012	$2s^2$
12	Mg	24.31	$3s^2$
20	Ca	40.08	$4s^2$
38	Sr	87.62	$5s^2$
56	Ba	137.34	$6s^2$
88	Ra	(226)	$7s^2$

Transition elements

Z	Symbol	Mass	Config
21	Sc	44.96	$3d^14s^2$
22	Ti	47.90	$3d^24s^2$
23	V	50.94	$3d^34s^2$
24	Cr	51.996	$3d^54s^1$
25	Mn	54.94	$3d^54s^2$
26	Fe	55.85	$3d^64s^2$
27	Co	58.93	$3d^74s^2$
28	Ni	58.71	$3d^84s^2$
29	Cu	63.54	$3d^{10}4s^1$
30	Zn	65.37	$3d^{10}4s^2$
39	Y	88.906	$4d^15s^2$
40	Zr	91.22	$4d^25s^2$
41	Nb	92.91	$4d^45s^1$
42	Mo	95.94	$4d^55s^1$
43	Tc	(99)	$4d^55s^2$
44	Ru	101.1	$4d^75s^1$
45	Rh	102.91	$4d^85s^1$
46	Pd	106.4	$4d^{10}$
47	Ag	107.87	$4d^{10}5s^1$
48	Cd	112.40	$4d^{10}5s^2$
72	Hf	178.49	$5d^26s^2$
73	Ta	180.95	$5d^36s^2$
74	W	183.85	$5d^46s^2$
75	Re	186.2	$5d^56s^2$
76	Os	190.2	$5d^66s^2$
77	Ir	192.2	$5d^76s^2$
78	Pt	195.09	$5d^96s^1$
79	Au	196.97	$5d^{10}6s^1$
80	Hg	200.59	$5d^{10}6s^2$
104	Rf†	(261)	$6d^27s^2$
105	Ha	(262)	$6d^37s^2$
106	Sg	(263)	
107	Ns	(262)	
108	Hs	(264)	
109	Mt	(266)	
110		Discovered Nov. 1994	
111		Discovered Nov. 1994	

Group III

Z	Symbol	Mass	Config
5	B	10.81	$2p^1$
13	Al	26.98	$3p^1$
31	Ga	69.72	$4p^1$
49	In	114.82	$5p^1$
81	Tl	204.37	$6p^1$

Group IV

Z	Symbol	Mass	Config
6	C	12.011	$2p^2$
14	Si	28.09	$3p^2$
32	Ge	72.59	$4p^2$
50	Sn	118.69	$5p^2$
82	Pb	207.2	$6p^2$

Group V

Z	Symbol	Mass	Config
7	N	14.007	$2p^3$
15	P	30.97	$3p^3$
33	As	74.92	$4p^3$
51	Sb	121.75	$5p^3$
83	Bi	208.98	$6p^3$

Group VI

Z	Symbol	Mass	Config
8	O	15.999	$2p^4$
16	S	32.06	$3p^4$
34	Se	78.96	$4p^4$
52	Te	127.60	$5p^4$
84	Po	(210)	$6p^4$

Group VII

Z	Symbol	Mass	Config
1	H	1.0080	$1s^1$
9	F	18.998	$2p^5$
17	Cl	35.453	$3p^5$
35	Br	79.91	$4p^5$
53	I	126.90	$5p^5$
85	At	(218)	$6p^5$

Group 0

Z	Symbol	Mass	Config
2	He	4.0026	$1s^2$
10	Ne	20.18	$2p^6$
18	Ar	39.948	$3p^6$
36	Kr	83.80	$4p^6$
54	Xe	131.30	$5p^6$
86	Rn	(222)	$6p^6$

*Lanthanide series

Z	Symbol	Mass	Config
57	La	138.91	$5d^16s^2$
58	Ce	140.12	$5d^14f^16s^2$
59	Pr	140.91	$4f^36s^2$
60	Nd	144.24	$4f^46s^2$
61	Pm	(147)	$4f^56s^2$
62	Sm	150.4	$4f^66s^2$
63	Eu	152.0	$4f^76s^2$
64	Gd	157.25	$5d^14f^76s^2$
65	Tb	158.92	$5d^14f^86s^2$
66	Dy	162.50	$4f^{10}6s^2$
67	Ho	164.93	$4f^{11}6s^2$
68	Er	167.26	$4f^{12}6s^2$
69	Tm	168.93	$4f^{13}6s^2$
70	Yb	173.04	$4f^{14}6s^2$
71	Lu	174.97	$5d^14f^{14}6s^2$

**Actinide series

Z	Symbol	Mass	Config
89	Ac	(227)	$6d^17s^2$
90	Th	(232)	$6d^27s^2$
91	Pa	(231)	$5f^26d^17s^2$
92	U	(238)	$5f^36d^17s^2$
93	Np	(239)	$5f^46d^17s^2$
94	Pu	(239)	$5f^67s^2$
95	Am	(243)	$5f^77s^2$
96	Cm	(245)	$5f^76d^17s^2$
97	Bk	(247)	$5f^86d^17s^2$
98	Cf	(249)	$5f^{10}6d^07s^2$
99	Es	(254)	$5f^{11}6d^07s^2$
100	Fm	(253)	$5f^{12}6d^07s^2$
101	Md	(255)	$5f^{13}6d^07s^2$
102	No	(255)	$5f^{14}6d^07s^2$
103	Lr	(257)	$6d^17s^2$

□ Atomic mass values given are averaged over isotopes in the percentages in which they exist in nature.

† For an unstable element, mass number of the most stable known isotope is given in parentheses.

‡ The names for elements 104–109 are controversial. The names given here are those recommended by the discoverers.